Herbert Reichl (Editor)

MICRO SYSTEM Technologies 90

1st International Conference on Micro Electro, Opto, Mechanic Systems and Components
Berlin, 10-13 September 1990

With 642 Figures

Springer-Verlag Berlin Heidelberg GmbH

Prof. Dr.-Ing. Herbert Reichl

Technische Universität Berlin
Forschungsschwerpunkt
Technologie der Mikroperipherik
Einsteinufer 17
1000 Berlin 10

Organized by

AMK Berlin
Ausstellungs-Messe-Kongreß GmbH

MESAGO
Messe & Kongreß GmbH Stuttgart

This event is supported by the
West-Berlin Senat, Dept. for Economics

ISBN 978-3-642-45680-0 ISBN 978-3-642-45678-7 (eBook)
DOI 10.1007/978-3-642-45678-7

Softcover reprint of the hardcover 1st edition 1990

2161/3020-543210 - Printed on acid-free paper

Preface

On September 10-13, 1990, the first international meeting on Microsystem Technologies takes place at the Berlin International Congress Center.

Most of the traditional congresses deal with themes that become more and more specific, and only a small part of the scientific world is reflected.

The Micro System Technologies is attempting to take the opposite direction: During the last two decades the development of microelectronics was characterized by a tremendous increase of complexity of integrated circuits. At the same time the fields of microoptics and micromechanics have been developed to an advanced state of the art by the application of thin film and semiconductor technologies.

The trend of the future development is to increase the integration density by combining the microelectronic, microoptic, and micromechanic aspects to new complex multifunctional systems, which are able to comprise sensors, actuators, analogue and digital circuits on the same chip or on multichip-modules.

Microsystems will lead to extensions of the field of microelectronic applications with important technical alterations and can open new considerable markets.

For the realization of economical solutions for microsystems a lot of interdisciplinary cooperation and know-how has to be developed.

New materials for sensitive layers, substrates, conducting, semiconducting, or isolating thin films are the basis for the development of new technologies.

The increasing complexity leads to increasing interaction among electrical and non-electrical quantities.

New CAD and simulation tools are to be developed, which are able to take into account thermal, electrical and mechanical properties of the device. Design for reliability and testing becomes an important part of the technological evolution.

I thank all the authors from 15 countries in Europe, USA and Asia for their contributions, the members of the steering committee for their cooperation, the organizers, AMK Berlin and MESAGO Europe, for their engagement and the local organizing committee for doing the hard work of organizing the conference. In the name of all of them I wish you a successful congress.

September 1990 Herbert Reichl

General Chairman

H. Reichl	Technische Universität Berlin	W. Germany

Technical Programme Chairmen

C. Baack	Heinrich-Hertz-Institut Berlin	W. Germany
A. Heuberger	FhG IMT Berlin	W. Germany
E. Obermeier	Technische Universität Berlin	W. Germany

Steering Committee

H. DeMan	IMEC Leuven	Belgium
M. Esashi	Tohoku University	Japan
B. Freund	Siemens AG Munich	W. Germany
W. Kroy	MBB GmbH Munich	W. Germany
W.H. Ko	Case Western Reserve University	USA
K. Kurzweil	Bull Les-Clayes-Sous-Bois	France
J.P. Lazarri	LETI Grenoble	France
C.R. Lowe	University of Cambridge	England
H. Luginbuhl	CSEM Neuchatel	Switzerland
H. Melchior	ETH Zürich	Switzerland
S. Middelhoek	Delft University of Technology	Netherlands
H. Ohnsorge	SEL AG Stuttgart	W. Germany
A. Paoletti	Universita degli Studi di Rana	Italy
H.-J. Queisser	MPI Stuttgart	W. Germany
M. Sage	BPA Ltd Surrey	England
D. Seipler	Robert Bosch GmbH Reutlingen	W. Germany
H. Smith	MIT Cambridge	USA
S. Sugiyama	Toyota Inc. Aichi-ken	Japan
K.D. Wise	University of Michigan, Ann Arbor	USA
H. Wurmus	Technische Hochschule Ilmenau	E. Germany

Local Organizing Committee

H. Reichl	Technische Universität Berlin
W.Tornow	Technische Universität Berlin
H.J. Griese	Technische Universität Berlin
G. Zimmer	Technische Universität Berlin
H.-P. Nolting	Heinrich-Hertz-Institut Berlin
W. Döldissen	Heinrich-Hertz-Institut Berlin
W. Benecke	FhG IMT Berlin
S. Jakob	FhG IMT Berlin

Table of Contents

Simulation and Design

Design of Micro Systems

CAD of Optical Devices

Simulation and Design - Electrical and Thermal Properties

Material Characterization - Optical Methods

Materials

Materials for Packaging and Passivation

Integrated Optics Materials and Technologies

Technologies

Pattern Generation and Image Transfer

Laser Supported Deposition

Reactive Etching and Deposition

Contact Methods

Packaging

Optical Assembly and Interconnection Technologies

Sensor and Actuator Technologies

Basic Components and Applications

Sensors for Physical Parameters

Gas and Ionic Sensors

Sensors for Biological and Medical Application

Optical Sensors

Optical Components

Actuators

Simulation and Design

Design of Micro Systems

Microsystems: A Challenge for CAD Development

Hugo De Man
Prof. Katholieke Universiteit Leuven
IMEC vzw. , Kapeldreef 75, B-3030 Heverlee, Belgium
Ph. 32/16/281 201/ Fax : 32/16/229 400
Email : deman@imec.be

1. Microsystems : an emerging discipline.

Micro systems have become possible as a result of the fast evolution of VLSI technology. Modern (BI)CMOS technology allows complete systems to be integrated on a single chip while new packaging technologies allow the use of micropattern techniques also on multichip carriers. The same micropattern techniques can also be used for designing micromechanical structures as well as silicon sensors possibly with inclusion of signal conditioning, conversion and interpretation circuitry.
For the discussion on CAD, microsystems can be partitioned into 3 categories :

1. Autonomous signal and data processing systems on a single chip. In almost all cases such micro systems contain analog and real time digital signal processing as well as data interpretation circuits on the same chip. We will call them micro system in restricted sense.

2. Smart sensor / actuator systems. Silicon sensor / actuator together with above signal processing on the same chip. Microsystem in extended sense.

3. Multi chip microsystems interconnected on multi chip modules as carriers. These systems are mixtures of analog and digital SP as well as microcontrolers, host CPU's and smart sensors.

2. Actual CAD situation and problem areas

Although in principle all these microsystems are technologically feasible, it is clear that designing them may be the true bottleneck for the exploitation of their potential.
CAD can obviously be of help to shorten the design cost of these complex systems. In the sequel we willl investigate what todays CAD means in this respect.
It is this authors opinion that the design of microsystems goes far beyond the capabilities of actual CAD systems. Therefore we have to ponder in what direction future CAD must be developed in order to cope with this important new challenge.
It will be discussed how the new CAD required will be much more design knowledge oriented than todays tool oriented CAD.
Therefore such *CAD* systems will have *to be developed by designers themselves* and although this sounds exciting, designers are usually *not the best software people* and that may be one of the biggest obstacles to the future of design. Traditional CAD may only be restricted to the delivery of the framework into which tools and methodologies are to be embedded. The succesfullness of future CAD systems will be measured on their friendly openess to creative designers. That leaves still quite some way to go !
When evaluating todays available CAD systems with respect to the above three categories we can make the following statements :

1. Best served is category 1. , little is available for 2. and virtually nothing for 3. No total integration for any of these categories is available.

2. CAD for category 1. is best developed to support *perspirational* design of digital systems and analog components (opamps, comparators, convertors). Perspirational means low level design starting from schematics entry (digital) respectively polygons (analog).
This contrasts sharply with *inspirational design* i.e. design whereby *conceptual specifications,* after validation are interactively *synthesised* into silicon (subchips).
Key issues here are requirements capturing, concept formalisation, and interactive synthesis rather than analysis.
In the **digital area** of synchronous systems there is now a gradual tendency and acceptance of logic and register transfer level *synthesis* of small controller parts and of *silicon assembly* of data path components. Most digital designers *accept* a limited set of parameterisable, precharacterised library components and the use of synchronous sequential logic synthesis tools to create synchronous digital subsystems in much the same way as PCB's were designed before .
Impressive results in microcontrollers and signal processors designed by traditional system designers using library based commercial CAD systems have been reported.
Problem areas in digital CAD have therefore moved upwards on the abstraction scale to architectural synthesis from algorithmic specification of signal processing or from state transition diagrams of micro controlled systems.
A very often forgotten area is *automated synthesis of (a)synchronous interfaces* to provide the on-chip race and hazard free link to bus oriented external systems, which invariably show up in most microsystem designs. Some recent work has been reported here but it is amazing *how little attention the computer science oriented CAD developers pay to real life system designers problems....*
Other problem areas, more related to catagorie 2. and 3. will be discussed below.
CAD for **analog subsystem design** is far behind the digital counterpart on the abstraction scale. Where subsystem level synthesis at micro architectural level has made its entry in the digital CAD world, analog design is barely breaking away from the transistor level.
In daily practice, the most intensively used tool set consists of polygon pushers, SPICE like circuit simulation and DRC. Recently also some switched capacitor simulation and filter synthesis tools have emerged.
This has several reasons. First there is the *belief* that *analog design is a black art* and that the use of analog libraries is not possible " because there is always a spec that can be made better or is not satisfied (*at that level of abstraction*)" . Obviously the analog specs are much more complex than digital specifications. *Catching, managing, transferring and validating these specifications of mixed analog digital systems is a real bottleneck* to microsystem design, even for category 1. This sounds like digital 10 years ago.
The major (*US based*) CAD vendors do*not* have enough *in house design experience* to drive adequate developments in this area. A chance for Europe (to cooperate with them if they want to listen) ?
The obvious trend therefore is to go high performance digital as soon as possible and limit analog signal processing to the interface between the sensor and the DSP - MCU part.
That is however not alleviating the requirements to analog CAD as most sensors require high precision, low noise electronics in a louzy digital CMOS process where *no* real *semi-custom analog design style exits (yet ?).*

In the research area of analog CAD two trends are emerging :

- Analog module generators for so called *fixed topologies* of analog functional building blocks such as op-amps, comparators, oversampled convertors, flash converters, switched capacitor filters etc...

These are somewhat similar to the *silicon assemblers* now well accepted by digital designers. Analog design then must make use of this *partly parameterisable library*. Module layout is generated for a given technology file from a specification sheet by a *module specific program* that *encapsulates* the design equations and *the design expertise* for that module.
The analog subsystem designer then must accept the limitations imposed by these fixed topologies, just as he/she had to live with good old catalogi of discrete analog components.

A particular problem here is the lack of good automated place and route tools that take care of the special requirements of analog design (matching, balancing, crosstalk...).

- Today, the *analog silicon wizzard* seldom agrees with the solutions generated by these expert programs and therefore prefers *non-fixed topology* oriented systems.

Such systems are now at the prototype level in the CAD research community.
The non fixed topology systems can be described as *open declarative programming environments* which allow a designer to enter his / her design knowledge in a user friendly way such that next time the problem reappears, all design expertise can be reused.

It is an open question what will happen in the future.
This author believes that, as much more design knowledge will be captured in these future CAD systems and as more and more problems will arise at the *analog systems level rather than the analog building block (module) level*, more acceptance will occur in the use of the library modules *without too much concern about their content.* That may indeed serve quite a substantial part of the system designer community and is similar to the actual situation in the digital world.
The remaining *hard problems, requiring solutions at the border of technological possibilities* will then be tackled by the non-fixed topology CAD systems by expert designers and their accumulated experience then will become available later to the system level designer at the higher level of abstraction.

This evolution is also taking place in the digital world where actual research is leading to *open frameworks for synthesis into which* design expertise for broad classes of applications and their corresponding architectural design scenarios are stored into an expert designer *scenario shell.* This scenario shell will be embedded into a *specification and planning shell* for system design.
The scenario shell itself operates on a *set of synthesis, analysis and optimisation tools* to be called from the scenario shell.
Finally the tools must operate on the inner kernel which is the *object oriented data model of the design paradigm under consideration.*
This author believes that this kind of *design centered CAD environments* as opposed to *tool centered CAD frameworks* must be the answer to the emerging design crisis that will hamper the development of microsystems.
This kind of CAD will require a *rethinking of the CAD industry.* This author indeed believes that the traditional CAD industry will become *the supplier of the frameworks of design centered CAD systems* as well as of a set of *generic perspirational tools.* However the inspirational tools will have to be developed by the expert designer community. How this expertise will be made available to the many potential design houses for microsystems remains an open question and may give rise to new industrial opportunities.
In any case, this observation may be of great importance to the academic world which in this paradigm is offered a unique opportunity of creativity in the *in the formalisation of design knowledge rather than building the nth perpirational tool which will never make it to the CAD market, dominated by only two major US players. Therefore it is of utmost importance that commercial CAD is freely available to the academic community, not only for learning how to use it but to add value to these systems by implementing design formalisation when these systems will be open enough to accept it easily (see below).*
In the same light, there is a definite need to abolish exclusive CAD research or exclusive CAD conferences. CAD research and conferences must be merged with design formalisation itself. The confrontation of design and the tools for it can only lead to more meaningful results.
If this happens, there will be a better way to identify a number of new tools needed in order to satisfy the microsystems need. Indeed most of the *existing tools* are semi-custom logic control system design oriented but *leave a lot to be desired if real time mixed analog-digital signal processing is envisioned.*
A lot of this is caused by the fact that CAD development is too much dominated by pure computer science engineers who obviously understand how a computer works but have little

understanding of the intricacies of signal processing in the broad sense. A clear case for breadth of education !

Some of the missing tools are :

- High level specification tools allowing multi-paradigm descriptions with well defined semantics.

- Based on the above : multi-paradigm simulation, including frequency, noise, distortion and digital-analog interaction. It is believed that this kind of simulation will only be possible on some special hardware architecture which will be much closer to multi-DSP than to traditional workstation architectures.

- Merger of the analog and digital CAD systems into a unified CAD framework (in the sense indicated above). Today indeed analog and digital CAD live a separate life, also as software systems. In the microsystem world however both are intimately mixed and this mixture must also show up naturally to the microsystem designer .

3. When looking at category 2. where sensors are integrated together with signal processing electronics, much more sophisticated problems arise for which no integrated CAD tools nor CAD environments do exist today. Only some experimental systems emerge in some research institutes.
Clearly, if analog design is emerging from the era of black art and more and more design formalisation is occuring, this is certainly not yet the case for silicon sensor design. This is indeed a relatively young engineering branch, where design is based on the use of *physical effects* not necessarily modeled by any of the existing traditional CAD tools.
CAD tools able to help create micromechanical and electrochemical sensors must be able to *model both the processing steps and the physico chemical, thermal and opto electronic effects in the silicon (or GaAs?) medium.* They must also provide the means to create *electrical models linking the sensor or actuator signal to the properties of the electrical signal going to the interface electronics* envisioned in category 1.
Therefore these tools, if they emerge, *must be integrated in the same CAD framework as the one used for category 1.*
It is clear that the discipline from which these tools must emerge is from the *device and process modeling CAD tools.*
These tools are at present mainly developed to study effects in advanced submicron structures for the few major semiconductor manufacturers still remaining on the globe who can afford this technology. *However there might be a much greater challenge and a much greater market potential* if a lot of this research were oriented towards the effects occurring in the microsystems structures. The challenges are at least as great due to the much higher variety of physical, chemical , optical and electromagnetic interactions !
On the other hand it would be nice if the designers of these programs tried to implement their CAD tools in the same *designer oriented environment as the traditional CAD systems.*
In summary we hope to see more device research in the academic world to turn to this sort of device and process modeling which ultimately might address *a much wider customer base as the three remaining players in the mega memory business.*
Ultimately this might be of great importance for Europe, in view of its strength in high quality consumer electronics (audio, video), intelligent machine construction as well as automotive and telecom industry.
The challenges are certainly not smaller !

4. Finally, one might wonder what to do with the multi chip modules (MCM) which in the future will offer the opportunity for thight integration of an heterogeneous system with an economically most optimum use of ASIC's, (category 1 and 2) as well as firmware (PLD's), microcontrollers, DSP processors (including CSIC's) and host microprocessors. The latter categories include the famous hardware / software trade-off in real time system implementations. All of these naked chips are then thightly integrated on the MCM carrier.

Such systems are very likely to occur in portable telecom (personal communication system), automotive systems, consumer multimedia systems etc...
What is the CAD available here ? Unfortunately virtually nothing...
There are multiple reasons for this :

- MCM is an emerging technology which is not just another way of doing PCB's. Still a lot of understanding of the interaction of chip I/O with the new interconnection medium is necessary.

- There is no formalised design methodology for these complex systems. Industry just tries to meet thight schedules to the market and universities have too small, too individualistic groups to understand all aspects of the problem and to be able to generalise concepts in design to be implemented into a new CAD system (not only tools !).

- CAD vendors are too far remote from the day to day system design problems and it is not clear whether there will be a return on investment for them in this highly service oriented activity.

- A formal technique can only be successful if it results in some form of standard knowledge encapsulation and standardisation on component specification ...will all component vendors adhere to JTAG testability, make VHDL (??) or Verilog (??) models available, provide computer executable timing specs etc...etc...

- Programmable DSP and micro controller programming is in the state microprocessor programming was 15 years ago...efficient compilers tuned to the particularity of these processor architectures are necessary.

- Virtually no CAD is available to design analog systems (above the module level). Especially tools are necessary to decompose system level specs into optimised subspecs for analog module generators as are arising from the actual CAD research mentioned for category 1.

The above may be interpreted as a rather pessimistic statement or as a stimulus for searching creative solutions. The task is enormous. Executing it is as vital to the application of VLSI submicron technology as the technology itself...maybe even more...who else than the systems people are going to absorb the enormous number of transistors technology is able to produce by the end of the decade?
System oriented CAD will have to face the following tasks :

- Full system specification of multi-paradigm signal types and their attributes in a formally well defined semantics allowing execution (emulation is preferrable over simulation) , formal verification and synthesis.

- Automated interactive partitioning in ASIC's, CSIC's, PLD, MCU, DSP, µP including hardware software trade-off.

- Full design management and estimators of design alternatives. Automatic subspec derivation.

- Highly interactive high level architectural synthesis techniques for ASIC design by the system designer.

- Automated test assembly at ASIC level as well as the MCM level

- Full performance characterisation of ASIC components and performance evaluation tools coupled to MCM place and route. Notice that the latter two aspects are not possible without vendor standardisation on commodity component specification *amenable to CAD frameworks...*

- Integration of the sensor and actuator design CAD to emerge from the design and process modeling into the CAD framework. A serious question arising here is whether there will be silicon "foundries" (expensive places!) available to the highly specialised micro patterning for silicon sensors and actuators, especially if the latter have to contain also advanced (BI)CMOS processing for adding the "smart" part of the components of category 2. Is this a chance and justification for specialised technology oriented institutes ?

Tackling the above problems is not just another Ph.D. project. It requires a big concerted effort which needs the willingness to join forces of both academia and industry around major themes rather than scattered activities. Maybe EC or JESSI projects can stimulate this although unfortunately from an outside observation there seems to be quite a lack of unified centralised vision on the design aspects...It is not because a miriad of applications are done once that a formal design methodology and its associated CAD system will emerge !

3. Is CAD the only salvation for microsystem design ?

Obviously CAD is important for alleviating the design crisis to be expected in the microsystems area. However, lack of CAD tools is all too often seen as an explanation why not more of the VLSI technology is exploited into systems.
It is this authors opinion that the key problem is often not the lack of tools *but the lack of multidisciplinary education in systematic design.*
Education in design is indeed a lot more than having access to a workstation having commercial CAD on it (which takes 3 months to master anyway)...let's not make computer hackers but designers understanding systematic formalised design of microsystems (even without tools...).
Most universities have system design, computer science and chip design oriented activities without any joint projects or communication.In CAD research, many universities are still trying to invent perspirational CAD tools often already available from the CAD vendors. The emergence of commercial frameworks, professional object oriented programming environments (C++), the universal use of the UNIX programming environment and a more open free exchange of research grade software (cfr US) must be established in Europe.
Then universities should be funded *only* for projects involving the use of available tools *and* where at least three *different* subdisciplines propose research around a theme involving the *formalisation* of *design activity* based on challenging problems from the systems industry.
By doing so, universities are not only creating students highly specialised in CAD algorithms, but individuals who have learned how to function within a *multidisciplinary team* and who are *familiar with the aspects of design itself.* This will be of utmost importance if indeed CAD systems evolve to design centered CAD systems rather than tool centered CAD systems.

4. Conclusions.

Microsystem design is an emerging new discipline combining all exciting aspects of electronics system design and implemeting them in advanced silicon technology.
Microsystems are of extreme importance for the European systems industry and its semiconductor industry since Europe does not have the outlet of the memory market nor the computer market. Launching major activities in the latter areas is too little and too late.
However Europe still has a good position in consumer, telecom, machine construction and automotive systems and has a good tradition of practical system theory and design.
Clearly it is in this direction that we should complement the perspirational CAD developments supplied by the major US CAD vendors. This should be done by installing our proprietary system design knowledge on these systems but that requires the willingnes to create multidisciplinary research activities which are design centered and lead to a competitive edge by being able to exploit the VLSI technology of the nineties.

A Specification Driven Environment for Microsystem Design

K.D. Müller-Glaser, H. Rauch, W. Wolz, J. Bortolazzi, C. Kuntzsch, R. Zippelius

University of Erlangen-Nürnberg
West Germany

Summary
The design of microsystems will not be realizable without new computer aided design tools. Special emphasis is necessary on CAD for early system design phases, namely feasibility study, design management and specification, as well as on new requirements for simulation tools and modelling. Furthermore, test strategy planning and development of microsystem specific test equipment mandates particular efforts. The DEBYS system, currently under development at the University of Erlangen-Nürnberg, provides an approach to a specification-driven environment for microsystem design.

Introduction

Some of the main markets of microelectronic components, e.g. consumer electronics, communication equipment, automotive and industrial control, show an increasing demand for true system integration. True system integration means integrating a complete system (e.g. an internal cardiac pacemaker) or a large part of a system in a single integrated circuit, a small ceramic substrate or printed circuit board incorporating not only digital functions, but also analog functions and optoelectronic components as well as "intelligent" sensors and actuators and in the near future even micromechanical components (Fig.1).

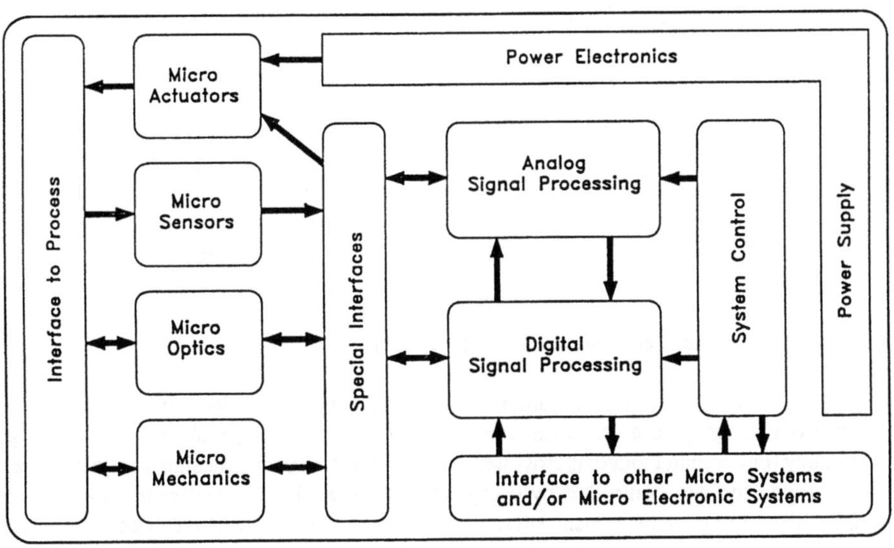

Fig.1 Principal Structure of a Microsystem

Although the design of complete systems has been accomplished for several years, the miniaturization and integration of system components with the resulting decrease of device dimensions and increase of compaction mandates the consideration of parasitic effects and device interactions. These microsystem specific effects include environment dependencies of actuator or sensor characteristics, thermal coupling of these components and digital or analog blocks with a high power dissipation on one silicon substrate or ceramic carrier, electromagnetic compatibility and reliability of safety-related systems such as pacemakers or automotive ABS. These interactions and the complexity of both system and system components mandates an exhaustive use of CAE/CAD/CAD tools. Even today a variety of specific computer-aided design, analysis and test tools provide support for the development of components from the digital, analog, or mechanical domain. Nevertheless, the design of microsystems mandates an integrated design environment that provides common data management, system level design support and interface description as well as a "common look and feel" user interface to provide effective man-machine interaction. Furthermore, an essential extension of computer support towards computer aided assistance of the early design phases of feasibility study, requirement specification, environment description, and design management is necessary to increase first silicon success rates, design productivity and quality. The goal is to support not only domain specific, e.g. electronic system design, but to provide an environment for general system design (Fig.2).

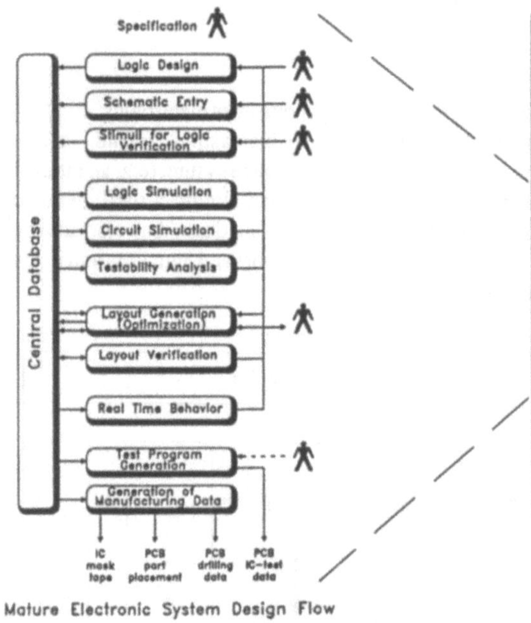

Phase			Content	Result
1	Study		Definition of the requirements as target	Requirement catalogue
2	Specification		Development of a solution concept and the test method	Specification - object-specification - test-specification
3	Systemdesign		Architecture development functional definition module structure test environment	Object structure: architecture structure of functions module structure Test conditions: test system test environment
4	Implementation		Construction of modules and test profiles (coding, compilation)	Tested module
5	Integration and test		Module integration and design verification of the complete object	Prototype
6	Production	HW	Transfer to product management and production	Released product
		SW	Transfer to product management	

General System Design Flow
(Design of VLSI Circuits
E. Hoerbst et al, 1987)

Fig.2 Design Flow for Electronic and General System

The DEBYS (DEsign BY Specification) system /1/, currently under development at the University of Erlangen, represents an approach to the realization of a specification driven design methodology for computer aided microsystem design. Domain specific design, analysis and test tools are supported and controlled by computer aided specification and estimation and are embedded in a design management environment (Fig.3). DEBYS includes commercial and proprietary tools as well as a combined test system for internal and external test.

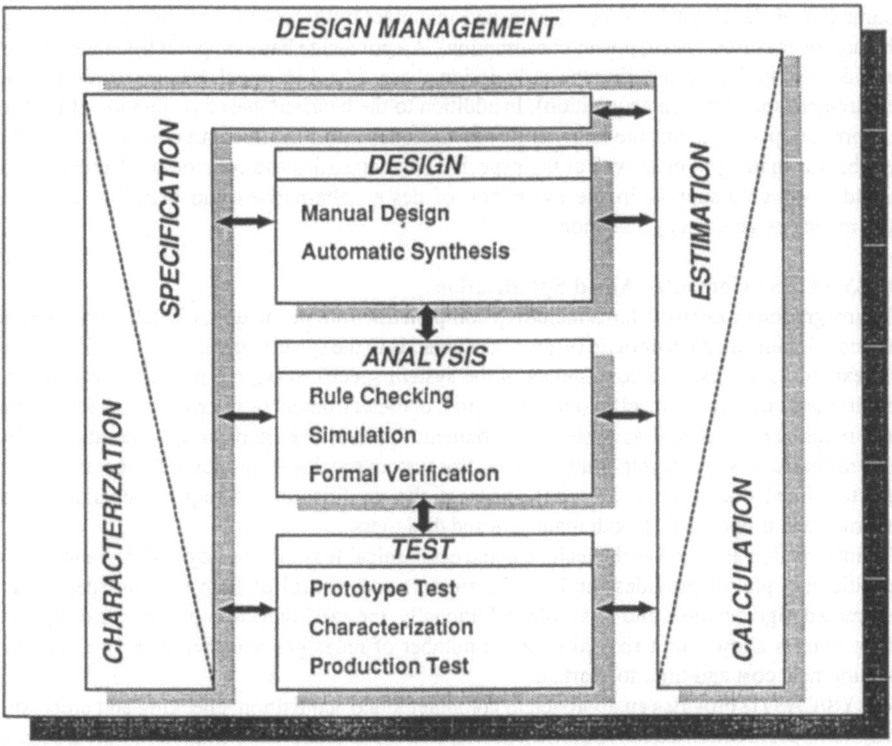

Fig.3 The DEBYS Concept for an Integrated Microsystem Design Environment

DEBYS - DEsign BY Specification

The DEBYS project is intended to develop an integrated design, analysis and test system for microsystem design based on a specification-driven design methodology. Design, analysis and test are controlled and supported by computer aided specification and estimation of essential design characteristics and embedded in a design management environment (Fig. 1). DEBYS combines commercial tools (e.g. Silvar Lisco SDS, CASS, CAL-MP, PRINCESS; ZYCAD System VHDL; Silicon Compiler Systems GDT; Isdata LOG/iC; HHB Cadat and Intelligen; GenRad System Hilo and HiTest; TSSI TDS; Analogy Inc. SABER; SPICE) and proprietary tools where necessary as well as an integrated test environment consisting of a TEK LT 1001 VLSI test system and a ICT 9030 electron beam tester. The key aspects of the DEBYS development are described in the following sections.

Design Management and Computer Aided Estimation

An essential criterion for the effectivity and quality of a design process is project planning and control of design phase coordination. The DEBYS design management includes project management (resource planning and scheduling of design steps), design flow control and cost estimation. An integrated knowledge base includes information about resources, personel, heuristics about design time and an automatic comparison of actual and desired project data. Based on the results of computer aided estimation and specification, alternative project plans are generated and evaluated. The early evaluation of technical and economical feasibility requires an estimation of development time, design and manufacturing cost as well as essential design characteristics (area, speed, power consumption). A knowledge based approach enables computer-aided estimation even during the early design phases (i.e. functional system partitioning and requirement specification acquisition). In addition to the heuristic-based evaluation of more or less precise specifications, the knowledge base includes information about components that are suitable for an integration as well as their specific characteristics and constraints. The estimation provides valuable support in the evaluation of design alternatives and therefore enables a systematic design strategy selection.

DEBYS/CAS - Computer Aided Specification

The integration of microsystems including components from the analog, digital, optoelectronical and micromechanical domain offers high demands to the system specification quality. In this context, completeness and consistency of the system specification, the propagation of system requirements to domain specific (analog, digital, optoelectronical or micromechanical) component requirements as well as system environment dependencies and the consideration of test system characteristics are important aspects. The system specification should be accessible, well structured and documented. Last but not least, the modification of requirements should be recorded and propagated to both managers and designers.

The information acquired in the technical and economical feasibility study and the requirement specification phases provides the base for support and control of both domain specific and common design, analysis and test tools. Additionally, the early detection of possible design and test problems allows for a reduction of the number of redesigns and therefore a reduction of development cost and time-to-market.

DEBYS/CAS /1/ provides an approach to computer aided acquisition, checking and processing of system specification and environment description for microsystems and microsystem components. The necessity to cope with highly related, partwise uncertain and incomplete specifications as well as to provide consistency checking based on heuristic methods mandates a knowledge based approach. To provide effective man-machine interaction, adequate concepts with respect to acceptance and ease of use have to be developed. The main features of DEBYS/CAS are:

Acquisition of environment description and requirement specification. The acquisition is based on the six specification areas shown in table 1. The specification data is interactively acquired using a comfortable graphical interface, that offers effective interaction concepts for certain parts of the specification (system partitioning, timing specification). The specification is represented as a semantic network of predefined objects and relationships that includes all relevant elements, properties and relationships. Due to different domain specific description forms, DEBYS/CAS provides the representation of min-typ-max specifications, tolerance bands, time-frequency domain specifications and dependencies as well as default values, natural language comments, parameter definitions and a description on different levels of certainty (hard and soft requirements) /2/.

System level design includes the successive partitioning of the design, the classification of components (software, hardware, analog, digital, optoelectronical, etc.), the connectivity and a unified description of interfaces, i.e. the definition and classification of module ports. The knowledge base contains information about existing modules, for example a library of actuators/ sensors for the acquisition of specific physical, chemical, or optical parameters, the conditions and constraints for their integration and available macromodels. Last but not least, the acquisition and management of natural language descriptions provides the base for the automatic generation of structured and detailed documentations.

Interface description	Behavioral description	Body description	System environment	Test environment	Design management
Port definition Maximum ratings Operating conditions Power constraints Package Interchip critical paths Static/dynamic/ transfer specifications Tolerances	Function definition Input/output relationships High level simulation model: Logic expressions Transfer functions VHDL Schematic Entry Analog HDL Templates	Partitioning Hardware Software Component definition Hierarchy Component and interface connection Testability strategy Stimuli generation strategy	Temperature Humidity Vibration Acceleration Shock Radiation ESD, EMC Reliability	ATE equipment Measurement accuracy Tester correlations Safety margins Production test Quality assurance AQL, PPM,ZDS	Timing and manpower schedules Design team management and responsibilities Price performance estimation

Tab.1 The DEBYS/CAS Specification Areas

Early and simulation independent checking of specification data. A rule-based representation of expert knowledge enables the evaluation of specification data with respect to completeness and consistency. This includes rule based checking of feasibility under consideration of technical and economical aspects as well as the consideration of technology and manufacturer-specific data. The rule based approach provides the possibility to describe interaction between modules, such as temperature dependency of characteristic curves or supply voltage sensitivity (influence of digital building blocks). A detailed description of the available test system allows for the early detection of possible test problems and provides support for the development of additional test and measurement systems. The rule base is flexible and extensible to receive application, technology or manufacturer specific data.
The implementation of DEBYS/CAS is based on the object-oriented data management system IDM /3/ and the programming languages C and Prolog. The rule base and essential parts of the system control are implemented in Prolog. The user interface is based on X/Window/Motif /1/.

Support of Design, Analysis and Test

Specification-Driven Support and Control

The structured representation of complete and consistent specifications provides the base for support and control of design, analysis and test. Based on the detection of possible test problems, the concurrent development of test and measurement systems is supported. Design tools, especially for automatic synthesis are supplied with specific input data and are controlled by the constraints included in the requirement specification and the environment description. The partitioning of the system into functional blocks enables the identification of frequently used, standard functions with a given set of specification parameters, such as digital counters, registers, ALU's or analog bulding blocks such as operational amplifiers, A/D and D/A converters. As analog and digital building blocks used within microsystems typically do not stress the limits of technology and complexity, available synthesis tools as described in /4/ for the digital domain and /5/ for the analog domain can be used to increase design productivity.
The simulation of microsystems and microsystem components is supported by an automatic setup of simulation environments including environmental conditions (temperature, load, etc., system and test specific) and by an automatic comparison of simulation results with given requirements

based on the SIMUEVA evaluation tool /6/. Further essential relationships are recognized through specification-based estimation and project management.

Simulation of Microsystems

The consideration of complex interactions not only in the electrical, but also the thermal, optical and mechanical domain provides one of the major problems for the simulation of microsystems. The modeling of such interactions demands interdomain expert knowledge and powerful modeling languages. However, as microsystem prototypes and the necessary test and measurement equipment are very expensive, efforts towards the development of sophisticated microsystem simulation approaches are mandatory. Today there are two major strategies to cope with the increasing system complexity: usage of more powerful computers, especially dedicated hardware (accelerators) or simulation restricted to certain aspects or views of a system. Typically, complex designs are described on different levels of abstraction. This approach enables an evident decrease of the system description complexity by considering a restricted set of aspects on each level of abstraction. Accordingly, the simulation of microsystems necessitates a reduction of model complexity and therefore demands simulation on different levels of abstraction, each focussing on a specific aspect of system behavior. For this reason, level specific modeling approaches have to be used to provide only the necessary accuracy of a model. Furthermore, system simulation should include the possibility to simulate the interaction of system components that are described on different levels of abstraction (mixed level simulation). An example of a hierarchical description of digital components and the corresponding simulation levels is illustrated in figure 4.

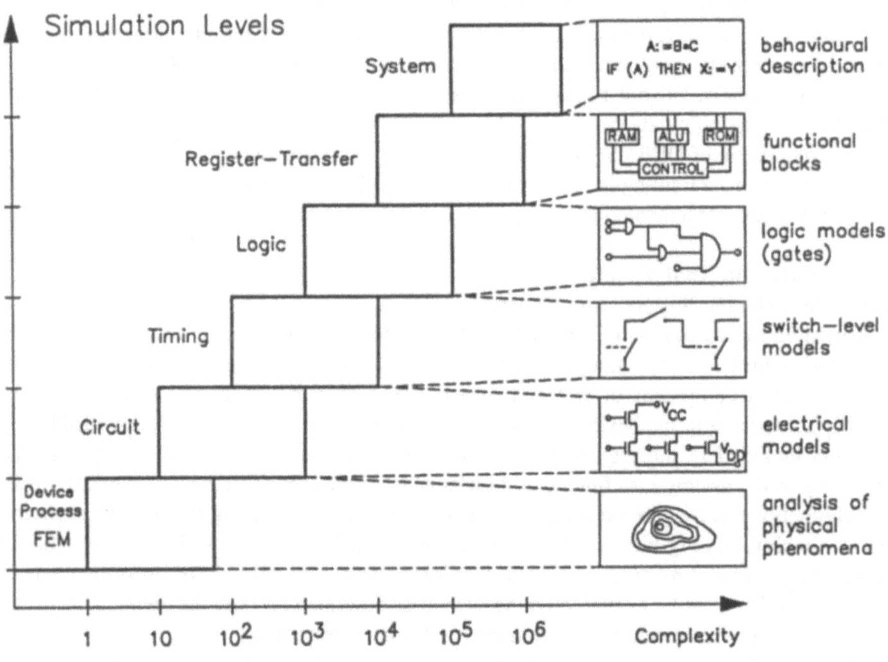

Fig.4 Simulation Hierarchy and Appropriate Simulators

The system level simulation concentrates on the functional simulation without consideration of structural information. Register transfer (RT) and logic level simulation both uses functional and structural information, whereby functional units on different levels of abstraction (RT level, gate level) are used. As the timing behavior gains more and more importance, the timing simulation allows for a detailed timing analysis. Whereas the former levels of simulation are based on discrete quantities (time, voltage), the transition to circuit level simulation leads to continous quantities and therfore enables the simulation of both digital and analog components. Finally, the device level simulation includes three dimensional models and field theory. On this level, FEMs (Finite Element Methods) provide a description of complex relationships with a tolerable effort and are currently used especially in the mechanical domain. This multi-level approch allows the simulation of electronical systems with more than a million devices as well as the detailed analysis of passive and active devices in a reasonable time.
The practical problems to be solved for microsystem simulation are essentially based on differences of domain specific descriptions, the complexity of interactions and the lack of experience, especially concerning interdomain relationships in this area. Obviously, the time range reaching from nanoseconds for digital to minutes or hours for thermal or mechanical processes represents an important requirement to the capabilities of the simulation environment. Due to restrictions of the hierarchical approach to system simulation on different levels of abstraction, fault simulation and timing analysis including back annotation of data extracted from layout (such as precise line delay) mandates extended simulation runs and therefore the use of dedicated hardware. A unified approach to the description, verification and maintenance of models and model libraries is necessary. Incremental compilation of models should be used to allow for a reduction of compilation time. The description of the interaction of system components belonging to different domains such as the quantisation of analog signals within digital components, demands a unified interface description. For high frequency applications, models for physical lines including for example crosstalk or reflections are necessary. The setup of microsystem simulation environments includes the consideration of system and test environment characteristics, a unified stimuli and expected response description to allow the use of different simulators as well as computer support for the evaluation of simulation results.

Test of Microsystems

The capacity of microsystem technologies to integrate micromechanics as well as analog and digital circuitry on a single chip dramatically increases the test problem. As microsystem test cannot be covered using standard test equipment, new methods and tools are necessary to generate analog/digital electrical, optical or mechanical/acoustical test stimuli and expected responses for microsystems, and to allow for the measurement of test coverage in a given mixed-technology case. To keep pace with the soaring capabilities of microsystem technology, microsystem test aspects have to be addressed early during the design cycle in order to circumvent the test problem bottleneck (i.e. 'design for testability' rules have a major impact on microsystem designs). The design process has to be a concurrent, hierarchical development of both the microsystem and the appropriate test system. In addition, highly sophisticated diagnostic tools need to be developed for prototype test, quality assurance, production test, field maintenance test and fault localization purposes. Partitioning of microsystems solves the complexity problem partly, as it eases the test of embedded modules. There exist only preliminary concepts to describe the different types of signals and their availability during the test (e.g. by using electron/optical beam testers in combination with close links to appropriate simulators and standard analog/digital testers).
An extended database that includes test-oriented information and knowledge on the production process supports the design of test-procedures for all phases of the production cycle. Tester

hardware, to date aimed at the test of pure electrical signals, has to be upgraded such that optical and acoustical/mechanical stimuli can be generated and device responses can be acquired and compared against expected responses. In addition, test systems have to deal with cycle times and clock periods with differences up to ten orders of magnitude (ps to hours or days).

Experimental Results

First valuable experimental results where gained during the design of a 7400 gate integrated processor control unit, a Forth RISC microprocessor core and a high speed gallium arsenide A to D converter.

Acknowledgements

This research has been partly supported by the Fraunhofer-Gesellschaft München and the Bayerische Staatsministerium für Wirtschaft und Verkehr.

References

/1/ K.D. Müller-Glaser, J. Bortolazzi: "An Approach to Computer Aided Specification", in IEEE Journal of Solid-State Circuits Vol. 25 NO. 2, April 1990 pp.335-345

/2/ J. Bortolazzi, K.D. Müller-Glaser: "Towards Computer Aided Specification of Analog Components", in Proc. IEEE Custom Integrated Circuits Conference 1990, pp.10.7.1-10.7.4

/3/ "CADLAB Object Management System Release 2.1 - IDM - An Overview of the Data Model", Internal report, CADLAB, Bahnhofstr.32, D-4790 Paderborn

/4/ R. Zavala, K.D. Müller-Glaser: "GENSYS - A Hierarchical VHDL Generator System", to be published 1990

/5/ M.G.R. Degrauwe e.a.: "Towards an Analog System Design Environment", in IEEE Journal of Solid-State Circuits Vol. 24 NO. 3 June 1989 pp.659-671

/6/ R. Buschke, K.Lagemann: "An Approach to Understanding Evaluation of Simulation Results as an Integrated Task", in Proc. IEEE Custom Integrated Circuits Conference 1989, pp.13.6.1-13.6.4

Conceptual Design of Micro-Electro-Mechanical Systems

Selden Crary* and Sridhar Kota†

*Center for Integrated Sensors and Circuits
Electrical Engineering and Computer Science Department
†Mechanical Engineering Design Laboratory
Department of Mechanical Engineering and Applied Mechanics
The University of Michigan, Ann Arbor, Michigan 48109-2122, USA

ABSTRACT

We introduce a new conceptual-design methodology for the qualitative-motion synthesis of micro-electro-mechanical systems (MEMS). The approach abstracts the functionality of MEMS building blocks as matrices and allows for the symbolic synthesis of systems via matrix operations. This work is an extension of research on conceptual design of machines and mechanisms using qualitative reasoning and artificial intelligence. The fundamental motivation is to extract the logical steps involved in the creative design so that it is possible to trace the design process using computers.

INTRODUCTION

Micro-electro-mechanical devices that have been fabricated using silicon processing techniques include micro-scale pins, joints, sliders, cantilevers, beams, diaphragms, springs, bearings, rotors, racks, pinions, tweezers, torsion bars, linear and rotary motors, and even binary logic devices. As this MEMS-device menagerie has grown, so has the need for appropriate CAD tools that will allow engineers to design micro-electro-mechanical systems accurately and efficiently. In response to this need, a few centers have recently announced research programs directed toward the development of integrated CAD tools for MEMS. In particular, Senturia [1] in 1987 outlined an approach to this area based on an analogy with integrated-circuit CAD. The emphasis of this approach, and of its subsequent early stages of implementation [2], is on predicting the mechanical properties of microsensors based upon knowledge of the process sequences. More recently, Crary and Zhang [3] introduced a CAD architecture, known as CAEMEMS, that is similar in many respects to Senturia's. However, in contrast to the Senturia approach, the first implementation of the architecture, CAEMEMS-D [4], takes device geometries, rather than process sequences, as the inputs for predicting mechanical properties of MEMS diaphragm structures. One advantage of this approach is that it skips the difficult process-modeling step in order to provide a useful CAD module in the short term.

In this paper we take another starting point for our discussion of CAD for MEMS, one provided by the engineering design hierarchy advanced by Pahl and Beitz [5] and others, as shown diagrammatically in Figure 1. In this

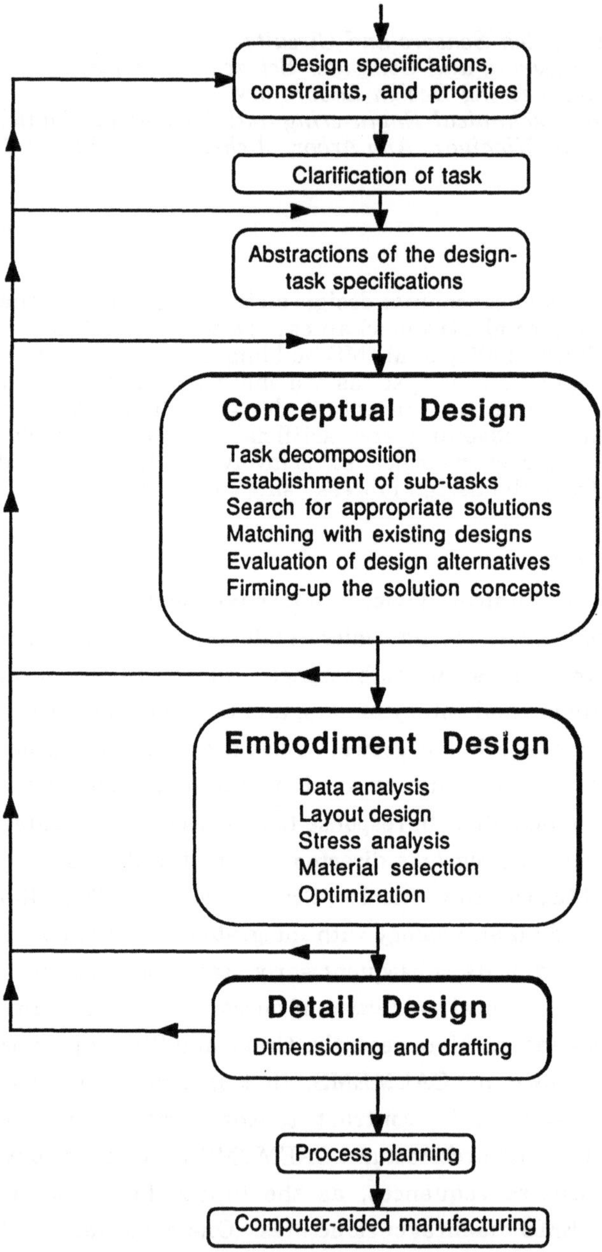

Figure 1. Flowchart of the engineering-design process (after Pahl and Beitz [5]

scheme the previously mentioned CAD architectures and early implementations of Senturia et al. and Crary et al. are difficult to classify simply, but are essentially examples of CAD tools performing tasks principally in the *detail design* and *embodiment design* phases. We believe it is now important to consider also how CAD tools may be developed for the tasks in the *conceptual design* phase, and we begin by establishing a rational basis for conceptual design for MEMS, demonstrated with simple examples. The extension to more complex systems is straightforward.

CONCEPTUAL DESIGN

Conceptual design is based on an appropriate *means of abstracting functionality,* whether it be the relatively simple functionality of an individual micro-mechanical widget or the more complex functionality of a micro-electro-mechanical system. Of several alternative approaches, we have chosen the parsimonious matrix formulation of Kota [6] and have applied it to the micro-mechanical realm. In this approach, the functionality of a MEMS primitive mechanism is expressed as a concatenation of matrices, including a motion transformation matrix (MTM) and a sequence of constraint matrices. We have now catalogued most of the published micro-electro-mechanical primitive mechanisms as matrices using this formulation. These abstractions are perhaps best understood with two simple examples.

EXAMPLE 1 - Micro-tongs

The micro-tongs reported by Mehregany, Gabriel, and Trimmer [7], which are shown in Figure 2, are a primitive mechanism that linearly transforms translation to rotation. The corresponding MTM and constraint matrices for this mechanism are also given in Figure 2. The first 6x6 matrix is the MTM and is constructed in such a way that it abstracts the couplings between the output rotations and translations (the R_o's and T_o's) and the respective input rotations and translations (the R_i's and T_i's) for three cartesian coordinates (subscripted by x, y, and z) through the usual rules of matrix multiplication. The 1's represent all the motion transformations that are possible with this class of devices. The transformation from $T_y \rightarrow R_z$ happens to be just one of them. The other two 6x6 matrices are constraint matrices. The first of these indicates the reversibility or irreversibility of the allowed couplings, which in this simple case are all reversible. The other indicates whether the responses to uni-directional inputs result in uni-directional (or in the case of rotation, uni-chiral) or bi-directional (bi-chiral) outputs, as indicated by the presence of singly- or doubly-signed matrix elements corresponding to the appropriate couplings. In terms of MEMS, a comb-structured rectilinear motor could, in principle, be used to drive the micro-tongs. In this case, the comb structure with its integral rack represents the linear, reciprocating actuator. In reality, as indicated by the 1's in the MTM, a generalized rack-and-pinion can be used to convert a translational motion along any direction within the x-y plane into rotational motion about the z-axis.

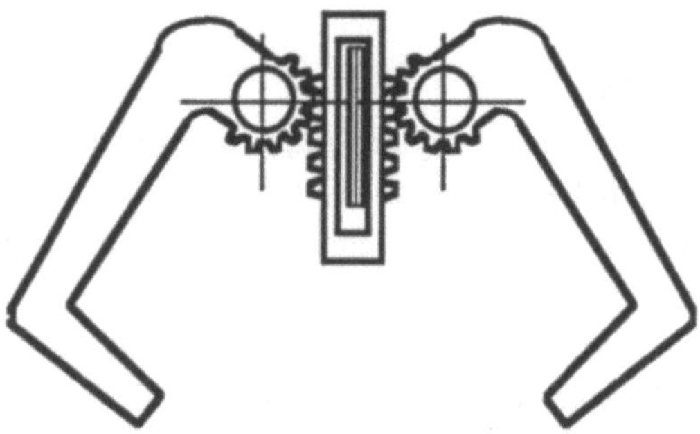

$$\begin{Bmatrix} R_{xo} \\ R_{yo} \\ R_{zo} \\ T_{xo} \\ T_{yo} \\ T_{zo} \end{Bmatrix} = \left\{ \begin{bmatrix} 0 & 0 & 0 & 0 & 1 & 1 \\ 0 & 0 & 0 & 1 & 0 & 1 \\ 0 & 0 & 0 & 1 & 1 & 0 \\ 0 & 1 & 1 & 0 & 0 & 0 \\ 1 & 0 & 1 & 0 & 0 & 0 \\ 1 & 1 & 0 & 0 & 0 & 0 \end{bmatrix} + \begin{bmatrix} 0 & 0 & 0 & 0 & \leftrightarrow & \leftrightarrow \\ 0 & 0 & 0 & \leftrightarrow & 0 & \leftrightarrow \\ 0 & 0 & 0 & \leftrightarrow & \leftrightarrow & 0 \\ 0 & \leftrightarrow & \leftrightarrow & 0 & 0 & 0 \\ \leftrightarrow & 0 & \leftrightarrow & 0 & 0 & 0 \\ \leftrightarrow & \leftrightarrow & 0 & 0 & 0 & 0 \end{bmatrix} + \begin{bmatrix} 0 & 0 & 0 & 0 & + & + \\ 0 & 0 & 0 & + & 0 & + \\ 0 & 0 & 0 & + & + & 0 \\ 0 & + & + & 0 & 0 & 0 \\ + & 0 & + & 0 & 0 & 0 \\ + & + & 0 & 0 & 0 & 0 \end{bmatrix} \right\} \begin{Bmatrix} R_{xi} \\ R_{yi} \\ R_{zi} \\ T_{xi} \\ T_{yi} \\ T_{zi} \end{Bmatrix}$$

Figure 2. A micro-mechanical mechanism [7] and its associated matrix representation. The gripping motion, at near-contact of the micro-tongs, is along the x-axis.

EXAMPLE 2 - Alternative gripper

As a means of contrasting different mechanisms, the alternative micro-gripper design of Kim et al. [8] is shown in Figure 3, along with its MTM and constraint matrices. In this case the matrices are simpler, due to the fact that the motion transformation is of the type translation ↔ translation, rather than rotation ↔ translation. In addition, a column vector representing an instance of a linear micro-motor is shown at the far right. Notice that vector instantiation of the micro-motor restricts the micro-motor motion to a single axis, whereas MTM's and their constraint matrices are axis and instance independent.

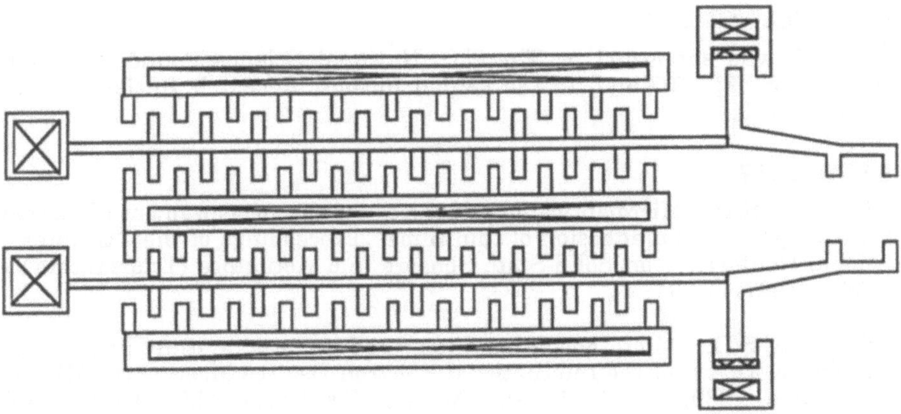

$$\begin{pmatrix} R_{xo} \\ R_{yo} \\ R_{zo} \\ T_{xo} \\ T_{yo} \\ T_{zo} \end{pmatrix} = \left\{ \begin{bmatrix} 0&0&0&0&0&0 \\ 0&0&0&0&0&0 \\ 0&0&0&0&0&0 \\ 0&0&0&1&0&0 \\ 0&0&0&0&1&0 \\ 0&0&0&0&0&1 \end{bmatrix} + \begin{bmatrix} 0&0&0&0&0&0 \\ 0&0&0&0&0&0 \\ 0&0&0&0&0&0 \\ 0&0&0&\leftrightarrow&0&0 \\ 0&0&0&0&\leftrightarrow&0 \\ 0&0&0&0&0&\leftrightarrow \end{bmatrix} + \begin{bmatrix} 0&0&0&0&0&0 \\ 0&0&0&0&0&0 \\ 0&0&0&0&0&0 \\ 0&0&0&+&0&0 \\ 0&0&0&0&+&0 \\ 0&0&0&0&0&+ \end{bmatrix} \right\} \begin{pmatrix} 0 \\ 0 \\ 0 \\ 1 \\ 0 \\ 0 \end{pmatrix}$$

Figure 3. An alternative mechanism for gripping [8] and its matrix representation that includes an instance of translational micromotor actuation.

CAD

In a CAD system, the conceptual design process is initiated with the statement of the design task that describes the input signal and desired output motion. The prescribed task is then decomposed into a series of input and motion transformation matrices. The process of decomposition is based on generating coordinate transformation matrices (CTM's) and is described in detail in [10]. Each mode of decomposition leads to a unique *design configuration.* The decomposed CTM's are then matched against a library of matrices (MTM's and constraints) representing existing devices (or working principles) in the knowledge-base. Thus, the computable processes of systematic decomposition and matrix-matching constitute the "reasoning process" involved in conceptual design. At the lower levels of abstraction, the symbolic elements of candidate matrices are replaced by the actual design parameters, before performing *qualitative simulation* of the designed artifact. Concatenation of motion matrices has the same effect as physically connecting the building blocks of the designed system.

ACKNOWLEDGMENTS

This work was supported by The U.S. National Science Foundation under Grant #ECS-8915215 and, to the second author, Research Initiation Grant #DDM-8910160.

REFERENCES

1. Senturia, S.D.: Microfabricated structures for the measurement of mechanical properties and adhesion of thin films. Proceedings of the 4'th International Conference on Solid-State Sensors and Actuators (Transducers'87), Tokyo (1987) 11-16.

2. Senturia, S.D.: Microsensors vs. integrated circuits: a study in contrasts. Technical Digest of the International Electron Devices Meeting (IEDM), San Francisco (1989) 1.1.1-1.1.5. Also, Maseeh, F.; Harris, R.M.; Senturia, S.D.: A CAD architecture for microelectromechanical systems. Proceedings of the IEEE Micro Electro Mechanical Systems Workshop, Napa Valley, CA (1990) 44-49. And also, Harris, R.M.; Maseeh, F.; Senturia, S.D.: Automatic generation of a 3-D solid model of a microfabricated structure. Technical Digest of the IEEE Solid-State Sensor and Actuator Workshop, Hilton Head Island, SC (1990) 36-41.

3. Crary, S.; Zhang, Y.: CAEMEMS: An integrated computer-aided engineering workbench for micro-electro-mechanical systems.

4. Zhang, Y.; Crary, S.B.; Wise, K.D.: Pressure sensor design and simulation using the CAEMEMS-D module. Technical Digest of the IEEE Solid-State Sensor and Actuator Workshop, Hilton Head Island, SC (1990) 32-35.

5. Pahl, G.; Beitz, W.: Engineering design. Berlin, Heidelberg, New York, Tokyo: Springer-Verlag (first English edition) 1984.

6. Kota, S.: Qualitative motion synthesis. Proc. of the First National Conference on Applied Mechanisms and Robotics, Cincinnati, OH (1989).

7. Mehregany, M.; Gabriel, K.J.; Trimmer, W.S.N.: Integrated fabrication of polysilicon mechanisms. IEEE Transactions on Electron Devices ED-35 (1988) 719-730.

8. Kim, C.-J.; Pisano, A.P.; Muller, R.S.; Lim, M.G.: Polysilicon Microgripper. Technical Digest of the IEEE Solid-State Sensor and Actuator Workshop, Hilton Head Island, SC (1990) 48-51.

9. Kota, S.: Automatic selection of mechanism designs from a three-dimensional design map. To be published in the ASME Journal of Mechanical Design. Also, Kota, S.; Chiou, S.-J.: Taguchi methods in mechanisms design. Presented at the First National Applied Mechanisms and Robotics Conference, Cincinnati, OH (1989).

10. Kota S.: A Qualitative matrix representation scheme for the conceptual design of mechanisms. Proc. of the 1990 ASME Design Technical Conferences (21'st Biennial Mechanisms Conference), Chicago, (September 16-19, 1990).

Intelligent CAD for Micro Mechanics

Richard Amster[1], Lee S. Tavrow[2] and Anita M. Flynn[3]

[1]ICAD Inc., Cambridge, MA 02139, USA
[2]Microsystems Technologies Laboratory, MIT, Cambridge, MA 02139, USA
[3]Artificial Intelligence Laboratory, MIT, Cambridge, MA 02139, USA

Abstract

The new field of micro mechanics, including both micro sensors and micro actuators, is progressing by taking advantage of a majority of the fabrication steps, lithography equipment and design tools of silicon integrated circuit production. Many of the design issues in micro mechanics are very different, however, and require tools which capture the essentials of mechanical design while maintaining the representations required to interface to silicon processing equipment.

This paper describes a new set of tools for designing microfabricated parts. The underlying structure is provided by a commercially available *design language*, ICAD, which is based on artificial intelligence techniques. While most computer aided design systems accept geometrical constraints as inputs, ICAD instead manipulates relationships between parts and subassemblies formulated as higher-level rules. Thus the designer can concentrate on the intrinsic features of the design, capturing spatial relationships and materials parameters, while letting the software generate graphical output, engineering drawings, costs, and the like.

Here we illustrate the power of such intelligent tools in the design of micromotors. The system accepts descriptions of classes of types of motors and can be interfaced easily to generate outputs such as mask layouts, input formats for finite element simulations and process planning for fabrication equipment.

1 Introduction

Increasingly, the progress of advanced technological development depends upon engineering and design software tools. One reason is that further progress in mature fields relies upon incremental improvements in design. Another reason is that the broad application of a given technology requires minor adaptations and modifications which would be prohibitively time-consuming and expensive without appropriate software tools. Thus, a wide variety of software tools have been, and are being developed to support engineering design, engineering analysis and the subsequent activities related to the manufacture of those designs. Typically, each engineering field develops its own specialized tools as, for example, the integrated circuit (IC) industry which has separate tools for circuit layout, circuit simulation, device simulation, process simulation, manufacturing, and testing. Other useful tools include Finite Element Method (FEM) packages for the analysis of mechanical, fluid, and electric-field properties.

The unique challenge of "micro-electrical-mechanical design" is implicit in its very name. By definition a Micro Electro-Mechanical System (MEMS) must successfully combine a functional mechanical design with a functional electrical design in a system or device which cannot be manufactured by conventional "macro scale" processes. While it may still be possible to create certain device prototypes with minimal software tools, the day will soon come when the scale and complexity of state-of-the-art MEMS absolutely require computer models for their design and manufacture (such as [1]). Our focus has been an exploration of what form the future software tools must take, what functions the software must provide, and the knowledge representation methods which will be needed.

In particular, the familiar problem of variable-capacitance electric micromotor design [2, 3, 4, 5] (see Fig. 1) was approached by building a rule and feature based model which can be used to generate design geometries according to non-geometric input parameters. The model is capable of providing input to an FEM package for electric-field analysis. The model is also compatible with Process Planning tools which can be used to analyze the motor for fabrication by various manufacturing methods.

2 Problems with conventional CAD tools: the representations

Conventional design and analysis software tends to rely upon specialized *data representation* schemes to function. The MIT experience in designing micromotors illustrates some of the

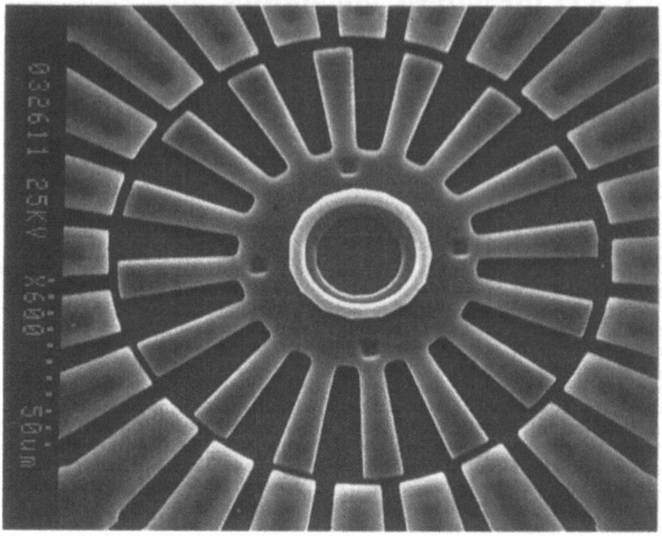

Figure 1: A scanning-electron micrograph of a 24 stator-pole 16 rotor-pole LOCOS-based microfabricated variable-capacitance motor.

problems encountered when attempting to use some typical conventional tools for MEMS design. The initial design ideas were layed out on paper, with the parameters of the device jotted onto the paper as the ideas developed. Thus the first representation of the *concepts* was the written page.

When enough of the necessary details had been worked out, the design was transferred into an interactive low-level graphical IC layout tool called KIC. The KIC data representation is oriented toward driving a shutter-based mask generator to produce the photolithography masks for manufacturing the micromotor. Although much more sophisticated tools exist for IC layout, none available were capable of readily handling the complex curves and structures necessary for the micromotor. Other than niche applications such as microwave circuits and power devices, not to mention the proliferation of intricate mask logos, complicated curved shapes are almost never required in digital circuit design. Consequently, most software tools for digital circuit design are limited to "Manhattan geometries," i.e., only boxes oriented at right angles to each other. In addition, KIC has been greatly expanded to provide the means of generating complex geometries, but the results must still be painstakingly fine-tuned by hand. The KIC *layout* representation amounts to a list of box locations, sizes, and rotation angles for each of the different masking layers.

Because of the shutter-based mask generation system, curves were achieved by superimposing rotated rectangles. This method causes the mask generator to overexpose the fringe around the approximated curve, which has the benefit of smoothing the curve. Further smoothing occurs during the etching in the manufacturing process, resulting in an acceptable approximation to the desired curve. Even though the result is satisfactory, considerable human intervention is required in each step of the design process. Consequently, even minute changes in the micromotor design such as the rotor/stator spacing, size, or number require a laborious session with KIC.

The electrical properties of the motor designs were also explored using FEM techniques. Unfortunately, a separate model is needed to drive the FEM software. For this purpose, a representative portion of the air gap between the rotor to stator was broken into a set of wedge-shaped volume elements compatible with the FEM code. The representation used in this *analysis* is then a set of vertices representing the extrema of the volume elements. These elements are then combined to produce the boundary conditions of the three-dimensional solution to Laplace's equation [6].

Neither the KIC model nor the FEM representation were suitable for generating a graphical preview of the design. For this purpose a graphics routine running under X-windows on a UNIX workstation was implemented. The routines were capable of presenting the rotor at varying angles with respect to the stators, providing a graphical *simulation* of the motor. The representation in this model was a set of lines and arcs.

Note the vastly different representations needed for *concepts*, *layout*, *analysis* and *simulation*. Also note that any significant change to the motor design must be manually propagated by the designer through several separate software systems.

2.1 The issue of integration

Integrated systems for MEMS design are already in development. Crary and Zhang [7] at the University of Michigan are developing an integrated system called CAEMEMS which will include FEM device modeling, mechanical simulation, and process modeling in addition to conventional mechanical design. A system developed at MIT [8] includes mechanical CAD, material property simulation, and structure simulation. The MIT work utilizes an object-oriented approach to modeling material properties.

The emphasis in these projects has been to build interfaces between various mature software components in order to create an integrated system. In these systems, an interactive user-interface is the primary means of capturing design information from the user. This contrasts with the approach taken in this work, in which the primary means of capturing design knowledge is through the creation of a highly structured set of rules, which subsequently become capable of generating the design geometry as required.

The knowledge based approach also differs in that (ideally) a single representation is used to capture the differing kinds of knowledge needed. This implies that most of the information flow is channeled through this central representation, rather than flowing from module to module. This has the advantage that the results of analysis can easily be fed back into the design process.

3 Knowledge based design

A new class of *knowledge based* design tools are beginning to emerge as an alternative to the scenario described above. A knowledge based system should have the potential to provide a single representation for conceptual, mechanical, analytical, simulation and manufacturing development.

Our work was done using the ICAD Design Language (IDL) [9]. IDL is an object oriented, declarative language implemented on top of a Lisp programming environment.[1] IDL has been implemented using *delayed evaluation* techniques [10]. Because it runs under Lisp, IDL is also incrementally compilable, which means that small changes in the design can be made with little effort. In developing our prototype of the micromotor model we took a top-down approach, starting by determining which engineering parameters would drive the design of the motor, and eventually leading to the creation of individual component models (such as rotor vanes, hubs, etc.) which where assembled into a complete motor (Fig. 2).

The essential activity in using a knowledge based tool is to define rules for computing a set of characteristic attributes which model the behavior, function and form of a given design. Thus a designer will ask: "What electrical parameters and relationships dictate the size of this rotor structure?," rather than asking "How big was this particular rotor structure supposed to be?"

To illustrate how these techniques are used, we now present selected portions of our actual IDL micromotor model.

3.1 A 30 second introduction to IDL

The fundamental syntactic unit of IDL is the `defpart`. The meaning of the word "defpart" can be taken to mean that the following set of declarations is the Definition of a certain Part. A `defpart` is typically created by combining one or more primitive geometric parts (such as boxes, cones, etc.), and then explicitly defining rules for computing various categories of attributes. Typical attribute categories are `:inputs`, `:attributes` and `:parts`. To define a simple part, such as a `cylinder`, which has a computed attribute of `:aspect-ratio` (which we define as the length divided by two times the radius), we would write:

[1] The Lisp programming language has been in use for over two decades, and has been the favored language of Artificial Intelligence.

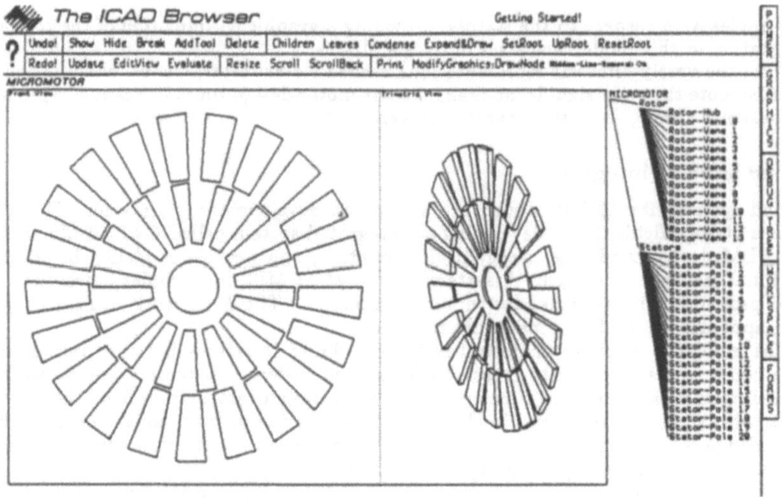

Figure 2: The ICAD browser displaying two views of a sample electric motor and the components which comprise their construction.

```
(defpart example-cylinder (cylinder)
   :inputs
     (:length :radius)
   :attributes
     (:aspect-ratio (div (the :length)
                          (* 2 (the :radius)))))
```

This definition would read as follows: Define a part named `example-cylinder`, which is made from the previously defined part `cylinder` (which happens to be a primitive, but any defined part, or set of parts, could be used or combined instead). The characteristic inputs to `example-cylinder` are `:length` and `:radius`, and a computed attribute is the `:aspect-ratio` which is defined by the expression that follows it. Note that attributes within a `defpart` are referenced by using "`(the <attribute name>)`", which is a simple example of a *referencing chain*.

IDL is a fully object-oriented system. To utilize a model as defined above, one must *instantiate* an object which can be done through the ICAD User Interface, or by a LISP form such as:

```
(make-part 'example-cylinder :length 3 :radius .125)
```

The result of this LISP form is a specific instance of `example-cylinder`, which can be queried for the value of its attributes such as the instance's aspect-ratio.

3.2 The micromotor model

The first task in creating a useful representation of a micromotor is to determine the set of input parameters which are most important to the design. We have broken these parameters into two groups, namely the parameters relating to fabrication constraints [4] and the parameters pertaining to motor torque optimization [5].

```
(defpart field-properties-mixin ()
  :optional-inputs
  (:stator-fraction 0.6     ; dimensionless quantities
   :stator-rotor-size-ratio 1
   :width-gap-ratio 5
   :phases 3
   :rotors-per-phase 2))
```

```
(defpart fab-parameters-mixin ()
  :optional-inputs
  (:gap 2                    ; values in microns
   :rotor-thickness 2
   :stator-thickness 2
   :rotor-outer-radius 60
   :rotor-inner-radius 15
   :rotor-annular-radius 25
   :stator-outer-radius 100)
  :attributes
  (:aspect-ratio (div (the :rotor-thickness) (the :gap))
   :rotor-vane-length (- (the :rotor-outer-radius)(the :rotor-annular-radius))))
```

These **defpart**'s contain no geometrical primitives but instead only group properties. They will never be used on their own (i.e., separately instantiated) so they are named according to a convention with a **-mixin** suffix. Finally, their separation draws attention to the different kinds of information needed to design a micromotors. This modularity also makes the code easier to read, understand, and maintain.

Motors can now be instantiated using the default values given by the **field-properties-mixin**, the **fab-parameters-mixin** and the construction rules in the following micromotor **defpart**:

```
(defpart micromotor (field-properties-mixin fab-parameters-mixin assembled-solid)
  :optional-inputs (:rotor-angle 0)
  :attributes
  (:solids (the :children)
   :preliminary-stator-poles (floor (div (* 2 pi
                                          (+ (the :rotor-outer-radius)(the :gap))
                                          (the :stator-fraction))
                                          (* (the :gap)(the :width-gap-ratio))))
   :stator-poles (* (the :phases)
                    (floor (div (the :preliminary-stator-poles)
                                (the :phases))))
   :rotor-poles (div (* (the :stator-poles)(the :rotors-per-phase))
                     (the :phases))
   :stator-pole-width (div (* 2 pi (the :stator-fraction))       ;In radians
                           (the :stator-poles))
   :rotor-pole-width (* (the :stator-rotor-size-ratio)
                        (the :stator-pole-width)))               ;Also in radians
  :descendent-attributes (:rotor-poles :stator-poles
                          :stator-pole-width :rotor-pole-width)
  :parts
  ((rotor :orientation (:numeric (roll :longitudinal (the :rotor-angle))))
   (stators)))
```

Changes in any of these three **defpart**'s will then propagate through into different instances of motors with differing shapes, sizes, and even topologies. The default motor is shown in Fig. 2 along with a micrograph of the actual implementation of that design in Fig. 1, for comparison. In addition, these motor instances can also be used to drive other design tools as in the case of FEM analysis or to create plans, drawings, etc.

3.3 Relationship to expert systems technology

It should be clearly understood that the approach described above is not an Expert System. Although an Expert System provides a single repository of knowledge which could be used for a number of purposes, it does not structure the knowledge in analogy with the design. This means that an Expert System must search a database of rules to determine whether a rule applies to the problem of interest. The technique embodied by IDL puts the knowledge exactly where it is needed to define the attribute of interest. No inferencing engine is needed because each rule is contained in the appropriate **defpart**.

4 Simulation analysis

In order to demonstrate the usefulness of the micromotor model developed in the previous section, in addition to visualization, simulation, and layout, we will now briefly describe its interface to a FEM package. Automatic mesh generation is generally quite difficult, in part

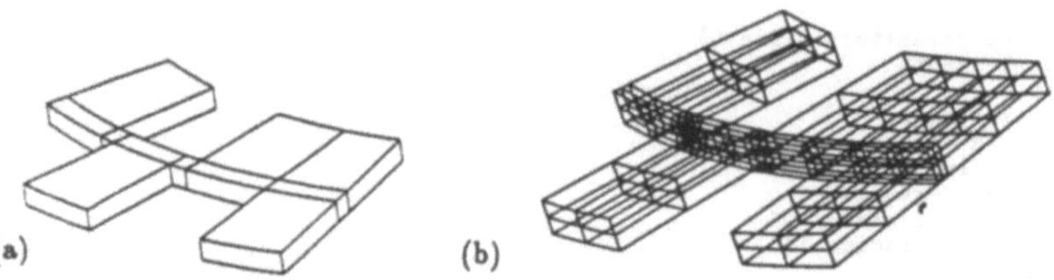

Figure 3: (a) An inverted section of a micromotor produced by ICAD suitable for FEM analysis and (b) the meshed section.

because the relevant features of the design cannot easily be extracted from the graphical representation of the design alone. The use of a feature based language relieves this situation, because information which defines how the volumes are constructed is available. In order to automatically generated the Finite Element volumes for the critical rotor/stator air-gap, inverse-volumes of the micromotor in the region of the air-gap were "assembled" by ICAD as shown in Fig. 3a. These volumes can then be meshed by the FEM package (Fig. 3b) and the electric drive torque computed. The complete drive torque curve for all rotations of the rotor of any motor variation can thus be easily computed.

5 Conclusion

We have described a knowledge-based design tool, ICAD, which is applicable to micro mechanics. In addition, we have compared and contrasted this approach with other techniques and implementations. Using an example of a micromotor, we have then demonstrated its operation and proven the design advantages of such a system.

6 Acknowledgments

This work was supported in part by the U.S. National Science Foundation under grant ECS-8614328 (Tavrow) and the Gnat Robot Corp. (Flynn).

References

[1] A.M. Flynn, R.A. Brooks and L.S. Tavrow, "Twilight Zones and Cornerstones: A Gnat Robot Double Feature," *MIT AI Memo 1126*, July 1989.

[2] S.F. Bart, T.A. Lober, R.T. Howe, J.H. Lang and M.F. Schlecht, "Design Considerations for Microfabricated Electric Actuators," *Sensors and Actuators*, Vol. 14, No. 3, 1988, pp. 269–292.

[3] L.-S. Fan, Y.-C. Tai and R.S. Muller, "IC–processed Electrostatic Micro-motors," *IEEE Int. Electron Devices Meeting*, December 11–14, 1988, pp. 666–669.

[4] L.S. Tavrow, S.F. Bart, M.F. Schlecht and J.H. Lang, "A LOCOS Process for an Electrostatic Microfabricated Motor," *Sensors and Actuators*, Vol. A, No. 23, April 1990, pp. 893–898.

[5] M. Mehregany, S.F. Bart, L.S. Tavrow, J.H. Lang and S.D. Senturia, "Principles in Design and Microfabrication of Variable–capacitance Side–drive Motors," *Journal of Vacuum Science and Technology*, July–Aug. 1990.

[6] L.S. Tavrow, "Microfabricated Electric Motors," *Electrical Manufacturing*, May 1990, pp. 23–28.

[7] S. Crary and Y. Zhang, "CAEMEMS: An Integrated Computer–Aided Engineering Workbench for Micro–Electro–Mechanical Systems," *Proc. of IEEE MEMS 1990*, Feb. 11–14, 1990, pp. 113–114.

[8] F. Maseeh, R.M. Harris and S.D. Senturia, "A CAD Architecture for Microelectromechanical Systems," *Proc. of IEEE MEMS 1990*, Feb. 11–14, 1990, pp. 44–49.

[9] *The ICAD System User's Manual*, ICAD Inc., Cambridge, MA, Dec. 1989.

[10] H. Abelson and G.J. Sussman, *Structure and Interpretation of Computer Programs*, MIT Press, 1985.

CAD of Optical Devices

Computer Aided Engineering for Microsystems - Status and Trends

Reinhard März

Siemens AG, Research Laboratories, Otto-Hahn Ring 6, 8000 München 83, FRG

Introduction : During the last decades, microelectronics has revolutionized the most branches of engineering. Parallel to this, other chip technologies such as integrated optics, micromechanics and acoustoelectronics have been developed to an advanced state of the art. During the next few years, integrated components of different chip technologies will converge to form microsystems. For their development interdisciplinary know-how and cooperation is required. Increasing reliability and decreasing costs are the expected benefits of microsystems in comparison to hybrid solutions. Typical examples are sensors and actuators integrated with their analog electronics as well as optoelectronic integrated chips for optical transmission lines [1]. This paper attempts to summarize the current activities on computer aided engineering for the various disciplines of microsystems technology and to discuss some trends.

Fig. 1 shows the CAE scenario for a typical chip technology. The diagram distinguishes between two types of paths, which correspond to the main tasks of computer aided engineering; support for the development of technology on the one hand and CAD, i.e. layout for fabrication, on the other. The CAE path of technology development ($\Longrightarrow$) for a certain component starts with its physical description, i.e. with the layout for a set of masks required for fabrication. The "process modeling" delivers a software model of the device that describes the geometry and some other relevant properties, e.g. profiles of doping or refractive index. The subsequent "device modeling" investigates the functional behavior of the device starting from the software model. It delivers a set of parameters (e.g. SPICE parameters of microelectronics), which describe the operation of the device in a circuit or another form of composite structure. It should be clearly pointed out that both "process" and "device modeling" are standalone techniques used mainly for process and device optimization. Their interfacing only makes sense if the corresponding modeling tools offer precise predictions. The CAD path ($\longrightarrow$) starts with the description of the chip or a bigger system, which is then worked out as far as the physical layout of the chip using one or more circuit descriptions.

The physical layout is the interface to fabrication, circuit descriptions that to behavioral simulations.

Only microelectronics, the most well established discipline within microsystems technology, shows a fully expanded CAE scenario. The field of VLSI, which is characterized by great complexity, exhibits several levels of circuit description. Micromechanics, in contrast, has no circuit level at all and the circuits of integrated optics are of such low complexity ($<$ 100 devices/circuit) that no circuit level description has been established until now. Since micromechanics and integrated optics are "young" technologies undergoing rapid changes, process modeling is rarely applied. Layout editors of microelectronics serve as CAD tools of micromechanics, whereas a special layout editor is available for integrated optics .

Device Modeling : Originally, the development of new devices was accompanied by theoretical investigations to obtain a deeper understanding of their operation and thus to derive strategies for their layout. With the growing technological outlays during recent years, a rapidly increasing demand for quantitative predictions can be observed, since the precise layout of devices reduces the number of prototypes required for the development of a new device and thus time and costs. This evolution was accompanied in the field of CAE by a trend from more qualitative analytical modeling to numerically oriented predictions of device properties. The numerical investigation of fabrication tolerances in particular is a topic that has become more and more important. These efforts were successful and are being continued.

Behavioral simulation is one important method of device modeling. It investigates the response of a device to a certain stimulus (more or less realistic) on a computer. All simulation tools used for device modeling are field oriented. Most of them calculate time dependent variations of a field, e.g. the response of the electric field, particle or current density to an exciting pulse, some of them such as the beam propagation method of integrated optics calculate the stationary or time harmonic field, which is observed after the onset of the field has faded away. Device simulations act on the one hand as a test bed for analytical or numerical models, while on the other hand they provide answers to questions, which are impossible or difficult to find by experiment. Rigorous design methods such as the eigenmode analysis of integrated optics and the stress and displacement analysis of micromechanics are the other important category of numerical methods for microsystems technology. Knowledge of these results is crucial for the layout of nearly all components of integrated optics, micromechanics and acoustoelectronics.

Device modeling for microelectronics is dominated by simulation tools, which solve a system of partial differential equations for the transistor consisting of the continuity and transport equations for electrons and holes and of Poisson's equation [2]. The calculated SPICE parameters fit well with the measured results (deviation of the currents $< 20\%$), i.e. device modeling can be used to predict the potential of a new generation of chips. Another highlight during recent years was the demonstration and subsequent avoidance of peaks of the electric field inside the MOS transistor, which gave rise to aging of the device. The software tools used for simulation such as the general package PISCES or the MOS package MINIMOS are now well established; some of them are commercially available.

During the last decade, the device modeling for integrated optics has made great strides. The beam propagation method, which solves the paraxial Helmholtz equation, is now widely used to simulate the behavior of nearly all integrated optical devices. Using more advanced calculation schemes such as the z-transient variational principle [3], it was possible to overcome its restrictions to small refractive index changes. Finite element, finite differences, mode matching techniques and similar methods for the calculation of eigenmodes [4] act as rigorous design tools for integrated optics (e.g. typical deviations for the beat length of an InGaAsP directional coupler $< 10\%$). Although the tools are now well established, commercial software is not available.

Steady state and eigenmode analysis as well as the simulation of acoustic waves running through sensors or filters are the main tasks of micromechanical device modeling. Commercial general purpose finite element packages for stress and displacement analysis, such as ANSYS or ADINA, provide solutions of the corresponding partial differential equations. The interaction between electrical and mechanical phenomena via capacitive, piezoresistive and piezoelectrical effects as well as the propagation of acoustic waves and stress and displacements caused by changes of the temperature can be included [5]. The results of device modeling in general fit the experiments well (e.g. typical deviations for the eigenmodes of a telephone transducer are $< 10\%$).

Process Modeling : Process modeling offers the most direct support of technology development provided by computer aided engineering. It has to provide simulation tools for two largely different tasks. On the one hand, the software model of a device, which is fabricated using a given sequence of process steps, should be delivered. On the other hand, tools for the optimization of the single processes themselves are required. Nowadays process modeling is mainly applied to further improve the technology of silicon microelectronics. The simulation tools are partly based on rigorous calculations and partly on heuristic models, which have to be fitted to equipment and process conditions. The software for the simulation of lithography and dry etching (e.g. SAMPLE) as well as for implantation, diffusion and oxidation (e.g. MIMAS or SUPREM) is based on refined models [6], which give a good description of the processes. Commercial packages are available for these tasks. For the optimization of processes the physics of reactors must be simulated [7]. These problems are tackled using commercial simulators of hydrodynamic systems such as PHOENICS.

For the other disciplines of microsystems technology, integrated optics and micromechanics, process modeling is rarely applied nowadays.

Computer Aided Design : Computer Aided Design (CAD) translates the system description of a chip into its physical layout. i.e. into a set of masks used for fabrication. CAD-systems have to speed up the design process and to offer "correctness by construction" or checkers to verify a certain design level. Circuit level simulators should allow the chip operation to be previewed.

The CAD systems of microelectronics are developed to handle the extremely complex VLSI circuits. They provide placement, floorplanning and routing and the required components within cell libraries [8]. Several levels of circuit description allow the simulation of logical and electrical behavior and of timing. Semicustom CAD systems, which are now commercially available, expect a netlist of standard and macro cells as a system description. High level synthesis [9] currently under development starts from a behavioral description, i.e. from a procedural programming language (e.g. VHDL, DSDL).

The CAD-problems of integrated optics are completely different. The optical circuit is formed by a small number (typically < 100) of basic structures (e.g. Y-branches, Bragg gratings, optical switches,..). These complex shaped basic structures ("primitives") must be mutually adjusted without kinks, since even kink angles of $1°$ can significantly distort the operation of the optical circuit. A first layout editor, Topcad-Optik, was published recently [10]. It stores the design specifications, and not the geometry and position of the individual primitives, to generate an optical circuit. To support the continuous joining of basic structures, an auxilary primitive called a "connector" was introduced (see Fig. 2). This connector - in fact a waveguide of zero length - is described by its width and by its relative position and orientation. It connects two adjacent primitives and controls the position and width of their interfaces. Due to the low complexity of optical circuits, circuit level simulations are not available for integrated optics.

The CAD problems of micromechanics are similar to those of integrated optics, i.e. the layouts consist of a few complicated structures, but there are no requirements on adjusting primitives to each other. Usually, these CAD problems are solved by using layout editors from microelectronics.

Interchange Formats : It can be expected in the future too that the software tools for microsystems technology will be developed by independent groups in a multisite environment with mainframes, workstations and personal computers. To connect the CAD/CAE tools under such circumstances powerful, flexible and user-friendly standard interface definitions are required. The EDIF (**E**lectronic **D**esign **I**nterchange **F**ormat), which describes VLSI circuits from the structural, behavioral and physical points of view, has become an IEEE standard, it is becoming increasingly accepted by the software vendors. In the scenario of process and device modeling nowadays data are not exchanged at all or via input and output formats, which are specific to a single software package. The PIF (**P**rofile **I**nterchange **F**ormat) [11] is a first attempt to establish a standard interface format for the interchange of data inside the design level of microelectronics. Its blockoriented syntax is EDIF-like, the format itself is defined in an ASCII-mode (Fig. 3) for intersite transmission as well as in an equivalent binary mode for in site use. Another type of interchange format is realized inside the integrated optics group at Siemens research laboratories. The data transmission here is based on a family of directory oriented low-level formats, which consist of the primitive items "line", "section of a surface" for the storage of the profiles and "attribute" for the storage of additional information. The direct access to these formats is effected by using a software library, which provides the operations "add line", "read line", "delete line", "list entries" for working with profiles and "read attribute" and "write attribute" for handling other information. The procedural interface allows new software tools to be connected within hours.

Conclusions : It was demonstrated in the previous sections of this paper that nowadays process modeling is applied to silicon microelectronics and that device modeling has attained an advanced state of the art for all disciplines of microsystems technology. For both process and device modeling an overall trend to 2d- and 3d-simulators and modeling tools, which is driven by the increasing performance of workstations and mainframes, can be observed. With the increasing technological activities on microsystems it can be expected that the corresponding modeling activities will move together, and the importance of process modeling will grow with the increasing expense of technology.

The CAD systems of microelectronics are now fully developed, and for integrated optics a layout editor is demonstrated. What future CAD or even CAD/CAE systems will look like is difficult to answer. The huge and rapidly varying knowledge base of microsystems technology favorizes, at least for the near future, a scenario of tools which are weakly coupled by a user-friendly standard interchange format.

Acknowledgements : The author gratefully acknowledges the valuable discussions with T. Grandke, W. Hänsch, H. Jacobs and S. März from the Research Laboratories of the Siemens AG.

References

[1] A. Heuberger (ed.): "Mikromechanik", Springer(1989)

[2] S. Selberherr: "Analysis and Simulation of Semiconductor Devices", Springer(1984)

[3] T.B. Koch, R. März, J.B. Davies: "Beam-Propagation Method Using Z-Transient Variational Principle", ECOC'90 (submitted)

[4] COST 216 (WG 1): "Comparison of Different Modeling Techniques for Longitudinally Invariant Integrated Optical Waveguides", IEE Proc. **136J**, 273-280(1989)

[5] H. Baltes, A. Nathan: "Sensor Modeling", in "Sensors" Verlag Chemie Weinheim **1**,46-77(1989)

[6] C.P.Ho, S.E. Hansen: "SUPREM III - A Program for Integrated Circuit Process Modeling and Simulation", Stanford University, Techn. Rep. SEL83-001 (1983)

[7] J.I. Ulacia, S. Howell, H. Körner, C. Werner: "Flow and Reaction Simulation of a Tungsten CVD Reactor" Appl. Surf. Sci. **38**, 370-385 (1989)

[8] E. Hörbst, M. Nett, H. Schwärtzel: "VENUS: Entwurf von VLSI-Schaltungen", Springer(1986)

[9] J. Biesenack et al.: "Synthese - Die Entwurfsmethode der Zukunft" Elektronik **23**(1989)

[10] R. Müller-Nawrath, R. März: "A Layout Editor for Integrated Optics", Proc. IOOC'89 **2**, 152-153(1989)

[11] S.G.Duvall: "An Interchange Format for Process and Device Simulation", IEEE Trans. CAD **7**, 741-754(1988)

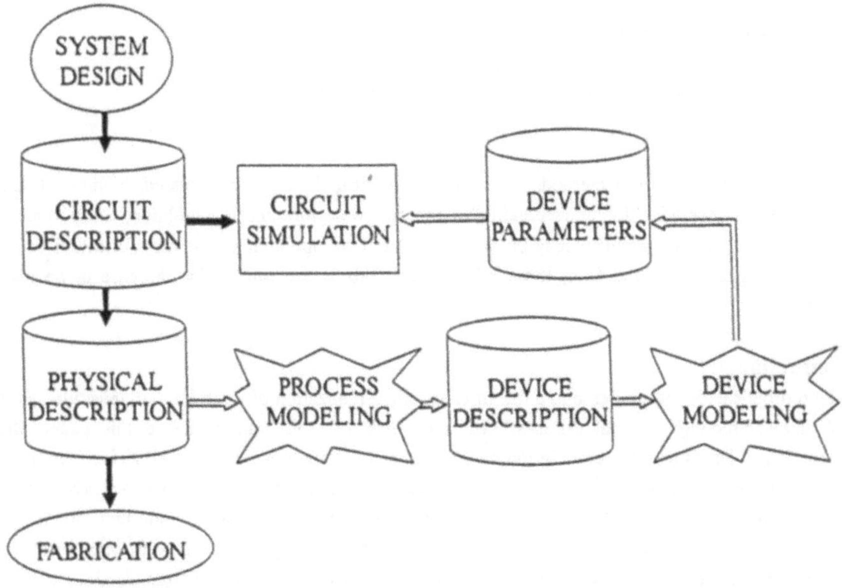

Fig. 1: CAE scenario of a chip technology.

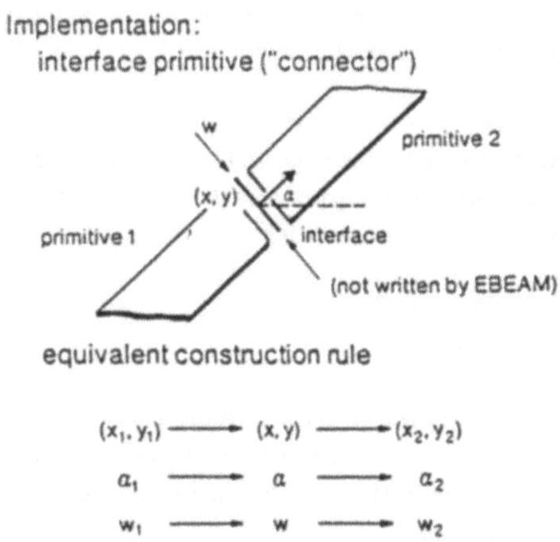

Fig. 2: Adjustment strategy of Topcad-Optik.

```
(PIF
********* geometry from Section IV.C.5        *********
********* grid from Section IV.C.5            *********
********* boron profile from Section IV.E.1   *********
  (materialType
    (specificationName "MaterialTypeList")
    (table
      (objectName "Segment" )
      (data "silicon" "siliconDioxide" "polysilicon")
      (dielectricConstant 11.9 3.9 11.9)
      (electronMobility 1000. ? 1000.)
      (holeMobility 500. ? 500.))
  (snapshot
    (snapshotName "TimeStep1")
    (snapshotDimension 2)
    (units "Micron" )
    (usesGeometry "FloatingGateStructure")
    (usesGrids "CoarseTriangularGrid")
    (usesAttributes "BoronProfile" "MaterialTypeList")
    (attribute
      (attributeName "TimeStep")
      (valueType shortReal)
      (units "ns")
      (value .1))
    (TerminalBias (value (objectName "SourceContact") (data 0.0)))
    (TerminalBias (value (objectName "DrainContact")  (data 1.0)))
    (TerminalBias (value (objectName "GateContact")   (data 1.0)))
    (TerminalBias (value (objectName "BaseContact")   (data 0.0)))
    (TerminalCurrent 1.E-7))
```

Fig. 3: PIF example (from [11]).

Numerical Simulation of Optical Devices

C. M. Weinert and H.-P. Nolting

Heinrich-Hertz-Institut für Nachrichtentechnik Berlin GmbH., Berlin, Germany

Abstract

Some of the main numerical methods for simulating optical devices are discussed. Numerical simulations yield essential information for understanding of the device performance. In addition, the parameters most sensitive to processing tolerances may be determined and an optimum design for fabrication of integrated optical devices can be found from these calculations. This is exemplified for the case of an optical 3dB coupler.

Introduction

Integrated optical devices are essential for an economical realization of optical communication systems. The III-V semiconducting material GaInAsP/InP on which we will concentrate in this paper allows the construction of lasers and detectors in the 1.3μm/1.5μm wavelength window as well as fabrication of passive optical and electrooptical devices such as waveguides, couplers and polarization sensitive devices. The common feature of all these devices is that the optical wave propagates along some kind of waveguide and thereby can be manipulated in the desired fashion. Therefore waveguides do not only interconnect different components but are themselves the main part of each optical component. Interconnections of waveguides with different types generally give rise to large losses and reflections. These junctions between different waveguides can be smoothed out by waveguide tapers which, however, are difficult to fabricate and increase the size of the integrated device. Therefore one of the main aims of the design for integrated optical components is to find structures with a common waveguide [1]. Since the development of the epitaxial techniques and processing steps are time consuming and costly it is important to get a thorough understanding of the component by numerical simulation already before fabrication.

Numerical simulation has to be done in close connection with measurements of the essential device parameters which are the device geometry and the refractive indices of the epitaxial layers. These data serve as input for the calculation and the results of the simulation are then compared with the experimental results. In case of disagreement between experiment and theory two main questions arise:

- is the theory valid and accurate enough to describe the device?
- are the measured input data accurately given?

In order to answer these questions profound knowledge is needed of the theoretical approaches, of the tolerances during fabrication, and of the accuracy of the measurement techniques. We therefore will discuss some of the main numerical simulation techniques derived from Maxwell's equations and then treat an optical 3 dB coupler as a specific example of numerical simulation.

Methods of Numerical Simulation for Optical Components

The propagation of an electromagnetic wave with harmonic time dependence is expressed by the solutions of the the three-dimensional Helmholtz Equation

$$(\Delta + (n^2 \omega^2/c^2))\, E(x,y,z) = 0 \qquad (1)$$

Even if very efficient computing facilities are available, numerical solutions of this partial differential equation for the three-dimensional case are very time consuming and the stability of the numerical solution is not always given. Therefore only some special cases have been treated in the past.

For most waveguiding devices, however, the structure in the direction of light propagation (z-direction) remains constant. This symmetry can be used to separate the z-part of the fields by using the ansatz

$$E(x,y,z) = E(x,y)\, \exp(-i\,\beta\, z) \qquad (2)$$

whereby (1) is reduced to a two-dimensional equation in the cross-section of the waveguide

$$(\Delta_t + (n(x,y)^2\, \omega^2/c^2 - \beta_i^2))\, E_i(x,y) = 0 \qquad (3).$$

The eigenvalues $E_i(x,y)$ are the electrical fields and β_i the propagation constants for the i-th guided mode. This two-dimensional eigenvalue problem has been solved by a number of different methods such as [2]:

- Finite Element Method (FEM)
- Finite Difference Method (FDM)
- Mode Matching Method
- Method of Lines

These two-dimensional calculation methods still have a high degree of numerical sophistication. For a number of problems, however, equation (3) can be approximately solved by successively solving the one-dimensional slab waveguide problem in x- and y-direction. This " Effective-Index Method" (EIM) is a very good approximation for waveguides with low lateral mode confinement such as rib waveguides with small rib heights. Another advantage of the EIM is that it can be applied to calculate structures with complex refractive indices (lossy media) whereas a two dimensional algorithm treating the general case of refractive indices has not been developed up to now.

For structures which vary in z-direction such as waveguide tapers, bends and discontinuities, ansatz (2) cannot be applied any longer. In this case the changes of wave propagation in z-direction can be numerically calculated by the beam-propagation method (BPM). The three-dimensional BPM which computes the two-dimensional field of the x-y-cross-section along the z-axis needs very large computer storage and computing time. Therefore, a two-dimensional BPM is often used in which the refractive index profile in the cross-section are approximated by the one dimensional Effective-Index Method.

Another method of calculating the wave propagation in z-direction is the Coupled-Mode Theory (CPT). This method allows to calculate the coupling between weakly coupled modes, e. g. waveguide couplers, and the effects of gratings on guided waves.

The methods listed above are complementary, the first set of methods treating the modal fields in x-y-cross-section and the second set calculating the wave propagation in z-direction. For a number of components, e. g. for couplers, both types of methods can be applied and a cross check is possible. This is important to verify the validity of the different theories and the approximations used.

Numerical methods have been applied by us for the following important optical waveguide problems:

- Calculation of the propagation constants, field distribution and cutoff properties of arbitrarily shaped waveguides for TE- and TM-polarization.
- Coupling behaviour of symmetrical and asymmetrical couplers also including TE/TM mode splitter structures [3].

Wavelength dispersion used for wavelength division multiplexing devices like couplers, interferometers and meander couplers [4].

For devices utilizing the electrooptic effects additional efforts have to be made to solve for the electrostatic fields. Such a program has been developed to simulate a TE/TM mode converter [5]. For laser structures the optical power confinement in the active region can be calculated as well as the influence of Bragg gratings and coupling between a laser and a waveguide. Numerical simulation (with complex refractive indices) of coupling between waveguide and detector for different device structures gives a deep understanding of the mode behaviour and leads to optimized coupling configurations [6].

Simulation of a 3-dB Coupler

A wave which is coupled into one of two adjacent parallel waveguides of identical shape will gradually couple over into the neighboring waveguide. This effect may be utilized to build a 3 dB coupler (figure 1) which splits the optical power into two equal amounts.

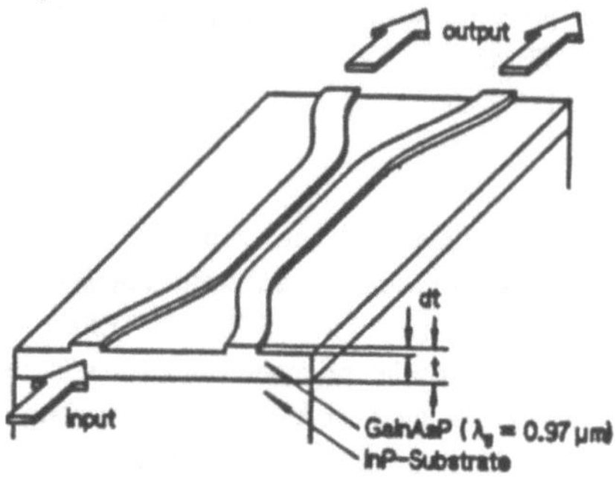

Figure 1: Design of an optical coupler on GaInAsP/InP

The cross-section of the coupler consists of rib waveguides fabricated on GaInAsP grown on InP. The wave is guided in the high refractive index region below the ribs. The interaction between the waves below the adjacent ribs leads to a symmetric and antisymmetric mode of the coupled structure. The field distribution of theses modes have been calculated by two dimensional FDM. Figure 2 shows the isolines of the field distribution of the symmetric mode.

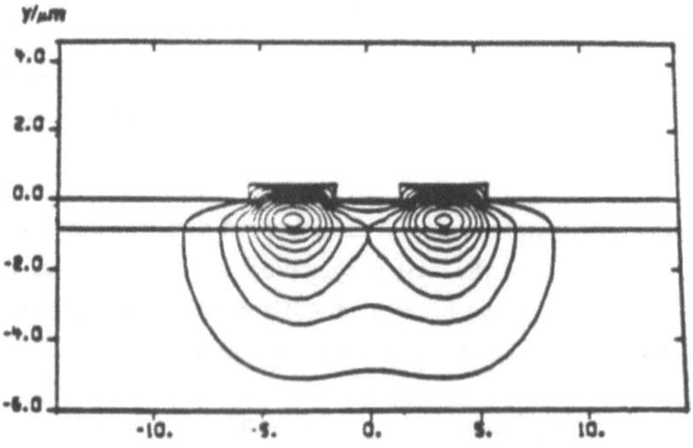

Figure 2: Isolines of the field distribution for the symmetric mode of the coupler.

The coupling length L_C is evaluated from the difference of the propagation constant of the symmetric mode β_S and of the antisymmetric β_A mode by

$$L_C = \pi/(\beta_S - \beta_A).$$

This coupling length for which half of the wave intensity has been coupled over depends sensitively on variations of the waveguide geometry and the gap between both waveguides. Therefore already small variations due to processing tolerances will largely influence the performance of a 3-dB-coupler.

To analyse the sensitive parameters of this structure, the dependence of the coupling length L_C on the layer thickness, rib height and coupler gap has been calculated and is plotted in figure 3. These calculations can be done with sufficient accuracy with the simple EIM as long as the rib height is small. For large rib heights the coupling length has been found to be wrong by 30 % and more. In this case, the two-dimensional FDM was applied.

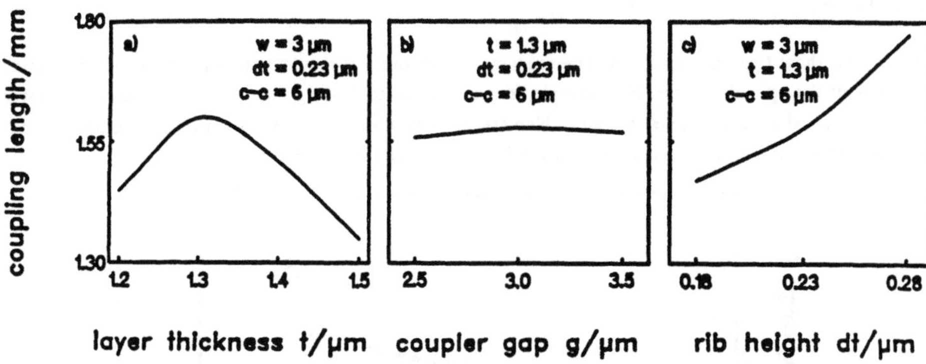

Figure 3: Calculated coupling length to evaluate the parameters most sensitive to fabrication tolerances

Figure 3a shows the dependence of L_C on the epitaxial layer thickness of the quaternary layer t. L_C has a maximum around t= 1.3 μm which means that at this thickness the coupling length is insensitive with respect to layer thickness variations. Next, we look at variations of the coupling length with coupler gap. In this case, deviations of the coupling gap from the nominal value are due to variations during the photolithographic exposure. It is known that these variations can change the coupler gap as much as 10% but leave the distance c-c between the center of both waveguides constant. This means that if the gap increases the width of the waveguide decreases and vice versa. The numerical simulation for this case is shown in figure 3b. Fortunately, changes of L_C due to gap variations are compensated for by changes of the width variation. We see that in this case fabrication tolerances are optimized because of a specific process used.

Finally, figure 3c shows that L_C always increases with the rib height dt. Therefore dt is the sensitive parameter of this structure. A possible way to avoid the strong sensitivity of L_C on dt is to find other coupler structures. By numerical simulations of different coupler structures (figure 4)we have found that rib waveguides having an InP rib on top of the quaternary layer are much less sensitive to rib height tolerances compared to the quaternary rib waveguide couplers discussed here.

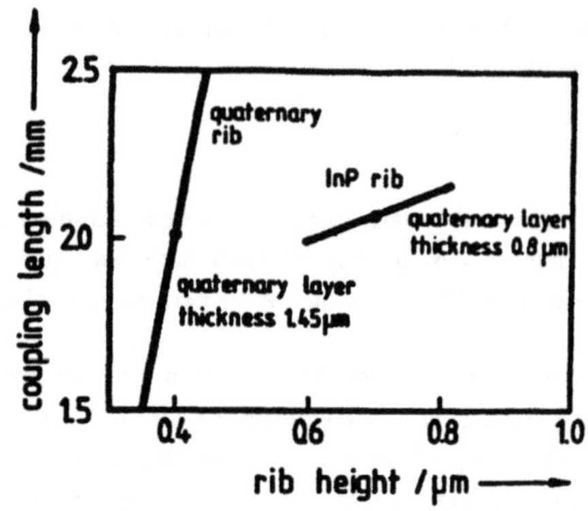

Figure 4: Variation of the coupling length with the rib height for a coupler with a quaternary rib and a coupler with an InP-rib.

We will next show that comparison of experimental and theoretical results can also be used to determine properties of the epitaxial layers. For example, it is important to know if there are variations of the layer thickness over the semiconductor wafer. Numerical simulations have shown that L_C only depends exponentially on the coupler gap if the quaternary layer thickness remains constant.

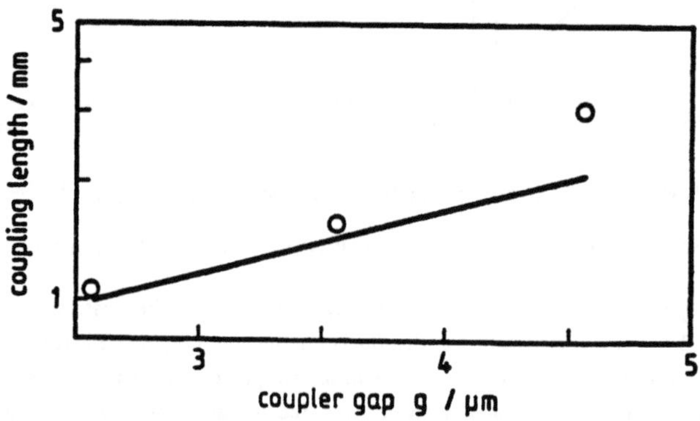

Figure 5: Deviation of the measured coupling length (circles) from the theoretical curve (solid line) calculated for constant layer thickness.

For a specific coupler grown by liquid phase epitaxy measurements of L_C showed a nonexponential behavior (figure 5). The calculated values could only be fitted to the experimentally found L_C if a varying quaternary layer thickness over the wafer was assumed. After cleaving the wafer, measurements by scanning electron microscopy confirmed our assumption that a gradient in the layer thickness existed.

Finally, since the couplers depend sensitively on the refractive indices of the epitaxial layers refractive indices can be determined by coupler measurements if the geometrical parameters are known. Usually, to determine the refractive indices, the luminescence peak position of the band to band transition is measured. The refractive index is then evaluated from a formula fitted to experiment relating the luminescence peak position to the refractive index. Several such formulas exist [7] and the calculated refractive indices can differ considerably with respect to the fitting formula used. From a great number of comparisons between calculations and experiments we have found that the formula by Henry et al. [8] yielded the best agreement with measured waveguides with quarternary layers having an energy band gap equivalent wavelength between 0.95µm and 1.06µm .

Conclusion

Theoretical approaches to simulate integrated optical devices lead to a number of numerical simulation methods which differ in their degree of sophistication. Input data for the numerical simulation are the device geometry and the refractive indices. Both validity of the theory and accuracy of the measurements can be checked by comparing the simulated optical performance with the experimental results. Main aims of numerical simulation are:

- understanding of the device performance
- quantitative determination of the device characteristics
- selection of a best set of parameters which are least sensitive with respect to variations of the fabrication processes
- choice of a common waveguide suitable for integration

References:

[1] A paper dealing with the integration of different waveguide devices entitled "Elements for Integrated Optics Coherent Receiver Networks", will be given by H. Heidrich et al. at this conference.

[2] Working Group I, COST 216: "Comparison of different modelling techniques for longitudinally invariant integrated optical waveguides", IEE Proceedings, 136 (1989), pp. 273-280 and references therein.

[3] P. Albrecht, M. Hamacher, H. Heidrich, D. Hoffmann, H.-P. Nolting, C. M. Weinert, IEEE Photon. Technol. Lett. 2 pp. 114-115

[4] R. März and H.-P. Nolting: "Spectral properties of asymmetrical optical directional couplers with periodic structures", Opt. Quantum Electron. 19 (1987), pp. 273-287

[5] M. Schlak, H.-P. Nolting, P. Albrecht, W. Döldissen, D. Franke, U. Niggebrügge, F. Schmitt: "Integrated-optic polarization convertor on (001)-InP substrate", Electron. Lett. 22 (1986), pp. 883-885.

[6] H.-P. Nolting and M. Gravert:in proceedings of "Numerical Simulation and Analysis in Guided-Wave Optics and Optoelectronics", Houston (1989) pp. 24-27.

[7] For a discussion of the different formulas see: F. Fiedler and A. Schlachetzki, Solid-State Electronics 30 (1987) pp. 73-83

[8] C. H. Henry, L. F. Johnson, R. A. Logan and D. P. Clarke: IEEE J. Quantum Electron. QE-21 (1985) pp. 1887-1892.

CAD-Tool for Integrated Optic Coupling Gratings

V.A. Sychugov, A.V. Tishchenko, M.S. Klimov
Institute of General Physics, Academy of Sciences of the USSR, Moscow.

O. Parriaux
Swiss Center for Electronics and Microtechnology, CH-2000 Neuchâtel

ABSTRACT

A CAD tool allowing the analysis and synthesis of integrated optic coupling gratings is presented. Finite size gratings of arbitrary groove shape with finite incident beam cross-section of any polarization and direction are considered.

I. INTRODUCTION

One generally agrees that the industrial future of integrated optics strongly depends on the solution given to a plumbing problem. When fibers are concerned, the plumbing problem is that of the waveguide edge preparation and of the best matching of the incident and coupled fields. However, a number of integrated photonics applications do not rely on fibers; this is particularly the case in the first important non-communication industrial application of integrated optics: interferometric distance measurement(1). In such case, grating coupling is an attractive solution as it is compatible with batch planar technologies and avoids the problem of lengthy preparation of the integrated optic chip edges. Grating coupling structures have not received much attention recently. However two major recent achievements are about to give a new impulse to their development.

The first achievement concerns the writing technique. It is now possible to write any type of groove line one by one with sufficiently short pitch by means of electron beam pattern generators(2). This means that a corrugation grating can achieve its conventional coupling function together with a wave front transformation, e.g., focusing.

The second achievement concerns the coupling efficiency. Careless grating coupling generally leads to a coupling efficiency lower than one percent. Careful adjustment leads to 20 - 40 % mainly because of the inherent existence of more than one diffraction order. It is now possible by simple technological means avoiding the difficult fabrication of blazed gratings to achieve coupling to a single diffraction order(3).

It has now become important to have a design method allowing the analysis of actual coupling structures and their optimization within the limits imposed by a practical opto-geometric configuration. This is what the present contribution does.

II. THE CAD - TOOL

The program deals with corrugation gratings realized on top of, or within planar optical waveguides. It provides the optimum opto-geometrical coupling configuration between an incoming Gaussian beam and a finite area waveguide grating and it calculates the coupling coefficient to TE and TM propagation modes under Rayleigh approximation[(4)].

The index profile of the planar waveguide can be a step, a Gaussian or an ERFC function. It can also be any function that the user may want to define. The grating is located at the waveguide - superstrate interface. A grating period is composed of a groove and a ridge whose mark-space ratio can be arbitrary. The grating groove shape can be sinusoidal, rectangular, triangular, trapezoidal. The user can define his own groove shape provided the latter has no overhang. The grating can be etched into the waveguide. It can also consist of a deposited thin film in which the grooves will be etched down to the waveguide surface or only partially. Figure 1 sketches the type of structure with the used parameters.

The incident beam has a circular Gaussian amplitude distribution. The incidence is from the superstrate side with arbitrary direction. The beam polarization can be linear, circular or have any elliptical polarization state. This has important consequences in the case of non-colinear incidence.

The analysis of the wave coupling problem rests on the essential hypothesis that the Rayleigh criterium $d/\Lambda < 0.144$ must be satisfied where d is the grating depth and Λ is the grating period. The Rayleigh criterium is the only fundamental assumption in the underlying theoretical development. There is no limitation in the index difference between the various materials involved provided reasonable index values are taken, i.e., $n > 0$. There is no fundamental limitation on the grating depth / waveguide width ratio ($d/h < 1$) but the program will deliver correct coupling values only for $d/h \ll 1$.

The possibilities that this CAD tool offers are: calculation of the coupling efficiency in a given configuration imposed by practical conditions and calculation of the optimum coupling configuration. The optimization procedure is performed in two steps: the phase matching condition is first set by adjusting the free parameters; then the optimal coupling efficiency is searched for in terms of the beam radius, the grating size and the impact of the beam center on the waveguide surface.

III. EXAMPLES

The program will be applied hereafter to three examples of application:

1) Saw-tooth groove profile

As discussed by Tamir[5], saw-tooth profiles are expected to provide a means of directing the coupled power preferentially into the substrate or into the superstrate. This problem is addressed again here in the configuration described in figure 2a) where input coupling of a finite size Gaussian beam via a finite size grating is considered in the colinear case.

The waveguide has a complementary error function profile given by

$$n(x) = 1.52 + 0.01 \cdot \mathrm{ERFC}\,(x/h)$$

with h = 2 microns. The guide is excited at 780 nm wavelength. The grating period for phase matched excitation is ca 0.35 micron for both TE_o and TM_o modes at 45° incidence in the contradirectional coupling case.

A grating period consists of a 50 nm height tooth of TiO_2 (index 2.3). Its symmetry is varied from $a/\Lambda = 0$ to 1 where a is the apex coordinate. Figures 2b) and 2c) show the dependence of the optimum input coupling efficiency η and of the total radiation loss from the waveguide in cm^{-1}, α versus the normalized apex.

The results on the coupling efficiency for a TM wave are qualitatively in agreement with Tamir's expectations[5]. TM efficiency goes to a maximum where the apex heads opposite to the incident beam direction. The TE mode efficiency remains remarkably constant. The total radiation loss coefficient exhibits a maximum for both polarizations at 0.5 normalized apex where the fundamental Fourier component is maximum.

2) Incident beam parameter values for optimum coupling

Future applications in the field of microtechnology and interferometric movement measurement will use planar waveguides and flat optical antennea with non-colinear incidence. In such case the incidence parameters are four in number: beam radius; impact - grating front edge distance; angle between the incident beam projection and the edge of the grating area; polarization.

Figure 3a), 3b) and 3c) represent the dependence of the first three versus the free parameters: angle of incidence θ and the angle of mode propagation under the assumption that the phase matching condition is satisfied for optimum coupling. These relationships between configuration parameters are true whatever waveguide and grating groove shape and material.

Figure 4a) and b) shows the influence of the input beam polarization on the input coupling efficiency in a non-colinear case with a step index waveguide of substrate, guide index 1.52, 1.66 respectively and film thickness 500 nm. The grating is a sinusoidal corrugation of period 618 nm, thickness 50 nm. For a given incidence angle of 45° and an angle between the plane of incidence and the mode propagation direction of 45° and for the condition of optimum coupling (beam radius, impact location and size of grating area), the solid line of figure 4a) and b) represents the direction of the principal axis of a given elliptical input polarization ensuring maximum coupling. In figure 4a), for TE mode coupling,the ratio between the polarization ellipse axes was set to 0.125 whereas in figure 4b), for TM mode coupling, the ratio is 0.382. The dashed line represents the coupling efficiency obtained if, instead of an elliptical polarization, a linearly polarized input beam is used with the same orientation as the principal axis of the ellipse. This shows that the effect is more pronounced in the case of TM mode excitation than for TE modes.

REFERENCES

(1) P. Roth, O. Parriaux, "Integrated optic interferometer with phase diversity", IOOC'89, Kobe, Japan, July 18-21, 1989, pp. 120-121.

(2) T. Suhara, H. Nishihara, J. Koyama, "High-performance focusing grating coupler fabricated by electron-beam writing", IGWO'84, Kissimmee, USA, April 24-26, 1984, paper ThD4-1.

(3) I.A. Avrutsky, A.S. Svakhin, V.A. Sychugov, "Interference phenomena in waveguides with two corrugated boundaries", J. Modern Optics, 36(1989), pp. 1303-1320.

(4) V.A. Sychugov, A.V. Tishchenko, "Light emission in a corrugated dielectric waveguide", Kvantovaia Electronica, 7(1980), pp. 326-331.

I.A. Avrutsky, V.A. Sychugov, A.V. Tishchenko, IOFAN report, to be published.

(5) S.T. Peng, T.Tamir, "Directional blazing of waves guided by asymmetrical dielectric gratings", Optics Commun., 11(1974), pp. 405-409.

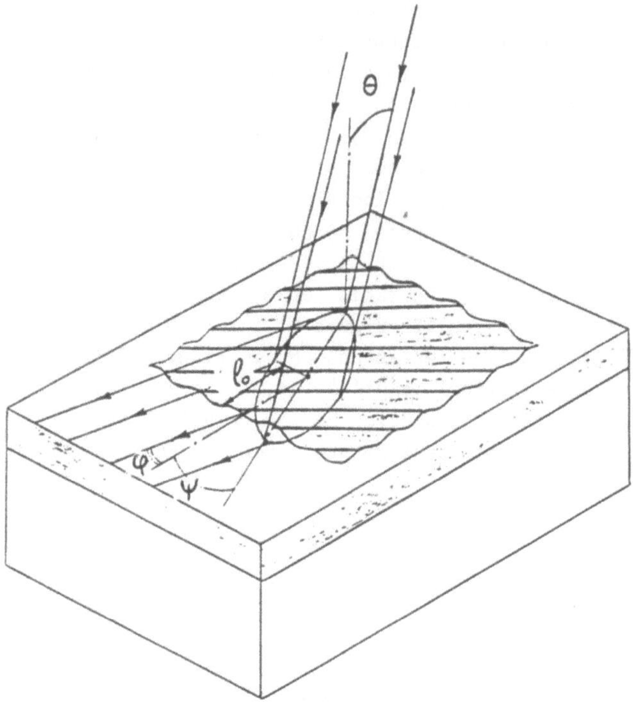

Fig. 1

General configuration of the waveguide coupling of an input Gaussian beam via a finite size corrugation grating.

Input parameters are:

* incidence angle
* angle between incidence plane and grating grooves
* angle between guided beam and grating grooves
* Polarization state and wavelength.

Optimization parameters are:

* beam radius
* normal distance of impact to grating front edge
* angle between grating grooves and front edge.

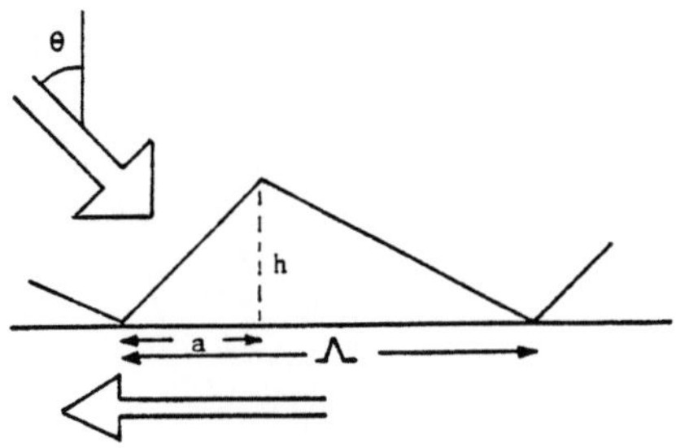

Fig. 2a)

Saw-tooth groove profile of a high index grating with contra-directional waveguide coupling.

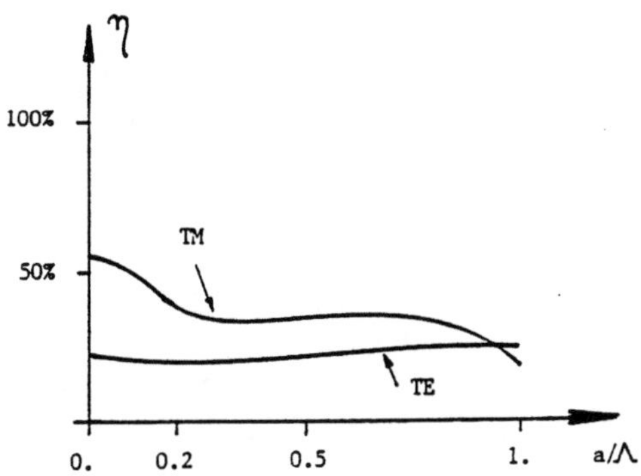

Fig. 2b)

Coupling efficiency versus normalized apex

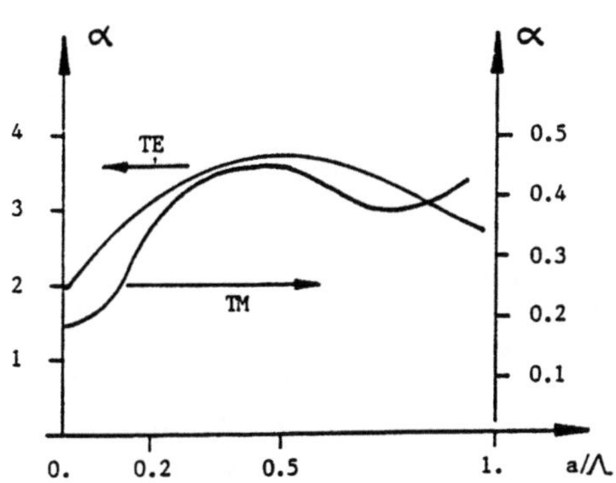

Fig. 2c)

Total radiation loss term in cm^{-1} versus normalized apex.

Fig. 3a)

Dependence of the beam radius normalised to the radiation loss coefficient under phase matching and optimum coupling conditions versus angle of incidence and mode propagation angle.

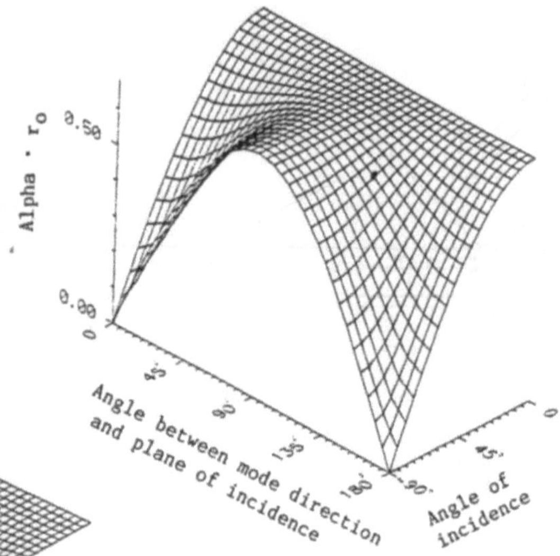

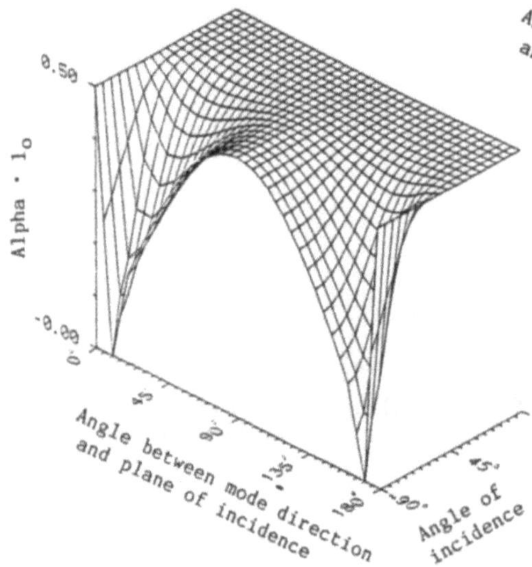

Fig. 3b)

Dependence of the impact distance from grating edge normalised to the radiation loss coefficient versus same parameters under same conditions.

Fig. 3c)

Dependence of the angle between the plane of incidence and the boundary of the grating area versus same parameters and same conditions.

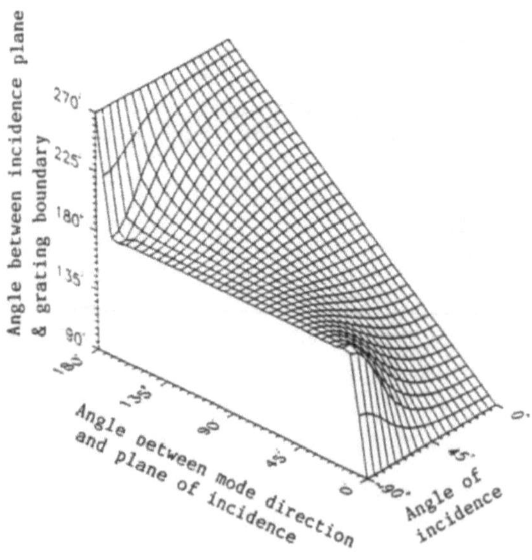

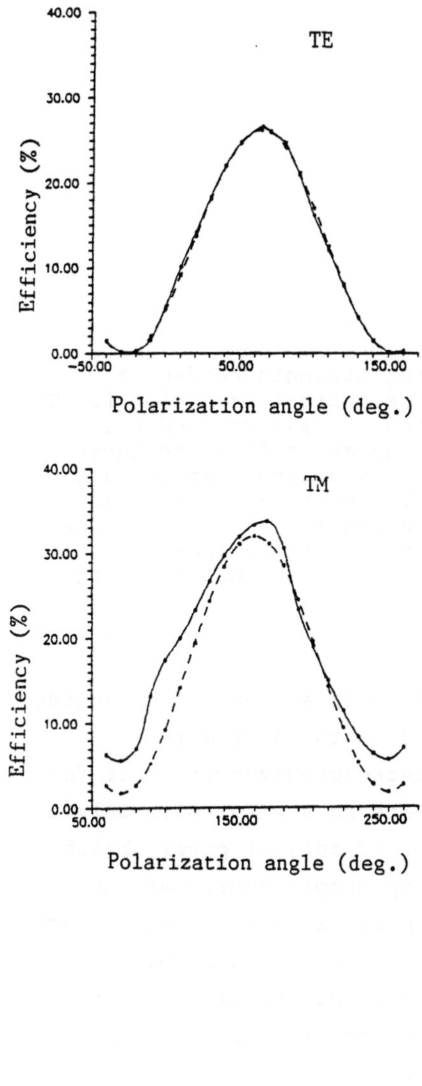

Fig. 4

Coupling efficiency in a non-colinear configuration versus the polarization angle of the input Gaussian beam.
Solid line : elliptical polarization,
Dashed line: linear polarization.

a) Excitation of TE mode
b) Excitation of TM mode

Simulation on Lateral Detector – Optical Waveguide Coupling

K.Kapser, P.P.Deimel

Messerschmitt-Bölkow-Blohm
ZTA 13, PO Box 80 11 09
8000 München 80

Summary
We have calculated the optical coupling strength between a lateral photodetector and a single-mode optical waveguide for TE and TM modes, respectively. In addition we have studied the influence on the coupling strength of an antireflection layer between waveguide and photodetector. In a second example we have looked at the optical interaction of two parallel waveguides using even and odd supermodes. We have studied the case of one of the two waveguides having a complex refractive index and its influence on the coupling length compared to the nonabsorbing case.

Introduction

The integration of an optical detector with an optical waveguide offers a number of advantages compared to the customary fiber-detector coupling scheme. The detector-waveguide pair for example can be batch-produced in numbers on a wafer with only a few alignment steps needed. The amount of optical power absorbed from the waveguide can be determined by proper choice of the detector length and thickness. Both, polarization and wavelength of the optical modes have a large influence on the coupling strength between waveguide and detector. This provides a scheme to selectively look at the information contained in various modes differing in wavelength in the guide.
In connection with silicon as a substrate material a number of new concepts and possibilities open up to combine micromechanical sensor ideas with optical readout techniques. In Fig. 1 a typical example is shown which combines an anisotropically etched trough with a freestanding beam carrying an optical waveguide.[1]

In our preliminary work we have concentrated on the device parameters to obtain a feeling for the processing parameters needed to fabricate a waveguide-detector pair.

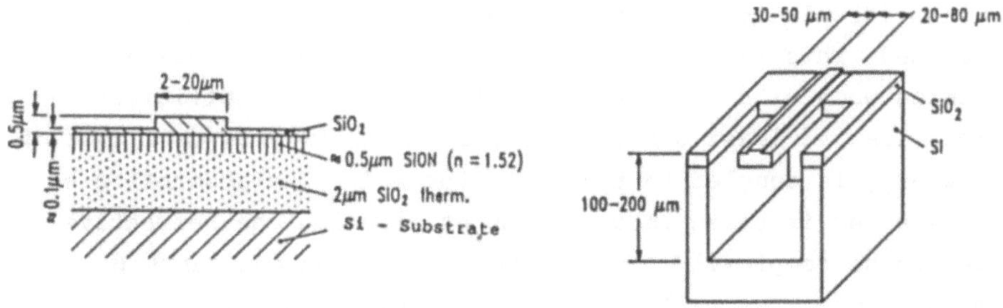

Fig. 1: SiON rib-waveguides and anisotropically etched Si

Model

The principle structure of the waveguide-detector combination is shown in Fig. 2. Light is propagating in the z-direction. At z=0 the incoming waveguide mode is transferred to the detector modes.Solving the Maxwell-equations for this problem with numerical techniques leads to the propagation constants β_j for the various eigenmodes of the system.[2]

At z=0 the boundary condition for the tangential fields of the incoming mode and the fields of the waveguide/detector-system (i.e. detector modes) must be satisfied. In general, the excitation of radiation modes at the discontinuity at z=0 is possible. Also, a part of the incoming wave is reflected.

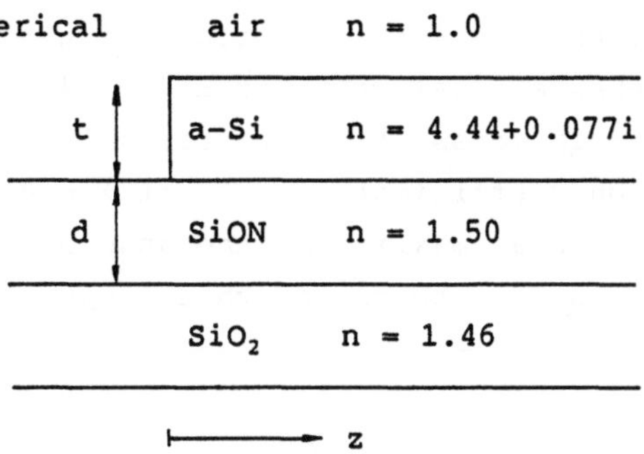

(Fig.2)

$$E_0 + r * E_0 = \Sigma\, a_j E_j + E_{rad} \Big|_{z=0}$$

E_0 : incoming mode

$r*E_0$: reflected mode

E_j : eigenmodes

E_{rad} : radiation modes

In this calculation, reflection and radiation modes are neglected. This is only reasonable, if the overlap of the incoming mode with at least one detector mode is sufficiently large (i.e. at least one a_j is close to 1). In this case most of the energy is coupled to that detector mode. The absorption of the detector is taken into account by allowing the refractive index to be complex.[3,4,5] Thus, the propagation constants β_j are complex numbers as well. The real part gives the effective refractive index of the mode, whereas the imaginary part is related to the absorption coefficient α.

$$\beta = \beta_r + i*\beta_i \qquad n_{eff} = \beta_r/k_0 \qquad \alpha = 2*\beta_i$$

Calculation

We have calculated the absorption coefficient of the coupled waveguide/detector system as a functon of detector thickness (Fig.3). The absorption for TM modes is about one order of magnitude larger than that of the TE modes. With increasing detector thickness the amplitude of the oscillations decreases and reaches a final value of α = 0.04 dB/μm for TE and α = 0.36 dB/μm for TM modes at a detector thickness t of about 3 μm. (The absorption coefficient of a-Si at λ = 0.6328 μm is α = 6.64 dB/μm). A local maximum of the absorption occurs when the detector is just thick enough to support a new mode.

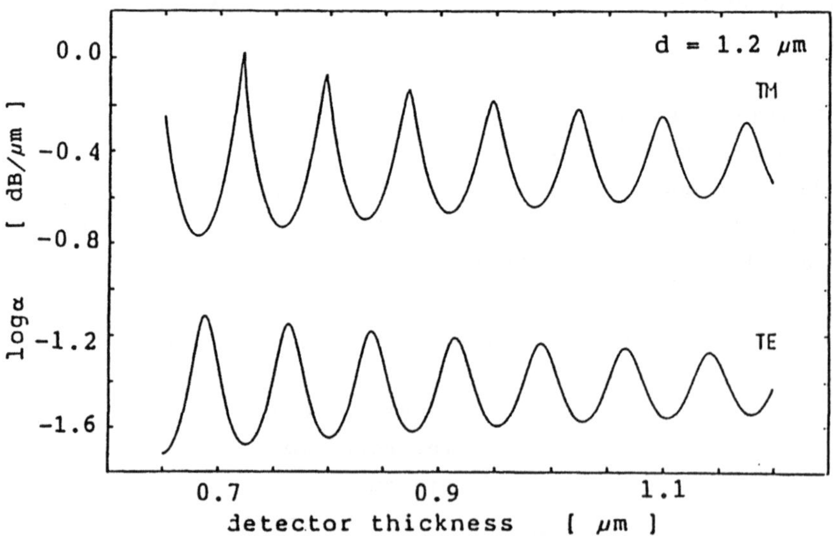

Fig. 3: Optical absorption α as a function of detector thickness

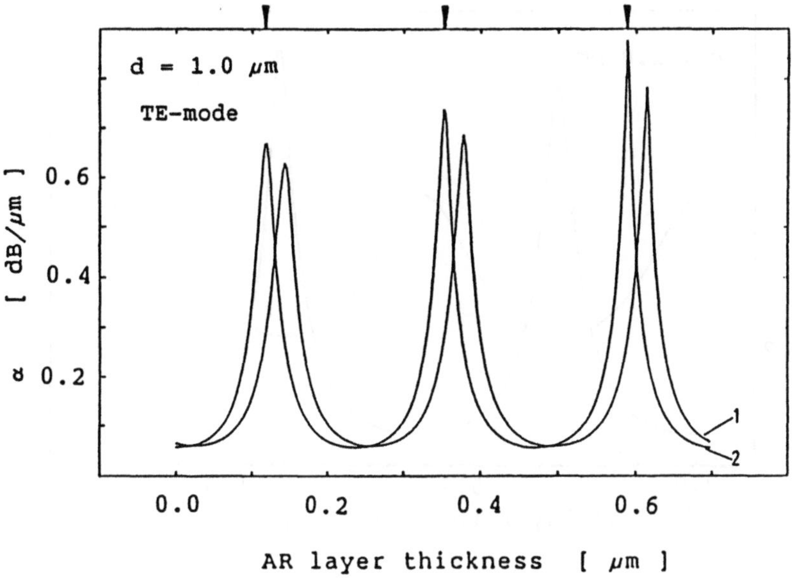

Fig. 4: Optical absorption as a function of AR layer thickness for a thin (0.7μm,"1") and thick (1.5μm,"2") detector layer

To enhance the absorption in the detector we introduced an additional layer between the waveguide and the detector.[6] This layer works as an antireflective (AR) layer, if its refractive index is larger than the index of the waveguide (Fig.4). The results show that a classical analysis holds only for a thick detector (The arrows show the classicaly expected values). Using a thin detector the reflection at the detector/air interface can not be neglected and the simple 3-layer model for an AR layer is incorrect. Fig.5 shows the absorption for TE and TM modes as a function of the AR layer thickness at a detector thickness of 1.5 μm. The refractive index of the AR layer is taken as n_b = 2.0 .A proper choice of the AR layer thickness can greatly enhance the absorption coefficient of either the TE mode or the TM mode. A further optimization may be possible by adjusting the refractive index of the AR layer.

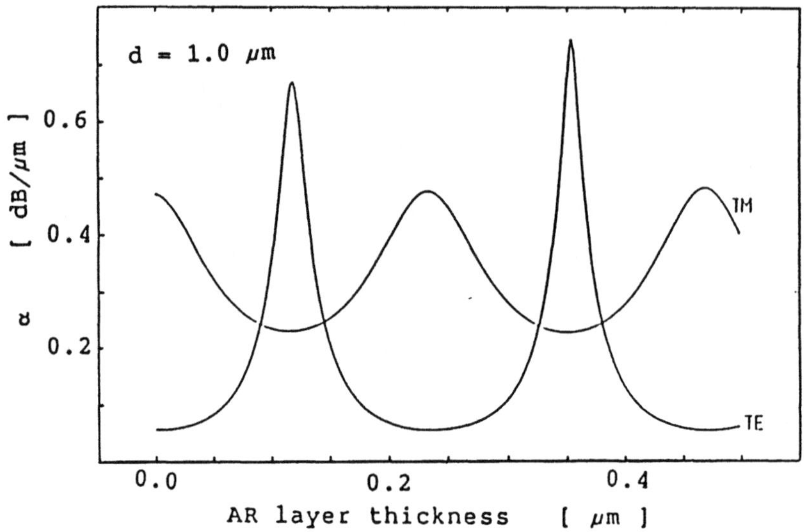

Fig. 5: Optical absorption as a function of AR layer thickness

Additionally, we have calculated the coupling length of two parallel waveguides seperated by a buffer layer.[7] One of the waveguides is assumed to be absorbing. In the same way as in the first example, the eigenmodes of the system can be calculated. They are called the supermodes of the system. If the two waveguides are sufficiently thin only two supermodes, the even and the odd one with propagation constants β_e and β_o, exist. If light is coupled at z=0 into waveguide 1, both supermodes will be excited and the interference of the two modes causes the energy being transfered to waveguide 2. After the coupling length l_c the energy in the second waveguide reaches a maximum and then swings back to the first. In the nonabsorbing case the coupling length can be calculated as

$$l_c = \pi/(\beta_e-\beta_o)$$

where β_e and β_o are the propagation constants of the even and odd supermodes. If one of the waveguides is absorbing, however, this simple relation does not hold. The coupling length l_c, defined as the length at which the energy in waveguide 2 reaches a maximum if light is coupled into the absorbing guide 1 at z=0, must be calculated numerically. Fig. 6 shows the intensity of light in waveguide 2 with various absorption constants of waveguide 1. With increasing imaginary part of the refractive index the peak of the intensity shifts towards smaller values. Fig.7 shows the calculated coupling length compared to that one would get by using the simple formula for nonabsorbing waveguides.

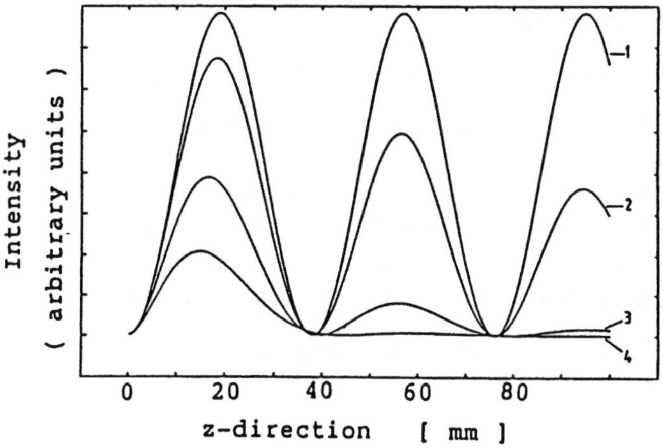

waveguide thickness : 0.4 μm
waveguide seperation : 1.7 μm
waveguide ref.index : 1.60
cladding and buffer layer ref.index : 1.46
λ = 0.6328 μm

Fig. 6: Optical intensity in the second nonabsorbing waveguide as a function of the propagation in z-direction. (parameter $Im(n_1)$: "1" = 0, "2" = 10^{-6}, "3" = $5*10^{-6}$, "4" = 10^{-5})

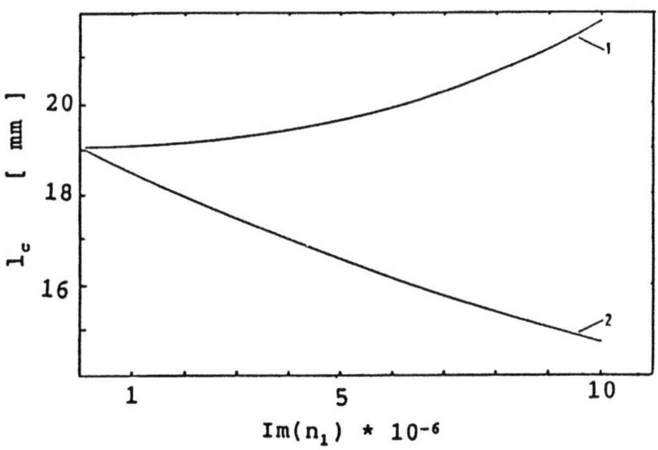

Fig.7: Coupling length l_c as a function of $Im(n_1)$
"1" : $l_c = \pi/(\beta_e-\beta_o)$, "2" : exact numerical calculation

Conclusion

The combination of lateral detector and optical waveguide using silicon as a substrate offers a wide range of possibilities in the area of optically addressed micromechanical sensors.

References

1. Private communication from Prof. E. Voges, Universität Dortmund

2. A.W.Snyder, J.D.Love, "Optical Waveguide Theory",Chapman and Hall Ltd, 1983, New York

3. M.C.Amann, Electr.Lett., 23(17), 1987, p.895

4. J.F.Vinchant, J.P.Vilcot, J.L.Lorriaux, D.Decoster, Appl.Phys.Lett., 55(19), 1989, p.1966

5. F.Mallecot, J.F.Vinchant, M.Razeghi, D.Vandermoere, J.P.Vilcot, D.Decoster, Appl.Phys.Lett., 53(25), 1988, p.2522

6. R.J.Deri, O.Wada, Appl.Phys.Lett., 55(26), 1989, p.2712

7. B.Broberg, B.S.Lindgren, M.G.Öberg, H.Jiang, J.Lightwave Techn., LT-4(2), 1986, p. 196

Simulation and Design – Electrical and Thermal Properties

Electrical Connections Through Silicon Wafer: Application to 3D-Multichip Modules

G. NICOLAS LETI-CENG-CEA Grenoble (F)
T. DUPEUX THOMSON-LCC Courbevoie (F)

Summary
Electrical connections through silicon wafers could have a direct application in 3D packaging. We discuss in this paper the way to use a "Blind Hole Process", developed in Magnetic Thin Film Head Technology, to achieve improvement over conventionnal Dual in Line Packages.

INTRODUCTION

Through Hole connections have already been proposed to solve different problems in microwaves devices, especially to improve connection to ground, through Alumine or Gallium Arsenide substrates [1, 2].
Nowadays, connecting both sides of a silicon wafer has an application in Thin Film Heads fabrication [3, 4]. The Blind Hole concept [6] that has been studied for such an application, could be extend to 3D packaging.
In a first part, we recall the main processes investigated in order to connect faces of a substrate.
The second part is focussed on the LETI Blind Hole Process applied to Magnetic Heads. Endly, an analyse of integrating this basic process in packaging is given.

1. BACK-GROUND TECHNIQUES

In order to form an electrical connection through silicon wafer, we must consider 3 main steps : drill the hole, insulate the walls, deposit the conductive material.

Forming the Hole

Silicon is a material of first interest in packaging considering its mechanical properties, thermal expansion coefficient and thermal conductivity.

At first sight, chemical etching is the simpliest way to form holes through silicon. In fact, a great choice of bathes is available [7, 8] ; etch rate and anisotropy depend on cristallographic orientation, doping type and level of the substrate.
Unfortunately, the aspect ratio, length versus diameter is low, less than 1. So integration is limited by space between holes [12].
Usually the alternative to chemical etching, in microelectronics, is plasma dry etching. Etching through a wafer requires high etch rate and excellent selectivity over the mask. Conventional 13.56 MHz Reactive Ion Etching does not match these requirements. With a Microwave Propagative Ion Etching system [13] etch rate greater than 10 μm/min and selectivity Si versus SiO_2 over 100 have been performed in LETI. However aspect ratio has not been significantly improved.
The only way to achieve a breakthrough in aspect ratio is a laser beam, combined or not with reactive gazes, photochemical [14-21] processing or laser drilling. Most of power laser interact with silicon and the dimensions of holes will depend on the laser choice.
As far as we can see, solid state YAG laser at 1.06 μm or frequency doubled at 532 nm is the most attractive. A YAG laser run in Q-switch mode provides enough power to drill holes less than 10 μm in diameter, with high speed 200 holes/s. The drilling speed is then limited by positionning the wafer.

Insulation

For most applications, the connections have to be insulated each from the others. Thermal SiO_2 is well-know in microelectronics for its good insulating properties. The problem is in this case the high temperature, near 1000°C required by oxydation processes. This leads to handle to process at the beginning of the technology.
An other technique that have been use is to polarize a cylindrical junction formed around the hole [22]. But diffusion of impurities is also a high temperature process. A major issue, is to develop a low temperature oxyde deposition process, conformal in holes of 500 μm depth and 20 μm diameter.

Deposition of conductor

Most of the method described in papers combines sputtering and electroless or electroplating [23, 24]. These techniques appear to be limited by aspect ratio of 6. In fact the deposition process in limited by the diffusion of species in the hole.

To get round this problem chemical, vapor deposition is a solution when the process is limited by the surface reaction and when surface diffusion is high. This is the case for Polysilicon below 650°C and Tungsten.

2. THE BLIND HOLE PROCESS (5)

In previous studies several ways to realize holes have been analysed : chemical etching, dry etching, ultrasonic drilling and laser drilling. Among all ot these techniques, laser drilling seems to provide the best compromise between high aspect ratio, diameter versus depth hole (1:50), relatively high drilling speed (10 holes/sec), accuracy in hole placement with double-sided alignement (± 2.5 μm).

A new cassette to cassette laser drilling machine have been designed for our needs by Microcontrole (located in Evry, France) to achieve above requirements. The machine combines YAG laser drilling and fast and accurate placement with an interferometric stage.

The laser delivers 10 Watts output in the continuous wave mode. To drill silicon wafers, the laser is operated in the Q-switch TEM00 mode and the power density used is about 10^{11}W/cm^2.
The wavelength of 1.063 μm is particularly suitable for absorption of photon (1.2 eV) by electronic band to band transition through silicon gap (1.1 eV). On the contrary, the photon energy is lower than the gap energy of silicon dioxyde (8 eV) and then only intraband or phonon absorption is possible. Coupling in the regime is weak [15] fig 1-2. Then it is possible to drill silicon without damage on a silicon dioxyde layer, located on the laser exit face. This provides a "blind hole" in silicon, covered by a thin SiO_2 layer.

The principal source of damage to the wafer undergoing laser drilling is the debris ejected from the hole during the drilling. Molten debris and vaporized material fall and condense on the surface of the wafer around the entrance to the laser drilled hole and also on the inner walls of the hole. The surface of the wafer can easily be protected by as sacrificial layer of resist or SiO_2, but the debris on inner walls forbide proper insulation by thermal SiO_2. So, we use a silicon chemical etchant (KOH), to remove these debris and to obtain flat and smooth inner surface, which permits proper growth of thermal SiO_2.

To reduce the capacitance of the connection we are using thick SiO_2 layers (about 2 μm). In this case, the leakage current measured under 100 Volts DC, is 10^{-11} A.

As the aspect ratio of holes is high (1:50) it is difficult to use metallization processes as evaporation, sputtering, electroless or electroplating usually used by other authors [1, 2, 24]. So we deposit CVD-Tungsten layer which seems to be a very promising material in microelectronic for metal levels and contact holes. Unfortunatelly CVD-Tungsten has a poor adhesion on SiO_2, so our conductive layer is a double layer of polysilicon (0.15 μm) and Tungsten (1 to 2 μm).
Photo 1 shows the good step coverage of the CVD-Tungsten. Photo 2 and Photo 3 show the contact obtained after drilling with the "blind hole process", (fig. 3) after deposition of tungsten and etching of the SiO_2 sacrificial layer. The roughness of the surface has been transfered from the tungsten during dry etching (the resisitivity of tungsten layer is about 10^{-5} Ω/cm).

3. APPLICATION TO 3D PACKAGING

This through hole connection process on silicon wafers is now developed for future packaging application in three dimensional integration.

Stacked silicon frames carrying chips, associated to multilayers interconnection systems need vertical connections for electrical vias and for thermal paths.

For Texas Instruments [26], 3D packaging promises a 250 to 300 times improvement in memory density over conventional Dual In line Packages. If power dissipation does not pose problems it is the most dense of any packaging technique to date. System modules toward the mid 1990s will be based on the junction card concept, utilizing silicon substrates and incorporating this third dimension.

In Europe, 3D integration and via forming technology, are included in the Basic Long Term Research Packaging JESSI project. The goals are to characterize, in 1996, these new technologies on high speed high density demonstrators.
The main studies proposed in this program are :

- The measurements of potential laser drilling effects on active and passive circuits surrounding the holes.
- The deposition of dielectric at low temperature for hole insulation (CVD-Metal Oxydes)
- Laser system improvements : pulse energy and focus control throughput and repeatability increase of the drilling process.
- Reliability studies and measurements constitute an important part of the program.

CONCLUSION

A through hole connection process using laser drilling has been developed. We have shown that among different possible technologies the laser drilling was the one which gave the best depth / diameter ratio.
We think that after some improvements 10 to 20 micron diameters are feasible and can establish reliable electrical connections through a wafer with a resistance less than 1 Ω.
3D silicon Multichips Modules is one very promising application of this new technology. Future process developement are part of the JESSI BLR PACKAGING Project.

REFERENCES

[1] C.B. COOPER III and al, JES vol B4, n° 10
Reactive ion etching of through the wafer via connections for contacts to AsGA FET'S.

[2] R.T. THOMSON and B.W. WHITACKER,
32nd Elect. Comp. Conf. Proceedings San Diego 1982
Conductive via Processing utilizing RF Sputtering and Selective electroposition for Hybrid Thin Film Circuits.

[3] J.P. LAZZARI and P. DEROUX-DAUPHIN,
INTERMAG Fall Meeting, Washington (march 1989).

[4] P. DEROUX-DAUPHIN and H. SIBUET,
5th annual Head/Media Technology Review
Comdex Las Vegas, november 12th 1989.

[5] T. DUPEUX, H. SIBUET, P. DEROUX-DAUPHIN,
Electrical Connection Through Silicon Wafer
JES Fall Meeting, Chicago (october 1988).

[6] Y. AKASAKA and al, Trends in Three Dimensional Integration. Solid State Technology, Feb. 1988.

[7] K. PETERSEN, Proceedings on the IEEE, V° 70, N° 5, may 1982.

[8] J.B. ANGELL, S.C. TERRY, P.W. BARTH, "Pour la Science", June 1983.

[9] L. SCEPREGI, Microelectronic Engineering 3, 1985, 221-234.

[10] E. BASSOUS, IEEE Trans. on Elect. Devices n° 10, oct. 1978.

[11] E. BASSOUS, E.F. BARAN, "The fabrication of high precision nozzles by the anisotropic etching of <100> silicon, J. Electrochem. Soc. 125, p. 1321, 1978.

[12] T. DUPEUX, G. NICOLAS, P. DEROUX-DAUPHIN, NATO Workshop "in situ Technology", may 1987, Cargèse (Fr).

[13] B. CHARLET, T. DUPEUX, L. PECCOUD, "6th Intern. Colloquium on Plasma and Sputttering", June 1/5th 1987, Antibes (Fr).

[14] T.J. CHUANG, J. Vac. Science Tech. B5, 1985.

[15] D.J. EHRLICH, R.M. OSGOOD, T.F. DEUTSCH, "Laser Chemical technique for rapid writing on surface relief in silicon", App. Phys. Letters v° 38, p. 1081, 1981.

[16] D.J. EHRLICH, J.Y. TSAO, A review of laser-microchemical processing, J. Vac. Sci. Techno. B1(4), Oct/Dec. 1983.

[17] D.J. EHRLICH, D.J. SILVERSMITH, R.W. MOUNTAIN, J. TSAO, IEEE Trans. on Comp. Hybrids and Man. Techn. v° 5, n° 4, December 1982.

[18] PODLESNICK, GILGEN, OSGOOD, "Deep UV induced wet etching of AsGa", Appl. Phys. Lett. 45(5), Sept. 1984.

[19] OSGOOD, SANCHEZ, RUBIO, EHRLICH, DANEU, Localized laser etching of compound S.C. in aqueous solution. Appl. Phys. Lett. 40(5), March 1982.

[20] BREWER, McCLURE, OSGOOD, "Dye laser assisted rapid HBr etching of GaAs", Appl. Phys. Lett. 47(3), August 1985.

[21] MAEGELE, "A better way to connect optical fiber and chips". Electronics, January 6th 1986.

[22] T.R. ANTHONY, "Diodes formed by laser drilling and diffusion", J. Appl. Phys. 53(12,) Dec. 1982.

[23] T.R. ANTHONY, "The random walk of a drilling laser beam", J. Appl. Phys. 51(2), Feb. 1980.

[24] T.R. ANTHONY, "Forming electrical interconnections through semiconductor wafers", J.A.P. 52(8), August 1981.

[25] M.F. von ALLMEN, "Coupling of laser radiation to materials and semiconductors" Physical processes in laser-materials interaction, Note ASI 1980.

[26] M.L. BUSCHBON, Hybrid circuit technology jan. 90, p. 6-10, Multichips modules help sidestep costly fine pitch problems.

photo 1 : cross sectionnal view of THC. (2cm = 5 μm)

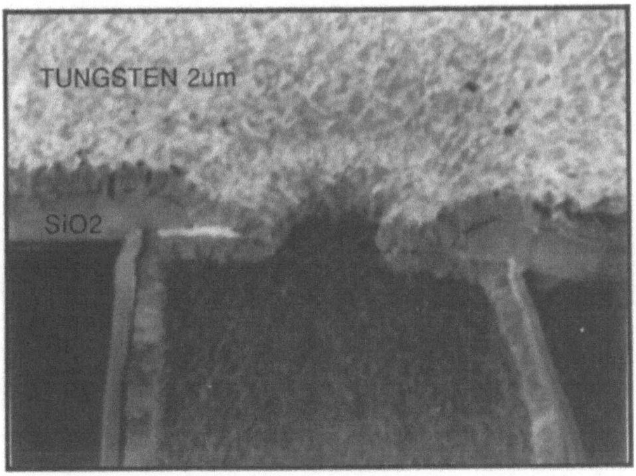

photo 2 : cross sectionnal view THC in the blind hole process (1cm = 2 μm)

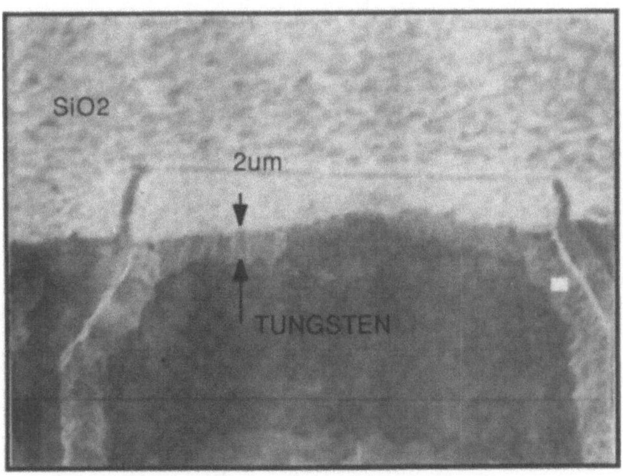

photo 3 : contact and planarized surface obtained with the blind hole process .(1cm = 4μm)

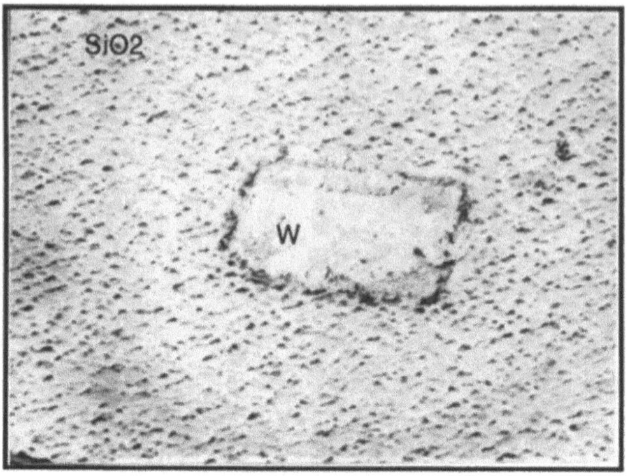

fig 1 : absorption coefficient and reflectivity of cristalline Silicon

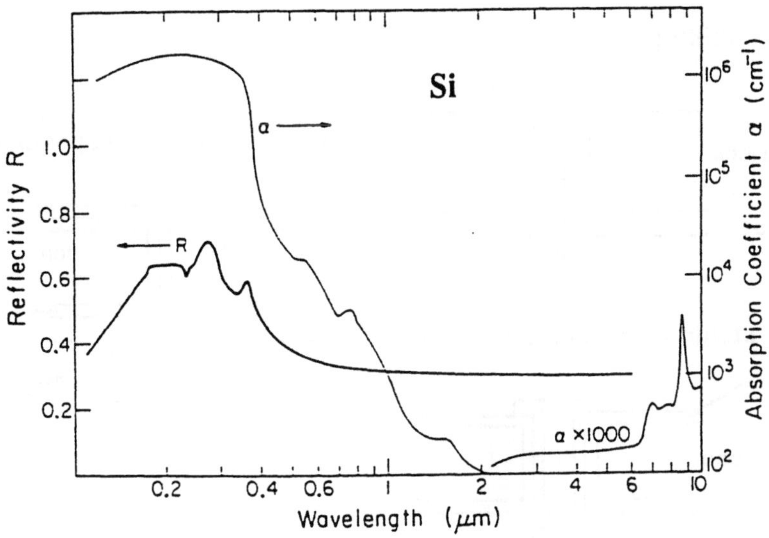

fig2 : absorption coefficient and reflectivity of cristalline Silicon Dioxyde

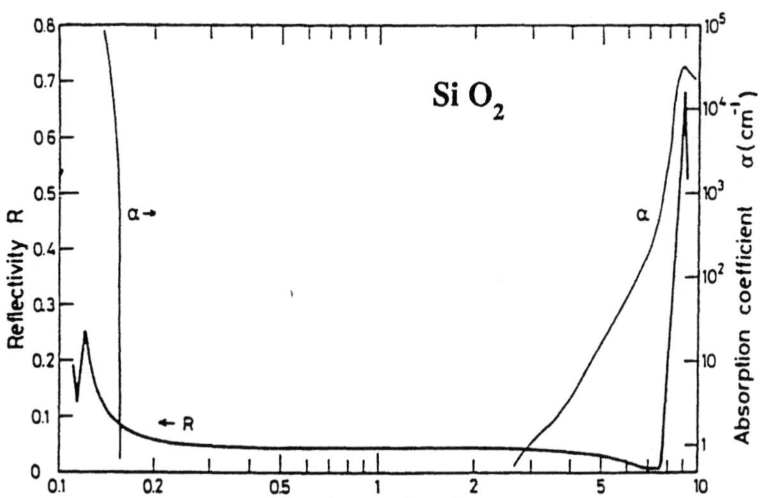

fig 3 : the 3 steps of the blind hole process

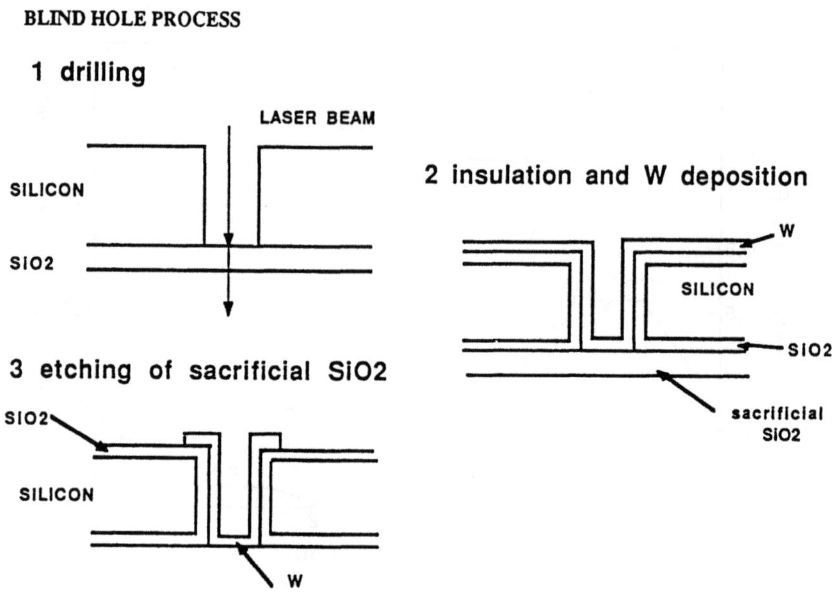

Design Methods for Multichip Systems

R. Dümcke

Technical University of Berlin
Forschungsschwerpunkt Technologien der Mikroperipherik

Summary

In the design of multichip systems the geometrical and physical properties of the materials and components play a major role. Therefore new design and simulation methods have to be introduced to handle the evaluation of the electrical performance, the thermal thermal management and to predict mechanical stresses. The integration of numerical simulation methods in the development of optimal technologies for multichip systems and in the design process is discussed.

Introduction

The introduction of Multichip modules as a first level of packaging is driven by two sources: VLSI integration trends and applications.

Multichip modules provide a solution to the packaging problems arising from the VLSI integration trends. The exponential growth of the number of transistors integrated on a single die is accompanied by an increase in chip area, pin count, clock frequency and power dissipation. Especially for high end applications, the delay times for chip to chip signal transmission determine the total system performance, therefore a close packaging of chips is required.

The other driving factor for multichip modules are applications which require compact system solutions as for example portable electronic equipment (computer, telecommunications) or in automobile and medical electronics.

Due to the higher integration levels on multichip modules the physical properties of the system have much more influence on the function than in conventional microelectronic systems. The high density of the interconnect system and the short rise times require a control of transmission line impedances, reflections at discontinuities and crosstalk. With increasing power densities and tighter packaging of components the thermal properties of the package have to be determined. The minimisation of thermally induced stresses is crucial for the reliability of the systems.

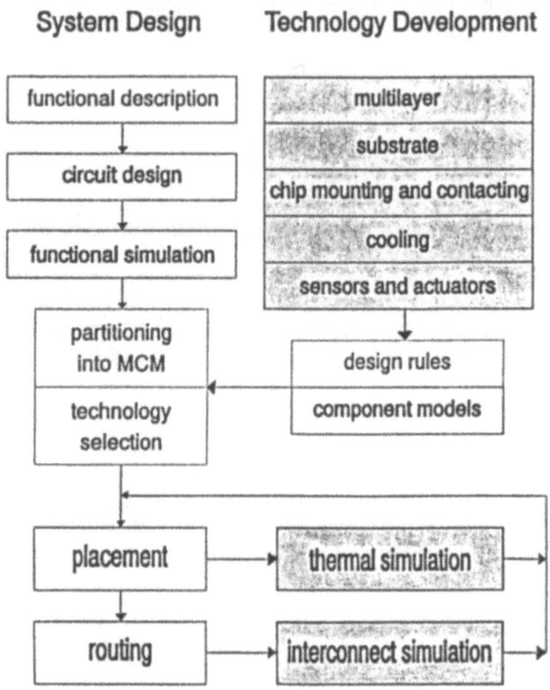

Fig. 1: Design Process for Multi Chip Systems

These new aspects have to be taken into account in the design processs of multichip systems. In Fig. 1 a scematic of this process is shown. The design of a multichip system has to rely on a description of the electrical, thermal and mechanical properties of the systems components and their interactions. Therefore together with the technology development a model of the function of a component should be developed. Here the use of simulation methods speeds up the development process by reducing the number of technological cycles and faciliates the development of design rules and component models.

The availability of this information is necessary to chose among different technological alternatives the one which leads to the optimal system solution and not to the optimisation of a particular property of the system.

In general the function of the system can not completely be guaranteed by design rules. Therefore in the layout phase simulation tools to control the interactions of the system components have to be used. Therefore efficient tools for the simulation of thermal distributions on modules and the transmission characteristics for critical signals are important.

Electrical Interconnect

When the signal rise time in a chip to chip interconnect line is shorter than twice the line delay, a transmission line layout is neccessary. The typical transmission line structures in multilayer packages are microstrip and triplate lines (Fig. 2a and 2b).

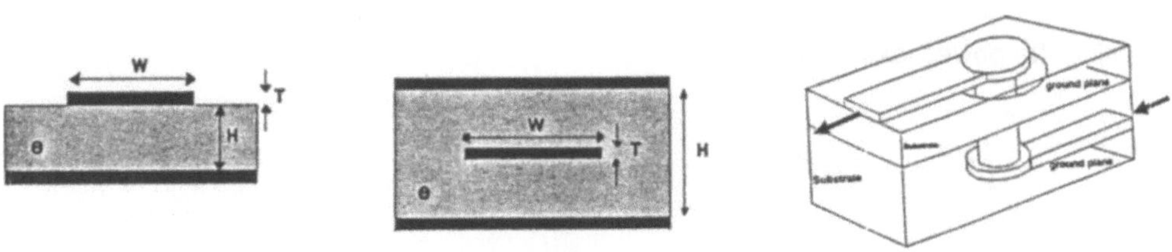

(2a) microstrip line (2b) triplate line (2c) via

Fig 2: Transmission Line Elements

Coupled straight line segments are characterized by the matrices of capacitance C', inductance L', serial resistance R' and transverse conductivity G' per unit line length. From these the line impedances and coupling coefficients can be calculated. For the conductor geometries in multi chip packages only the quasi TEM mode is propagating for frequencies up to about 20 GHz, and C' is obtained for ideal conductors by solving Laplace's equation

$$\Delta \Phi = 0 \tag{1}$$

and calculating the capacitance matrix from the energy relation:

$$\sum_{i,j} C_{ij}\Phi_i\Phi_j = \int \varepsilon\,|\nabla\Phi|^2 d^3x \tag{2}$$

The inductance matrix is obtained from the capacitance matrix for ε_r=1 by the well known relation

$$L'\,C' = c_0^{-2}\,1 \tag{3}$$

for TEM modes.

Transmission line losses become important for the small conductor cross sections (e.g. 20 µm x 10 µm) in thin film packages. The frequency dependent resistance and inductance can be calculated from the solution of a Helmholtz equation for the z-component A_z of the electromagnetic vector potential:

$$\Delta A_z + (\omega^2\mu\varepsilon - i\omega\mu\sigma)A_z = j_0 \;, \tag{4}$$

using the power loss integral

$$\frac{1}{2}RI^2 = \int \sigma j_z^2 d^2x \tag{5}$$

where j_z is the current density and I is the total current.

Except for very simple geometries no analytical solutions to these equations are known and one has to use numerical methods. A number of techniques has been used for the solution of these partial differential equations. The method of lines /1/ is computationally rather efficient, but due to the discretisation sceme the geometric fexibiltity is limited. Finite Difference methods are rather easy to implement, but are also limited in the choice of geometries. The most flexible methods are the Finite Element Method (FEM) and the Boundary Element Method (BEM) (implemented e.g. in /2/). For the calculation of the skin-effect variational methods /3/, the FE-Method /4/ and the BEM /5/ have been used.

Transmission line discontinuities like the via structure shown in fig. (2c) require three dimensional calculations. As long as the size of the discontinuity is short compared to the wavelength of the signal, a static approach may be used. One calculates the capacitance from the solution of eq. (1) and the inductance from the solution of the Laplace equation for the magnetic potential /6/. In /7/ a mixed Finite Element and Integral Equation technique is used for frequency dependent resistance and inductance calculations in 3D structures.

As shown in fig. 3 the numerical analysis results in a library of characterized transmission line elements which can be used as building blocks for the modeling of complex line structures. As lumped element models of longer transmission lines are computationally inefficient, appropriate models for coupled transmission lines have to be designed for the time domain simulation including nonlinear drivers and input stages. A number of models have been proposed for coupled transmission lines /8 - 11/. In most cases the basic technique is to use transformation networks at the beginning and at the end of the line to go to the eigenmode representation of a set of coupled lines and to use single line models to describe the propagation of each mode. In the case of coupled lossy lines this leads to quite complex models /8/.

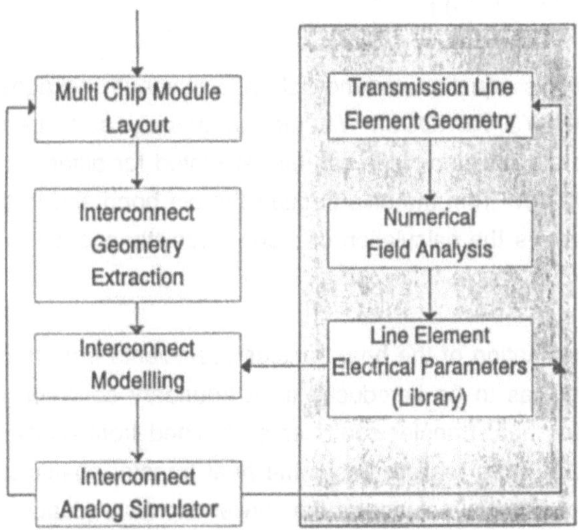

Fig. 3: Interconnect simulation

During the technology development the analog simulation of transmission lines is used to derive design rules for the layout of interconnects. In the design process after the completion of the layout the function of the interconnect system is verified by the extraction of models for critical lines from the layout geometry, and detecting possible timing or crosstalk problems in the analog simulation.

Thermal Simulation

With increasing power densities and closer packaging new methods for the cooling of multi chip modules have to be designed. The first step in heat removal from the chip into the environment is to enlarge the effective cooling surface by thermal conduction. In the second step the heat is transferred to the cooling medium, which in most cases is air for low and medium power densities and water for high performance systems.

The simulation of the conductive heat transfer involves the solution of the heat equation

$$div(k \nabla T) = \rho c \frac{\partial T}{\partial t} + Q \tag{6}$$

For the solution of this equation commercial finite element packages are available (e.g. NASTRAN, ANSYS, ABAQUS). With the use of these software tools 2- and 3-dimensional temperature distributions can be calculated for different boundary conditions and the heat flow from the chip through the die bond and in the subtrate be determined. This allows the calculation of thermal resistances for complex structures.

However, the determination of the heat transfer coefficients from the substrate into the cooling medium has to be introduced as a boundary condition in the calculation. In simple cases heat transfer coefficients obtained from anatytical considerations can be used. In more realistic cases the heat transfer coefficients have to be determined by the numerical solution of the Navier-Stokes equations in the cooling medium. In some cases the heat transfer coefficients obtained from two dimensional simulations (e.g. the fluid flow capabilities of ANSYS) can be used as boundary conditions for 3D-problems with sufficient accuracy /12/. For more complex problems as the design of cooling fins or microchannel cooling modules 3D fluid flow packages are available (e.g. Phoenics, Fidap).

Due to the rather large effort for the construction of finite element models and the high computation times of commercially available programs, the advantage of finite element simulations is mainly in the detailed analysis of complex geometries and in parametric studies for the development of design rules.

To determine the temperature distribution on a multi chip system with a large number of chips more efficient methods than FE-methods are available for this essentially two dimensional problem. One such method is the Fast Fourier Transform for the solution of the heat equation on rectangular substrates. With this method also layered media may be treated and to some extent spatially varying and temperature dependent boundary conditions may be handled by iterative solution.

Mechanical Stress

As the FEM method was originally developed for the simulation of mechanical stress distributions all the mentioned FEM software packages are capable of performing linear and nonlinear mechanical analysis. Thermally induced stress is an important failure mechanism in packages. Although a quantitative relation of stress distributions to lifetime predictions is very difficult, a qualitative analysis of the stress concentrations gives an indication where reliability problems may occur. FEM analysis was applied to stress distributions in plastic packages (e.g. /13/) and to the fatigue of solder joints(e.g. /14/).

Another topic in this field is the calculation of stress distributions during chip contacting processes. For the TAB thermocompression bond process the elastic-plastic deformations in the bump-lead system and the stress concentrations at the edge of the bump in the passivation are determined /15/.

Conclusion

Physically orientated simulation methods become an important tool for the design of multichip systems. However, the present methods are limited to the investigation of components or subsystems and for parametric studies during the technology development. The present tools may be characterized as analysis tools. In future, design tools are needed which allow for a fast optimisation of the simulated structures. Furthermore a close connection to measuring and characterisation techniques is necessary to obtain validated simulation models.

References

1. W. Pascher; R. Pregla: Full wave analysis of complex microwave structures. Radio Science 22 (1987) 999 -1002

2. A. T. Murhphy et al.: ICONSIM, Computer Aided Interconnection simulation. 7th International EMC Symposium, Zürich 1987

3. P. Waldow; I. Wolff: The Skin-Effect at High Frequencies. Trans. Microwave Theory Techn. **33** (1985) 1076 -1081

4. O.-K. Kwon; R. F. W. Pease: Closely Packed Microstrip Lines as Very High-Speed Chip-to-Chip Interconnects. Trans. Components, Hybrids, and Manuf. Technology **10** (1987) 314 -319

5. W. Wlodarczyk: Modeling of interconnects by boundary-element methods. This volume.

6. R.Beck; M. Borst; E. Schaffer: Electrical Simulation of Chip-to-Chip Interconnections. VLSI Packaging Workshop, Garmisch-Patenkirchen, 1989

7. A. C. Cangellaris; J. L. Prince; L. Vakanas: Frequency-Dependent Inductance and resistance Calculation for Three-Dimensional Structures in High-Speed Interconnect systems. Proceedings of the 39th Electronic Components Conference (1989) 398 - 403

8. A. J. Gruodis; C. S. Chang: Coupled Lossy Transmission Line Characterisation and Simulation. IBM J. Res. Develop. **25** (1981) 25 - 41

9. J. Siegl: PLANAR - A Circuit Simulator Including Distributed Transmission-Line-Structures in High-Speed Circuits. Frequenz **8** (1986) 209 - 214

10. F. Romeo; M. Santomauro: Time-Domain Simulation of n Coupled Transmission Lines. IEEE Trans. Microwave Theory Tech. 35 (1987) 131 - 137

11. O. A. Palusinski; J. C. Liao; J. L. Prince; A. C. Cangellaris: Simulation of the Transients in VLSI Packaging Interconnects. Proceedings of the 40th Electronic Components and Technology Conference (1990) 173 - 179

12. A. Hobl: Simulation der Kühlung von Substraten mit freier Konvektion. SMT/ASIC/Hybrid Int. Conf. (1990) 275 -282

13. A. Nishimura; S. Kawai; G. Murakami: Effect of Lead Frame Material on Plastic-Encapsulated IC Package Cracking under Temperature Cycling. Proceedings of the 39th Electronic Components Conference (1989) 524 -530

14. J. H. Lau; D. W. Rice: Thermal Stress/Strain Analysis of Ceramic Quad Flat Pack Packages and Interconnections. Proceedings of the 40th Electronic Components Conference (1990) 824 - 834

15. A. Eberle: Simulation of the inner lead bonding process. This volume.

A Fully-Automatic Macromodeling Procedure for the Electrical Simulation of Integrated Circuits

C. Turchetti*, Zs. M. V.-Kovács**, G. Masetti**, M. Conti *

* Dept of Electronics, University of Ancona, v. Brecce Bianche, 60131 Ancona, ITALY
Tel. 071/5893848, Telex 561838 Univan I, Telefax 071/5893835

** DEIS - University of Bologna, v.le Risorgimento 2, 40136 Bologna, ITALY

Summary

A new fully automatic macromodeling procedure for the simulation of large integrated circuits at device level is presented. The proposed approach has been implemented in SPICE and enables to obtain for both the DC and transient analyses exactly the same results achieved with the original version of SPICE but with a reduction of a factor four in the CPU time and much more in memory occupation.

Outline of the macromodeling approach

The starting point of the approach was based on the following observations:

i) Often (also for integrated circuits of intermediate complexity) circuit topologies are quite regular in structure, i. e., it is possible to identify a set of subcircuits (cells), that are repeated in the design, allowing the hierarchical description of the circuit itself through this cells, rather than at the device level. This means also that there are a lot of devices in the circuit described by the same parameters [1, 4].

ii) Circuit designers are interested to the voltage and current behaviours at the ports (or external nodes) of these cells and not to the voltage behaviours at the internal nodes.

Of course, the behaviour of the circuit does not change if the above cells are replaced by macromodels, having at the external nodes the same current-voltage characteristics of the original ones. In general, due to the non-linearity of the cells, it is not possible to obtain a macromodel using closed form analytical expressions.

The development of the method presented in this work is based on the fact that in circuit simulators like SPICE the non-linear equations are solved by using the Newton-Raphson algorithm. This is equivalent to solve the non-linear equations by iteration, starting from an initial guess, substituting a linear companion model to each non-linear device and solving the resulting linear network with some sparse matrix technique. Thus, at each step of this algorithm, explicit relationships relating the currents and the voltages of each cell at the external nodes only can be derived; in turn, this is equivalent to the "elimination" of the internal nodes. This way the dimension of matrices involved in the analysis are reduced and, as a consequence, both a gain in CPU time and memory occupation can be achieved. A further significant reduction in memory occupation can be obtained considering that in the data structure of SPICE each device occupies a complete record, containing all the parameters needed to describe it, but several devices have the same model and the same parameter values. Besides, the reduction of memory usage can avoid memory swapping during the simulation of a big circuit, reducing significantly the simulation time.

With reference to a transient analysis, the proposed approach was implemented as follows:

a) First, the complete circuit to be analyzed is described using a SPICE description, as a connection of standard components and macrocells. Besides, the topology of each macrocell is described using the standard SPICE subcircuit description, if not yet present in the macrocell library. Notice that macrocells which differ only in transistor size are considered equal as the geometrical parameters appear in the macrocell description.

b) Then, the symbolic macrocell compiler derives a macromodel for each new macrocell, starting just from topology and device models, thanks to the substitution of linear companion models for non-linear devices. The macromodel of a cell is defined as the analytical expressions of the current at the external terminals as a function of the voltages at the external terminals only. It seems worth to notice that, formally, the adopted procedure eliminates the analysis of the internal nodes of the macrocells.

c) At this point, the new macromodels are added to the cell library for future use.

d) At the end, a suitably implemented code generates for the new macromodels the calls and modifications to the DCTRAN and ITER8 routines of SPICE. As a result, a modified version of the SPICE code, called MMSPICE from now on, is obtained.

Results

Figures 1-6 show some benchmarks and results achieved with MMSPICE in comparison with the results obtained with the original version of SPICE 2G. The simulations were performed using macromodels which eliminate 1 and 4 internal nodes (i) of the circuit only. Data refer to transient analyses of NMOS chains of inverters, cascades of NANDs, EX-OR circuits and full-adders. Even by using these simple structures, gains up to a factor of about 4 in the CPU time, of about 5 in the circuit matrix dimensions and more than 10 in the memory occupation were found. It seems worth to notice that, as expected, the results achieved with MMSPICE were exactly coincident with those obtained using the original version of SPICE for both the DC and transient analyses, when the same set of tolerances VNTOL, RELTOL, ABSTOL, and CHGTOL was used. As an example, Fig. 6 shows the time dependence of the voltages in some relevant nodes of the 4-bit full adder circuit of Fig. 4 predicted by the original version of SPICE (dotted lines) and by the MMSPICE version with macromodels (continuous lines) for the chosen input waveform A_i, B_i, C_i. Each subcircuit in Fig. 4, implemented as shown in Fig. 5, was macromodeled as cells of NANDs and EX-NORs.

Future work

Trying methods to improve the speed of the simulation, a possibility is to use a parallel architecture in place of the serial one. It is difficult and time consuming to convert it globally into a vectorial code, since SPICE was not written for a multi-processor environment; but, using macrocells, it comes natural to introduce the computation of each of them on a separate processor, when available. Very interesting preliminary results have been found implementing the code on a 4-processor system.

One of the most difficult problems which arise during circuit simulation is the convergence of the algorithm. Observing the voltages at the nodes of the circuits during the calculation, it is possible to notice, that the simulation aborts very often for the occurrence of a numerical overflow, what is not the same as a non-convergence. One possibility to overcome this drawback is to increase the range of the variables in the code, reducing the occurrence of the numerical overflow condition, an other is to try to reduce the ringing observed in the voltage (or current) values. Research is in progress to find, from a numerical point of view, the convergence properties of a hierarchical simulator, where the whole circuit is obtained by the connection of macrocells. Each macrocell is treated as a non linear circuit, being simulated solving the non-linear equations by iteration, using as boundary conditions the voltage (or current) values given by the remaining part of the whole circuit. In this way, dealing with smaller circuits at a time, improvement of the behaviour of the voltages should be obtained.

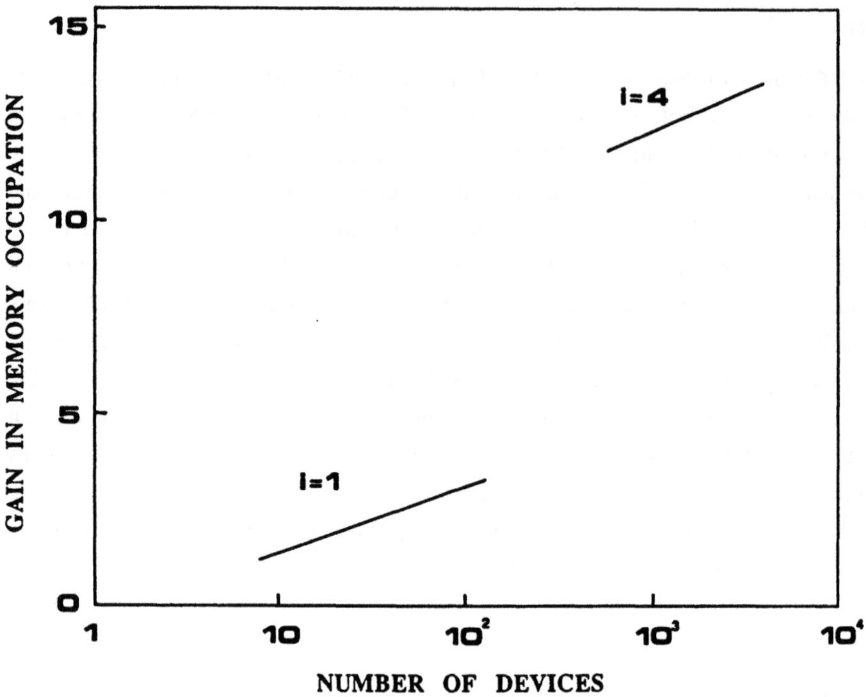
15
10
5
0
1
10
10^2
10^3
10^4
GAIN IN MEMORY OCCUPATION
NUMBER OF DEVICES
i=4
i=1

Fig. 1

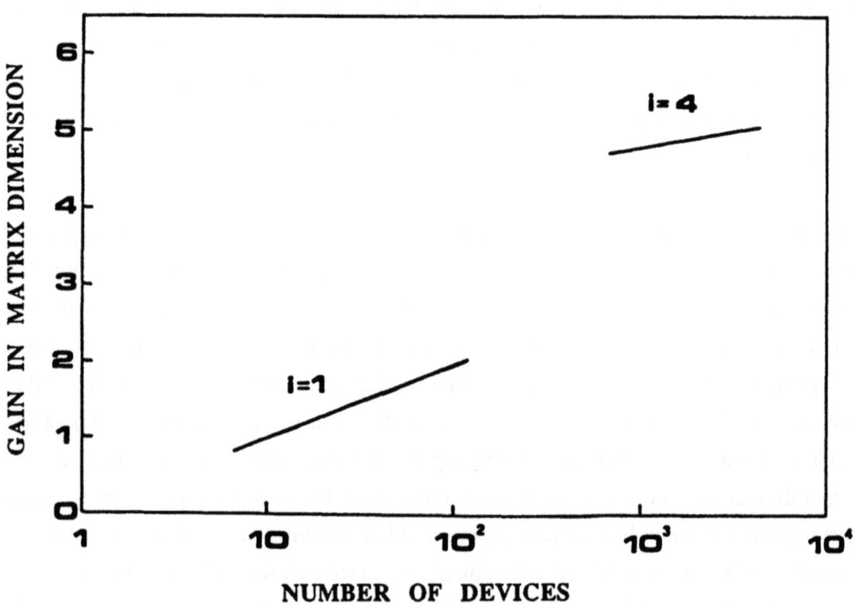
6
5
4
3
2
1
0
1
10
10^2
10^3
10^4
GAIN IN MATRIX DIMENSION
NUMBER OF DEVICES
i=4
i=1

Fig. 2

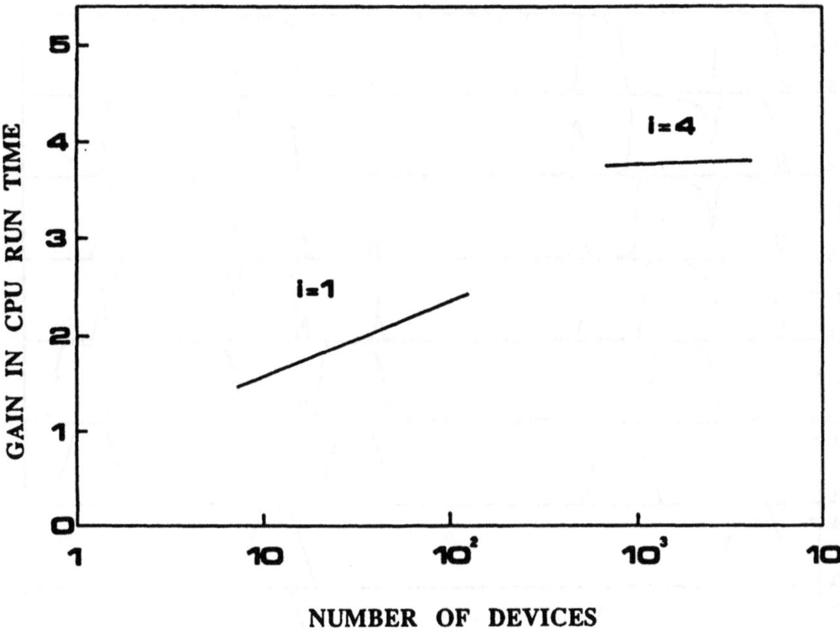

Fig. 3

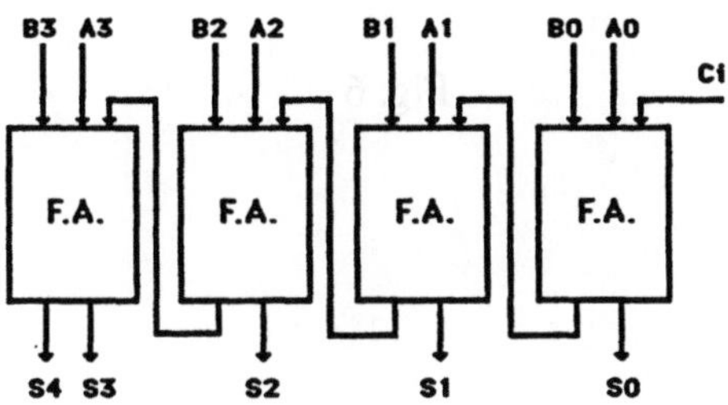

Fig. 4

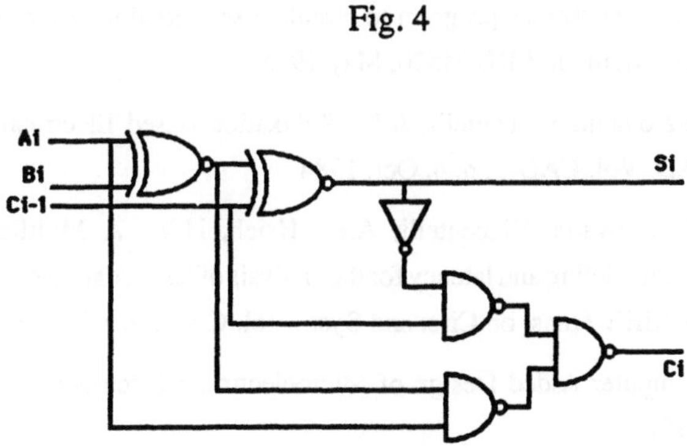

Fig. 5

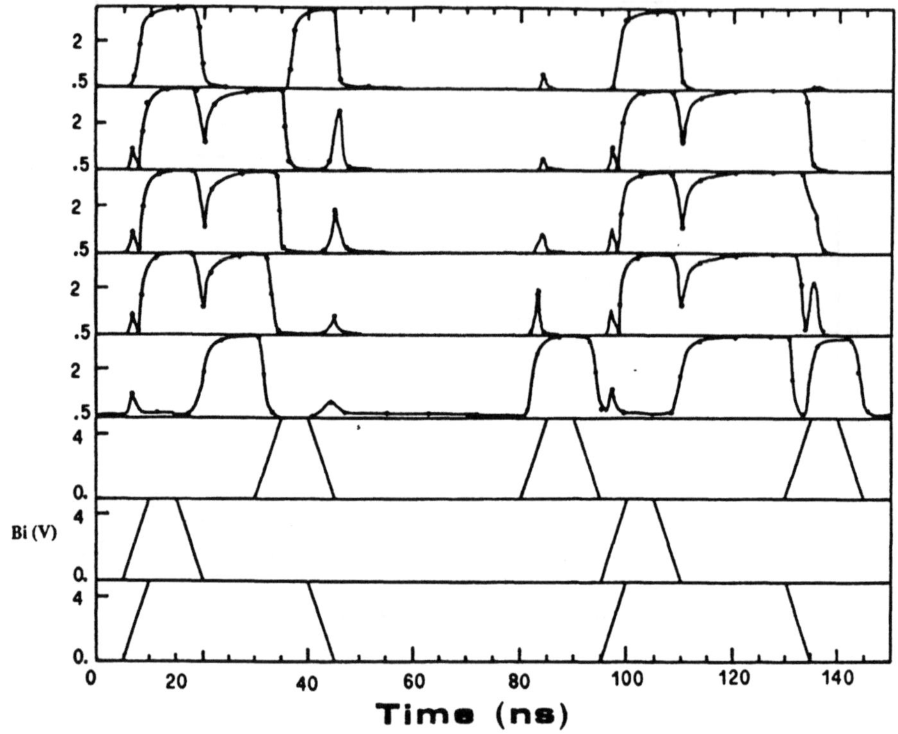

Fig. 6

References

1. Nagel, W.: SPICE2, A computer program to simulate semiconductor circuits, Univ. of California, Berkeley, Memo n. ERL-M520, May 1975

2. Newton, A.R.; Sangiovanni-Vincentelli, A.L.: Relaxation-Based Electrical Simulation, IEEE Trans. on CAD, Vol. CAD-3, n. 4, Oct. 1984

3. Rabbat, N.B.G.; Sangiovanni-Vincentelli, A.L.; Hsieh, H.Y.: A Multilevel Newton algorithm with macromodeling and latency for the analysis of large scale non-linear circuits in the time domain, IEEE Trans. on Circ. and Syst., Vol. CAS-26, n.9, Sept. 1979

4. Schwartz, A.F.: Computer Aided Design of Microelectronic Circuits and Systems, Ed. Academic Press, 1987

KOSIM - A Multi-Level Simulator for the Design of Integrated Circuits

P. Schwarz

Academy of Sciences of the German Democratic Republic
Central Institute of Cybernetics and Information Processes
GDR-8027 Dresden, Haeckelstraße 20

Summary

KOSIM is a multi-level simulator covering a wide range of modelling levels: behavioural simulation, PETRI-net simulation, register-transfer simulation, dynamic logic simulation, and electrical simulation may by carried out concurrently. Originally developed for the design of digital MOS circuits, the program may be used also for the simulation of computer structures, mixed analog-digital systems as well as nonelectrical systems.

Multi-Level Simulation

Simulation programs are becoming more and more important tools whithin the design process of integrated circuits. Multi-level description and simulation proved to be good for the design of VLSI circuits. Multi-level simulation means the simultaneous modelling and simulation of subsystems on different levels of design and with several simulation algorithms. DIANA /1/ and SPLICE /2/ have been the first programs of the kind during the past.

KOSIM /3/, /4/ is an interactive multi-level simulation program directed to following levels of design:

logic block level (behavioural simulation):
complex logic blocks whose behaviour is defined by user-written subroutines (FORTRAN, PASCAL, C) /5/.

PETRI nets: as a useful tool for modelling general nets, control parts of computers and integrated circuits, parallel processes.

register-transfer level: the RT language DDL /6/ is being implemented, with some extensions (three-valued logic, tristate behaviour, embedding different DDL blocks in a multi-level environment).

logic gate level: dynamic logic simulation, for a variety of elements (NAND, NOR, flip-flops, RAM, ROM, PLA, bidirectional gates, tristate elements,...)

electrical level: MOS and CMOS transistors, bipolar transistors, resistances, capacitances, inductances, transformers and gyrators,... Two simulation algorithms are implemented:

- electrical network analysis
- timing simulation.

KOSIM is written in FORTAN77 and can be implemented on VAX computers with VMS operating system. Systems with up to 100 000 elements and nodes may be simulated.

Simulation Algorithms and Signal Representation

All algorithms are event-oriented and controlled by a global event scheduler. Each event is a signal change at a node characterized by the tupel (event time, node number). Only those parts of the system being active at the same time are simulated. All events are scheduled into a normalized time scale with a smallest time step being 1 (corresponding to a real time step of 1 ns or less). The algorithm of electrical network analysis is based on an internal time scale, with time steps less than 1 (for step size control of the integration).

a) Logic block and gate simulation

A table-driven, event-oriented algorithm /7/ for dynamic logic simulation has been implemented. Each element output may have an individual delay time. Signals are characterized by tupels again:

signal = (state, value)

Logic gate level:

value $\in$ $\{0, 1, X\}$ X = undefined

state $\in$ { LOW_IMPEDANCE (R=0),
RESISTIVE (R=1,2,...,126),
HIGH_IMPEDANCE (R=127)
UNKNOWN }

By using different resistance levels (or signal strenght, inversely) bi-directional transfer gates may be simulated /8/, too.

Logic block level:

value $\in$ $\{0, 1\}^n$, bit string of the length n

state $\in$ { LOW_IMPEDANCE DEFINED
HIGH_IMPEDANCE DEFINED
LOW IMPEDANCE UNDEFINED
HIGH IMPEDANCE UNDEFINED }

The bit strings are representing signals which can be interpreted as a set of different types of data such as INTEGER, REAL, CHARACTER, ...

b) PETRI net simulation

The simulation of timed PETRI nets is based on a sligthly modified version of gate level simulation. Each transition may have a delay time using a transport delay model. Some kinds of firing rules are implemented. For coupling with subsystems modelled on the logic level a signal conversion is applied:

token on place - logic signal "1"
no token on place - logic signal "0"

In such a way the simulation of interacting the control part (modelled as PETRI nets) and the data path (modelled on logic gate or block level) can be done.

c) Register-transfer simulation

The multi-automaton model of a DDL description is transformed into a set of boolean equations, with signals represented by bit strings. Each "bit" has the value of 0, 1 or X (undefined). The system may be constructed by several subsystems (blocks), each of them with its own DDL description. The clock signals for registers and memories are generated by logic circuits outside of the DDL blocks.

d) Timing simulation

Timing simulation is an iterative network analysis method especially for digital MOS circuits, characterized by

- decoupling of nodes, solving scalar node equations instead of systems of equations,
- some few iterations (or only a single iteration) over all node equations,
- using the latency principle for each individual node,
- simple transistor models.

We are using a slightly modified iterated timing analysis /9/. To avoid numerical problems or to many iterations in the case of strongly coupled nodes (floating capacitances), small systems of equations are solved for these nodes instead of scalar equations.

e) Block-oriented network analysis

The electrical part of the system is to be partitioned into some subnetworks and a main network. The node voltages of the main network are the signal sources of the subnetworks and have to be calculated in such a way, that the Kirchhoff current law is fulfilled. Modified nodal analysis is applied for the formulation of the nonlinear differential equations. The trapezoidal rule is used for integration. For each subnetwork, individual step size control and exploitation of latency is carried out.
Electrical circuits with up to 20 000 transistors have been simulated.

User-defined models

The function of logic and electrical elements is defined by means of subroutines. The "behavioural procedures" of new elements have to be programmed, compiled, and linked into the simulator. Linking is not necessary if the subroutines are compiled as "shareable images" under the VMS operating system.There is a simmple interface definition for input and output signals and for the time point of next element activation.

Programming of new behavioural procedures specified by the user is important, above all on the logic block level and for new transistor models. But the same concept could be used for coupling KOSIM with other simulators, for instance with an ISPS simulator and an other stand-alone DDL simulator. In both cases, the "core" of the other simulator must be called in the "behavioural procedure" which corresponds to the subsystem which is modelled with the ISPS or DDL language. In this sense, KOSIM is an open simulator which can be coupled easily with different simulation algorithms.

User communication

KOSIM is an interactive simulation program but it can also operate in the batch mode. The simulation can be controlled by hierarchically structured commands. These commands can be collected in a file. Breakpoints may be defined at signal values and signal changes. The results of the simulation can be written on a display, a printer, or into a file (bit pattern, numbers, print-plot, ...). Testing and debugging of user-written behavioural procedures is supported by monitor commands. A path-follower helps to find circuit errors by backtracking the signal flow. By using the commands DUMP and RESTART the user can order the continuation of the simulation.

Application

The program has been successfully applied within the industry, during the design process of LSI and VLSI circuits (full custom, gate array and standard cell ICs, memory chips). KOSIM offers the possibility for modelling not only a single chip, but also the whole system environment. Analogue and digital functional units can be simulated simultaneously. Non-electrical problems can be solved by application of analogies /11/,/12/,/13/. The modelling with flow quantities (electrical current, mechanical force, thermal flow,...) and difference quantities (electrical voltage, mechanical pressure, temperature, ...), and the construction of networks with lumped elements are involved for the simulation of combined systems. Transformers and gyrators are connecting elements between different object domains. These possibilities are important for the simulation of microsystems.
Two other model description languages for KOSIM which are under development may be very useful for microsystem simulation:

- a subsystem can be described as a set of nonlinear differential equations which are solved with the algorithm of electrical network analysis;
- a subsystem can be modelled as a net of elements like summators, integrators, function formers, threshold elements,... (CSMP-like description).

For the net description on all modelling levels, the hierarchical net description language NBS-84 /14/ is used.

Acknowledgement

KOSIM was developed under contract with the Research Center of Microelectronics Dresden (ZMD). I would like to thank my colleagues Christoph Clauss, Ulrich Donath, Joachim Haase, Jürgen Haufe, Gunter Kurth, and Peter Trappe for their contributions in developing the program.

References

1. Arnout,G.; DeMan,H.: The use of threshold functions and boolean-controlled network elements for macromodelling of LSI circuits. IEEE J.Solid-State Circuits SC-13(1978)3,pp.326-332
2. Newton,R.A.: Techniques for the simulation of large-scale integrated circuits. IEEE Trans. CAS-26(1979)9, pp.741-749
3. Donath,U.;Schwarz,P.;Trappe,P.;Clauss,C.;Kurth,G.: A multi-level simulator for integrated circuit design. Proc. ECCTD'87, Paris 1987,pp.465-470
4. Clauss,C.;Donath,U.;Kurth,G.;Schwarz,P.;Trappe,P.;Wittwer,H.: KOSIM - ein Multi-Level-Simulator für den Schaltkreisentwurf Z. messen-steuern-regeln, 32(1989)5, pp.223-227
5. Donath,U.;Schwarz,P.;Trappe,P.: Dynamische Logiksimulation auf Bit- und Wort-Niveau. 1.Tagung Schaltkreisentwurf, TU Dresden 1986, pp.218-228
6. Duley,J.R.;Diethmeyer,D.L.: A Digital System Design Language (DDL). IEEE Trans.on Comp.,C-17(1968)9,pp.850-861
7. Szygenda,S.A.;Thompson,E.W.: Digital logic simulation in a time-based, table-driven environment. Computer(1975)3,pp.24-36
8. Donath,U.: Behandlung von MOS-Schaltungen in der dynamische Logiksimulation. 3. Tagung Schaltkreisentwurf,TU Dresden 1989, pp.201-205
9. Saleh,R.A.;Kleckner,J.E.;Newton,A.R.: Iterated timing analysis. Proc. ICCAD 1983, pp.139-140
10. Clauss,C.;Schwarz,P.: Ein Verfahren zur blockorientierten Analyse nichtlinearer Netzwerke. 3. Tagung Schaltkreisentwurf,TU Dresden 1989, pp.153-157
11. Reinschke,K.;Schwarz,P.: Verfahren zur rechnergestützten Analyse linearer Netzwerke. Akademie-Verlag Berlin 1976
12. Koenig,H.E.;Blackwell,W.A.: Electromechanical system theory. New York McGraw-Hill 1961
13. Thorn,D.C.: An introduction to generalized circuits. Belmont: Wadsworth 1965
14. Issel,W. et al.: NBS-84: A structural description languague for VLSI design. Proc. ECCDT'85,Prag 1985,pp.62-65

Waveform Estimation and Hazard Detection in Digital MOS Circuits

D. Sass, H. Warmers and E.-H. Horneber
Institut fuer Netzwerktheorie und Schaltungstechnik
Technische Universitaet Braunschweig, FRG

Abstract

In the design process of digital VLSI circuits it is necessary to detect hazards. In this approach node transients are calculated by using switch-level timing simulation which gives an estimation of signal waveforms and a worst-case approximation for the peak values of occuring hazards.

1. Introduction

In the design process of digital integrated MOS circuits, switch-level simulation - i.e. logic simulation of MOS circuits on transistor level - has found increasing acceptance. Based on principles developed by Bryant [1] and Hayes [2] about 1982, many switch-level simulators have been developed in industry and universities. In most switch-level logic simulators a circuit is decomposed into subcircuits which are logically evaluated and connected via unit delay signal propagation. Consequently, simulations are restricted to synchronous circuits with nonoverlapping multiphase clock cycles. Faults e.g. due to unsuitable aspect ratios in NMOS circuits which lead to unexpected timing behaviour like hazards cannot be detected.

In the past years switch-level timing simulators have been developed which additionally give an estimation of the signal waveforms. Due to the timing models used they often show insufficient accuracy if nonstandard network configurations or clocking schemes are used or in case of time skews which lead to hazards.

To receive reliable simulation results also in these cases a new switch-level timing model [3] has been developed at the University of Braunschweig and implemented in the switch-level simulator BRASIL. This paper reports the simulation approach and discusses the results.

2. The Simulation Objects in BRASIL

The event-driven simulator BRASIL includes algorithms for switch-level logic and switch-level timing simulation and can be applied to CMOS as well as NMOS circuits. The following circuit elements are admitted: p- and n-channel MOSFETs, linear resistances and capacitances (e.g. to model interconnections extracted from the layout) and functional blocks which can be understood as arbitrary macromodels whose terminal behaviour is described by a C-program. Three functions have to be defined to integrate such blocks into the surrounding circuitry: An initialization function for block variables is called when simulation starts, a second function monitors input node changes and controls event scheduling and a third function repeatedly starts the evaluation algorithm after user defined time intervals. The third function is optional and prevents the necessity of defining an event-driven macromodel.

3. The Simulation Approach

To evaluate, the timing behaviour of a MOS circuit the following steps are necessary:

A. Partitioning of the circuit into subcircuits and calculation of their steady-state responses by switch-level logic simulation

To decompose the circuit, input and output nodes of functional blocks are separated. Each transistor is statically cut between its gate and its drain-source region. For nonconducting transistors drain and source regions can dynamically be split apart. All transistors being still connected after that procedure form a subcircuit. The node signals of a subcircuit can change if a transistor of this subcircuit is switching due to a signal change at its gate. In this case an improved standard switch-level logic algorithm is applied to predict the steady-states of all nodes of the subcircuit. Subcircuits which are not driven have not to be concidered. Thus the approach is event-driven.

B. Modeling each subcircuit as linear RC network

In the next step a subcircuit is modeled as a linear RC network by summing up the parasitic transistor capacitances and the linear circuit capacitances to node and floating capacitances and by calculating the linear resistances of the drain-source regions of the conducting and switching transistors.

The resistance of a conducting transistor can easily be calculated from the Shichman-Hodges model using the final node voltages. For switching transistors their actual node states are changing. Therefore, average resistances for standard gate configurations are estimated using empirical table functions which are normalized to the input slope at the transistor's gate and to the output load capacitance of an inverter which transient behaviour is equivalent to the subcircuit. Normalized delay time tables are used to define the time interval between an event activating the subcircuit and the time when the RC network model becomes valid. For a detailed explanation of the generation of the table functions see [3].

C. Calculating the node voltages

After mapping the subcircuit into a linear RC network a system of first order differential equations is generated for the node voltages. It can be solved in closed form by using a linear Gauss-Seidel waveform relaxation algorithm which essentially is an extension of Zukowski's algorithm [4].

Advantages of this approach are that initial charges at the node capacitances can be handled as well as floating capacitances and resistive meshes. These features are the prerequisite for detecting and qualitatively calculating dynamical hazards which do not cross the logic level predicted in the first step.

4. Conflict Handling and Hazard Detection

If a new event appears at an input node of the subcircuit before the node voltages of that subcircuit have settled a conflict occurs. The time interval the RC network is valid for is limited by the moment when the new event affects a node

voltage. A new evaluation for the subcircuit is started causing the node voltages to be calculated once more using the actual node voltages as initial conditions.
Any conflict can cause a hazard which means that the sign of the waveform gradient will change. As the resistances corresponding to the switching transistors are estimated for complete switching, the empirically calculated resistance tables give underestimated resistance values for noncomplete switching. That means that the waveform gradient in case of an occuring hazard is steeper than in the real circuit, so that the estimated hazard peak value can be regarded as a worst case approximation. For the propagation of the hazard through the connected subcircuits it is important whether its peak value crosses the logic trigger level of the following subcircuit's input node or not. This trigger level is defined as the output voltage resulting from connecting the output and the input node which corresponds to the gate of the switching transistor. If a hazard at a gate input does not pass the trigger level the output does not respond in a critical way, i.e. the hazard at the output won't pass the level, too. Otherwise the hazard is probably critical depending on its peak value and the delay time of the output signal. This approach guarantees all occuring critical hazards to be found and declared as probably critical.

5. Simulation Results

The presented approach was implemented in the switch-level timing simulator BRASIL [3]. Results are discussed guided by the following typical example circuit for occuring hazards, a NOR-gate driven by the input signal IN and its delayed inverted signal NIN, as shown in figure 1.

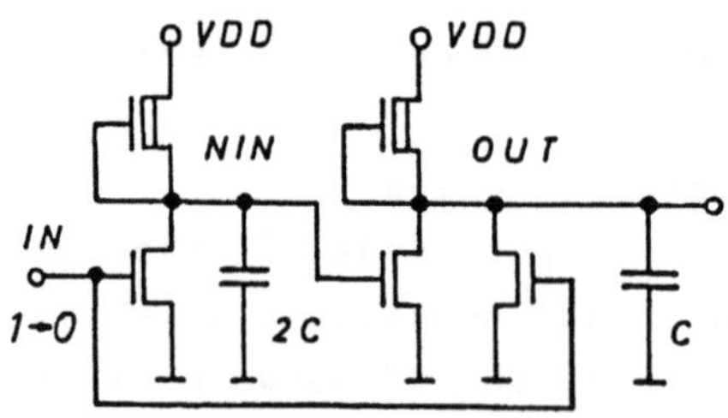

Fig.1: Example circuit for hazard detection

At the beginning the input node state (IN) is set to "High" which causes the output node states of the inverter (NIN) and the nor gate (OUT) to be "Low". Then signal IN falls to "Low" and consequently signal NIN rises to "High". Signal OUT rises as well, until the transistor which gate is connected to NIN switches. Then signal OUT changes to "Low" which produces a dynamic hazard at the output node of the circuit.
In figure 2 the simulation results for this circuit obtained with BRASIL are compared to the results from the circuit simulator SPICE.

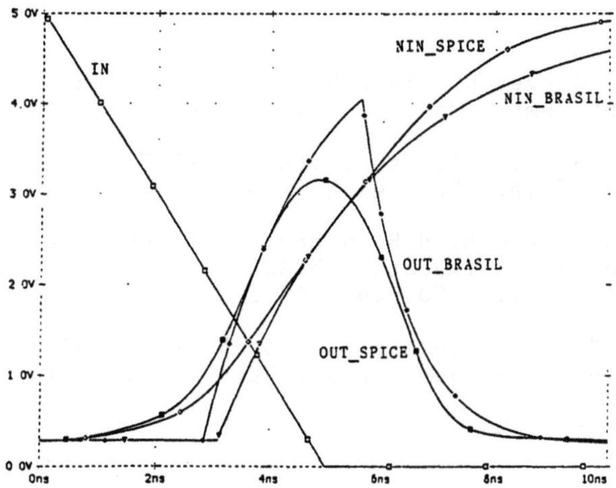

Fig.2: Comparison of the simulation results obtained with BRASIL and SPICE

This simple example shows that the hazard is detected in BRASIL and the signal waveforms are approximated by exponential functions which can be considered as a worst-case approximation in case of the hazard at the output node.
Also in more complex configurations all critical hazards which peak values are high enough to produce malfunctions of the connected gates are reliably detected. Compared to circuit simulations runtime is reduced by two to three orders of magnitude. Simulating circuits including pass transistors the accuracy of the waveform approximation is reduced. To solve this problem suitable pass transistor models are under development.

7. Conclusions

The presented approach was implemented in the switch-level simulator BRASIL. All hazards occuring in a digital MOS circuit are reliably detected and worst-case approximated by simulation with BRASIL. Hence, hazard propagation which can lead to a malfunction of the following gates can easily be observed. In spite of handling signal waveforms, run time reduction of BRASIL is two to three orders of magnitude compared to SPICE2.

References

1. R.E.Bryant: "A Survey of Switch-Level Algorithms", IEEE Design & Test of Comp., Vol. 4, No. 4, Aug. 1987, pp 26-40.

2. J.P.Hayes: "An Introduction to Switch-Level Modeling", IEEE Design & Test of Comp., Vol. 4, No. 4, Aug. 1987, pp 18-25.

3. H.Warmers, D.Sass and E.-H.Horneber: "Switch-Level Timing Models in the MOS Simulator BRASIL", European Design Automation Conf., Mar. 1990, pp. 568-572.

4. C.A.Zukowski: "The Bounding Approach to VLSI Circuit Simulation", Kluwer Academic Publishers Boston, 1986, pp. 161-171,187-189.

Modelling of Interconnects by Boundary Element Method

W.Wlodarczyk

Technische Universität Berlin
Institut für Hochfrequenztechnik
Einsteinufer 25, D-1000 Berlin 10, F.R.G.

Summary

Scaling down of chip-to-chip interconnections to finer geometries can limit the overall performance of a digital system and therefore better characterization of its electrical parameters is demanded. In order to extract the equivalent circuit parameters of interconnect structures, related 2-D electromagnetic field problems are solved by use of the boundary element method (BEM). Special attention is paid to the frequency dependent conductor losses due to the skin effect.

Introduction

One of the main limitations to very high speed digital system performance is the limited speed of chip-to-chip communications. The standard approach to decreasing the interconnect delay is by scaling down its physical size. On the other side, this is limited by the increasing line resistance because of the finite resistivity of the conductors.

Up to now, the loss calculations at low frequencies have been based on the dc definition of resistance and at high frequencies on the Wheeler's incremental inductance rule /1/. For resistance calculations for the whole frequency spectrum of speed pulses on micrometer interconnect structures, finite element techniques (FEM) have been used /2/.

In this contribution, the boundary element method (BEM) /3/ is applied to calculate the frequency dependent resistance as well as other line parameters. Over FEM, this method has the advantage of the reduction of one dimension, thus of CPU time and memory storage.

As a first step in designing complex interconnect structures, results for single conductor and conductor over ground plane are presented and verified. In addition, the effect of different loss assumptions on the step response of an interconnect line is shown.

Formulation of the problem

For 2-D capacitance calculations quasi-TEM analysis is sufficient because of cross-sectional dimensions which are negligibly small with regard to the wavelength. For broadband resistance calculations this is to be modified, because a significant axial electrical field component appears due to lossy conductor.

2-D eddy current problem is posed in terms of the axial magnetic vector potential (**A** - ϕ formulation, e.g. /4/). This formulation introduces a frequency dependent complex constant ($\mathbf{A}_0$ = grad $\phi / j\omega$) to modify the interlinkage flux in the conductor instead of splitting the solution of field equation into the static and dynamic parts /5/. When conductors embedded in homogeneous dielectric medium are parallel to the z-axis, as shown in Fig. 1, the current density **J** and the vector potential **A** have only a z-component. The Helmholtz equation

$$(\nabla^2 + k^2) A = 0 \tag{1}$$

holds in both regions, the conducting and the dielectric one. Merely the wavenumber k and the total vector potential A are dependent on the kind of medium. The vector potential in the conducting region consists of its two parts, the varying and the constant one.

Solution technique

Applying the weighted residual method to the Helmholtz equation (1) and using the Green's 2nd theorem, the differential description of the boundary value problem in 2-D domain Ω can be replaced by the integral description on the 1-D boundary Γ, reducing in this way CPU time and storage memory required for the numerical treatment.

For the sake of universality we set for the vector potential u = A and for its outward normal derivative q = $\partial A / \partial n$, and introducting the fundamental solution (Green's function)

$$u^* = -\frac{j}{4} H_0^{(2)}(kr) , \tag{2}$$

where $H_0^{(2)}$ is the Hankel function of the second kind and order zero, and its outward normal derivative

$$q^* = \frac{\partial u^*}{\partial n} , \tag{3}$$

we obtain the singular boundary integral equation (BIE) for the ith point as follows:

$$u_i + \int_\Gamma uq^* - \int_\Gamma qu^* = 0 . \tag{4}$$

In order to discretize the BIE, the boundary Γ is approximated by the connection of line elements, on each of these u and q are assumed to vary linearly due to linear basis functions. When the source point i does not belong to the considered element, contributions of this element to the integration are calculated using Gauss quadrature rules. Otherwise, the integrals can be determined in the sense of the Cauchy's principal value (smoothed boundary). After discretizing the BIE for all points and elements, a linear system of equations is obtained. In multi-media cases, as the eddy current problem is, the final system of equations for the whole region is obtained by adding a set of equations for the conductive and dielectric subregions together with compatibility and equilibrium conditions between their interfaces. For conductors with bounded cross section, nodal potentials and fluxes only at interfaces are to be considered. For structures with unbounded ground planes, both interface and external boundaries are combined. Finally, the total system of equations is completed by the discretized Ampere's law.

Results

As a first example, the line-to-ground capacitance of an interconnect transmission line for different width-to-height ratios with thickness-to-height ratio as a parameter is calculated (Fig. 2) and verified by one of available empirical formulas /6/.

The AC resistance of a single rectangular conductor over the normalized frequency for different width-to-thickness ratios is shown in Fig. 3. The BEM results are compared with those of the variational analysis /7/.

The numerical treatment of structures containing an unbounded ground plane is very cumbersome. The broadband resistance of the stripline (Cu at 77 K) over the frequency range from 10 MHz to 10 GHz is shown in Fig. 4. The losses calculated by the BEM are higher than those obtained from the FEM /2/, which are the only ones published until now. The BEM results, however, are more exact, because the losses of the strip alone calculated by the BEM and verified by the very exact results of the variational analysis /7/ are also higher than the losses of the total stripline obtained by the FEM.

In order to investigate the influence of losses on the pulse propagation, the frequency dependent transmission line model is transformed by the inverse FFT into the time domain /8/. The effect of the different loss models (lossless, DC, Wheeler, broadband) on the step response of the previous stripline structure of 20 cm length is shown in Fig. 5. The step edge affected by the exact broadband losses clearly differs from that, where losses have been idealized or approximated.

Conclusions

The boundary element method (BEM) has been used to solve 2-D electromagnetic field problems related to the chip-to-chip interconnects. Special attention has been paid to the frequency dependent losses. The BEM results for some simple structures, which are in agreement with those obtained by other methods, are considered as a start point for the future application of this approach to more complex interconnect structures.

References

/1/ Wheeler, H.A.: Formulas for the skin effect, Proc. IRE 30 (1942) 412-424.

/2/ Ghoshal, U.; Smith, L.N.: Finite element analysis of skin effect in copper interconnects at 77K and 300K, Proc. 1988 IEEE MTT-S Int. Microwave Symp. 773-776.

/3/ Brebbia, C.A.: The boundary element method for engineers. London: Pentech Press 1978.

/4/ Costache, G.I.: Finite elements method applied to skin-effect problems in strip transmission lines, IEEE Trans. Microw. Theory and Techn. MTT-35 (1987) 1009-1013.

/5/ Kwon, O.; Pease, R.F.W.: Closely packed microstrip lines as very high-speed chip-to-chip interconnects, IEEE Trans. Comp., Hybrids and Manuf. Techn. CHMT-10 (1987) 314-320.

/6/ Chang, W.H.: Analytic IC-metal-line capacitance formulas, IEEE Trans. Microw. Theory and Techn. MTT-25 (1977) 712.

/7/ Waldow, P.; Wolff, I.: The skin effect at high frequency, IEEE Trans. Microw. Theory and Techn. MTT-33 (1985) 1076-1082.

/8/ Schutt-Aine, J.E.; Mittra, R.: Scattering parameter transient analysis of transmission lines loaded with nonlinear terminations, IEEE Trans. Microw. Theory and Techn. MTT 36 (1988) 529-536.

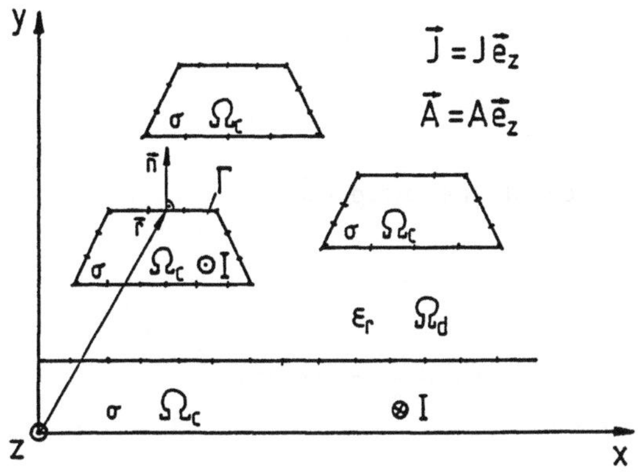

Fig.1. Multiconductor configuration consisting of the conducting regions Ω_c (including conductor carrying currents) surrounded by the homogeneous dielectric region Ω_d with interfaces divided into boundary elements

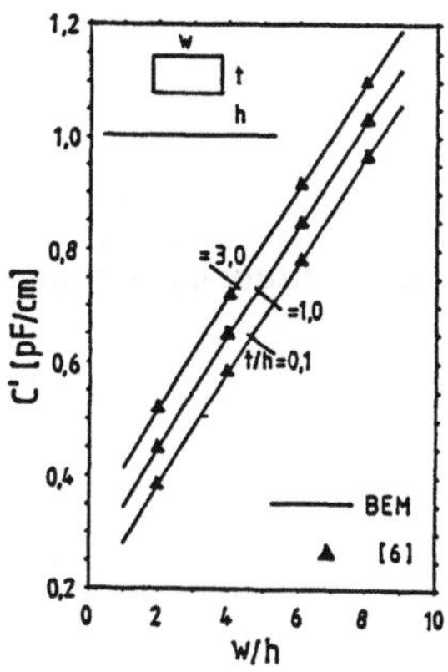

Fig.2. Length-specific line-to-ground capacitance for different geometries

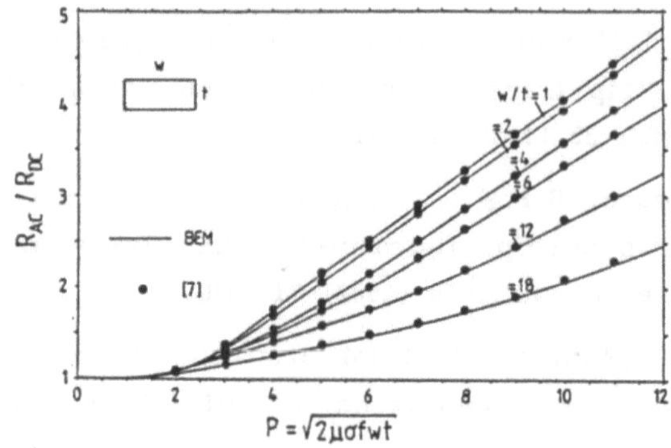

Fig.3. Normalized AC resistance of a single rectangular conductor versus normalized frequency P

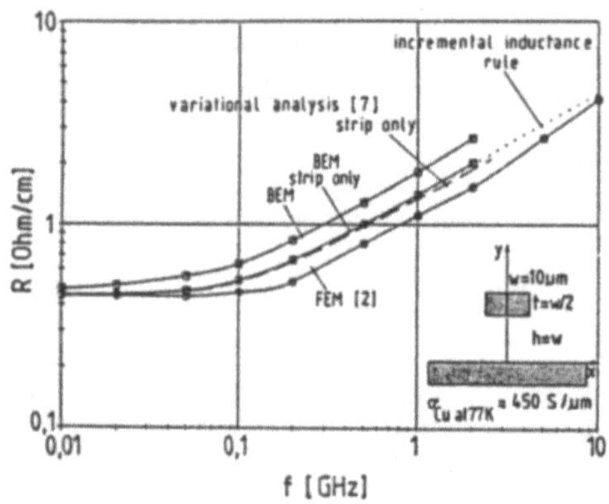

Fig.4. Frequency dependent resistance of the stripline

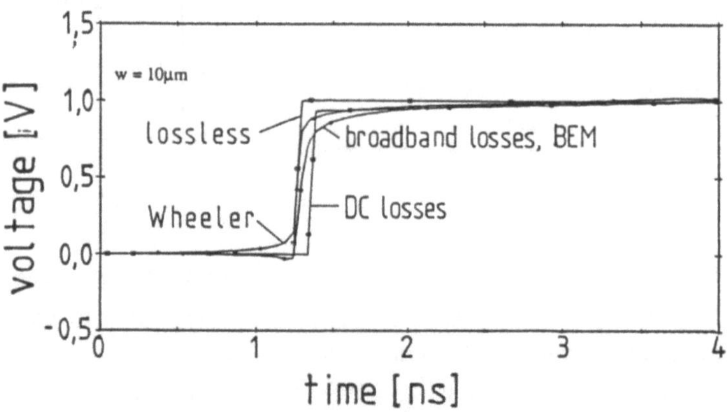

Fig.5. Step responses of a 20 cm long interconnect line depicted in Fig. 4 with different loss assumptions

Thermal Analysis of Integrated Circuits

H. Rudolph, D. Seitzer
Fraunhofer Gesellschaft
Institut für Integrierte Schaltungen
Am Wetterkreuz 13
D-8520 Erlangen-Tennenlohe

0. Introduction

With the increasing use of analog ASICs, especially in the instrumentation and sensor area, considerations about thermal effects on the circuit design become more and more important. This is because thermal time constants and the offset voltages due to them are in the same range as conversion time and resolution of the measurement instrumentation.

This paper describes investigations on the thermal analysis of bipolar integrated circuits within the design phase. In the first example the influence of self-heating and interaction thermals was simulated with special SPICE subcircuits. In a second example, we investigated the on chip temperature contribution with the finite-elements program ANSYS.

1. Basic Theory

Every electrical device changes its properties depending on the temperature. For example a bipolar transistor behaviour is represented by a change of -1.6 mV/deg of U_{BE}. The device's actual temperature is a result of a generally inhomogenous temperature contribution over the chip, which is defined by the ambient temperature and the power dissipation of the single devices. For a static temperature contribution and a constant thermal conductivity the following form of Poisson's equation is valid :

$$\text{div grad } \Theta(x,y,z) + 1/\lambda_0 \cdot W_q(x,y,z) = 0 \qquad (1)$$

with:

$\Theta(x,y,z)$: temperature field
λ_0 : thermal conductivity
$W_q(x,y,z)$: power density

Thus the temperature contribution of a point heat source with the power P_0 is radially symmetrical and decreases inversely proportional to the distance:

$$\Theta(r) = P_0 / 4\pi \cdot \lambda_0 \cdot r \qquad (2)$$

Refering to /1/, that result can be extrapolated to a square heat source with help of Fig.1:

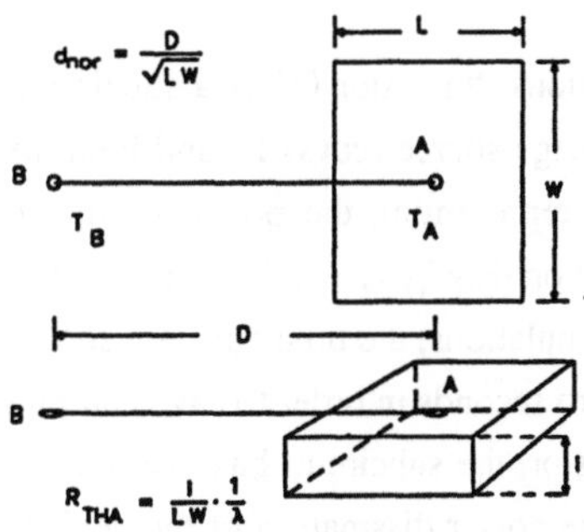

Fig.1: Temperature increase at point "B" due to a square heat source at point "A".

$$T_B = P_A \cdot R_{THA} \cdot 1/d_{nor} \tag{3}$$

Equations (2) and (3) allow simulations of the thermal behaviour of a circuit with SPICE as well as with a finite-elements program after layout.

While SPICE simulations with thermal models give an insight into the thermal sensitivity of the circuitry, the temperature profile yielded from ANSYS shows whether critical parts of the circuit, e.g. differential pairs, see a temperature gradient or not.

2. Thermal Models for SPICE

Due to a suggestion by P. Quinn (/2/), thermal models of bipolar transistors have been developed as subcircuits. The development is based on the use of additional controlled voltage and current sources to model the power dissipation and the resulting decrease of U_{BE}. The output of such sources is a function of a multi-dimensional polynomial of the controlling inputs, e.g. in the 2-dimensional case:

$$U_{out} = P_0 + P_1U_1 + P_2U_2 + P_3U_1 \; + P_4U_1U_2 + \ldots \tag{4}$$

In the model we have to distinguish between self-heating due to the power dissipation in the device itself and thermal interactions due to the power dissipation of devices in the near neighbourhood. A model for self-heating is shown in Fig. 2.

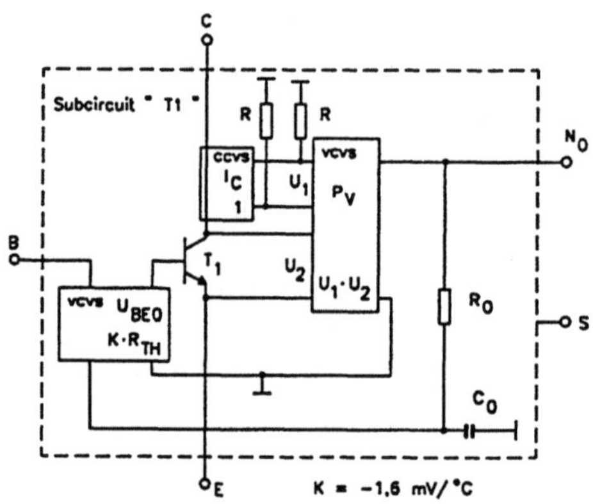

Fig.2: Subcircuit to model self-heating of bipolar transistors with SPICE.

According to the SPICE definitions, transistor Q1 is a subcircuit consisting of the transistor itself, the current controlled voltage source (ccvs) IC and both the voltage controlled voltage sources (vcvs) PV and UBE0, representing the power dissipation and the resulting dU_{BE} respectively. The RC-combination R_0C_0 is used to model the dynamic temperature behaviour. In practical transient simulations, the time constant should be choosen in the range of nano seconds rather than in micro seconds in order to save computing time.

Compared to the original transistor, the subcircuit has one more node, N_0. The voltage magnitude at that node represents the power dissipation of the device. Thus by implementing additional nodes N_μ, time constants $R_\mu C_\mu$ and controlled sources UBE_μ, adjacent devices can be taken into account. To simulate the thermal interactions of a single differential pair, every subcircuit must have at least two N-type nodes and node N_0 of the left transistor has to be connected to N_1 of the right hand one and vice versa.

So far, the thermal conductivity λ of silicon has been held constant; in fact itis decreasing between 20 K and 500 K (-253 °C - 773 °C) approximately inversely proportional to the temperature (/3/). That means that λ is changing 240 % within the temperature range between -55 °C and +125 °C . That effect is included in the following model (Fig. 3):

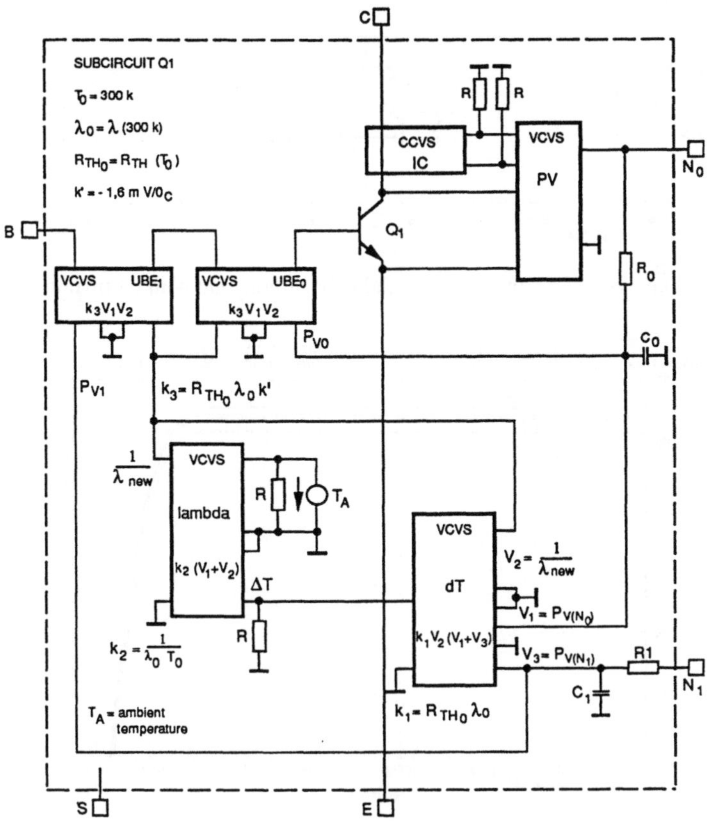

Fig.3: SPICE transistor model to simulate self-heating and interaction thermals with respect to the temperature dependence of the thermal conductivity.

Compared to the previous solution there are now two additional controlled sources "dT" and "lambda". These are used to recalulate the temperature increase and the resulting "new" thermal conductivity as a result of the ambient temperature T_A, self-heating ($P_V(N_0)$) and thermal interactions ($P_V(N_1)$).

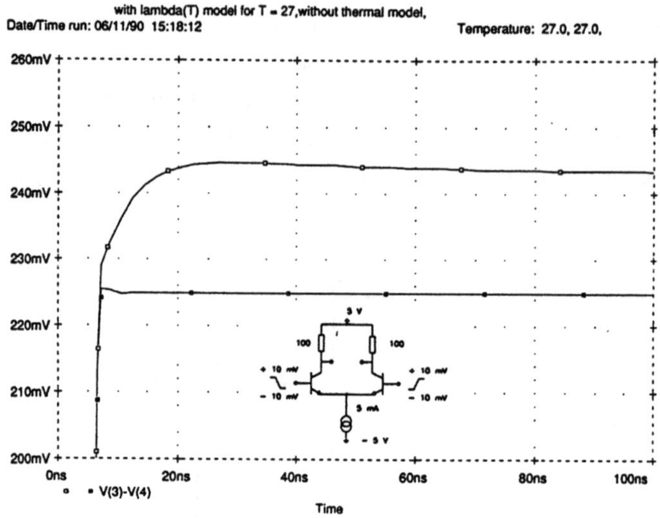

Fig.4: Transient analysis of a simple differential pair without and with thermal model (Fig.3).

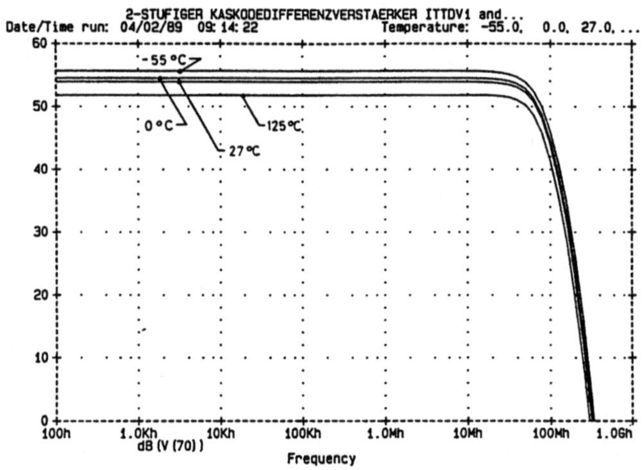

Fig.5: Transfer characteristics of a two stage cascode differential amplifier in the temperature range from -55 °C to 125 °C with thermal model of Fig.3.

3. Temperature Profile Analysis

The example here is a two stage cascode differential amplifier. The analysis was done with ANSYS on a PC running MSDOS. As a result of some memory restrictions the following considerations were done:

- Only 4 nodes are defined in the z - direction (depth) because the wafer thickness of 400 μm is considered to be very large compared to the lateral dimensions of the chip.
- The lateral dimensions of the net are choosen to be 550 · 850 μm^2 which means that the active area is surrounded by a 100 μm stripe.

- The horizontal definition of the net nodes was based on the prior result showing that the temperature increase of an even square heat source can be described with an 1/r - function. Thus the net was done as narrow as possible within a circle defined from the centre of the most important heat sources. The diameters of these circles are estimated under the condition that the increase in temperature caused by that particular heat source is reduced to less than 5 % of its maximum.

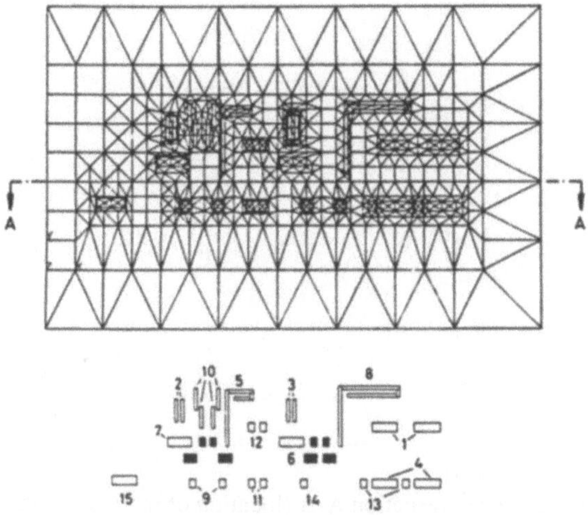

Fig.6: Net used for the thermal analysis of the two stage cascode amplifier and below the location of the considered circuit elements where:
1,4,13: 2nd stage emitter follower; 2,7: cascode 1 base potential; 3,6: cascode 2 base potential; 5,9: 1st stage current mirrors; 8,14: 2nd stage current mirrors; 10: cascode 1 load resistance; 11,12: 1st stage emitter follower; 15: reference for bias-current cancellation circuitry; Black boxes: 1st and 2nd cascode.

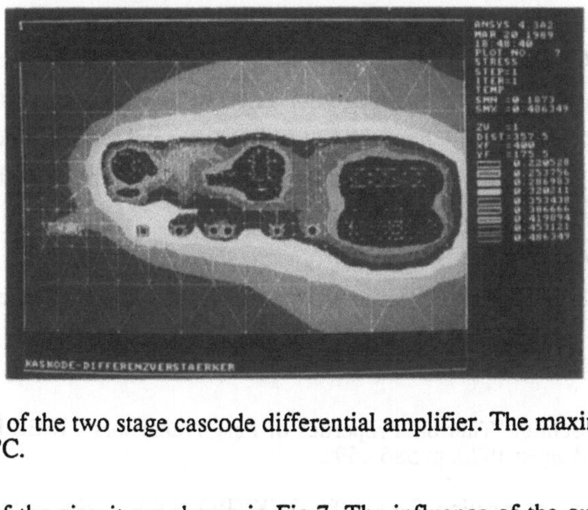

Fig.7: Isothermals of the two stage cascode differential amplifier. The maximum temperature rise is 0.5 °C.

The isothermals of the circuit are shown in Fig.7. The influence of the output stage and the low-impedance resistors used to define the base potential of the cascodes is clearly seen. Fig.8

shows a cross section along line A-A in Fig.6. This line correspondes to the location of the differential pair of both stages in the y-direction. The temperature difference between the left and right hand transistor is 0.0016 °C in the first stage and approximately zero in the second one. This difference leads to an input offset voltage of only 25 μV.

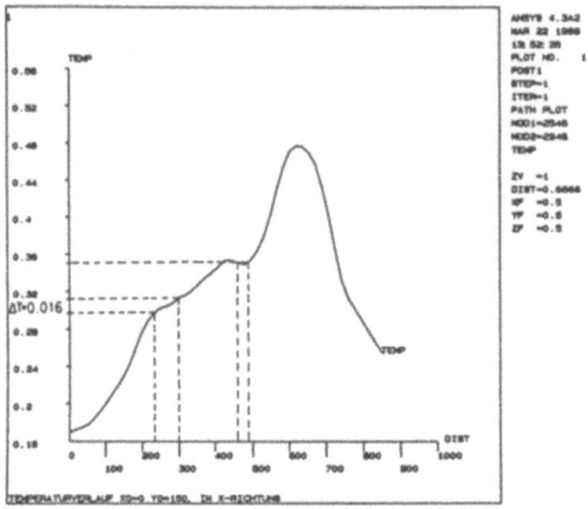

Fig.8: Temperature rise along cross section A-A (location of the input differential pairs of the cascodes).

4. Conclusions

Two examples of thermal simulations on integrated circuits have been presented. A finite element analysis can be seen as a final check for critical temperature contributions before fabrication, but requires great effort in terms of experience, compute time, etc. to get reliable results.

The reason for SPICE simulations in this case is to get an impression rather than exact data for the thermal sensitivity of the circuitry with as small an effort as possible. Results like those in Fig.4 show that designers have to keep in mind thermal settling times (actually in the range of micro seconds) on the one hand and on the other hand, the poor thermal modelling capability of "standard" SPICE transistor models.

5. Acknowledgements

Thanks are due to INTERMETALL, Freiburg, for the friendly cooperation in carrying out the mask layout for the amplifier and the bipolar fabrication process.

6. References

/1/ R. C. Joy, E. S. Schlig: "Thermal Properties of Fast Transistors", IEEE Electron Devices, Vol. ED-17, No. 8, August 1970, p. 586 - 593

/2/ P. Quinn: Informal information, Sept. 1988, TEKTRONIX Inc., Beaverton, Oregon

/3/ P. D. Madock: "Thermal Conductivity of Silicon, Germanium, III-V Compounds and III-V Alloys", Solid State Electronics, Vol. 10, p. 161 -168, Pergamon Press, 1967

Sensitivity Analysis in Thermal Wave Measurement

M. Wiedemann

TU Berlin, Institut für Luft- und Raumfahrt, Salzufer 17-19, D-1000 Berlin 10

This paper adds to the mathematical description of the technique of photothermal measurement by thermal waves. The technique itself is assumed to be known.
The method of finite elements (FEM) is applied to describe the simulation model of the examined structure. A following sensitivity analysis allows an effective and time-saving use of such FE-models for simultaneously determination of several thermal parameters by only few complete calculations.

Introduction

In many applications of microsystems an important question concerns the thermal behaviour of the structures and the transport of heat within them. The simulation for better thermal design to increase reliability and lifeduration needs a tool to determine the thermal parameters of the real microstructure quantitatively.

A well known and proved technique is the thermal wave measurement [1], [2]. Theoreticaly this technic is achieved mainly by analytical models, which allow the determination of one, sometimes of two thermal parameters. But often complexe structures cannot be described by analytical models. And due to the special conditions of manufactoring and their extreme small dimensions often more than one parameter is not known well enough.

The method presented in this paper is able to neutralize these restrictions partially. The examined structure is decribed by the FE-method. At the beginning, the used thermal parameters are estimated. On the whole, the theoretical results do not agree with the measured information. The differences can be used to correct the estimated parameters, provided that it is known, in which way each parameter influences the theoretical result.

To gain this knowledge one can calculate the model many times with different parameter sets. In case of FE-models with many degrees of freedom that means a waste of time and may lead to confusion if many parameters are in action.

Instead of that it is possible to examine the results of the FE-calculation by using the sensitivity analysis. This method supplies a functional relationship between the interesting parameters and the theoretical results. Sensitivity analyses are mainly used in structural dynamics to determine stable configurations of fast rotating systems [3], [4].

The determination of the really acting parameters now consists in the solution of a system of nonlinear equations. This solution can be achieved automatically including a least-square-fit and - what is the main advantage of this method - it can be achieved simultaneously for more than two parameters.

Two examples will demonstrate the possibilities of thermal wave measurement completed by the sensitivity analysis of a simulation model.

Simulation

The structure is discreticized by finite elements and the spatial field problem of instationary heat flux (1) is replaced by a system of I coupled differential equations for I temperature degrees of freedom, containing only partial derivations with respect to time (2).

$$\rho\,(x,y,z)\; c\,(x,y,z)\; \frac{\partial}{\partial t} T(x,y,z,t) \;-\; \nabla\{\lambda\,(x,y,z)\; \nabla T(x,y,z,t)\} \;=\; \kappa\,(x,y,z,t) \qquad (1)$$

$$\underline{\boldsymbol{M}}\, \frac{\partial}{\partial t} \boldsymbol{T}(t) \;+\; \underline{\boldsymbol{L}}\; \boldsymbol{T}(t) \;=\; \boldsymbol{Q}(t) \qquad (2)$$

T stands for temperature, ρ for the density, c for the heat capacity, λ for the heat conductivity, κ for the heat source, $\underline{\boldsymbol{M}}$ represents the heat-mass matrix, $\underline{\boldsymbol{L}}$ the matrix of conductivities, $\boldsymbol{T}$ the vector of the discrete temperatures and $\boldsymbol{Q}$ the load vector. For the chosen FE-model is to demand, that its accuracy is proved by a second model with double fine mesh, obtaining phase results that differ no more than 0.5% in comparision.

The homogenous part of equation (2) leads to the eigenvalueproblem. Its solution results in I eigenvalues and I corresponding eigenvectors. The latter can be normalized and are orthogonal to each other.

$$Det\,[\,\underline{\boldsymbol{L}} - \omega_i \underline{\boldsymbol{M}}\,] = 0 \quad \text{leads to i eigenvalues} \quad \omega_i;\; i=1 \text{ to } I \qquad (3a)$$

$$[\,\underline{\boldsymbol{L}} - \omega_i \underline{\boldsymbol{M}}\,]\; \boldsymbol{r}_i = 0 \quad \text{leads to i eigenvectors} \quad \boldsymbol{r}_i;\; i=1 \text{ to } I \qquad (3b)$$

Making use of the eigenvectors it is possible to describe the total solution of (2) for any form of $\boldsymbol{Q}(t)$ by a superposition of the single solutions of the eigenvalueproblem. In case of thermal waves with frequency $\Omega = 2\pi F$, the solution for the temperature in the k-th node of structure - for example a point on the surface - can be written in the following way:

$$\text{with} \quad \boldsymbol{Q}(t) = \boldsymbol{Q}_0\, e^{j\Omega t}; \qquad q_i = \boldsymbol{r}_i^{\,T} \boldsymbol{Q}_0; \qquad \eta_i = \frac{\Omega}{\omega_i}:$$

$$\overset{*}{T}_k(\Omega t) = \sum_{i=1}^{I} \frac{r_{k,i}\, q_i}{\omega_i\,(1+\eta_i^{\,2})}\,[1 - j\eta_i]\; e^{j\Omega t}. \qquad (4)$$

$r_{k,i}$ means the k-th element of the i-th eigenvector. The resulting phase in the k-th node of the structure is :

$$\tan \varphi_k = -\frac{T_k^{\,Im}}{T_k^{\,Re}}. \qquad (5)$$

It is to emphasize that in this formulation the whole physical behaviour of the structure is represented by the I eigenvalues and by the elements of the I eigenvectors!

Sensitivity analysis

Every layer of a microstructure is characterized mainly by three thermal parameters; the conductivity, the capacity and the thickness. The latter ist in fact a geometric parameter, but treated in this context as a thermal one too. Within N layers that means 3N thermal parameters.

$$\boldsymbol{p}^T = \{\lambda_1 ; c_1 ; d_1 ; \ldots \lambda_N ; c_N ; d_N\} = \{p_1 ; \ldots p_{3N}\}$$

The sensitivity analysis is based on the taylorisation of eigenvalues and eigenvectors with respect to the parameters of interest.

$$\omega_i(\boldsymbol{p}) = \omega_{i0} + \sum_v \frac{1}{v!}\left\{ \left.\frac{\partial}{\partial p_1}\right|_0 \Delta p_1 + \ldots \left.\frac{\partial}{\partial p_{3N}}\right|_0 \Delta p_{3N} \right\}^v \omega_i \tag{6a}$$

$$\boldsymbol{r}_i(\boldsymbol{p}) = \boldsymbol{r}_{i0} + \sum_v \frac{1}{v!}\left\{ \left.\frac{\partial}{\partial p_1}\right|_0 \Delta p_1 + \ldots \left.\frac{\partial}{\partial p_{3N}}\right|_0 \Delta p_{3N} \right\}^v \boldsymbol{r}_i . \tag{6b}$$

To determine the partial derivations, one has to develop the matrices $\underline{\boldsymbol{M}}(\boldsymbol{p})$ and $\underline{\boldsymbol{L}}(\boldsymbol{p})$ with respect to the interesting parameters. Introducing them into equation (3a) together with the statements (6a) and (6b) we get the modified eigenvalueproblem:

$$[\,\underline{\boldsymbol{L}}(\boldsymbol{p}) - \omega_i(\boldsymbol{p})\,\underline{\boldsymbol{M}}(\boldsymbol{p})\,]\;\boldsymbol{r}_i(\boldsymbol{p}) = \boldsymbol{0} . \tag{7}$$

The evaluation of (7) and the comparision of the resulting combinations of arguments Δp_i leads to equations which can be solved by taking advantage of the orthogonality of the eigenvectors. The resulting derivatives are:

$$\frac{\partial \omega_i}{\partial p_1} ; \frac{1}{2}\frac{\partial^2 \omega_i}{\partial p_1^2} ; \ldots \frac{\partial \omega_i}{\partial p_{3N}} \quad \text{e.t.c.} \qquad \frac{\partial \boldsymbol{r}_i}{\partial p_1} ; \frac{1}{2}\frac{\partial^2 \boldsymbol{r}_i}{\partial p_1^2} ; \ldots \frac{\partial \boldsymbol{r}_i}{\partial p_{3N}} \quad \text{e.t.c.}$$

These results applied to the statements (6a) and (6b) provide the wanted relationship between the theoretical results for the temperature phase and the interesting parameters. A fairly good approximation of the real behaviour in the range of ± 40% divergenze to the estimated parameters is reached by considering the following derivatives:

$$\frac{\partial}{\partial \lambda_n} ; \frac{\partial}{\partial c_n} ; \frac{\partial}{\partial d_n} ; \frac{\partial^2}{\partial \lambda_n \partial c_n} ; \frac{\partial^2}{\partial c_n \partial d_n} ; \frac{\partial^2}{\partial \lambda_n \partial d_n} ; \frac{1}{2}\frac{\partial^2}{\partial \lambda_n^2} ; \frac{1}{2}\frac{\partial^2}{\partial c_n^2} ; \frac{1}{2}\frac{\partial^2}{\partial d_n^2} \quad ; \; n = 1 \text{ to } N$$

Coupled partial derivations of parameters of different layers do not occure.

Identification

The aim of the identification procedure is to correct the estimated parameters $\boldsymbol{p}$ of the model in such a manner, that theoretical results reconcile with measured information. This demand leads to the system of equations (8) for frequencies Ω_m (m = 1 to M) of periodically heating. This equation is strongly nonlinear due to the thermal parameters, which are 'hidden' in functions of the eigenvalues and eigenvector elements. f_m represents the sum of the unavoidable errors of measurement, the discretisation failure and the approximation error of the sensitivity analysis.

$$g_m(\boldsymbol{p}) = \varphi_k^{Mod}(\Omega_m, \boldsymbol{p}) - \varphi_k^{Mes}(\Omega_m) = 0 + f_m ; \qquad \text{respectively} \qquad \boldsymbol{g}(\boldsymbol{p}) = \boldsymbol{0} + \boldsymbol{f} \tag{8}$$

To solve equation (8), the method by Newton-Kantorowitsch is used. In case of $\boldsymbol{p}^0$ being an approximation of the vector $\boldsymbol{p}$ of the corrected parameters and $\boldsymbol{g}(\boldsymbol{p})$ being C^0-continual with respect to $\boldsymbol{p}$, one may write a new linear equation system (9) by taylorisation of (8). $\boldsymbol{z} = \boldsymbol{p} - \boldsymbol{p}^0$ represents the new argument vector and $\underline{\boldsymbol{A}}$ the jacobian matrix of the partial derivations of $\boldsymbol{g}(\boldsymbol{p})$ with respect to $\boldsymbol{p}$ for all measurements. $\boldsymbol{v}$ contains the remainder of the taylorisation and the error-vector $\boldsymbol{f}$.

$$\underline{\boldsymbol{A}} = \frac{\partial (g_1 ; g_2 ; \ldots\ g_M ;)}{\partial (p_1 ; p_2 ; \ldots\ p_{3N} ;)}$$

$$\underline{\boldsymbol{A}}\, \boldsymbol{z} + \boldsymbol{g}(\boldsymbol{p}^0) = \boldsymbol{0} - \boldsymbol{v} \tag{9}$$

If there exist more data of measurement than correctable parameters, the overdetermination can be used for a least-square-fit.

$$\underline{\boldsymbol{A}}^T \underline{\boldsymbol{A}}\, \boldsymbol{z} + \underline{\boldsymbol{A}}^T \boldsymbol{g}(\boldsymbol{p}^0) = \boldsymbol{0} \tag{10}$$

In general this procedure shows good convergence qualities for this application.

Supplements

a) As already other authors pointed out, the chance to determine thermal parameters of different layers separately depends on their effussivities e_l. The reflection coefficient derived from them has to be bigger than 0.5.

$$e_l = \sqrt{(\rho c)_l\, \lambda_l} ; \qquad | \alpha_{l,l+1} | = \left| \frac{e_l / e_{l+1} - 1}{e_l / e_{l+1} + 1} \right| < 0.5 . \tag{11}$$

b) The depth in the structure which may be reached by a thermal wave depends on the frequency of heating. It is useful to start the procedure of identification at the first surface layer by using the corresponding frequencies and then to extend the examination to deeper layers. The minimum frequenzy allowed measuring effects from the first till the N-th layer may be calculated by:

$$F_{\min} = \frac{\bar{\lambda}}{\pi\, \overline{\rho c}\, \bar{d}^{\,2}} ; \quad \bar{d} = \sum_{l=1}^{N} d_l ; \quad \bar{\lambda} = \frac{\bar{d}}{\sum_{l=1}^{N} \frac{d_l}{\lambda_l}} ; \quad \overline{\rho c} = \frac{\sum_{l=1}^{N} \rho_l c_l d_l}{\bar{d}} \tag{12}$$

Results

First of all results achieved by Baumann and Beyfuss [5] were used to proof the developed algorithm. A thin glaze on a ceramic was measured by thermal waves. The spot of the laserbeam was chosen large enough to allow the description of the thermal behaviour by an one-dimensional FE-model (figure 1). The thickness of the glaze and the thickness and heat capacitiy of the ceramic are assumed to be known exactly. In this structure of two layers therefore three parameters remain, which need to be determined by thermal wave measurement. Curve 1 in diagram 1 shows the theoretical phase results of the chosen FE-model with estimated parameters at frequency F. Curve 2 represents the results of a simulationmodel with double fine mesh, showing no differences to the simpler one. Curve 3 shows the measured phases for different frequencies and curve 4 the results of the corrected simulationmodel. The first column of table 1 represents the starting values. After the first identification step the corrected values were used again and a second identification was done. The last column shows the determined values refering to [5]. Starting with estimated values which differ more than 40% to the finally determined ones, the method convergences after 4 steps.

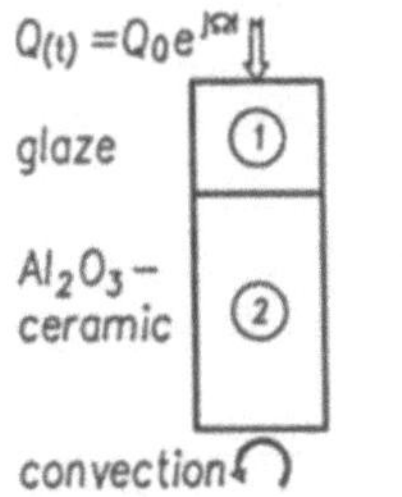

Figure1: 1D-modell of glaze / ceramic

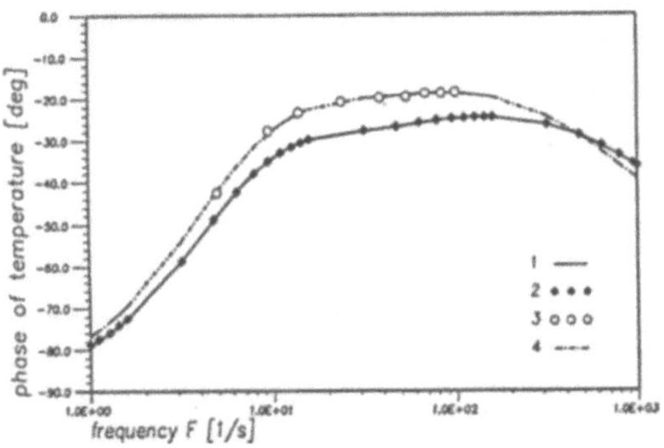

Diagram 1: (1) simulationsmodel, (2) simulationsmodel with double fine mesh, (3) measured values ref. [5], (4) corrected simulationsmodel

parameter	Step 1	Step 2	Step 3	Step 4	values ref. [5]
λ_1 [W/mK]	2.0	1.1224	1.1546	1.1690	1.16
ρc_1 [J/cm^{3}K]	2.5	2.6143	2.0658	2.1220	2.07
d_1 [µm]	15	fixed	fixed	fixed	15
λ_2 [W/mK]	20	24.681	24.900	24.900	unknown
ρc_2 [J/cm^{3}K]	2.9	fixed	fixed	fixed	unknown
d_2 [µm]	620	fixed	fixed	fixed	620

Table 1: parameter values in four steps of identification and ref. [5]

A theoretical examination shall demonstrate the extended possibilities of thermal wave measurement completed by the sensitivity analysis. A structure of three layers - chip / adhesive / substrate - was modelled by two-dimensional axisymmetric finite elements (figure 2) and the phase was calculated with given parameters for various frequencies (measurement-model). A model with different parameter values was examined by the sensitivity analysis (identification-model) and corrected by the results of the measurement-model. In two identification steps even a difference of 20% in the heat capacity of the third layer was detected with good accuracy. Diagram 2 shows the phase results of the identification-model and the measurement-model in two nodes on the surface, which were used in the identification process. Table 2 lists the starting values and the corrected parameters of two steps in comparision to the given values of the measurement-model.

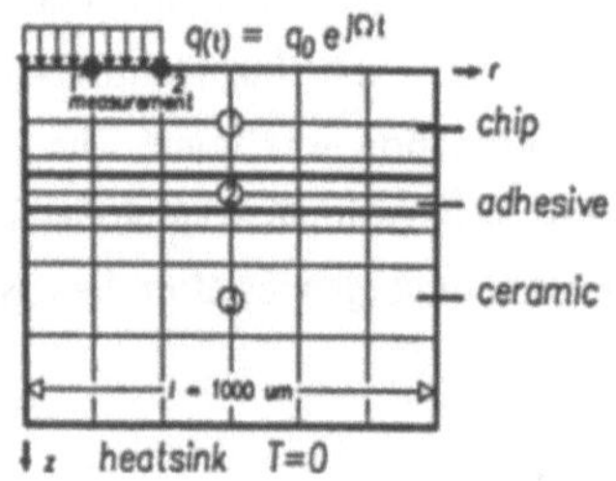

Figure 2: 2D-model; chip / adhesive / substrate

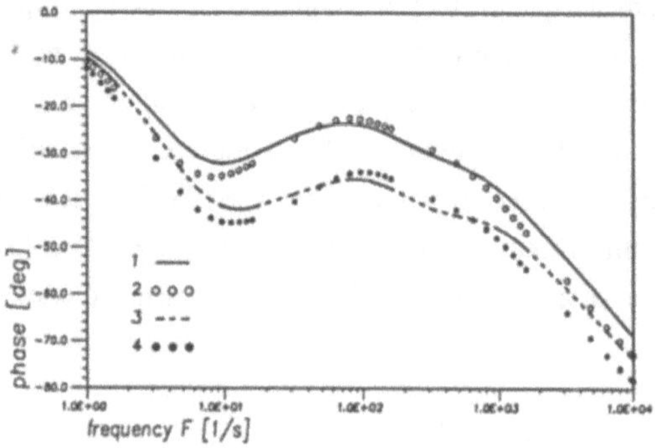

Diagram 2: (1) ident.-model, node 1; (2) meas.-model, node 1; (3) ident.-model, node 2; (4) meas.-model, node 2

parameter	ident.-model	Step 2	Step 3	meas.-model
λ_1 [W/mK]	141.0	fixed	fixed	141.0
ρc_1 [J/cm^{3}K]	1.708	1.7615	1.709	1.7079
d_1 [µm]	300	368.37	359.68	360.0
λ_2 [W/mK]	6.0	4.756	4.197	4.2
ρc_2 [J/cm^{3}K]	1.26	fixed	fixed	1.26
d_2 [µm]	75	fixed	fixed	75
λ_3 [W/mK]	23.0	fixed	fixed	23
ρc_3 [J/cm^{3}K]	2.5506	2.7864	3.029	3.0607
d_3 [µm]	600	fixed	fixed	600

Table 2: parameter values of identification-model and measurement-model

Considering the accuracy of 0.5° in phase measurement and the accuracy of FE-models which may be achieved it seems to be possible to extend the determination of thermal parameters by thermal wave measurement up to three layers of optical and X-ray opaque structures.

Literatur

[1] M. Mandelis (Ed.); Photoacoustic and thermal wave phenomena in semiconductors; North Holland, 1987

[2] P. Hess, J. Pelzl (Eds.); Photoacoustic and photothermal phenomena; Springer Series in Optical Science Vol. 58, 1987

[3] C. P. Fritzen, R. Nordmann; Influence of parameter changes to the stability behaviour of rotors; Workshop 'Rotordynamic Instability Problems in High Performance Turbomachinery', Texas, 1982, NASA Conference Publication

[4] R. Gasch, K. Knothe; Strukturdynamik, Band 1, Diskrete Systeme; Springer Verlag 1987

[5] M. Beyfuss, J. Baumann; Determination of the properties of thin layers by a photothermal technique; 3. Inern. Symposium Zerstörungsfreie Bestimmung von Materialkenngrößen, Saarbrücken; Springer Verlag, 1988

Simulation and Design of Sensors

Design and Simulation of Sensors

HENRY BALTES

Physical Electronics Laboratory
Institute of Quantum Electronics
Swiss Federal Institute of Technology Zurich
ETH-Hoenggerberg, CH-8093 Zurich, Switzerland

Summary

Computer-aided design and simulation of silicon microsensors and micromechanical structures in standard CMOS or bipolar IC technology with or without additional processing steps are reviewed. Examples include integrated magnetotransistors, humidity sensors and flow meters.

Introduction

The progress of silicon integrated circuit (IC) design and technology has led to a strong demand for sensors - devices that convert physical or chemical measurands into useful electronic signals - which are compatible with modern IC signal processing. The answer to this challenge is currently attempted by silicon microsensors fabricated with IC technology and related techniques. The goal is to transfer the advantages of microelectronics to sensor systems, namely to achieve inexpensive batch fabrication of highly reliable sensors with on-chip circuitry for the necessary signal conditioning.

Certain temperature, magnetic field, and radiation sensors can be obtained in standard IC technologies such as CMOS or bipolar by judicious design. Other sensors, notably mechanical and chemical sensors, require etching or thin film deposition as additional processing steps (post-processing) compatible with the IC process. An approach analogous to application specific integrated circuits (ASIC) leads to application specific integrated (or intelligent) sensors (ASIS). These are families of sensors that can be personalized to customer specifications and combine individual product development with the advantages of semiconductor batch fabrication [1].

Computer aided design (CAD) tools including device and circuit simulation are crucial for fast turn-around IC development. By the same token, design methodologies and numerical modeling for the sensor elements to be integrated on a sensor IC chip are required for microsensor development. These are the very topics of the present paper. In contrast to the mature fields of semiconductor IC design automation and device modeling, the art of sensor design and model-

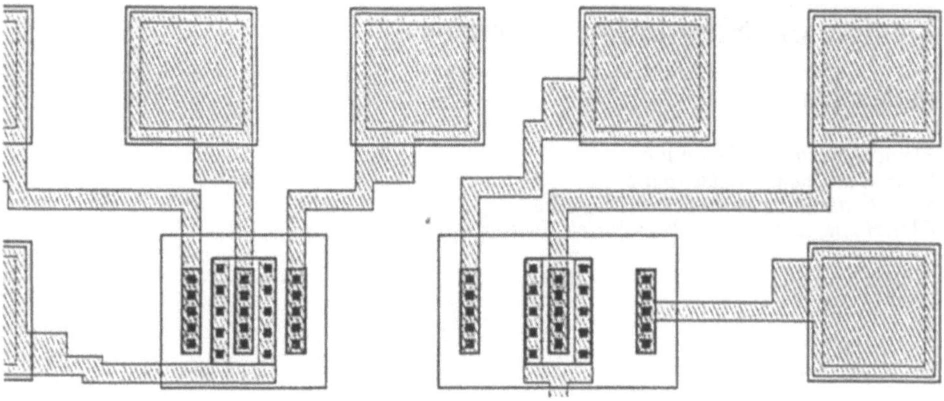

Fig. 1a Layout of vertical dual collector magnetotransistors

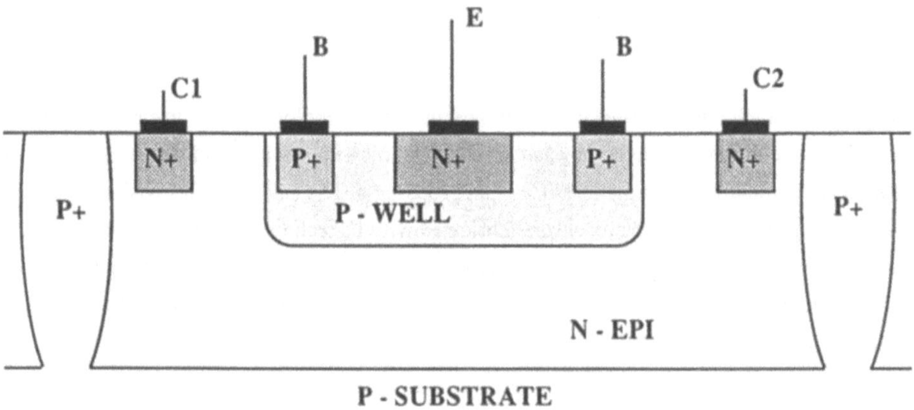

Fig. 1b Cross section of a vertical dual collector magnetotransistor

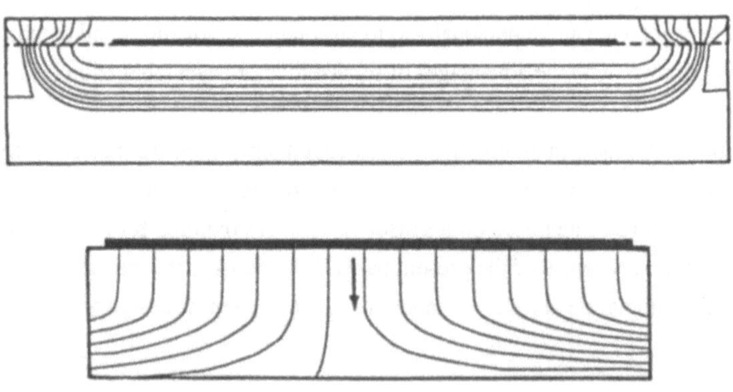

Fig. 2 Equipotential lines and minority carrier flow near the emitter

ling [2] are in their infancy. We therefore illustrate the intricacy of microsensor design and simulation by selected examples. Futher cases are presented in two companion papers [3, 4].

Magnetotransistors in bipolar and CMOS technology

Semiconductor magnetic field sensors exploit the Lorentz force on moving carriers (Hall effect, Lorentz deflection, magnetoconcentration, or magnetoresistance). A wide variety of integrated silicon magnetic field sensors have been realized such as bulk or MOSFET Hall plates,vertical or lateral magnetotransistors, and magnetodiodes [5]. In Figs.1a and 1b we show layout and cross section of a vertical dual collector magnetotransistor fabricated by the 20 V bipolar process on (111) or (100) silicon of *Ascom Microelectronics*. The magnetic field leads to an asymmetry in the carrier trajectories (Fig.2) and hence a current imbalance between the two collectors. The layout was generated using AUTOCAD on a SUN work station. Lateral magnetotransistors can be also designed for CMOS technology, which allows the more sensitive npn transistor. The purpose of our prototypes is to explore the degree of noise correlation between the two collector output signals and find geometry and operating conditions with high correlation and hence low noise and high magnetic resolution [6] .

Fig. 2 shows the potential and current distribution in a magnetotransistor obtained from the numerical solution of the carrier transport equations in the presence of the Lorentz force [2]. In the drift diffusion approximation with small magnetic induction and under isothermal condition, the physical model equations are the *Poisson, continuity* and *current transport equations*

$$div\,(\varepsilon\, grad\, \psi) = -q\,(p - n + N_D - N_A) \qquad (1)$$

$$div\, \boldsymbol{J_n} - q\, \partial n/\partial t = q\,(R - G) \qquad div\, \boldsymbol{J_p} + q\, \partial p/\partial t = -q\,(R - G) \qquad (2)$$

$$\boldsymbol{J_{nB}} + \mu_n{}^* \boldsymbol{J_{nB}} \times \boldsymbol{B} = qD_n\, grad\, n \qquad \boldsymbol{J_{pB}} - \mu_p{}^* \boldsymbol{J_{nB}} \times \boldsymbol{B} = -qD_p\, grad\, p \qquad (3)$$

with ε denoting the material permittivity, ψ the electrostatic potential, q the elementary charge, n and p the electron and hole concentrations, N_D and N_A the ionized donor and acceptor impurity distributions, $\boldsymbol{J_n}$ and $\boldsymbol{J_p}$ the electron and hole current densities, R and G the carrier recombination and generation rates, D_n and D_p the carrier diffusion coefficients, $\mu_n{}^*$ and $\mu_p{}^*$ the Hall mobilities and $\boldsymbol{B}$ the magnetic induction vector. The technicalities of the numerical solution of these equations under realistic boundary and operating conditions and further magnetotransistor results are reported in [7]. Extended modeling equations proposed recently [8] include thermal diffusion and will allow to model thermomagnetic effects.

Magnetic microsensors with on-chip circuitry are made in bipolar and CMOS technology. Magnetic sensor chips include amplifiers, filters, Schmitt triggers, emitter followers, or oscillators. Highly functional sensors are currently developed such as the *monolithic silicon magnetic com-*

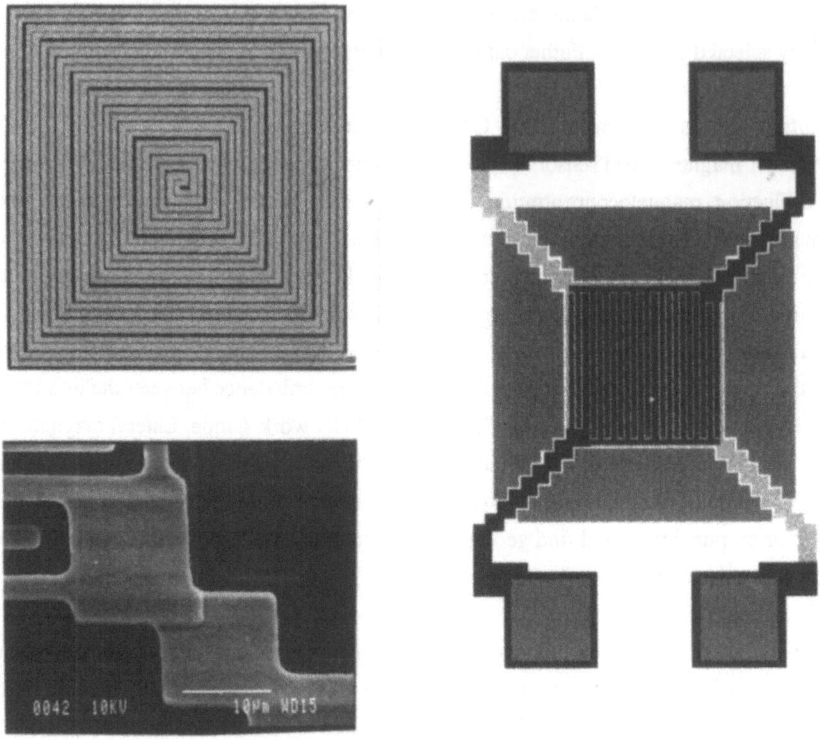

Fig. 3. CMOS designs of lateral capacitor with coil-shaped aluminium and poly electrodes (upper left) and comb-shaped aluminium electrodes separated by CVD oxide from poly ground plate on suspended membrane (right and photo of detail)

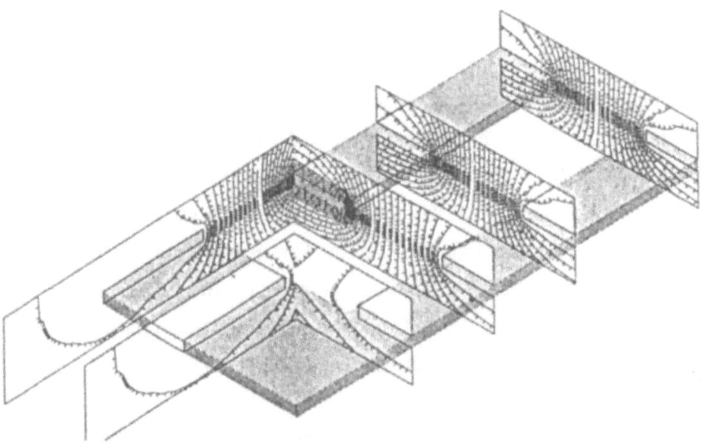

Fig. 4. Equipotential lines in selected sections through comb-shaped capacitor electrodes

pass [9] fabricated in bipolar analog IC technology based on two vertical Hall devices with circuitry (190 transistors and 40 resistors) providing the absolute value and the angle of $\boldsymbol{B}$. Circuits for temperature or offset reduction have been integrated as well.

Capacitive humidity sensors in CMOS technology

Lateral capacitors with aluminium or polysilicon electrodes can be designed for CMOS technology. Coated with a hygroscopic layer such as polyimide [10] they function as humidity sensors. The moisture modulates the dielectric environment of the electrodes and hence the capacity. Examples of capacitor layout for a CMOS process of *FASELEC* are shown in Fig.3. The interdigitated structure has aluminum electrodes separated by silicon oxide from a polysilicon ground plate. The structure is placed on a suspended membrane over an etch pit (see below) separating the capacitor from the bulk and providing access of moisture from below. Fig.4 shows the distribution of equipotential lines around the electrodes obtained from a finite element solution of the Laplace equation with appropriate boundary conditions [11].

As long as we work within the conventional IC design rules, the whole chip except the contact pads is passivated. Exposure of the electrodes (for post-processing coating without intermediate passivation layer) or silicon (for post-processing anisotropic etch) can be achieved by unconventional design [12,13], at variance with the design rules, but compatible with the process. This leads to the design rule check (DRC) problem of distinguishing between incidental design errors (to be corrected) and intentional ones (to be kept). A "window DRC" exempting the capacitor helps to solve this problem. The design of a capacitive moisture sensor for a CMOS process of *VTI* with on-chip circuitry is described in [3].

Micromechanical structures with CMOS and bipolar technology

Micromechanical silicon-oxide structures with or without sandwiched metal or poly layers on a *CMOS* chip can be obtained by maskless post-processing anisotropic etching of (100) silicon from the front of the wafer [13]. The silicon substrate area to be etched remains exposed after the standard CMOS process by virtue of an unconventional layout design: active area, contact cut, and pad opening placed one above the other define the area where the silicon substrate should be exposed. N^+ and n-guard exclusion masks surround the area in order to keep a low dopant concentration in the silicon to be exposed. Again this layout design is at variance with conventional IC design rules, but compatible with the CMOS fabrication process.

A typical layout for a cantilever beam is shown in Fig.5. The cross section (b) results after the CMOS process: the CVD oxide layers are removed in the active area, but doping is prevented by the exclusion masks. The anisotropic silicon etchant undercuts the oxide and thus forms the beam. Free-standing sandwiched layers consisting of CVD on polysilicon or aluminium on field oxide are designed in a similar fashion. This is how the capacitor on a membrane (Fig.3)

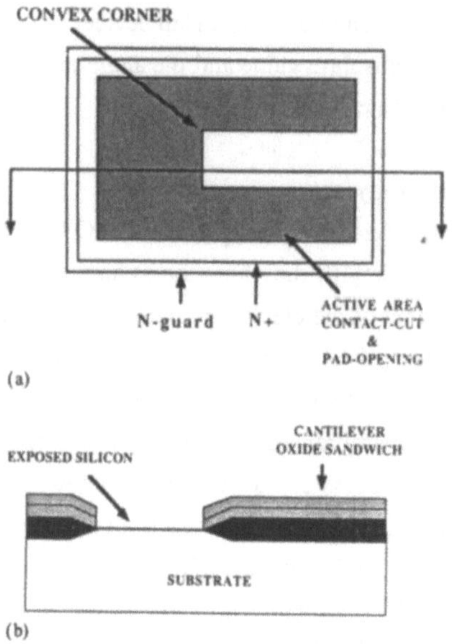

Fig. 5. (a) Layout design for fabrication of oxide beam with CMOS process
(b) Cross section before anisotropic etch post-processing

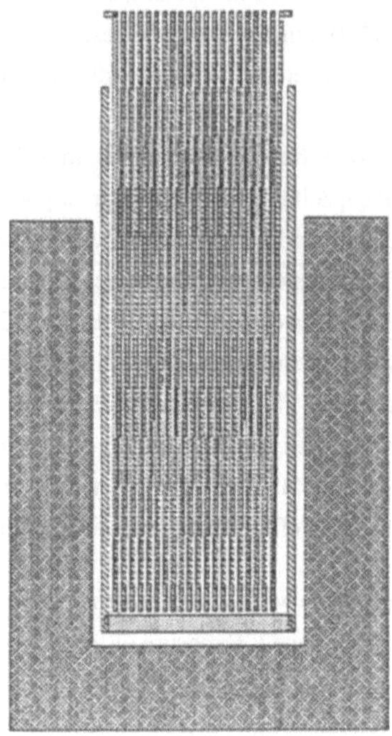

Fig. 6 a. Layout design of aluminium-poly thermopile on oxide beam

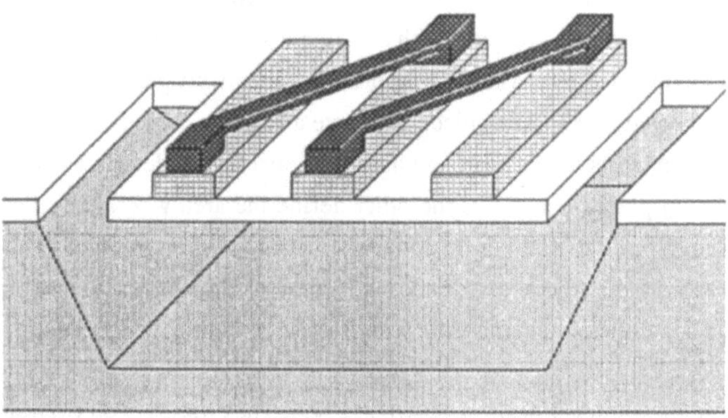

Fig. 6 b. Cross section of aluminium-poly thermopile on oxide beam over etch pit (CVD oxide between poly and metal is not shown)

is made. *Thermal gas flow sensors* can be made in that way: the design for a CMOS process of *FASELEC* shown in Fig.6 has a polysilicon resistive heater at the tip of the beam and a polysilicon/aluminium thermopile operating between the hot tip and the cold bulk silicon (the CVD oxide separating the metal and poly strips has been omitted in Fig.6b).

Anisotropic etching of silicon can be combined with a *bipolar* process on (100) silicon [1]. The simultaneous sequence of IC and sensor processing steps is outlined in Fig.7. The membrane with piezoresistors can serve as pressure sensor. Accelerometers can be obtained by a further etching step. The finite element modeling software ANSYS is quite useful for the simulation of three-dimensional micromechanical structures. The simulation of silicon accelerometers using ANSYS is reported in [4].

References

1. A. H. Kayal, N. Rauch: Anwendungsspezifische intelligente Sensoren ("ASIS"). *Elektronik* **9** (1998) 4-10.
2. H. Baltes, A. Nathan: Sensor modeling. ***Sensors*** **1** (1989) 45-77.
3. T. Boltshauser, E. Charbon, H. Baltes: Design of integrated humidity sensor with digital output in double metal CMOS technology. Microsystems 1990.
4. D. Moser, P. Sigg, S. Ansermet, H. Baltes: Finite element simulation of silicon accelerometers. Microsystems 1990.
5. H. Baltes, A. Nathan: Integrated magnetic sensors. ***Sensors*** **1** (1989) 195-215.
6. R. Castagnetti, H. Baltes: Noise correlations and operating conditions of dual-collector magnetotransistors. *Eurosensors IV*, Karlsruhe, Germany, October 1-3, 1990.
7. W. Allegretto, A. Nathan, H. Baltes. Numerical analysis of magnetic-field sensitive bipolar devices. *IEEE Trans. CAD of ICAS*, in press.
8. S. Rudin, H. Baltes. Treatment of thermomagnetic effects in semiconductor device modeling. ESSDERC 90, Nottingham, U.K., September 10-13, 1990.
9. K. Maenaka, M. Tsukahara, T. Nakamura. Monolithic silicon magnetic compass. *Sensors and Actuators* **A 21** (1990) 747-750.
10. M. C. Buncick, D. Denton: Effects of aging on polyimide: a study of bulk and interface chemistry. *1990 IEEE Solid State Sensor and Actuator Workshop Digest*, in press.
11. L. Chandran, J. Korvink, H. Baltes: Three-dimensional modelling of capacitive humidity sensors. *Eurosensors IV*, Karlsruhe, FRG, 1-3 Oct. 1990.
12. T. Boltshauser, E. Charbon, H. Baltes: Capacitive humidity sensors in SACMOS technology. *Eurosensors IV*, Karlsruhe, Germany, 1-3 Oct. 1990.
13. D. Moser, M. Parameswaran, H. Baltes: Field oxide microbridges, cantilever beams, coils and suspended membranes in SACMOS technology. *Sensors and Actuators* **A 23** (1990) 1019-1022.

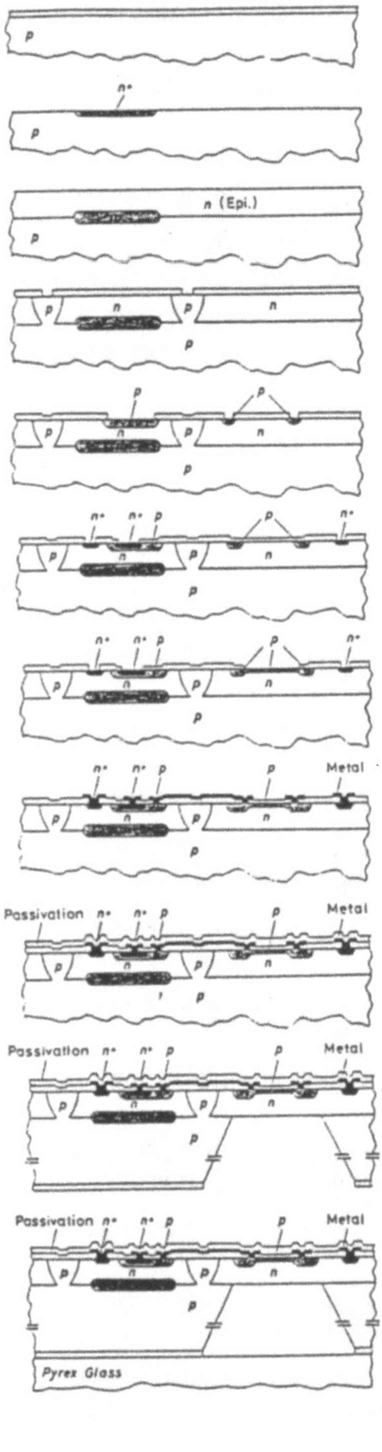

Thermal oxidation (IC & Sensor)

Buried layer diffusion (IC)

Epitaxial layer growth (IC & S)

Isolation and diffusion (IC)

Base diffusion (IC & S)

Emitter diffusion (IC & S)

Piezoresistor implantation (S)

Aluminium interconnection (IC & S)

Passivation layer deposition (IC & S)

Back side silicon micromachining (S)

Silicon to glass (or silicon) sealing (S)

Fig. 7. Process sequence of simultaneous IC (left) and sensor (right) fabrication [1] (Courtesy of *Ascom Microelectronics*)

Finite-Element Simulation of Silicon Accelerometers

David Moser, Paulo Sigg, Stephane Ansermet* and Henry Baltes

Physical Electronics Laboratory, Institute of Quantum Electronics
Swiss Federal Institute of Technology Zurich
ETH-Hoenggerberg, CH-8093 Zurich, Switzerland

**Ascom Microelectronics*
CH-2022 Bevaix, Switzerland

Summary

We report the simulation of integrated piezoresistive silicon accelerometers using the general purpose finite-element program ANSYS. Modelling of stress patterns and deflections due to applied forces as well as to the fabrication process (bipolar technology) are demonstrated. Predictions of resonant modes and resonant frequencies are also reported.

Introduction

The two typical types of integrated silicon accelerometers are shown in figures 1 and 2. A seismic mass is supported by thin beams. The beam-type sensor supports the mass only on one side, the symmetric-type on all sides. Two different sensing effects are commonly used. The first is based on the piezoresistive effect. The deflection of the seismic mass induces stress into implanted piezoresistors on the thin supports and therefore induces a change in resistivity [1]. The other kind of sensing effect measures the deflection of the seismic mass capacitively [2].

Besides design, fabrication and test, modelling is an important step of sensor production. For micromechanical devices not only the electrical circuits have to be simulated, but also the mechanical parts. The finite-element modelling (FEM) of micromechanical accelerometers helps to increase the sensitivity and resolution of the sensors by finding the optimal position of the sensitive elements such as piezoresistors.

Geometry

Using the university version of the general purpose finite-element code ANSYS version 4.4 on a SUN-4 Sparcstation we simulated silicon accelerometers of the beam-type as well as of the symmetric-type. The seismic mass of the beam-type sensor is supported by two beams on the same side. For the simulation example reported here we have chosen a beam thickness of about 10 µm, a length of about 200 µm, and an overall size of the sensor of 3.5 mm by 3.5 mm by 0.4 mm (Fig. 1). For the symmetric-type accelerometer we have chosen the same overall size as above with a square mass of 1.9 mg, size 1.6 mm by 1.6 mm, supported by four beams of a thickness of about 10 µm, one on each side (Fig. 2).

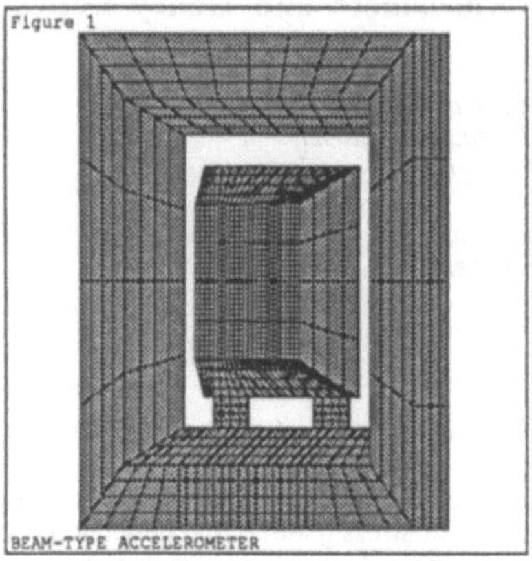

Fig. 1: Beam-type silicon accelerometer

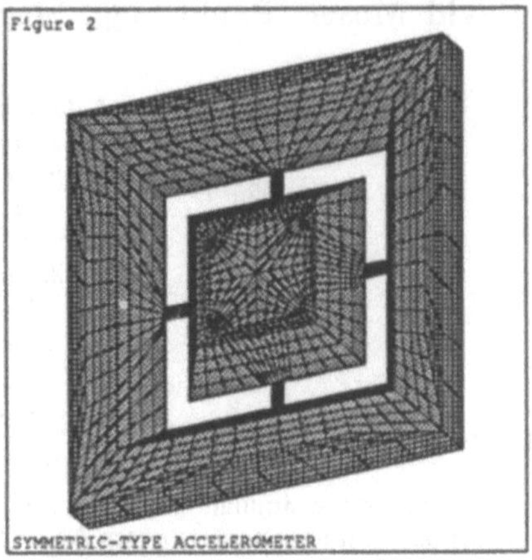

Fig. 2: Symmetric-type silicon accelerometer

Modelling with ANSYS

The first step of an ANSYS simulation is the preprocessing. This is done with the so-called PREP7 module. Here all the required information has to be given as input. The analysis type (e. g. static stress analysis or modal frequency analysis) has to be chosen followed by the material parameters and the geometry of the structure. After that the finite-element mesh is automatically generated. At this point a reordering of the elements is required to minimize the so-called wavefront and the required memory. The wavefront is the number of equations which are active after any element has been processed during the solution phase at that point. The computer time required for the solution procedure is proportional to the mean square of the wavefront size. The final step in the preprocessing is the determination of the boundary conditions and the load (e. g. acceleration in z-direction) applied to the sensor. The symmetric-boundary-condition option of ANSYS allows to reduce the simulation to half of the sensor in the case of the beam-type and to one eighth of the sensor in the symmetric case, with more than proportional reduction of calculation time. As a matter of fact it was not possible to simulate the whole structure with the university version of ANSYS (maximum wavefront allowed is 500) without using the symmetric boundary condition option.

The second step of an ANSYS simulation is the solution phase. Depending on the complexity of the structure and the number of users on the network the required time for the solution was between five minutes and two hours. For very complex problems the solution run is usually done in batch mode during the night.

The last step of an ANSYS simulation is the postprocessing phase. Mostly the so-called POST1 module is used to produce 3D-plots or 2D-graphs of the obtained results.

Modelling of stress

First we simulated the stress induced by the high temperature steps of the fabrication of the sensor. Typically, the passivation layer of such a sensor consists of a thermally grown oxide of approximately 1 µm thickness and a LPCVD nitride of about the same thickness. We assumed an oxidation temperature of 1000° C and a LPCVD temperature of 600° C. Due to the different thermal expansion coefficients of silicon, oxide and nitride, a thermal stress is induced by cooling the sensor down to room temperature. Figures 3 and 4 show plots of the stress s_y in y-direction (parallel to the beam) through a cross-section in z-direction in the center of the beam for both sensor types. No acceleration is applied. A large tensile stress of -160 MPa occurs for the beam-type sensor at the interface between the silicon and the dielectrics, where the piezoresistive elements are located (Fig. 3). This value has to be reduced by using thinner passivation layers. The deflection in z-direction at the end of the seismic mass due to the thermal stress is 9.4 µm. The temperature difference induces compressive stress into the center of the beams of the symmetric-type sensor. The values are much higher than above, because the symmetric supports allow only a deflection of 1.3 µm. The stress is very close to the point where fraction occurs (Fig. 4).

Figures 5 and 6 show plots of the stress s_y due to the applied acceleration at the interface along the beams. We applied accelerations of 1 g, 5 g, 10 g, 20 g and 50 g in the negative z-direction. This stress is compressive and cancels the thermal induced stress to a certain degree in the case of the beam-type accelerometer (Fig. 5). The net overall stress is still tensile and has its maximum at the base of the beams. Because the seismic mass of the symmetric structure can only move in z-direction, the stress s_y changes the sign in the middle of the beams (Fig. 6). The stress values are reduced by a factor of 6. Figure 7 shows the maximum stress s_y at the silicon/silicondioxide interface as a function of the applied acceleration. The maximum stress occurs in the beams at about 30 µm away from the frame. It is proportional to the acceleration for both sensor types.

Figure 8 shows the maximum deflection of the seismic mass for the applied accelerations (without thermal stress) in logarithmic scale. For the symmetric-type sensor this deflection is more than two orders of magnitude smaller than for the beam-type sensor. For both types it is proportional to the acceleration.

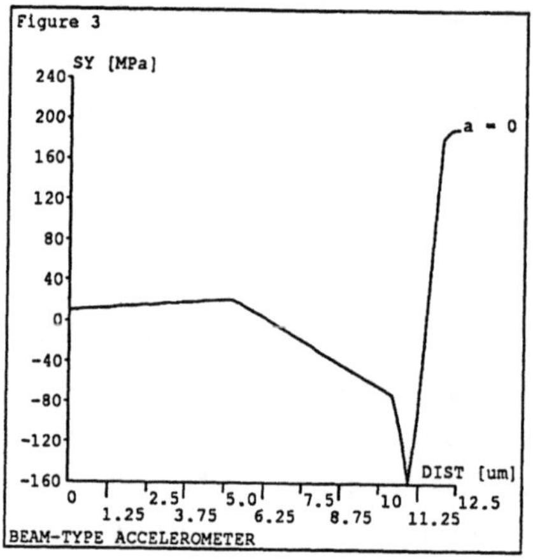

Fig. 3: Thermal stress across the beam (beam-type sensor)

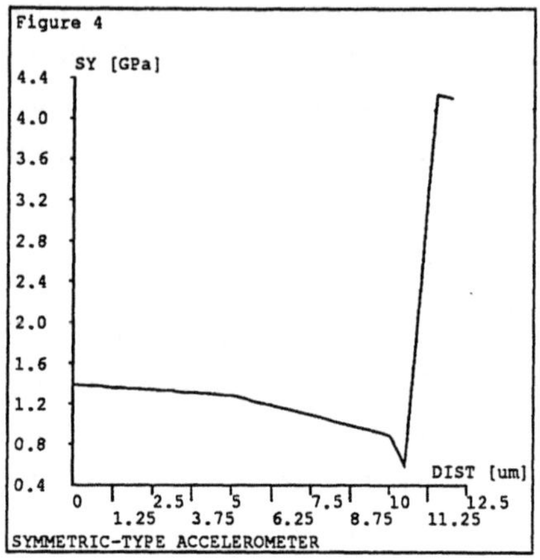

Fig. 4: Thermal stress across the beam (symmetric-type sensor)

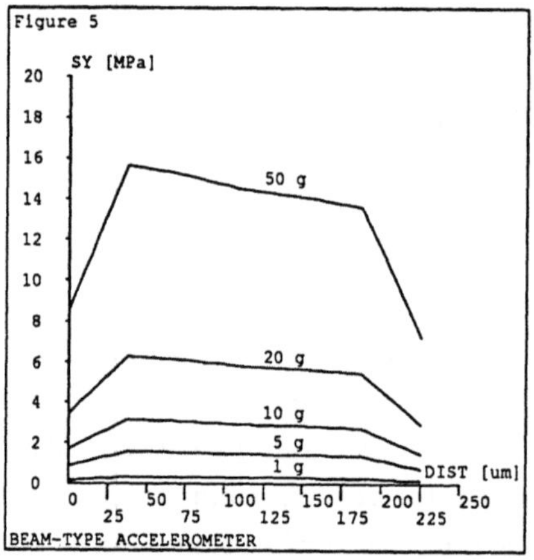

Fig. 5: Acceleration induced stress along the beams for beam-type sensor with tip of beam located at 225 μm

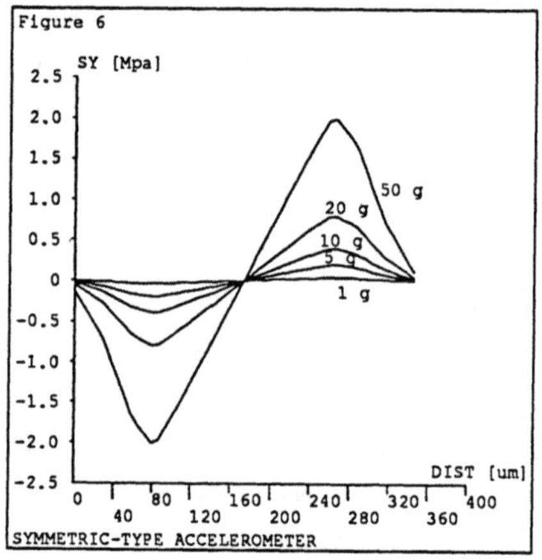

Fig. 6: Acceleration induced stress along the beams for symmetric-type sensor with seismic mass located between 60 μm and 285 μm

The cross-sensitivity, i. e. the sensitivity of the sensor to an acceleration in x- or y-direction instead of an acceleration in z-direction, was simulated by applying the same acceleration in all three directions. Figures 9 and 10 show the stress s_y for all three directions together with their sum. Acceleration in x-direction (perpendicular to the beam) has almost no influence. But the acceleration in y-direction parallel to the piezoresistors increases the stress s_y by 15 - 18 % for the beam-type sensor (Fig. 9). As expected, the cross-sensitivity for the symmetric sensor is much smaller than above. Only the acceleration in z-direction induces a significant stress into the beams. The cross-sensitivity is less than 1 %.

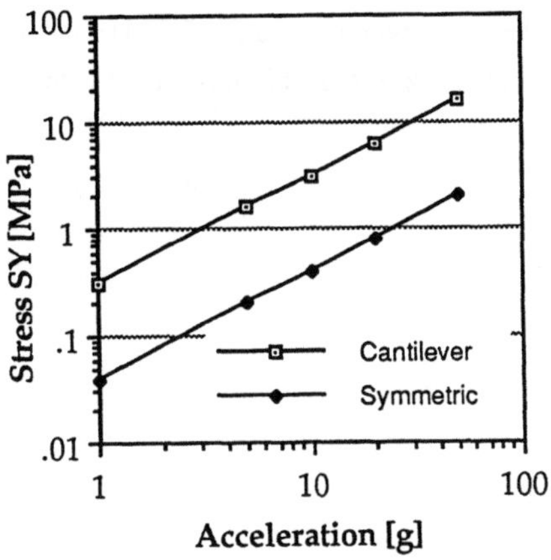

Fig. 7: Maximum stress in the beam

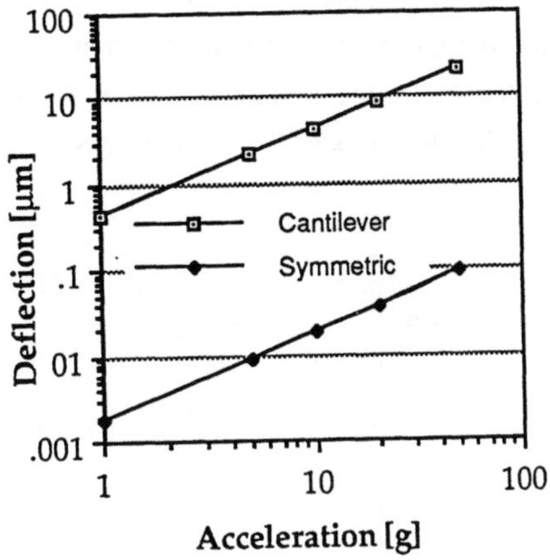

Fig. 8: Deflection of the seismic mass

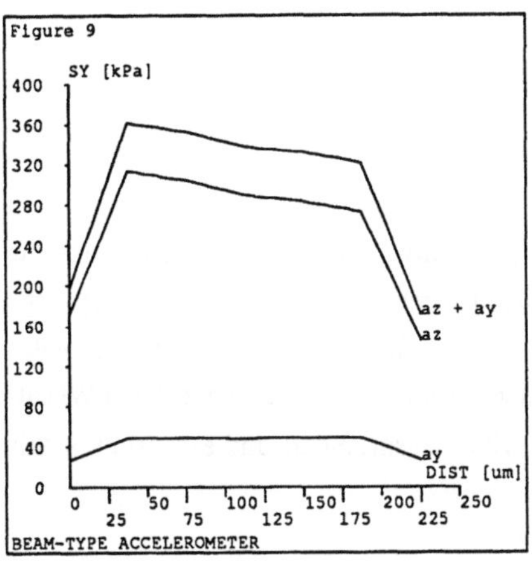

Fig. 9: Cross-sensitivity for beam-type sensor with tip of beam at 225 µm

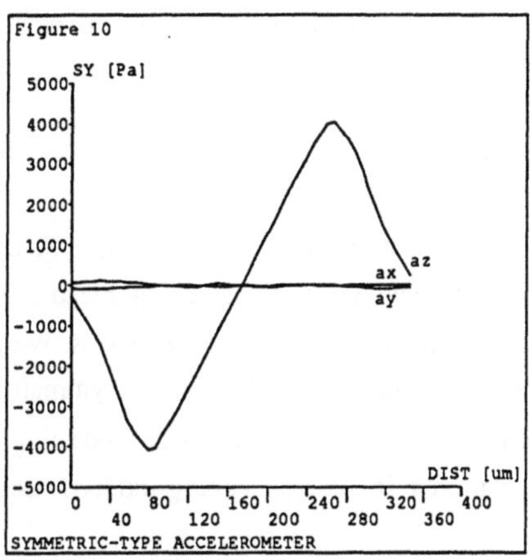

Fig. 10: Cross-sensitivity for symmetric -type sensor with seismic mass between 60 and 285 µm

Modelling of frequencies

To find out the bandwidth of the accelerometers we simulated their dynamic behavior. ANSYS allows not only to calculate the resonant frequencies, but also the shape of the modes. We calculated the resonant frequency to be 930 Hz for the beam-type sensor and 11.6 kHz for the symmetric-type. The next two higher modes of the beam-type vibrate at 32.5 kHz and at 226 kHz, for the symmetric-type at 740 kHz and at 1.15 MHz. Figure 11 shows the shape of the first three modes for the beam-type accelerometer.

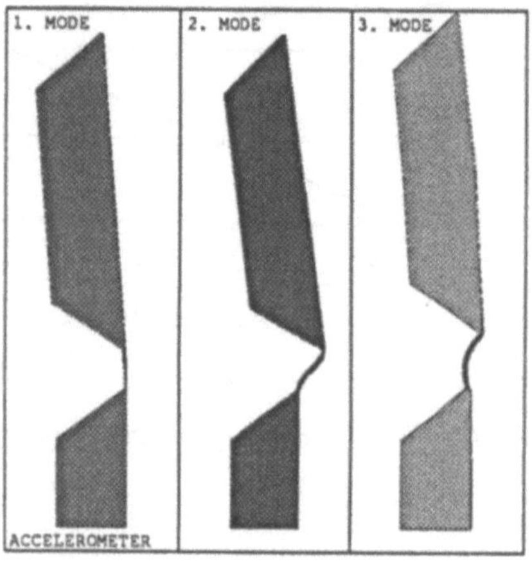

Fig. 11: Mode-shapes for beam-type accelerometer

Conclusion

Using the finite-element software ANSYS we have simulated stress, deflection, and the resonant frequencies of two typical piezoresistive silicon accelerometers. We obtain very large values of the fabrication-induced stress. We predict a sensitivity of 0.6 mV/V/g for the beam-type sensor and 120 µV/V/g for the symmetric-type accelerometer when connected to a Wheatstone bridge. Comparison of our modelling results with measurements and extension of this simulation to include packaging-induced stress are in progress.

References

[1] H. Sandmaier, K. Kühl, E. Obermeier, A Silicon Based Accelerometer with Cross Acceleration Sensitivity Compensation, *TRANSDUCERS '87, Digest of Technical Papers*, Tokyo, Japan, 1987, pp. 391-402.

[2] H. Seidel, H. Riedel, R. Kolbeck, G. Mück, W. Kupke and M. Königer, Capacitive Silicon Accelerometer with Highly Symmetrical Design, *Sensors and Actuators, A21* (1990) 312-315.

Accelerometer Design Considerations

By William C. Dunn
Semiconductor Research and Development Laboratories
Motorola Inc.
Phoenix

Introduction

Micromachining technology has been used in low cost pressure sensors for a number of years. However, in other structures such as accelerometers, the devices were fragile, suffered from cross axis sensitivity, and lacked overload protection. Low yields also made the devices very expensive, and not cost effective in many applications. But with the technology being driven by low cost, high volume applications, such as in the automotive industry, new cost effective, and rugged devices are being developed to meet these exacting requirements.

Technology

Micromachining technology can be subdivided into bulk, and surface micromachining. In bulk micromachining, shapes defined by masking are etched into the silicon wafer, using isotropic and anisotropic etch techniques, the lower shapes are defined by diffusions that are also used as etch stops, and/or by using wafer bonding techniques[1]. Where as with surface micromachining[2], layers of structural material (polysilicon), and sacrificial material (silicon dioxide) are deposited and patterned, and the sacrificial material then etched away, to give operational structures, these are shown in Figure 1. Either micromachining methology is suitable for accelerometers. Surface micromachining technology gives a more cost effective structure, however, stresses in the layers have to be annealed out, to prevent warping. Bulk

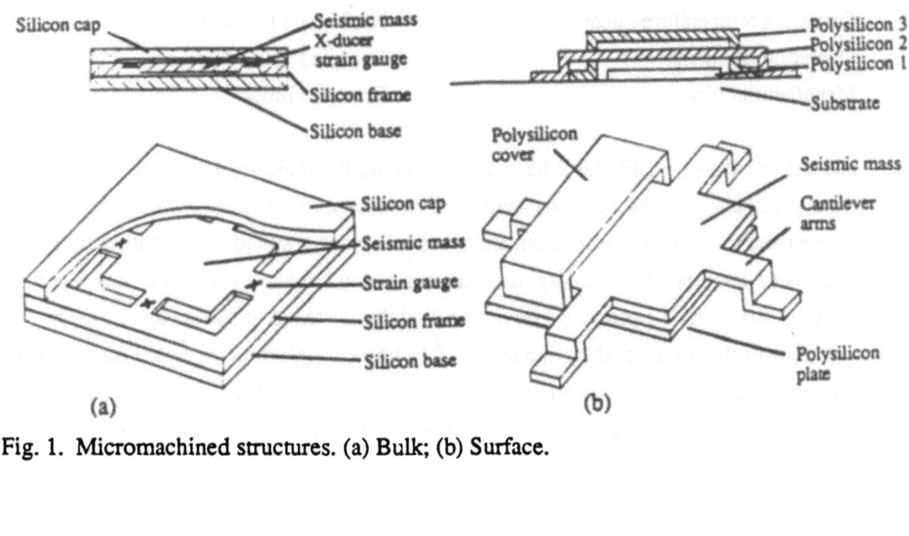

Fig. 1. Micromachined structures. (a) Bulk; (b) Surface.

material is stress free. The materials of choice in these structures are silicon, polysilicon, and silicon dioxide. Silicon has a number of unique properties which make it well suited for microstructures, these are

High modulus of elasticity	High tensile strength
Good elastic properties, no hystersis	Low cost defect free material
Microcircuit technology compatable	Machinable in a batch mode

Piezoresistive, piezoelectric, and capacitive are three methods of sensing that can be used in micromachined structures. The characteristics of these methods of sensing are shown below.

	Capacitive	Piezoelectric	Piezoresistive
Loading effects	high	high	low
Size	small	small	medium
Temperature range	very wide	wide	medium
DC response	yes	no	yes
Sensitivity	high	medium	medium
Shift due to shock	no	yes	no
Self test features	yes	no	capacitive
Damping available	yes	no	yes

Because of a dc response, no shift due to shock, and the requirements for self test features in many applications, piezoresistive or capacitive sensing are preferred.

A number of shapes, and configurations for accelerometers are possible. However, in considering the alternatives the following requirements must be taken into consideration.

Sensitivity	Cross axis sensitivity
Self test features	Stability
Linearity	Material fatigue
Operating temperature range	Resistance to shock
Operating frequency	Damping
Manufacturability	Reproduciblity

Three basic shapes shown in Figure 2 have been selected for their ability to meet the above requirements, particularly cross axis sensitivity, and resistance to shock. Shown are the double beam cantilever,the torsion bar suspension, and a seismic mass suspended by cantilevers from all four sides. As silicon is a low density material, mass may have to be added to improve the sensitivity of the structure. To produce a device with high resistance to shock (up to 5,000g), it is necessary to limit the travel of the seismic mass[3] to prevent deformation or breakage of the suspension members. Travel is limited in the vertical direction by the substrate, and by the addition of a top cap. Air damping (squeeze-film) is also used between the mass and travel

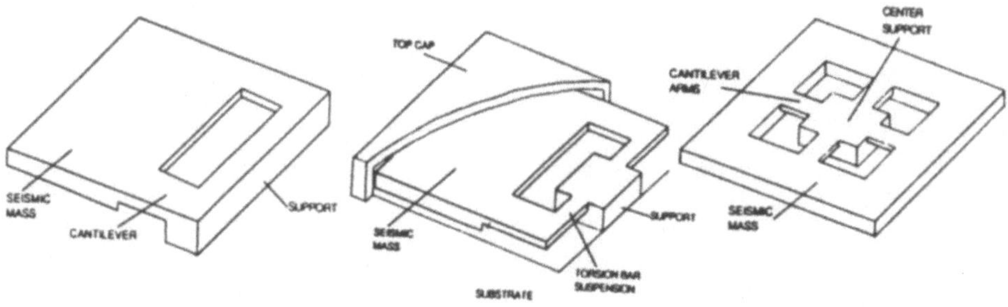

(a) (b) (c)

Fig. 2. Basic accelerometer shapes. (a) Cantilever; (b) Torsion bar; (c) Suspended mass.

stops. These also dampen any resonant frequency oscillations that can occur in the structure. In bulk and surface devices the travel is limited to about 10 μ, and 2 μ respectively. Full scale deflection being about 10 % of this value. The travel stops are also used to seal the accelerometer, to prevent particulates from entering the structure, and causing reliability problems, particularly during the sawing and assembly operations. Travel stops are also necessary in the lateral directions, however in these directions, some turning or twisting motion caused by the center of mass, not being in line with the point of suspension[4] can be experienced, here motion must also be limited to a few microns.

Piezoresistive Sensing

Piezoresistive sensing can either be made with a bridge circuit, or the X-ducer™ strain gauge, in either case the sensing resistors are P- ion implantations in a N type substrate. These resistors are diffused into the suspension arms for maximum sensitivity, this type of sensing can be used with either silicon or polysilicon systems. Trimming of control resistors is necessary to set the zero operating point, and the gain of the system. The maximum operating temperature is limited to about 100° C., due to the junction leakage current in the sensing elements (higher operating temperatures have been obtained using oxide isolated strain gauges[1]), temperature correction is also required. The X-ducer strain gauge is shown in Figure 3, together with the trimming, and compensation network.

Capacitive Sensing

Capacitive sensing[5] has two modes of operation, firstly open loop and secondly closed loop operation. Sensing in either case, is accomplished by differential capacitor measurements, using a charge redistribution technique[6]. The circuit and the cross section of a capacitive sensor is shown in Figure 4. The capacitors are alternately switched between a reference voltage (Vref),

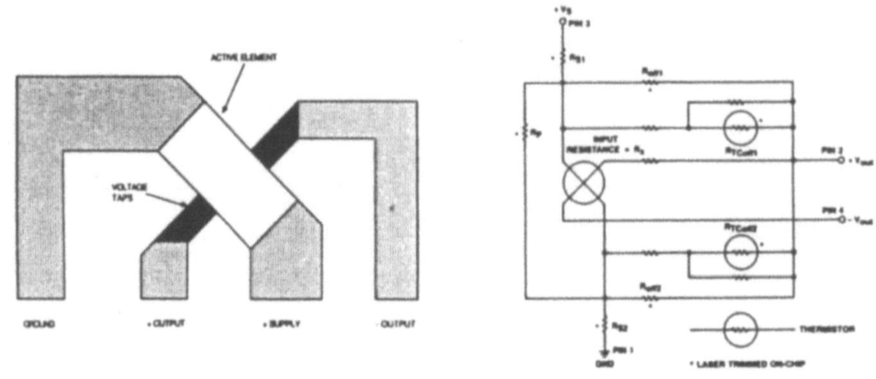

(a) (b)

Fig. 3 X-ducer. (a) Strain gauge; (b) Trimming, and compensation network.

and ground. The output of the integrator is also periodically shorted to the input, this circuit converts the differential capacitance to a voltage. The switching frequency is not critical (typically 250 kHz), and the output is independent of stray capacitances. Using switch capacitor techniques, it is also possible to set the band-width of the accelerometer to meet the requirements of the application. With this type of sensing, differential capacitive changes as low as 0.5 fF can be detected, which translates into sensing capacitors of 250 fF, or a capacitor size of 160 μ x 160 μ with 2 μ spacing. This assumes requirements of 1% accuracy, and 10% deflection for full scale output. In closed loop operation, electrostatic forces are used to counter balance the accelerating forces, and hold the seismic mass in its central position. This will reduce any tendency for fatigue in the suspension members. An accelerating force can be sensed with piezoresistors, or capacitive techniques, and the appropriate voltage applied to another set of capacitors to produce the electrostatic balance force. The block diagram and equation used for

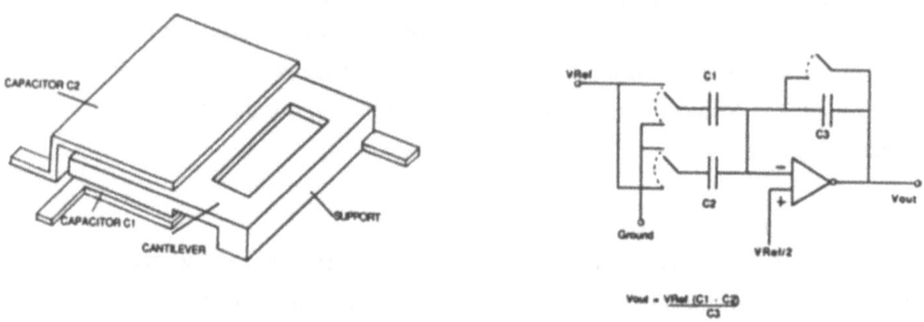

(a) (b)

Fig. 4. Capacitive sensing. (a) Capacitive layout; (b) Sensing circuit

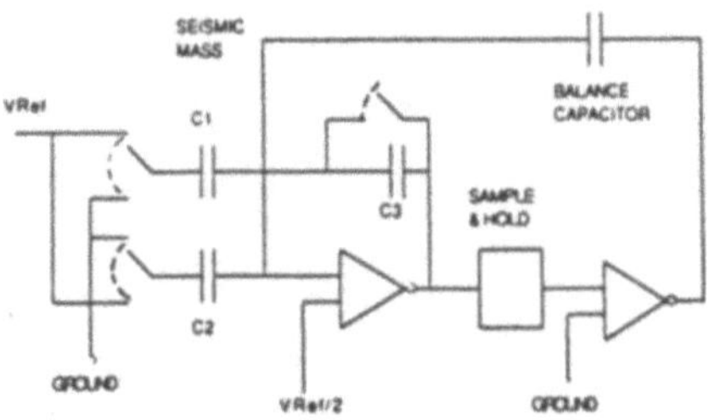

$$\text{Force} = \text{acc.x mass x } 9.8 = \frac{C\,V^2}{2\,d} = \frac{E\,A\,V^2}{2\,d^2}$$

Where acc = acceleration in g
C = capacitance
V = voltage applied
E = dielectric constant
A = area of capacitor
d = distance between capacitor plates

Fig. 5. Closed loop operation.

closed loop operation are shown in Figure 5. The same set of capacitors can be used for sensing deflection, and to develop the opposing electrostatic field. The electrostatic force [2] produced by 5 volts, using the above dimensions was equivalent to 20g acceleration. Capacitive sensing can be used at high temperatures (> 150° C.) as the capacitors are dielectrically isolated, temperature correction is also not required.

Self Test Features

In some applications self testing[4] is a requirement to check that the device, and also the system is operating correctly, and to check the calibration of the accelerometer. This can be achieved with an additional capacitor as shown in Figure 6. Self testing can be used with either piezoresistive or capacitive sensing. However in either case, the self test features use electrostatic forces to deflect the mass and verify operation. By applying a dc voltage between the plates of the capacitor, the mass can be deflected to simulate an externally applied force. The deflection can then be measured using the normal sensing circuits. Other advantages of the self test feature are the ability to test the device at probe and when assembled, and to exercise the device for drift and life time measurements without having to apply external forces.

Integration

For monolithic integration of sensors and microcircuits, the processes must be compatable, any extra masking steps required in one section must not affect the other section. Bulk micromachined structures are better suited for integration when using piezoresistive sensing. Because with capacitive sensing, the capacitors are on different layers, and inter-connections become a problem. The N substrate used with piezoresistive sensing is compatable with either bipolar N epi., or P well CMOS technology. Surface micromachined devices have better sensitivity using capacitive sensing techniques. The polysilicon used in the micromachined

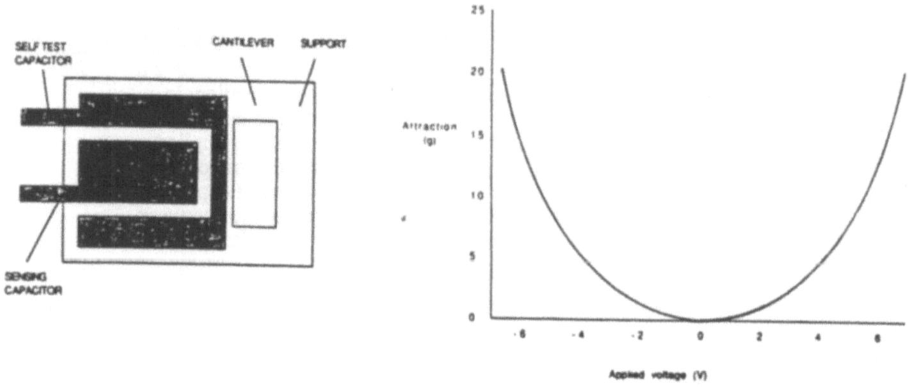

(a) (b)

Fig. 6. Self test. (a) Test structure; (b) Applied voltage vs attraction.

structures is compatable with CMOS gate polysilicon, and CMOS is well suited for sensing using switched capacitor techniques, and for low current operation. This applies to both closed loop and open loop operation.

Conclusion

The development of low cost batch producible accelerometers has been discussed. These devices can use either piezoresistive or capacitive sensing, and be made using bulk or surface micromachining techniques. However, self test features, resistance to shock, and monolithic integration are essential to meet cost goals, and for the future development of smart sensors. As the cost goals are meet, new applications will present themselves, for linear accelerometers as well as trigger devices.

X-ducer is a trade mark of Motorola, Inc.

References

1. Peterson, K.; Barth, P.; "Silicon Fusion Bonding: Revolutionary New Tool for Silicon Bonding and Microstructures," Wescon, (November 1989) 220-224.

2. Tang, W.C.; Nguyen, T.C.H.; Howe, R.T.; "Laterally Driven Polysilicon Resonant Microstructures," IEEE Micro Electro Mechanical Systems Workshop, February 1989.

3. Barth, P.W.; Pourahmadi, F.; Mayer, R.; Poydock, J.; Petersen, K.; "A Monolithic Silicon Accelerometer with Integeral Air Damping and Overange Protection," IEEE Solid-state Sensor and Actuator Workshop, June 1988.

4. Allen, H.V.; Terry, S.C.; deBruin, D.W.; "Accerometer Systems with Self-testable Features," Sensors and Actuators, Vol. 20 (1989) 153-161.

5. Watanabe, K.; Chung, W-S.; "A Switched Capacitor Interface for Intelligent Capacitive Transducers," IEEE Trans. on Inst. and Meas., Vol. IM35, No. 4, December 1986.

6. Kung, J.T.; Lee, H-S.; Howe, R.T.; "A Digital Readout Technique for Capacitive Sensor Applications," IEEE Journal of Solid-State Circuits, Vol. 23, No. 4, August 1988. 972-977

Measurement and Testing

Testing

Development of a Thermal Test Chip

J. HELM

Technical University of Berlin

Technologien der Mikroperipherik

Summary

A thermal test chip has been developped which can be used for the investigation of thermal properties of die attachment, bonds and substrates. It consists of an array of 32x32 cells, each equipped with a storage element, a heater and a temperature sensor. An arbitrary arrangement of heaters can be fixed and the temperature field evaluated in real time. The chip has been produced using a 3.5µ CMOS process. The layout and the electrical simulation have been carried out with software tools of the Mentor Idea System. A preliminary test and measurement has been made on a wafer prober and with wire-bonded substrates. An overview of a complete testing and measurement setup will be given.

Purpose

The final objective of the test chip and its measurement software is to obtain thermal images of a variable heater configuration either electrically (reading the sensors) or with thermography.

In static mode (thermal equilibrium) the chip can be used to measure the thermal resistance of the die-attach and the bonds in situ and for the examination of material and bond failures, also in connection with thermography. Another application is temperature measurement inside a multilayer structure.

In dynamic mode with the thermal time constant $\tau = d^2/a$ ($a = \kappa/\rho c$ diffusivity, τ is about 1ms for the cell distance d=300µm, c specific heat, κ thermal conductivity) the expansion of the heat front can be depicted. Finally the chip lends itself to an application in thermal wave imaging: modulating the input current generates a thermal wave thus replacing the modulated laser heating used normally. In this mode the thermal wave picture can be produced and the thermal diffusion constant $\mu = (2/a\omega)^{1/2}$ of different layers can be determined.

Internal Structure

The chip consists of an array of 32x32 cells, the row decoder ZE_DECODER, shift register TOP_SHIFT and current mirrors TOP_MIRROR.

Each cell contains a heater supplying about 8mA current (at 12V power), a temperature sensor and a storage element, which stores the heater state (on or off). As soon as the cell is activated, the temperature sensor is connected via analog switches with two of the 32 analog outputs VTHi, VTLi. The voltage difference VTHi-VTLi, proportional to the local temperature, is about 90mV at 27°C, the temperature, its slope is about 135µV/K. The cell array is partitioned in blocks of 2x2 cells. A physical row of 32 cells is subdivided logically in a left and a right half-row, which are activated separately.

The activation of rows is handled by the row decoder. The signals ZA4..ZA0 address the row; EZ0=L,EZ1=H activates the left and EZ0=H,EZ1=L the right half-row. The shift register supplies the heater control signals: the 16 heater bits IN are shifted serially in one burst into the register, timed by clock pulses PHI0.

The current mirrors distribute the regulated current IIN (10-20 µA) into the individual columns and via the the block current mirror to the 4 cells of the 2x2 block and to the next block.

The crucial parts of a cell are the temperature sensor and the heater. The sensor consists of two transistors circuited as diodes and supplied with current from a current mirror in relation 1:10. The resulting voltage difference $V = (kT/e_0)\ln 10$ depends on temperature only, provided that the two transistors are matched perfectly. The heater is a large PMOS transistor L=3.5µm, W=200µm. Its heating current is derived from the input current of the cell via two cascaded current mirrors and can thus be modulated when changing the input current.

Control Environment

A control cycle consists of two phases: loading the shift register and row activation.

During phase 1 the row activation signals are inactive (EZ0=EZ1=H) and therefore no row is activated. The burst enable signal in the external circuitry (BUREN) generates 16 clock pulses with corresponding 16 data bit IN, which switch on or off one of the 16 heaters of a half-row. After the burst the 16 heater control signals are applied to the inputs of the 16 cells of a half-row.

In the phase 2 the row address ZA4..ZA0 is set and EZ0,EZ1 activate the left or right half-row. At the same time the thermal voltages VTHi,VTLi will become valid: their magnitude is about 2V, VTHi-VTLi is about 90 mV at 27°C.

The input current IIN is supplied by a current source and controls the heater current; it depends approximately linearly on the input current.

Preliminary Test Results

The test has been carried out on a Rucker&Kolls wafer prober RK681 using a probe card and the preliminary test cicuit VTP. The control cicuit works in the line-scanning mode, where all rows are activated periodically or in the pulsed-enable mode,which activates one row only. The heater configuration can be fixed by jumpers. VTP can amplify and measure one of the 16 thermal voltages. Below the results for heater current and thermal voltage are given.

	test 5V	test 12V
heater current, mA	1.8	6
Vth 27 °C, mV	81	89

Measurement

The measurement system consists of a preamplifier circuit, the data acquisition cards DASH16 and PIO12 in a PC/AT and the measurement software ASYST running under DOS.. The preamplifier circuit contains 16 amplifier for the thermal voltages, a shift register with circuitry for control of the on-chip shift register, a current generator for the regulated current and a current measurement logic. The digital signals are passed via PIO2 to the preamplifier circuit and the analog outputs VTHi..VTLi are converted on DASH16 by a fast AD converter and processed by ASYST. The speed of the system is about 100000 measurements/sec.

References

1. Rosencwaig,A.: Application of thermal wave physics in microelectronics, VLSI electronics/ microstructure science vol.9. Academic Press 1985

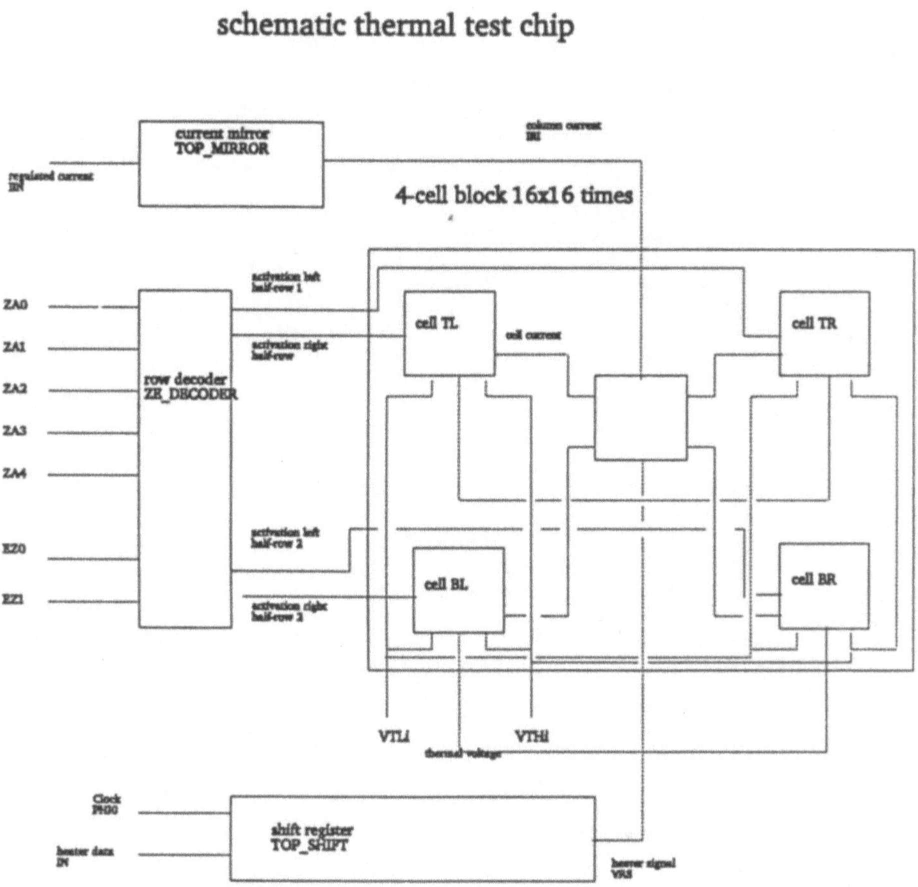

Fig.1. Schematic of the thermal test chip

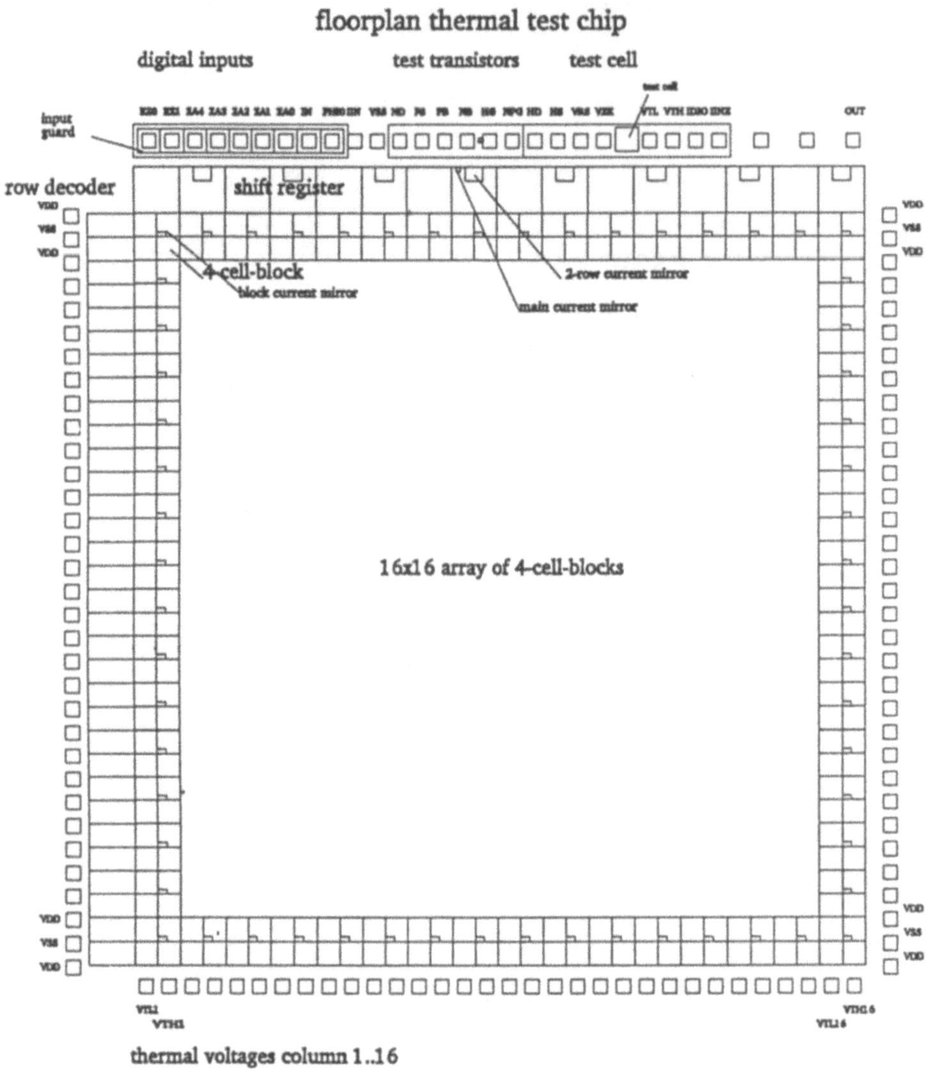

Fig.2. Floorplan of the thermal test chip

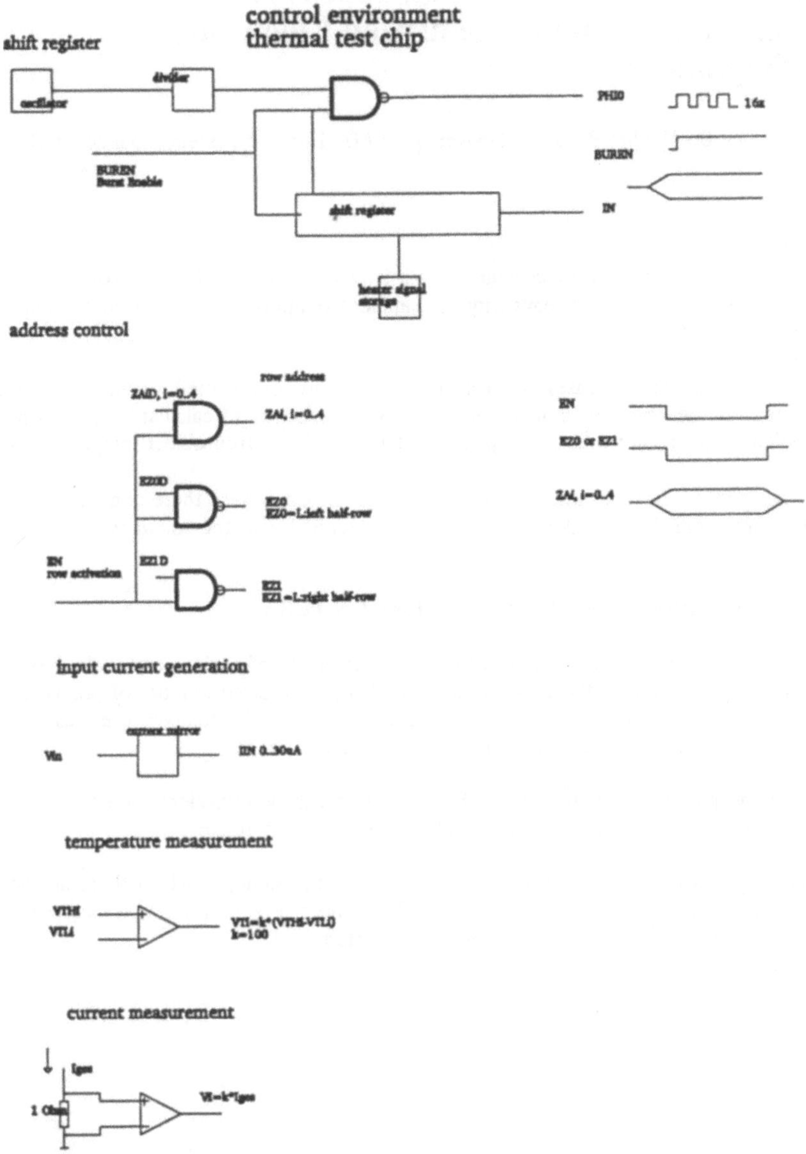

Fig.3. The control environment

Infrared Microscopy – Contactless Temperature Measurement on Miniaturized Electronic Components

by Bo Wallin, Product Manager, AGEMA Infrared Systems, Danderyd, Sweden

Summary:
In non-contact temperature measurement it is possible to calculate the correct object temperature provided the object emissivity, the ambient temperature and the atmosphere temperatures are known.

In many measurement cases on electronic components one should assume that the emissivity varies over the object. It is in special cases even necessary to calculate with varying reflected ambient temperature over the object in order to find the true object temperature.

In this paper different measurement cases are decribed, where these conditions are to be taken into consideration, but also how this is catered for in the software.

1. TEMPERATURE MEASUREMENT ON PCB:s

On a PCB we normally find a lot of components, made of various materials. From a thermografic point of view the PCB surface is built up by a large number of points, each of them with a different emissivity value. The calculation of the temperature must therefore consider the individual emissivity in each point of the image.

The atmospheric attenuation of the heat radiation can be considered the same for all points on the object. The distance is too small to cause any difference.

The reflected ambient radiation can be considered the same, provided there are no lamps or other strongly radiating objects behind the IR camera. Figure 1 shows a measurement situation with control of the reflected ambient radiation.

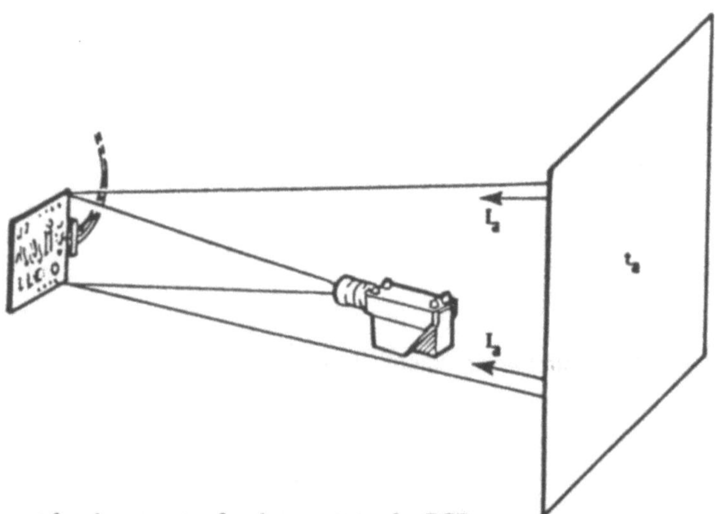

Fig. 1 Arrangement for the capture of a thermogram of a PCB.

With the arrangement, shown in figure 1, it is possible to capture a thermogram of a PCB. However, because of the unknown and sometimes strongly varying emissivities, the temperatures can be calculated only with limited accuracy.

1.2 How to measure true temperatures

The true temperatures can be calculated only if some type of equalization is used. Equalization means that the emissivity is known in each point of the object. This, in its turn makes it necessary to measure the emissivity in each point. The final temperature calculation will make use of these individual emissivity values. The emissivity is equalized, meaning that each point value is recalculated to a situation where the emissivity can be considered the same, i.e. equal, in the whole image.

The in this somewhat artificial way selected emissivity will then no more show the real emissivity in the image, but the temperatures will be correct. A normal artificial emissivity value can be 0.5.

It was for example found that the emissivity on an arbitrary PCB varied from 0.13 to 0.93.

1.3 Measurement set-up

To find the true temperatures on such complex objects needs some special measurement conditions. The individual emissivity in each point of the object can be calculated by means of a so-called reference image. This can be done using the method shown in figure 2.

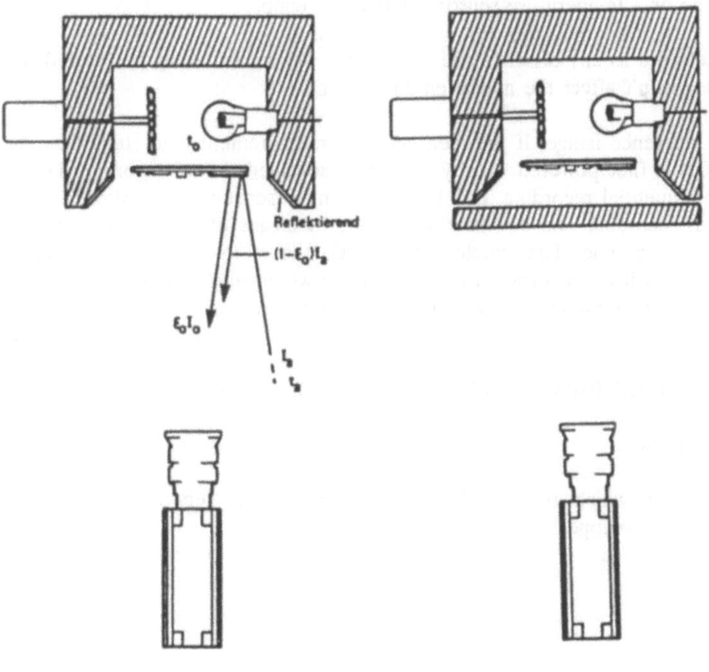

Fig. 2 Arrangement for capture of a reference image.

The objective is to heat up the whole object as equally as possible to a temperature of e.g. 50°C. The PCB is therefore put into an insulated box. The desired temperature is created by means of lamps and the heated air is circulated around the object by means of a fan. The temperature should be in the order of 20°C above the normal ambient temperature. It should preferably be in the order of the expected operating temperature of the PCB.

When the wanted reference temperature is reached, the lid is drawn off the box and the reference thermogram is captured.

As the reference thermogram is connected to the thermogram with the temperatures to be corrected, it is most important that the two thermograms be captured with the PCB and the IR camera in the same positions. The idea with this method is built on the fact that all points in the two thermograms correspond exactly to each other spatially.

1.4 Practical method for the calculation of the true temperatures

The equalization algorithm is implemented as the standard function EMEQ (Emissivity equalization) in AGEMAs image processing system TIC 8000 with the software CATS E. The calculation time on an AT is a few seconds.

In order to find the true temperature on the PCB after the heating there are mainly 2 possibilities

a) Paint a small part (e.g. 15 x 15 mm) with black paint of a known emissivity.

b) Use a probe or a temperature sensor for the box temp.

Method a) is to be preferred as any offset in the instrument will automatically be catered for. Any offset in the sensor won't affect the measurement either.

It is enough with one reference image. If the idea with the measurement is to study how the PCB heats up, the PCB is first powered up. By means of so-called "burst" recording i.e. 25 images per second, or sequential recording, e.g. 12, 6 or 1 image/second or any slower image frequency the process is recorded. The equalization follows by means of one single reference image, one "raw" image at a time. The simplest way to do this is to write a macro, i.e. a keystroke sequence, which will automatically run through the whole sequence and equalize it. The result is a new series of thermograms showing the true temperatures.

2. TEMPERATURE MEASUREMENT WITH AN IR MICROSCOPE

2.1 Measurement conditions

The measurement conditions which are described in chapter 1 are in general valid also for measurement with an IR microscope.

The object, though, is often built up of shiny materials. A power transistor has for example a very shiny conductor material. Between the conductors there is a much less shiny material. Hence the emissivity changes drastically over the object surface and can go as low as 0.05! On top of that the reflected ambient also changes very much as the very cold detector (-80°C or lower) is reflected in the reflecting material, which looks as if the object has a very low temperature. This will of course complicate the measurement situations!

However, todays computers give us a powerful tool for coping also with these problems.

2.2 Practical set-up for temperature measurement with an IR microscope

The microscope is equipped with an XYZ table on which the object is attached. As the depth of focus of this field of view, 1.6 x 1.6 mm, is quite small and sensitivity of the thermovision camera at this strong magnification is only some 8% of the normal sensitivity, there is a parallel video channel for easy location and focusing of the object.

As in this case the emissivity as well as the reflected ambient temperature both vary, we have to calculate with two unknowns. Hence two reference thermograms at different temperatures need be captured.

This can be achieved by means of a heating stage, which can heat up the object to known temperatures. The temperatures can be set to 40, 60, 80, 100 and 120°C.

2.3 Practical measurement method

In the CATS E there are functions for a comfortable use and study of the real-time thermogram, e.g. SHINT, i.e. SHow INTegration and ALIGN. The SHINT function displays beside the real-time image also an integrated image, in which the noise is strongly reduced. The ALIGN function makes it possible to check and adjust the object to perfect alignment with e.g. an earlier captured thermogram, in order to prepare for equalization. It is extremely important, that a reference image and the image to be equalized are exactly aligned.

The final calculation of the true temperatures is carried out by means of a special function, which is implemented as an option to CATS E. In this way it is possible to extend CATS E with new functions without changing the main CATS E program at all.

The calculation takes approximately 20 seconds. The result is an image, in which the emissivity as well as the reflected ambient temperature have been equalized. The temperatures, which are read out from the resulting thermogram are the true temperatures.

If the real-time recording or sequential recording possibilities are used, then it is possible also in this case to carry out the processing by means of a macro.

2.4 The practical method in steps

1. Put the object on the heating stage.

2. Localize the object with the video camera and bring it into focus.

3. Switch to the IR channel and adjust, if necessary, the focus.

4. Call up the SHINT function.

5. Select the necessary surrounding temperature to the object and power up the heating stage.

6. When the yellow lamp on the heating stage is flashing, the selected temperature is reached. The first reference image can now be acquired. It is usually practical to name this image e.g. REF60, i.e. the reference image was captured at 60°C.

7. Power up the object with the wanted voltage/current.

8. Capture a thermogram of the object under power. If a whole heating-up process is to be studied, then a whole sequence with the wanted speed should be captured, e.g. 6 images-/second during say 20 seconds (120 images!).

9. Now switch the heating stage to the next temperature e.g. 80°C.

10. When the flashing lamp announces that the set temperature is reached, capture the second reference image, and store it under, e.g. the name REF80.

11. Call up the option EQUAL2 i.e. "Equalization" using 2 reference images. The result is a thermogram where the true temperatures are displayed. The emissitivity and the reflected ambient will be artifical.

The complete measurement procedure might seem somewhat cumbersome. The correct - true - temperatures, however, demand this method. By means of macros and the sequential recording possibilities in CATS E the image processing can run completely automatically. It is possible to study the result of a whole sequence after a few minutes.

3. PRACTICAL MEASUREMENT ON A PCB, WHICH IS SITTING IN A RACK OR IN A CASE

There is one condition with non-contact temperature measurement by means of thermographic imaging which simply has to be fulfilled: The object must be "seen" by the IR camera!

This fact unfortunately makes if diffucult to study the temperature distribution on a PCB in its normal environment.

The following method is built up on the principle, that the temperature of an object adapts itself to the (new) ambient temperature along an exponential curve.

1. Let the PCB heat up and reach its normal operating temperature in the rack.

2. Switch off the power, take out the PCB and put it into a measurement box (see fig 2). Simultaneously (with power off), capture the first thermogram - which will show nothing of interest, but the power off time will be recorded.

3. When the PCB is fixed to its place in the box, capture image no 2 and after a similar time also image no 3, so that the images be captured with an interval of some 20 seconds. See figure 3.

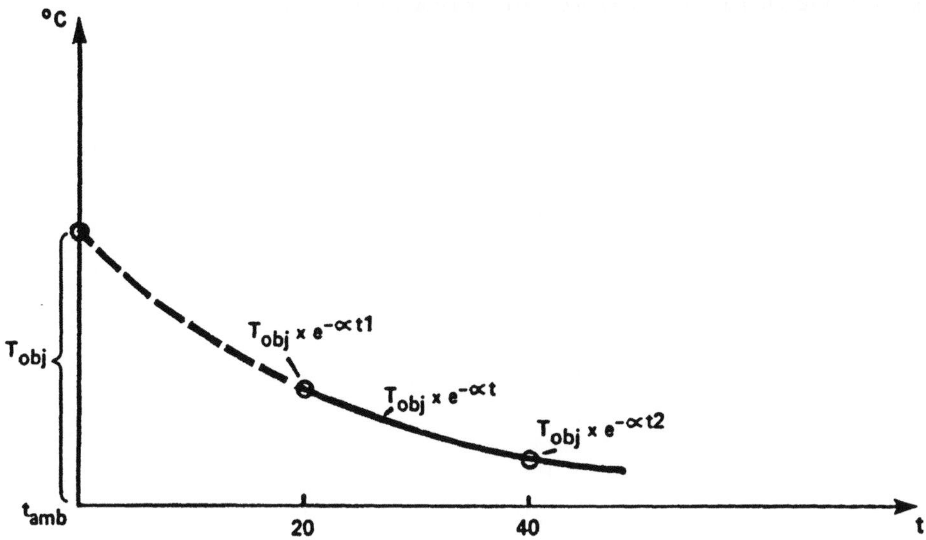

Figure 3. By means of extrapolation the temperature can be calculated backwards.

Different components have different cooling off speeds. This speed is therefore calculated individually and so is its extrapolated temperature.

The processing time is around 20 seconds (for a 12 MHz AT 286).

When the extrapolated image is calculated, the "normal" equalization procedure with a reference image will take place. Then, finally, there is an extrapolated, true temperature image available.

4. AN APPLICATION OF THE MICROSCOPE, WHICH HAS YET TO BE TESTED

4.1 Study of very fast processes

A combination of an IR microscope and fast line scanning might give a possibility to study the temperature development in connection with fast switching processes in a thyristor.

The so-called LINETIC function captures the temperatures along a line with a repetition frequency of 2500 Hz.

Normally the temperatures do not change that fast, but thyristor frequencies of a few hundred Hz in DC-DC converters can be studied.

The fast line scanning will in this case make it possible to study the object along the scanned line with a repetition time of 400 microseconds between the lines.

Highly Sensitive Strain Sensor for the Qualification of Epoxy Compounds Designed for the Encapsulation of Microsystems

H. Weidner, H. Borner[1)], and J. Villain[1)]

Siemens AG, Corporate Production and Logistics Department
and Data and Information System Group [[1)]];
Otto-Hahn-Ring 6, D-8000 München 83

Summary
The sensitivity of semiconductor strain gauges has been greatly increased by cutting a u-shaped groove into the back side of the sensor chips. With two different modes of deformation, these sensors can discriminate between molding compounds with high Young's modulus and low thermal expansion or vice versa.

1. Introduction

Potting electronic circuit boards with epoxy compounds may exert high mechanical stresses to individual components due to differences in the thermal expansion. The controversy is not yet settled whether a combination of a high Young's modulus E with a low coefficient α of the thermal expansion is to be preferred to the opposite pairing. Even the value of their combination to the integral (1) is disputed [/1/, /2/].

$$\int E * \alpha \, dT \qquad (1)$$

In this paper we offer an experimental approach to the selection of suitable molding compounds: we have developed a chip sensor which is highly sensitive to the strain characteristics of different molding compounds.

2. Description of the sensor

The sensor is a tiny (100)-silicon chip, bearing 10 boron doped polysilicon resistors in 4 different orientations as strain gauges on its surface, see figure 1. Each resistor consists of 6 parallel lines (length .12 mm, width .005 mm) connected in series, giving a total of about 17 kOhms. A u-shaped groove in the chip's back side greatly enhances its sensitivity, at the same time providing the sensor with two different modes of de-

formation in response to the E/α -combination of the molding compound, see figure 2.
The sensor is TAB-mounted within a polyimide frame, only suspended on its 17 copper leads.

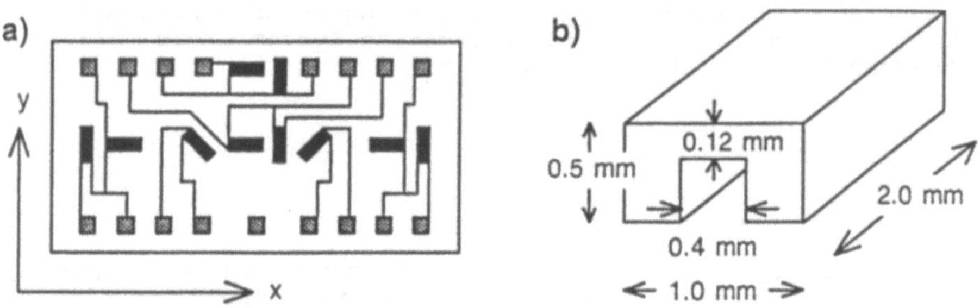

Figure 1: Geometry of the chip sensor
a) position and orientation of the resistors
b) geometric dimensions of the chip with its groove

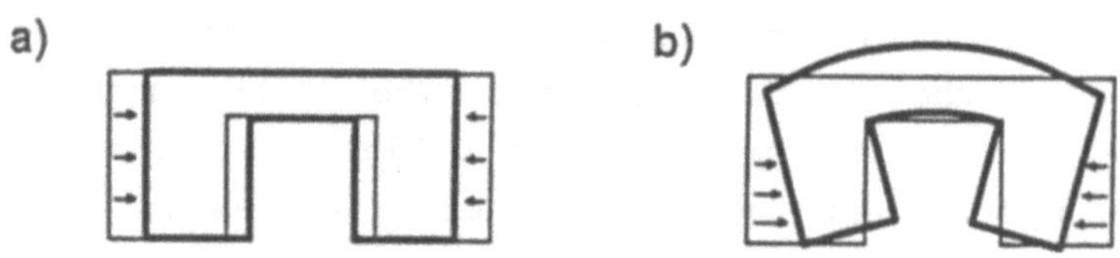

Figure 2: Two principal modes of deformation of the sensor. The epoxy exerts
a) high load and small displacements
b) large displacements under weak load

The *strain sensitivity* of the sensor is calibrated experimentally. A silicon bar ($15 * 3 * 0.45\ mm^3$) is glued under each of the sensor's sockets, thus providing a cantilever which we support upside down on two razor blades 20 mm apart. A well defined load suspended from the chip's groove will bend its bridge. The resulting surface strain in the y-direction (see fig. 1a) can be calculated from basic bending formulae. Comparing it to the resistance changes observed we determined the enhancement factors to be 33 for parallel and -2.7 for transversonal strains, respectively.

The *temperature dependence of the unstrained resistors* has been followed between -60 and +100 °C. It may be approximated by

$$\Delta R/R = 4.3 * 10^{-4} * T + 5.6 * 10^{-7} * T^2 . \qquad (2)$$

Albeit the deviation in the coefficients is less than 1 % for individual probes, at extreme temperatures this may still lead to errors in the strain interpretations in the order of the strains themselves unless statistics of large samples are applied. We therefore decided to calibrate the temperature dependence of each individual resistance before its use. This can be readily achieved within our computer-based experimental setup and provides us with accurate reference values for the sensors throughout the temperature range to be investigated.
The *temperature dependence of the enhancement factors* has not been determined for these sensor chips. According to our experience with various kinds of semiconductor strain gauges we expect a decrease of the sensor's sensitivity of about 0.1 % per Kelvin of temperature rise. In the present work, however, we neglect this effect.

3. Experimental Details

In this paper we report the use of the sensor for the characterization of 3 different epoxy compounds. The polyimide frame with the chip is suspended about 3 mm above the bottom of a rubber pot (30 mm * 30 mm). Electrical contacts to a connector outside the pot are provided by 0.1 mm thin copper wires. The pots are filled up to 10 mm with the epoxy under vacuum. After curing according to the due specifications we remove the pots and measure the sensor resistances at 30, 60, 90, and 120 °C.

The properties of the epoxy compounds used in our tests are listed in table 1. In addition, the influence of the following parameters on the chip strain were investigated:

a) stiff walls of the pot (glass pot instead of rubber)
b) inhibition of the adhesion of the epoxy to the chip
   (by spraying the chip with Tyreno Fluid B)
c) aging the epoxy by temperature cycles (60 times -40/+125 °C)

	Epoxy A	Epoxy B	Epoxy C
curing conditions	1h / 110°C	3h / 120°C	24h / 80°C
density (g/cm^3)	2.26	1.80	1.43
glass temp. T_g *(°C)*	120	146	47
therm.exp.$*10^6 K$ *[temp.range (°C)]*	28 / 103 $T<T_g$ / $T>T_g$	54 / 131 $T<T_g$ / $T>T_g$	150 $50<T<150$
E-modulus (N/mm^2) *[temp. (°C)]*	5000 / 4000 25 / 70	4000 / 400 25 / 170	2500 / 1600 25 / 70

Table 1: Properties of the epoxy compounds investigated

4. Results

4.1. The directions of the *principal strains* are sufficiently close to the chip axes (x and y in fig. 1a) that the readings of the diagonal resistors are redundant.

4.2. As expected, the *most sensitive readings* are provided by the resistors across the bridge. However, the other resistors have to be considered as well in order to fully understand the deformation profile which, in general, is a superposition of the extremes depicted in fig. 2.

4.3. *Accuracy* of our measurements and *reproducibility* of the potting conditions are within 0.01 % absolute strain.

4.4. There are *large differences* in the deformation profiles due to the *individual molding compounds*. This is demonstrated in fig. 3a by the strains across the bridge. An interpretation will be given in the discussion, sect. 5.1.

4.5. *Molding conditions and epoxy aging* are discernible, too, as is shown in diagram 3b. However, the characteristic strain profile of the individual epoxy remains dominant.

5. Discussion

5.1. Differences due to the individual compounds, see fig. 3a. Let us first look at epoxy A with its exceptionally high E modulus and low α-value due to its high filler content. This epoxy works on the sensor chip almost as ideally as sketched in

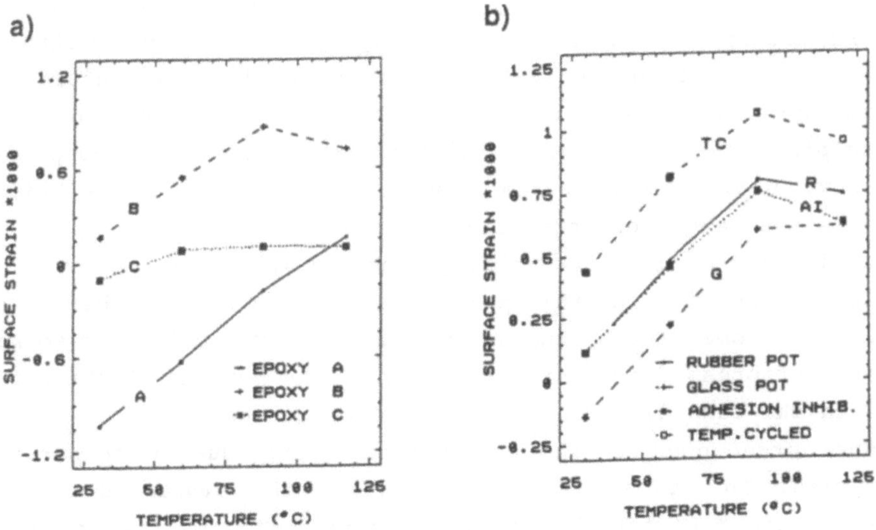

Figure 3: a) Temperature dependent surface strain for the epoxy molding compounds investigated
b) influence of potting conditions and epoxy aging on the surface strain (epoxy B of diagram a)

figure 2a: the lateral compression of the bridge increases nearly linearly with decreasing temperature without any hint to a surface tension due to chip bending.

Epoxy B behaves quite different. Its shrinkage during the curing process forces the chip's bridge to bend (a tension of $0.5*10^{-3}$ corresponds to a radius of curvature in the range of 100 mm, pressing the sockets of the chip against each other by 1.8 μm!). The bending is enhanced upon cooling until - below 90°C - a lateral compression is superposed to it. The longitudinal strains of the chip are compressive throughout the temperature range investigated (data not shown).

Epoxy C is characterized by its low glass transition temperature of about 47 °C. In spite of its high thermal expansion above T_g it does not stress the chip greatly (note the low curing temperature), the effects of shrinkage during curing and of the thermal expansion seem to nearly compensate at least up to 120 °C. Below T_g, the increase of the compressional strain seems to be comparable to that exerted by the other compounds.

5.2. Influence of potting parameters and aging, see fig. 3b.
Inhibiting the adhesion of the epoxy changes the mechanical stress only slightly at and above the curing temperature: due to the missing adhesion, the shrinking epoxy cannot draw the chip's sockets together. The compression from the surrounding epoxy still persists and causes the bending of the bridge, albeit somehow less pronounced.
Curing and cooling the molding compound in a *glass pot* with its hard walls will put the epoxy under tensile stress, thereby releasing some of the compressive stress to the chip. The sockets are not pressed together as strongly, the bending of the bridge with the tensile strain of its surface is reduced.
Most pronounced is the deformation change due to *temperature cycles*. In the plane of its bridge, the sensor exerts the strongest resistance against the compression by the epoxy. In this plane, therefore, the aging effect of the molding compound is at its maximum. In turn, then, the planar compressive strain to the bridge will relax first whereas the bending forces are reduced by a lesser amount: the tensile surface strain increases as is apparent from fig. 3b.

6. Conclusions

- Cutting a u-shaped groove into the back of a sensor chip greatly enhances its sensitivity for mechanical stress.
- Due to the groove the sensor reacts differently for high E / low α and low α/ high E molding compounds.
- In many applications there will be a need for molding compounds that do not cause large geometric distortions of the electronic assembly while homogeneous compressive loads might be tolerated. These compounds can be identified by simply potting our sensor chip with them. However, for practical applications, the tests should cover the full range of specified temperatures, e.g. -50 to +125 °C.

8. References

/1/ D. Kuwata, K. Ito, H. Tabata
IEEE Trans.Compon.Hybr.Manufact.Technol. 8 (1985), 486-489
/2/ G. Draeger, M. Maier, F. Quella
IEEE/IEMT Europ.Electron.Manufact.Technol.Symp.1988, 37-40

Material Characterization – Physical Properties

Mechanical Reliability of Electronic Devices

S. Kawai, A. Nishimura, M. Kitano and T. Shimizu

Mechanical Engineering Research Laboratory, Hitachi, Ltd. Japan

SUMMARY

Stress simulation techniques in semiconductor fabricating processes and strength evaluation techniques for electronic devices and equipment are described. Semiconductor fabricating process simulators are developed for stress analysis of thermal oxidation and film deposition processes. Strengths of plastic encapsulants, silicon chips, bonding wires and adhesive interfaces in IC plastic packages are evaluated using fracture mechanics and other techniques. Thermal fatigue strength of solder joints is evaluated using fatigue test data of solder specimens and simplified modeling of IC packages.

INTRODUCTION

Semiconductor integrated circuits (IC's), and electronic equipment have become so widely used as to now be regarded as necessities in daily life. This trend will be accelerated in the future, requiring electronic devices and electronic equipment to become more dense and more reliable. Thus, considerable attention has recently been paid to mechanical factors, such as stress and fracture, in electronic devices.

Stress causes not only mechanical failures but also creates troubles in semiconductor fabrication processes. Hence, stress is a very important factor for designers to take into account. From the viewpoint of a practical objective, total structure reliability has to be investigated in terms of stress and fracture. Therefore, related academic fields from materials, including mechanical properties, microstructures and chemical composition, to mechanics, including theories of elasticity and plasticity, numerical analysis techniques and experimental techniques should all undergo an expansion in research.

In this paper, stress simulation techniques in semiconductor fabrication, and strength evaluation methods for electronic devices and equipment are described.

STRESS SIMULATION IN SEMICONDUCTOR FABRICATING PROCESSES

Semiconductor devices such as LSI's (large scale integrated circuits) are formed as multilayered structures of thin films, for example, silicon dioxide layers or aluminum lines. In their fabricating processes, consisting of film deposition, etching, oxidation of silicon and annealing at high temperature, it is likely that rather high stress is locally generated in the devices. With the current trend toward high density integration and size reduction (Fig. 1), this local stress has come to influence device characteristics. There have been some reports on stress induced defects: dislocations generated in silicon wafers through thermal oxidation (1) or aluminum line breakage (2). Therefore, it is necessary to estimate the stress state in the devices by computer simulation. From the viewpoint of local stress generation, fabrication processes are divided into three parts: thermal oxidation of silicon, thin film deposition and etching, as shown in Fig. 2. Two simulators (OXSIM (3) and SIMUS (4)) have been developed to analyze the stress state in devices corresponding to thermal oxidation and film deposition. Only an outline of OXSIM will be given here.

In the thermal oxidation process, oxidant species diffuse through an already existing dioxide layer on the silicon substrate. Oxidation reaction occurs when the oxidant species reach the interface between the silicon substrate and the dioxide layer and a new dioxide layer is formed. As the silicon is changed into silicon dioxide, its volume expands and stress is generated in both the silicon substrate and silicon dioxide layer.

Oxidant species diffusion and stress generation are simulated in OXSIM using the finite element method. Finite element meshes are refined by the Body-fit method according to the movement of the interface. Results of an example analysis of a LOCOS (local oxidation of silicon) structure are shown in Fig. 3. The dioxide layer swells on the silicon substrate, because of the body expansion by oxidation reaction. Near the edges of oxidation masks made of silicon nitride, so called "bird's beaks" are formed and stress is concentrated in these areas. Local stress distributions obtained by Raman spectrum analysis are shown in Fig. 4. It can be seen that the formation of W thin film on SiO_2 film causes compressive stresses in the Si substrate near the edge of W thin film. Further, these stresses decrease sharply as the distance from the edge of W thin film increases.

STRENGTH EVALUATION FOR PLASTIC PACKAGE DESIGN

Most semiconductor devices are mounted in packages to prevent mechanical failures and corrosion, as well as for easy handling. Although various kinds of materials and configurations are used for packages, discussion will be confined here to plastic packages, the most commonly used type of package.

Recently, the size of IC chips has been rapidly increasing due to demands for large scale integration, while the package dimensions have been restricted or even made smaller and thinner as shown in Fig. 5. This trend increases thermal stresses in the packages during the molding process or temperature cycling tests, causing a variety of failures. Hence, the development of sophisticated technologies has become necessary for the structural design of plastic packages to prevent these failures and to improve reliability (5), (6).

Typical failure modes of IC packages are shown in Fig. 6 (5). Chip cracks occur mainly during die bonding steps in which the silicon chip is bonded to the lead frame before transfer molding in an epoxy compound. Discrepancy between the thermal expansion coefficients of the silicon chip, the lead frame, and the epoxy molding compound causes residual stresses in the IC packages after molding. These residual stresses fluctuate in temperature cycling tests, and cause resin fatigue cracks originating from edges of the chip and the chip pad. Bonding wires are also fractured by cyclic displacement during temperature cycling. Hence, preventing fractures is the main problem in IC plastic package structure design.

Estimation of temperature cycling life (7)

Fracture mechanics, originally developed for metallic materials, has been applied to estimation of the package cracking life of plastic packages under temperature cycling.

Fatigue crack propagation rate data of a plastic encapsulant are shown in Fig. 7. Fatigue crack propagation rate da/dN is plotted against stress intensity factor range ΔK. The results show that a linear relationship exists between log da/dN and log ΔK at each temperature, and that da/dN increases with the rise in temperature. The linear relationships in Fig. 7 indicate that the crack propagation rates of the encapsulant can be described by the well-known Paris relationship

$$\frac{da}{dN} = C(\Delta K)^m \tag{1}$$

where a is crack length, N is number of cycles, and C and m are constants.

Stress intensity factor K for packages under temperature cycling was analyzed by means of the finite element method. The analysis was performed for a package with resin crack(s) originating from the lower edge(s) of the chip pad, as shown in Fig. 6. The crack propagation life of packages can be estimated by substituting the relationship between K and crack length *a* into Eq. (1) and performing the integration with respect to crack length *a*.

The estimated results of crack propagation behavior in a package are shown in Fig. 8. Crack propagation behavior was estimated for both a single crack from one side of the chip pad and a pair of cracks symmetrically extending from both sides of the chip pad. Experimental data are also plotted in this figure. It is shown that the temperature cycling life for the pair of cracks is far longer than that for the single crack. The crack propagation in most of the test packages was observed to be considerably asymmetric and, therefore, was close to the single crack model. The tendency of the estimated crack propagation behavior for the single crack correlates fairly well with the experimental data.

The symmetry effect of crack growth on temperature cycling life is confirmed by the experimental data, as shown in Fig. 9. The crack from one side of the chip pad was found to have reached the bottom surface of the package at a number of temperature cycles N. The length of the embedded crack increases with the number of cycles N. Extended life of the package can be observed when cracks on the opposing sides of the chip pad propagate at almost equal rates during temperature cycles. The result points to the importance of symmetry in the design and assembling of packages.

Strength of silicon chips (5)

A typical fracture mode of a silicon chip is shown in Fig. 6. This type of horizontal cracking is caused by high local stress at the chip edge. Another typical type, vertical cracking, is caused by tensile stress at the chip surface due to bending after die bonding.

Chip bending strengths are shown in Fig. 10. Because of the extreme brittleness of silicon, the bending strength varies greatly for each test condition. Bending strength for surface tensile stress is far larger than that for back surface tensile stress. This is because the back surface roughness is greater than that of the chip surface. Fracture toughness K_{IC}'s for (100), (110) and (111) planes are 0.74, 0.66 and 0.62 MPa$\sqrt{m}$, respectively. Since the K_{IC} value of silicon is very low, careful surface finishing and dicing are required to prevent cracks.

Strength of bonding wires (5)

Since the diameter of the bonding wire is very small (20-40μm), special apparatuses are required for wire fatigue testing. Therefore, a special microwire fatigue test system was developed. Bending fatigue test data of 32μm diameter gold wires are shown in Fig. 11(a). There is no difference between fatigue strength at room temperature and that at 150°C. Figure 11(b) shows the relationship between strain range at the fracture point and cycles to failure. Strain ranges were calculated from the displacement range by assuming a deflection curve. Gold wire has a higher fatigue strength than solder.

Adhesive strength evaluation using stress singularity parameters (8)

A typical model of plastic encapsulated LSI devices is shown in Fig. 12. Adhesive interfaces in these LSI devices have various bonding edges, as shown by the circles in this figure, and the stress fields near these bonding edges show singularity behavior. Since in these regions, the maximum stresses calculated by numerical stress analysis, such as the finite element method, depend on the size of the element meshes, overall strength evaluation cannot be made using only these calculated maximum stress values. Therefore, a new adhesion strength evaluation method using two stress singularity parameters, K and λ, is presented, which can express stress distribution near a bonding edge as follows:

$$\tau(r)=K/r^{\lambda} \tag{2}$$

where $\tau(r)$ is shearing stress (MPa), r is distance from the singularity point (mm), K is intensity of stress singularity, and λ is order of stress singularity.

Order of stress singularity λ can be calculated using edge angles, Young's moduli and Poisson's ratios of the bonded materials. The intensity of stress singularity K is calculated by fitting Eq. (2) to the stress distribution near a bonding edge, calculated by numerical stress analysis.

To obtain the relation between the critical intensity value of stress singularity, K_{crit}, where delamination starts from the bonding edges, and the order of stress singularity λ, we bonded these two materials at 170°C and post-cured them in a furnace at the same temperature for 5 hours. Delamination in processes was measured using two strain gauges mounted on both surfaces.

During the cooling process, the specimens were first warped by thermal stresses. Then, with the delamination, these thermal stresses were released, as shown in Fig. 13. The values of K_{crit} for various orders of stress singularity λ are shown in Fig. 14. The delamination occurrence can be estimated by this criterion curve. These delamination criteria are now being applied to the structural design of plastic encapsulated LSI devices.

THERMAL FATIGUE STRENGTH OF SOLDER JOINTS

Fatigue failure of the solder joint, caused by thermal expansion mismatch between the device and the printed circuit board (PCB), is the most critical issue with surface mounted devices. This is because the solder joint is the only mechanical means of attaching the surface mounted device to the PCB and also establishes the electrical path. Hence, structural design of solder joints has become increasingly important. Many studies have been carried out to improve the reliability of solder joints (9)-(11). The strength evaluation method for solder joints will be briefly described in the following section.

Stress-strain behavior and fatigue strength of solder (10),(11)

The relationship between stress range $\Delta\tau$ and strain range $\Delta\gamma$ of the solder was obtained by cyclic torsion tests of hollow cylindrical specimens 15 mm in diameter. The solder tested was type 60 Sn/40 Pb solder used most commonly for surface mount devices. Test results showed that the relationship between $\Delta\tau$ and $\Delta\gamma$ can be expressed as

$$\Delta\gamma = \frac{\Delta\tau}{G} + \left(\frac{\Delta\tau}{Ff^m}\right)^{1/n} \tag{3}$$

where G is modulus of elasticity in shear, F is plastic coefficient, f is frequency, m is sensitivity to strain rate, and n is strain hardening exponent. If the effect of temperature is incorporated in Eq. (3), it gives a final relationship between $\Delta\tau$ and $\Delta\gamma$, including the effects of both temperature and frequency. Hence, the values of each constant were determined as functions of temperature from the test results run at 0.01 Hz and 0.3 Hz at various temperatures:

$G = 1.11 \times 104 - 29.8T$ $\quad$ $F = 106 - 0.444T$

$n = 0.168 + 5.51 \times 10^{-5}T$ $\quad$ $m = 0.0468 + 5.85 \times 10^{-4}T$

where the unit of G, F and T is MPa, MPa and °C, respectively.

Two kinds of specimens, the bulk specimens and the solder joint specimens, were used in fatigue test. The bulk specimens were employed to obtain fundamental fatigue strength of the solder and the solder joint specimens were employed to obtain actual fatigue strength of joints. Fatigue tests were carried out in torsion under strain control in a triangular wave at 1 Hz. Figure 15 correlates the number of cycles to failure N_f versus applied shear strain range $\Delta\gamma$. Fatigue life of bulk solder specimens has little scatter, while that of solder joint specimens has considerable scatter. Also, the fatigue life of solder joint specimens with the longest life at each strain level approximately agrees with that of bulk solder specimens. It was found from observation of the fractured surface of solder joint specimens that specimens with longer fatigue life fractured at the solder, while those having shorter fatigue life initiated cracks at defects in the boundary between the solder and copper and propagated along the boundary. It is concluded from the above discussion that, when there are no defects, fatigue life of solder joints agrees with that of bulk solder.

Fatigue life of solder joints of actual IC packages (11)

Strain on the solder joint, induced by the thermal expansion mismatch between the package and the substrate, can be calculated using the stress-strain behavior of the solder expressed by Eq. (3) and also by using a simplified model in which leads are treated as rigid frames.

An example of the fatigue strength analysis of the solder joints is shown in Fig. 16. The solid and dotted lines show the fatigue life of the solder joint specimen and the bulk specimen at a 50% probability of failure, respectively. Other symbols show the fatigue test results of solder joints of actual IC packages of which the strain range was calculated by the above method. The figure shows that the fatigue life of the solder joints can be accurately estimated by the above method.

CONCLUSIONS

It is certain that mechanical reliability issues will increase as electronic devices and electronic equipment are increasingly integrated, miniaturized, and applied to broader fields. As has been described , studies on materials and mechanics have already been playing significant roles in the design of electronic devices, where only functional design had been needed before. Studies on migration and corrosion in devices also fall under this category. Structural design has even exerted some influence on functional design (12). However, real challenges to more complicated issues are now beginning.

REFERENCES

(1) —, Nikkei Microdevices, No.17 (1986), p.42-43 (in Japanese).

(2) Klema, J. et al., Proc. 22nd Int. Reliability Physics Symp.(1984), p.1-5.

(3) Saito, N. et al., Proc. 2nd Conf. on Computational Mechanics (in Japanese), JSME (1989).

(4) Saito, N. et al., IEICE Technical Report, SDM88-83, Inst. of Electronics, Information and Communication Engineers (in Japanese), (1988).

(5) Kawai, S., JSME Int. J., Ser.I, Vol.32, No.3 (1989), p.320-330.

(6) Edwards, D. R. et al., Tutorial Notes, Int. Reliability Physics Symp. (1988), p.8.1-8.39.

(7) Nishimura, A. et al., IEEE Trans. Comp., Hybrids, Manuf. Technol., Vol.CHMT-10, No.4 (1987), p.637-642.

(8) Hattori, T. et al., JSME Int. J., Ser.I, Vol.31, No.4 (1988), p.718-723.

(9) Lau, J. H. and Rice, D. W., Solid State Technology, Vol.28, No.10 (1985), p.91-104.

(10) Kitano, M. et al., "Current Japanese Materials Research", Soc. of Mater. Sci., Japan., Vol.2 (1987), p.235-250.

(11) Kitano, M. et al., Soldering and Surface Mount Technology, No.2 (1989), p.4-8.

(12) Tsubosaki, K. et al., Nikkei Microdevices (in Japanese), No.51 (1989), p.109-114.

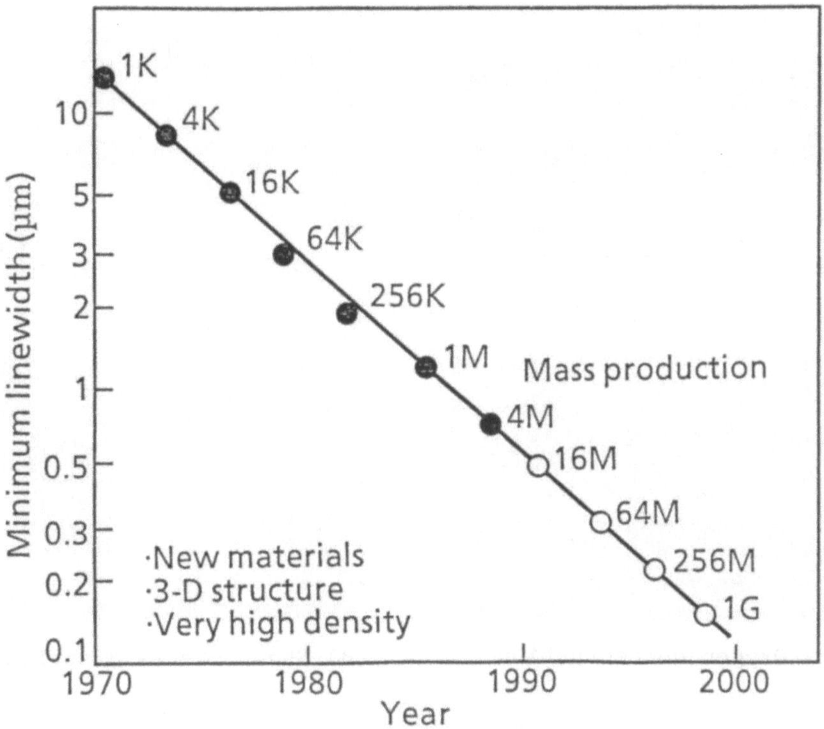

Fig. 1 Evolution of memory cell density in a DRAM chip

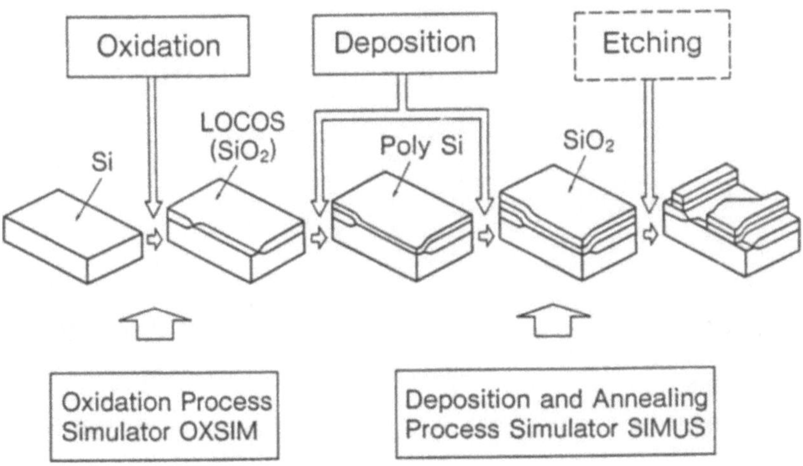

Fig. 2 LSI fabricating processes and simulators

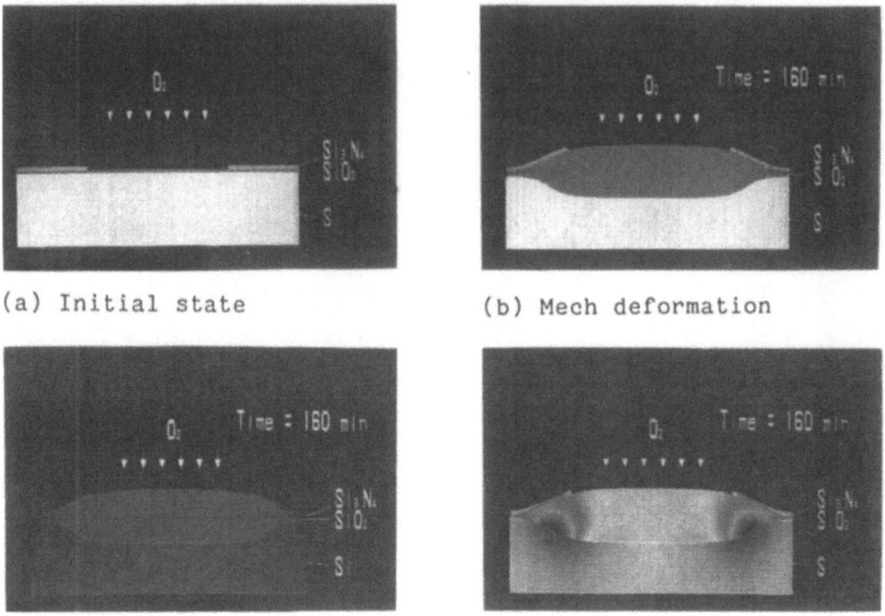

(a) Initial state (b) Mech deformation

(c) Oxygen concentration (d) Shear stress distribution

Fig. 3 Example analysis of LOCOS structure

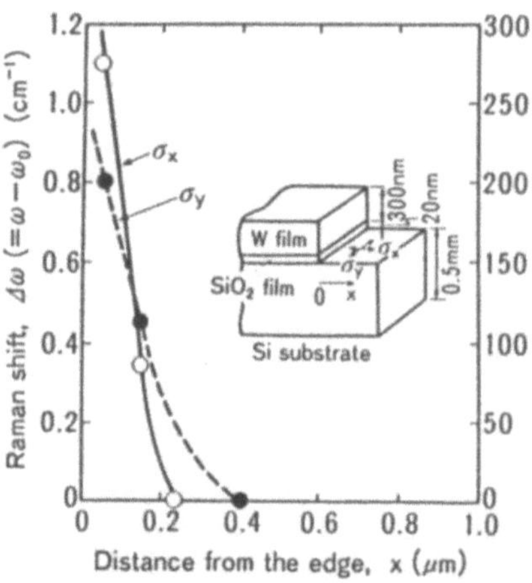

Fig. 4 Plot of Raman shift and compressive stress vs. distance from edge of W thin film using Raman spectrum analysis

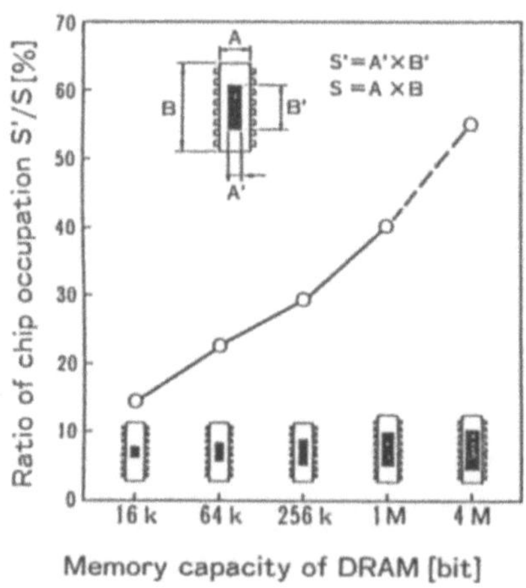

Fig. 5 Trends for ratio of chip occupation

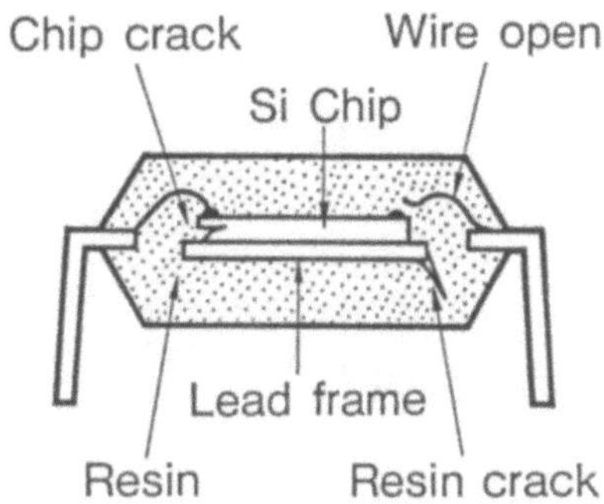

Fig. 6 Typical failure modes of IC plastic package

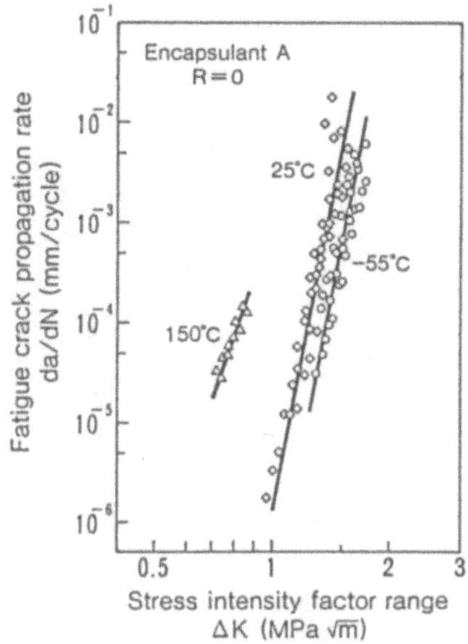

Fig. 7 Fatigue crack propagation rates of encapsulant

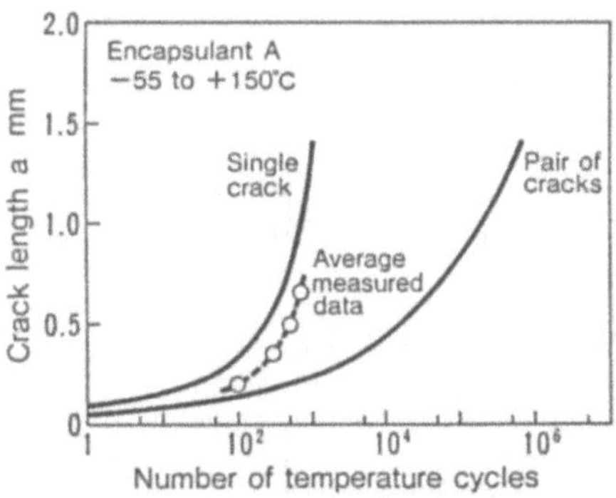

Fig. 8 Estimation of crack propagation behavior in a package

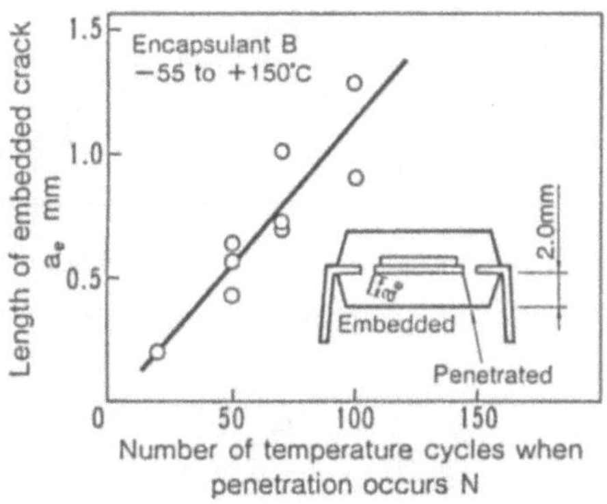

Fig. 9 Symmetry effect of crack propagation on temperature cycling life

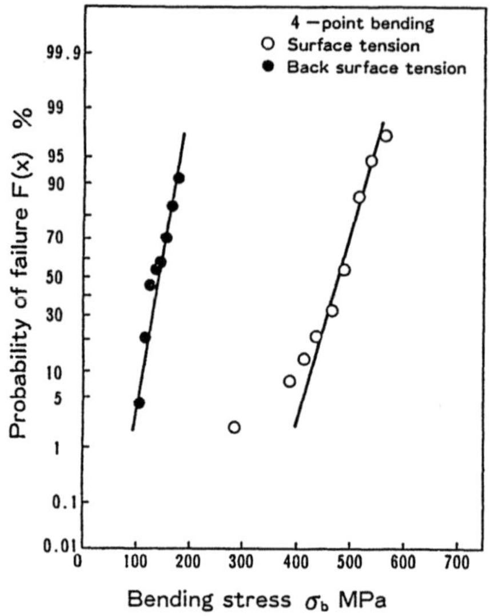

Fig. 10 Bending strength of silicon chip

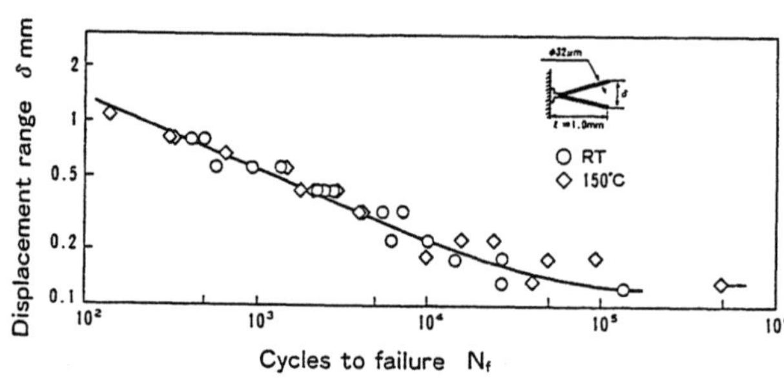

(a) Displacement range vs. fatigue life

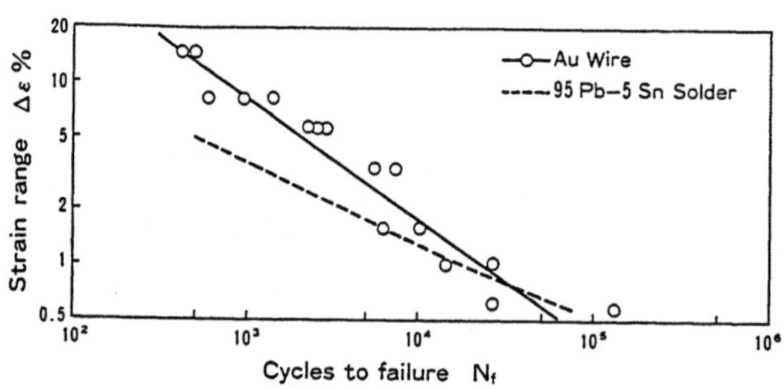

(b) Strain range vs. fatigue life

Fig. 11 Bending fatigue strength of gold wire

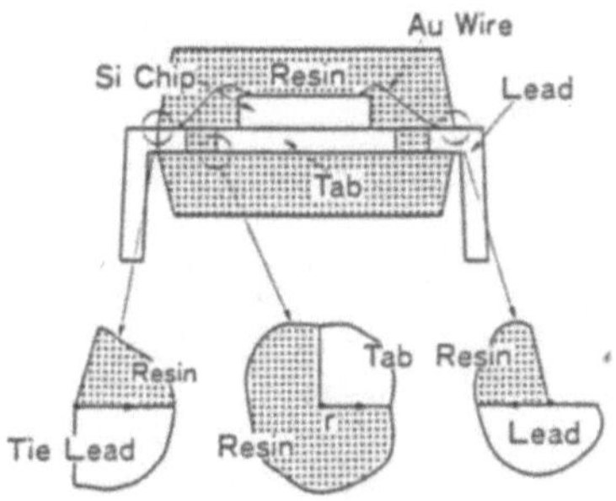

Fig. 12 Plastic encapsulated LSI device and various kinds of bonding edges

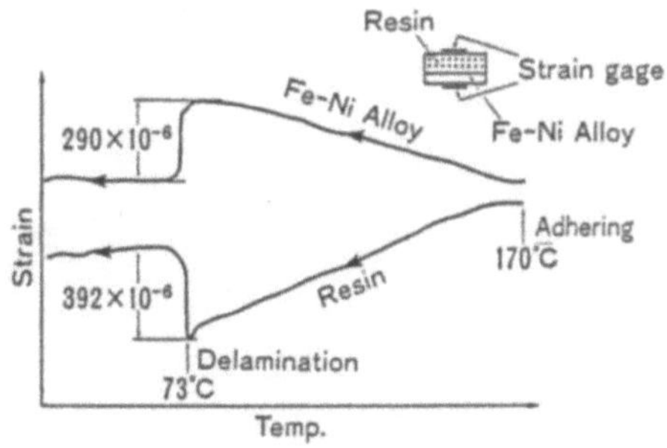

Fig. 13 Measured result of delamination in the cooling process

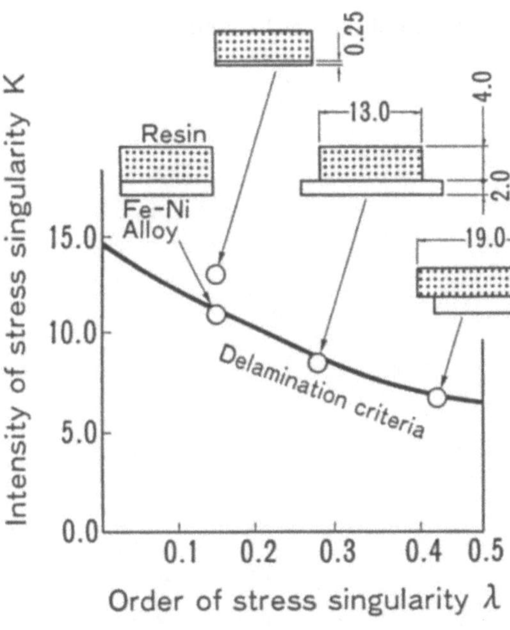

Fig. 14 Delamination criteria using stress singularity parameters K and λ

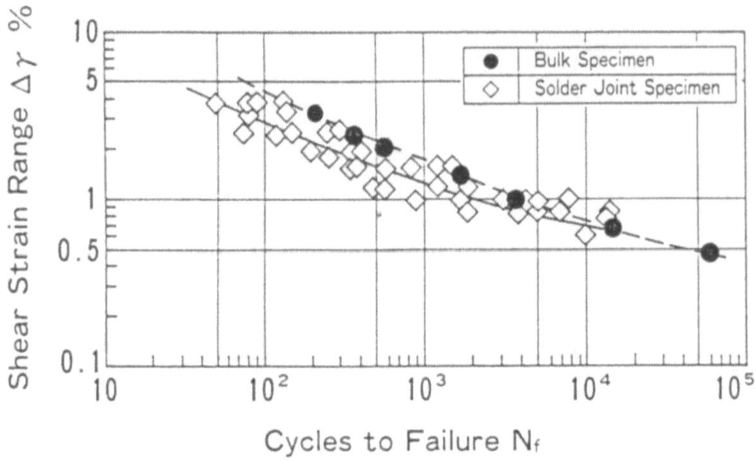

Fig. 15 Fatigue strength of bulk and solder joint specimens

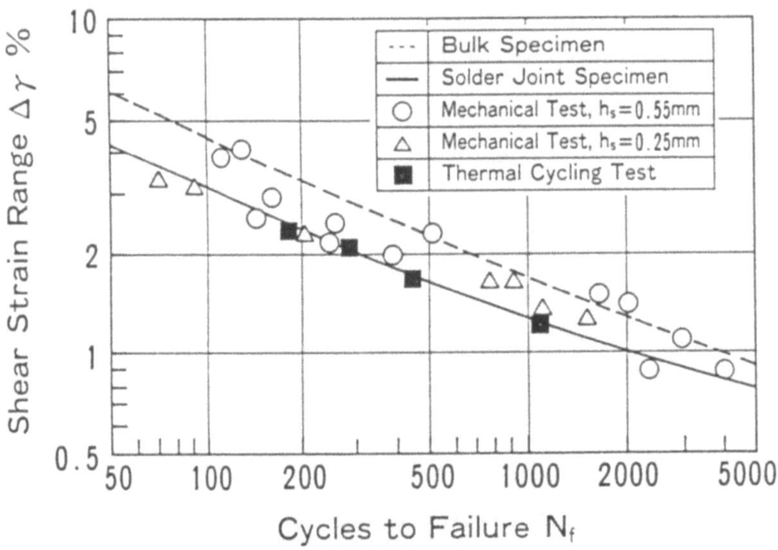

Fig. 16 Fatigue strength of solder joints of IC packages
(h_s : height of solder fillet)

Measurement of Internal Stress in Thin Silicon Membranes

S. BÜTTGENBACH[a], W. ENGELHARDT[a], W. KULCKE[b]

a) Hahn-Schickard-Gesellschaft für angewandte Forschung e.V., D-7000 Stuttgart
b) IBM Deutschland GmbH, D-7032 Sindelfingen

Abstract

This paper reports on four different methods to determine the internal stress in thin membranes as they are used for shadow lithography with electron beams or X-rays. Of the four methods the "Electrostatic Bulging Method" is described in detail. The paper also reports the measurement of the Young's modulus of Boron doped Silicon for two crystal directions.

1. Introduction

Thin Silicon membranes are used as mask substrates in electron beam and X-ray lithography [1]. For their fabrication a Silicon wafer is, within the pattern window, etched down to a thickness of about 3 μm. A uniform thickness of the membrane is achieved by a highly Boron-doped layer implanted into the surface of the wafer. The chemical etching stops when it reaches this layer. The Boron atoms have a smaller radius than the substituted Silicon atoms and, thus tend to contract the membrane, i.e. they introduce an internal stress. On one hand, this stress is desired as it keeps the membrane stretched flat. On the other hand, the internal stress causes mask pattern distortions if the pattern introduces a non-uniform elasticity in the membrane. In order to control these pattern distortions one needs to precisely know the tensile stress of the membrane. In this paper we will describe four different methods to determine the tensile stress of membranes. The investigated membranes had thicknesses of about 3 μm and were about 20 mm x 20 mm in size.

2. Measuring methods

2.1 Resonance method

A membrane is the two-dimensional analogon of a string. The resonance frequency of a string depends on its length and its tensile stress. Consequently, as for the string, the tensile stress of a membrane can be determined from its resonance frequency and its dimensions. This method can be employed in vacuum or in air [2].

a) Resonance method in air

In air the resonance method is aggravated by damping of the membrane oscillations. This damping causes a shift of the resonance frequency, which must be known in order to calculate the membrane's stress.

We made a few experiments with this method and used the following setup (Fig.1): A laser beam was reflected at the membrane toward a photo detector. The membrane was stimulated to oscillate by a loudspeaker or a piezo-ceramic actuator. The sensor signal was probed by an oscilloscope. This showed maximum amplitudes at the resonance frequencies of the membrane.

It turned out to be rather difficult to precisely locate the first resonance frequency. Inconstant air pressure and humidity as well as an accounting for the radiation losses lead into difficult correction calculations.

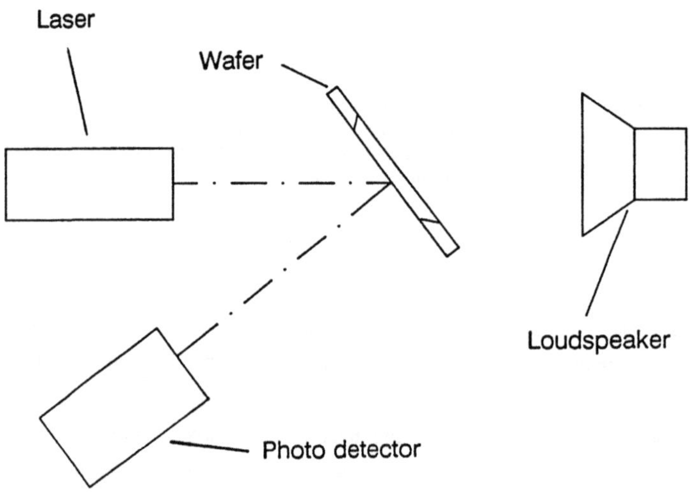

Fig.1. Measurement of membrane stress by determining the resonance frequency

b) Resonance method in vacuum

The problems of taking into account damping of the membrane oscillations are avoided if the membrane is placed into vacuum. In these experiments the first resonance frequency could be precisely determined. The resulting measurement error was about 1.5 %. The main drawback of this method ist, however, the need for substantial equipment.

2.2 Thermal method

The tensile stress of the membrane can be considered as originating from a cooling of the membrane. While the surrounding Silicon frame is kept at constant temperature the membrane, in trying to contract, develops the tensile stress. Inversely, if the membrane is warmed above the temperature of the surrounding Silicon frame the tensile stress ist reduced until, at the relief temperature, the membrane stress is compensated and the membrane starts to buckle. When knowing Young's modulus and the coefficient of thermal expansion of the membrane material the relief temperature allows to determine the tensile stress of the membrane for equal temperature of membrane and surrounding frame.

The problems of this method lie in the very fast dissipation of the heat from the membrane into the frame. One can calculate a temperature decrease of 2.5 K within 100 ms. To keep the error of the measured stress under 5 % one needs to measure the temperature with an accuracy of at least 1 K.

2.3 Measurement structures in the membrane

The tensile stress of the membrane can also be measured by measuring the deformation of special structures in the membrane. Special structures have been developed [3, 4] which magnify the deformations and allow to read the tensile stress from observations under a microscope.

As such structures impact the usefulness of the membrane for lithography applications this method can be used only to compare the results of other measuring methods.

2.4 Electrostatic bulging method

In the electrostatic bulging method the membrane is used as one plate of a condensor. Under the influence of an electric field the membrane bulges, the amount of bulging depending on the tensile stress of the membrane.

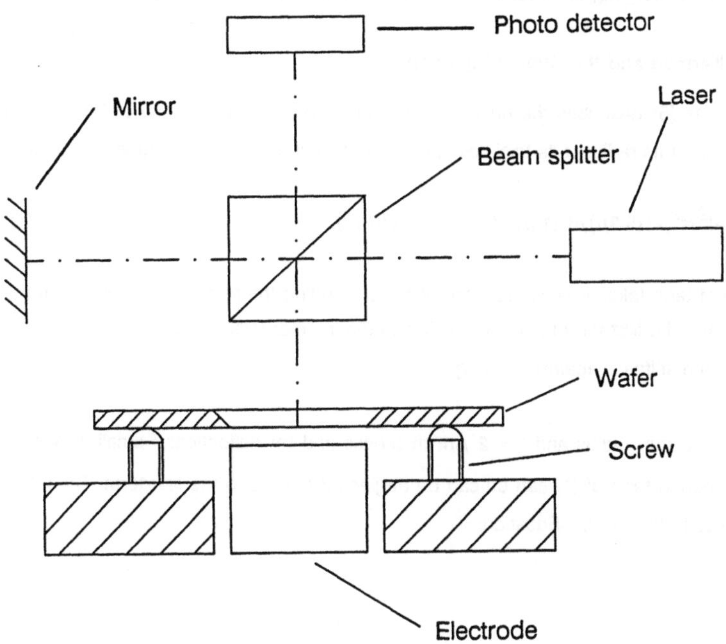

Fig.2. Setup for electrostatic bulging method

In our experimental setup (Fig.2) the wafer with the membrane window rests on three screws over the other electrode of the condensor. With the aid of the screws the distance between wafer and electrode is adjusted to 0.5 mm. The three screws also connect the ground to the wafer. The voltage between wafer and electrode bulges the membrane. The deflection of the membrane ist measured by a laser interferometer.

The advantages of this method are:

- Simple equipment,
- Measurement error of tensile stress less than 5 %,
- Time required for a breadboard measurement less than 5 minutes,
- Automation easily possible.

With these advantages the bulging method appeared to be superior to the alternatives. Therefore it will be detailed in the following section.

3. Measurement of the tensile stress by the electrostatic bulging method

3.1 The membrane in the Theory of Elasticity

In the Theory of Elasticity membranes can be considered as a special kind of plates. They differ from plates by their infinitesimal thickness. Consequently the plate's thickness collapses into the neutral plane with the effect that, at bulging, the forces required to bend the thin plate disappear.

3.2 The Equation of Karman and the Bubble-Equation

Karman's theory of plates [5] describes the bulge of a round plate under a pressure p. With σ_i the internal stress in the plate, r its radius, t its thickness and f the deflection, Karman's equation states:

$$\sigma_i = p\cdot r^2/4f\cdot t - (Y\cdot f^2/r^2)\cdot\{(5-3\mu)/6\cdot(1-\mu)+2/7\} - 16\cdot K/(t\cdot r^2). \qquad (1)$$

In this equation the third term takes into account the effect of bending the plate, the second term the stretching of the plate and the first term the conter-action to the internal stress. The factor $K = Yt^3/[12\cdot(1-\mu^2)]$ is the stiffness against bending.

For membranes of $r = 10$ mm radius and $t = 2$ μm thickness and for deflections f smaller than 10 μm the third and the second term of (1) can be can be neglected relative to the first term. Then Karman's equation simplifies to the Bubble-Equation:

$$\sigma_i = p\cdot r^2/4f\cdot t. \qquad (2)$$

For square membranes of side length 'a' equation (2) becomes:

$$\sigma_i = 0.082\cdot p\cdot a^2\,/(f\cdot t). \qquad (3)$$

3.3 Experimental results

Square membranes of side lengths 'a' between 10 and 20 mm and thicknesses t between 2 and 3 μm were measured. Fig.3 shows a typical deflection plot through the center of a membrane.
The calculated stress values of the membranes resulted between 5 and 30 N/mm^2. In the same membranes they corresponded with stress values measured by the vaccum resonance method, to about 4 %.

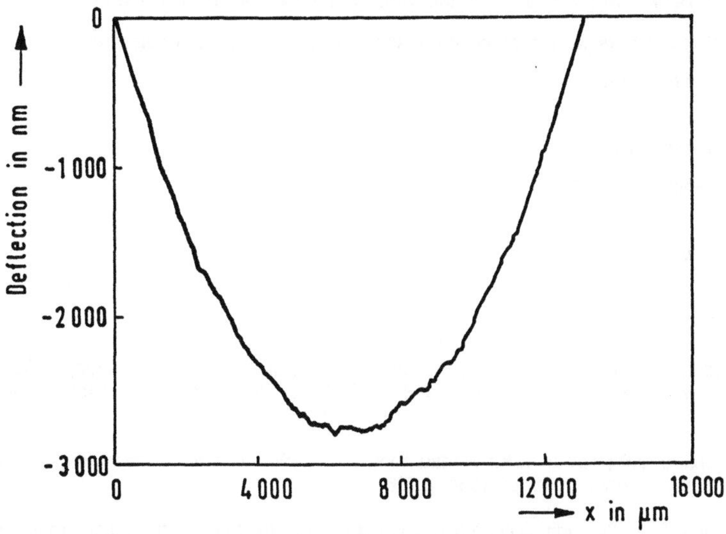

Fig.3. Deflection plot through the center of a square membrane

4. Young's modulus of Boron-doped Silicon

For several methods to measure the internal stress of membranes the knowledge of Young's modulus is required. Young's modulus has been reported for pure Silicon [6] as:

$$Y(Si,<010>) = 130\ 100\ N/mm^2,$$
$$Y(Si,<011>) = 169\ 100\ N/mm^2.$$

As the membranes used in our investigations all were highly doped with Boron (10^{20} B atoms/cm^3), Young's modulus had to be determined extra. Petersen and Guarnieri [7] describe a method and formula to determine Young's modulus by measuring the resonance frequency of little bars.

4.1 Resonance frequency of bars

The resonance frequency of bars can be calculated:

$$f_{vac} = 0.162 \cdot t \cdot (Y_s/d)^{1/2}/l^2, \tag{4}$$

where t is the thickness, l the length of the bars, and d the density, Y Young's modulus of the material. All bars tested in our experiments were at least ten times wider than tick. Therefore the areal modulus Y_s instead of the unidirectional modulus Y had to be used:

$$Y_s = Y/(1-\mu^2) \tag{5}$$

Here μ is the Poisson ratio ($\mu = 0.3$).

4.2 Experimental results

The lengths of the used bars were l = 1 to 2 mm, the withs 35, 85 and 180 μm and the thicknesses t = 3 μm. With these different bars we measured the resonance frequencies and calculated Young's modulus for two crystal directions:

$$Y(Si/B,\langle 010\rangle) = 107400\ N/mm^2$$
$$Y(Si/B,\langle 011\rangle) = 146200\ N/mm^2.$$

The measurement error was 8 %.

References

1. Bohlen, H.; Behringer, U.; Keyser, J.; Nehmiz, P.; Zapka, W.; Kulcke, W.: High Throughput Submicron Lithography with Electron Beam Proximity Printing. Solid State Technology (Sept. 1984) p.210

2. Acosta, R.E.; Maldonado, J.R.; Towart, L.K.; Warlaumont, J.M.: B-Si Masks for Storage Ring X-Ray Lithography. Solid State Technology (Sept. 1984) p.205

3. Mehregany, M., Howe, R.T., Senturia, S.D.: Novel Microstructures for the in situ Measurement of Mechanical Properties of Thin Films. J.Appl.Phys. 62 (1987) 3579

4. Asch, A.; Behringer, U.; Keyser, J.; Zapka, W.: Measuring Local Stress in a Membrane. IBM Technical Disclosure Bulletin 29 (1986) 3212

5. Göldner, H.: Höhere Festigkeitslehre, Vol.2. Weinheim: Physik Verlag 1985

6. Landolt-Börnstein: Zahlenwerte und Funktionen aus Naturwissenschaft und Technik. Neue Serie, Gruppe III, Band 17a. Berlin: Springer 1982

7. Petersen, K.E.; Guarnieri, C.R.: Young's Modulus Measurements of Thin Films using Micromechanics. J.Appl.Phys. 50 (1979) 6761

Mesurement of the Thermal Diffusivity of Electrically Nonconducting Substrates

A.Drost, D.Hemmetzberger, J.Emmer
Fraunhofer Institut für Festkörpertechnologie, München

Summary
An apparatus was built up for measuring the thermal diffusivity of thin ($\geq$ 500 μm), electrically nonconducting substrates. The apparatus is based on the method of Angström. The ratio of amplitudes and the phase difference of temperature waves at the upper and bottom side of a periodically heated substrate are measured. The detection of temperature is done by sputtered Ni– temperature sensors. Results are shown from measurements on Al_2O_3 and AlN substrates.

Introduction

The thermal conductivity of substrate materials is an important property in electronic packaging. As the complexity of integrated circuits is increasing the dissipated power is rising. Part of the heat has to be removed over the substrate.
The methods for determing the thermal conductivity can generally be divided in static and nonstatic methods. With the static methods the thermal conductivity λ is obtained by measuring the heat flow and the temperature gradient. In the nonstatic methods the propagation time of temperature signals is determined. As a result the thermal diffusivity α is obtained, which is connected to the thermal conductivity by the density ρ and the specific heat c.

$$\lambda = \alpha \rho c \tag{1}$$

Two of the most common nonstatic methods are the flash method and the Angström method. In the flash method a heat pulse (laser flash) is given to the front side of the sample and the rise of the temperature at the backside is measured as a function of time. In the Angström method the sample is periodically heated. For this method several implementations exist. The generation of the temperature modulation (mostly sinus waves) can be done by a modulated laser beam /1/, electron bombardment /2/ or electrically /3/; the measurement of the temperature waves can be done with the help of radiation (IR), temperature sensors, the photoacustic effect /4/ or the reflectivity of the surface. The phase difference between the creating wave and the temperature wave at the surface of the sample as well as between temperature waves at two places of the sample can be used for determing the thermal diffusivity.

Theoretical treatment

Fig. 1 shows a schematic sketch of the arrangement, used in the experiment. The sample P is fixed mechanically between a heater H and a heat sink S. To ensure, that most of the heat flows across the sample, a heat resistance R has to be put at the back of the heater H. Moreover every interface (eg. between sample P and the heat sink S) can be regarded as a heat resistance.

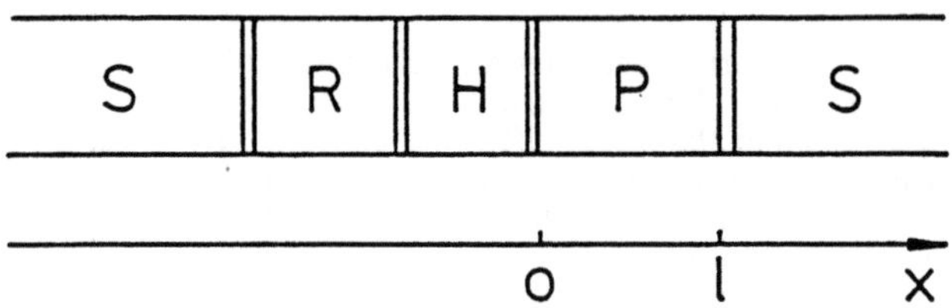

Fig. 1: Scheme of the arrangement

The description of the arrangement concerning its thermal properties can be found in /2/. When the sample is heated with a sinus wave modulation, beyond the general equation of heat flow (for one dimension):

$$\frac{\delta^2 T}{\delta x^2} = \frac{1}{\alpha}\frac{\delta T}{\delta t} \qquad \text{with } 0 \leq x \leq l$$

the following boundary conditions are valid (see /2/):

$$T = T_0 + V \sin(\omega t) \qquad \text{at } x = 0$$

$$\lambda \frac{\delta T}{\delta x} = q \qquad \text{at } x = l$$

The system of equations has been solved in /2/. The result is given by:

$$T(x,t) = T_0(x) + V(x) \sin(\omega t + \phi(x)) \qquad \text{with } 0 \leq x \leq l$$

Here $T_0(x)$ is the static solution of the system of equations (for $V(x=0)=0$); $V(x)$ and $\phi(x)$ stand for an amplitude and a phase respectively of the temperature wave. The ratio of amplitudes $A = V(x=l)/V(x=0)$ and the phase difference $\phi = \phi(x=l) - \phi(x=0)$ are given by:

$$A = \{ (\cosh^2(\beta l) - \sin^2(\beta l)) + \frac{2K}{\beta l}(\cosh(\beta l)\sinh(\beta l) + \cos(\beta l)\sin(\beta l)) + \frac{2K^2}{(\beta l)^2}(\cosh^2(\beta l) - \cos^2(\beta l)) \}^{-1/2} \qquad (2)$$

$$\phi = \{\tanh(\beta l)\tan(\beta l) - \frac{K}{\beta l}(\tanh(\beta l) - \tan(\beta l)) \} / \{1 + \frac{K}{\beta l}(\tanh(\beta l) + \tan(\beta l)) \} \qquad (3)$$

with $1/\beta = (2\alpha/\omega)^{1/2}$ thermal diffusion length
K heat coupling factor between sample P and heat sink S

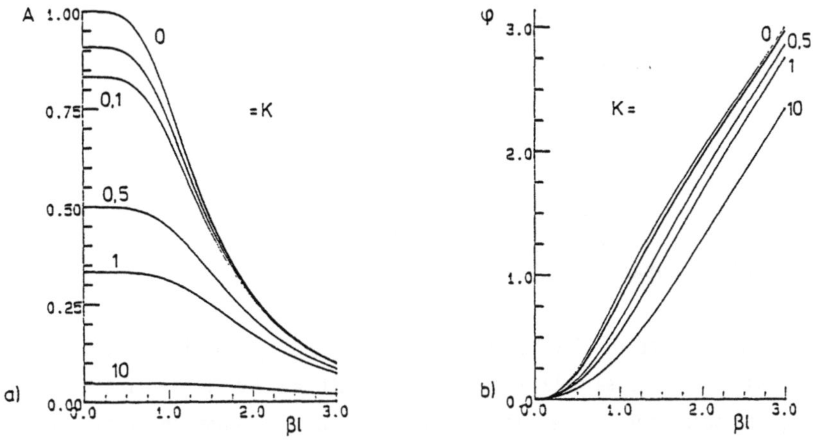

Fig. 2: a) Ratio of amplitudes and b) phase difference as a function of βl

Fig.2 a and b illustrate the equations given above. The ratio of amplitudes is decreasing with increasing thickness l of the sample or decreasing thermal diffusion length $1/\beta$. In the case the heat resistance is low between the sample and the heat sink ($K \rightarrow \infty$), the ratio of amplitudes is low, because the amplitude of the temperature wave at $x = l$ is getting reduced by the almost constant temperature of the heat sink. The phase difference is increasing with increasing propagation time (increasing thickness l, decreasing thermal diffusion length $1/\beta$).

Experimental setup

Fig.3 shows a schematic drawing of the apparatus. The heater was realised as a thick film resistor. To yield an amplitude of the power wave of about 50 W the voltage, provided by a sinus wave generator, had to be amplified. The sample is fixed mechanically between the heater and the heat sink. As it can be seen from fig.2, a reasonable range for βl is from about 0.2 to 3 (for $\beta l < 0.2$ the phase difference and for $\beta l > 3$ the ratio of amplitudes are quite small). Therefore the thickness l of the samples, which can be measured by this method, can be estimated as $l \approx (0.1 - 1.7)\,(\alpha/f)^{1/2}$. In the experiments described here the frequencies ranged from 1 Hz to 150 Hz; the thickness l varied from 0.5 mm to 1 mm. At the upper and bottom side of the sample Ni– resistors are realised in thin film technology. The Ni layers have a thickness of about 0.2 μm and are in very good thermal contact to the surface of the sample. The thermal coefficient of resistance of the Ni– films is something less than the TCR of bulk Ni (TCR $\approx 6.5\ 10^{-3}$ 1/K).

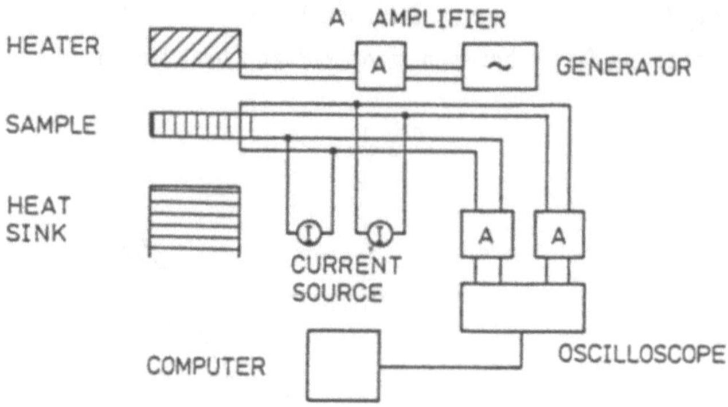

Fig.3: Scheme of the apparatus

Depending on the power of the heater and the thermal properties of the sample temperature waves at the sample were produced with an amplitude of about 1 K. Using four point technique the temperature waves lead to changes in voltage at the Ni– resistors. These changes have to be amplified to allow a detection on an oscilloscope. The curves, measured at the oscilloscope, are transfered to a computer. Fig.4 shows typical curves. The extraction of the amplitudes and the phases from the data is done by the computer too.

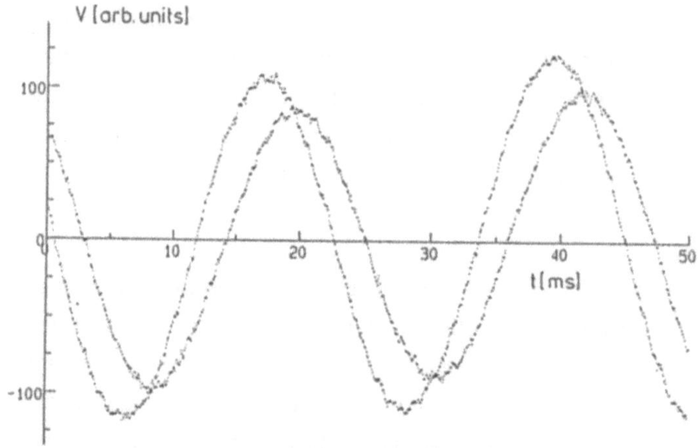

Fig.4: Signals, measured at the upper (a) and bottom (b) side of the sample. The amplitudes are proportional to the changes in temperature.

With the help of equation (2) and (3) the quantity βl can be determined, when the ratio of amplitudes A and the phase difference ϕ are known. The thermal diffusivity α is then given by the following equation:

$$\alpha = \frac{l^2 \ \omega}{2\,(\beta l)^2} \tag{5}$$

Materials

As the Ni– temperature sensors are in direct contact with the sample, only electrically nonconducting substrates can be measured (eg. ceramics). Following materials were available: Al_2O_3– and AlN– ceramics and sapphire. Some properties of the samples are listed in Tab.1. The accuracy of the density data is about 3%. The value for the specific heat of the sapphire is taken from /5/. The specific heat of the Al_2O_3 and AlN samples was measured at ComTech by DSC.

material	manufacturer	thickness	density	spec. heat
		mm	gr/cm^3	Ws/kgK
Al_2O_3	Hoechst CeramTec	0.640	3.86	800
AlN	Tokuyama Soda	0.640	3.25	
	Standard Oil	0.700	3.22	705
	Keramont	0.635	3.13	
	ANCeram	1.035	3.26	
sapphire	Goodfellow[1]	0.895	3.98	765

Tab.1: Properties of the samples, used in the experiments, [1] supplier

Results

Fig.5 shows the phase difference ϕ and ratio of amplitudes A, which had been measured, as well as the quantities βl and K, which had been extracted from the data, as a function of the square root of the frequency $f = \omega/2\pi$. (The data were measured at an AlN sample by Tokuyama Soda.) The ratio of amplitudes and the phase difference show the decrease respectively increase with increasing frequency as predicted by theory.

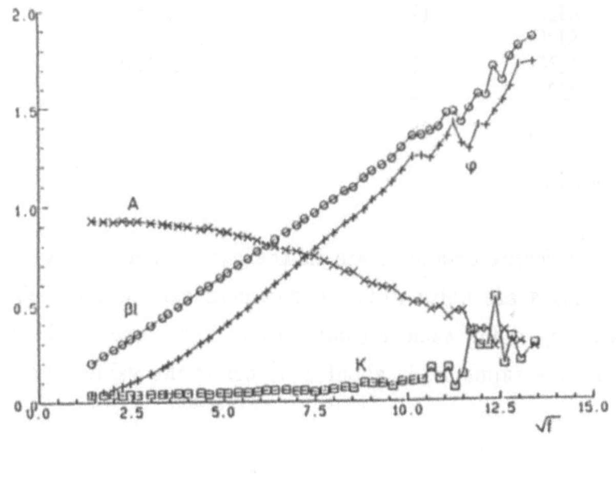

Fig.5: Ratio of amplitudes A, phase difference ϕ, βl and K as a function of $f^{1/2}$

In order to compare the results with theory (see fig. 2 and 3) the quantities ϕ and A, which had been measured, are shown as a function of βl, which had been extracted from the data. As it can be seen in fig.6, the measurement agrees excellent with theory.

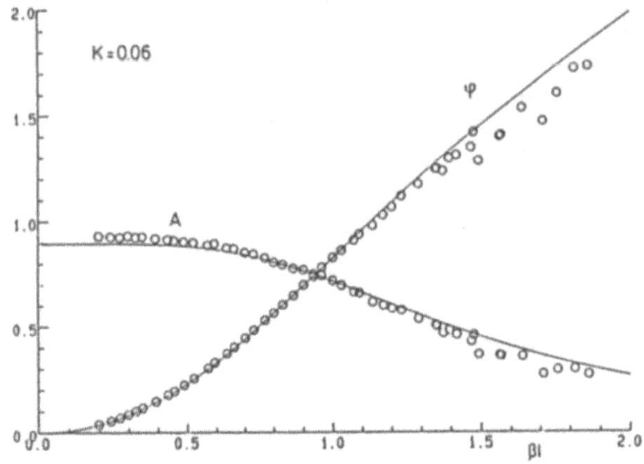

Fig.6: Comparision of theory and measurement,
ratio of amplitudes phase difference

In Tab.2 values of the thermal conductivity, measured with this apparatus, are compared with manufacturers data.

manufacturer	material	λ_{meas} (W/mK)	λ_{manuf} (W/mK)
Hoechst CeramTec	Al_2O_3	27	25–30
Tokuyama Soda	AlN	175	170–180
Standard Oil	AlN	159	140–220
Keramont	AlN	126	100–160
ANCeram	AlN	90	100–140
	sapphire	38	42

Tab.2: Thermal conductivities

The values, measured at different ceramics, are in the range of data, given by the manufacturer. As these ranges are quite wide, no comment can be given about the accuracy of the apparatus by doing such a comparision. The value of the thermal conductivity, measured at the sapphire, is about 10% below the value given by the supplier.

Conclusion

The apparatus, described above, allows to measure the thermal diffusivity of electrically nonconducting substrates. It is simple to use and an unexpensive construction. The use of the Angström method is possible because of the very good thermal contact of the Ni– temperature sensors with the samples. The accuracy of the data for the thermal diffusivity is about 10%, but it can be increased by further improvements.

Acknowledgement

The authors thank Mr. D.Bonfert for his support and Mrs. M.Träger for the realisation of the temperature sensors on the substrates.

References

/1/ R.Tilgner, J.Baumann, M.Beyfuss, 'On the influence of thin absorber layers in thermal–wave applications', Can.J.Phys. Vol.64 1986, p1287

/2/ M.Cerceo, H.Childers, 'Thermal Diffusivity by Electron Bombardment Heating', J.of Appl.Phys. Vol.34 1963, p1445

/3/ N.Savides, W.Murray, 'Apparatus for the measurement of thermal diffusivity featuring a low frequency sine wave generator and a digital phase meter', J.Phys.E Vol.11 1978, p941

/4/ A.Rosencwaig, A.Gersho, 'Theory of the photoacoustic effect with solids', J.of Appl.Phys. V01.47 1976, p64

/5/ Y.S.Touloukian et al., 'Thermophysical Properties of Matter, Specific Heat', Vol.5 1970, p24

Characterization of Supersmooth Surfaces with Roughness Below 0.1 nm

Dirk-Roger Schmitt

Deutsche Forschungsanstalt für Luft- und Raumfahrt (DLR)
Institut für Flugführung
Flughafen, D-3300 Braunschweig, Germany

Summary

Supersmooth surfaces with a roughness of 0.05 nm have been developed. Nomarski differential interference contrast microscopy was used to characterize the quality of the surfaces. Using an especially selected microscope objective, the microscope system resolves an rms roughness of lower than 0.05 nm. Photomicrographs of these supersmooth surfaces of BK7, Zerodur, and silicon are being presented. To obtain quantitative informations, the roughness of the surfaces was determined using light scattering, phase-shifting interferometry, heterodyne interferometry and stylus profilometry.

Introduction

With an increasing degree of integration of microsystems the dimensions of the manufactured structures will decrease. To produce very tiny structures in the submicron range, supersmooth surfaces with roughnesses of lower than 0.1 nm are required. We developed and characterized supersmooth surfaces made of BK7, Zerodur, and silicon.

Experimental Setup

We used Nomarski differential interference contrast microscopy (DIC) to get an overview of the surfaces. The high resolution of this microscope system has been demonstrated in previous work [1, 2, 3]. Using a Leitz Metallux-3 microscope with an especially developed and selected objective-Normarski-prism unit (aperture 0.45), a depth resolution of better than 0.05 nm was achieved. To get quantitative informations of the surface parameters, the samples were measured with a Zygo 5500 optical heterodyne profiler (OHP) [4], with the Wyko Topo-2D phase-shift interferometer (PI) [5], with the Rank Taylor Hobson Talystep system with 12,5 µm stylus (M) [6], and with an integrating Ulbricht sphere [1].

Experimental Results on Glass Surfaces

Figure 1 shows 4 typical results of supersmooth samples. The photos were taken with the Nomarski microscope. The rms roughnesses are given as δ for each sample. The index relates to the used measurement system.

Figure 1(a) presents a surface which is being obtained using an improved conventional polishing process with cerium oxide. The roughness values obtained with the phase-shift interferometer and the Talystep system are in the range of 0.2 nm and are in rather good agreement. In Figure 1(b) and 1(c) photomicrographs are shown of samples which were polished with an especially developed polishing process [3]. The improvement of the surface roughness can clearly be seen on that photos. In Figure 1(b) the OHP system as well as the PI system measures a roughness of 0.11 nm. The Talystep system gives a value of 0.44 nm. This can be explained by the different lateral resolution the profiling systems. The OHP and PI system resolve surface spatial wavelengths down to 1 µm - 2 µm. So it might occur that the scratches, which can be identified in the photomicrograph, cannot be resolved with these optical profilers. The Talystep has a lateral resolution better than 1 µm and resolves the scratches, and therefore measures a much higher roughness.

Figure 1(c) shows a much better sample. The OHP surface profile is presented in Figure 2. At this sample a roughness of 0.05 nm is obtained. We want to point out, that even on that supersmooth surface the Nomarski microscope resolves a lot of details. Another sample with an OHP roughness of 0.05 nm is shown in Figure 1(d). Although the OHP roughnesses of the samples in Figure 1(c) and Figure 1(d) are equal, the photomicrographs show that the sample of Figure 1(d) seems to be much smoother than the other one. This demonstrates the different lateral resolutions of the Nomarski system (1 μm) and the OHP system (2 μm).

Results on Silicon Wafers

Silicon wafers, which are used in the semiconductor industry, are known to have an extremely smooth surface. Therefore also substrates made of this material were investigated. Figure 3 presents the photos of two different wafers, which were manufactured by Wacker-Chemitronic GmbH, Burghausen, Germany.

The light scattering of the wafer in Figure 3(a) was measured in the Ulbricht sphere [1]. From the scattering a roughness of δ = 0.19 nm was calculated. Other measurements using a Coblentz hemisphere [7, 8] gave a result of δ = 0.10 nm. These differences may be explained by the slightly different spatial wavelengths collected in the two spheres. The photomicrographs show a long range spatial wavelength ("waviness") of about 15 μm which dominates the surface topography. This wavelength is rather close to the cutoff wavelengths of both spheres. If the cutoff wavelengths differ slightly so that our Ulbricht sphere collects only a small amount of longer spatial wavelengths than the Coblentz sphere, our measurements will give a higher scattering level and thus a higher roughness. So the cutoff wavelengths of both spheres have to be adjusted to the same values to give identical results. The PI system resolves completely the 15 μm waviness, and therefore a result of δ = 0,59 nm is obtained.

Figure 3(b) shows a better polished sample which has a roughness of about 0.05 nm [7, 8]. The long range surface spatial wavelengths of 20 μm on the surface in Figure 3(b) are not collected in the TIS measurements. So the measured roughness of 0.05 nm corresponds mainly to the wavelengths shorter than about 5 μm which can be identified on the microphotograph.

Conclusions

We want to point out that the DIC system has a very high depth resolution. The microphotographs show a large number of details on supersmooth surfaces. Topographic height differences of 0.05 nm can be resolved on glass (Fig. 1(d)), and on silicon (Fig. 3(b)). Further work is requested to get well correlated measurement results within the 0.05 nm range.

The cooperation with G. Staps of Rollei Fototechnic GmbH, Braunschweig, for the development of the supersmooth samples is gratefully acknowledged.

References

1. Schmitt, D.-R.: Characterization of mirror surfaces for laser-gyro applications. In: Surface Measurement and Characterization, J.M. Bennett, ed., Proc. SPIE **1009**, 155 (1989).

2. Schmitt, D.-R.: High reflectance low scatter mirrors. In: Thin Films in Optics, T. Tschudi, ed., Proc. SPIE 1125, 94 (1990).

3. Schmitt, D.-R.: Characterization of high quality surfaces by Nomarski microscopy and light scattering. In: Proc. 5th International Precision Engineering Seminar IPES5, Monterey (1989), p. 221.

4. Sommargren, G.E.: Optical heterodyne profilometry. Appl. Opt. **20**, 610 (1981).

5. Bhushan, B., Wyant, J.G., Koliopoulos, C.L.: Measurement of surface topography of magnetic tapes by Mireau interferometry. Appl. Opt. **24**, 1489 (1985).

6. Bennett, J.M., Dancy, J.H.: Stylus profiling instrument for measuring statistical properties of smooth optical surfaces. Appl. Opt. **20**, 1785 (1985).

7. Hahn, P.O., Kersten, M.: Elastic lightscattering measurements of Si-surfaces. In: Surface Measurement and Characterization, J.M. Bennett, ed., Proc. SPIE **1009**, pp. 172 - 181 (1989).

8. Mattson, L.: Characterization of supersmooth surfaces by light-scattering techniques. In: Surface Measurement and Characterization, J.M. Bennett, ed., Proc. SPIE **1009**, pp. 165 - 171 (1989).

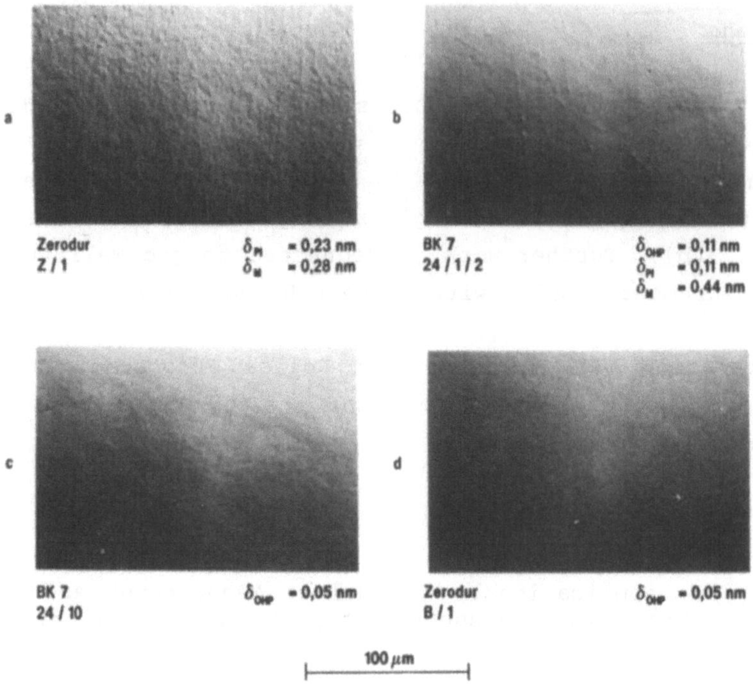

Fig.1. DIC photomicrographs of different superpolished optical surfaces.
OHP = Optical heterodyne profiler
PI = phase-shift interferometer
M = Talystep system

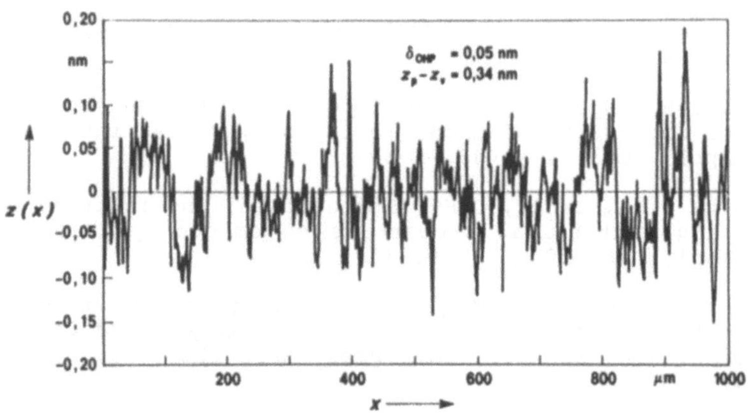

Fig.2. OHP surface profile of the ultrasmooth sample 24/10 in Fig. 1(c).

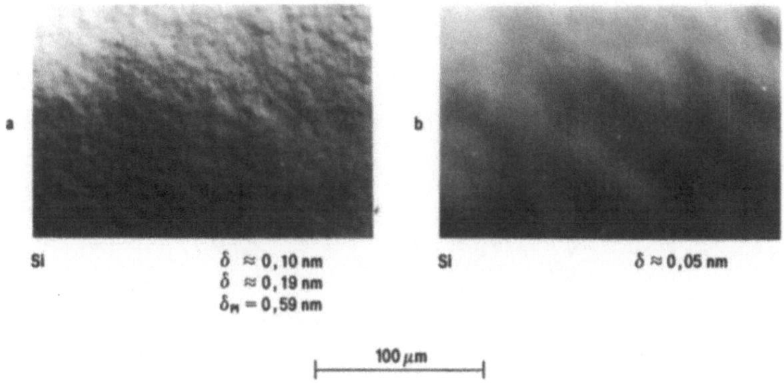

Fig.3. DIC photomicrographs of two supersmooth silicon wafers. The roughness δ was determined using scattering measurements.

Material Characterisation – Semiconductor Charateristics

Non-Destructive Testing of Semiconductor Materials Using Microwave Photoconductivity

G.BETZ, R.PLOSCHIES, CH.WOLK, CH.WINTER, J.VALLDORF

Phoenicon Photoaktive Systeme GmbH
Bismarckstr.58
1000 Berlin 12

Summary

For semiconductor research and production the ability to test and measure important parameters of materials and devices is highly desirable. The usual methods necessitate the formation of ohmic contacts and special test structures, which may not only change the parameters to be tested but may also cause damage and add contaminations. A convenient, non-contact method is presented which is capable of measuring the free carrier´s lifetime. The method is based on the measurement of reflected microwave power from the semiconductor after optical excitation.

Carrier Lifetime as a Material Parameter

The kinetics of excess charge carrier is directly correlated with the defect structure of the semiconductor material via generation, recombination (R) and trapping processes (T). This is shown in a simplified energy band diagram of a crystalline semiconductor (Fig.1).

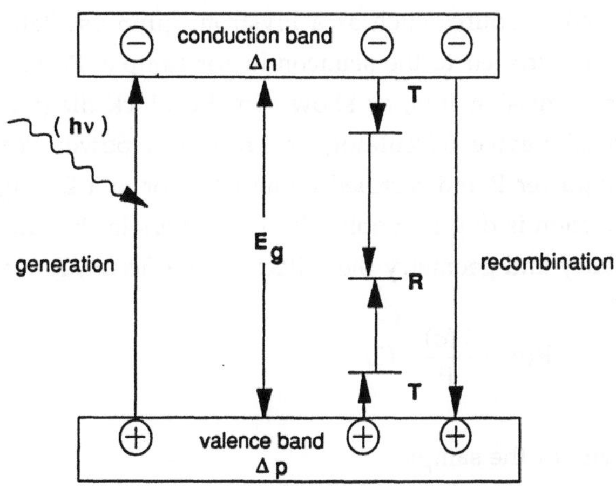

Fig.1: Optoelectronic processes visualized in an energy band diagram

The excess carrier concentrations, Δn, Δp, give rise to an excess conductivity, the photoconductivity, $\Delta\sigma$.

$$\Delta\sigma = q\,(\mu_n \Delta n + \mu_p \Delta p) \quad (1)$$

with q as the elementary charge and μ_n, μ_p as the mobilities of electron and holes respectively. Thus, the defect structure can be analysed by measuring the decay of the photoconductivity. Such an analysis yields the lifetime of minority carriers, τ, from the time constant τ_p of the photoconductive decay. For silicon already defect densities of 10^{10} cm^{-3} change the lifetime τ drastically. Since the defect density in highly pure silicon is 10^{15} cm^{-3}, lifetime measurements can detect deviations of 10 ppb from the ideal material. Lifetime measurements are therefore much more sensitive to defects than measurements of resistivity or mobility. The minority carrier lifetime is also a functional parameter for transport processes in solar cells, photodetectors or bipolar circuits. An effective process control for the manufacturing of these devices necessitates the monitoring of τ.

Microwave Photoconductivity

The photoconductivity can be either directly measured using electrical contacts or by non-contact methods. Such methods using the interaction of electromagnetic waves with the semiconductor material.

The use of microwaves for the detection offers some advantages such as sensitivity and reproducibility if compared to the low frequency and the IR range. Microwave radiation generated in source such as a klystron, gunn oscillator or dielectric resonator oscillator is directed to the semiconductor sample. Partially it is reflected back into the transmission line, as shown in the block diagram of Fig.2. Using an non-reciprocal device (circulator) the reflected power $P(\sigma)$ is separated from the incoming power P and rectified by an detector to a DC-signal giving $P(\sigma)$. Microwave reflection is due to mobile charge carriers in the sample. At a given microwave frequency and geometry the reflection coefficient, $R(\sigma)$:

$$R(\sigma) = \frac{P(\sigma)}{P} \quad (2)$$

is a function of the conductivity of the sample.

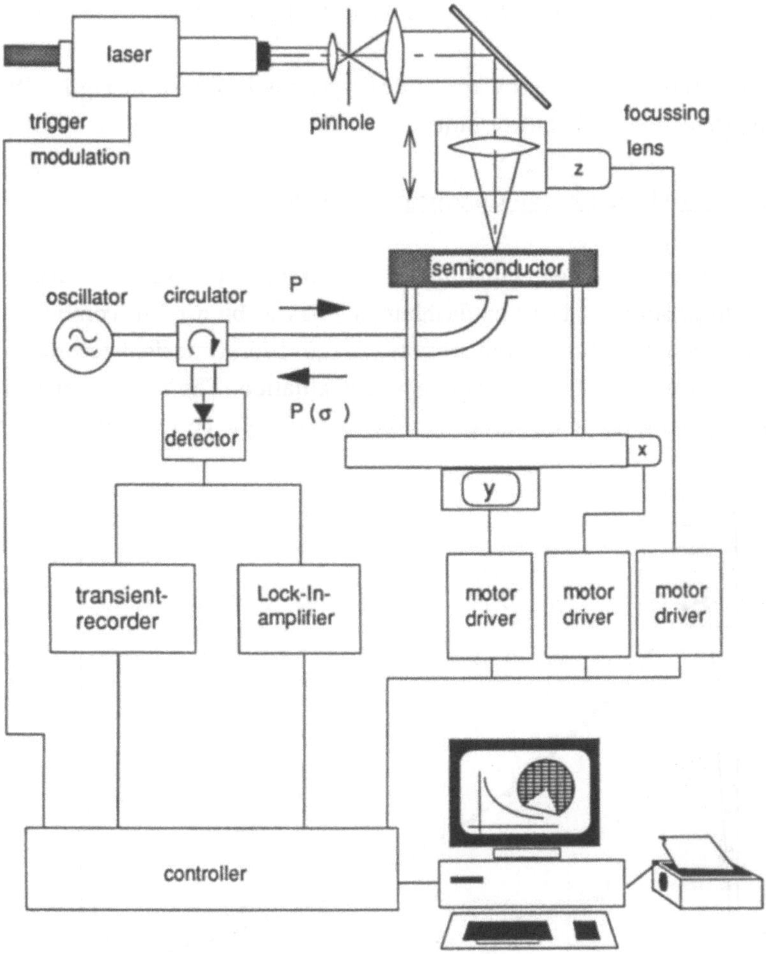

Fig.2 Block diagram of the lifetime and defect imaging system MRM with the microwave reflectometer consisting of oscillator, circulator, detector and transmission line.

Upon illumination of the sample the conductivity increases resulting in a change of the reflected power ΔP. For small changes ΔP the relative change ΔP/P is proportional to Δσ with A as the proportional constant:

$$\frac{\Delta P}{P(\sigma)} = \frac{1}{R(\sigma)} \left(\frac{dR(\sigma)}{ds}\right)_\sigma \Delta\sigma = A\,\Delta\sigma \quad (3)$$

Since no currents but charge densities are measured, no interfaces are needed for charge separation as with OBIC or EBIC.

Measurement Techniques, Instruments and Applications

The basic set-up of a microwave optical systems is shown in Fig.2 with the light source, detection and control units.

Time Resolved Microwave Conductivity (TRMC)

For the quantative determination of the lifetime of excited charge carriers a pulsed light source (laser or flashlamps) has to be used in combination with a time-resolved detection of the microwave reflection (transient recorder). The data are transferred to a PC for display and evaluation. The measurement procedure including sample positioning can be completely controlled via a graphical interface.

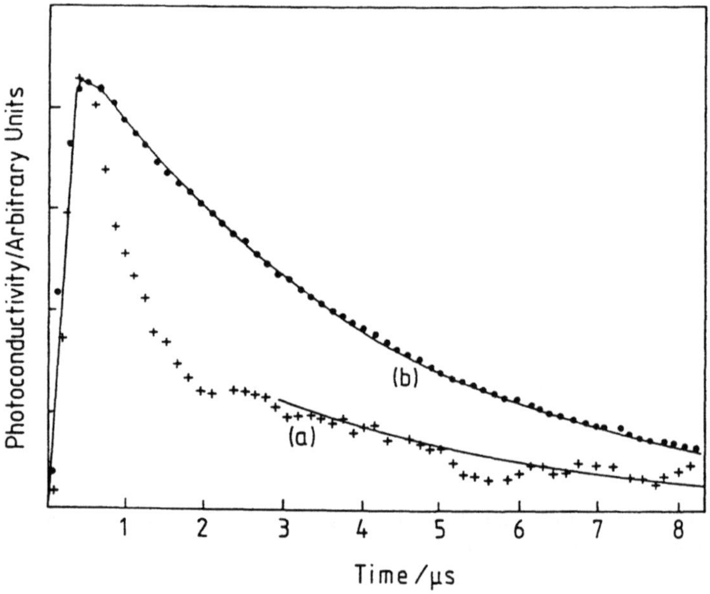

Fig.3. Photoconductivity transients of a polished Si-wafer at a) 532 nm and b) 1064 nm excitation. The solid lines are calculated with $s = 10^5$ cm/sec and $\tau_b = 4$ µsec /2/.

In Fig.2 the microwave power and the light pulse are incident from opposite sides This allows a simultaneous optical monitoring of the sample surface either by a CCD camera or by measuring the reflected laser light confocally. If one side of the sample is metallically conducting light and microwave power have to be irradiated both from the top /1/.

If the surface recombination velocity, s, of the semiconductor material is not dominating ($s < 10^4$ cm/sec) excitation with several wavelengths (i.e. 1064 nm and 532 nm of a Nd:YAG - laser) allows one to separate the recombination in the bulk ("true" minority carrier lifetime τ) from the surface recombination term. In this case the differential equation describing generation, recombination and diffusion is fitted into the measured data as shown in Fig.3. This procedure is necessary when transport of carriers across the bulk is vital for the device. By etching the sample or covering it with a passivating layer the condition of low surface recombination is generally fulfilled. These processes occur anyhow during the first steps of device fabrication. In many applications such as the control of Si/SiO_2 - interfaces the surface recombination on the other hand indicates the defects that have to be avoided /2/.

Scanning Microwave Photoconductivity (SMPC)

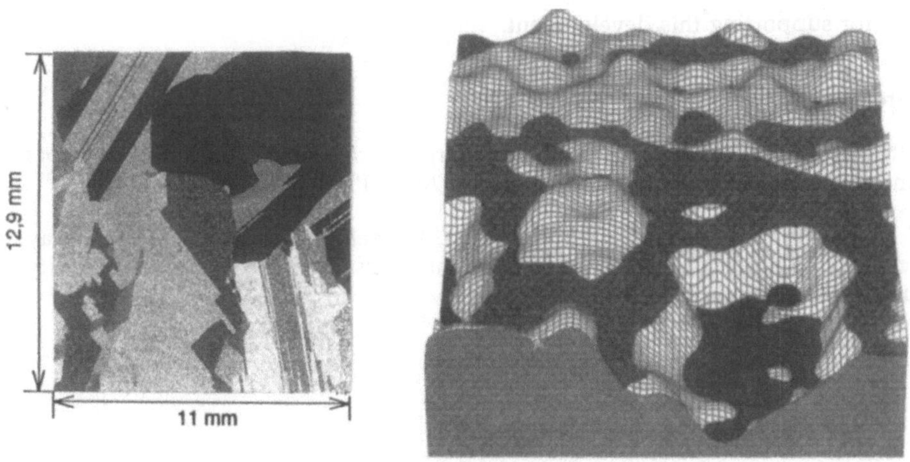

Fig.4 Distribution of diffusion lengths of a polycrystalline solar grade Si sample recorded with a SMPC -system (Phoenicon instrument "MRM"). The diffusion lengths vary between 90 and 94 µm.

In order to map the defect distribution with maximum lateral resolution the laser beam is focussed on the sample to a diffraction limited spot size (ca. 1 µm). To obtain a maximum detection sensitivity a modulated cw - laser is used for excitation.

The microwave photoconductivity signal, which is proportional to the steady state photoconductivity, $(\Delta\sigma)_s$, as measured via a lock-in amplifier. The $(\Delta\sigma)_s$ is directly related to the photoconductive decay time τ_p and thus the carrier lifetime:

$$\left(\frac{\Delta P}{P(\sigma)}\right)_s \approx (\Delta\sigma)_s = q(\mu_n+\mu_p)g\,\tau_p \quad (4)$$

The optical generation rate, g, is proportional to the light intensity and the optical absorption. For a material of homogenous optical properties the steady state signal gives the lifetime when calibrated at one point with a TRMC measurement. The semiconductor sample is scanned by an x-y-stage between the fixed laser beam and transmission line yielding a point-by-point image of the defect distribution or the diffusion length as shown in Fig.4 for solar grade polycrystalline silicon.

Acknowledgements

The authors thank Dr. Kunst, Hahn-Meitner-Institut Berlin and the BMFT (TOU 1557) for supporting this development.

References

/1/ M. Kunst and G. Beck; The study of charge carrier kinetics in semiconductors by microwave conductivity measurements. J. Appl. Phys. 60, 3558 (1986).

/2/ M. Kunst, G. Müller, R. Schmidt and H. Wetzel: Surface and Volume Decay Processes in Semiconductors Studied by Contactless Transient Photoconductivity Measurements.Appl. Phys. A 46, 77 (1988).

Electrical Characterization of III/V Materials for Heterostructure Devices

H. G. BACH

Heinrich-Hertz-Institut für Nachrichtentechnik Berlin GmbH
Einsteinufer 37, D-1000 Berlin 10, FRG

1. Summary

The quality of optoelectronic devices is highly influenced by the semiconductor material bulk properties as well as by the heterojunction interface parameters. Design requirements for heterostructure devices like lasers, photodiodes, bipolar transistors, and High Electron Mobility Transistors (HEMTs) are related to material and heterointerface properties. Current-Voltage (I-V) analyses, equivalent circuit modelling and free carrier profile simulation tools are applied to optimize the design parameters. Test structures suitable for parameter extraction are discussed.

2. Introduction

Design of heterostructure optoelectronic devices comprises a careful analysis of semiconductor bulk properties like doping, mobility and deep traps. If, in addition, effects of band discontinuities are to be exploited, the heterointerfaces involved have to be optimized and characterized in detail.

Different devices often rely on common design principles. The wide-gap emitter, for example, is applied for efficient carrier injection into an active region as is the case for heterostructure bipolar transistors /1/ as well as for double heterostructure lasers. The principles of the I-V analysis for such structures are described in section 3.

A second study (section 4) is dedicated to transmission line networks. Applications are the investigation of the increase of channel resistance in FETs near the pinch-off point in case of channel doping capacitance voltage (C-V) profiling /2/, distributed gate electrodes in high-speed (> 10 GHz) electrooptical modulators /3/ or the high-frequency current confinement in the active stripe of an oxide stripe or ridge waveguide (RW) laser.

Quantum well devices and HEMTs exploite the confinement of carriers in a two-dimensional electron gas (2DEG) existing in the narrow-gap zone at an isotype heterointerface. The suppression of substrate currents and the requirement of high transconductance call for a high effective barrier between channel and substrate in the case of the heterostructure FET (HFET). A measurement of the conduction band discontinuity and the interface charge in a HEMT structure is described in section 5.

Section 6 presents examples of additional test structures to be applied to parameter extraction in the frame work of device simulation.

3. Current injection analyses of heterostructure bipolar transistors

In bipolar devices the injection across an anisotype heterojunction boundary has to be separated into currents injected into the active region and recombination currents occuring on either side of the junction. In a bipolar transistor these current components may be separated easily using the Gummel measurement scheme depicted in the inset of fig. 1. The structure to be analyzed here is an invertible double-heterostructure bipolar transistor (DHBT) fabricated in InP comprising a quaternary base layer /4/ which is accessed by a deep Zn-diffusion. The equivalent circuit of the transistor is completed by emitter and base series resistances and an emitter-collector leakage path G_{ec} (simulating surface effects due to passivation).

Fig. 1

Gummel plot of invertible heterostructure bipolar transistor

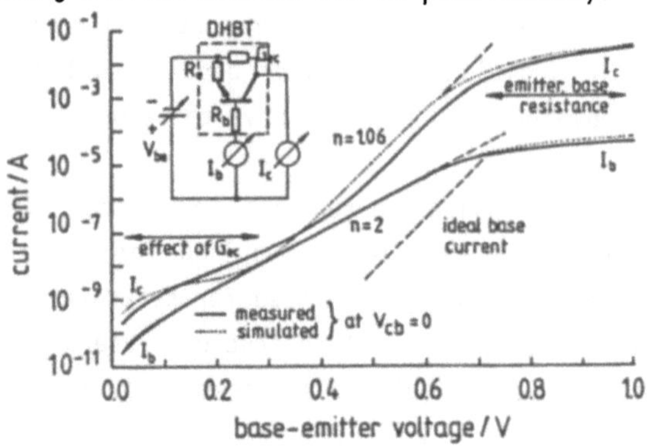

Fig. 1 shows measured and simulated (dotted line) collector and base current curves of a DHBT. The simulation is based on the well-known Gummel-Poon model. The upper curves I_c gives the transferred electron current across the base layer. The ideality factor (n = 1.06) indicates a thermal current injection mechanism for base-emitter voltages higher than 0.4 V. In the same voltage region this analysis yields a dominant recombination contribution for the base current which explains the increasing current gain vs. collector current often found in III/V HBTs. The ideal base current is shown in dashed line which indicates a very high intrinsic current gain. The high current end can be modelled by emitter and base series resistance effects. In the low collector current regime surface leakage explains the backed up collector curve, which otherwise would still decrease exponentially towards lower currents.

4. Equivalent circuit modelling of laser admittance

In this section an example of a transmission line analysis will be given by means of the frequency (and forward bias) dependent current distribution inside an oxide stripe laser the principle structure of which is given in fig. 2.

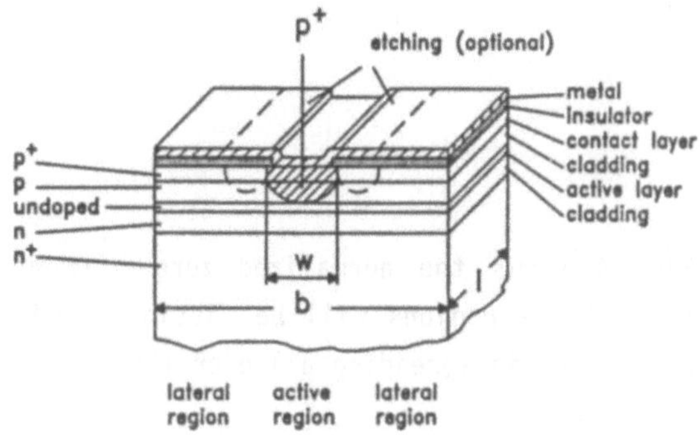

Fig. 2

Scheme of an oxide stripe laser structure

(the optional etching grooves are intended to reduce further lateral conduction)

The active region is defined laterally by the width w of the diffused p-contact. However, ac and dc current flow may penetrate into the adjacent lateral conducting p-channel regions isolated by the pin-junction between the p-cladding and the n-type buffer. The frequency dependence of this lateral current spread can be elucidated by an equivalent network using distributed elements outside the active layer as shown in fig. 3.

The capacitance and conductance of the active region are lumped together to discrete elements C_a and G_a. The p-type layer embracing the highly doped contact layer and the p-cladding layer is characterized by its mean square resistance $dR_{\blacksquare}$. The distributed pin-capacitance is represented by the elements dC_j and dG_j. The latter three distributed components form a low-pass channel. Thus the effective area of the laser structure is frequency (and bias) dependent, which affects its high-frequency operation and also the analysis of doping and deep levels in the active pn-junction by C-V methods. The equivalent circuit is completed by the elements C_{it} and R_{it}

representing rechargeable traps in the active layer and by the distributed oxide capacitance elements dC_i.

Fig. 3

Equivalent circuit for oxide stripe and ridge waveguide lasers

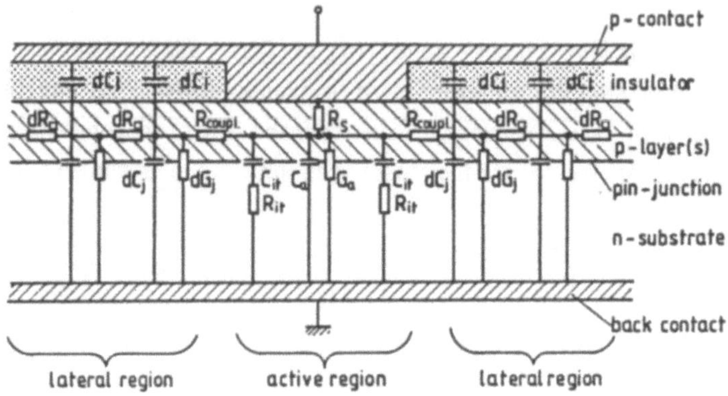

Fig. 4 shows the normalized zero bias admittance over a wide frequency span. Three regions will be distinguished for discussion. In region I the ac current is spreading all over the sample of width b and only a constant

Fig. 4

Zero bias admittance of a 1.3 µm quaternary planar oxide stripe laser on InP

crosses, circles: measured
full lines: simulated, applying network of fig. 3

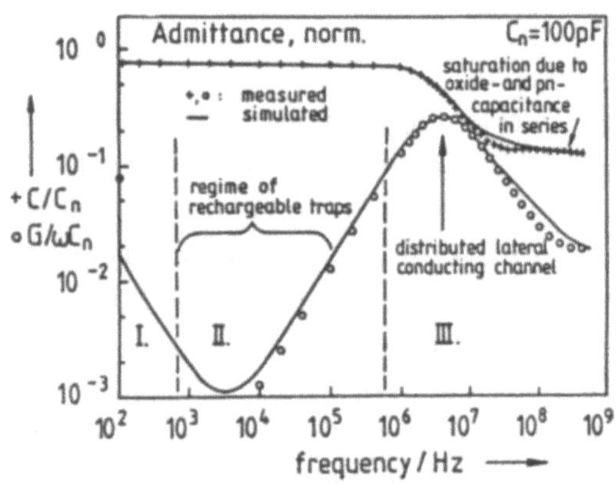

leakage pn-conductance is noted. The conductance in region II is sensitive to trap charging and discharging processes /5/, none of which can be noticed for this sample. In region III the frequency is high enough for the vertical pin-conductance to exceed the lateral p-channel conductance. Consequently the lossy lateral channel provokes an increasing normalized conductance and the total capacitance of the structure decreases. This has to be considered with care in case of doping profiling by C-V methods normally operating at 1 MHz. Towards higher frequencies the ac current is confined to the active region of width w. Due to the distributed oxide capacitance the total sample capacitance saturates above 100 MHz to a considerably higher value than defined by the geometry of the active area w•l. Thus this equivalent circuit analysis provides information about all sheet resistances and evenly distributed capacitances. If the bias is pushed into strong forward conduction, some of the elements will change due to high injection effects.

5. Band discontinuity and interface charge extraction in HEMT structures

To give an example with wider application range this section treats the characterization of the quality of the interface between an AlInAs barrier layer and an GaInAs channel layer which forms the most important part in a HEMT device. The samples investigated were grown by MBE. The AlInAs barrier layer was Si-doped ($6 \cdot 10^{17}$ cm^{-3}). An undoped AlInAs spacer of 3.4 nm separates the intentionally undoped channel from the highly doped barrier material.

The conduction band discontinuity leads to an electron redistribution at the isotype heterojunction somewhat distorted by residual (fixed) interface charges. The resulting potential well carries the channel charge which should be as high as possible. The electron distribution can be measured accurately by the C-V technique. The heterojunction properties may be extracted by Kroemer's method /6/ if the shape of the free carrier distribution indicates a sufficient barrier quality. In this work this integrating evaluation procedure was complemented by the numerical modelling of the free electron distribution /7/ to demonstrate the high junction quality achieved by a close fit of a simulation with the measurement.

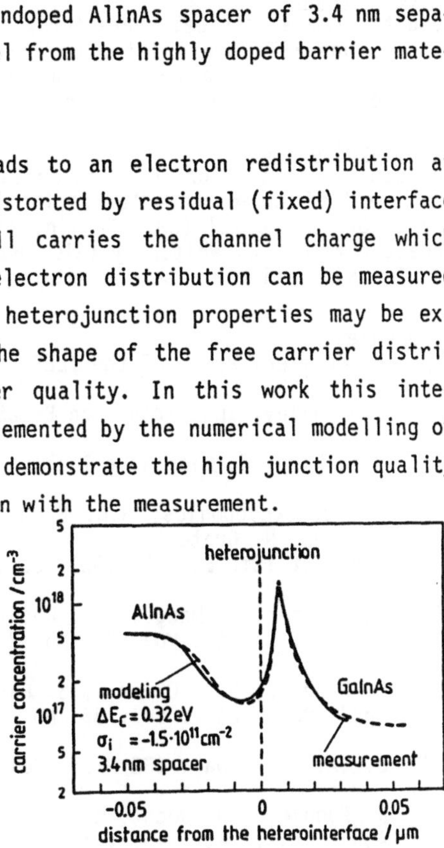

Fig. 5

Numerical fit of free electron distribution to the measured data at an AlInAs/GaInAs heterojunction with a 3.4 nm spacer

The conduction band discontinuity was evaluated to be 0.32 eV and the fixed interface charge density amounted to $-1.5 \cdot 10^{11}$ cm^{-2}. The discontinuity is somewhat lower compared to that determined in /8/ where the same method was used. A more detailed analysis will be given elsewhere.

6. Needs for extra test structures in device optimization

Besides direct device analyses such as demonstrated above, extra test structures are needed for nearly all devices for parameter extraction before device modelling. These are test structures for contact resistance measurements (transmission line model (TLM), cross-bridge Kelvin resistor (CBKR)), unconnected pads for the determination of stray capacitances, guard-ring structures for passivation leakage, separate homo- and hetero-diodes for current suppression modelling, van der Pauw structures for sheet resistances and, apart from this, fine-line structures for lithography resolution and yield assessment /9/.

7. Conclusion

Results from electrical characterization techniques applied to transistors, lasers, and FETs are related to device features and optimization. Besides device analysis, measurements carried out on test structures provide unequivocal information needed for device modelling.

8. Acknowledgement

The contribution of Dr. G. Beister (Akademie der Wissenschaften der DDR) in the field of laser characterization is greatfully acknowledged. The author thanks Dr. H. Künzel (HHI, Berlin) for supplying AlInAs MBE samples and is indepted to Dr. J. Whiteaway (STC, Harlow) for a program for numerical free carrier simulation at isotype heterojunctions.

References

/1/ H. Kroemer: Proc. of the IEEE, Vol. 70, No. 1 (1982) 13-25

/2/ J.D. Wiley: IEEE Trans. on ED-25 (1978) 1317 ff

/3/ S.Y. Wang: J. Lightwave Techn., Vol. 6 (1988) 758-771

/4/ A. Paraskevopoulos, H.G. Bach, G. Mekonnen, H. Schroeter-Janßen, F. Fiedler, N. Grote:
Proc. of 19th ESSDERC 1989, Springer Verlag, 385-388

/5/ G. Beister, J. Maege, H.G. Bach:
Sol. St. Electr. Vol. 33, No. 2 (1990) 227-233

/6/ H. Kroemer, W.Y. Chien, J.S. Harris, D.D. Edwall:
Appl. Phys. Lett. Vol. 36, No. 4 (1980) 295-297

/7/ J.E.A. Whiteaway: IEE Proc. Vol. 130, Pt. 1, No. 4 (1983) 165-170

/8/ R. People, K.W. Wecht, K. Alavi, A.Y. Cho:
Appl. Phys. Lett. 43, Vol. 1 (1983) 118-120

/9/ Proceedings of ICMTS, 13.-14. March 1989, Edinburgh, Scottland

Material Characterization – Optical Methods

Assessment of III-V Epitaxial Layers by Optical Methods

B. Sartorius

Heinrich-Hertz-Institut für Nachrichtentechnik Berlin GmbH, Berlin (West), FRG

Summary

Optical methods for measuring thickness and bandgap of epitaxial layers are presented. Procedures for nondestructive quality assessment of material and processing as well as methods for defect analysis are described. Examples demonstrate the potential for selective investigations of buried layers and for measuring in sub-micron dimensions.

Introduction

Optoelectronic devices consist frequently of sequences of epitaxial layers. The bandgaps and thicknesses of these layers are critical device parameters which must be controlled precisely. The crystalline perfection of III-V compounds and the quality of their processing are still inferior to the high standard achieved in the silicon technology. Consequently quality assessment of materials and processing are important. Optical methods are nondestructive and thus well suited for testing the actual material at any stage of processing. In this paper the versatility of optical methods for the investigation of a variety of different material parameters will be shown.

Thickness of epitaxial layers

A very important effect in optoelectronic devices is the dependence of the effective index of refraction upon the thickness of waveguiding layers. As an example we mention distributed feedback (DFB) lasers, where the Bragg wavelength depends on the corrugation period and the thickness of the waveguiding layer as well. In order to measure nondestructively the thickness of InGaAsP/InP epitaxial layers we developed a new optical method. It is based on light absorption measurements at a wavelength where the layer is absorbing and the substrate is transparent. If the absorption coefficient is known the thickness can be calculated. In our two-wavelength technique [1] the (high) reflection and scattering losses are experimentally compensated by using a reference beam at a wavelength where no absorption occurs. High thickness resolution is obtained since the intensity difference of the two beams after they have passed the sample is directly proportional to the (low) absorption. The method provides spatial resolution, it is very fast, and thus it is possible to get thickness images of epitaxial layers as shown in Fig.1.

In optoelectronic heterostructures often the thicknesses of buried layers has to be determined. By proper choice of the two wavelengths the selective investigation of a buried layer is possible. The layers in Fig.1 provide an example: They are grown one on top of the other in one run in order to give a laser heterostructure. Similar structures in both layers show the spatial matching, while inhomogeneities visible in only one layer (e.g. bowing in the lower part of the λg= 1.3 μm layer) demonstrate the selectivity of the method. The thickness of the layers is about 0.2 μm each and variations of 10 nm are clearly resolved. This demonstrates that optical methods have the potential for measuring structures in submicron dimensions.

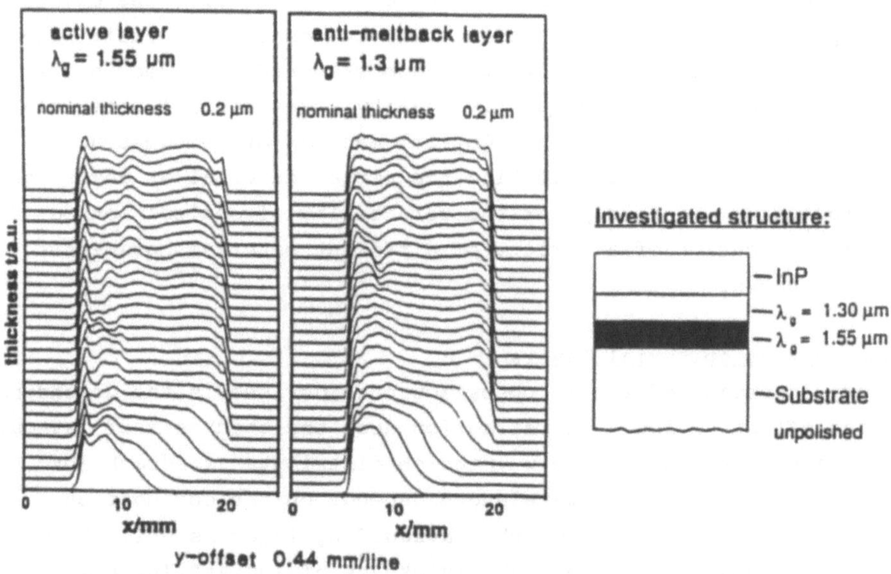

Fig.1. Thickness maps of buried InGaAsP/InP epitaxial layers (bandgap equivalent wavelengths λg= 1.3 μm and λg=1.55 μm) measured by two-wavelengths transmission.
Right: Layer composition of investigated laser heterostructure

Composition and bandgap of epitaxial layers

In the InGaAsP/InP material system the bandgap of the lattice matched epitaxial layers can be varied between 0.75 eV (1650 nm, InGaAs) and 1.35 eV (920 nm, InP) by adjusting the composition of the quaternary material. The bandgap - and by this the composition - can be determined by luminescence spectroscopy. Free carriers are generated by excitation with a laser at a photon energy above the bandgap of the layer. The carriers relax down to the band minima and the emission spectrum of the following radiative recombination processes (direct transitions in most III-V compounds) reveals the density of states and the occupation probability near the bandedge. If the ideal model holds the bandgap is 0.5 kT (k Boltzmann constant, T temperature) below the luminescence maximum. Fig.2 shows, for example, the luminescence spectrum of a layer from which we derive a bandgap of 0.807 eV. The emission wavelength of a laser made from this material is estimated to be at 0.82 eV (1.512 µm) and will shift to shorter wavelengths with increasing injection current as the Burstein effect gets more pronounced. The layer characterized in Fig.2 is a buried active layer in an actual laser heterostructure (similar to Fig.1). The buried layer can be investigated because the surrounding materials have a higher bandgap and therefore they are transparent at the luminescence wavelengths. Bandgap and composition of epitaxially grown layers are not necessarily homogeneous over a wafer. Thus spatially resolved measurements are needed, and for characterization of many growth processes luminescence spectra mapping has become standard.

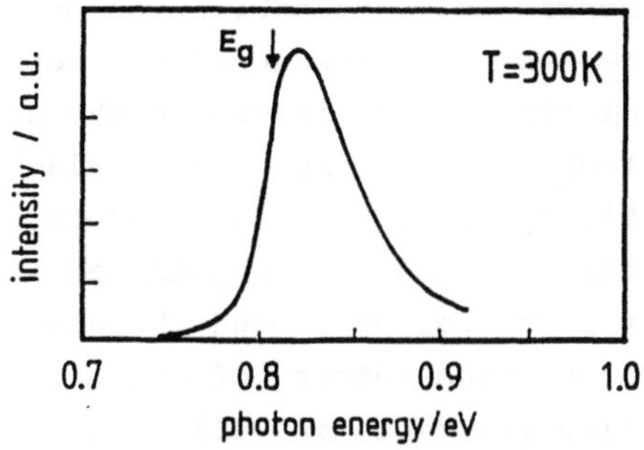

Fig.2. Luminescence spectrum of the active layer inside a laser hetero-structure (similar to Fig.1)

Quality check of materials and processing

Crystalline defects act as nonradiative recombination centres which reduce the luminescence intensity. Thus the intensity can be taken as a means to monitor the crystal quality. This enables one to check a sample after a critical process in order to detect malfunctions. Another application of luminescence intensity analysis is the optimization of process parameters. Fig.3 illustrates the investigation of the damage introduced by a dry etching process. We tried to remove the damaged material by a soft post-etching process. Surprisingly an increase of luminescence intensity was obtained only for short post etching times. The signal dropped for longer times, which we ascribed to heating of the sample by the process itself. Completing the process by wet chemical etching, however, resulted in improved crystal quality, as indicated by the high luminescence intensity. The strong increase of intensity after removing only 100 Å of material shows that this method has a better depth resolution than estimated from the penetration depth (about 5000 Å) of the exciting laser light. This is due to the high efficiency of nonradiative recombination processes. Carriers generated sufficiently close diffuse to the defect site, recombine nonradiatively, and, as a consequence, the luminescence is quenched in a large volume around the defect. Thus defects of sub-micron size (in depth as well as in width) can be detected by luminescence investigations.

The high sensitivity of luminescence can be used to identify modifications of the surface chemistry since surface states act as recombination centers. The efficiency of surface recombination depends on the density and the energetic position of the surface states, which can both be changed by chemical treat-

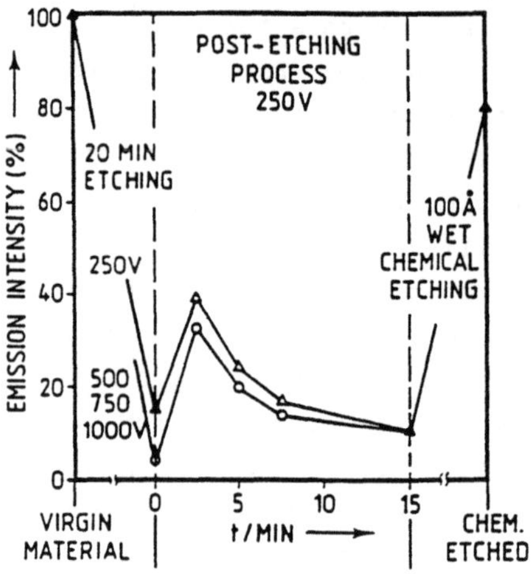

Fig.3. Luminescence intensity as a monitor for crystal damage introduced by ion beam etching with different voltages.

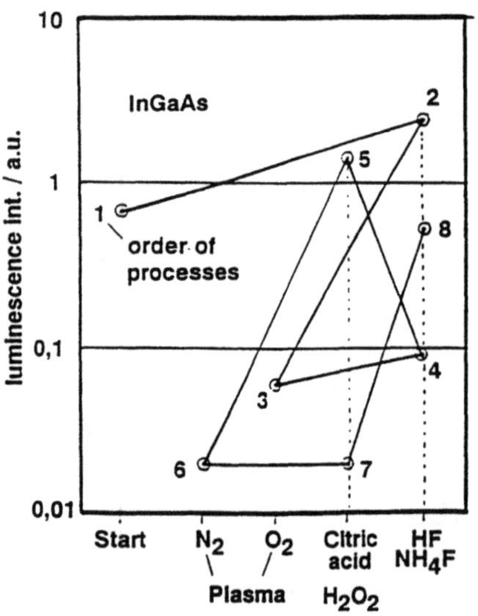

Fig.4. Modifications of the surface quality in a sequence of chemical treatments monitored by luminescence intensity

ments. Fig.4 shows the luminescence intensity of an InGaAs layer after a series of surface treatments. The intensity changes by about two orders of magnitude. The different intensities corresponding to one kind of treatment (e.g. HF/NH_4F) show that similar final treatments do not guarantee similar surface characteristics. The surface quality, however, is important: any growth and deposition start on the surface and the function of many devices is based on junction effects. With luminescence we have a tool to recognize modifications of the surface.

Analysis of type and origin of defects

Crystallographic defects exhibit a characteristic geometry in most cases. This geometry can be made visible by taking luminescence images. The characteristic shape of different defect types then allows their identification. As an example, Fig.5b shows luminescence images of an InP wafer, taken with an optical microscope [2]. All dark features correspond to defects. Dark points are caused by dislocations pointing through the surface. Straight lines correspond to a sub-surface damage generated during sawing or polishing. The jagged lines are significant for micro-cracks. It s an important advantage if various imaging methods can be superimposed. In our microscope the quasi-simultaneous observation of luminescence intensity, Nomarsky interference contrast, infra-red transmission and polarization is possible. For example, the interference-contrast images in Fig.5a show that in this case the defects are not detectable by any surface structures. In Fig.6 the comparison of different images of the identical sample section enables one to correlate defects in the substrate and in the layer. In the luminescence image (Fig.6b) of the epitaxial layer dislo-

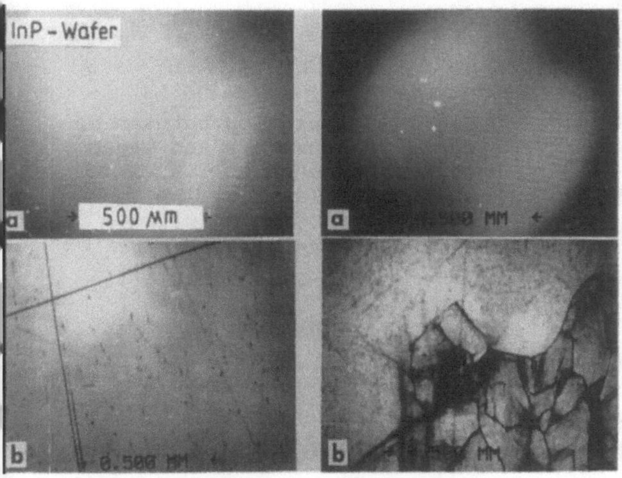

Fig.5. Images of an InP wafer
a. interference contrast: flat surface
b. luminescence (sample part identical to a): defects are visible and can be identified by their characteristic shape

cation clusters are detected. With the transmission image (Fig.6a) the existence of an inclusion in the substrate just below each cluster is revealed. This suggests that the defects are correlated. In Fig.7 the different emission wavelengths of the active layer and of the substrate are used to establish a correlation between defects (dark section) in a laser stripe and the substrate quality. Scratches in the substrate are clearly visible and can be identified as sources for the reduced luminescence in the laser stripe.

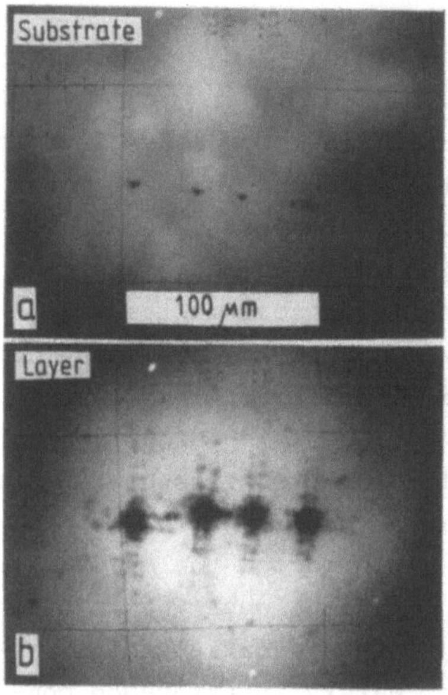

Fig.6. Correlation between defects in substrate and epitaxial InGaAsP layer.
a. IR transmission: inclusions in substrate
b. Luminescence: dislocation cluster in layer

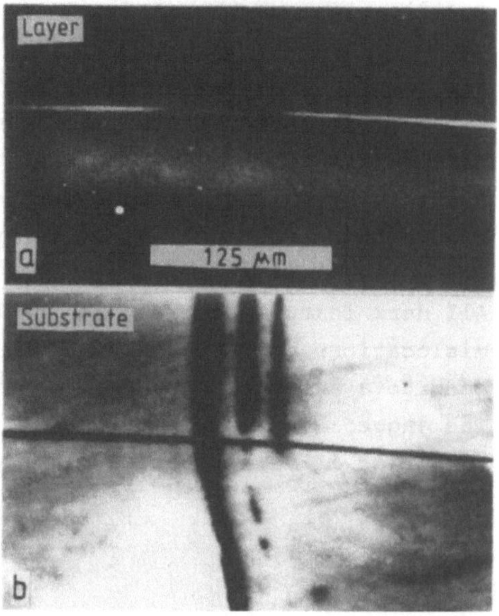

Fig.7. Correlation between defects in substrate and layer
a. Luminescence at 1.5 μm of an InGaAsP laser stripe
b. Luminescence at 0.92 μm of the InP substrate

Conclusion

Optical methods are well suited for the nondestructive and spatially resolved determination of thickness, bandgap and quality of epitaxial layers. They have the potential for selective investigations of buried layers and for measurements in sub-micron dimensions.

References

1. Sartorius, B,; Brandstättner, M.: Nondestructive thickness mapping of epitaxial InGaAsP/InP layers. Proc. "2. Int. Conf. InP and Related Materials", Denver, USA, 23.-25 April 1990.
2. Sartorius, B.; Franke, D.; Schlak, M.: Luminescence microscopy for quality control of material and processing. J. Cryst. Growth 83 (1987) 238-245.

Ellipsometry for Process Control in Reactive Ion Etching

Roland Müller

Heinrich-Hertz-Institut für Nachrichtentechnik Berlin GmbH
Einsteinufer 37
D-1000 Berlin 10 , FRG

Summary

Ellipsometry in the near infra red is shown to be a powerful tool for the control of dry etch processing of InP and its related compounds. Both, etch depth monitoring and endpoint detection can be performed with a high degree of precision even for refractive index differences as low as 0.1. Furthermore the method allows the examination of surface modifications caused by different etching chemistries and the measurement of the wafer temperature during the process.

Introduction

Reactive ion etching (RIE) using a CH_4/H_2 plasma has been demonstrated to be a well suited method for the transfer of fine patterns into In(GaAs)P layers [1]. However, the achievement of a nanometer accuracy concerning the etch depth, indispensable for the fabrication of various integrated optics and optoelectronic devices, requires an in situ monitoring of the process, because fluctuations of the etch rate are unavoidable. Furthermore, as CH_4/H_2-RIE exhibits a very low selectivity with respect to differently composed In(GaAs)P layers, an endpoint detection technique has to be employed to stop the etching precisely at heterojunctions. In this contribution an ellipsometric method is presented which fulfills both requirements. A surface contamination caused by the etching mechanism may have consequences on the subsequent processing steps. Its in situ examination yields helpful information on appropiate cleaning procedures to be applied after etching. The real wafer temperature is an important parameter for the adjustment of the rf power. In situ ellipsometry offers a simple and accurate way to measure the increase of the wafer temperature due to plasma exposure.

Preliminary Considerations

The ellipsometric angles Ψ and Δ, representing the polarisation state of the reflected light, depend on the angle of incidence α, the wavelength of the ellipsometer λ_E, the complex refractive indices n_0, n_s, n_i of the ambient medium, the substrate and the layers, respectively, and the thicknesses of the layers. Ideally the diminishing thickness of a layer leads to a periodic alteration of Ψ and Δ. For constant α and λ_E the thickness equivalent to one period depends only on the refractive index of the layer [2] and ranges from 183 nm for InGaAs to 213 nm for InP (α=76.85°; λ_E=1300 nm). In the worst case considered here (n_s=3.21 for InP, n_i=3.30 for InGaAsP with λ_G=1.06 µm) the oscillation amplitudes exceed the ellipsometric resolution limit achievable under process conditions by a factor of 20. As a sufficiently large optical response from the lower layer boundary is required, an upper limit for the measurable film thickness exists for materials which absorb at the wavelength of 1300 nm. In the case of InGaAs this amounts to 2 µm. On the other hand the InP substrate is transparent at this wavelength and reflections from the wafer backside contribute to the ellipsometric signal. Therefore the thermal expansion of the InP substrate due to heating by the plasma results in additional oscillations of Ψ and Δ, which can be exploited for the measurement of the sample temperature.

Experimental

A rotating analyser ellipsometer [3] is attached to a parallel plate RIE reactor. The light source consists of an electronically chopped semiconductor laser emitting at 1300 nm. Through collimating lenses, a quarter-wave plate and a stress-free optical window the polarized beam is focused onto the wafer with a spot size of about 0.5 mm^2. For light detection a Si-Ge diode is employed; the use of the lock-in technique avoids disturbances from the burning plasma. A computer system allows the on-line display of the ellipsometric angles and their fitting to a theoretical model previously defined by the user.

Etching of In(GaAs)P materials was performed in a CH_4/H_2 plasma with 16 vol.-% CH_4 at 3 Pa total pressure and 0.3 W/cm² rf power density; for the etching of SiO_2 a mixture of 95 % CHF_3 / 5 % H_2 by volume was used at a total pressure of 2 Pa and an rf power density of 0.3 W/cm².

Etch Depth Monitoring

The graph of Ψ vs. Δ, recorded during RIE of an InGaAsP / InP heterostructure is expected to have a closed elliptical shape because of the periodicity of both variables with respect to the layer thickness (solid curve in Fig.1). The comparison with the experimentally determined curve reveals two significant differences (cf. Fig.1): Firstly, the ellipsometric angles exhibit a rapidly evolving deviation from the virgin

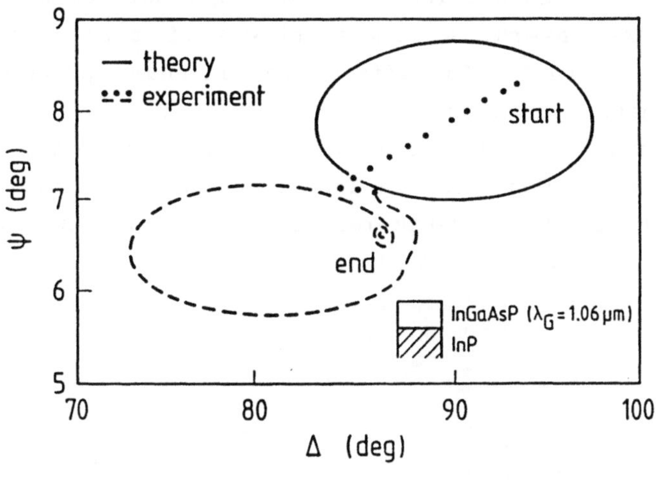

Fig.1: Theoretically expected and experimentally recorded Ψ vs. Δ graph for reactive ion etching into an InGaAsP layer on an InP substrate.

state (dotted line in Fig.1), before a substantial removal of semiconductor material has occured. This behaviour is attributed to the formation of a thin layer of surface contamination, presumedly involved in the etching mechanism. Secondly, small oscillations are superimposed on the Ψ vs. Δ ellipse in the early stage of etching and after switching off the plasma. As explained above they are generated by the thermal expansion and contraction respectively of the InP wafer. The RIE process was interrupted after completion of one period in Ψ and Δ (dashed line in Fig.1), theoretically corresponding to a thickness decrease of 207 nm. This agrees very well with an etch depth of 215 nm as determined by stylus measurements.

Endpoint Detection

The change of the refractive index, occuring when the etching front passes the boundary between differently composed In(GaAs)P layers, produces a discontinuity in the relation between Ψ and Δ. This effect enables the detection of heterojunctions during the etching as exemplarily depicted in Fig.2 for the case of InGaAs quantum wells embedded in InP. Each of the indentations in the graph corresponds to one of the InGaAs layers. A film thickness of only 1.5 nm (second indentation in Fig.2) suffices for an unambiguous recognition of the heterojunction.

Fig.3 shows the evolution of Δ in time, recorded during the etching of 280 nm SiO_2 on InP. Especially for very thin SiO_2 layers (< 30 nm) a depth resolution of 1 nm is obtained (cf. markers on the curve in Fig.3).

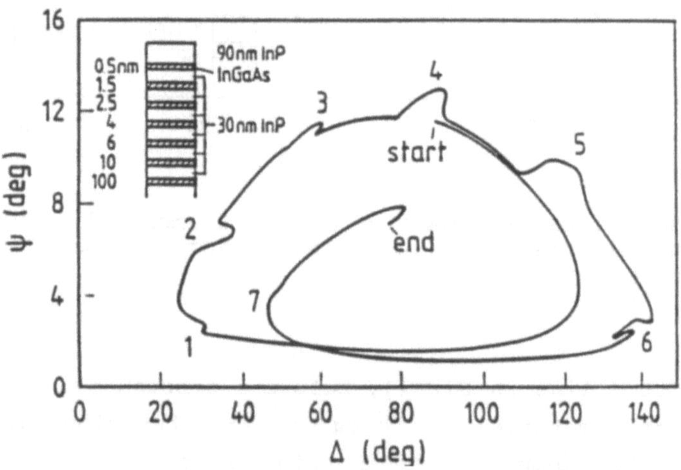

Fig.2: Ellipsometric monitoring of reactive ion etching through InGaAs quantum wells embedded in 30 nm thick InP barrier layers.

After the removal of the SiO_2 Δ increases by approximately five degrees, indicating the formation of a layer on the InP surface, whose consistency obviously reaches a steady state after about one minute of plasma exposure.

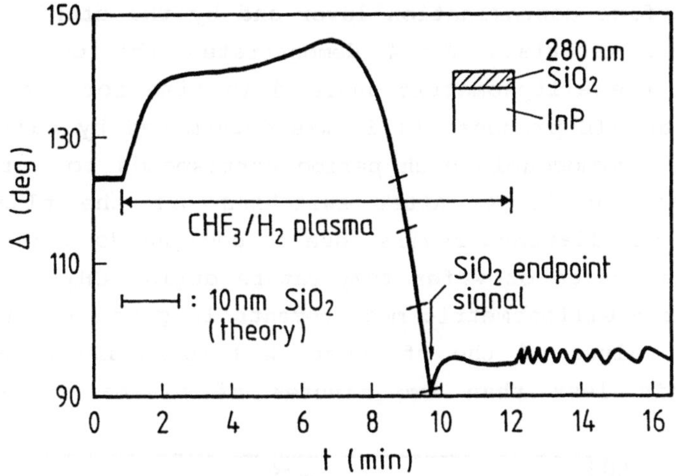

Fig.3: Evolution or Δ in time during the etching of 280 nm SiO_2 on InP

<u>Measurement of Wafer Temperature</u>

As mentioned before the InP substrate acts as an optical resonator whose thermal expansion or contraction causes oscillations of the ellipsometric angles Ψ and Δ.

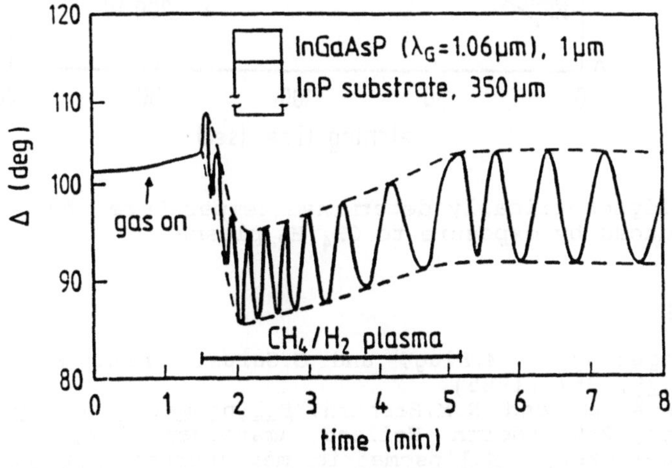

Fig.4:'Thermal' oscillations of Δ during CH_4/H_2-RIE of an InGaAsP layer grown on a backside polished InP wafer

The use of backside polished wafers leads to strongly increased amplitudes of these oscillations. Under these circumstances they are quantitatively resolvable in addition to those variations of Ψ and Δ which are caused by the formation of the surface contamination layer and by the etching of semiconductor material. Fig.4 demonstrates the overall evolution of the ellipsometric angle Δ in time for the initial stage of an RIE process. As it was determined by calibration against a thermocouple each period corresponds to a temperature change of 12 K. After switching off the plasma the 'thermal' oscillations reverse due to cooling down of the InP wafer. The change of wafer temperature during CH_4/H_2-RIE extracted from ellipsometric measurements is given in Fig.5 for different values of the rf power. A temperature rise up to 60 K within less than two minutes of plasma exposure is observed.

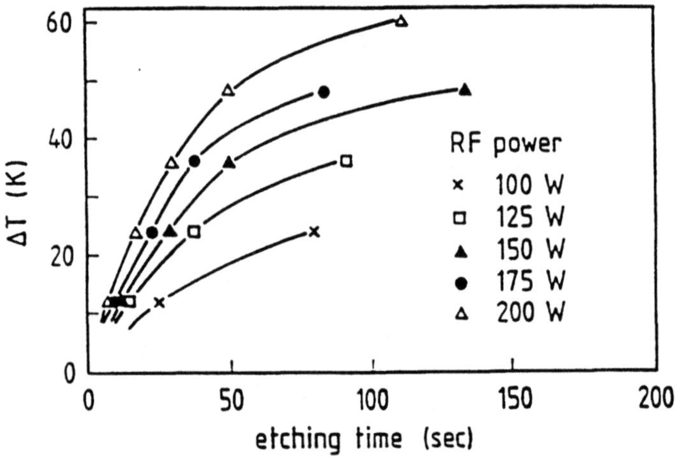

Fig.5: Ellipsometrically determined temperature change of InP wafers induced by exposure to CH_4/H_2 plasma

References

[1] U.Niggebrügge, M.Klug, and G.Garus, Inst. Phys. Conf. Ser. 79, 367 (1985)

[2] R.M.A.Azzam and N.M.Bashara,"Ellipsometry and Polarized Light", 290, (North Holland, Amsterdam 1977)

[3] H.Schwiecker, "Ellipsometric measurement for thin film technology", SPIE conference 11/12.01.1988, Los Angeles

Thin Film Ellipsometric Measurements - A Tool for Research and Production

M. TAMME, R. KAMILLI, P. PADUSCHEK, PLASMOS GmbH, Munich
M.STADTMÜLLER, Siemens AG, Munich

1. Introduction

In research and development as well as production, surface and thin film examinations are frequently needed.

Although ellipsometry has been applied in optics, semiconductor industry, chemistry and medicine for a long time, it remained the "complicated" tool for the scientist due to the complexity of the scientific interests.

Now with the availability of fast computers, the ease of use, and design of modern ellipsometers, these measurements became widely used in research as well as production.

2. Principles of Ellipsometry

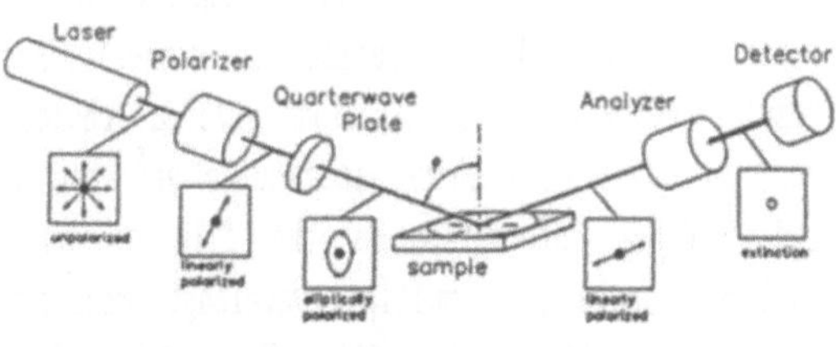

Figure 1: The main ellipsometer components

The measurement principles of ellipsometry is based upon the fact of reflection of polarized light as described in the Fresnel equations (1). Fig. 1 shows the principle.

A polarized collimated monochrome light beam passes a polarizer and a Lambda/4 plate and is incident to the sample surface at an incidence angle Phi. The reflected polarized light is analyzed in the detector unit.

Rotating Analyzer (RAE)	Nulling Setup
Very short analysis time (<1 sec)	Measurement time 20 - 30 sec
Complicated mathematics and software necessary	Only simple software needed
Only one rotating element	Two rotating elements
Complicated electronics	Simple electronics
Critical alignment	Easier alignment

Table 1: Comparison between the rotating analyzer ellipsometer (RAE) and the nulling ellipsometer principles

In the rotating analyzer principle, the analyzer in front of the detector is rotated, while the output of the detector unit is recorded with the azimuth of the analyzer. With Fourier transformation, the characteristic angles of ellipsometry, Psi and Delta, can be determined. Once Psi and Delta are established the Fresnel equations are to be used. In a different arrangement of ellipsometry called the nulling principle both polarizer and analyzer must be rotated to minimize the readout at the detector unit. At least two measurements are made to eliminate optics imperfections. In tab. 1, both principles are compared (1,2,3). The rotating analyzer arrangement used by PLASMOS is much faster in measurements, while the nulling principle requires less quality electronics and optic components. Measurement accuracies are about the same. In fig. 2, the Psi and Delta curves of an ellipsometric arrangement (silicon substrate, incidence angle 70°, wavelength 632,8 nm) are shown. The curves show lines of constant refractive index; the film thicknesses increase from the origin of the curves at Psi = 12°, Delta = 180° counterclockwise. The curves are repeated (nonabsorbing film) after one period (revolution of curves). As a result the Fresnel calculations give a choice of very accurate film thickness results and refractive index. Other information on the film or multiple incidence angle like multiple wavelength measurements give the exact results (1). Thus, substrates as well as thin films or multiple layers can be evaluated.

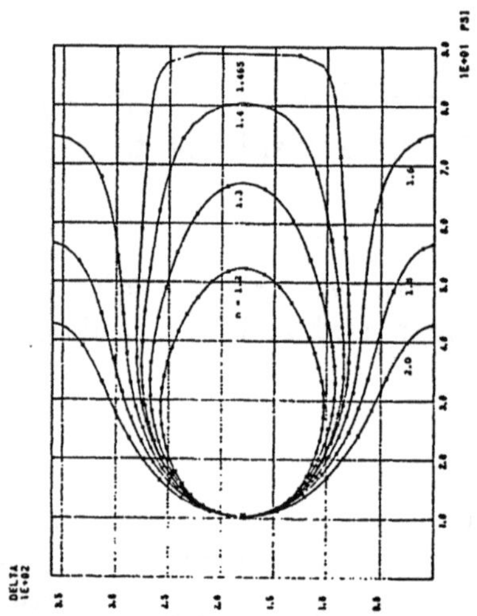

Figure 2: The ellipsometer curves for silicon substrate: Psi and Delta vs. refractive index and thickness of transparent film on silicon substrates. Lambda = 632.8 nm and Phi = 70°

	Ellipsometer	Interferometer
Measurement Speed	< 1 sec	4 - 10 sec
Refractive index	yes	inaccurate
Accuracy	< 0.1 nm	> 1 %
Microspot diameter	10 - 30 μ	2 - 20 μ
Multiple layers	yes	no
Absorbing layers	yes	no
Reference needed	no	yes
Measurement range	0 nm - 10 μm	> 30 nm - 20 μm
Thin film evaluation	yes	no

Table 2: Comparison between ellipsometry and interferometry

3. Comparison Ellipsometry - Interferometry

The most challenging competitor to ellipsometry is interferometry amongst stylus measurements, IR-transmission techniques or reflectrometry. A comparison of interferometry to ellipsometry is shown in table 2 (1,3).

As can be seen from above table, ellipsometry is superior to interferometry with respect to measurement speed, accuracy, multiple layer capability, thin film evaluations, absorbing films, and refractive index. The only drawback is μ spot size. Especially the lack of interferometers for refractive index determinations often leads to vast erroneous results.

4. Precision of Ellipsometry

As shown in chapter 3 and table 1, ellipsometry is very precise. Precision though depends on various parameters such as incidence angle of light beam, wavelength and of course optical alignments.

Calculations of the errors introduced by an average aligned ellipsometric instrument (1) show that the refractive index can be well determined down to 25 nm to the period. (For measurements nearer the period a different incidence angle or wavelength must be used.) The accuracy of film thickness is very high across the measurement range even for very thin films below 5 nm at fixed refractive index. Thus also natural oxides can be evaluated, although incidence angles and wavelengths must be carefully selected and established.

5. Evaluation of Very Thin Films

In semiconductor industry especially the thickness determination for very thin films below 100

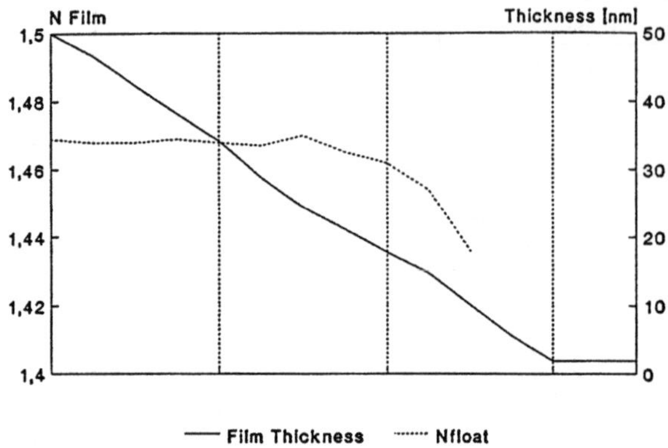

Figure 3: Wedged etched SiO_2 film 50 nm -> 2 nm with refractive index measurement

A becomes more and more important. With ellipsometry it is possible to determine these thicknesses at measurement spot sizes of 10 to 30 µ very accurately. This is shown in fig. 3. A thermal oxide on a silicon wafer was linearly etched down to the substrate. The linearity can be controlled very accurately. Shown is the result profile of such a wedged film together with the also determined refractive index. As can be seen the profile is very linear - the refractive index stays constant down to 20 nm.

6. In Situ Ellipsometry

Due to their short measurement time rotating analyzer ellipsometers can also be used for in situ thin film evaluation (4). Therefore the ellipsometric heads are attached to the specific process chamber via optical windows. Depending on growth/etch rates thicknesses can be measured (endpoint) with accuracies in the

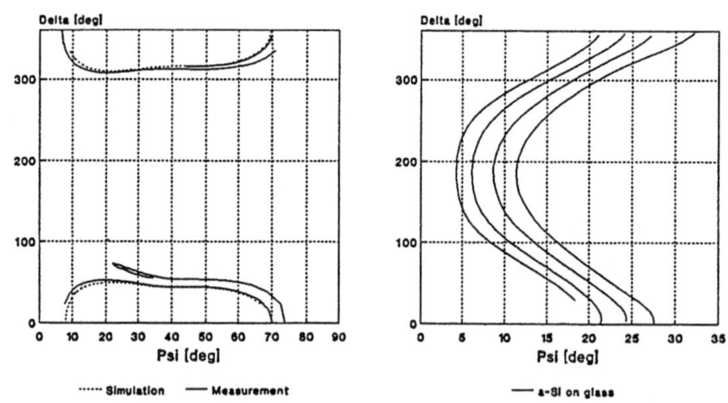

Figure 4: a) Theoretical Psi/Delta curves of an etched SiO_2 film on silicon b) growth of a-Si on glass substate

range of Angstroms. An example for Psi/Delta curves calculated theoretically and measured in situ during a plasma etch/growth process is shown in fig. 4a. While the thickness etch rate should be constant down to the substrate other phenomena such as plasma polymerization, sputtering effects, or surface roughening occur. The solid line shows the theoretical simulation of the etch process. Fig. 4b shows Psi/Delta curves of an in situ growth of amorphous silicon on a glass substrate. With the runaway in Psi the absorption constant and refractive index and thickness can be determined.

7. µ-Spot Ellipsometry

Typically the measurement beam diameter for ellipsometric measurements are 0.5 to 1 mm. However, a well designed projection optics unit can be used to focus the beam onto the substrate down to 10 - 30 µm (elliptic due to non normal incidence angles). With such a µ-spot lateral highresolution in circuit measurements can be made. Fig. 5 shows a surface scan across a USAF 1951 test target consisting of chromium etched windows on top of a glass substrate. The inverted number 6 clearly identifies the group number 6 with 10 µm wide windows.

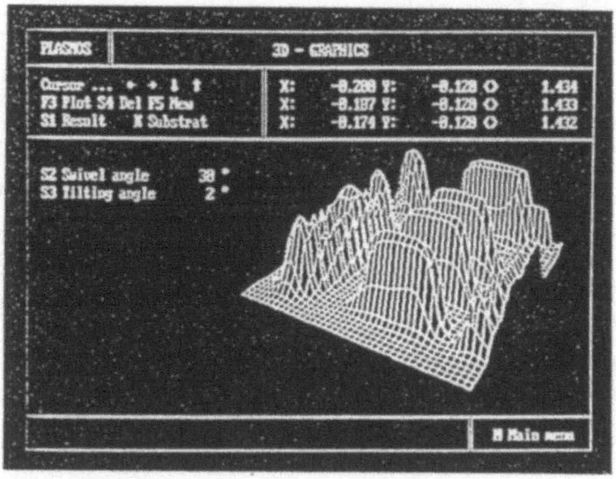

Figure 5: Mapping across USAF target structure

8. Polysilicon on Oxide Measurements

In this example a polysilicon film on oxide on silicon was examined. Fig. 6 shows the Psi-, Delta-results of a surface scan across a 200 mm Ø wafer together with theoretical curves. With the measurement data we determined thickness of oxide, thickness of polysilicon as well as absorption constant of polysilicon. The next important result however was the fact that a 20 nm thick SiO_2 on top of the polysilicon was present.

9. Discussion and Conclusion

Ellipsometry is the only non destroying measurement technique for thin film evaluation for research and development and production.

Very thin films can be precisely valuated by ellipsometry. Micro-beams for measurement areas down to 10 x 30 µm are possible. With theoretical simulations multiple films can be evaluated; however enhanced software is needed for instant determinations of multilayers or absorbing films. In situ thickness and refractive index measurements can be used as endpoint detection or growth mechanism evaluations. Implantation doses can be measured as a function of the ellipsometric angle Psi and Delta (2). The multiple choice thickness problem can be solved by multiple angle incidence measurements or multiple wavelength measurements. Selection of measurement wavelength allows determination of IR-material films (highly absorbing films in visible region).

Due to its fast measurement cycles ellipsometry can be reliably used in production for qualifying thin films or optical materials.

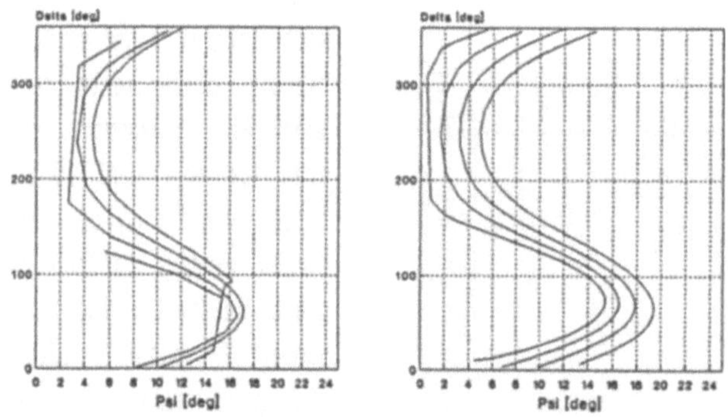

Figure 6: Psi/Delta curves for polysilicon on oxide/c-silicon a) measurement across 200 mm wafer, b) theoretical approximation

References

(1) R. M. A Azzam and N. M. Bashara: "Ellipsometry and Polarized Light", North Holland, personal library
(2) R. Kallweit, J. P. Biersack: "Radiation Effects", REDS, 1990, to be published
(3) P. Paduschek, R. Kamilli: "Ellipsometry for Thin Film Characterization", MMT, July 1988
(4) M. Tamme: "In Situ Ellipsometry in PECVD Production", diploma thesis, 1990

High Precision Non-Contacting Metrology Based on Short Coherence Interferometry

R. JONES, M. S. HAZELL, R. WELFORD

Cambridge Consultants Ltd,
Optics Group,
Science Park, Cambridge CB4 4DW

Summary
New sensing techniques based on short coherence interferometry are demonstrated to form the generic basis for a range of absolute, passive sensors.

1. Introduction

This paper describes new measurement techniques which rely upon short coherence source interferometry. These methods provide the basis for various configuration of passive, non-contacting sensors capable of the measurement of either the absolute co-ordinate of a surface or phase changes in a phase object. The measurement techniques have been developed with particular reference to the requirements that the sensor be compact and have:-

* High dynamic range (10^5 to 10^6) and repeatability. (Typical operating ranges 10 - 10^2μm, 1 - 10cm)
* line of sight operation
* immunity to the perturbation of the transmission medium (either free space or fibre optic)
* the ability to operate on both specular and scattering (ie. non-cooperative) surfaces

The decision to use interferometry based on sources having short coherence lengths as a basis for the measurement technique emerged from a detailed consideration of all of the above requirements. In this class of interferometry the output fringe field is modulated as a result of the source coherence (spectral) characteristics in a way which enables absolute data relating to the change in measureand to be determined. Such a modulation will be negligible in conventional interferometry based on a source with long coherence (ie. narrow spectral width).

Specific features of short coherence source interferometry may be exploited in a way which has a significant impact on the design and performance characteristics of the sensor. These are described in detail below.

2. Principles of operation and system design

The sensing system relies for its operation on:
(a) Coherence modulated interferometry
and
(b) Coherence discrimination

The principle of the above may be understood from Figures 1 and 2 respectively. Coherence modulation of the fringe field at the output of the Michelson interferometer (Figure 1) derives from the coherence function of the source. The latter is a function of the absolute path difference (l_1 - l_2) in the interferometer. This thereby enables the absolute fringe order number to be determined from the fringe envelope modulation provided that the differences in visibility between adjacent fringes: e.g. V_A, V_B is resolveable above system noise.

Coherence discrimination (Figure 2) enables the interference between the reference and measurement arms in the remote sensor to be detected at a transceiver mounted tracking interferometer. Coherence disrimination is achieved by adjusting the relative path lengths of the beams to satisfy the condition:-

$$| (a+c) - (b+d) | << l_c \qquad 1(a)$$

$$| (a+d) - (b+c) | >> l_c \qquad 1(b)$$

An important feature of coherence discrimination is that it allows the remote sensor to be connected to a transceiver by a common path optical link in which the interfering beams are separated only in the time domain. Such a link (either free space or fibre optic) is inherently immune to random perturbations. It is important to note also the sensor need not have the geometry of a conventional Michelson interferometer. Figure 3(a) and (b) show alternative geometrics of robust common path probes which may be engineered into compact, passive devices. An interesting feature of the tracking interferometer shown in Figure 3(a) is the geometry that generates the nomalised difference V_{12} between fringe contrasts.
Where

$$V_{12} = \frac{V_1 - V_2}{V_1 + V_2} \qquad - (2)$$

and

V_1 = the visibility of fringes in channel 1 of interferometer.

V_2 = visibility of fringes in channel 2 of interferometer.

V_{12} may be measured as a mean value derived from a set of samples at the fringe field and is extremely robust to differential variation in light level and component reflectivity.

3. Experimental Results

Experiments have been conducted using a laser diode source having the coherence characteristics shown in Figure 4. The coherence length, $l_c \sim 100\mu m$. The normalised differential fringe contrast, V_{12} observed over the fringe fields for an increase in fringe order of unity is shown in Figure 5. The change in V_{12} is of the order 50 times the system RMS noise level and demonstrate clearly that fringe order may be determined from the normalised absolute parameter V_{12}. Figure 6 shows the absolute parameter $\langle V_{12} \rangle$ plotted as a function of phase.

4. Discussion

The experimental results demonstrate that robust, absolute phase measurement using coherence modulated interferometry is feasible. This is based on a new configuartion of tracking interferometer which enables differential fringe constraint to be measured and requires phase modulation over only a few multiples of 2π. Current work is centred on the harmonisation of absolute integral fringe and sub-fringe measurement with the aim of realising maximum dynamic range. In parallel with this activity the geometry of various probes is being investigated with a view to the verification of their application in:-

Metrology: (e.g. surface profile and finish measurement)

Aerospace: (e.g. engine condition monitoring at high temperatures)

Healthcare: (e.g. en-vivo pressure measurement)

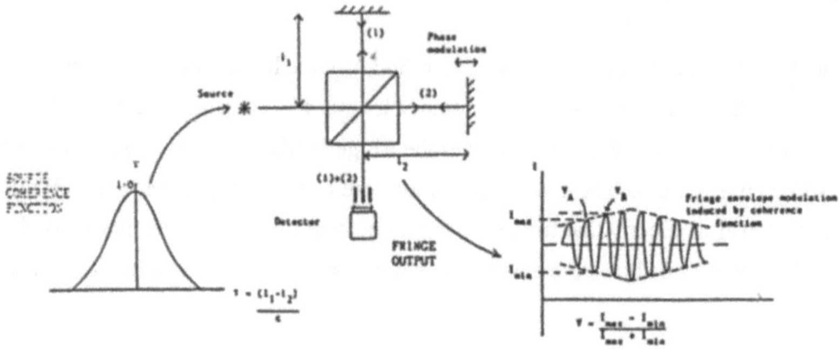

Coherence modulated interferometry

FIGURE 1

CAMBRIDGE CONSULTANTS

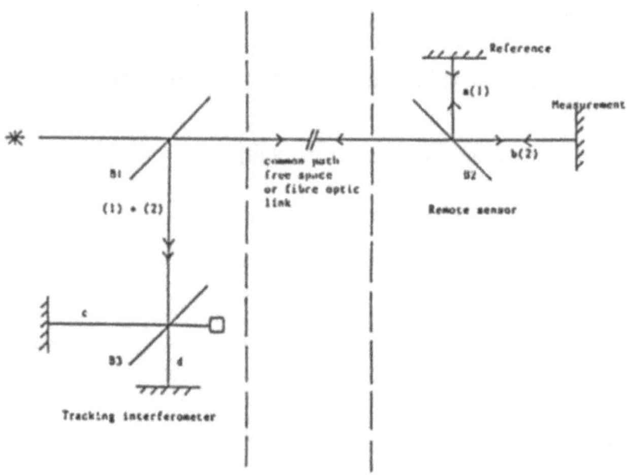

Coherence discrimination

FIGURE 2

CAMBRIDGE CONSULTANTS

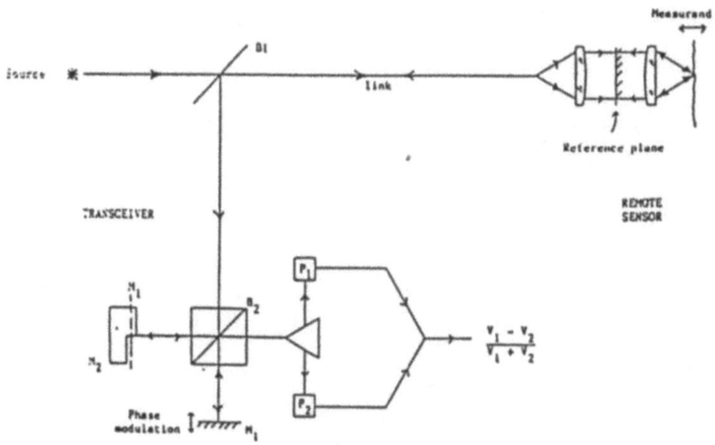

Practical system: combines normalised coherence modulated interferometry and discrimination

FIGURE 3.a

CAMBRIDGE CONSULTANTS

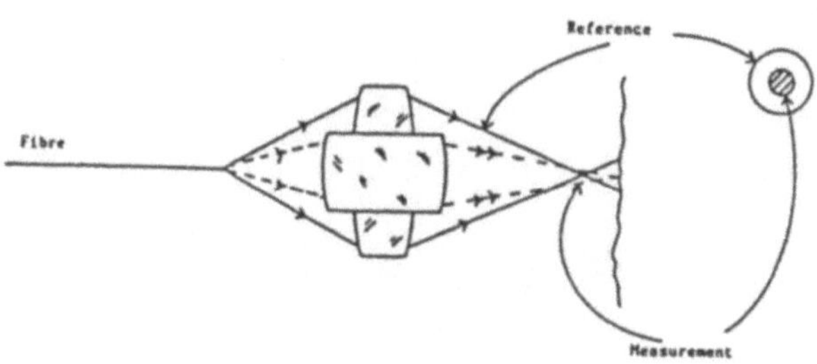

Alternative probe geometry for surface finish/profile measurement

FIGURE 3.b

CAMBRIDGE CONSULTANTS

Source Coherence Function

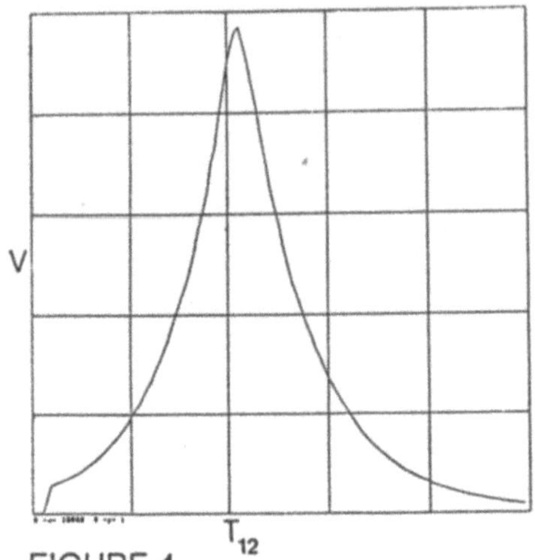

FIGURE 4

CAMBRIDGE CONSULTANTS

Normalised differential fringe contrast V_{12}
as a function at absolute fringe order n

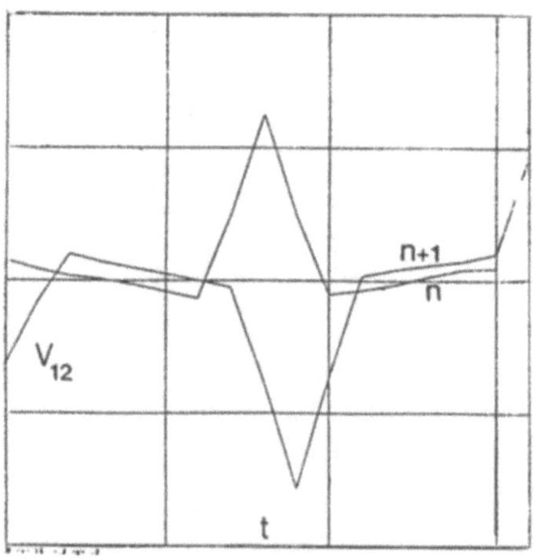

FIGURE 5

CAMBRIDGE CONSULTANTS

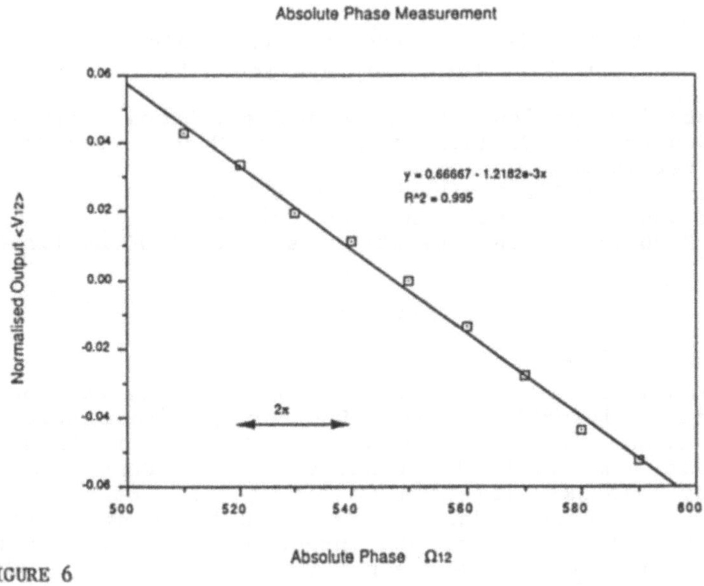

Absolute Phase Measurement
y = 0.66667 - 1.2182e-3x
R^2 = 0.995
2π
Normalised Output <V12>
Absolute Phase Ω12
0.06
0.04
0.02
0.00
-0.02
-0.04
-0.06
500
520
540
560
580
600

FIGURE 6

VIS-IR-Ellipsometry for Control and Characterization of Thin Film Processes

B. D. Doi[1], K. Hammann[2], A. Röseler[3]
U. Schneider[2], H. Schwiecker[4].

1. Institute for Physics, Nghia Do Tu Liem, Hanoi Vietnam
2. Firma Dünnschicht-Technik GmbH, Gustav Meyer Allee 25 D-1000 Berlin 653
3. Zentralinstitut für Optik undSpektroskopie AdW der DDR ,1199 Berlin, Rudower Chaussee 6
4. Techn. Fachhochschule, Luxemburgerstr.10, 1000 Berlin 65

ABSTRACT

IR-ellipsometry (4000-400 cm-1) is realized by the coupling of a photometric ellipsometer with a Fourier transform spectrometer. Two typical results in characterization of bulk and film material are given. In next part the results in measurement of surface roughness by VIS-IR-Ellispsometry using the Effective Medium Theory and a roughness model with a variable volume fraction are presented. In last part a computer program for model investigation at magnetooptic layer systems is demonstrated. By this magnetooptic contants could be determin from ellipsometric measurements .

I. SPECTROSCOPIC IR-ELLIPSOMETRY BY MEANS OF FOURIER TRANSFORM SPECTROSCOPY

In the ellipsometry a reflection is made with polarized light /1/. The complex ratio of reflection coefficients in the direction parallel and perpendicular to the plane of incidence Rp/Rs respectively, is measured:

$$\rho = \frac{Rp}{Rs} = \tan\Psi \, e^{i\Delta}$$

From these ellipsometric parameters PSI (Ψ),DELTA (Δ) the optical properties of the sample can be calculated. For the measurement a combination of a FTS with a photometric ellipsometer is used (Fig.1)/2/. The polarizer P1 in front of the sample is changed in steps of 45° (α_1 = 0°, 45°, 90° and in the polarimetric mode 135°). The observed intensity is the second harmonic of α_1 with the Stokes parameters as the Fourier-coefficients.

The fourth Stokes parameter S3 is measured by a retarder, which introduces an additionally known phase-difference. In this polarimetric mode the unambiguous determination of DELTA in the range of o° < DELTA < 360° is possible and the total degree of polarization P or the degree of phase-polarization Pph of the light after the reflection at the sample can be calculated :

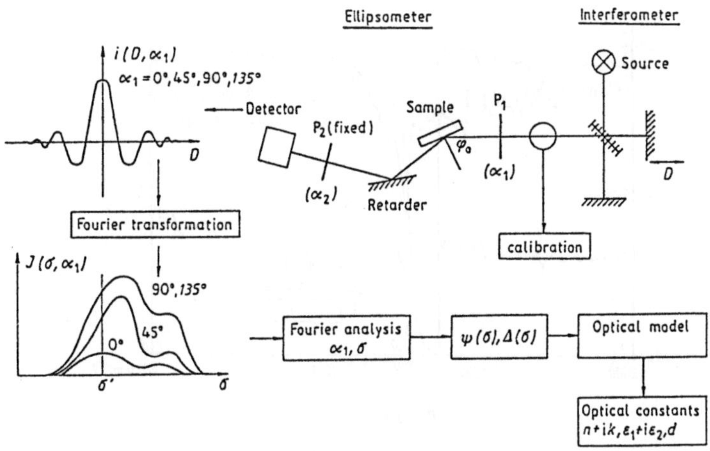

Fig.1 Shema of the spectroscopic IR-Ellipsometer

$$P = \frac{\left(S_1^2 + S_2^2 + S_3^2\right)^{1/2}}{S_0} \quad ; \quad P_{Ph} = \left(\left(\overline{\cos\Delta}\right)^2 + \left(\overline{\sin\Delta}\right)^2\right)^{1/2} = \left(\frac{S_2^2 + S_3^2}{S_0^2 - S_1^2}\right)^{1/2}$$

P and Pph is in many cases less than the unit and shows a pronounced spectral dependence. In the investigation of semiconducting samples the reflection of the rough backside can be eliminated by using measured Pph with a suitable computing and correction procedure. The polarizing properties of the interferometer are regarded by calibration measurements. The optical constants ε or n are calculated for a non-layerd surface by the relation of the metal optics:

$$\varepsilon_1 = n^2 - k^2 = (n_0 \sin\varphi_0)^2 \left(1 + \frac{\tan^2\varphi_0 (\cos^2 2\psi - \sin^2 2\psi \sin^2\Delta)}{(1 + \sin 2\psi \cos\Delta)^2}\right),$$

$$\varepsilon_2 = 2nk = (\sin\varphi_0 \tan\varphi_0)^2 \frac{n_0^2 \sin 4\psi \sin\Delta}{(1 + \sin 2\psi \cos\Delta)^2},$$

$$\hat{\varepsilon} = \varepsilon_1 + i\varepsilon_2, \qquad \hat{\varepsilon} = \hat{n}^2, \qquad \hat{n} = n + ik.$$

PHI = angle of incidence, n_0 = refractive index of the incident medium. An example is n of SiO2-glass in Fig.2 .

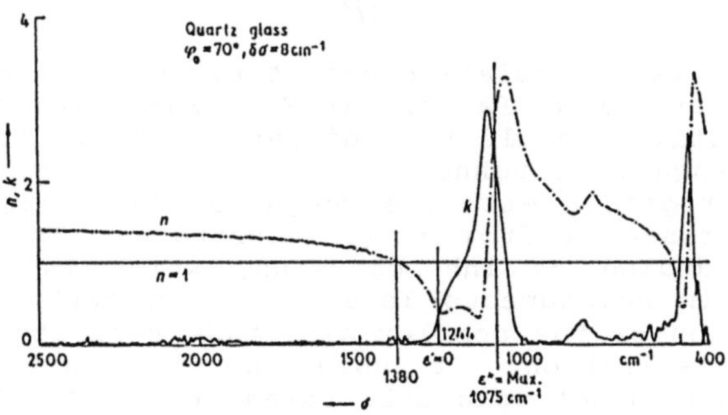

Fig.2

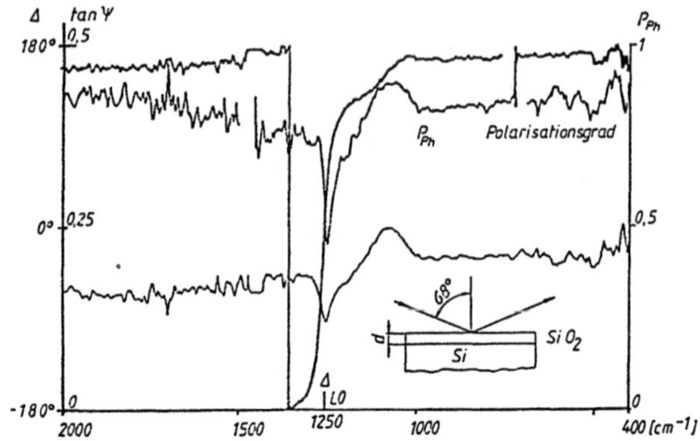

Fig.3

The optical properties of a film on a substrate are computed by a suitable iteration procedure. An example of a film-substrate combinatio is given in Fig.3, which shows a SiO2-film at a Si substrate. The degree of phase-polarization is represented too.

II. MEASUREMENT OF SURFACE ROUGHNESS BY VIS-IR-ELLIPSOMETRY

Assumed surface roughness σ is small compared to the measuring wavelenght and has a Gaussian distribution, Effective Medium Theory (EMT) can be used to interpret ellipsometric parameter from rough substrates. The roughness is to represent by an effective surface layer with a thickness corresponding to roughness Ra (average of valley to peak, Ra= $2\sqrt{2}\sigma$) and an effectiv complex refractive index Ne which is given by /6/ :

$$\frac{Ne^2 - Na^2}{Ne^2 + 2Na^2} = q\,\frac{Ns^2 - Na^2}{2Ns^2 + Na^2}$$

Na is the refractive index of the medium, Ns that of the substrate and q the volume fraction. For better correlation with the real surface a roughness model is used with the volume fraction q described by the equation :

$$q = \left(\frac{x}{Ra}\right)^{æ}$$

Where x/Ra is the relative height of the roughness and æ is the exponential fraction factor (EFF). Substrates with different roughness profiles should have different EFF (for example æ = 1.0 for the triangular roughness profil). The exponential fraction factor æ is together with the roughness Ra an important parameter for characterisation from rough substrates. The effective layer has to be handled as inhomogeneous, which is substituted by a number of thin homogeneous layers. A computer program is written for calculation of the complex amplitude reflectance Rp and Rs for P- and S-polarization. It used the 4x4 matrix method to do evaluations on anisotropic substrates too. The ellipsometric datas PSI and DELTA can be calculated from Rp and Rs

Fig.4 shows the calculated volume fraction q versus relative ight of roughness x/RA with æ as parameter. Figs.5 shows the lculated PSI-DELTA curve with variating roughness Ra (æ as rameter) for Gold substrate.. It is seen that the ellipsometric rameter PSI and DELTA are quite sensitive to the roughness Ra d to the exponential fraction factor æ

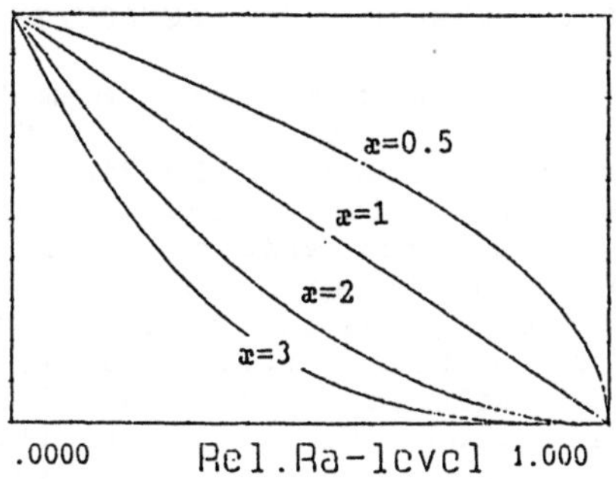

Fig.4

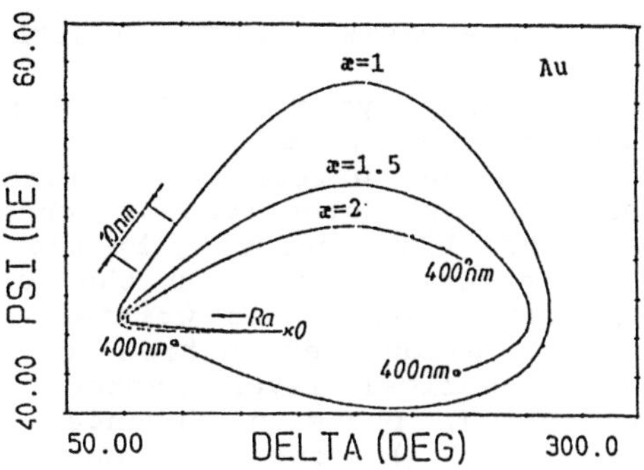

Fig.5

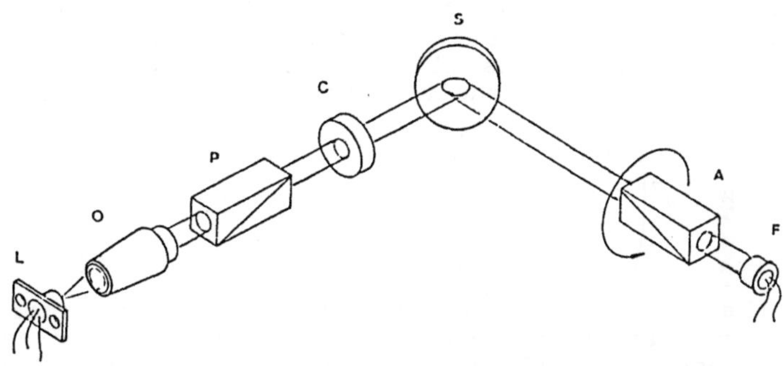

Fig.6 Optical arrangement of VIS-IR-Ellipsometer RAE

PSI and DELTA of rough substrate are measured by a Rotating ıalysator Ellipsometer (RAE) in Fig.6 /4/. The rouhgness Ra and ıe exponential fraction factor æ (or the refractive index Ns of ıbstrate) are determined from measured PSI and DELTA by a ıitable iteration procedure. Results for measurement on a Glas ıd a Gold substrate are the examples :

	Au	Bk7
æ	0.923	-
Ns	-	1.489
Ra (ellips)	980 ± 35 nm	18 ± 1 nm
Ra (stylus)	1000 nm	20 nm

For comparasion the roughness Ra measured by a stylus nstrument are also given.

The roughness measured by the ellipsometer correlate well with hose measured by the stylus instrument on different sustrate aterials. The minimal detectable roughness, limited by the ncertainty in measurement of PSI and DELTA, is about 1 nm.

The measurable range of roughness Ra is variable by the change of the measuring wavelenght Lamda (semicondutor laser 670, 780, 1300, 1500nm) and is about 600 nm for Lamda = 1300 nm. The measurement is fast, non destructive, non contac and could be an excellent tool for quality control or in-situ verification of processes for production of optic and electronic components.

III. ELLIPSOMETRIC MODEL INVESTIGATION AT MAGNETOOPTIC LAYERSYTEM

Thin film systems with incorporated magnetooptic layers seem to hold the most promise in the aera of optical data storage . The data can be written in a thermo-magnetic process. Reading is possible by detection of change of the polarization in the reflected light occuring by the magnetooptic Kerr effect. Based on the scattering theory on the magnetooptic multilayer structure an extensive program was written using the 4x4 matrix method. The program allows evaluation of all the optical responses from magneto-optical layer systems (i.e. reflection, transmission, absorption, phase difference, Kerr rotation, Kerr ellipticity, PSI, DELTA) depending on the system parameters (i.e.layer thickness, refractive index, magneto-optical constants, birefringence) and on the mode parameters (polarization, angle of incidence, wavelength). An additional subroutine provides the calculation of field intensity and absorption in the layers of the system. The program handles all anisotropic active layer materials which can be described by the following dielectric tensor :

$$\hat{\varepsilon} = N^2 \begin{pmatrix} 1 & -iQ & 0 \\ iQ & 1 & 0 \\ 0 & 0 & 1 \end{pmatrix} \quad \text{or} \quad \hat{\varepsilon} = \begin{pmatrix} \varepsilon x & -i\varepsilon a & 0 \\ i\varepsilon a & \varepsilon y & 0 \\ 0 & 0 & \varepsilon z \end{pmatrix}$$

With the aid of the computer program various model investigations have been done at magneto-optic layer systems. For example we show some calculations for study of the Kerr rotation and Kerr ellipticity in dependance on the other parameters for the well known quadrilayer structure outlined in Fig.7 /8/.

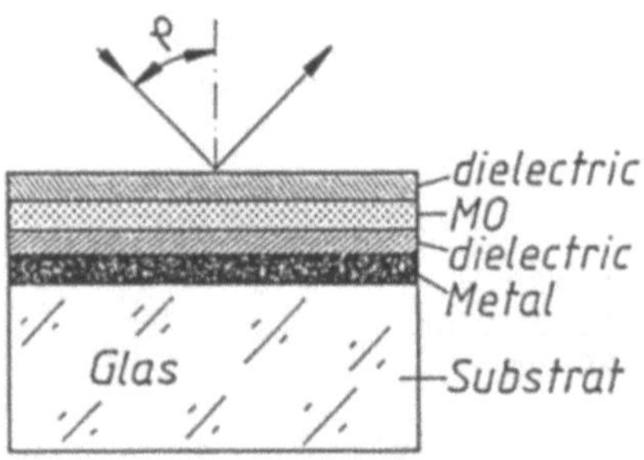

Fig.7

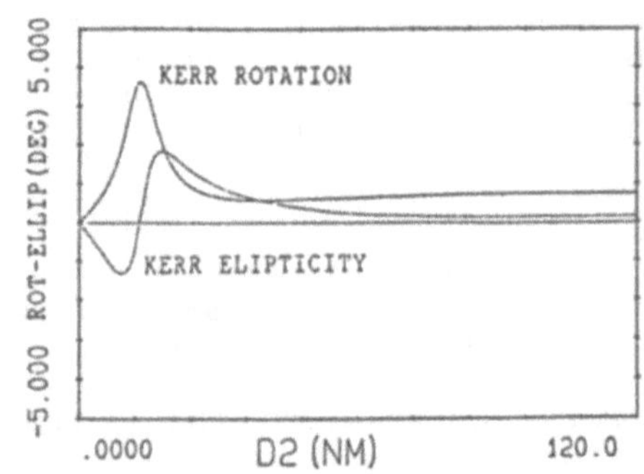

Fig.8

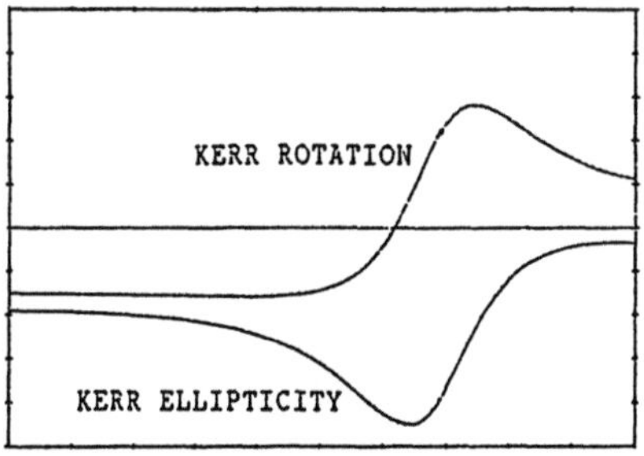

Fig.9

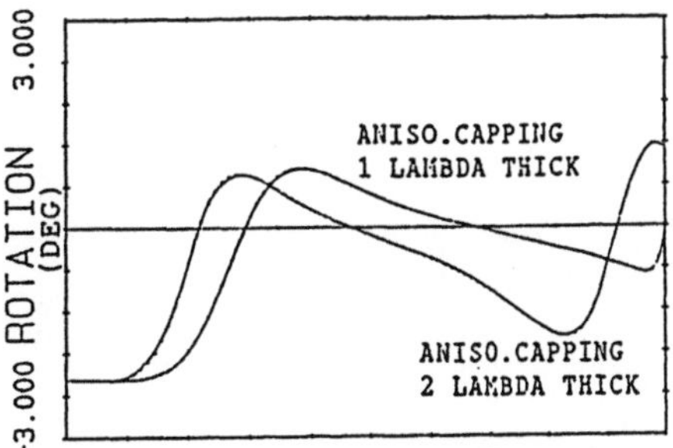

Fig.10

Fig.8 shows the Kerr rotation and Kerr ellipticity versus the ickness of the magneto-optical layer MO. In Fig.9 is a plot of s dependance on the angle of incidence. The effect of the .isotropic capping layer (1λ and 2λ thickness) to Kerr rotation is presented in Fig.10. The calculation above s done for S-polarisation .

The program is a powerful tool for study and design of .gnetooptic layer system, combined with an ellipsometer .gnetooptic constants could be determined.

:FERENCES

/1/ R.M.A. Azzam and N.M. Bashara Ellipsometry and Polarized Light North-Holland, Amsterdam 1977

/2/ A. Röseler , Infrared Phys. 21 (1981),349

/3/ A. Röseler , Preprint 85-84,1985, Zentralinstitut für Optik und Spektroskopie , Berlin GDR

/4/ H. Schwiecker, K. Hammann and U. Schneider SPIE Vol.(1988) 280-291

/5/ B. D. Doi, Schichtdickenkontrolle von Interferenzschicht-System mittels Ellipsometrischer Method , a Dissertation submitted to the Academy of Science of GDR , Berlin 1982.

/6/ D.E.Aspnes Handbook of Optical Constants of Solids, 89-110 Editor E.O.Palik , Academic Press, Inc.,1985

/7/ S.F. Nee and H.E. Benett,SPIE Vol 675 (1986) 260-289

/8/ K. Balasubramanian, A.S. Marathay and H.A. Macleod Thin Solid Film, 164 (1988) 391-403

Materials

Materials for Packaging and Passivation

Deposition, Patterning, Properties and Applications of Al_2O_3 Films

H. BURGHARDT, H. CEBULLA, W. LEIMBROCK, W. LEYFFER, D. THESS
Technical University Chemnitz

K. DRESCHER
Technical University Dresden

Summary

Alumina is an interesting material for different applications in microelectronics and sensor design. Remarkable properties are the excellent efficiency as a barrier against diffusion of alkali ions and water molecules, the chemical resistivity, and the high dielectric constant. Al_2O_3 films were deposited by NPCVD and by LPCVD using hydrolysis of $AlCl_3$. Wet and gas phase etching was applied for patterning the Al_2O_3 films. The stacked high capacitance dRAM cell as well as the ion sensitive FET (ISFET) are discussed as two special applications.

1. Deposition

Al_2O_3 films were deposited employing gas mixtures containing $AlCl_3$, CO_2, and H_2 /1, 2, 3, 4/. The scheme of the experimental apparatus /3/ is shown in fig. 1. The deposition system consists of a cold-walled silica tube, a graphite susceptor, a rf induction heating coil, and an $AlCl_3$ vapor source. $AlCl_3$ vapor is generated by sublimation from a solid source heated to about 100 °C. The hydrogen by-pass stream flowing through the source vessel becomes saturated with this $AlCl_3$ vapor. All depositions took place on 3" wafers at temperatures in the range from 900 °C to 1 000 °C. To avoid condensation of $AlCl_3$ before entering the reaction chamber all critical parts of the gas handling system are heated up to more than 100 °C. Preferred flow rates were:

H_2 (total):	2.3 mol/min
CO_2:	4×10^{-3} mol/min
$AlCl_3$:	10^{-4} ... 10^{-3} mol/min

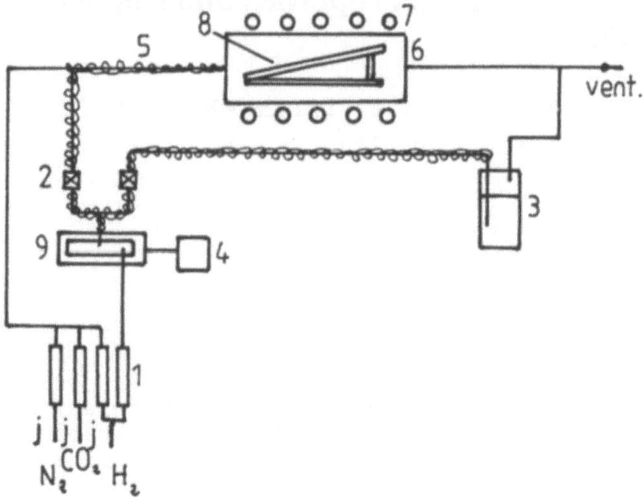

Fig. 1: Apparatus for chemical vapor deposition of Al_2O_3 films
1 flow meter 4 thermostat 7 rf coil
2 needle valve 5 heated pipe 8 susceptor
3 $AlCl_3$ trap 6 reaction chamber 9 evaporator

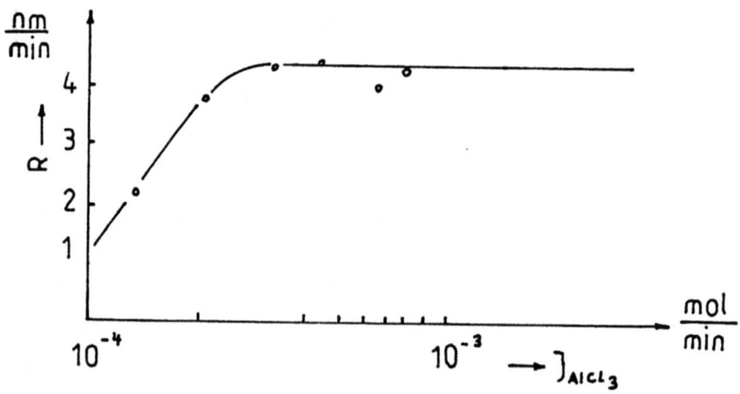

Fig. 2: Effect of the $AlCl_3$ flow rate on the Al_2O_3 deposition rate at 980 °C.
——— calculated deposition rates
o o o measured deposition rates

The dependence of deposition rate vs $AlCl_3$ flow rate at 980 °C is shown in fig. 2. The line was calculated using a model based on the thermodynamic equilibrium of this multicomponent chemical system /5/. Calculated and measured rates are in very good agreement.

2. Patterning of Al_2O_3 films

Al_2O_3 films deposited at temperatures higher than 750 °C are insoluble in hydrofluoric acid. Wet etching is possible only using phosphoric acid at elevated temperatures of at least 160 °C. Consequently standard resist can not be used for patterning. But masks consisting of SiO_2, Mo, or polysilicon films are applicable.

Etch rates of Al_2O_3 and SiO_2 are strongly dependent on the condensation behavior of the phosphoric acid. High selectivity between etching of SiO_2 and Al_2O_3 as well as good repeatability can be obtained using a closed reflux etch system /6, 7/ as shown in table 1.

Material	Etch rate ($nm\ min^{-1}$)	
	open etch system	reflux etch system
Al_2O_3	3.75	12.8
CVD-SiO_2	1.5	0.35
SiO_2 therm.	0.75	0.18
Si	0.3	< 0.3

Table 1: Etch rates in phosphoric acid at 180 °C /7/

For wet etching processes essentially a high excess of acid compared to the amount of material being etched is used. The mass of Al_2O_3 reacting with phosphoric acid is about 4×10^{-4} g/cm^2 for a 150 nm film thickness. Therefore we replaced the etch vessel by a thin film of phosphoric acid spun-on to the wafer like coating the surface with resist. When coating at room temperature the viscosity of phosphoric acid is remarkably high and suitable for film formation /7, 8/. Etching takes place only when temperature arises to more than

160 °. Heating up by a halogen lamp is a convenient method for accellerating the etch rate. Because of low heat capacity of wafers the etching can be started very quickly and can be stopped simply by switching off the lamp. Consumption of etchant is much less than for using an etch vessel.

Dry etching presumes the formation of volatile reaction products. Therefore $AlCl_3$ or $AlBr_3$ must be formed. In comparison with these AlF_3 is not volatile. At temperatures of at least 680 °C Al_2O_3 films react with sufficient rate in CCl_4 vapor forming $AlCl_3$. SiO_2 does not react under these conditions and can be used for masking /7, 9/.

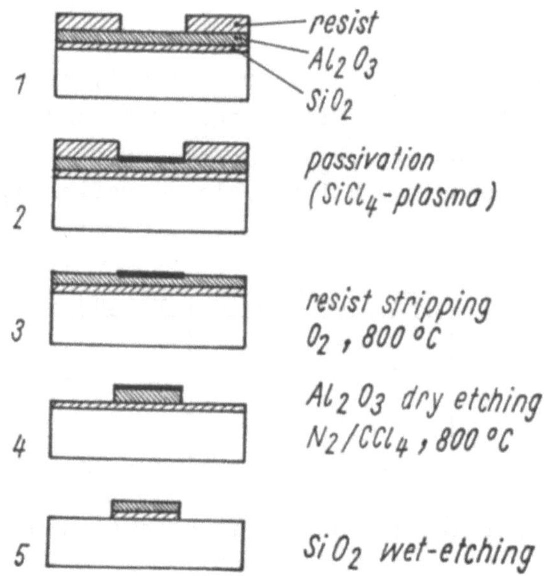

Fig. 3: Patterning an Al_2O_3 film by gas-phase etching with CCl_4

The steps of patterning are shown in fig. 3.

Step 1: Standard resist masking

Step 2: Covering Al_2O_3 with SiO_2 using a $SiCl_4$ plasma (barrel reactor)

Step 3: Decomposition of the resist mask by oxygen at 800 °C

Step 4: Gas-phase etching of Al_2O_3 with CCl_4 at 800 °C

Step 5: Removing the SiO_2 mask by wet etching

Typical etch conditions are:

- temperature: 820 °C
- CCl_4 concentration: 5 ... 10 Vol% in N_2
- flow rate: $3 \cdot 10^3$ sccm/min
- etch rate: 8 ... 16 nm/min

Plasma etching is not an acceptable method for patterning of Al_2O_3 films.

3. Applications

CVD-Al_2O_3 was tested as an insulator for stacked high capacitance dRAM cells as well as a gate membrane for ion-sensitive field-effect transistors (ISFET).

3.1. dRAM application /10/

The behavior of Al_2O_3 films was examinated using test pattern arrays containing stacked capacitance cells and standard one-transistor cells, too. The smallest verified capacitor area was 6 x 6 μm^2. Film thickness was 60 nm, the dielectric constant about 9.

We obtained a signal voltage of 150 ... 250 mV. This is nearly 2.5 x higher than for SiO_2 isolation layers. Further we found a better refresh behavior, a better radiation hardness, and an access time comparable with the one-transistor cell. Despite of those excellent properties Al_2O_3 films are not used in dRAM manufacturing.

3.2. ISFET application /11, 12/

ISFET's with high sensitivity against hydrogen-ions for pH-measurements have been designed using Al_2O_3 membrane on thermally grown SiO_2 gate insulator. Results of extensive investigations can be summarized by:

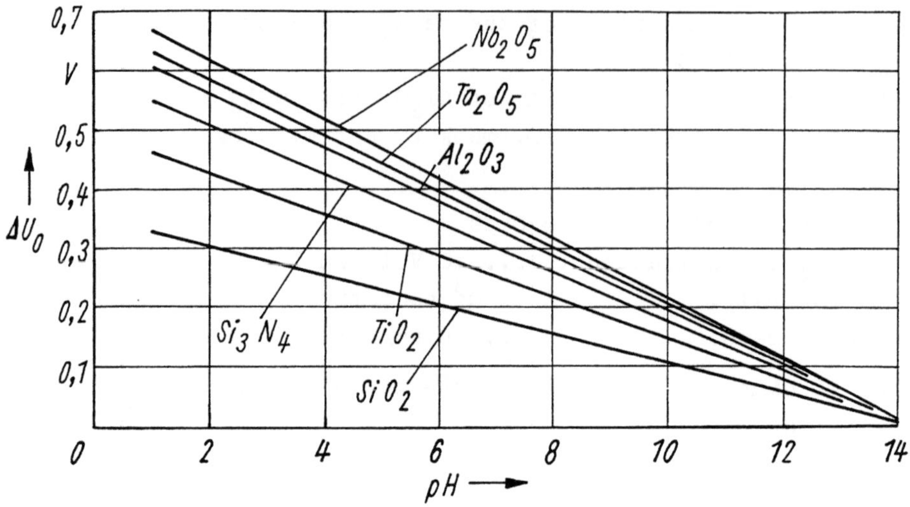

Fig. 4: Signal voltage of ISFET's vs. pH-value for different membrane materials /12/

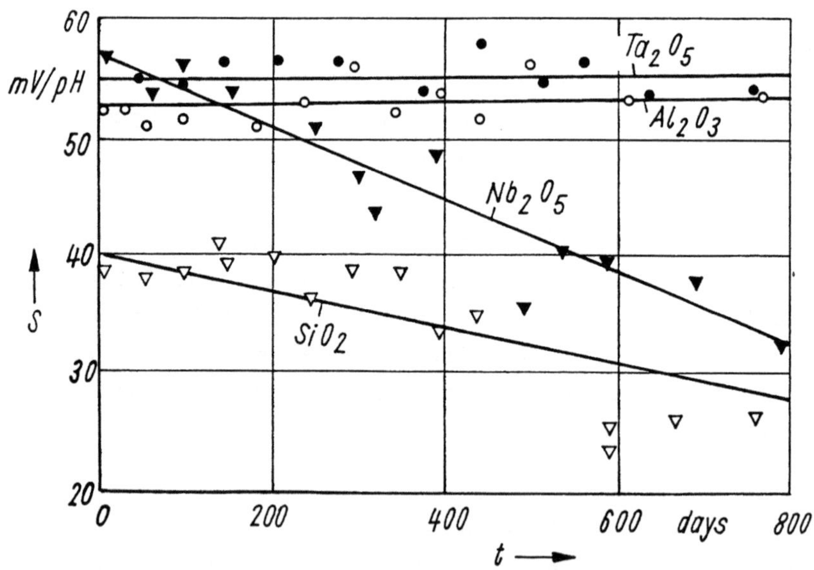

Fig. 5: Long-term stability of ISFET sensitivity for different membrane materials /12/

1. ISFET's with Al_2O_3 membrane have a sensitivity of 55 mV/pH in the range of pH = 1 ... 11 (fig. 4).
2. Al_2O_3 membranes show an excellent long-term stability compared to other dielectric materials (fig. 5). Signal voltage drift is lower than 10^{-3} mV/h or 0.02 pH/h.
3. The response time is much smaller than that of other membrane materials (fig. 6).

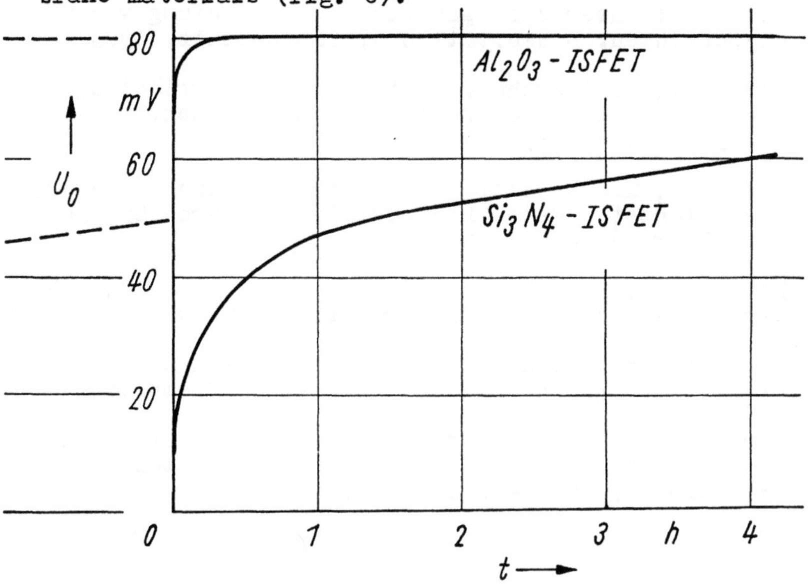

Fig. 6: Comparison of signal response of ISFET's with Al_2O_3 and Si_3N_4 membranes after keeping in dry ambient for 12 hours /12/

References

/1/ Leimbrock, W.: Untersuchungen zur Herstellung dünner Aluminiumoxidschichten für die Mikroelektronik durch CVD. Dissertation A, TU Chemnitz, 1977

/2/ Günther, St.: Untersuchungen zum Einsatz von Aluminiumoxidschichten in der MIS-Technik. Dissertation A, TU Chemnitz, 1982

/3/ Cebulla, H., u.a.: Normaldruck-CVD-Abscheidung von Al_2O_3-Isolatorschichten für die Mikroelektronik. Wiss. Zeitschr. TU Chemnitz 28 (1986) H. 4, 462

/4/ Röber, J., u.a.: Herstellungsprozeß dünner Al_2O_3-Schichten mit Hilfe des Niederdruck-CVD-Verfahrens. Wiss. Zeitschr. TU Chemnitz 31 (1989) H. 3, 425

/5/ Cebulla, H., u.a.: A thermodynamical model for CVD-systems. Thin solid Films (in press)

/6/ van Gelder, W.: The etching of silicon nitride in phosphoric acid with silicon dioxide as mask. J. Electrochem. Soc. 114 (1967) 869

/7/ Leyffer, W.: Untersuchungen zur Strukturierung von CVD-Al_2O_3-Schichten. Dissertation A, TU Chemnitz, 1983

/8/ Leyffer, W.; K. Drescher: Verfahren und Vorrichtung zum naßchemischen Ätzen von Festkörperoberflächen. DD-WP 13 88 51 (1978)

/9/ Leyffer, W.; K. Drescher: Verfahren zur Strukturierung von beschichteten Festkörperoberflächen durch Gasphasenätzen. DD-WP 13 88 50 (1978)

/10/ Burghardt, H.: Zur Physik und Anwendung von Al_2O_3-Schichten in MIS-Anordnungen. Dissertation B, TU Chemnitz, 1985

/11/ Stöhr, R.: Herstellung und Untersuchung ionensensitiver Halbleiterstrukturen. Dissertation A, TU Chemnitz, 1986

/12/ Leimbrock, W.: Ionensensitive Feldeffekttransistoren-Untersuchung der Membran- und Sensoreigenschaften. Dissertation B, TU Chemnitz, 1990

PECVD-Passivation Layers for Printheads in Micro-System-Technology

E. Unger

Siemens AG
Corporate Production and Logistics
Otto-Hahn-Ring 6
D-8000 München 83 FRG

Summary

Thin film layers deposited by plasma enhanced chemical vapor deposition (PECVD) are utilized in the construction of bubble-jet- and thermal transfer-printheads for the protection of the heater arrays. Particularly hard material like silicon carbide improves wear resistance and lifetime of such components. Deposition parameters and properties of suitable PECVD-layers are reported and discussed.

Introduction

In modern concepts of the construction of printheads a thin film control circuit is used in order to achieve high performance and resolution at moderate costs. If thin film ceramics or silicon wafers are used for the substrate, many of the widely available and well designed production systems for semiconductor fabrication can serve for photolithography, deposition, etching and other purposes. Therefore, printhead development has proceeded with these processes.

Printhead-modules consist of a number of structured thin film layers on the substrate, and cover plates and other materials may also be present. Figure 1 shows the cross section of a thermal printhead built on a ceramic substrate.

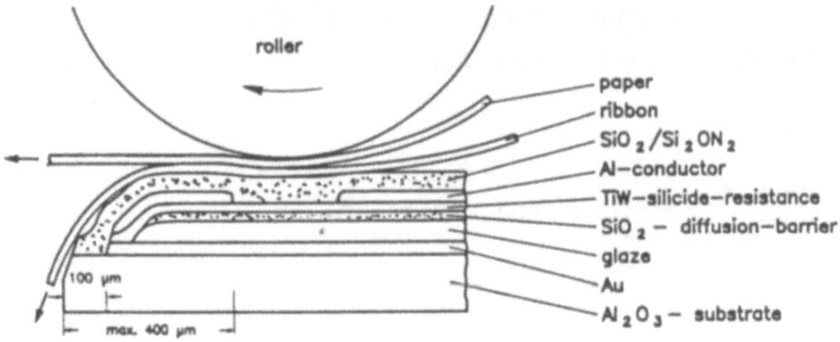

Fig. 1:

Thermal Printhead cross section

An array of silicide resistors is placed over a SiO_2 diffusion barrier, followed by Aluminium conductors and a passivation layer. During printing short electric pulses heat up the resistor melting a small amount of dye on the ribbon; this is transferred to the paper by pressing it between the ribbon and the roller. In this case, not only is a passivation needed to protect the resistor material from oxygen and moisture, but also a shield is needed against the abrasive forces from the paper running through the printer. For that reason it was suggested to deposit another layer of a hard material like silicon carbide or boron carbide over the first passivation layer.

In the "Bubble Jet" printhead device the resistor array produces vapor bubbles in the liquid ink in channels above the resistors. The shock waves of expanding bubbles eject ink droplets out of the nozzles to the paper in front of the printhead (Fig. 2). Here again a passivation layer is needed to protect the resistor material (HfB_2). In addition Tantalum is used to avoid cavitation damage.

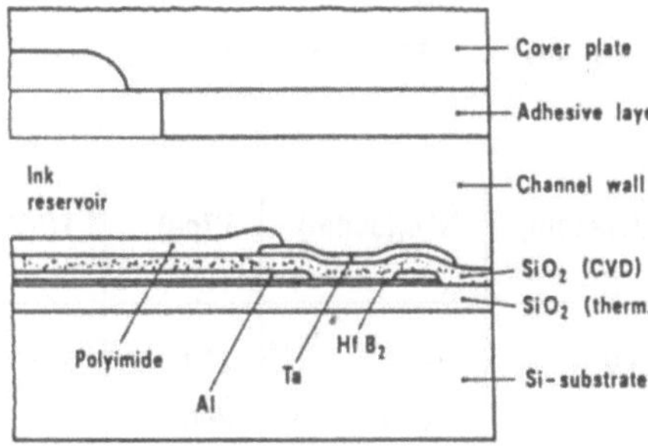

Fig. 2: Bubble-jet printhead cross section

The requirements of the passivation with respect to mechanical resistance and chemical inertness, combined with the need of a low temperature process, since aluminium is used in the circuits, leads to the use of PECVD with its pinhole-free layers, good step-coverage and high deposition rates.

Experimental Details

In all printhead systems SiO_2 or Si_3N_4 was used as first passivation layer. Both dielectric materials were deposited in a commercial hot-wall tube reactor (PWS 450) at 325°C (SiO_2) or 380°C (Si_3N_4) using silane (SiH_4) and N_2O or NH_3 respectively.

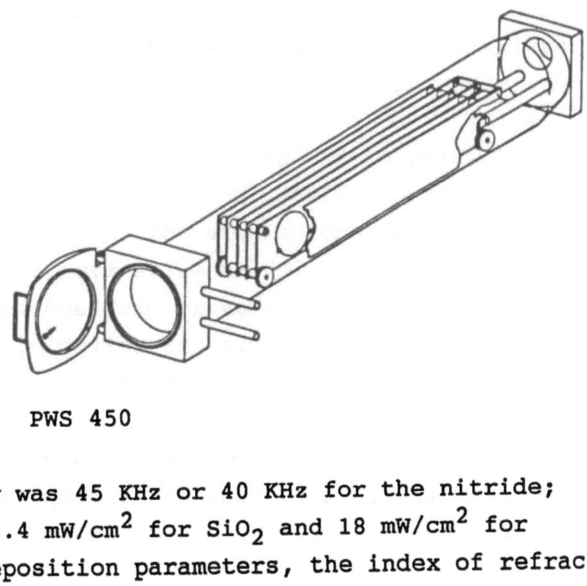

Fig. 3 Tube reactor PWS 450

The generator frequency was 45 KHz or 40 KHz for the nitride; the power density was 4.4 mW/cm^2 for SiO_2 and 18 mW/cm^2 for Si_3N_4. Table 1 shows deposition parameters, the index of refraction and the etchrates in "buffered hydrofluoric acid" (BHF 1:7). The thicknesses were 1500 or 2000 nm for SiO_2 and 700 nm for Si_3N_4.

	SiH_4(sccm)	N_2O(sccm)	NH_3(sccm)	P(pa)	T(°C)	n	etchrate nm/min
SiO_2	50	2400	-	132	325	1,5	180
Si_3N_4	200	-	1000	200	380	2,0	5

Table 1: Deposition Parameters

Carbon-rich, hard materials - silicon carbide, boron carbide and hard carbon were also examined. These were deposited in a commercial parallel plate reactor (Plasmox 2 supp. by Plasmos, Munich) with a plate diameter of 45 cm, so 12 wafers could be processed at one time. The uniformity of deposition was 5 % or better for all wafers within a run. The RF-generator (Plasma-loc 1 HF) was tuned to 85 KHz and coupled to the reactor via a capacitor in order to achive self-biasing of the substrate plate. The bombardment of positively charged ions was exploited to increase the hardness and density of the deposited films. The possible damage in the substrate lattice was ignored since no active elements were incorporated. The bias voltage reached several hundreds volts depending on pressure and RF-power. The excessive ion bombardment resulted in compressively stressed layers with relatively low hydrogen content containing only about 10 %.

The adhesion of the hard layers on Si-wafers which were not precleaned was excellent. We never observed a film peeling off of the substrate, and we were not able to remove a film or parts of it from the wafer with adhesive tape. The thicknesses were 100 to 750 nm. In order to avoid problems with gas mixtures, and to keep the processes simple and reliable, only one organic precursor was used for each deposition.

For hard carbon films, one can simply use methane -CH_4- as the source of the carbon. Many other compounds like benzene, ethylene, acetylene can also be used, however the pressure regime may be different. In our work the minimum deposition pressure was 14 pa as it was not possible to match the generator to the reactor at lower values. If one could do so, even harder films could be obtained at lower deposition rates. Process temperature was 50°C for all runs.

Boron carbide was deposited by the plasma-chemical decomposition of Triethylborane-$(C_2H_5)_3B$-, a liquid compound with a bp. of 95°C. This starting material must be handled under inert gas because of its poisonous and pyrophoric properties. The precursor was fed into the system by a stream of argon bubbling through a duran-glass-bubbler filled with about 50 ml of the compound. At room temperature Triethylborane is sufficiently volatile to provide a reasonable deposition rate. At the same time, the heat of evaporation is sufficient to avoid excessive adiabatic cooling of the supply.

The same technique was utilized for the deposition of silicon carbide. Many compounds can be used following the formula:

R_3SiX where $R = CH_3$ or C_2H_5 and
$X = H, Cl, CH_3, R_3Si$, etc.

In most cases these compounds are liquids, stable in air and can be handled without special precaution.

Results

Table 2 presents the data obtained from the depositions and some properties of the films. As observed by many authors, a-C:H exhibits excellent hardness but with very high compressive stress. So only thin layers, several 100 nm, should be used. The deposited boron carbide or better referred to as a hard carbon-boron alloy, has almost the same hardness as a-C:H, but only 25 % of the film stress. Furthermore this material can be produced at much higher pressure (46 pa) causing less technical problems. Silicon carbide can be deposited at high rates however it has intermediate properties. SiC should be a semiconductor, but further applications were not investigated in the present work.

The hard thin films were stable in all aqueous reagents, acidic or basic. Only a slight attack was observed in a mixture of 65 % HNO_3 and 50 % HF at room or higher temperature. But this mixture rapidly dissolved the Si-wafer from the backside forming brown clouds of nitrogen oxides. However the layers are easily etched in an oxygen/tetrafluoromethane-plasma. a-C:H can be etched in a oxygen plasma alone. It should be noted that the described processes with organic precursors are sensitive to oxygen and can yield the oxides B_2O_3 or SiO_2. This was found when several runs were processed with residual air in the Ar-manifold. A significant lower index of refraction was observed in an oxygen-containing layer (1.7 instead of 2.5 for SiC). It seems to be possible to get more than one compound out of the same starting material.

Compos.	hardness (HK) kgf/mm^2	density g/cm^3	stress x 10 pa	n	dep.-rate nm/min	bias $-V_B$ (V)	RF mW/cm^2	press. pa
a-C:H[a)]	1900	1,26		2,50	26	220	220	37
a-C:H	2900	1,55	150	2,50	16	360	220	22
a-C:H (10 pH)	-4300	1,82		2,50	10	500	220	14
SiC_2[a)]	3000	2,79	86	2,48	51	260	156	46
BC_6[b)]	3580	2,13	43	2,27	36	220	156	46

a) SNMS b) RBS

Table 2: Carbide film properties

Conclusion

Using commercial deposition systems we were able to produce a number of dielectric films with high hardness, low hydrogen content and excellent chemical and mechanical stability. In microelectronic devices those films can be used to protect functional layers and to improve the abrasion resistance of sensitive surfaces. The use of organic, mostly liquid, starting materials for hard coatings simplifies development and fabrication with this modern plasma-chemical tools.

Hybrid Integrated Circuits on Polycrystalline Aluminium Nitride

PHILIP PHILIPPOV

N.NICOLOV

VMEI - Sofia, dept. KTPPME, Bulgaria

IPM, BAN branch Razgrad

Abstract

By means of suitable technologic procedure substrates with high temperature conductivity have been prepared from pure aluminium nitride (AlN) with additives. The substrate sutface after polishing has been assessed and the possibility for holes drilling in AlN has been investigated. Finally, the metallization with direct joint of molybdenum (Mo) has been discussed.

Introduction

The semiconductor technology development brought to an increasing of thermal power dissipation in Hybrid Integrated Circuits (HIC's). The necessity for dissipation of a vast amount of heat from semiconductor crystal requires perfect thermal properties of materials designed for HIC - substrates, utilizing large size of bear chips.

In technologies, utilizing largr size bear chips the thermal extension matching of crystal holder material and silicon is extremely important. For high - frequency and microwave HIC's the low dielectric constant is important, too. The high volume resistance and sufficient mechanical strength are obligatory for every kind of substrates.

In table 1 are summarized the basic trends of modern microelectronics as well as the requirements towards substrate properties.

The best combination of all these properties possesses polycrystalline AlN (Table 2): high thermal conductivity, linear extension close to that of silicon, high volume resistivity, low dielectric constant, high bending strength, not toxic.

Table 1: Trends in microelectronics & substrate properties

TRENDS IN MICROELECTRONICS	SUBSTRATE PROPERTIES
1. Power dissipation increasing	1. High thermal conductivity
2. Chip size increasing	2. Linear extension - close to that of silicon
3. Working frequency increasing	3. Low dielectric factor
4. High amplitude signals utilization	4. High volume resistance
5. Larger size HIC's ceramic boards	5. High mechanical strength

Table 2: Ceramic substrate properties

Properties	AlN	SiC	BeO	Al_2O_3
Consistency (g/cm^3)	3,25	3,1	2,9	3,78
Bending strength (kg/mm^2)	40	49	20	30
Thermal conductivity (W/mK)	120	270	250	20
Linear extension ($x10^{-6}/^oC$)	4,4	3,7	8,0	7,2
Volume resistance	10^{12}	10^{10}	10^{14}	10^{14}
Dielectric constant	8,9	40	7	9,4

AlN substrate characteristics

In terms of suitable technologic procedure a high consistency polycrystalline AlN was obtained from ultra-pure AlN-powder with 0,5 μm + 0,8 μm size of grains. By using Ca and additives the mechanism of sintering proved to be the dissocietion (2). The polycristalline substrate obtained was transparent with throughput (maximum) of 4 + 6 wavelength (figure 1).
Figure 1 shows light-transperancy throughput of AlN-ceramic. The temperature conductivity of polycrystalline AlN is limited by phonon-dissipation from microstructural deffects, pores,

contaminant phases along the grains boundaries and by the grain aluminium vacances, engendered from AlN oxidation.

The closest thermal conductivity of 150 W/mK possess substrates, obtained with CaC_2 additives. Main factor for thermal conductivity is the creation of a balanced and maximum consistent microcapsule of polycrystalline material.

The lack of defects in grains and elimination of O_2 - impurities alond the grain boundaries is presented in fig. 2.

The polished AlN substrates have 35 x 0,635 sizes and two surface properties: Ra = 0,025 μm and maximum value of microunflatness Ra = 0,25 μm. During another kind of surface treatment the substrate properties were Ra = 0 μm and Ra_{max}= 1,40 μm.

The substrate flatness was measured by Talysurf and was 0,12 and 0,05 μm/mm, correspondingly.

Figure 4 shows the flatness of the two substrate types. For both of them the flatness was less than 1 μm/mm. The second type substrates are feasible for utilizing in thinfilm circuits (2).

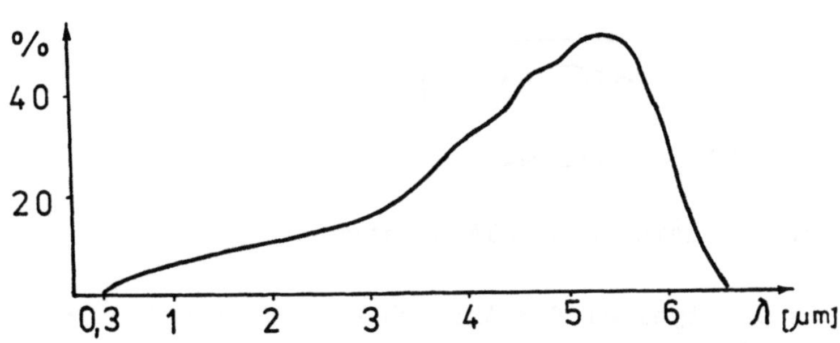

Fig. 1. Throughput spectrum of AlN - ceramic

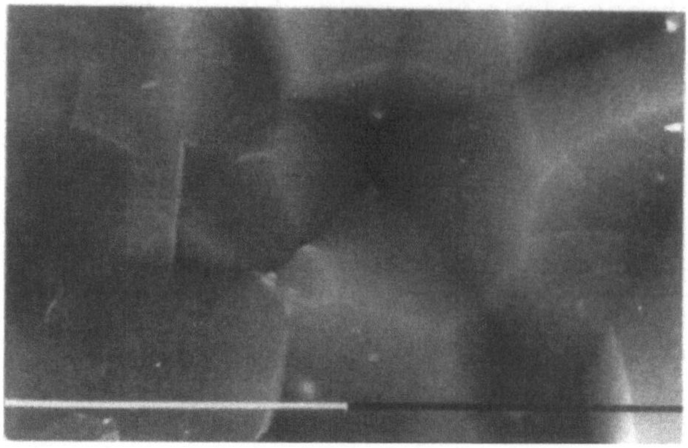

Fig. 2. A photo - picture of Marker 5 μm AlN - ceramic

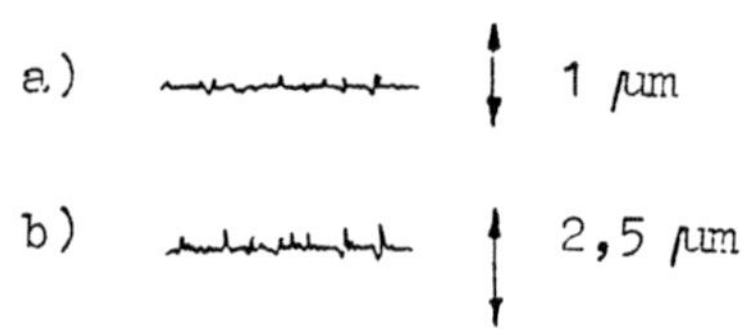

Fig. 3a, b. AlN substrate roughess

Laser treatment of aluminium nitride substrates

Through - holes creation in AlN ceramic is of great importance for its utilization both as a crystal - holder and as a substrate for high - frequency and microwave hybrids.

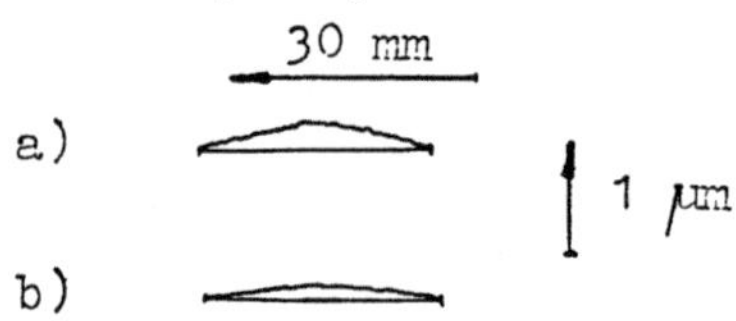

Fig. 4. Flatness of AlN substrates

Two methods are typical for through - hole creation laser and ultrasonic (4). The former method is suitable for AlN - ceramic since treatment velocity and exactness is much higher. The scribing and marking are easily carried out by laser and so a YAG-laser with 10,6 μm wavelength, 0,5 mm beam - diameter and power of 5,5 J/impulse was used. For cylindricity assessment of the drilled holes was utilized the diameters'correlation between laser irradiated facet diameter and reverse side diameter.

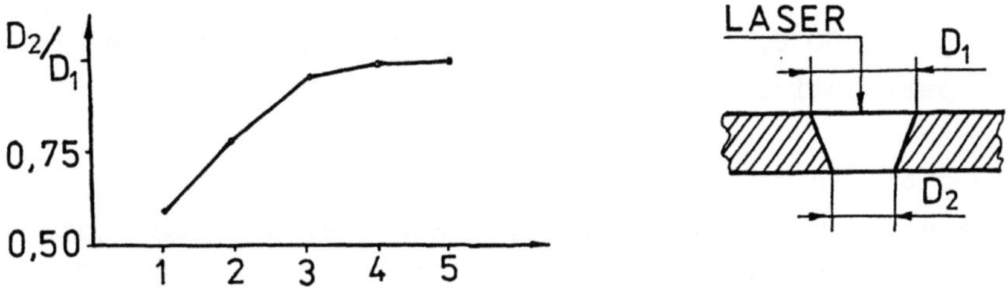

Fig. 5. Laser drilling characteristics

Figure 5 shows correlation between the two diameters.
The holes were not cylindrical and vertical until fourth irradiation had been accomplished. The shape of the hole was nor perfectly round until the fourth irradiation had been carried out either.

<u>AlN metallization</u>

The hole - shaping mechanism consists of AlN dissolvation in aluminium and nitrogen due to local heating at the laser spots; the two ingradients were evaporated without leaving craters on melted metal.

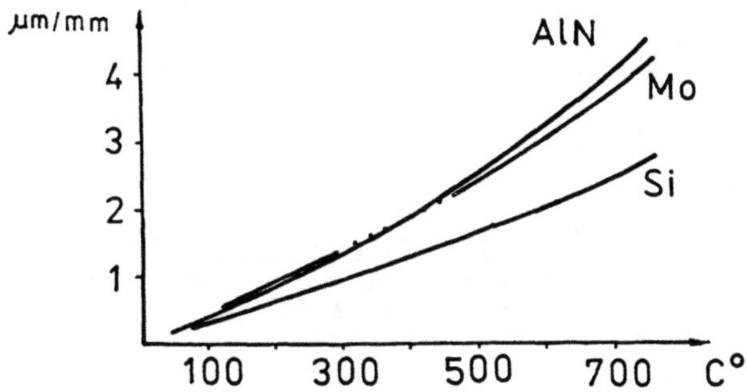

Fig. 6. Thermal-linear extension of Mo, Si and AlN

A feature of AlN - substrate metallization was the endeavoure to save the combination of AlN qualities and to assure all necessary properties of the metal layer at the same time. The most important thing concerning the metallization is the coincidence of layer and substrate linear extension coefficients. Figure 6 shows thermal linear extension of Mo, Si and AlN. Due to AlN and Mo linear extension close coincidence at temperatures up to 200°C, the Mo - metallization is concedered to be quite perspective.

It is possible to obtain metallization of AlN substrate by special treated molybdenum foil in a sole technological process. The layer adhesion was better than every other metallization.

Discussion

With such properties combination AlN ceramic seems to be rather suitable for HIC's substrates. The substrate surface quality can be easily reached which makes it convenient both for thick and for thin metallization.

AlN ceramic is suitable for laser treatment. However, the most interesting metallization, that saves the substrate properties, is that of directly joined molybdenum.

References

1. E.S.Detter and M.K.Charles Jr.:"AlN and SiC Substrate Properties and processing characteristics" - Advances in ceramics, V.26 ed. M.F. Yon at All, 1989.
2. L.J.Bostlaar, D.Vander Auwera and F.Gys.: "Thin Film Hybrid Circuits on AlN Substrates", Hybrid Circuits Number 21 january'90.
3. Lepkova D., Dimitriev J., Joleva A.,NIcolov N.:"Untersuchung des Einflusses von Zugaben und der Struktur des Aluminium'Nitrides" Vortrage der Internationalen Tagung "Glas und Keramik", Budapest, Mai 1989.
4. Ivase Noboy and Al."VHF - Range Power Amplifier Module on AlN substrate" Symposium on microelectronics, Tokio, 1985.

High Performance Resin System for Printed Circuit Boards

Siemens AG
R & D Center ZFE ME AC 51
P.O. Box 3220
D-8520 Erlangen
Federal Republic of Germany

Summary

Printed circuit boards from aromatic polyepoxide resins and newly developed aromatic polyamines with heterocyclic high temperature resistant structural units possess outstanding properties. According to the special chemical structure of the crosslinking components, the prepregs show good storage stability at room temperature. The cured material has high glass transition temperatures (> 200°C), a reduced flammability and good mechanical strength. In contrast to other high temperature resistant materials like polyimide, this resin can be processed by common methods. Testing of the basic materials according to MIL 13949 shows that all requirements for epoxies are fulfilled. An account is given of the synthesis of the crosslinking components, the manufacture and the properties of printed circuit boards produced therefrom.

Introduction

Polymer materials based on epoxy resins are widely used in electrical and electronic industry. The reasons are above all the excellent characteristics of the end products and the possibility to vary them as required. An important field of application is the use of epoxies as resin matrix for printed circuit boards.

In 1987, the worldwide production of printed circuit boards reached a total value of approximately 26 billion DM. 30 % of this is due to multilayer boards. The predicted growth rate of this market up to 1995 lies around 12 % per annum (1).

So-called FR 4-material is the resin matrix that is most used for the manufacture of laminates for printed circuit boards. However, apart from advantages like the low price, easy processability and low combustability, these basic materials have also several disadvantages. Because of the low glass transition temperature of 130-135°C, there is a strong decline in the mechanical strength at temperatures > 130°C. Furthermore these FR 4-materials show poor dimensional stability and insufficient soldering characteristics. In the case of fire corrosive products are also produced and under certain conditions toxic compounds may be released.

Technically available printed circuit boards with high glass transition temperature (e.g. polyimide) do not have the disadvantage of poor mechanical strength between 130°C and 200°C. On the other hand these laminates cannot be processed using normal techniques. Processing and treating ability are worse, demanding higher moulding pressures and curing temperatures and longer curing times. Additionally, drilling tools are worn out fast during the preparation of mounting holes. The price of these laminates exceeds that of FR 4 by a factor of 6 to 10.

We now report the development of a nonexpensive resin system, which possesses the advantages of laminates with high glass temperature and which can be produced with traditional techniques.

<u>Synthesis of crosslinking components for epoxies</u>
It is well known that polymers with isocyanurate structures give products with high glass transition temperatures and improved dimensional stability (2). We therefore developed aminetype crosslinking components containing these heterocyclic structures of high heat resistance (3). The synthesis of such components is shown in Fig. 1.

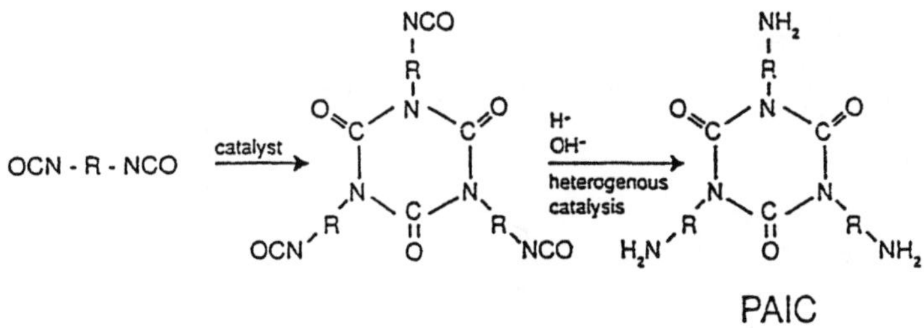

Fig. 1 Synthesis of crosslinking components

Starting from diisocyanates, for instance from toluylenediisocyanate, isocyanurate rings can be achieved with various catalysts. Depending on the viscosity of the starting materials, this oligomerization can be carried out with or without solvent. The reaction proceeds at room temperature and is readily controllable. It can be followed by determining the NCO value. At the particular desired degree of oligomerization, the reaction is stopped by deactivating the catalyst. These isocyanurate compounds with NCO groups can then be converted into the corresponding isocyanurates with NH_2 functional groups. There are a number of possibilities for doing this. In addition to alkaline or acidic hydrolysis, preparation by means of heterogeneous catalysts is also possible. Depending on the reaction conditions of oligomerization and hydrolsis, products with different NH_2 values can be obtained. The PAIC obtained from toluylenediisocyanate is a yellow powder which does not melt but decomposes at temperatures around 200°C. The storage stability at room temperature is more than 1 year without changes in solubility and amino group concentration.

These isocyanurates can be combined with any common epoxide. Mixtures can be used as well as aliphatic, aromatic, heterocyclic or halogenated epoxy resins.

<u>Preparation of prepregs</u>

A typical composition of an impregnating solution for glass fabric for preparation of prepregs is shown in Fig. 2.

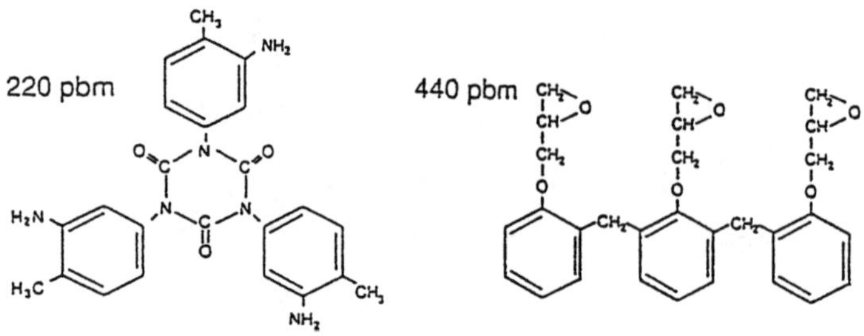

340 pbm Methyl-ethyl-Ketone

Fig. 2 Composition of an impregnation resin solution

The resin solution consists of 220 parts by mass of an isocyanurate based on toluylene-2,4-diisocyanate with a NH_2-value of 8.6 %, 440 parts by mass of an epoxidised novolak having a functionality of 3.6 and 340 parts by mass of methylethylketone.

The reaction of the epoxide group is slowed down by the steric shielding of the amine function of the isocyanurate by the ortho methyl group. This enables us to prepare the B-stage, which is important for the processing of prepregs.

Glass fabric was impregnated with this solution in a vertical lamination apparatus. By right choose of the temperatures in the drying channels we were thus able to prepare non-sticking prepregs with a resin content of 44 % and a residual solvent content of 0.2 %. If these prepregs are stored without control of temperature and humidity at room temperature the residual gelation time falls off by only 10 % after three months. Prepregs stored at 5°C show no change in the residual gelation time after 5 months.

Preparation and testing of basic material

The prepregs were processed in a preheated press to eight layered, double-sided copper-clad (18 µm foil thickness) basic material. After as little as half an hour in the press a material with a glass transition temperature of 210°C was obtained. By postcuring in an oven or in the press, the glass transition temperature could be raised to about 240°C.

The testing was done according to MIL 13949 under the stringent conditions which are applied to materials with high glass transition temperatures such as polyimide. Some results are shown in table 1.

Table 1: Properties of the basic material

- Thermal stress	145 s
- Peel strength after 10 s at 288°C	1.3 N/mm
- Flexural strength	540 N/mm^2
- Water adsorption	16 mg
- Dissipation factor (1 MHz)	0.017
- Dielectric constant (1MHz)	4.86
- Coefficient of linear thermal expansion	$90 \cdot 10^{-6} K^{-1}$ up to 220°C

The material fulfills all requirements of MIL 13949 for epoxide basic materials. Apart from the flexural strength at high temperatures, it also fulfills the requirements for materials with higher glass transition temperatures.

Furthermore the cured basic material has the advantage of a low coefficient of linear thermal expansion over a wide temperature range. This results in a particularly good dimensional stability up to high temperatures.

Because of the partially heterocyclic structure in the cured resin, our basic material has self-extinguishing properties even without the addition of ring-halogenated components. The stage of combustability is UL 94HB, which is just achieved by polyimide. Measurements on 1.6 mm thick test specimens gave a mean burning time of less than 5 seconds, so that flame retarding additives are not necessary for suitability under this standard. These materials have the advantage that they do not form any corrosive or particularly toxic decomposition products when thermally overloaded. For a further reduction of the combustability to the stage UL 94V-1 or UL 94V-0, it suffices to add only small quantities of halogenated compounds, so that the total halogen content of the basic material is below 4 %, compared to about 10 % with FR 4-material.

Another field of application for these newly developed crosslinking components is their use as low pressure moulding compounds for encapsulation of electronic elements or for the fabrication of plastic parts with special properties.

Many properties of polymers produced from epoxies and these crosslinking components can be suitably adjusted by varying and combining the diisocyanates for the crosslinking components, so that polymers containing these crosslinking components are suitable for many applications.

References

1. Galvanotechnik 79, 2031 (1988).
2. H. Markert et al, Siemens Forsch.-u. Entwickl.-Ber. 16, 105 (1987).
3. US Pat. 4,783345

Z-Axis Conductive, Stress Free, Thermally Conductive Adhesive Films and Pastes for Die- and Component Attach

Kevin K. T. Chung, Ph.D.
Garrett Dreier
Eldon Avery
Andy Boyle
AI TECHNOLOGY, INC.
P.O. BOX 3081
PRINCETON, NJ, USA

William Koehn
Guido Govaert
Dirk Theunissen
AIT EUROPE, NV
54 BOSDEL
GENK, BELGIUM

With the more massive production of very large scale integrated circuits (VLSI), and the developmental efforts in ultra-large scale integration (ULSI) and wafer scale integration (WSI), there are many changes that might be required in functional packaging of these devices at both the die and component levels of electronic packaging. In fact, much more critical consideration in terms of packaging materials and their effects on the performance of the finished products must be in place to assure the same and similar level of overall product reliability.

Some of the obvious aspects that are involved in putting the IC's into usable forms, are the facts that the area of the silicon dice is much larger, up to 1 cm by 1 cm, the pin counts for input and output also increased drastically to the range of 400 or close to 2-5 mils spacings, and most of the microprocessor-based and power dice are also generating more heat while consuming more power per unit area, in some cases over 30 watts per square centimeter.

It is clear that conventional die-attach and component attach materials and methods might not be able to accomodate all these changes. When the area of dice increases, die to substrate/ carrier/header stress also increases to such levels that dice are cracking or delamianting. Thus stress-free adhesives are required. When pin counts increase, it becomes more and more difficult to access the functional areas with solder-bumps (flip chips) and wires. It also becomes difficult for solder-paste to handle the stringent requirements of closely pitched packages in the circuit board levels. It will be desirable to have an adhesive that can assure electrical contacts without worrying about bridges between functional areas.

A high definition Z-axis conductor/adhesive may be one of the useful solutions. Also, because of the power dissipation, it is important that such an adhesive be thermally conductive to channel away the heat generated at both the die and component levels.

In this paper we will discuss some of these materials in both paste and film formats for use in die and component applications. Some tests by both our laboratory and by some IC manufacturers will also be presented.

Integrated Optics Materials and Technologies

Integrated Optics in Silicon

B.Schüppert, J.Schmidtchen, A.Splett, K.Petermann

Technische Universität Berlin
Institut für Hochfrequenztechnik
Einsteinufer 25, D-1000 Berlin 10, F.R.G.

Summary

A technique for fabricating low loss channel waveguides in silicon is reported. The waveguides are obtained by an indiffusion of germanium diffused from a SiGe-alloy. The fabrication process is described in detail and typical fabrication parameters are given. The waveguides are characterized optically by loss- and spot-size-measurements, both, theoretically and experimentally.

Introduction

Even though silicon is the most important material for electronics, it is yet an unexplored material for integrated optics. Up to now, integrated optics research interests are concentrated on III/V-compounds, which require large technological efforts. Even though silicon is no direct semiconductor, hence not allowing a laser fabrication, the realization of photodiodes /1/, optical waveguides /2/, modulators and switches is possible.

The major advantages of silicon are

well established material for electronics,
inexpensive material of high quality,
transparent for wavelength > 1.2 µm,
compatibility of electronic and photonic devices,
potential for various microstructuring techniques.

As an example of a possible future application of integrated optics in silicon, Fig. 1 shows an integrated coherent-optical receiver. This device contains a coupling between the input single-mode fiber and the channel waveguide, an integrated optical directional coupler, and two integrated optical photodetectors which can be easily combined with the complete electronic circuitry. A precise positioning of the incoupling single-mode fiber with respect to the integrated optical channel waveguide is simply achieved by using well known anisotropic etching techniques in <100>-silicon. The precision of the v-grooves fabricated with this technique, hence the relative positioning between fiber and channel waveguide is only limited by the precision of the photomask, being usually better than 1 μm. The local semiconductor laser, which is necessary for a coherent receiver is hybrid mounted upside-down on the silicon chip as shown in Fig. 1. Etching a v-groove in the substrate and a suitable rib at the semiconductor laser offers a potentially simple preadjusted mounting technique. By further investigation in the field of heteroepitaxy of III/V-compounds on silicon a monolithic integration of the laser diode seems to be possible, thus yielding a complete monolithically integrated optoelectronic device in silicon. However, for realizing any optoelectronic devices in silicon, the key-component is a low-loss optical waveguide in silicon.

Basic principles of optical waveguides in silicon

For optical applications in the IR-region silicon is a material with very low losses. Fig. 2 /3/ shows the losses of single-crystal silicon dependent on the optical wavelength. It can be seen that the intrinsic optical absorption of silicon yields losses as low as a few hundreds dB/cm for a wavelength of 1.3 μm, which is a wavelength of great interest for optical communication systems. So it should be possible to fabricate integrated optical waveguides in silicon. In order to realize such an optical waveguide it is necessary to enhance the refractive-index locally along the waveguide. The simplest way is to indiffuse a material with a larger refractive-index than silicon into the silicon substrate by using a stripe configuration of the diffusion source as shown in Fig. 3. Additionally, low excess loss should be introduced by the indiffused material. In our experiments we found a diffusion source containing germanium to be useful. On one hand, germanium has a refractive-index of approximately $n = 4.3$ at $\lambda = 1.3$ μm, hence being significantly larger than the refractive-index of silicon at this wavelength ($n = 3.5$), thus providing an index-enhancement by indiffusion. On the other hand, germanium belongs to group IV of the periodic system, hence not yielding additional absorption due to free carriers.

Waveguide fabrication

In practice, the diffusion source has been realized by means of a SiGe-alloy. The fabrication process is shown schematically in Figs. 4a-4f. The SiGe-alloy is deposited through high-vacuum E-beam evaporation by using standard lift-off techniques for defining the diffusion-source stripes (Figs. 4a-4c). Then a SiO_2-layer is sputtered onto the heated substrate (Fig. 4d) to avoid evaporation during the following diffusion process, which is carried out at 1200 °C for 69 h (Fig. 4e), yielding the indiffused Si:Ge-waveguide (Fig. 4f). These diffusion data correspond to a diffusion depth of 2.1 µm.

Waveguide characterization

With this technique, single-mode channel waveguides can be fabricated, yielding optical spot-sizes as depicted in Fig. 5 for both the vertical and horizontal spot-sizes. These waveguides, for example, have been fabricated by diffusing an alloy with variable stripe-width W_{Ge}, a layer thickness of $\tau = 290$ nm and a germanium content of 27 %. The measured spot-sizes are in the range of 4...8 µm and are well matched to spot-sizes of typical single-mode fibers. Comparing the measured spot-sizes with a theoretical analysis carried out by solving the scalar wave-equation numerically we get an estimate about the maximum refractive-index enhancement in the indiffused region. This refractive-index enhancement is shown in Fig. 6 as a function of the germanium content yielding an almost linear relation between germanium-content and index enhancement. As an example, a germanium content of 4.5 % yields a refractive-index enhancement of approximately $\Delta n_{max} = 1.0\ 10^{-2}$ corresponding to a waveguide with a relatively good confinement. Therefore we expect that small bend-radii in integrated-optical devices could be realized.

In order to determine the waveguide-losses, we measured the loss of a 11 mm long waveguide-chip which is inserted between two single-mode fibers. The losses have been measured from fiber to fiber (Fig. 7) to be approximately 8.4 dB. Subtracting the reflection losses and field-mismatch losses we found pure waveguide losses of 3.6 dB/cm. These values are encouraging results even though further investigations are necessary to reduce these preliminary waveguide-losses.

Conclusions

We have demonstrated the feasibility of low-loss integrated optical waveguides in silicon, even though not much attention has been payed to this material for integrated optical applications up to now.

This enables the use of the same material for electronics and photonics hence being potentially suitable for monolithic integration. In comparison with those integrated-optical waveguides on silicon /4/ the advantage of our proposal is to use a common material for waveguiding and electronic circuitry. It is expected that the techniques of microstructuring, such as anisotropic etching techniques, can be applied as long as the germanium content is small.

/1/ T.P.Pearsall, H.Temkin, J.C.Bean, S.Luryi: "Avalanche gain in Ge_xSi_{1-x}/Si infrared waveguide detectors"; IEEE Electron Device Lett., 1986, EDL-7, pp. 330-332

/2/ B.Schüppert, J.Schmidtchen, K.Petermann: "Optical channel waveguides in silicon diffued from GeSi alloy"; Electron. Lett., 1989, 25, pp. 1500-1502

/3/ R.A.Soref, J.P.Lorenzo: "All-silicon active and passive guided-wave components for λ = 1.3 and 1.6 µm"; IEEE J. Quant. Electron., 1986, QE-22, pp. 873-879

/4/ C.H.Henry, R.F.Kazarinov, H.J.Lee, K.J.Orlowsky, L.E.Katz: "Low loss Si_3N_4-SiO_2 optical waveguides on Si"; Appl. Optics, 1987, 26, pp. 2621-2624

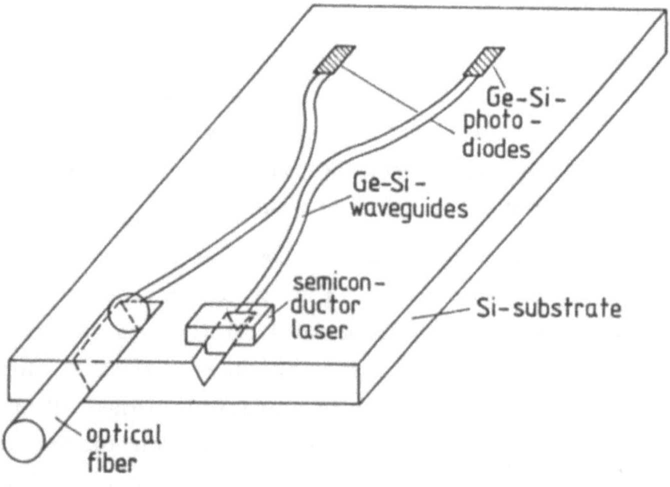

Fig. 1: Integrated coherent-optical receiver in silicon

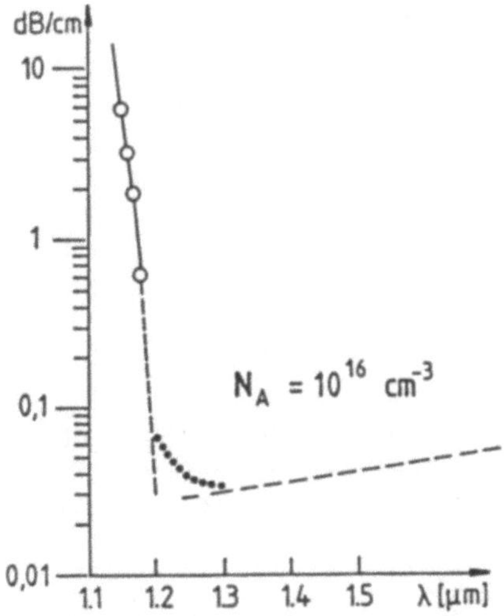

Fig. 2:
Intrinsic losses of single-crystal silicon against the wavelength

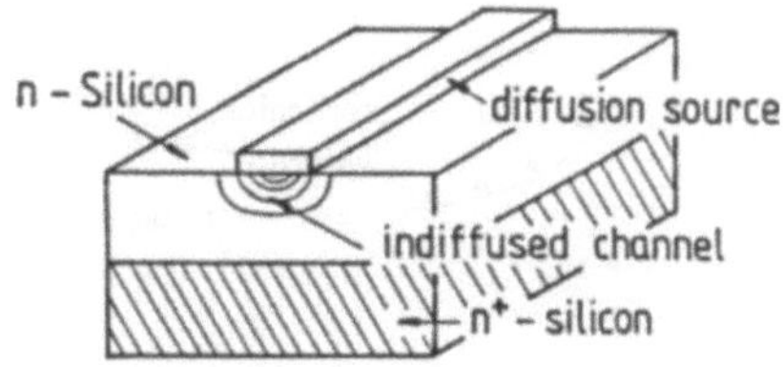

Fig. 3:
Indiffused Si:Ge-waveguide

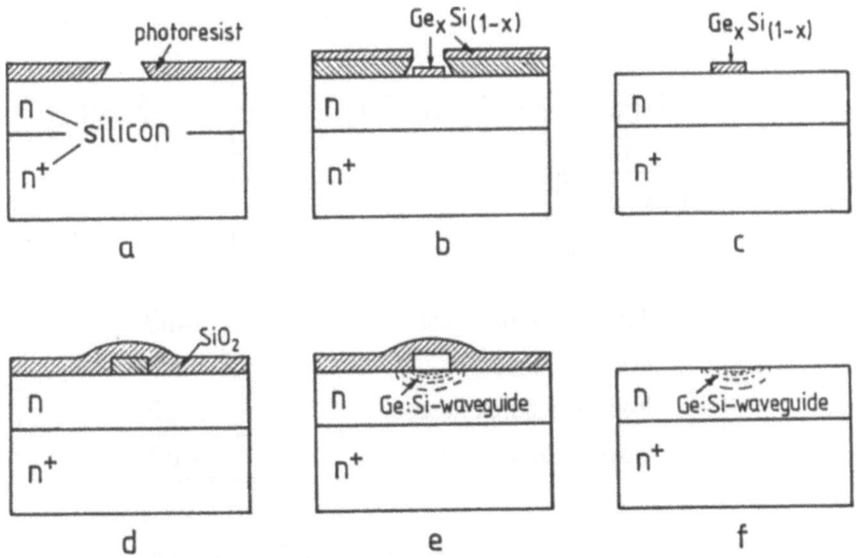

Fig. 4: Fabrication process
a) photoresist: deposition, exposure, development
b) evaporation of a SiGe-layer
c) after "lift-off"
d) deposition of a sputtered SiO_2-overlay
e) indiffusion
f) SiO_2-overlay removed by etching

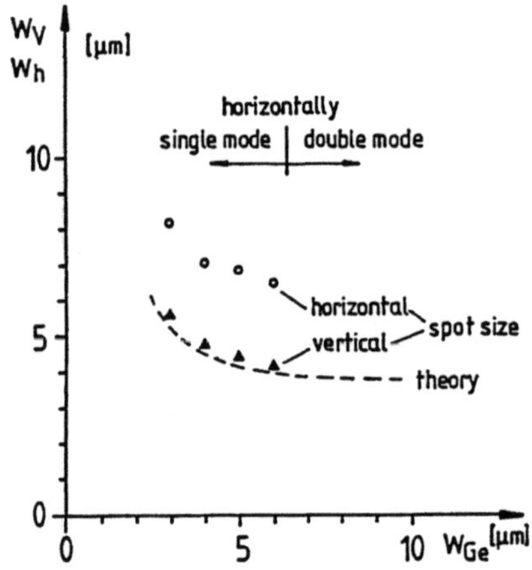

Fig. 5:

Horizontal and vertical spot-sizes of Si:Ge-waveguides

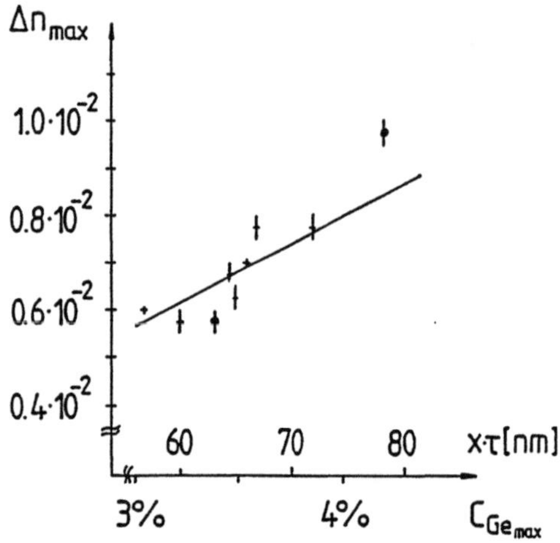

Fig. 6:

Refractive index enhancement against germanium content

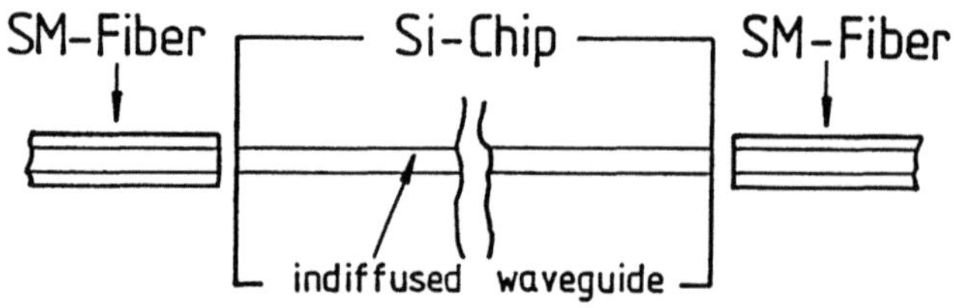

Total insertion loss	8.4dB
Fresnel losses	− 3.2dB
Mode mismatch losses	− 1.3dB
Waveguide loss 11mm	3.9dB

Waveguide losses 3.6 dB/cm

Fig. 7: Measurement of the insertion-loss of integrated-optical Si:Ge-waveguides

Integrated Optics on Silicon with PECVD-Fabricated Waveguides

H. Bezzaoui, A. Baus, E. Voges

Universität Dortmund, Lehrstuhl für Hochfrequenztechnik
Postfach 500500, D-4600 Dortmund - 50 FRG
Tel.: FRG (231) 7553909, Fax: FRG (231) 7554631

Abstract

Silicon provides as a particular advantage the monolithic integration of optical circuits, photodiodes, microelectronic circuits and micromechanics for electronical-optical-micromechanical microsystems. The optical waveguiding layers are fabricated by Plasma Enhanced Chemical Vapor Deposition (PECVD) for a compatibility with CMOS-processes and optical strip waveguides are delineated by Reactive Ion Etching (RIE). The technology, the waveguides and their characterization are demonstrated.

Introduction

The fabrication of low loss optical waveguides by means of dielectrical thin films requires specific methods of deposition and structuring. In [1] and [2] silica based embedded single-mode channel waveguides are formed by a combination of flame hydrolysis and RIE. Optical waveguide devices on silicon fabricated by combined PECVD- and LPCVD-processes are reported in [3], [4]. Other approaches to optical waveguide devices on silicon have been summarized by Hall [5] and Hickernell [6].

With regard to a monolithic integration of optical and microelectronical components we use PECVD-processes for the fabrication of waveguides on the base of SiO_2- and nitrogene doped silicon oxide (SiON)-layers since they are low temperature processes (T = 200 - 400 °C). The PECVD-process ensures adjustable refractive indices and thickness of layers with high homogeneity and reproducibility, a high transparency of the light-guiding films for wavelengths between 0.6 to 0.9 μm and a sufficiently high deposition rate.

PECVD-process

PECVD-processes operate with gases as reactances. The gases are ionized by a high-frequency plasma discharge and react on the hot surface of substrate. Silicon dioxide (SiO_2) is fabricated with a gaseous mixture of silane (SiH_4) and N_2O, the nitrogen-doped siliconoxide (SiON)-films are synthesized with SiH_4, N_2O and NH_4.

The chemical reactions in the ideal case are shown in equation (1) and (2).

SiO_2-films:

$$SiH_4 + 2\,N_2O \xrightarrow{\text{Plasma}} SiO_2\downarrow + 2\,N_2\uparrow + 2\,H_2\uparrow \tag{1}$$

SiON-films:

$$SiH_4 + x\,N_2O + y\,NH_4 \xrightarrow{\text{Plasma}} SiO_xN_y\downarrow + x\,N_2\uparrow + (2+2y)\,H_2\uparrow \tag{2}$$

Since a PECVD-process is not in thermodynamical equilibrium because of the low process temperatures, the fabricated layers have no stoichiometrical composition. Especially the undesirable inclusion of hydrogen (H^+) into the layer causes a strong absorption of light due to exitation of molecular (O-H- resp. N-H-) resonances in the infrared wavelength region ($\lambda > 1.3\ \mu m$).
The PECVD-depositions of SiO_2- and SiON-layers were performed on thermically oxidized silicon substrates. These thermically oxidized layers have a refractive index of $n = 1.46$ at $\lambda = 0.633\ \mu m$ and a thickness homogeneity better than 0.5 % for 4"-wafer size. Optimum PECVD parameters for homogeneous SiO_2-layers are substrate temperatures of 300 °C, a process pressure of 200 mTorr and a RF-power of 10 W. We achieved a homogenity better than 2 % (refered to 4"-wafer size).
Fig. 1 demonstrates the dependence of the refractive index from the ratio of gas flux. A refractive index of $n = 1.46$ is achieved for ratios $N_2O/SiH_4 > 40$. Smaller gas-ratios produce an increase of the refractive index due to inclusion of hydrogen and uncombined silicon within the layer.
In consideration of wave guidance by total reflection low-loss layers with a higher refractive index than SiO_2 are fabricated in form of nitrogen doped silicon oxide films. Fig. 2 shows that the refractive index is controlled by the gas ratio of N_2O and NH_4. With a increasing flow rate of NH_4 the combination of silicon with hydrogene instead of nitrogene increases with the consequence that there is no deposition of pure Si_3N_4. Therefore the refactive index $n = 2$ of Si_3N_4 is not reached.

SiO_2/SiON - waveguides

The light-guiding structures are fabricated in form of strip loaded rib-waveguides. Fig. 3 shows the cross section of such a waveguide.
The rib in the SiO_2 top layer has been structured in a RIE-process. The etched ribs exhibit vertical and graded sidewalls. Waveguides with oblique sidewalls as demonstrated by the SEM-photograph in Fig. 4 are fabricated by variation of RIE-parameters. The high quality waveguide faces are achieved by sawing and cleaving the silicon substrates along a crystal axis.

We found as a result of loss meassurements by means of the cut back method (Fig. 5) that waveguides with oblique sidewalls of the ribs have lower loss than waveguides with vertical sidewalls. The waveguide loss of 0.5 - 1 dB/cm for the wavelength λ = 0.633 μm is 0.5 dB/cm smaller compared to the loss for the wavelength λ = 0.83 μm. This additional loss is caused by the unsufficient isolation of the lightwave by the 2 μm thick thermically oxidized SiO_2-layer which produces a partial radiation of the wave into the silicon substrate.

Design rules for the strip loaded waveguides are obtained from numerical calculations of the guided modes by a finite difference algorithm (the corresponding computer program has been obtained from Prof. Dr. Lagasse, Univ. Gent). A calculated electrical field distribution of a rib-waveguide is shown in Fig. 6.

Near-field measurements (Fig. 7) confirm the theoretical considerations and show that the waveguides are very good qualified for the direct coupling with laserdiodes.

Rib-waveguides with a width of the rib smaller than 4 microns guide the fundamental wave only.

Figures

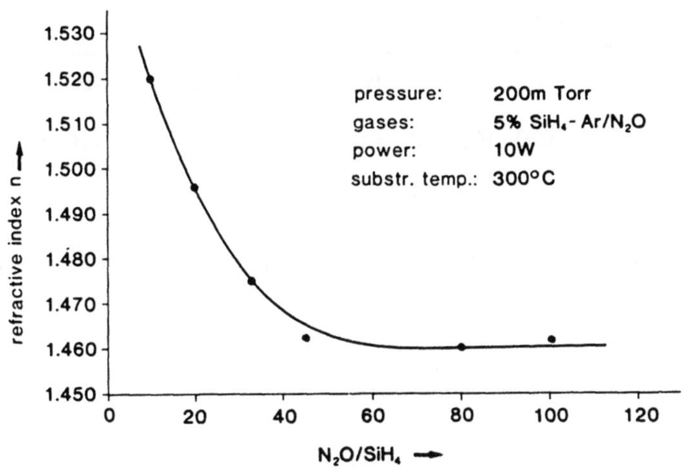

Fig. 1. Refractive index of PECVD-SiO_2 in dependence of the gas-ratio

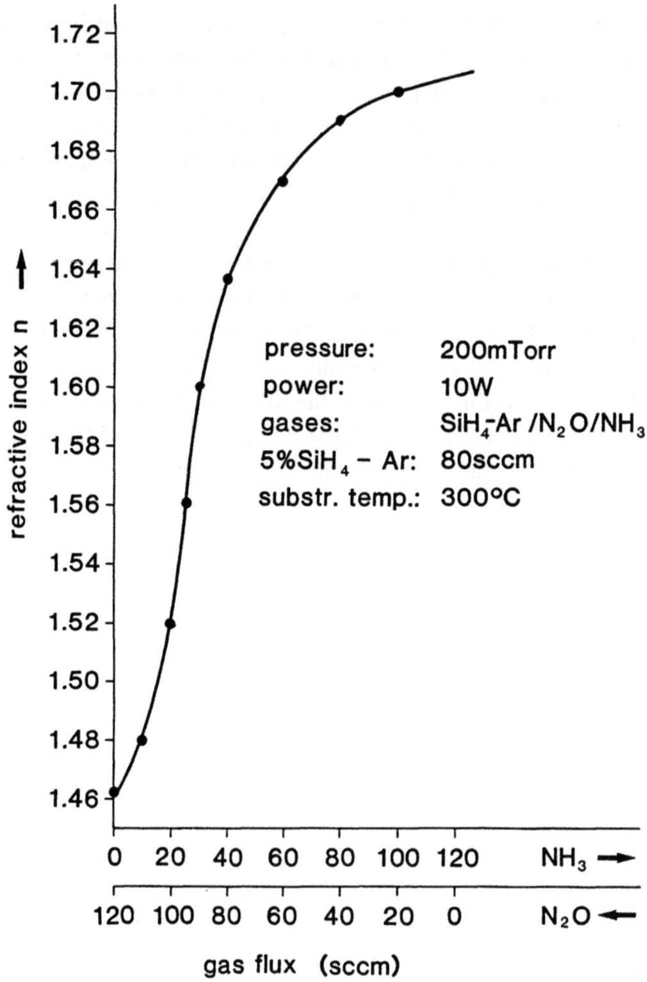

Fig. 2. Refractive index of PECVD-SiON in dependence of the gas-ratio

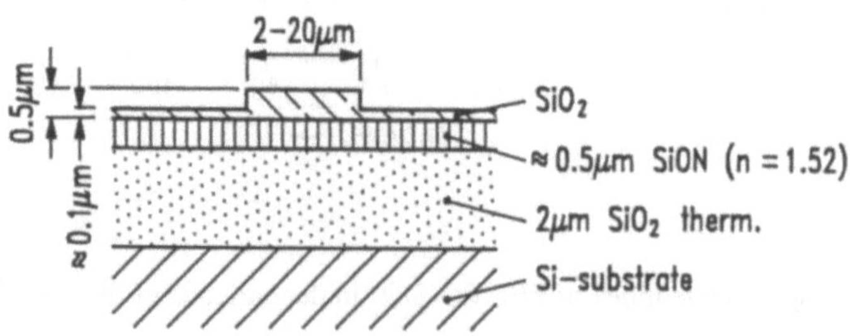

Fig. 3. Cross section of a SiO_2/SiON-rib-waveguide

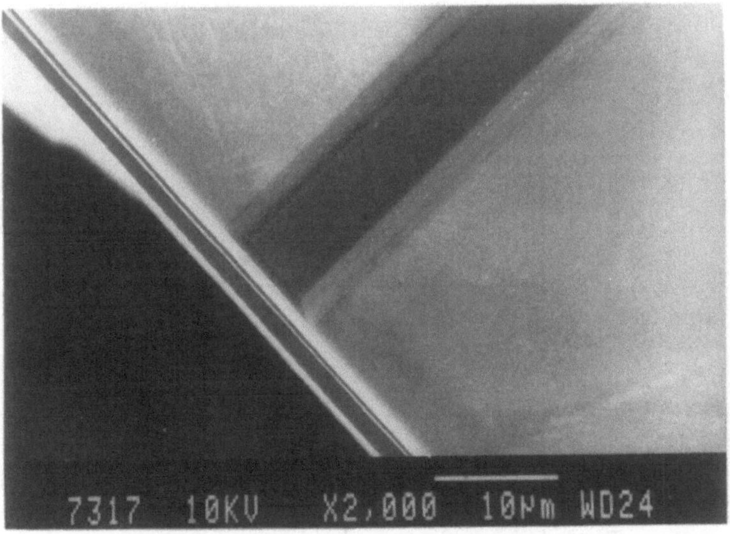

Fig. 4. SEM picture of a rib-waveguide

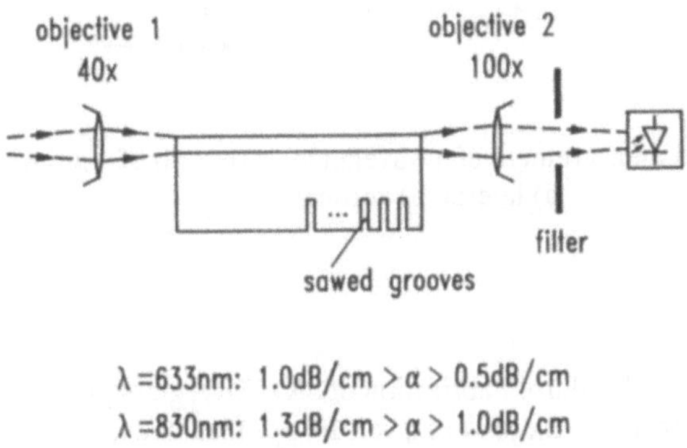

Fig. 5. Loss measurement with the cut back method

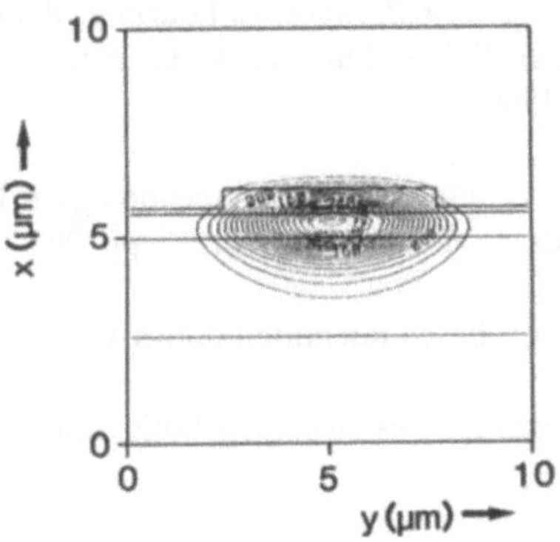

Fig. 6. Calculated electrical field distribution of a waveguide with a rib of 4 µm width

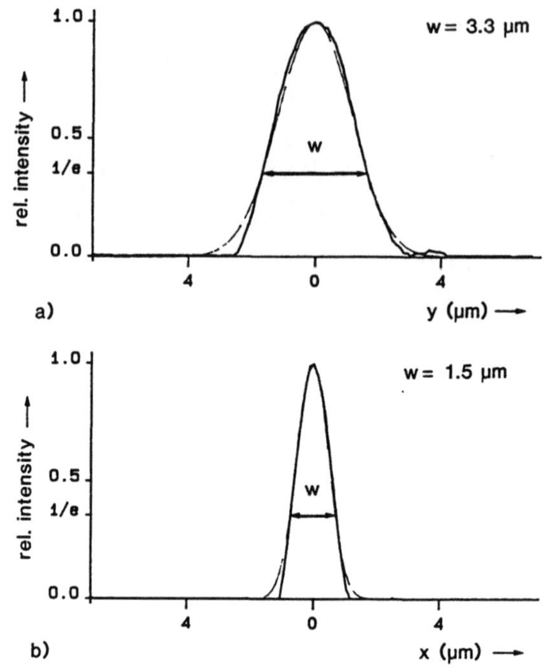

Fig. 7. Near field measurement of a waveguide with a rib of 4 μm width
a) lateral b) vertical

References

1. M. Kawachi, M. Yasu, T. Edahiro: Fabrication of SiO_2-TiO_2 glass planar waveguides by flame hydrolysis deposition. Electr. Lett., vol. 19, pp. 1883-1584, 1983.

2. N. Takato, M. Yasu, M. Kawachi: Low-loss high-silica embedded channel waveguides. Electr. Lett., vol. 22, pp. 321-322, 1986.

3. S. Valette, S. Renard, H. Denis, J. P. Jadot, A. Fournier, P. Philippe, P. Gidou, A. M. Grouillet, E. Desgranges: Si-based integrated optics technologies. Solid State Technol., pp. 69-74, Feb. 1989.

4. S. Valette: State of the art of integrated optics technology at LETI for achieving passive optical components. J. Modern Opt., vol. 35, pp. 993-1005, 1988.

5. D. G. Hall: Survey on silicon-based integrated optics. Computer, p. 25-32, Dec.1987.

6. H. S. Hickernell: Optical waveguides on silicon. Solid State Technol., pp. 83-87, Nov. 1988.

Polymer Waveguides in Integrated Optics

N. Keil, B. Strebel, H. H. Yao

Heinrich-Hertz-Institut für Nachrichtentechnik Berlin GmbH
Einsteinufer 37, D-1000 Berlin 10, FRG

Abstract

Future optical communication and optical signal processing systems will require a variety of passive and active optical components. Recently, some polymers with attractive processing properties have been developed and there has been a significant increase in interest in using such organics as optical waveguide materials to form different waveguide devices. In this paper, the possible applications of polymer waveguides in integrated optics, including passive, dynamic passive and active devices will be discussed.

Introduction

There is a growing interest in the research and development of organic polymers as optical waveguide materials for the application in the future optical communication and optical signal processing systems. Polymer fibers made from some organic materials exhibit very low optical loss (0.05 dB/m; λ = 650 - 660 nm) [1], quite high optical nonlinear coefficients (r_{33} = 45 pm/V) [2] and could be easily fabricated by low cost technologies [3-4]. Also, it is possible to synthesize rare earth metal-containing polymers [5]. This opens up the field for optical passive, dynamic passive and active waveguide devices, such as power deviders and combiners, optical couplers, optical filters, modulators, switches, wavelength converters and optical amplifiers, which can be integrated on a variety of substrate materials. Also, organic photodetectors have already been demonstrated [6].

The monolithic integration of a GaAs/AlGaAs laserdiode and a polymeric multilayer waveguide has been demonstrated [7]. This integration scheme can also be used for the fabrication of other components, such as laserdiodes with external cavities or integrated coherent receivers as shown conceptually in Fig. 1. It may be possible to integrate polymer passive devices and dynamic passive devices with laserdiodes and photodetectors to form very complicated coherent diversity receivers or transmission systems. Such polymer devices will be very interesting for future optical communication techniques.

In this paper, our research work on polymer waveguides in integrated optics, including passive, dynamic passive devices and especially the Nd^{3+} and Er^{3+} doped polymers, is presented.

Passive polymer waveguide devices

The polymethylmethacrylate (PMMA) and a photoinitiator have been used for fabricating monomode strip waveguides as the basic elements for passive devices [8]. The fabrication process is shown in Fig. 2 and it includes: mixing of waveguide materials, spin-coating the material on a substrate, UV-exposure and finally an annealing treatment. Different substrates, such as quartz, optical glasses and Si-wafers with SiO_2 buffer layers have been used. The passive monomode waveguide structures are fabricated by using filtered UV-light (λ = 366 nm) through a negative mask at about 6-8 mW/cm^2. During UV-exposure there is a photo-chemical process in the illuminated area of the doped polymer film and the refractive index is changed. In order to develop and fix this refractive index change, an annealing treatment is followed.

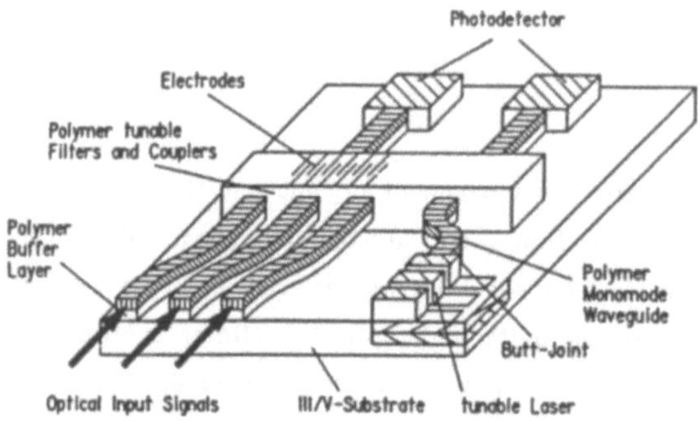

Fig.1: Integration concept of a future coherent receiver with polymer waveguide devices.

Fig.2: Fabrication process of polymer waveguides:

a) mixing of waveguide materials
b) spin-coating
c) exposure (8 mW/cm^2; λ = 366 nm)
d) annealing (60 - 70 °C).

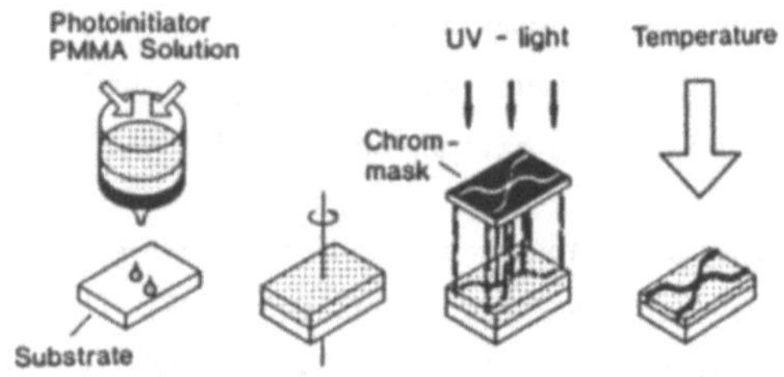

The measured relationship of the refractive index change with exposure and annealing time for a polymer waveguide with 40% benzildimethylketal in PMMA is shown in Fig. 3. It can be seen that the refractive index is increased in the exposured area and reached the saturated value after about 60 minutes exposure time, however it has a rather low change in the unilluminated area. During annealing, the index in the exposured area shows rather low change while in the unilluminated area it decreases rapidly to the value near the index of PMMA. After annealing, an index difference of 0.025 is obtained for 40% of photoinitiator concentration. The measured relationship between photo-induced refractive index change and concentration after the annealing treatment is shown in Fig. 4.

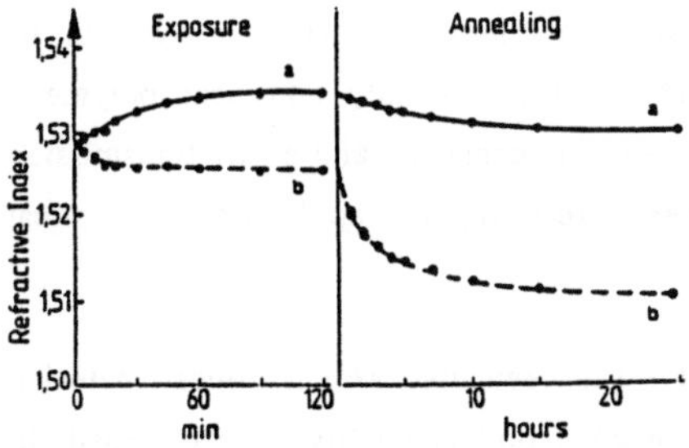

Fig.3: Refractive index change vs. exposure and annealing time (40% Benzildimethylketal).
a: illuminated area
b: unilluminated area.

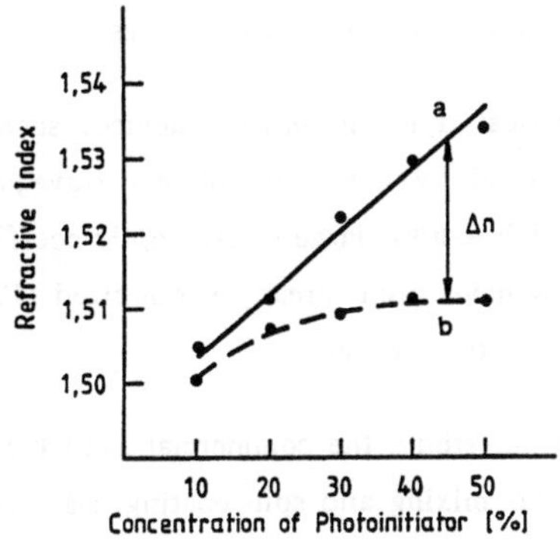

Fig.4: Refractive index change vs. photoinitiator concentration.

a: illuminated area
b: unilluminated area.

The polymer monomode strip waveguides and related passive devices such as couplers and Mach-Zehnder interferometers have been fabricated. The microscopic picture of such devices is shown in Fig. 5. In order to measure the transmission loss of polymer slab waveguides, the scattering-detection method is used. The measured transmission loss for our polymer slab waveguides is about 0.6 dB/cm. The near-field pattern of a monomode polymer strip waveguide operated at λ = 1.55 μm is shown in Fig. 6.

Fig 5: Microscopic picture of polymer Mach-Zehnder interferometers.

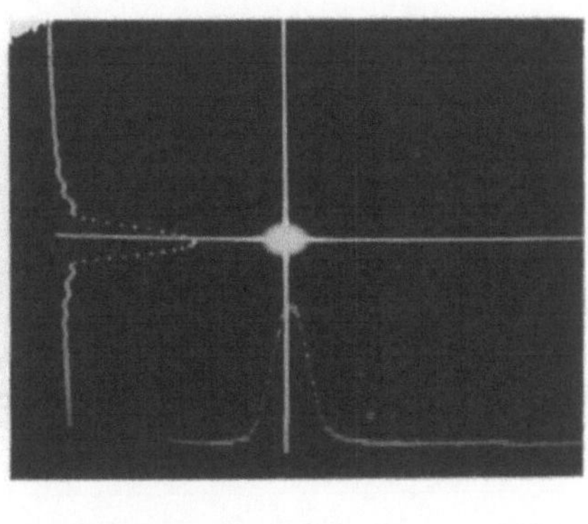

Fig.6: Near-field pattern of a monomode polymer strip waveguide (λ = 1.55 μm).

Optical dynamic passive devices

Optical dynamic passive devices such as modulators, switches and switch arrays can be realized by low cost polymer waveguide technology [9-11]. Optically nonlinear polymers, which exhibit large electro-optic coefficients, low dielectric constants and a low transmission loss have been already synthesized [12] and are very promising for applications in optical signal processing.

In our study, the commercial available disperse red 1 has been used as a nonlinear material. After mixing and spin-coating, slab waveguides with good optical quality were obtained. In order to form nonlinear strip waveguides, first the photoinitiator is added and the standard photolithiographic technique has been applied. The index increase is observed in the exposured strip area and the nonlinear properties still remain.

After the spin-coating process, the optical polymer waveguides containing nonlinear molecules do not show second order nonlinear effects because of the random orientation of the organic molecules. In order to achieve molecular orientation, secondly a corona poling process is followed [13]. After poling, the electro-optic phase modulation was measured in a Mach-Zehnder interferometer as shown in Fig. 7.

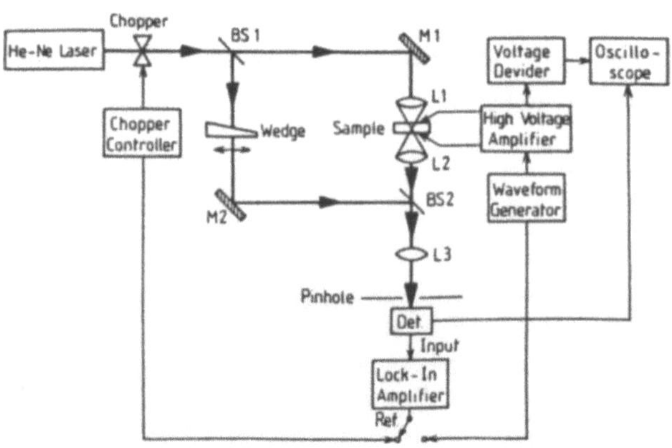

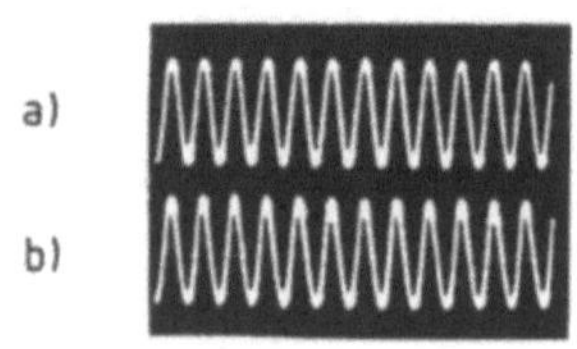

a: Modulation signal (V_{pp} = 500 V; f = 1 KHz).

b: Interferometer output signal.

Fig.7: Experimental layout for E/O-coefficient measurement.

To build practical devices, a multilayer structure is necessary. The multilayer structures, including substrate, electrode, buffer layer and nonlinear polymer layer have been fabricated.

Active devices

Rare-earth doped single-mode fibers have led to a new class of both active and passive fiber devices including lasers, optical amplifiers, filters and distributed sensors [14]. On the other hand, the study of polymers containing rare-earth metals has been initiated [5]. Very recently, the study of Er^{3+} doped single-mode $LiNbO_3$ waveguides, which could be a promising candidate for optical amplifiers and lasers in the 1.5 - 1.6 μm wavelength range, has been reported [15]. From these, especially for rare-earth doped polymers, the fluorescence intensities reach a maximum at much higher rare-earth metal concentrations [16], therefore it can be expected that the rare-earth containing polymer waveguides may be also promising candidates for optical amplifers and lasers.
The Er^{3+} and Nd^{3+} containing PMMA has been synthesized and the absorption spectra have been measured as shown in Fig. 8 and Fig. 9. The concentration of rare-earth metal is about a few percent. The measured absorption bands are quite similar to that of Er^{3+} and Nd^{3+} doped fibers. The first Er^{3+} doped polymer film waveguide has been developed. As next steps, the material synthesis will be further improved and the fluorescence spectra and the lifetime will be measured.

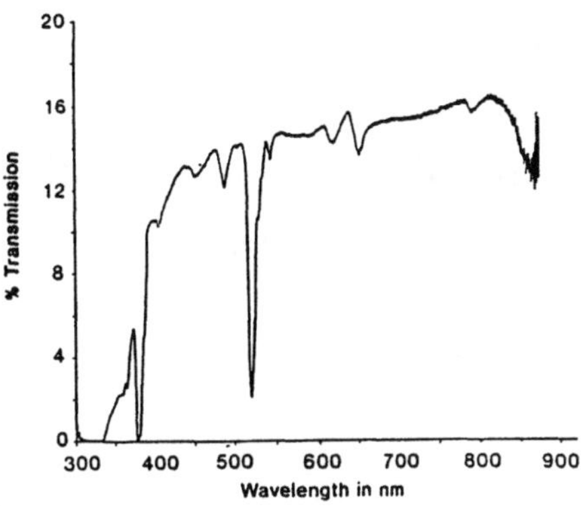

Fig.8: Absorption spectrum of Er^{3+} doped PMMA.

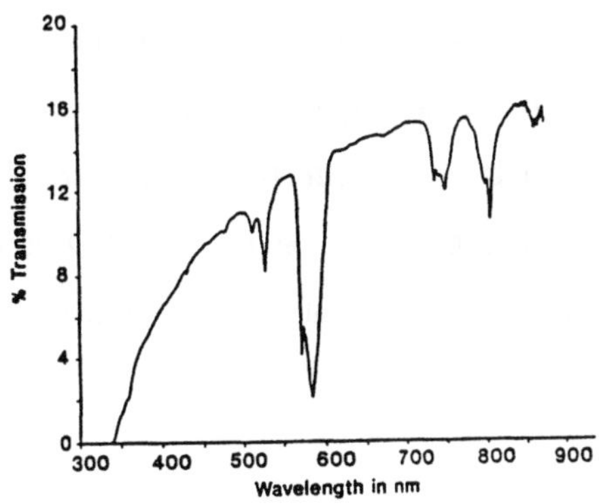

Fig.9: Absorption spectrum of Nd^{3+} doped PMMA.

Acknowledgement

This work is part of a project supported by the Deutsche Bundespost Telekom. The authors wish also to thank J. Krauser for performing the near-field pattern measurements and C. Baack and G. Heydt for valuable discussions.

References

[1]: W. Groh, et. al.; Elektronikpraxis (1988), 26-31.

[2]: G. H. Cross, et. al.; SPIE Proceedings, Int. Soc. Opt. Eng., Integrated Optics and Optoelectronics, Boston, 5-8 Sept. 1989.

[3]: P. Pantelis, et. al.; British Telecom Techn. Journ. 6 (1988), 5-17.

[4]: B. L. Booth; J. Lightwave Technol. 7 (1989), 1445-1453.

[5]: Y. Okamoto; J. Macromol. Sci. Chem. A24 (3&4) (1987), 455-477.

[6]: R. Reuter and H. Franke; Appl. Phys. B48 (1989), 219-224.

[7]: P. Van Daele, et. al.; in Proceedings Vol.3, post-deadline papers, 15th Eur. Conf. Optical Commun.(Gothenburg, Sweden) 10-14 Sept. (1989), 58-61.

[8]: H. Franke and W. Heuer; SPIE Proceedings, Vol. 651, (1986), Integrated Optical Circuit Eng., 120-125.

[9]: M.B.J. Diemeer, et. al.; in Proceedings Vol.1, regular papers, 15th Eur. Conf. Optical Commun.(Gothenburg, Sweden) 10-14 Sept. (1989), 425-428.

[10]: M.B.J. Diemeer, et.al.; J. Lightwave Technol. 7 (1989), 449-453.

[11]: P. Kaczmarski, et. al.; IEE Proceedings, 136 Pt. J, (1989), 152-158.

[12]: D. Hass, et. al; Photonics Spectra 4, (1989), 169-174.

[13]: K.D. Singer, et.al.; Appl. Phys. Lett. 53 (1988), 1800-1802.

[14]: D. N. Payne; In Techn. Digest Vol. 5, IOOC'89, Kobe, Japan, 18-21 July, (1989), 2-3.

[15]: R. Brinkmann, et. al.; OFC'90 22-26 Jan. 1990, San Francisco, USA, paper PD 1.

[16]: E. Banks, et. al; J. Appl. Polymer Sci. 25 (1980), 359-368.

Technologies

Pattern Generation and Image Transfer

Aspects of Patterning Inorganic-Organic Copolymers (ORMOCERs)

M. POPALL, H. MEYER, J. SCHULZ

Fraunhofer-Institut für Silicatforschung, Neunerplatz 2, D-8700 Würzburg, F.R.G.

Summary

Inorganic-organic copolymers (ORMOCERs ≡ ORganically MOdified CERamics), transparent for optical purposes and synthesized by sol-gel processing have been developed for interconnection technologies in microelectronics. Unsaturated substituents like methacrylates or epoxides and UV-sensitive initiators enable a photopolymerization. Using a frequency doubled Argon-ion laser at 257 nm for direct laser writing, patterned layers with rectangular profile have been realized. The influence of different parameters (age of lacquer, writing velocity, initiator concentration and laser intensity) on the height, width and visual quality of patterns are studied. In combination with high breakthrough voltages, low permittivity constants and high bulk resistivities they open interesting aspects for very large system integration techniques (VSLI). Investigations to establish their potential for optoelectronic applications are in progress.

Introduction

Complex future applications of microelectronic systems demand intelligent products with maximum integration of different components like sensors, memories, processors, actuators and an optimized interconnection technology for transducing and transmitting the electronical, optical and mechanical information. Micro system technologies and better materials to meet these demands are under investigation.

Microelectronics require tailormade materials and new techniques for interconnection, especially for enhanced integration and multilayer technology [1]. Patternable dielectric materials with high electrical resistivity for high density of conducting paths, low permittivity constants for minimum signal deformation are desired. Also good adhesion to substrate material and to active and passive components is necessary. Dielectric polymers like polyimide and polyimide/polyamide blends are well established and the state of the art [2]. For higher integration, the permittivity constant should be as low as possible, water take up has to be minimized.
Presently, for integrated optics, organic polymers like polyacrylic acid are patterned to achieve devices for multiplexing or for expanding diode arrays. Some applications require better thermal stability or light transmission (CH stretch overtones) [3].

ORMOCERs are materials whose properties can be varied between those of pure inorganic materials and organic polymers. They consist of an inorganic oxidic backbone (Si, Al, Ti, Zr, ...) chemically bonded to an organic network, e. g. by use of (functionalized) organosilanes and are prepared by sol-gel processing. Due to the functionalization properties relevant to various applications could be established [4,5,6,7].

For electronic purposes, ORMOCER layers were synthesized with low ϵ values, good adhesion to different substrates, rather low H_2O take-up and moderate curing temperatures [8]. Preliminary aspects of patterning ORMOCER dielectrics with photolithographic methods like proximity printing or laser direct writing are reported in ref. [9] and [10], while the potential of ORMOCERs for optoelectronic applications is discussed in ref. [11].

During current investigations we are developing ORMOCERs which can be patterned by direct laser writing techniques. This technique provides a flexible tool for microelectronic and optoelectronic design. In this article we will discuss various aspects relevant to direct laser writing of two types of ORMOCERs.

Experimental

The compositions of the ORMOCERs are:
1: 0.39 mol 3-(glycidoxypropyl)trimethoxysilane, 0.39 mol 3-methacryloxypropyltrimethoxysilane, 0.18 mol vinyltrimethoxysilane, 0.04 mol tetraethoxysilane.
2: 0.39 mol 3-(glycidoxypropyl)trimethoxysilane, 0.39 mol 3-methacryloxypropyltrimethoxysilane, 0.18 mol diphenylsilanediol, 0.04 mol tetraethoxysilane.
The components are mixed together without solvent and refluxed under air up to 6 hours. 50 % of the water necessary for a stoichiometric hydrolysis is added successively. The clear, viscous liquid (subsequently called lacquer) is cooled to room temperature and stored in closed bottles. Shelf life of both systems is in the order of about 4 months.
The lacquer is applied to glass substrates by spin-on coating. Further information about processing and characterization are given in [10].
For laser writing a frequency doubled Ar-ion laser yielding up to 100 mW at 257 nm was used.

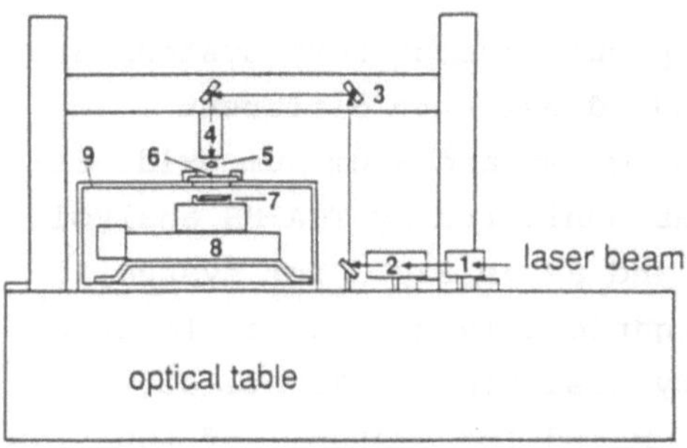

1. electro-optic modulator
2. polarizer
3. mirrors
4. Gallilei telescope
5. lens (fused silica)
6. window (fused silica)
7. target holder and target
8. motor driven stages
9. reactor

Fig.1. Scheme of the laser writing equipment

The set-up is shown in fig. 1. The laser beam is directed towards the horizontal target via several mirrors. To achieve a minimum laser spot size, the beam passes through a gallilean telescope expanding the beam diameter by a factor of 2. A mechanical shutter or an electro-optic modulator are used to switch the beam by computer control. Finally the laser beam is

focussed by a quartz lens onto the target. The target itself is placed onto two motor driven stages providing the horizontal xy-movement. To adjust the focus on the target the lens position can be varied. The positioning system can be placed completely inside a vacuum vessel to operate under oxygen-free conditions. For flexibility of the patterning process, the positioning system as well as the lens position are controlled by a computer. The resolution of the system is 100 nm at a maxium velocity of 20 mm/s.
For controlling focus diameter and profile, the laser spot radius is measured using the scanning knife edge method [13]. A razor blade is placed on the target. While scanning the edge through the laser spot the intensity is monitored.
The laser spot diameter is determined by differentiating the intensity profile. So far the smallest spot size thus obtained is 8 μm.
The patterns are developed either by organic solvents like acetone or by 0.1 N NaOH.

Results and discussion

Details of the general material properties of both systems are described in [10]. They show good adhesion to different substrate materials like silicon, glass and aluminum oxide ceramics. Thermal decomposition (as monitored by TGA-MS analysis) begins above $200^{\circ}C$ for system 1 and above $270^{\circ}C$ for system 2. Both systems are resistent to organic solvents for at least a few minutes and pass the humidity test DIN 50017. Dielectric strengths for system 1 of 100 V/μm and for system 2 of 400 V/μm are measured. Both systems show a bulk resistivity better than 10^{14} $\Omega\cdot$cm and permittivity constants in the range of 3-6, depending on sample preparation and curing history. The first optical data are estimated for system 1: The refractive index n_D20 is 1.504, the absorption spectrum is given in fig.2. In the range of 400-1100 nm only very low absorption can be seen.

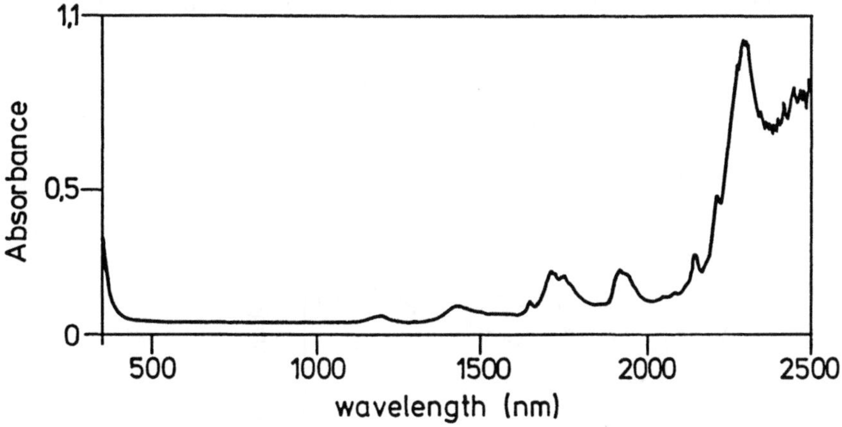

Fig.2. Absorption spectrum of the cured system 1

As the photochemical curing of both systems depends on the same patterning parameters system 1 has been chosen for further investigations.

The patterning process is based on the organic crosslinking (polymerization of methacrylic groups) of the polysiloxanes obtained by hydrolysis and condensation.During the photopolymerization more than 80 % of the C=C bonds are consumed [10].

The quality of the patterned layer depends on the material properties (especially the degree of prepolymerization), the polymerization conditions (mainly the type of initiator and the laser parameters) and the development conditions. Particulary important are parameters like exposure dose, initiator concentration and the age of the lacquer. For laser direct writing, radiation density, exposure time, absorption of the matrix, quantum yield and radical life time are also important parameters. Therefore we determined the influence of these processing parameters on the width, height and visual appearance of the structures.

The influence of the initiator concentration and the laser intensity for direct laser writing on structure width has been studied previously [10]. A linear dependence on the laser intensity and a logarithmic one on the initiator concentration is found.

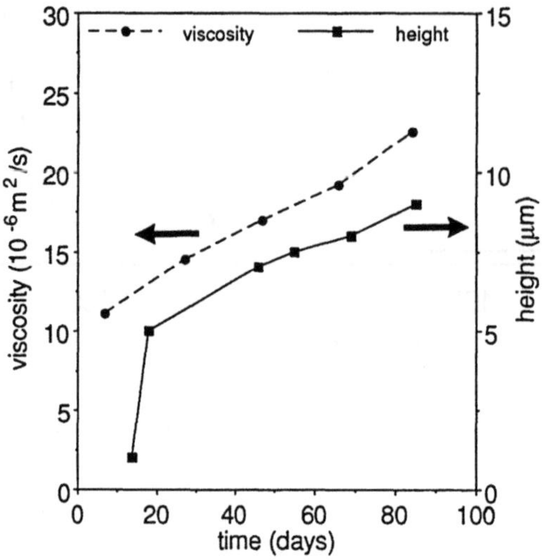

Fig.3. Dependence of height and viscosity of system 1 on the age of the lacquer

The influence of the age of the lacquer on the height of structure is shown in Fig.3. The structure was created using 0.1 % starter concentration, 80 W/cm^2 laser intensity and a writing velocity of 0.2 cm/s. Under these conditions structures could only be produced starting 16 days after the synthesis of the lacquer. Within several days the achieved height of the structure on glass increases from 1 μm to 5 μm. Thereafter the photocured structures exhibit heights which show a linear dependence on the age of the used lacquer. Although no values for the thickness of the uncured layer have been measured, a step-like change of the thickness is unlikely, since the viscosity (Fig.3) shows a strict linear time dependence over the total time range studied. By raising the laser power and the initiator concentration the height of the structures can be increased (in the time range of 16 to 20 days). A possible explanation is the formation of only partly polymerized oligomers, which are still soluble in the development solutions. The subsequently linear increase of height will probably be due to the higher thickness of the uncured layer caused by the rise of the viscosity.

Another important parameter is the writing velocity. Their influence is studied for two different initiator concentrations. At the low initiator concentration the height of the structure passes through a maximum at a velocity of about 13 mm/s. At the high one the height increases monotonically with the velocity. The maximum might be explained by two competing processes: The degree of photopolymerization throughout the lacquer is determined by the quantum yield of radical formation from the photoinitiator. We assume that the higher the degree of polymerization the higher the density of the photocured material will be. Therefore a high photon flux (achieved at low velocities) results in smaller heights. On the other hand at small photon fluxes the degree of photopolymerization might not be sufficient to ensure a reduction of solubility during the development process which is high enough. This assumption also explains the dependence of structure height on initiator concentration and laser intensity as shown in figure 4.

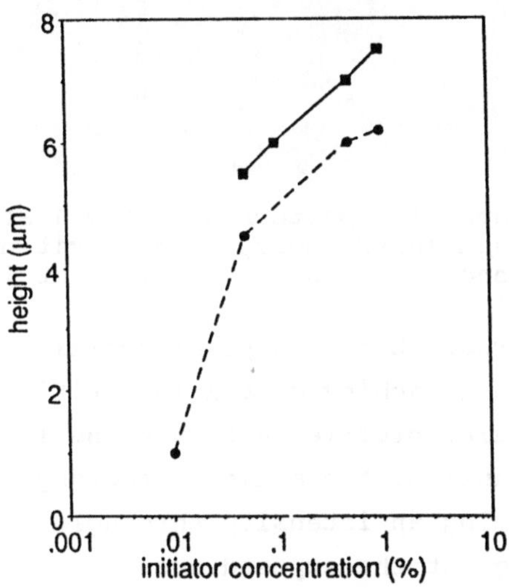

Fig.4. Dependence of height of structure on initiator concentrations, -▪- at 300 W/cm^2 , -•- at 600 W/cm^2

The structure height increases logarithmically with the initiator concentration in the range of 0.05 to 1.00 mass %. At smaller concentrations the height decreases dramatically. Furthermore, at 0.01 mass %, structures could only be achieved at high laser intensity.

Also the dependence of structure height and width on spinning frequency has been investigated. The width decreases slightly with increasing writing velocity but is independent of the spin on frequency. The optimum width is given by the minimum spot size of the laser beam. The structure height shows a small increase with the writing velocity. Its value can be varied between 8 and 15 μm by changing the spin-on frequency from 400 to 600 rpm for a given viscosity (here 14 mm^2/s).

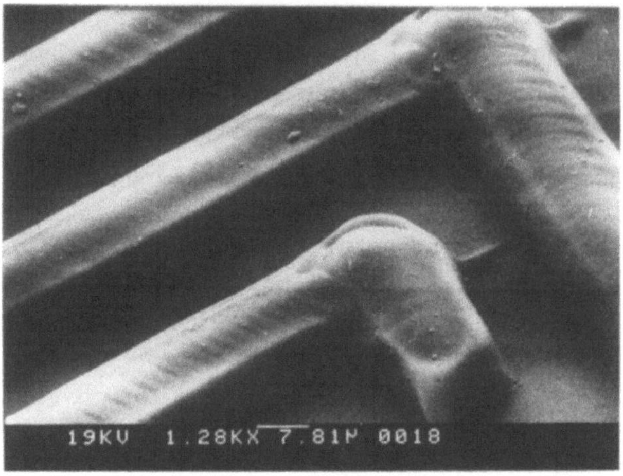

Fig.5. SEM micrograph of a patterned system 1 at 80 W/cm^2 laser intensity, 8μm laser focus, 2 mm/s writing speed, 0.1% initiator concentration and development in acetone.

In fig.5 a SEM micrograph of a typical optimized pattern is shown. This pattern is achieved at a low writing speed (2 mm/s) yielding a rectangular profile. Although the laser beam intensity profile is gaussian, the achieved rectangular profile can be understood assuming an intensity threshold for complete polymerization throughout the exposed area.

Conclusions

Patterned ORMOCERs are developed using UV-radiation and UV-sensitive initiators. A process to pattern ORMOCERs via laser direct writing could be established. Very interesting materials for opto- and microelectronic purposes and good quality of structure are achieved. Further investigation of the correlations between system composition, exposure to UV-radiation for polymerization and etching process have to be carried out to optimize patternable ORMOCER layers for electronic and opto-electronic applications.

Acknowledgement

We wish to thank the Bundesministerium für Forschung und Technologie of the Federal Republic of Germany for their financial support. We also like to acknowledge the work of the technical staff members.

References

[1] M. Terasawa, S. Minami and J. Rubin, Proceedings of ISHM Symposium on Microelectronics, 607 - 615 (1983).
C. A. Neugebauer and R. O. Carlson, IEEE Transactions on Components, Hybrids and Manufacturing Technology, 10 (2), 184 - 189 (1987).
Y. S. Lin, H.S. Cole in: Electronic Packaging Materials, Materials Science IV, eds. R. Jaccodine, K. A. Jackson, E. D. Lillie, R. C. Sundahl, MRS Vol. 154, Pittsburg 11 (1989).

[2] Plastics for Electronics, ed. M. T. Goosey, Elsevier Applied Science Publishers, London (1985).

[3] W. Groh, D. Lupo and H. Sixl, Adv. Mater., No. 11 (1989) 366

[4] H.Schmidt, B.Seiferling, G.Philipp and K.Deichmann in : Proc.Int.Congress on Ultrastructure Processing of Ceramics, Glasses and Composites, San Diego, California, USA, February 1987, J.D.Mackenzie and D.R.Ulrich eds..

[5] H.Schmidt and B.Seiferling, Mat.Res.Soc.Symp.Proc.73 (1986) 739.

[6] H.Schmidt, Mat.Res.Soc.Symp.Proc.32 (1984) 327.

[7] H.Schmidt in: ACS Symposium Series No. 360, Inorganic and Organometallic Polymers, M.Zeldin, K.J.Wynne and H.R.Allcock. eds. (1988) 333.
- J. Non-Cryst. Solids 73 (1985) 681.
- J. Non-Cryst. Solids 100 (1988) 51.

[8] H. Schmidt and H. Wolter, Fifth International Workshop on Glasses and Ceramics from Gels, Rio de Janeiro (1989).
M.Popall und H.Schmidt: Electrical and electrochemical applications of ORMOCERs. In: Proceedings TEC 88 - Materiaux, Grenoble, Oktober 1988.
H.Schmidt, M.Popall und H.Wolter: Sol-Gel-Techniken für neue Materialien der Mikrotechnik. VDI Berichte 796 (1989) 51.

[9] H. Schmidt, M. Popall and J. Schulz in: Proceedings Second International Symposium on New Glass, The Association of New Glass Industries,Tokyo, 134 - 135 (1989).
D. B. Bailey, R. C. Daly, T. B. Brust and J. M. Pearson in: Polymers for Advanced Technologies, IUPAC International Symposium, ed.: Menachem Lewin, Jerusalem, 18 - 33 (1987).

[10] M. Popall, H. Meyer, H. Schmidt and J. Schulz, MRS Spring Meeting, San Francisco (1990)

[11] B. Lintner, N. Arfsten, H. Dislich, H. Schmidt, G. Philipp and B. Seiferling, J. Non-Cryst. Solids 100 (1988) 378

[12] Y. Suzaki, A. Tachibana, Appl. Opt. 14, 2809 (1971).

Photopatternable Layers from Amorphous, Diamond-Like Carbon – An Innovative Material for Microsystems Technology

A. Hammerschmidt, S. Birkle, J. Kammermaier, R. Leuschner, W. Schmidt, R. Schulte

Siemens AG, Corporate Research and Technology, Paul-Gossen-Str. 100, D-8520 Erlangen, Federal Republic of Germany

Summary

A class of material, promising for micro system technology are amorphous, diamond-like carbon films deposited in low pressure plasmas. They can be produced with controllable electrical and optical properties. They exhibit outstanding mechanical properties and an excellent stability against acids, bases and organic solvents. These films can be patterned via silicon containing photoresists with a subsequent O_2-RIE-etching process down to the sub-micrometer range. They are excellent barriers against moisture.

Introduction

Amorphous hydrogenated diamond like carbon films (DLC-films) deposited in a plasma process are finding increasing interest in the recent past because of their outstanding mechanical, optical and electrical properties. They form hard films (1) with low friction coefficients down to 0.08 for DLC/Al or 0.02 for DLC/DLC (2). They exhibit a low infrared absorption and, in the visible range, the optical properties as absorption coefficient (optical band gap) or refractive index can be varied in a wide range (3). The same is true for the electrical parameters as the dielectric constant or the specific resistance (3) which can be changed between 2.3 to 12 and 10^2 to 10^{17} Ωcm respectively.

This great variety of DLC film properties can be explained by the molecular structure. In contrary to diamond films in which sp^3-hybridized carbon is bonded tetrahedrally to adjacent carbon atoms, in DLC films we find carbon atoms with sp^1-, sp^2- and sp^3-hybrid orbitals (4) with a typical ratio of 1:2 for sp^2/sp^3, and, this is important, a significant amount of hydrogen with a H/C-ratio up to 0.6 which may be bonded to the different kinds of carbon atoms. Because these ratios can vary the layer properties can vary too.

As far as the structure is concerned different structural models can be proposed. With respect to the amorphous behaviour a random network of sp^1-, sp^2- and sp^3-carbons saturated at the edges with hydrogen can be assumed for the most simple case. The structure can be explained in a more complex way assuming a heterogeneous structure consisting of polycyclic aromatic hydrocarbons interconnected with tetrahedral carbon atoms (5). Until now the structure of DLC-films is not yet completely understood in detail.

We will report on layer deposition, characterization, measurements of barrier properties and photolithographic production of DLC patterns.

<u>Layer deposition</u>

DLC films can be produced using chemical vapor deposition or sputtering CH-targets. In our case we used low pressure plasmas with a characteristic pressure between 0.1 to 5 mbar.

Fig. 1 shows a schematic view of the apparatus used in the experiments .

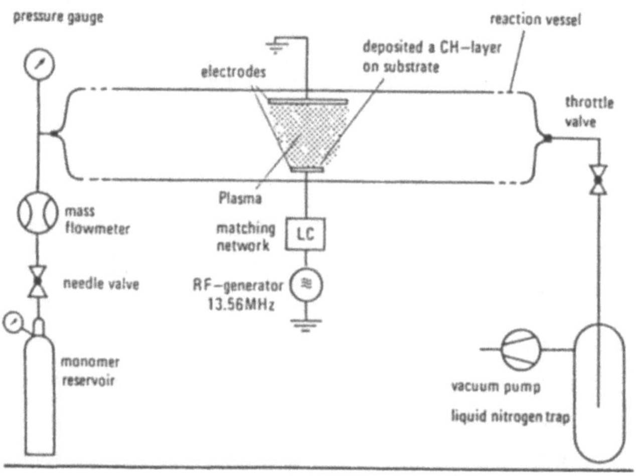

Fig. 1: Radiofrequency-plasmadeposition of methane – schematic of the used apparatus

A radiofrequency excited plasma was used for the deposition. For this purpose the power produced by a 13.56 MHz generator (output max. 500 W) is transduced via a tunable matchbox capacitively to the reaction vessel. The electrodes are dimensioned differently (area ratio anode/cathode = 6.2:1) to get a DC self bias potential superimposed on the high frequency voltage. This effect is important for DLC deposition: the smaller electrode is the cathode, at which usually the DLC deposition is performed.

As process gas methane was used. As substrate Si-wafers or quartz plates were usually taken. They were cleaned prior to deposition in an

ultrasonic bath with an commercial alkaline cleaner, ethanol, methylenchloride and benzene and then dried and stored at 200° C. Kapton films for the permeation measurements were cleaned with ethanol.
Typical deposition parameters are:

- pressure : 0.2 - 2.5 mbar
- power : 0.7 - 2.5 W/cm^2
- resulting self bias : -900 - - 70 V

Layer characterization

The characterization of DLC layers was only possible via samples on substrates (layer thickness between 0.1 and 1.5 µm). For infra-red spectroscopy we deposited DLC layers on non-doped silicon wafers on which they adhere very well. Thus direct measurements were possible by Fourier Transform IR-spectroscopy.
A sensitive representative indicator of layer properties is the optical band gap. For determining this quantity depositions were made on quartz plates under different conditions. The absorption in the near UV and visible range was measured with a conventional UV/Vis spectrometer. The Tauc plot valid for amorphous semiconductors (6) according eq.(1)

$$(\alpha E)^{1/2} = A\,(E - E_{opt}) \qquad (1)$$

wherein α is the absorption coefficient, A a constant, E the incident energy and E_{opt} the optical gap provides an easy way to determine the optical gap.
The electrical data were determined in capacitance measurements. In Tab. 1 a correlation between optical band gap and electrical parameters is given. As shown it is possible to tailor the electrical properties by means of the deposition parameters.

Measurement of barrier properties

The barrier properties DLC films concerning water permeation through DLC films were determined. Therefore a 128 µm thick Kapton film was coated with 1.3 µm DLC and placed within the measurement apparatus shown in Fig. 2. The leakage rate Q can be calculated either from the temporal pressure increase p_0 within the constant volume V_0 or from the amount of gas pumped off at constant pressure p_0. In volume V_1 we have the interesting gas with the pressure p_1. The permeation coefficient can be calculated from equ. (2)

deposition parameters	optical gap (eV)	specific resistance (Ω cm)	dielectric constant
2 mbar methane -150 V self bias	2	10^{17}	2.3
1 mbar methane -420 V self bias	1.2	10^{12}	4
0.2 mbar methane -880 V self bias	0.9	$5*10^{7}$	10.8

Table 1: Correlation between deposition parameters, optical band gap and electrical properties

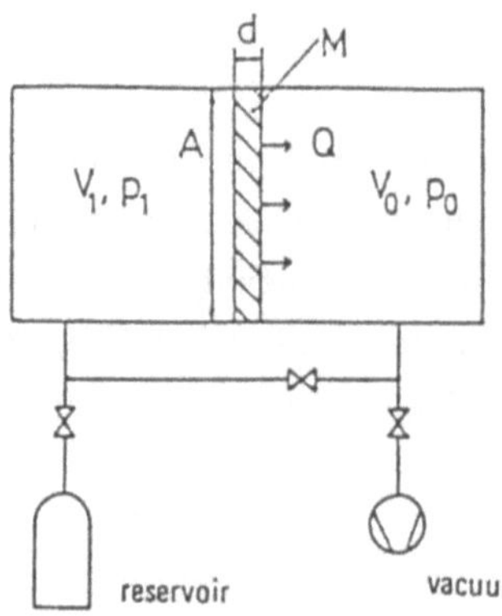

Fig. 2: Measurement of the permeation coefficient: sample M with the thickness d, area A

$$\mathcal{P} = (d/a)*(\dot{p}_0 V_0/p_1) \qquad (2)$$

wherein $\mathcal{P}$ is the permeation coefficient, d the thickness and a the area of the sample. In the case of a bilayer sample (DLC on Kapton) we get for the permeation coefficient of DLC $\mathcal{P}_{DLC}$ from the overall permeation coefficient $\mathcal{P}$ and the coefficient of Kapton $\mathcal{P}_{Kapton}$:

$$\mathcal{P}_{DLC} = \frac{d_{DLC}}{d_{Kapton}} \left[\frac{1}{\mathcal{P}} - \frac{1}{\mathcal{P}_{Kapton}} \right]^{-1} \qquad (3)$$

In Tab. 2 some results of the measurements are presented. DLC films show significantly lower permeation rates of water than the polyimide Kapton; that means DLC-films have good barrier properties against moisture. It is

deposition parameter	permeation coefficient (m^2/s)
(Kapton)	$1.5*10^{-10}$
1 mbar, - 450 V selfbias	$5*10^{-13}$
0.2 mbar, - 880 V selfbias	$2*10^{-13}$

Table 2: Deposition parameters and water permeation coefficient of Kapton and DLC

remarkable that the permeation coefficient of DLC depends clearly on the deposition parameters.

Photopatterning of DLC films

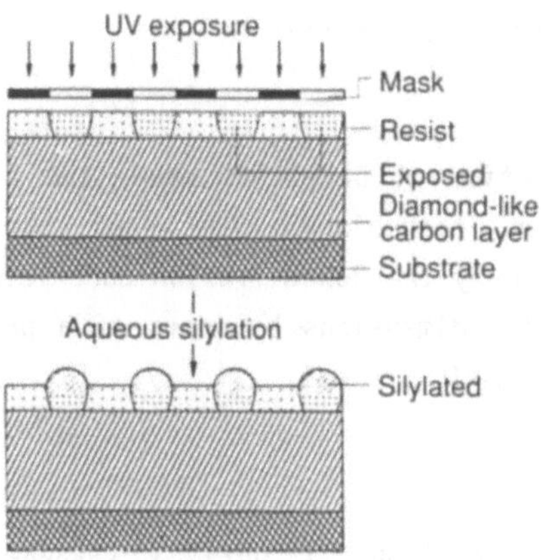

Fig. 3: TOP CARL process

Photopatterning of DLC films is done by a silicon containing top imaging system in which pattern transfer is achieved by a subsequent O_2-reactive ion etching (O_2-RIE-) process.
For the option to make even very fine patterns the TOP-CARL process developed by Siemens (7, 8) was chosen (Fig. 3). The photoresist coated on the DLC-film is exposed through a corresponding mask and treated with a the aqueous silylation reagent. This reagent reacts only on the exposed areas with the photoresist polymer resulting in an incorporation of Si atoms. In the subsequent reactive ion etching process these Si atoms act as an etch mask against the oxygen plasma.
Thus the non-silylated regions are etched and the resist patterns are transfered to the underlying layer. After etching the Si-containing auxiliary layer is removed by an appropriate solvent. Fig. 4 shows 1 µm

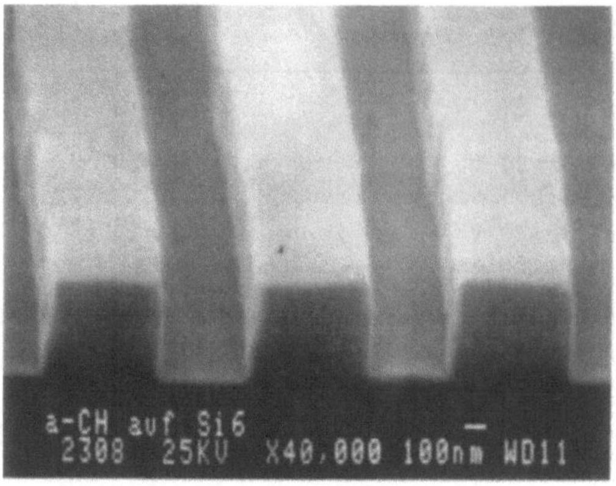

Fig. 4: SEM picture of 0.7 um lines and 0.3 um spaces in photopatterned DLC-layers made by contact printing

test patterns made from DLC-films in the described way. In the same manner larger and smaller relief patterns can be obtained.

Conclusions

The electrical and mechanical properties of DLC-films make them interesting canditates for microsystem technology. The good barrier properties against water are described by the low diffusion coefficient. They can be photopatterned even in low dimensions by means of a photolithographic technique followed by a O_2-RIE process.

References:

1) R. Kleber, W. Dworschak, J. Gerber, A. Fuchs, T. Pütz, J. Scherer, K. Jung, H. Ehrhardt, to be published in Vacuum.
2) R. S. Bonetti, M. Tobler, Metalloberfläche 44, 209, (1990).
3) J. C. Angus, P. Koidl, S. Domitz, Carbon Thin Films in Plasma Deposited Thin Films, (Editors: J. Mort, F. Jansen), CRC Press, Boca Raton, Florida, USA 1986.
4) J. Robertson, Advances in Physics 35, 317, (1986).
5) D. E. Polk, J. Non-Cryst. Solids 5, 365, (1971).
6) J. Tauc, Amorphous Semiconductors, (Editor: M. H. Brodsky), Plenum Press, New York, USA 1973.
7) R. Sezi, R. Leuschner, M. Sebald, H. Ahne, S. Birkle, H. Borndörfer, Microcircuit Engineering 89, Cambridge, England, September 1989.
8) R. Sezi. M. Sebald, R. Leuschner, H. Ahne, S. Birkle, H. Borndörfer, SPIE Symposium on Microlithography, San Jose, California, USA, March 1990.

Submicrometer Photolithography by Surface Imaging – Experiments and Simulation

L. Bauch*, U. Jagdhold*, W. Höppner*, J. Bauer*, R. Klabes**, G.Mehliß***, W. Abraham***

* AdW Institut für Halbleiterphysik, Walter-Korsing-Str. 2 1200 Frankfurt(Oder), PSF 409
** AdW Zentralinstitut für Kernforschung, Rossendorf
*** Fotochemische Werke Berlin

Summary
Results obtained by investigation of the silylation and dry develoment process in photolithography will be presented. An improved depth resolution of 0.65/um structures was obtained (exposure wavelength 436 nm, aperture 0.3). The degree of silylating was obtained by RBS measurements. A first way of process simulation will be outlined and results are discussed.

Introduction

An extension of the field of application of wafer steppers can be effected by surface imaging, that means by transfering the original pattern into the surface layer of a thick planarising resist layer. This surface layer, which is modified due to the action of the incident ligth, is used for the formation of an etching mask. The complete resist structure is obtained by anisotropic etching utilizing this mask (dry development). A practicable variant of the outlined procedure is given by the DESIRE-process /1/. Here the etching mask is produced by a selective diffusion of a silicon-organic compound (most HMDS = hexamethyldisilazan) within the illuminated resist region. Afterwards the resist structure is developed by dry etching (O_2-RIE). For a futher process optimization and for its later application a process simulation is of advantage. Therefore one goal of our investigations consists in the numerical simulation of that process and in the derivation of the necessary input parameters.

Experiments and results

A resist based on ORWO-photoresist FK2070 (binder: novolac, photoactive component (PAC): naphtoquinonediazid) was used. The deposition was made in the conventional manner by spinning (layer thickness 1.8/um at 5000 rpm, well planarizing). For exposure with the wafer stepper AUR (Carl- Zeiss Jena, wavelength 436 nm, numerical aperture NA=0.3) an amount of radiation was necessary, which turned out to be up to 50 % smaller than those commonly used standard techniques (AZ 1370). After exposure the resist is silylated by HMDS (chemical vapour process on a hot plate at 165°C for 180 s with HMDS). The degree of silylating was obtained by RBS measurements for different resist compositions (e.g. concentration of the PAC) and process parameters. Fig.1 shows the concentration of Si-atoms in dependence of the distance to the surface measured by RBS (1.2 MeV He^+). With higher exposure increased the depth of silylating. From the RBS-data the diffusion coefficient D needed for modelling was derived. D increases from 3.5 nm^2/s (nonexposed) up to 4125 nm^2/s (exposed)).

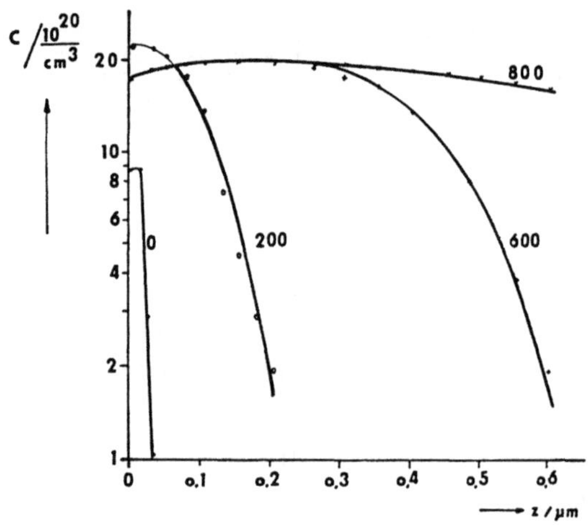

Fig.1: Dependence of the concentration of Si-atoms (c) from the depth to the resist surfaces (z). Parameter: exposure (arbitrary units)

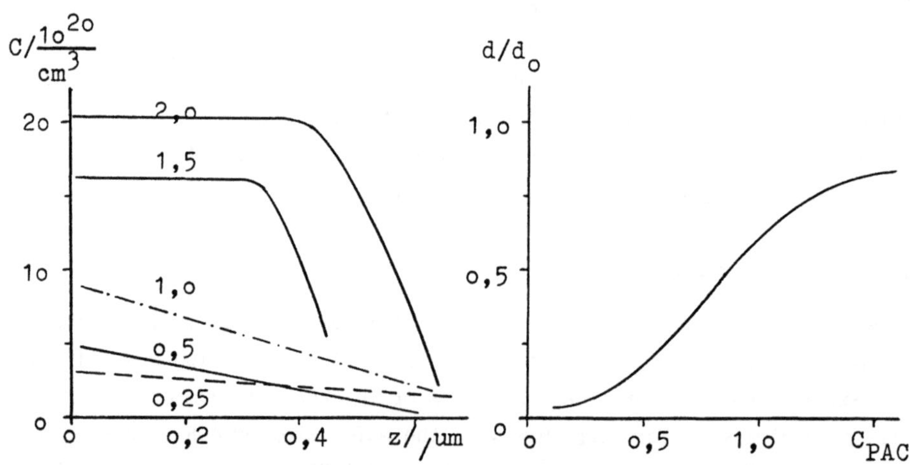

Fig.2: Dependence of the Si-concentration (c) from the distance to the surface of the resist (z) by variation of the PAC-conzentration (exposure 600 Skt) (left) and dependence of the relative resist thickness (d/d_o) after dry development from the relative PAC-concentration (C_{PAC})

The concentration of the PAC has a great influence to the selectivity by the silylating and dry development. For higher concentrations increased the depth of silylation and the concentration of the Si-atoms (fig.2). The si- lylating on the surface of the nonexposed resist decreased with higher PAC concentration. The diffusion from HMDS in the resin and their reaction is acomplised by increasing the layer thickness (fig.3). The degree of silylating, the increasing of the layer thickness and the resistance of the silylated resist by the dry development are corresponding.

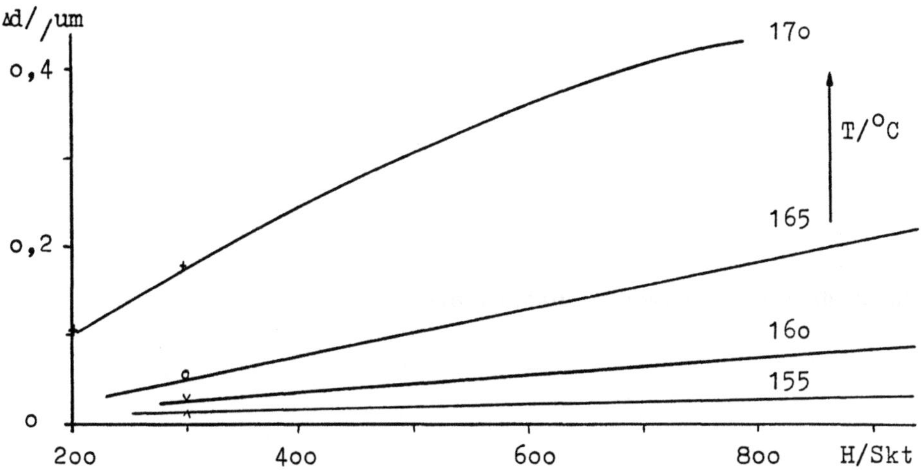

Fig.3: Dependence of thickness increasing (△d) by silylation from the exposure (H) by variation of the silylating temperature (T in °C)

In dry-etching a two-step-process was applied which consists of a nonselective etching of the resist surface layer by CF_4 (removal from the silylated surface layer in nonexposed regions; contrast enhanced) and following O_2-RIE for structure development. RBS measurements showed, that by O_2-RIE the silylated layer of the resist is converted in a SiO_x-layer (Fig.4). The SiO_x-layer act as a mask for the future etching of the resist. The dry development is maked by TU Dresden (see paper on this conference). Fig.5 shows the SEM-micrograph of a resist with 0.65/um structures processed in this way. In comparison to standard techniques this value represents an improvement of the limiting resolution by 0.2/um. Fig.6 shows the contrast curve. A high gradation caused by dry development is observed. The dependence of the line width and limiting resolution from the exposure is flate (fig. 6 and 7). By analyzing the dependence of the limiting resolution on focusing an improvement of the utilizable depth of focus latitude by 2/um was deduced.

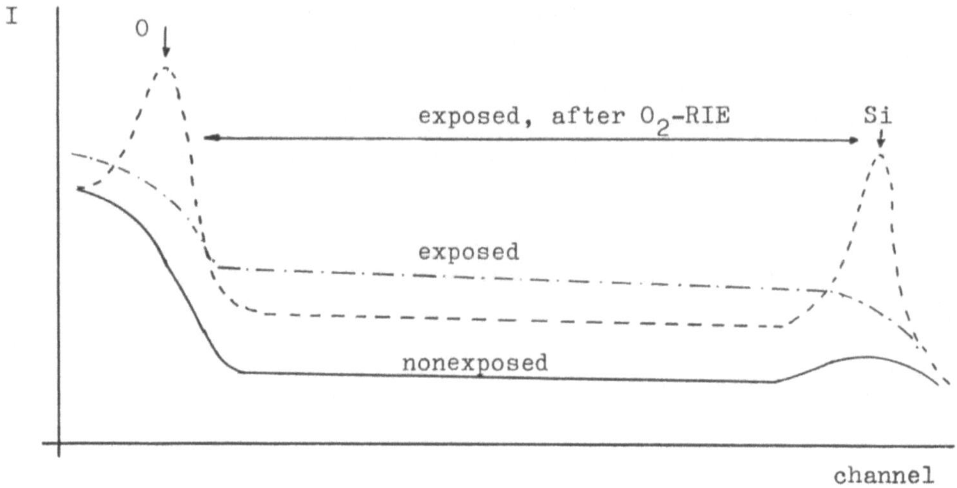

Fig.4: RBS-signal from a silylated resist

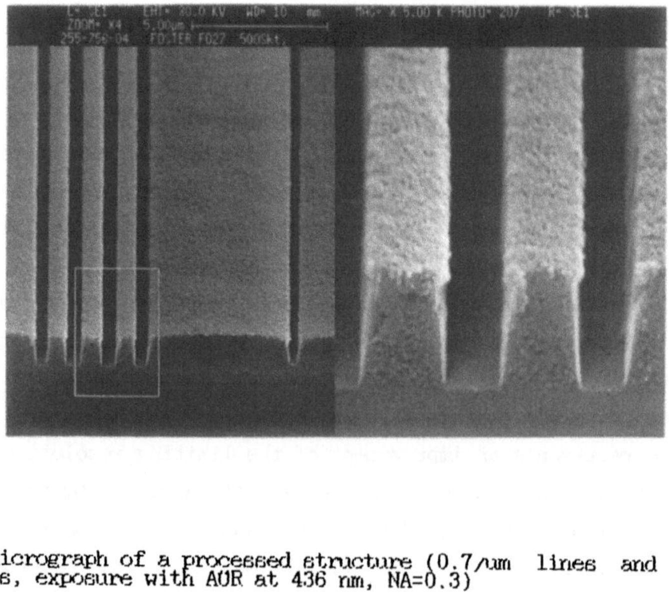

Fig.5: SEM-micrograph of a processed structure (0.7/um lines and 0.6/um spaces, exposure with AUR at 436 nm, NA=0.3)

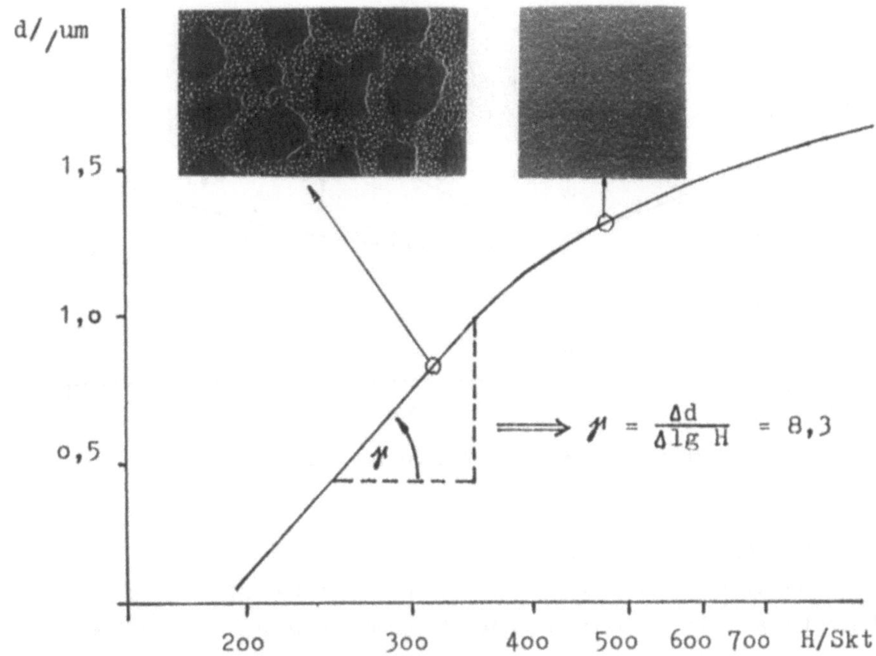

Fig.6: Contrast curve by silylated and dry developed resist (the SEM-micrographs show the resist surface)

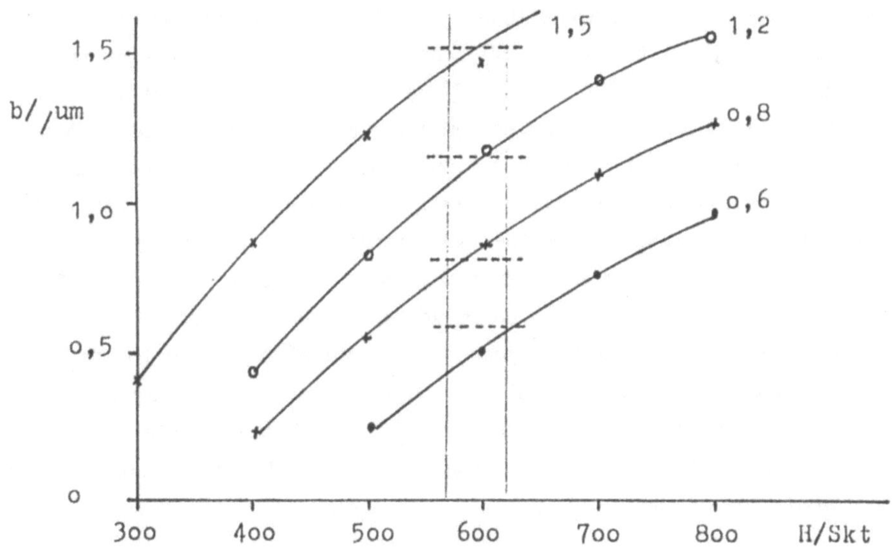

Fig.7: Dependence of the resist-line-width from the exposure by variation of the structure-width (in μm)

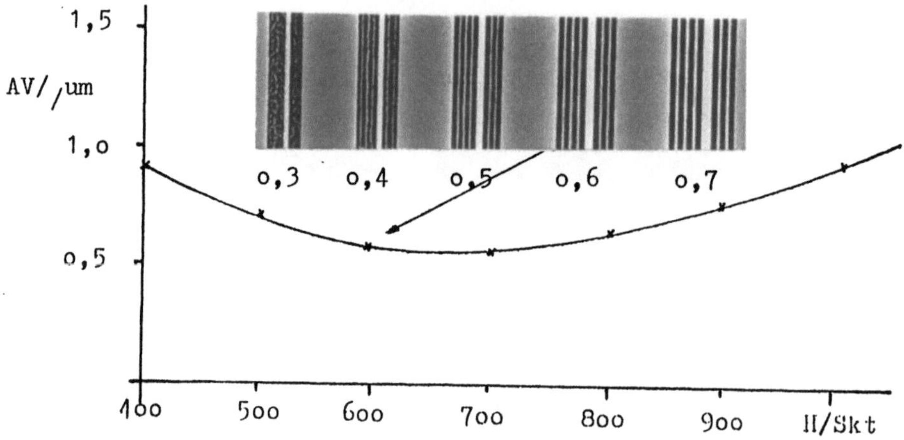

Fig.8: Dependence of the limiting resolution from exposure (line = space in µm)

Simulation of the silylation

The concentration distribution of the PAC after the exposure of the resist was calculated by means of a photolithography simulation programme /2/. We assumed a chemical reaction between PAC and the resin (C_{res}) to a product like ester (C_{ester}). This means, there is a connection between the molecules of the chains from the resin matrix. These connections block the diffusion of the silylating agent (C_{HMDS}). By RBS measurements we found the dependence of the diffusion coefficient D_{HMDS} on the ester concentration. The incorporation of HMDS on top of the surface of the resist is a sorption reaction. There is assumed an equilibrium between the vapour absorption flux (Henry´s law), the desorption flux and the flux of the transport into the resist (Fick´s first law). The transport of HMDS within the resist will be discribed by the two dimensional diffusion equation. The HMDS within the resist reacts with the rest of the resin to a resin bounded silicon, this is unmovable. We solve the diffusion equation with the sorption condition and the kinetics of the chemical reactions for the different components. The silicon distribution after the silylation is than the input for the simulation of the conversion of the silylated resist layer in a SiO_x-mask in the first stage of the O_2-RIE. After that follows the calculation of the resist profile with a simulation programme /3/.

References

/1/ Coopmans,F.; Roland,B.: SPIE **633** Optical Microlithography V (1986) 262

/2/ Bauer,J.; Mehr,W.; Glaubitz,U; Baborski,H.; Haase,N.: SPIE **1264** Optical/Laser Microlithography III (1990)

/3/ Erzgräber,H.J.; Glück,B.; Bauer,J.:V.Symposium "Physikalische Grundlagen zu Bauelementetechnologien der Mikroelektronik" Frankfurt(O) 1988

X-Ray Imaging System

H. Joseph Weaver and Gary E. Sommargren (consultant)

Engineering Sciences Division
Mechanical Engineering

We are developing a new type of x-ray imaging system in which an opaque sphere is used to image x rays by diffraction. We have successfully described the diffraction and imaging processes analytically and have shown excellent agreement with experimental results taken in the visible region of the spectrum. (We are presently preparing the x-ray experiments.) Because of the simplicity of the system, the potential for high spatial resolution, and the extremely large depth of field, this technique could have a significant impact on x-ray lithography (a field involved with the printing of high-density integrated circuits), the imaging of living cells with the aid of high-resolution microscopes, material studies, and diagnostics in the Weapons Programs.

Introduction

The diffraction of light (or any part of the electromagnetic spectrum including x rays) by an opaque sphere such that a point source of illumination is focused to a bright spot behind the sphere - at the center of the shadow cast by the sphere - has been a curiosity ever since Arago demonstrated the phenomenon experimentally in 1818. Figure 1 shows the basic arrangement of the source, opaque sphere, and observation plane. The distribution of light in the observation plane was derived by Mie in 1908 using rigorous electromagnetic theory. The resultant formulas, however, are quite involved and cannot easily be used to study the characteristics of the diffraction pattern. Some authors have described the pattern using scalar diffraction theory. Most have concentrated on the axial intensity of the central bright spot, with Osterberg and Smith solving the Rayleigh diffraction integral exactly. Their results showed that the bright spot exists at all axial points behind the sphere, but that the intensity is not increased beyond that which would be measured if the sphere were removed. Others have derived the radial intensity of the central bright spot using various levels of approximation.

In 1914 Hufford realized that this diffraction phenomenon could also be used to form an image. He demonstrated this using a carbon arc lamp, but never pursued the problem analytically or with invisible radiation. Since then, this phenomenon has received little attention, particularly for visible imaging applications where high quality lenses and mirrors are readily available.

However, for short, invisible wavelengths such as x rays, there are no transmissive materials with refractive indices high enough to be used to fabricate lenses. Thus, x rays are typically focused and imaged using grazing incidence mirrors, multilayer mirrors, or zone plates, each of which is formidibly difficult to fabricate. It is, however, relatively easy to fabricate a sphere, and if it could be shown that such a sphere could be used to image x rays, it would prove to be a useful addition to x-ray technology.

Our Approach

The goal of this project is to describe analytically and verify experimentally that an opaque sphere

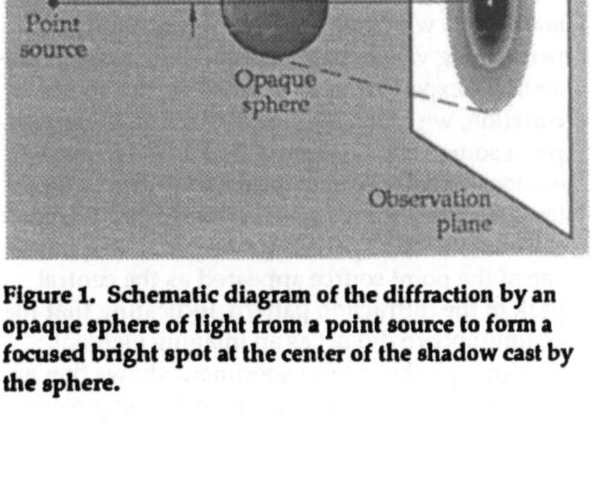

Figure 1. Schematic diagram of the diffraction by an opaque sphere of light from a point source to form a focused bright spot at the center of the shadow cast by the sphere.

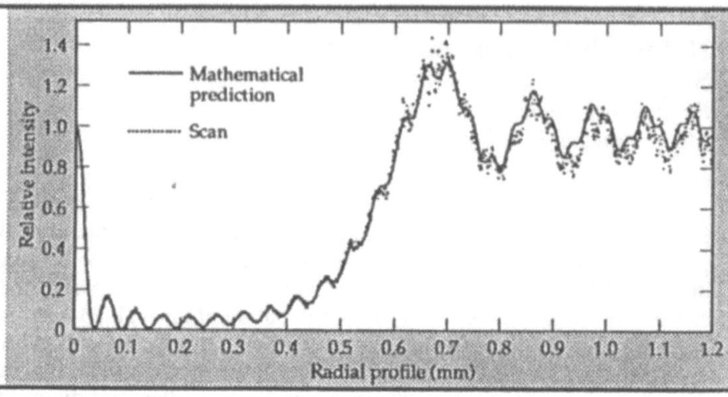

Figure 2. Comparison of the diffraction pattern of a point source with a microdensitometer scan of an actual diffraction pattern.

can be used as an imaging element for x rays. To do this we started with scalar diffraction theory, using minimal approximations, and derived analytic expressions that describe the diffraction of light from a point source by an opaque sphere. Using the geometry in Fig. 1, we found that the intensity, $I(s)$, of the diffraction pattern[1] is given by

$$I(s) = \frac{1}{(z+d)^2}\left[V_0(u,v)^2 + V_1(u,v)^2\right]$$

where

s = radial distance from center of spot in observation plane

d = distance from source to center of sphere

z = distance from the center of sphere to observation plane

$V_0(u,v)$ and $V_1(u,v)$ are Lommel functions

$$u = ka^2\left(\frac{z+d}{zd}\right)$$

$$v = \left(\frac{kas}{z}\right)$$

a = radius of sphere

$k = 2\pi, \lambda$

λ = wavelength of the source radiation.

This equation can be applied to both visible light and x rays, which means that we can verify it by performing visible light experiments (a much easier set of experiments to do). Using the above equation, we calculated the diffraction pattern of a point source and compared the results to a microdensitometer scan of an actual diffraction pattern. **Figure 2** shows excellent agreement between calculation and experiment. As predicted, the image of the point source appeared as the central peak of the diffraction pattern, indicating that the opaque sphere did act as an imaging element.[2]

Although the above experiment shows that an opaque sphere can be used an an imaging element, it does not indicate its image quality. One way to measure how well an opaque sphere images is to determine the minimum separation of two point sources, the images of which appear distinct and separate. (This is called spatial resolution.) Theoretically, we found the resolution of the opaque sphere to be given by

$$\text{Resolution} = \frac{0.38\lambda z}{a}.$$

This was verified by imaging two incoherent point sources that were separated by a distance equal to the resolution limit given by the above equation. **Figure 3** shows both the calculated and experimental results. As can be seen, the model is in good agreement with the experiment.

We then looked at how accurately the sphere could image extended objects (as opposed to points). Extended objects may be considered an aggregate of point sources, each one producing a corresponding diffraction pattern in the observation plane. The sum of these diffraction patterns is the resultant image. **Figure 4** shows the image of a standard three-bar target using visible illumination and the corresponding calculated image. Again, there is good agreement between theory and experiment; however, there is a loss of contrast in the image. This is related to the efficiency by which the imaging system passes different spatial frequencies. It is described mathematically by the modulation transfer function of the opaque sphere, which we are presently studying.

Discussion

Although imaging with an opaque sphere is not a new idea, it has never been fully developed, either analytically or experimentally. This is because it is not a practical alternative to imaging elements, such as lenses and mirrors, in the visible part of the spectrum. However, at x-ray wavelengths, where grazing incidence mirrors, multilayer mirrors, and zone plates are difficult to fabricate to tolerances required to achieve their theoretical resolution, an imaging opaque sphere may be a viable addition to x-ray technology. The art of manufacturing extremely round, smooth spheres (attributes that enhance their imaging performance) has improved considerably over the last several years, primarily due to the need for accurate spheres for general relativity experiments.

The imaging system described here has several unique characteristics that set it apart from conventional imaging systems. First, it has an almost infinite depth of field; i.e., the image exists at all distances, z, behind the sphere. This is advantageous for x-ray lithography because we are working to very close tolerances. In conventional systems the image is localized within a small interval. The dimension of this interval is related to the spatial resolution. Increasing the resolution by a given factor decreases the depth of field by that same factor squared. For high resolution conventional systems the depth of field is on the order of the wavelength of the illumination. For x rays this presents significant alignment and stability problems.

Secondly, imaging with an opaque sphere is independent of the wavelength. The source of illumination can be monochromatic, broad band or can be varied during its use without affecting the image or the alignment. Conventional systems using multilayer mirrors and zone plates must be designed for a specific wavelength for optimum performance.

Finally, the opaque sphere images at different magnifications by simply changing the ratio of z to d (see Fig. 1). When $z > d$, the system gives a magnified image. This is the geometry used when imaging living cells with a microscope. When $z < d$, the image is reduced, a geometry useful in x-ray lithography where the printing of integrated circuits in highly dense arrays calls for the exact

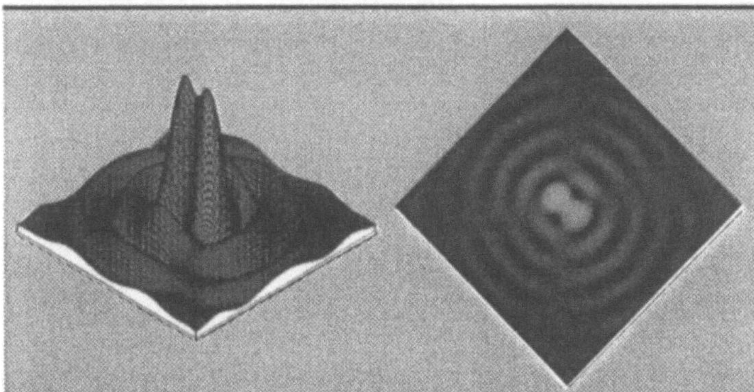

Figure 3. Comparison of a model and experiment of two incoherent point sources where the points are separated by a distance equal to the minimum predicted resolution.

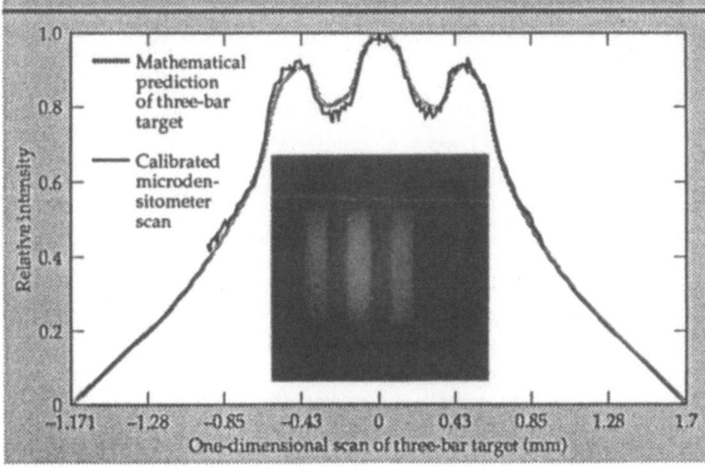

Figure 4. Comparison of the model and experiment of a three-bar target.

Table 1. Our approach to the problem of developing an x-ray imaging scheme based on the diffractive focusing of radiation by an opaque sphere.

- ■ Mathematical description of the diffraction problem
- ■ Experimental confirmation of mathematical analysis
- ■ Two-point resolution of imaging system
  - ■ Analytical studies
  - ■ Experimental studies
- ■ Mathematical description of sphere as imaging device
- ◪ Experimental demonstration of imaging
  - ■ Optical domain
  - □ X-ray domain
- ◪ Transfer function resolution of imaging system
  - ■ Analytical studies
  - □ Experimental studies
- □ Possible applications
  - □ X-ray lithography
  - □ Biological/medical research
  - □ LLNL Weapons Program

■ Completed work
◪ Started work
□ Work planned but not yet started

placement of extremely small lines. Conventional systems, on the other hand, must be designed for a particular magnification and application to keep aberrations to a minimum.

Future Work

Up to now, all our experiments have been done with visible illumination, primarily because it was a simple way of verifying our calculations. Now that we believe that our analysis is sound, we have started setting up an imaging system with an x-ray source. We expect to have preliminary results within the next few months. **Table 1** summarizes what has been accomplished and what future work needs to be done.

1. G. E. Sommargren and H. J. Weaver, "Diffraction of Light by a Circular Obstruction, Part 1: Description and Properties of the Diffraction Pattern," *Appl. Opt.* (to be published).
2. G. E. Sommargren and H. J. Weaver, "Diffraction of Light by a Circular Obstruction, Part II: Image Formation and Resolution Considerations," *Appl. Opt.* (submitted for publication).

Laser Supported Deposition

Laser Induced Gold Deposition on Polyimide Layers

D. Metzger, A. Paredes, Th. Kruck* and H. Reichl

TU Berlin, Research Center for Microperipheric Technologies
Sekr. EN 4, Einsteinufer 17, 1000 Berlin 10

* Universität zu Köln, Institut für Anorganische Chemie, Greinstr. 6, 5000 Köln 4

ABSTRACT

An apparatus was set up for laser induced chemical vapor deposition of gold on polyimide. Electrical interconnects were written by thermal decomposition of an organometallic gold compound in an argon ion laser focus. The low decomposition temperature (approx. 200 °C) of a volatile gold precusor and the selectivity of the direct writing technique offer advantages for hybrid systems.
The dependence of gold deposition on different process parameters is studied. The electrical resistivity is determined as a funktion of scan rate and laser power.

INTRODUCTION

The application of laser techniques for interconnection writing on multi chip systems has received great interest for use in packaging technology (1). Especially laser induced chemical vapor deposition (LICVD) can be a powerful tool in interconnect technology (2). LICVD of metals with high electrical conductivity like gold and copper has been reported (3-6). For thermal conductivity consideration in most cases SiO_2 covered substrates have been used (3-6). In our experiments polyimide was chosen as a low thermal conductivity and low ϵ dielectric layer for hybrid systems. In recent years LICVD on polyimide layers has found increasing interest as a new approach to maskless and low temperature processes (2). Tungsten interconnects written by LICVD on polyimide have been reported (2). Organometallic compounds with low decomposition temperature and high vapor pressure are used for deposition on polyimide. Dimethylgold (III) hexaflouroacetylacetonate (DMGHF) has proved to be a precursor with exellent features.

EXPERIMENTAL

In all experiments reported here an argon ion laser (Spectra Physics model 2025) was used for the deposition experiments. The laser works at the wavelenght of 514 nm with an output power of 2,5 W.

Fig. 1 LICVD deposition chamber with objectiv lens

The laser beam was directed by mirrors into a stainless steel chamber. Figure 1 shows the stainless steel chamber with the objective lens. The focal distance was about 30 mm. The focal diameter of the lens was 5 μm. The writing system, a model 4455 ESI laser trimmer, was modified for deposition applications. The system was controlled by a HP 9000 computer. Gold strips are written by moving the laser focus across the polyimide surface. The maximum writing area was 135 * 135 mm^2. To adjust the polyimide surface to the laser focus, the chamber is mounted on a computer controlled z-adjustment stage.

In most experiments 4" Si - Wafers or 2" * 2" alumina ceramics spin coated with Pyralin 2611 or Probimide 348 were used as substrates. Both polyimides were cured 2h at 350 °C . The thickness of the polyimide was in the range of 10 μm. Before carrying out the deposition experiments the chamber was evacuated to a base pressure of 10^{-6} mbar. The static pressure of DMGHF during deposition was measured by type 221A MKS Baratron capacitance manometer. DMGHF was supplied by Prof. Kruck, Universität of Cologne. It was stored in a temperature controlled stainless steel bubbler and was admitted into the deposition chamber right above the polyimide surface through cylindrical nozzels.

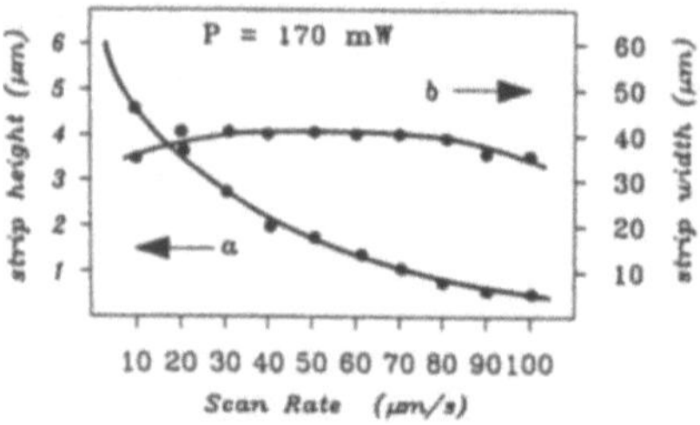

Fig. 2. Strip height and strip width as a function of laser power

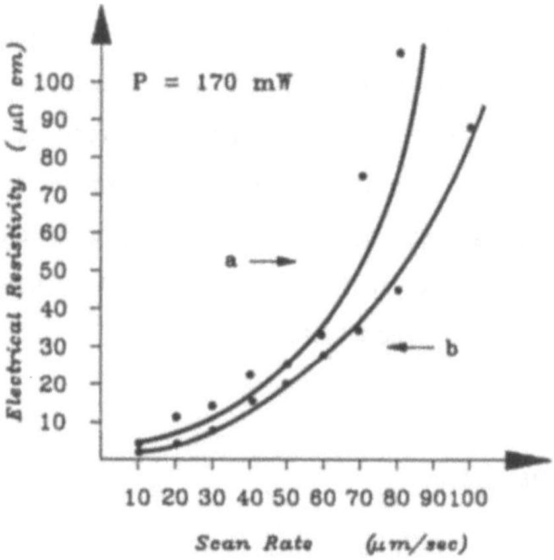

Fig. 3 Electrical resistivity as function of writing velocity; as deposited (a), after annealing at 350 °C by 45 min. (b)

All deposition experiments were performed at room temperature. The vapor pressure of the organometallic goid compound was about 0,40 mbar. During the gold deposition a slight increase of pressure was observed.

RESULTS

Figure 2 shows the dependence of the strip height and the strip width on writing velocity at a laser power of 170 mW. The strip height was measured with a Sloon Dektak 3030. The strip width remains constant (approx. 40 μm) nearly over the whole range of writing velocities. Only at low and high velocity the strip is decreasing by 20 %. Easily a thickness of some μm can be achieved . By variation of the laser power a linear dependence of the strip width is observed. Fig. 4 shows deposited gold lines written on polyimide. The lines are very homogeneous in shape. Without further preparation of the polyimide surface the cross section is reflecting the intensity profile of the laser beam (Gaussian). The grain size of the gold deposit is a function of laser intensity. At high laser intensity, surface temperature is increased, resulting in a large grain size in the centre of the focus (Fig. 4a). After increasing the surface roughness of the polyimide by O_2 plasma etching the scattering of the incident radiation results in a modification of the intensity profile. Now a rectangular cross section and a homogeneous grain size are achieved (Fig. 4b). These results were attained without further process optimization.

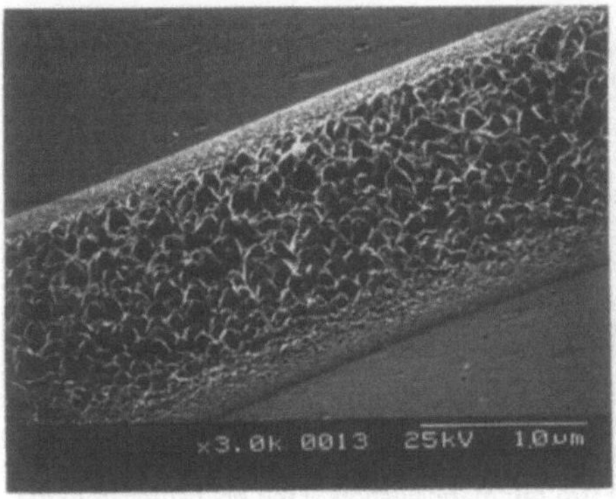

(a)

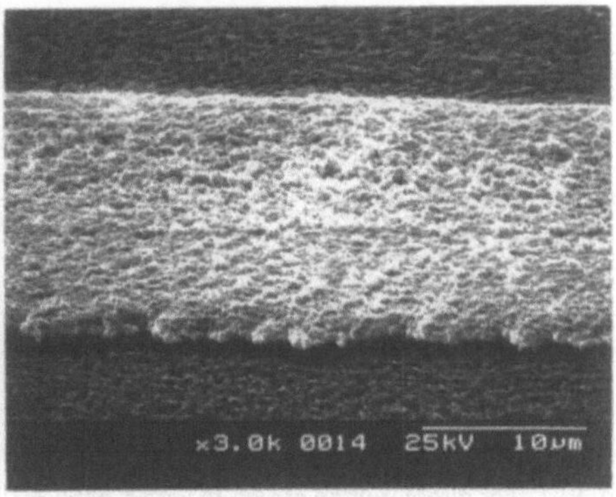

(b)

Fig. 4 Deposition of gold strips; with (b), without (a) modification of the poyimide surface

ELECTRICAL RESISTIVITY

To measure the electrical resistivity with the four point probe method (8) a special line geometry was laser written on polyimide. Current-voltage characteristics were measured with type 220 Keithley current source and type 196 Keithley digital multi meter. The cross section of the gold lines was calculated by inspection of cutted substrates with a microscope. Figure 3 shows the electrical resistivity as function of the writing velocity at a laser power of 170 mW. Low writing velocities yielded best results. At a velocity of 10 $\mu m/s$ the electrical resistivity was 4,38 $\mu\Omega$ cm (bulk gold 2,35 $\mu\Omega$ cm). After annealing at 350 °C, 45 min, 2,87 $\mu\Omega$ cm was measured. Little change was observed upon annealing (curve b). This can be attributed to the grain size distribution of the thermal deposit, which is not altered by further annealing (7). Nevertheless the overall dependence of the resistivity on the scan rate remained largely unaltered after annealing.

Figure 5a shows a test structure for electrical interconnect application, the lines having a pitch of 200 μm. A further decrease in pitch can be achieved easily . Details of the test structure are shown in Fig. 5b.

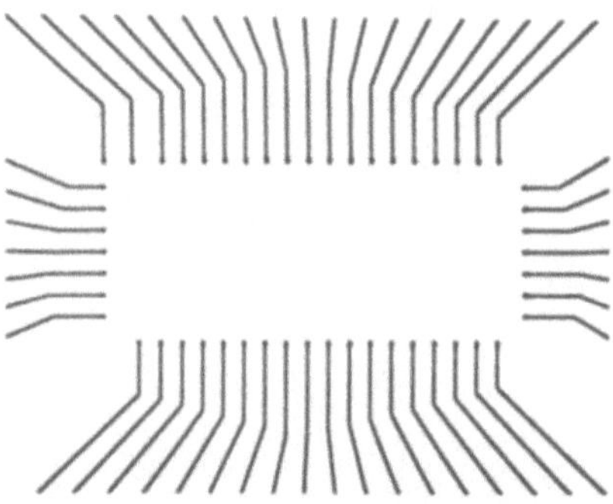

(a)

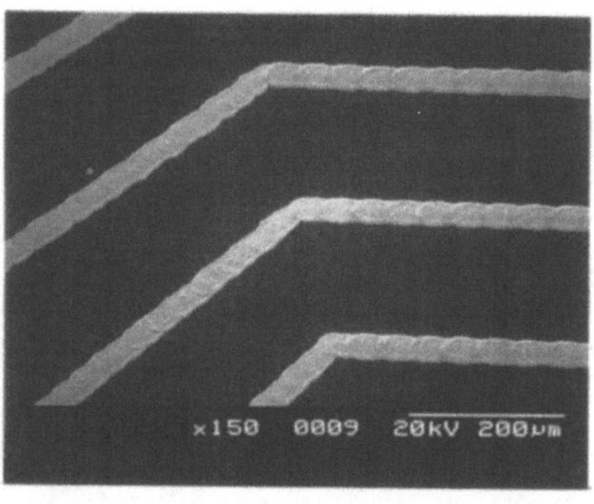

(b)

Fig. 5 Laser direct written structure with a pitch of 200 μm (a); magnification of the gold strips from the structure (b)

CONCLUSION

It has been demonstrated that DMGHF is a suitable precursor for the deposition of gold on polyimide by thermal decomposition with a 514 nm argon ion laser. Because of its high vapor pressure and low decomposition temperature, electrical interconnection lines can be written in situ. The measured electrical resistivity is close to that of bulk material. A special preparation of the polyimide surface can be used to smooth the morphology of gold lines. With the demonstrated spatial resolution of the computer controlled writing system, the LICVD technique can be applied to produce electrical interconnect systems directly from CAD layout.

REFERENCES

1 A. T. Barfknecht, D. B. Tuckerman, J. L. Kaschmitter, and B. M. McWilliams, IEEE Trans. Comp., Hybrids, Manuf. Technol. Vol. 12, No.4, 646 (1989)

2 R. F. Miracky, in Laser- and Particle-Beam Chemical Processes on Surface, edited by A. W. Johnson, G. L. Loper, T. W. Sigmon, Mater. Res. Soc. Proc. Vol. 129, 547 (1989)

3 T. H. Baum, C. R. Jones, J. Vac. Sci. Technol., B 4 (5),1187 (1986)

4 C. R. Moylan, T. H. Baum, C. R. Jones, Appl. Phys. A, 40, 1-5 (1986)

5 B. Markwalder, M. Widmer, D. Braichotte, and H. van den Bergh, J. Appl. Phys. 65 (6), 2470 (1989)

6 F. A. Houle, C. R. Jones, T. H. Baum, C. Pico and C. A. Kovac, Appl. Phys. Lett. 46 (2), 204 (1985)

7 C. E. Larsen, T. H. Baum, and R. L. Jackson, in Photon, Beam, and Plasma Stimulated Chemical Processes at Surfaces, ed. by V. M. Donnelly, I. P. Herman, M. Hirose, Mater. Res. Soc. Proc., Vol. 75, 721 (1987)

8 S. Pfüller, Halbleitermeßtechnik, Dr. Alfred Hüthig Verlag Heidelberg, p. 78, 1977

Selective Pattern Deposition of SnO_2 Thin Film by Laser Pyrolysis

BINGLIN ZHANG, WEI WENLOU

Physics Department and
Fundamental & Applied Science Research Institute
Zhengzhou University, Zhengzhou, Henan 450052
The People's Republic of China

Abstract

The laser induced chemical mist deposition (LCMD) of SnO_2 thin film is reported. The pyrolysis of $SnCl_4 \cdot 5H_2O$ mist is carried out on a small area of glass substrates by Co_2 laser, and the selective deposition of SnO_2 thin film is performed. The selective pattern deposition is obtained by moving X—Y table. The resistivity of the films is about $1 \times 10^{-2}\Omega$-cm. The transmision of the films remains a constant of 75% — 80% over a large wavelengthrange from 400nm to 700nm approximately.

1. Introduction

There has been a great deal of interest in the preparation of SnO_2 films for a long period, because of its great potential for applications in solar cells, optoelectronic devices and gas sensing elements [1 — 3]. A variety of deposition techniques have been employed for preparation of SnO_2 films, such as sputtering, spray pyrolysis, chemical vapor deposition and evaporation [4]. But these require high processing temperature or post deposition annealing to obtain the desired properties. Last year R. R. Kunz *et al* reported the laser photochemical selective deposition of transparent conductive SnO_2 films without annealing [5]. This paper reports the selective pattern deposition of SnO_2 films by laser pyrolysis. High optical transparency and electrical condoctivity of SnO_2 films were obtained without annealing.

2. Experimental Details

The experimental method is based on the pyrolytic decomposition of $SnCl_4 \cdot 5H_2O$ compond disolved in ethanol liquid sprayed onto a glass substrate. The pyrolytic reaction is

$$SnCl_4 + 2H_2O \rightarrow SnO_2 + 4HCl$$

The pyrolysis and deposition are carried out on a small area of substrates by a focussed CO_2 laser irradiation. A typical spray mixture concentration of $SnCl_4 \cdot 5H_2O$ is 30wt. %. The fine mist of very small droplets containing the reactants is formed by a sprayer. The flow velocity of the carrier gas N_2 is typically 4L/Min. The laser power density is varied from $1.6 \times 10^2 W/cm^2$ to $2.4 \times 10^2 W/cm^2$. The deposition rate as high as 400Å /Sec can be obtained, which is calculated by the measurement of the film thickness. The substrate is placed on a X—Y moving table, which is controlled by a computer. The selective pattern deposition has been obtained by moving X—Y table.

3. Results and Discussion

since the deposition of the SnO_2 requres laser irradiation on the selective area of substrates, the grating pattern of transparent and conductive films has been formed. For the purpose of optical and electrical measurement, the 1mm-wide lines of SnO_2 were diposited. The resistance measurement has been made with a standard four-point probe method. Under the condition of $2 \times 10^2 w/cm^2$ laser power density and 10mm/min scan speed, the resistivity is about $1 \times 10^{-2} \Omega$ — cm, the thickness of the film is 1200Å. The variation of conductivity with laser power density is shown in figure 1, It represents the deposition temperature dependence of conductivity indirectly. It is basically similar to Sundaram result [6]. probably, the increase in conductivity at high laser power density could be attributed to increased oxygen vacancies or due to the phase change.

Transmission measurement has been performed using a spectrophotometer. The thansmittance spectrum of SnO_2 film is shown in Figure 2, which is recorded over the UV—visible region. In Figure 2, it shows that the optical transmition of these films remains a constant of 75% — 80% over a large wavelength range from 400nm to 700nm approximately. The transmitance spectrum in UV region shows that the absorption transition falls in this region of the spectrum. According to the transmitance spectrum, the absorption coefficient can be calculated. A discrimination between direct and indirect transitions is possible [7] [8].

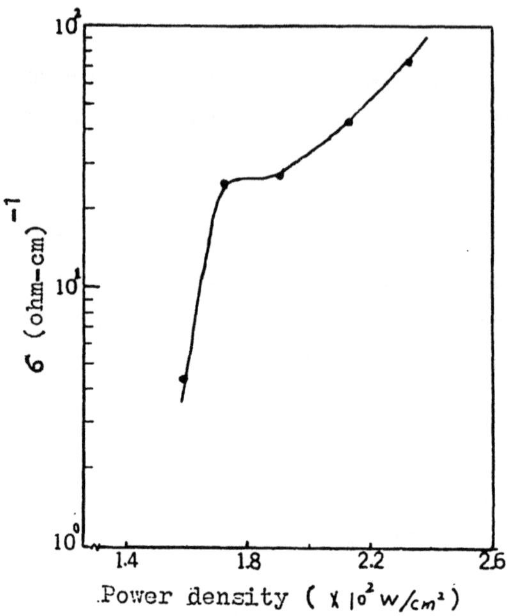

Figure 1. Electrical conductivity of SnO_2 films as function of laser power density.

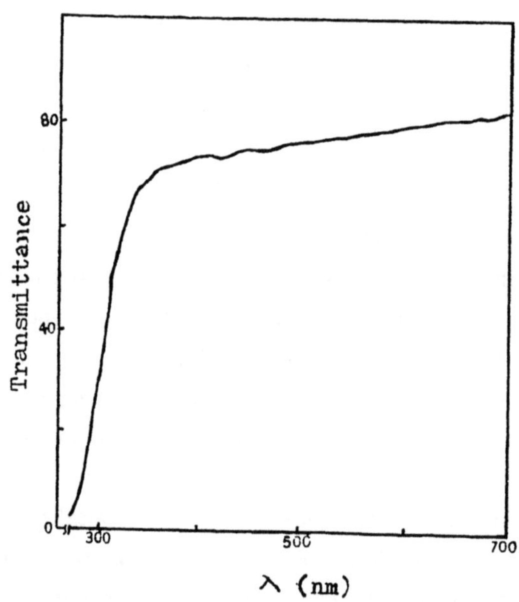

Figure 2 Transmittanc spectrum of SnO_2 film under the condition: $2 \times 10^2 w/cm^2$ — laser power density, 10 mm/min —scan speed, 4 L/min—flow velosity.

The variation in the absorption coefficient α with the absorbed photon energy is shown in Figure 3. When the linear portion is extrapolated to give zero absorption, it is determined that the absorption edge lies in 3. 85ev. Figure 4 shows the variation in α^2 with photon energy, indicating that the value of 4. 05ev corresponds to direct transition. Figure 5 shows the variation of $\alpha^{1/2}$ with photon energy. when the linear portion is extrapolated to zero absorption, the value for indirect transition is determined to lie in 3. 56ev. The values of absorption edge, direct transition, and indirect transition are closed to Spence´ s results [7].

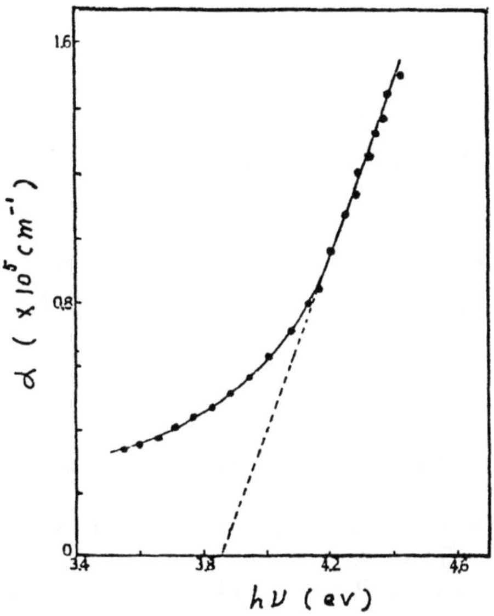

Figure 3. Plot of absorption coefficient α versus photon energy for SnO_2 film

The structural properties of the films are investigated by an X—ray diffraction using a X—ray diffractometer. The diffraction pattern indicates that the films are polycrystalline. It can also be seen that the direction of preferential orientation of the films is found to be (110) plane.

The diffraction patterns between anneal and unanneal samples are compared. We have observed that the preferential orientation of the films are same.

Acknowledgement

We are grateful to Mr. Han Xinmin for designing X—Y moving table.

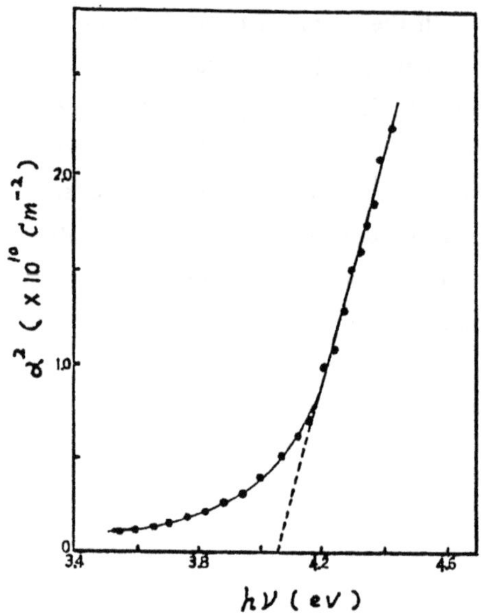

Figure 4. Plot of α^2 versus photon energy for SnO_2 film

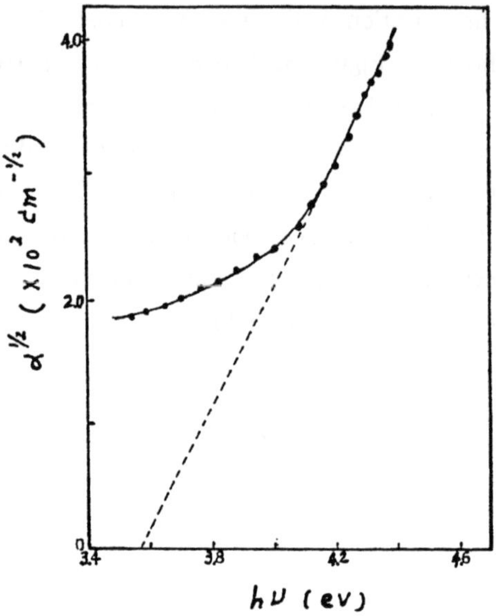

Figure 5. Plot of $\alpha^{1/2}$ versus photon energy for SnO_2 film

References

1. N. S. Murthy and S. R. Jawalekar, J. Inst. Electron. Teleco mmun. Eng. 30 (1984) 97.
2. C. Tatsuyama and S. Ichimura, Jpn. J. Appl. phys. 15 (1976) 843.
3. T. Oyabu, J. Appl. phys. 53 (1982) 2785.
4. K. L. Chopra, S. Major and D. K. Pandya, Thin Solid films, 102 (1983) 1.
5. R. R. Kunz, M. Rothschild, and D. J. Ehrlich, Appl. Phys. Lett. 54 (1989) 1632.
6. K. B. Sundaram and G. K. Bhagavat, J. Phys. D: Appl. Phys. ,14 (1981). 333.
7. Wendell Spence, J. of Appl. Phys. , 38 (1967) 3767.
8. K. B. Sundaram and G. K. Bhagavat, J. Phys. D: Appl. Phys. , 14 (1981) 921.

Plane Surface Embedding Technique for Thin Film Hybrids

Heinrich G. Müller, K. Jassim, K. Buschick, and H. Reichl
Forschungsschwerpunkt Mikroperipherik
TU Berlin, Sekretariat TIB 3.1, Gustav-Meyer-Allee 25
D-1000 Berlin 65

Abstract

Openings have been laser cut into Si and alumina substrates. Chips are placed into these openings with a vertical precision of better than 3 μm and lateral tolerances smaller than 30 μm. After filling the gaps between chip and substrate with an epoxy resin, a thin film quality, plane surface is achieved. This allows for interconnection of the integrated chips with thin film methods like sputtering or laser induced deposition techniques. By subsequent formation of further metallizations and dielectric films, multilayers can be formed. Preliminary results show very good high frequency properties of these assemblies.

Introduction

The development of high density, thin film multichip modules is one of the main technologic demands of the forthcoming years [1]. The special requirements of high pin count and low electrical reflection coefficients for high speed devices can be met best with flat thin film metallization methods. To achieve appropriate surfaces for those processes, special micromechanics are substantial. Johnson et al. [2,3] developed a process for embedding the chips into a thin film quality substrate and thin film interconnection with individual masks; Tuckerman et al. [4,5] used a microgrinder to provide a smooth 3-D structured surface of chip and substrate for the same reason.

Our aim has been to achieve flat surfaces by a simple embedding technique, which may easily be automated. The interconnections were to be performed with laser induced direct writing methods, where no masking or sputtering processes were required.

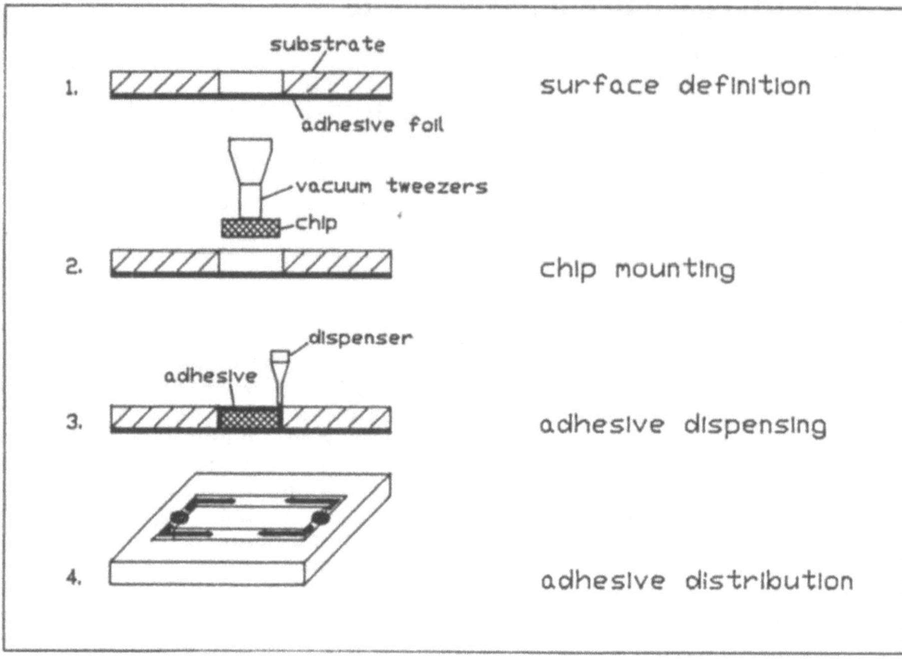

Fig.1: Scheme for the high precision embedding process

Process

A brief process scheme is given in fig 1. The alumina substrate is laser cut with a CO_2-laser at the required opening positions. As the openings do not have exactly vertical walls, care has to be taken that the future connection layer side of the substrate is not facing the laser beam during cutting operation. Also the corners must be cut rounded (some 50 micron radius) and slightly bigger than for an ideally shaped rectangle. This provides a better stability of the assembly against thermal stresses than very precise rectangular cuts.
Next, an adhesive foil is spanned with care over the entire substrate surface. Then the chips are introduced face down into the substrate and fixed onto the adhesive foil. Using vacuum tweezers and a pressure sensor for the embedding manipulator, even chips of very different thicknesses can be positioned with equal accuracy.

The gaps between chip and substrate are filled with an epoxy resin. This material has to be adjusted for very widespread demands. First, the shrinking factor must not exceed one percent, to avoid rupture during bake out. Second, temperatures of at least 350 °C must be sustained by the resin during subsequent polyimide processing of the module. Later, for heat removal during device operation, the thermal conductivity must be maximum. All of these conditions are met by an alumina filled epoxy; but the size of the filler particles must not exceed 15 µm, in order to fit into the very small gaps. Also, to make the adhesive creep into these gaps, the viscosity has to be quite small (<20000cps).

Surprisingly enough, the glass transition temperature must not be high (to withstand high temperature processes) but low (<100°C), to minimize build up of thermal induced stresses. The mechanical fixing of the embedded chips during high temperature cycles is achieved by a polyimide layer, which is spun on after bake out of the resin.

Fig.2: Vertical cross section of the chip - substrate interface

Finally, the Epotek H77S (Epoxy Technoloy) proved best to meet all of these requirements. A cross section of an embedded chip, the substrate and the 20 µm wide gap is given in fig.2.

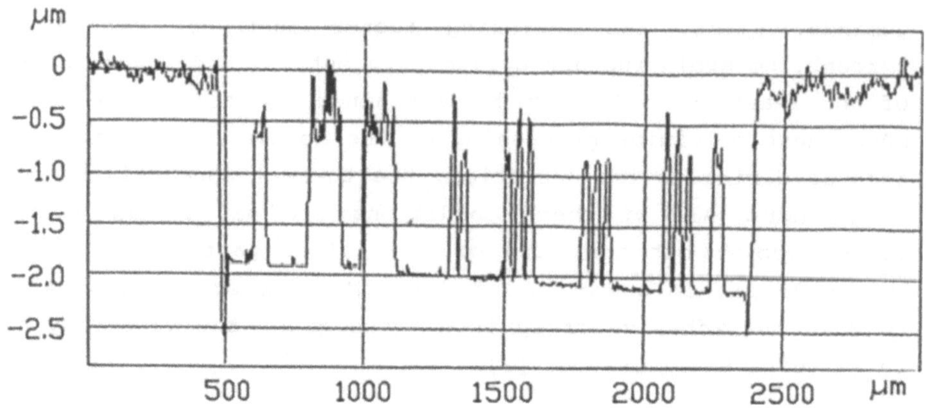

Fig.3: Profilometer scan across an embedded chip and the surrounding substrate area. The structures on chip (center area) stem from the conductor metallizations.

Fig.3 gives a profilometer scan across the front side of an assembly. The height of the conductor lines on chip is comparable or bigger than the steps at the substrate - chip interface.

A layer of about 10 µm polyimide (pyraline 2611 from Du Pont Electronics) was spun onto the module surface, to decrease the dielectric constant and to planarize the remaining surface steps. To achieve contact openings in this layer without needs for a mask illumination, a laser illumination technique was applied. A photoresist (AZ 4562, Kalle Hoechst) was spun onto the prebaked polyimide and illuminated with a frequency-tripled Nd:YAG laser at the positions of underlying pads. The alkaline developer (MIF 312, Kalle Hoechst) dissolved the prebaked polyimide together with the illuminated photoresist. Then the photoresist was stripped and the remaining polyimide was hard baked at 350 °C. A top view of the integrated chip with the polyimide overlayer and opened via holes to the contact pads is given in fig.4.

Fig.4: View upon integrated chip with polyimide cover layer and contact openings

Application

To prove the viability of this technique for practical purposes, a demonstrator system was produced. An entirely additive metallization process was developed, without making use of sputtering processes or masks. Instead, laser induced deposition techniques were applied, to allow for layout corrections if any geometrical misfits should have occured during the embedding operation.

First, the chip pads were metallized with deposition of copper from copper formate in a direct write process with an Nd:YAG-laser. Then, to seal the pads, a thin layer of chemical nickel was added. Third, again from copper formate, the interconnection pattern was directly written onto the polyimide layer and the nickel coated contact pads. Finally, the entire metallization was improved to a thickness of about 7 µm by a chemical copper deposition process. More details about this process are given in another paper [6].

A picture of the final demonstrator is given in fig. 5. The laser direct writing time for the connection lines was 45 seconds. The contact resistivity was less than 1 Ohm, and preliminary high frequency tests of the assembly showed a reflection factor of this interconnection below 5 percent at frequencies up to 26 Ghz.

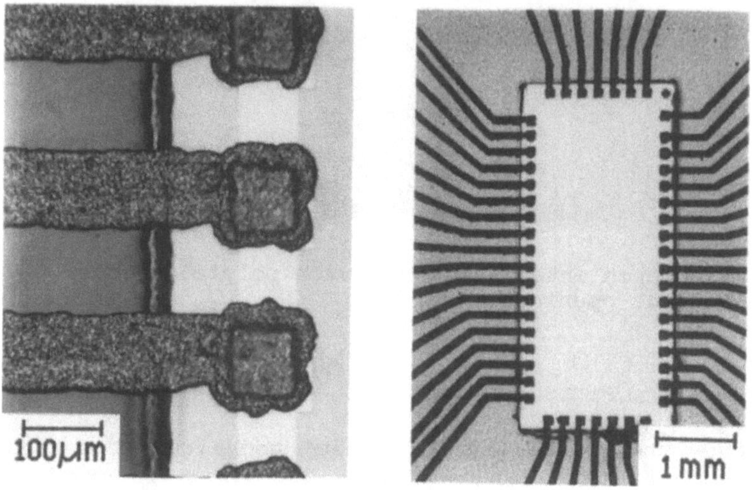

Fig.5: Demonstration system with laser written conductor lines across the chip - substrate interface.

References

1. M.G. Sage, Hybrid Circuits 21, p.36, (1990)

2. R.W. Johnson, J.L. Davidson, R.C. Jaeger, and D.V. Kerns, IEEE J. Sol. State Circ. SC-21, (1986)

3. R.W. Johnson, T.L. Phillips, R.C. Jaeger, S.F. Hahn, and D.C. Burdeaux, IEEE Trans. Comp. Hybr. 12, 185 (1989)

4. D.B. Tuckerman, IEEE Elec. Dev. Lett., EDL-8, 540 (1987)

5. A.T. Barfknecht, D.B. Tuckerman, J.L. Kaschmitter, and B.M. McWilliams, IEEE Trans. Comp. Hybr. 12, 646 (1989)

6. H. Müller, K. Buschick, S. Schuler, and A. Paredes, to be published in "Proceedings of the E.-MRS spring meeting, Straßburg, 1990"

Pulsed Laser-Plasma Deposition of Polycomponent Thin-Film Structures – A Review

S. Metev

BIAS, Bremen Institute of Applied Beam Technology,
2 Klagenfurter Str., 2800 Bremen 33, FRG

Summary

The pulsed laser-plasma deposition method is introduced. This relatively new but very prospective deposition technique combines the capabilities of some other methods as flash evaporation, ion-beam sputtering and molecular-beam epitaxy. The specific characteristics of the deposition process and their influence on the film properties are discussed. The results of the investigations obtained show that this method can be used for synthesis of thin polycomponent films and film structures of unique properties.

1. Introduction

The pulsed laser-plasma method is a fast developing technique for synthesis of thin-film materials [1-3]. This deposition method is based on physical processes, arrising from the impact of high-power pulsed laser radiation on solid targets and leading to removal of partially ionized material (plasma) from the impact zone which condense on the substrate surface in the form of a thin film [1] (Fig.1). In the last several years, very interesting and sometimes unique results have been obtained due to the possibility of varying over a wide range the process parameters such as plasma flux density, ionization degree, ion energy, substrate temperature and geometrical arrangement in the deposition chamber [1-3].

2. Characteristics of the deposition process

The laser-plasma deposition parameters are characterized by some specific features which determine to a great extend the unique capabilities of the method.

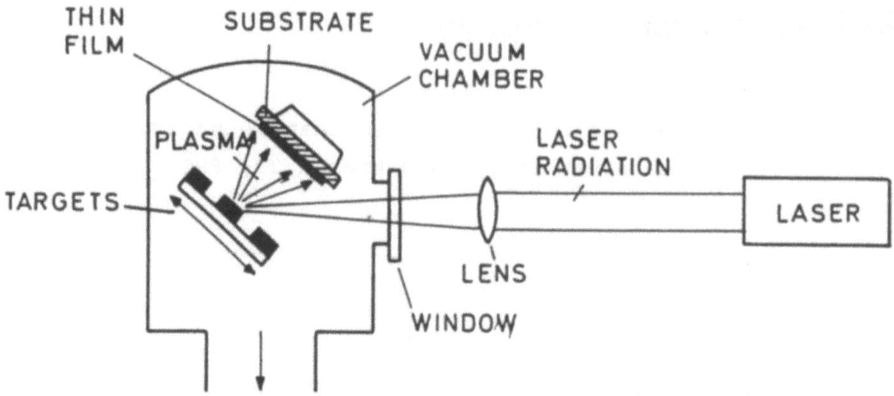

Fig.1 A principle set-up for laser-plasma deposition method

Congruent evaporation of polycomponent targets

The laser-plasma deposition process beginns with laser radiation action on the target resulting in heating up its surface and evaporation of target material from the irradiated zone. At sufficient light flux densities and short pulse durations heating rates of 10^{10} K.s^{-1} can be achieved. This ensures practically the instantaneous reaching the highest evaporation temperature and the simultaneous evaporation of all elements in the target. Even when, due to the high temperature, dissociation of the target material occurs, all elements in the vapour are in the same stoichiometric ratio as in the target. This is a precondition for synthesis of polycomponent films on the substrate surface at suitable other conditions [2].

Controllable particle energy and degree of ionization

An important peculiarity of the deposition process is the possibility to control the energetic parameters of the evaporated particles and the degree of their ionization by varying over the irradiation conditions.

With laser pulse duration $t_l=10^{-3}$s a significant part of the evaporated material is in the form of fine dispersive fraction and droplets of maximum radius $10^{-2}\mu m$ [1]. The initial particle velocity is $\approx 10^4$ cm/s and due to the high temperature they are partially ionized. At high light flux densities q 10^9 W/cm^2 and pulse duration $t_l=10^{-8}$s the vapor is ionized to a very high degree (up to 90% [1]). The energy of the main ionic component is in the region of 50 to 1000 eV with mean energy of about 350 eV. The neutral component of the plasma flow has relatively low mean energy (about 10 eV) and it comprises 10 to 70% of the evaporated material depending on the thermophysical properties of the target and on the irradiation conditions [1,3].

High deposition rate and supersaturation

Another specific feature of the pulsed laser-plasma deposition method is the high deposition rate during one laser pulse. For normally used experimental conditions in laser-plasma deposition experiments ($q=10^5$-10^8W.cm^{-2} and distance target-substrate l=2-15cm) the deposition rates could be controllably varied through the experimental conditions over a wide range from 10 to 10^5nm.s^{-1} [2], exceeding in some cases those available with conventional methods by many orders of magnitude. This circumstance leads to a very high degree of plasma flux supersaturation (>10^5cal.mol^{-1} [2]) resulting in specific nucleation and growth kinetics of the deposited film and in its specific properties respectively [4].

Deposition of ultrathin films of controlable structure

A detailed theoretical analysis of the film growth mechanism under the specific conditions of pulsed laser-plasma deposition [4] has shown that due to the high supersaturation (several orders of magnitude higher than that one of the conventional deposition techniques) it is possible to deposit dense and continuous films of monoatomic thickness. Depending on the experimental conditions such as density of the plasma fluxes (which determines the deposition rate R) and the substrate temperature, thin films of single crystalline, polycrystalline or amorphous structures can be synthesized. This conclusion is in agreement with published experimental results [2].

3. Deposition of polycomponent thin-film structures

The mentioned specific peculiarities of the laser-plasma deposition technique make it possible the synthesis of thin films and multilayer structures of complicated stoichiometric content, determined structure and precisely controllable thickness (down to several monoatomic layers) [1,4]. Depending on the energy of the particles several specific cases of application can be distinguished [1].

1. The thermal energy of the evaporated particles is much less than the dissociation energy of the molecules of the target material ($E_{pt} << E_d$). This regime may be realized with relatively long laser pulses ($t_l \sim 10^{-3}$s) and light flux densities $q \sim 10^5 W.cm^{-2}$. This conditions are suitable in obtaining stoichiometric films of polycomponent materials with high energy of dissociation, for example, oxides. This has been demonstrated with the deposition of dielectric films of SnO_2, TiO_2, ZrO_2, Al_2O_3, Nb_2O_5, BeO and their mixtures [1].

Varying over the process parameters polycomponent films of specific optical, electrical and mechanical properties have been synthesized. So for example films consisting of CdO and SnO_2 in weight ratios from 10:1 to 1:1 possess (depending on the experimental conditions) specific electrical surface resistance from $R_s=15\Omega/\Box$ up to $R_s=100k\Omega/\Box$ and high transmittance in the IR region from 2μm up to 15μm [1]. On the basis of BeO thin films X-ray exodetectors have been prepared characterized by a long storage time (2 months) of the radiative information [1]. Recently high quality high-Tc-superconducting films of (Y,Eu,Ho)$Ba_2Cu_3O_7$ and $Bi_2Sr_{3-x}Ca_xCu_2O_y$ have been synthesized by the laser-plasma method with critical current density for some of them better than $10^6 A.cm^{-2}$ [5].

2. The thermal energy of the particles is of the order of the dissociation energy ($E_{pt} \sim E_d$). In this case the target material is

partially or completely dissociated and the dissociation energy is released on the substrate surface in the process of atom-to-molecule association [1]. This regime concerns the most of the polycomponent semiconductors and may be realized also with long laser pulses ($\sim 10^{-3}$s) and flux densities $q \sim 10^5$-$10^6 W.cm^{-2}$. In this regime stoichiometric polycomponent semiconductor films of perfect structure, smooth surface and high carrier mobility can be obtained. The properties of thin films of GaAs, CdS, PbS, PbSe, PbTe and $Pb_{1-x}Cd_xSe$ deposited on this way have been intensively investigated [2]. The monocrystalline CdS films are highly photoconductive. The Hall mobility of PbSe on KCl substrate at 300K has a value of about $1000 cm^2 \cdot V^{-1}.s^{-1}$ which is of the order of the highest ones reported in the literature for monocrystalline PbSe. The high carrier mobility indicates perfect crystal structure.

3. The kinetic energy of the particles is of the order of the defect formation energy ($E_{pk} \sim E_{df} \sim 20eV$). Irradiation of monocrystalline substrates with particles of this energy produces a network of radiative defects (vacancies), which are additional crystallization centers. This allows the epitaxial growth of films at moderate mobility of the adsorbed atoms, i.e. at lower temperatures as compared to other methods [1].

This peculiarity has been used to prepare multilayer structures (up to 60 layers) of consecutively following heteroepitaxial semiconductor pairs of InSb(50Å)-CdTe(20Å), InSb(50Å)-PbTe(50Å) and Bi(50Å)-CdTe(20Å) [1]. Such structures, called superlattices, possess very interesting electrical and optical properties. The synthesis of high quality superlattice structures of HgTe and CdTe has also been demonstrated [3]. These showed high electron mobility and two-dimentional transport behavior including the quantum Hall effect.

4. Laser-plasma fluxes containing particles of kinetic energy higher than 200eV can be used for preparation of very interesting layer structures, consisting of monocrystalline films on amorphous or polycrystalline layers. As known location of adatoms in vapor condensation is influenced by the charge image of the electric field on the substrate surface. Under certain conditions

adatoms may form on an amorphous surface a single-crystal film, provided there is a field component with spatial frequency corresponding to the crystal lattice spacing.

A system of orienting potentials can be produced artificially by means of an ion stream. If the energy of the ions is high enough, they can pass through a thin unoriented layer deposited on a monocrystalline substrate and create vacancy-type defects in the interface of the single-crystal. If the thickness of the layer is not bigger than the Debye radius the surface of the structure will have orienting properties during the life time of the defects ($\sim 10^{-5}$s). In this case a film which is epitaxial to the single-crystal substrate can be grown on the surface of the unoriented layer, provided the amount of the condensate arriving during the vacancy life time is sufficient for the formation of a monolayer. This situation arises when the target is irradiated with laser beams of intensity $q > 2.10^{8}W.cm^{-2}$ [1].

On the bases of this effect laser-plasma deposition method has been used for synthesis of new type of superlattices, consisting of periodic multilayer structures of 50-100Å thick epitaxial semiconductor films of InSb and PbTe and 10-20Å thick nonoriented barier layers of GaAs, C, Ge [].

The possibility to deposit ultrathin films by the pulsed laser-plasma method has been used to prepare interference mirrors for soft X-rays consisting of up to 100 alternatively deposited layers of Cr(16Å) and C(34Å). Such mirror have for the carbon line (λ=44,7Å) 28% reflectivity for unpolarized radiation and 35% for the S-polarization [1].

4. Conclusion

The pulsed laser-plasma deposition technique is already well developed alternative to some conventional deposition techniques as flash evaporation, ion-beam sputtering and molucular-beam epitaxy. The results obtained have shown that this method is a powerful tool for synthesis of thin polycomponent films and film structures of unique properties, many of which can not be obtained any other way.

Literature

1. S.Metev, "Laser-plasma synthesis of thin-film polycomponent materials", in Laser Processing and Diagnostics (II), D. Bäuerle, K.L. Kompa, L. Laude eds., Proc. E-MRS, p. 143, (1986), and references therin

2. S.Metev, M.Sendova, "Thin-film compound formation with pulsed laser-plasma fluxes", in Trends in Quantum Electronics, (SPIE, Bellingham), Vol.1033, p.260, (1988), and ref. therein

3. H. Sankur, J.T. Cheung, "Formation of dielectric and semiconductor thin films by laser-assisted evaporation, Appl. Phys. A47, 271 (1988), and ref. therein

4. S.Metev, K.Meteva, "Nucleation and growth of laser-plasma deposited thin films", Appl. Surf. Sci., 43, 402 (1989)

5. A.I. Golovashkin,E.V. Ekimov et al., "Single-crystal films of high-temperature superconductors with the perovskite structure", JETP Lett.,47, No3, 191 (1988)

Laser-Induced Chemical Vapour Deposition of Thin Films in Microelectronics

G. Reisse, F. Gänsicke, A. Fischer, K. Zimmer, G. Zscherpe
University of Engineering Mittweida, Platz der DSF 17,
Mittweida 9250, GDR

Summary

Lateral thin film structures of Si, Mo, W, Co and $TiSi_2$ can be generated by laser-induced pyrolytical and/or photolytical chemical vapour deposition techniques (LCVD) using perpendicular incident focused laser light. The process, and film parameters, e. g. growth rate, film width, crystalline structure, and specific resistance have been investigated and compared with a mathematical model.

Experimental

The experimental apparatus exhibits the following specifications: argon-ion-laser ILA 190, TEM_{oo}-mode, maximum power P_L = 10 W (multi line) with external SHG (P_L = 1 mW), gas flow rate Q = 0 ... 100 sccm, total pressure range p = 13.3 ... 2.7 · 10^4 Pa, gaussian radius r_g = 2.6 ... 15 µm, scan velocity v_t = 0.02 ... 0.2 mms^{-1}, reaction chamber /1/ with substrate heater T_s = 77 ... 800 K. The partially reflected laser beam is detected for the in situ control of the focus on the surface /2/ and for the evaluation of the film thickness during and after a deposition run /3/.

Model calculations

The numerical model of the laser-induced pyrolytical film deposition process involves a temperature field calculation within the substrate and the growing film using an element method /4/. The heat conduction equation was transformed into a system of first order differential equations by subdividing the space into hollow cylinders. This differential equation system was numerically solved /5/. Reactant gas transport and the reactant gas concentration in the reaction zone are computed by solving the diffusion equation using a a point source idealization and a subdivision of the gas space into hemispheres. A description of the deposition rate R_D by applying Arrhenius relation is shown in the following equation /6/:

$$R_D = R_o \cdot \exp(-E_A/k_B \cdot T)$$

(E_A - activation, or dissociation energy, k_B - Boltzmann's factor, T - surface temperature)
The factor R_o is influenced by kinetic processes.
It is based on experimental results.
The following predictions could be made with the help of the model:

- Passivation layers, e. g. 40 nm SiO_2 on silicon lead to self-limiting effects in film thickness in connection with the growing film (Fig. 1).
  This behaviour could be verified experimentally by selecting the matching oxide film thickness ($x \cdot 170$ nm)
- With input data for the Si-deposition by silane pyrolysis: E_A = 1.95 eV, R_o = 1600 cms^{-1} (p_{SiH_4} = 1.3 · 10^4 Pa), and the diffusion coefficient $D_{SiH_4} = 1\, cm^2s^{-1}$ a stationary temperature field occurs after a thermal time constant $\tau_{th} > 10\ \mu s$. For small v_t - values the characteristic deposition time constant is: $\tau_D = r_g/v_t > \tau_{th}$ and the temperature distribution is a function of the intensity. At $R_D > 100\ \mu ms^{-1}$, and $r_g = 4\ \mu m$ the diffusion limited case of the Si-deposition reaction resulting from SiH_4 begins.

An analytical solution was derived for the film thickness dependence perpendicular to the scan direction by integrating the intensity distribution of a moving laser beam with respect to time:

$$d_s(y, v_t) = R_D \exp(-y^2/r_{gD})\ (\pi\, k_B \cdot T_c/E_A)^{1/2} \cdot r_g/v_t$$

(y - coordinate perpendicular to the scan direction,
T_c - temperature at the spot centre)
Therefore the deposited line has a gaussian profile with the deposition radius:

$$r_{gD} = r_g\,(k_B \cdot T_c/E_A)^{1/2}$$

Because of the strongly exponential temperature dependent deposition rate film widths smaller than the gaussian radius of the laser beam can be obtained.

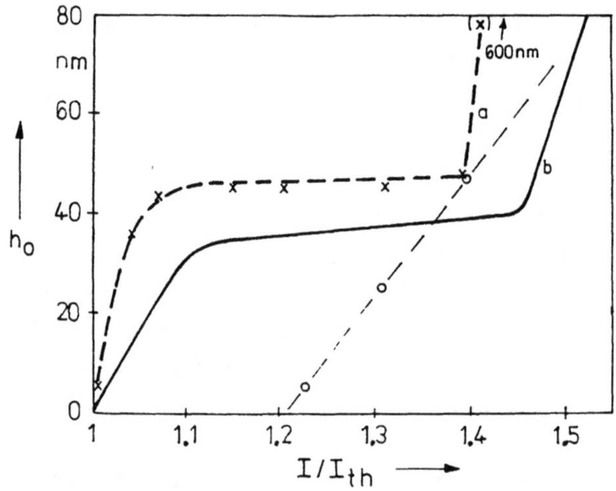

Fig. 1: Thickness dependence of a deposited Si-line on the normalized intensity.
(40 nm SiO_2/Si-substrate, r_g = 15 μm, p_{SiH_4} = 13 kPa, v_t = 0.1 mms^{-1}, T_s = 500 K)
a - experimental results, b - calculated values,
I_A - threshold intensity

Experimental results and discussion

The film width produced during pyrolytical Si-deposition from SiH_4 is preferably influenced by the gaussian radius within the focus, the laser power and the substrate structure /7/. To give one example the Si-film width deposited on c-Si was b = 2 μm with r_g = 4 μm, P_L = 1.5 W, T_s = 300 K, v_t = 0.09 mms^{-1}, p_{SiH_4} = 5.2 kPa and R_D = 5 μms^{-1}. The film width was enhanced with increasing laser power caused by shifting the deposition threshold temperature to larger distances from the focus centre.

After reaching the melting point of silicon a melting ditch was formed (Fig. 2)

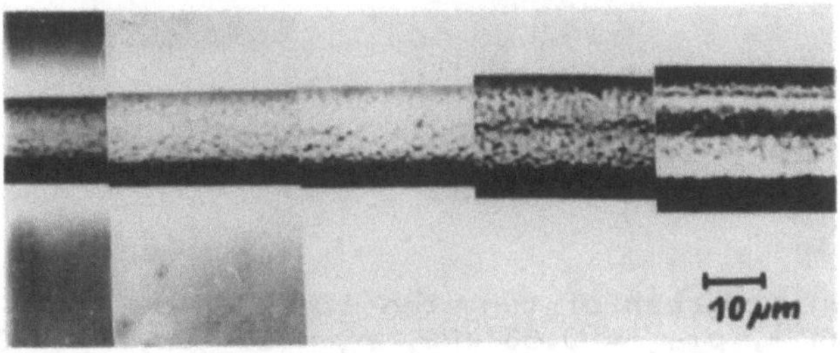

Fig. 2: Optical micrograph of Si-lines on 340 nm SiO_2/c-Si. (r_g = 15 µm, T_s = 580 K, v_t = 0.1 mms^{-1}, p_{SiH_4} = 13 kPa, t_s = 1.2 µm, P_L = 2.3; 2.4; 2.5; 2.6; 2.8 W)

The film thickness can be inreased by multiple scans without film width broadening. The polycrystalline structure of the Si-films is characterized by crystallites with the dimension 300 x 500 nm orientated toward the direction of film growth. The crystallite size decreases with the distance from focus centre.

The specific film resistance of about $\varrho_{Si} = 10^6$ Ohm cm, decreases to about $\varrho_{Si} = 10^4$ Ohm cm with inreasing laser from P_L = 2.3 to P_L = 2.7 W (r_g = 15 µm, T_s = 580 K, v_t=0.1 mms^{-1}, p_{SiH_4} = 13 kPa. One possible reason is the temperature dependent crystallite size.

The pyrolytical deposition of tungsten from tungsten hexafluoride WF_6 takes place through silicon substitution reaction at the Si-surface. Because of the Si-consumption the W-films grow into the surface. W-lines with 12 µm film width and a specific resistance of 170 µm Ohm cm could be deposited with the parameters: r_g = 15 µm, p = 0.53 kPa, v_t = 0.2 mms^{-1}, P_L = 3.5 W. The film thickness is limited by the inhibited diffusion of Si-atoms through the growing film (Fig. 3).

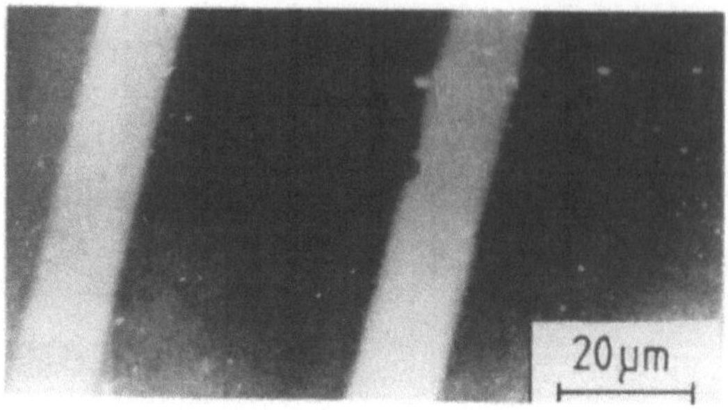

Fig. 3: SEM-micrograph of tungsten lines on c-Si.
(v_t = 0.1 mms^{-1}, p_{WF_6} = 0.53 kPa, r_g = 15 µm, P_L = 3.0 W, T_s = 300 K)

The film resistance of laser deposited tungsten could be decreased by replacing the WF_6 by tungsten hexacarbonyl W $(CO)_6$ as W-source. It drops to about 12 µOhm cm with Ar as carrier gas at v_t = 0.02 mms^{-1}, P_L = 2 W, p = 280 Pa, r_g = 15 µm, R_D = 0.25 $µms^{-1}$.

Further experiments were directed to molybdenum, and cobalt deposition from their respective carbonyls.
Mo-lines deposited at $p_{Mo(CO)_6}$ = 260 Pa, P_L = 2.3 W, r_g = 15 µm, v_t = 0.2 mms^{-1} show specific film resistances of 3 mOhm cm caused by high levels of O_2-contamination (50 %, AES). The deposition of Co from $CO_2(CO)_8$ takes place in two stages: a photolytical start reaction with a threshold power P_L = 60 mW and a pyrolytical film growth. The film thickness and film width increases with increasing laser power.

Some basic investigations were made for titanium silicide deposition from titanium tetrachloride $TiCl_4$, silane SiH_4 and H_2 or Ar as carrier gases /8/. Parameters used are: r_g = 4 µm, P_L = 1.5 W, v_t = 0.02 mms^{-1}, and p = 2.7 kPa. These $TiSi_x$-films have film widths of 10 µm and specific film resistances of ρ = 3 mOhm cm after deposition without post-annealing. High film resistances are attributed to the fact, that the film conductivity distribution is inhomogeneous (Fig. 4), and the films are contaminated by chlorine.
The films grow into the substrate material (Si) because of an etching reaction baused by Cl_2 and a Si-consumption by a substitution reaction.

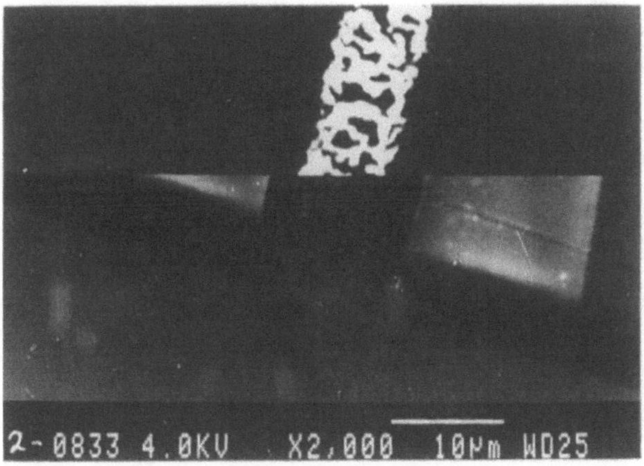

Fig. 4: EBIC-micrograph of $TiSi_x$-films,
(r_g = 4 µm, P_L = 1.5 W, v_t = 0.02 mms^{-1}, p = 2.7 kPa)

<u>References</u>

/1/ Reiße, G., H. Exner, G. Zscherpe, K. Zimmer, S. Weissmantel: DD-WP B 23 K 26/02 245 833 A 1, 1986

/2/ McWilliams, B. M., W. Chin, I. P. Herman, R. A. Hyde, F. Mitlitsky, I. C. Whitehead: Proc. of SPIE, Vol. 459 (1984) p. 49 n - 55

/3/ Zimmer, K., G. Reiße, F. Gänsicke, G. Zscherpe: DD-WP GO 1 B 11/30 270 365 A 1, 1988

/4/ Schwarzott, W.: Forsch.-Ing. Wes. 38, 6(1972) 165

/5/ Gear, C. W.: Numerical initial value problems in ordinary differential systems, Prentice Hall, Englewood Cliffs, New York 1971

/6/ Osgood, R. M., H. H. Gilgen, Ann. Rev. Sci. 15 (1985) 549

/7/ Reiße, G., K. Zimmer, F. Gänsicke, G. Zscherpe: Beiträge zur 10. Tagung Hochvakuum, Grenzflächen/Dünne Schichten, Dresden 1990, Bd. II, p 173 - 175

/8/ Gänsicke, F., K. Zimmer, G. Reiße, H. Johansen, B. Lämmel: Proc. of the EPM 89, Dresden, 1989

Reactive Etching and Deposition

Plasma-Assisted Etching

H. Mader

Fraunhofer-Institut für Mikrostrukturtechnik,
Dillenburger Str. 53, D-1000 Berlin 33 (F.R.G.)
Fachhochschule München, Fachbereich Elektrotechnik,
Dachauer Str. 98b, D-8000 München 2 (F.R.G.)

Summary

Plasma-assisted etching is an essential part of the microstructuring process in fabrication of integrated semiconductor circuits and micromechanic devices. Thereby structures down to the sub-µm-range are etched into layers of isolators, semiconductors, metals and metal-semiconductor-compounds. Such etch processes must show a high degree of anisotropy and a high selectivity in respect to the etch mask and the substrate. These demands can be satisfied by a chemical etch reaction which is induced by impinging energetic particles such as ions, electrons or photons. In the following article the properties of plasma-assisted etching will be described and applications in the microsystem technology will be discussed.

1. Introduction

Since the birth of microelectronics 30 years ago the complexity of integrated circuits (ICs) has grown vary rapidly. As an example, from 1970 the number of storage cells per IC chip has increased by a factor of 4 every 3 years [1, 2]. This progress had been achieved mainly by miniaturization of the chip components. The minimum linewidth within the 4 Mbit DRAM is about 0.8 µm. The demand for further increase in integration density will require etching techniques for features in the range of sub - 0.5 µm in the 1990s.
The actual state of plasma-assisted etching appropriate to the fabrication of such fine structures will be described below.

2. Basic methods of plasma-assisted etching

Plasma-assisted etching implies methods by which a solid surface is etched either physically by ion bombardement, chemically by a chemical reaction of reactive species at the surface or combined physically chemically by an ion-, electron-, or photoninduced chemical reaction at the surface, as shown in Fig. 1.

In physical etching, ions extracted from a plasma are accelerated in an electrical field towards the substrate to be etched. The impinging ions erode the surface by momentum transfer.

In chemical etching, neutral reactive species generated in a plasma diffuse to the substrate. Etching is affected by a chemical reaction of the reactive species with atoms of the surface to be etched. The reaction product has to be a volatile gas which can be removed by a vacuum pump. The isotropic velocity distribution of the reactive species leads to mask undercutting.

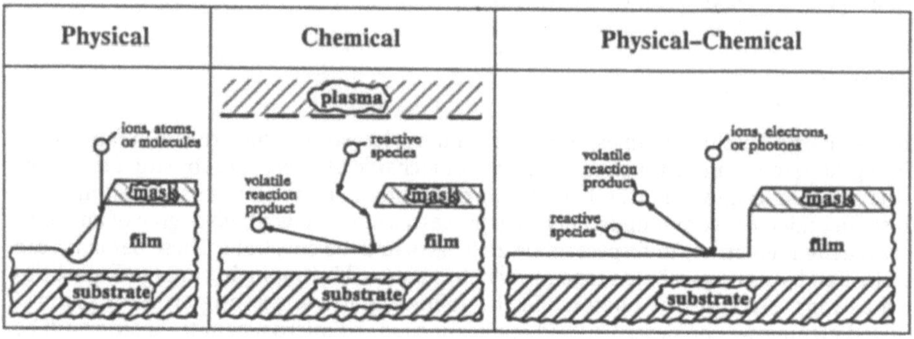

Fig. 1. Basic methods of plasma–assisted etching [3]

In physical–chemical etching, the chemical etch reaction is induced by impinging ions, electrons or photons. The reaction product also has to be a volatile gas. This etching method leads to an accurate transfer of the mask pattern to the film below, as required for the fabrication of fine patterns in integrated circuits and micromechanic devices.

3. Anisotropic plasma–assisted etching

As described before anisotropic plasma–assisted etching can be obtained by a combination of physical and chemical etching. This combination can be enhanced physical sputtering, enhanced chemical etching or chemical sputtering as shown in Fig. 2.

Enhanced physical sputtering is a two step process. In the first step a chemical etch reaction between incident neutral particles (A) and the substrate surface atoms (S) takes place. The reaction products (S poly) remain on the surface. Such reaction products can be polymers or adsorbed compounds. The second step of this etching process is physical sputtering of the reaction products (Ps) by impinging ions (B^+). A renewed etch reaction corresponding step one only can accour when the previous obtained etch products are sputtered away. This is not the case at the sidewalls. Therefore the etch process is anisotropic. The etch mask can be transfered accurate to the substrate below, as shown in Fig. 2.

Another anisotropic etch process is enhanced chemical etching which is also a two step process. In the first step an impingment of accelerated ions (B^+) on the substrate (S) leads to a damaged or modified surface (Sd). In a second step a chemical etch reaction between incident reactive species (A) and the damaged surface (Sd) takes place. This reaction leads to a volatile reaction product (P). Etching only takes place on the surface (Sd) which is bombarded by ions. This leads to an accurate transfer of the etch mask to the substrate below.

ANISOTROPIC PLASMA-ASSISTED ETCHING		
Etching Model	**Principle**	**Reactions**
Enhanced Physical Sputtering	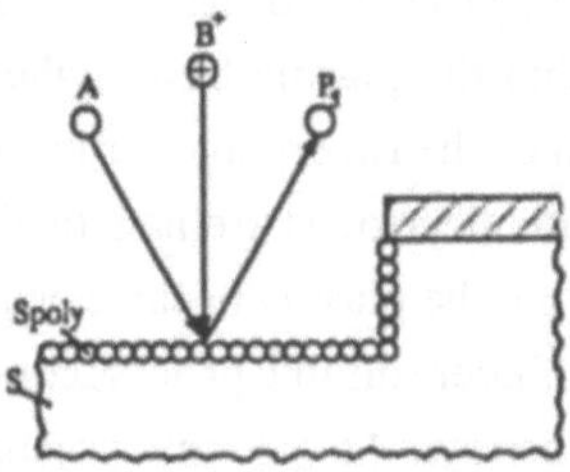	1. $A + S \rightarrow S\ poly$ A neutral reactive particles S substrate surface atoms S poly remaining reaction products on the surface 2. $B^+ + S\ poly \rightarrow P_s$ B^+ impinging ions Ps sputtered compounds
Enhanced Chemical Etching	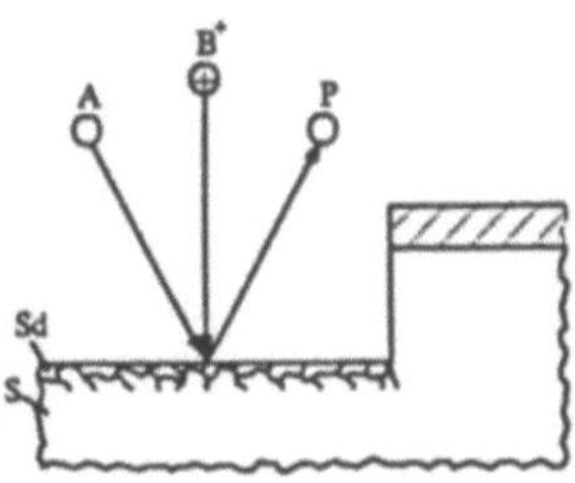	1. $B^+ + S \rightarrow Sd$ B^+ impinging ions S surface atoms Sd damaged or modified surface 2. $A + Sd \rightarrow P$ A neutral reactive particles P volatile reaction product
Chemical Sputtering	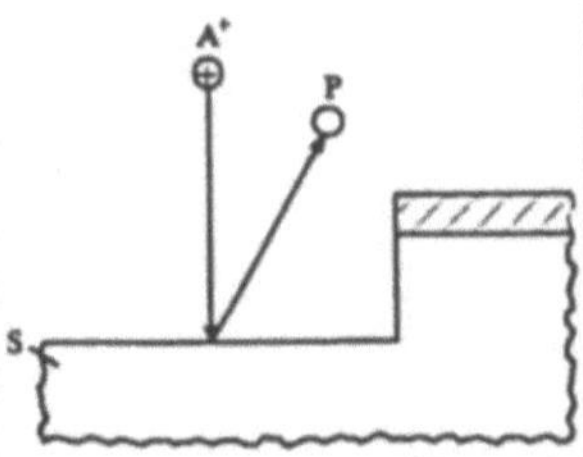	$A^+ + S \rightarrow P$ A^+ impinging <u>reactive</u> ions S surface atoms P volatile reaction product

Fig. 2. Mechanisms of anisotropic plasma-assisted etching [4]

A third variant on anisotropic etching is chemical sputtering. In this process a chemical etch reaction between incident reactive ions (A^+) and the substrate surface atoms (S) takes place resulting in a volatile reaction product (P). Chemical sputtering is a one step process and the best way to get anisotropic etch profiles.

In commercial etch reactors for physical–chemical etching, the chemical etch reaction is induced by impinging ions. The most used reactor, which is shown schematically in Fig. 3, consists of a recipient with a gas supply, a connection to a vacuum pump and two parallel electrodes. A reactive gas is fed to the previously evacuated recipient and held at constant flow rate and pressure. The gas between the electrodes is brought to a glow discharge by the application of an r.f. voltage, usually at a frequency of 13.56 MHz. The result is a low–pressure, low temperature plasma containing free electrons, ions and neutral radicals. On account of their low mass, electrons within the plasma have higher mobility than ions. Therefore immediately after switching on of the r.f. voltage far more electrons than ions reach the electrodes. In consequence, both electrodes are negatively charged relative to the plasma. The positive ions extracted from the plasma became accelerated in the direction of the negatively charged electrodes. Under suitable process conditions these ions initiate a chemically etch reaction on the substrate. An essential precondition for etching is the formation of volatile reaction products which can be removed by a vacuum pump.

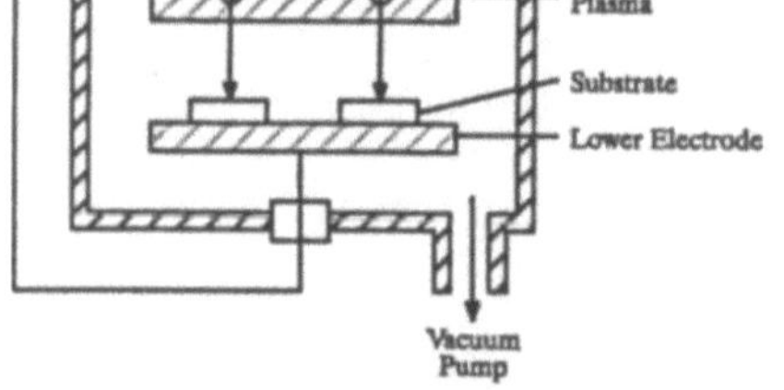

Fig. 3. Schematic drawing of a parallel plate reactor used for ion induced physical–chemical etching

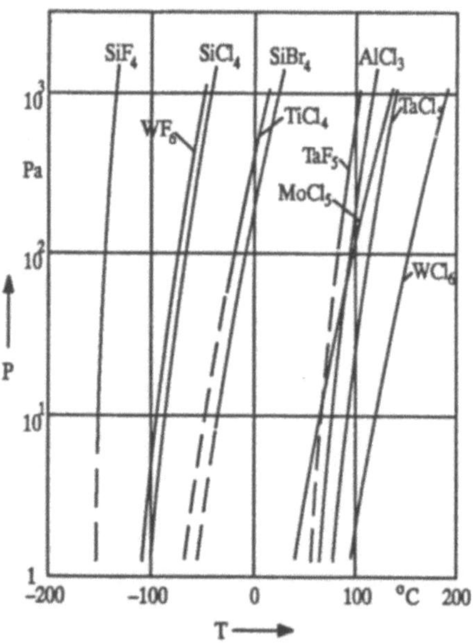

Fig. 4. Vapour pressure curves for some reaction products in plasma–assisted etching [3, 5]. A large number of additional curves can be found in [3].

Fig. 4 shows the vapor pressure curves of some florine, chlorine and bromine compounds with elements used in silicon technology. During the etching process the surface temperature ranges between 20 and 100° C and the gas pressure between 0.1 and 100 Pa. Therefore, reaction products of silicon, SiO_2 and Si_3N_4 with fluorine, chlorine and bromine are volatile under process conditions. Aluminium and refractory metals such as molybdenum, tungsten, titanium and tantalium deliver volatile reaction products partly only at higher temperature. While etching of such materials the etch reactor has to be warmed up to the temperature where the reaction products are volatile. Table I contains a selection of gases used for etching silicon, SiO_2, Si_3N_4, aluminium, refractory metals and organic materials. All these etching gases contain one or more elements which deliver volatile etch reaction products with the atoms of the substrate to be etched.

Layer	Etch gas
Si	CF_4, SF_6, Cl_2, $CBrF_3$ ...
SiO_2	CF_4, CF_4+H_2, CHF_3, C_2F_6 ...
Si_3N_4	CF_4, CF_4+O_2, CF_4+H_2, CHF_3 ...
Al	Cl_2, BCl_3, CCl_4, BCl_3+Cl_2 ...
$TaSi_2$, $MoSi_2$, $TiSi_2$, WSi_2	CF_4, Cl_2, BCl_3 ...
organic materials	O_2

Table 1. Selection of etch gases used in silicon technology [2, 3, 6]. A large number of additional etch gases can be found in [3]

4. Techniques for anisotropic plasma–assisted etching

At present anisotropic plasma assisted etching mostly is carried out in a parallel plate reactor shown in Fig. 3. Etching techniques using this reactor are reactive ion etching (RIE), magnetically enhanced reactive ion etching (MERIE) and triode plasma etching.

In reactive ion etching (RIE) the substrates to be etched are placed on the lower electrode to which the r.f. voltage is capacitively coupled. The upper electrode is grounded. The acceleration voltage for the impinging ions to the substrates is approximately the amplitude of the r.f. voltage, which is usually several hundred volts. The pressure of the reactive gas ranges typically between 1 and 10 Pa. In magnetically enhanced reactive ion etching (MERIE) the plasma is densified by an applied magnetic field. This results in an increased etch rate and a decreased ion energy compared to reactive ion etching.

In triode plasma etching systems, the discharge is sustained by an applied r.f. voltage on horizontally located parallel electrodes. The substrates to be etched are placed on a third electrode which can be biased by a second voltage. The energy and flux of the ions to the substrates are quasi–independendly controllable which is not possible in reactive ion etching.

The energy and flux of ions are independently controllable in plasma source based etching. Thereby the generation of plasma is separated from the place where the substrates are etched. Methods of plasma source based etching are reactive ion beam etching (RIBE) and chemically assisted ion beam etching (CAIBE), which are shown schematically in Fig. 5.

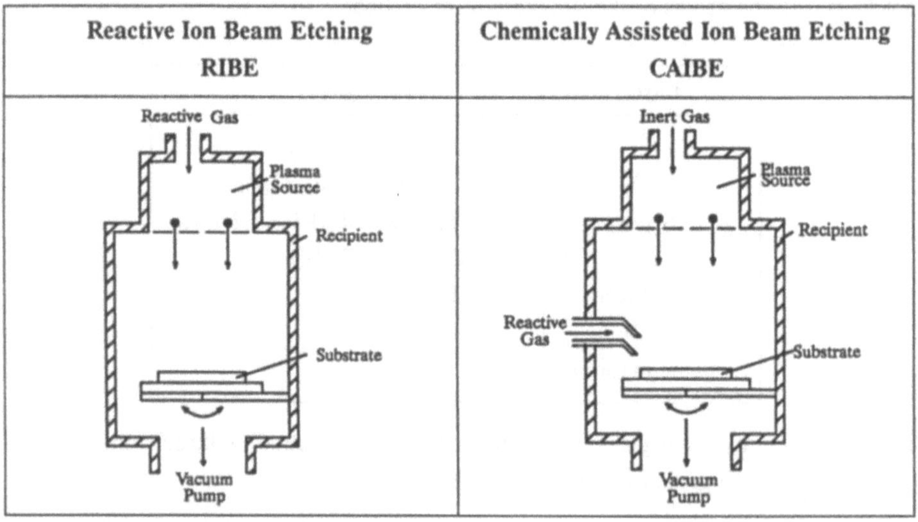

Fig. 5. Plasma source based etching:
reactive ion beam etching (RIBE) and chemically assisted ion beam etching (CAIBE)

In reactive ion beam etching reactive ions generated in a plasma source are extracted and accelerated in the direction of the substrate to be etched. Etching accours if the reaction product is volatile.

In chemically assisted ion beam etching ions from an inert gas are extracted from a plasma source and accelerated towards the substrate to be etched. The impinging ions initiate a chemical etch reaction between an reactive gas and the surface to be etched.

Frequently used plasma sources are based on the electron cyclotron resonance (ECR–sources) which produces high density plasmas.

5. Application in microelectronics

In the fabrication of integrated circuits anisotropic plasma–assisted etching is used for patterning $Si0_2$, Si_3N_3, silicon, silicides, metals and organic materials. Thereby the layers are etched in the following sequence [2, 6]:

1. Si_3N_4
   The etched silicon nitride pattern is used to produce the isolation regions by local oxidation. Oxidation is confined to the nitride–free regions.
2. Polysilicon and Silicide
   This etch step defines the length of MOS–transistor gates.
3. SiO_2
   Contact windows are etched through the isolation layer on silicon.
4. Metal film (Al/Si, Al/Si/Cu)
   This etch step serves for patterning interconnections.

All these etch steps must be performed with extreme pattern fidelity because they greatly influence the packing density and electrical behaviour of the fabricated integrated circuits. The complete demands on etch processes applied in fabrication of high density integrated circuits can be summarizes as follows:

1. high degree of anisotropy,
2. high selectivity to substrate and etch mask,
3. high etch rate,
4. high degree of etch rate uniformity across the wafer and from wafer to wafer,
5. high degree of reproducibility of etch results,
6. low heating of the substrate surface, because the resist etch mask lose its form stability at temperature higher than 100° C,
7. low surface damage,
8. low surface contamination,
9. low electric charge onto the substrate to avoid break down of thin isolator films,
10. environment compatibility concerning etch gases used and etch reaction products.

6. Application in micromechanics

In the fabrication of micromechanic devices plasma–assisted etching is used for patterning of thin films either directly or in combination with galvanic processes. Directly patterning etch processes are carried out mostly using a resist etch mask.

A micromechanic device patterned by directly plasma–assisted etching is the fluid amplifier [7, 8]. In the fabrication of this device 35 μm thick and 6 μm wide silicon structures are etched anisotropically using a gas mixture of SF_6 and C_2ClF_5.

The same gas composition is used for the fabrication of micromechanic silicon gears with a diameter of 300 μm [7, 9].

Another micromechanic devices, which are fabricated by plasma–assisted etching are ring nozzle sets for ink jet printers [7, 10]. The nozzles are etched using SF_6 gas.

Anisotropic plasma–assisted etching is also applied in the fabrication of hole masks used in ion– or electron–projection lithography. In this process CF_4 can be used for etching silicon nitrid layers and CCl_2F_2 for silicon [7, 11]. All etch processes have to be carried out with a high degree of anisotropy to get patterns down to 0.1 μm.

7. Conclusions

Increased integration density in microelectronic devices and decreased pattern dimensions of micromechanic structures need improved anisotropic etch processes. The demands can be accomplished using plasma–assisted etching which can be pure physical, pure chemical or combined physical–chemical. In commercial etch reactors for combined physical–chemical etching the chemical etch reaction is induced by impinging ions. The mostly used machine for such anisotropic etching is the parallel plate reactor. Etching techniques carried out in this reactor are reactive ion etching (RIE), magnetically enhanced reactive ion etching (MERIE) and triode plasma etching. Plasma source based etching, in which the plasma generation and the place of etching are separated, is a very interesting variant of anisotropic plasma assisted etching. This technique promises to solve some existing etch problems. Anisotropic plasma–assisted etching is absolutely necessary in the fabrication of integrated circuits and of increasing importance in the fabrication of micromechanic devices.

References

1. Arden, W.; Müller, K.–H.: Microelectronic Engineering, Vol. 6 (1987) 53 – 60
2. Widmann, D.; Mader, H.; Friedrich, H.: Technologie hochintegrierter Schaltungen, Berlin: Springer–Verlag 1988
3. Mader, H.: Etching processes, Landolt–Boernstein, Vol. 17c Berlin: Springer–Verlag 1984, 280 – 367
4. Zawaideh, E.; Kim, N. S.: A plasma etching model based on generalized transport approach. J. Appl. Phys. 62 (6), (1987) 2498 – 2507

5. Mader, H.: Microstructuring in semiconductor technology. Thin solid films, 175 (1989) 1 - 16.
6. Beinvogl, W.; Mader, H.: Reactive Dry Etching for Fabrication of Very–Large–Scale Integrated Circuits. Siemens Forsch.- u. Entwickl.-Ber. Bd. 11 (1982) Nr. 4, 180 -189
7. Heuberger, A.: Mikromechanik. Berlin: Springer–Verlag 1989
8. Zdeblick, M. J.; Barth, P. W.; Angell, J. B.: Microminiature fluidic amplifier. IEEE Solid–State Sensor Workshop. Hilton Head Island, South California 1986
9. Mehregany, M.; Gabriel, K. J.; Trimmer, W. S. N.: Micro gears and turbines etched from silicon. Sensors and Actuators
10. Bentin, H.; Doehring, M.; Koehler, I.: Mikromechanische Bauteile für Tintenstrahldrucker. Private communications
11. Behringer, V.: Silizium–Transmissionsmasken für die Sub–micron–Elektronenstrahllithographie. VDI–Berichte 584, Düsseldorf: VDI–Verlag (1985)
12. Aritome, H.; Aoki, H.; Namba, S.: X–ray zone plates fabricated using electron beam lithography and reactive ion etching J. Vac. Sci. Technol. 83 (1985) 1

Reactive Etching with ECR Plasma Source

G.CASTRISCHER, K.-H. KRETSCHMER, G. LORENZ, I. KESSLER

Leybold AG, Alzenau, FRG

Summary

An Electron Cyclotron Resonance (ECR) plasma source with an effective diameter of 9'' is described. The radially symmetric construction of the source as well as the chosen microwave input allows wafers with diameter up to 200 mm to be processed with high uniformity. Several etch processes are presented to demonstrate the application potential of the source.

Description of the ECR source

The Leybold ECR source is shown schematically in fig.1. The dome-shaped quartz glass plasma chamber is surrounded by permanent magnets in a configuration that results in a toroidal magnetostatic field. This magnetic field plus the 2.45 GHz microwave excitation produce an ECR effect in a toroidal-ring shaped region running along the wall of the plasma chamber. Thus a sufficient number of ions is created in the outer plasma region to compensate for the radial decay of ion density in the chamber. Microwave is fed to the chamber by a 2.45 GHz magnetron capable of generating up to 1 kW of microwave power. The rectangular waves excited by the magnetron are transmitted to a rectangular waveguide via a circulator, which keeps forward-propagating waves separated from reflected waves and thereby allows reflected power levels to be measured at a third port on the circulator. The rectangular waveguide section is terminated by a short-circuiting slide, which serves as an impedance adjustment and allows the reflected power to be minimized. To keep the plasma homogeneous a radially symmetric waveguide mode has to be generated within the plasma region. This is achieved with the microwave feed located between the rectangular waveguide and the horn-shaped radiator. This

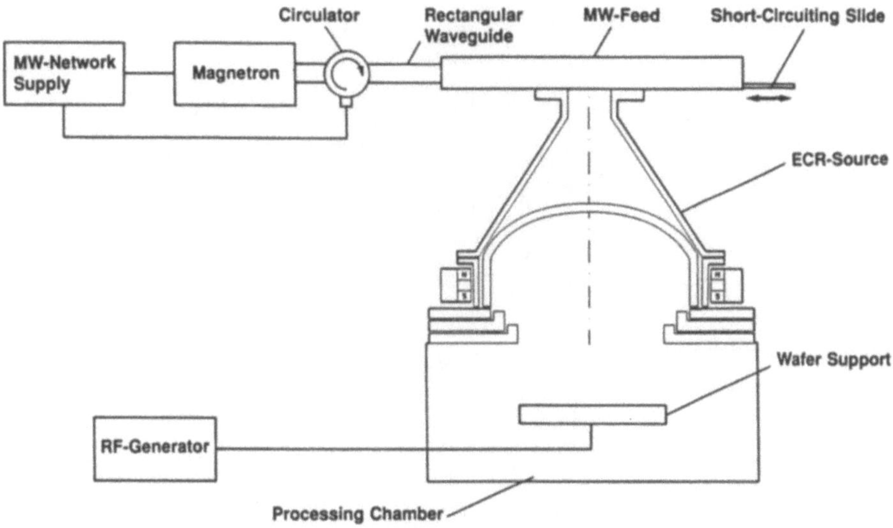

Fig.1 ECR system including microwave input

microwave feed transforms the waveguide excited by the generator into the radially symmetric E_{01} mode depicted in fig.2 [1]. The microwave input coupling is arranged such that only this fundamental mode can propagate into the circular waveguide section. As shown in fig.2, the maximal field strengths of this mode, and thus the most effective plasma generation, occur in the region where the ECR condition is met, that is, near the chamber wall in the magnetostatic toroidal ring.

The ECR source allows the generation of both isotropic and anisotropic etch profiles. If the wafer support shown in fig.1 is electrically grounded radicals and ions produced within the plasma region simply diffuse to the wafer surface. Since particles impacting the sample have energies less than 20 eV [2], purely chemical effects dominate leading to an isotropic etch profile. Anisotropic etch processes are possible if a 13.56 MHz rf voltage is applied to the wafer support. An rf-excited plasma is created near the wafer surface and a dc bias voltage is established at the wafer support. This rf-induced dc bias voltage accelerates ions from the plasma boundary layer to the wafer surface. The energy of the extracted ions and thus the physical etch component may be adjusted by variation of the rf power input.

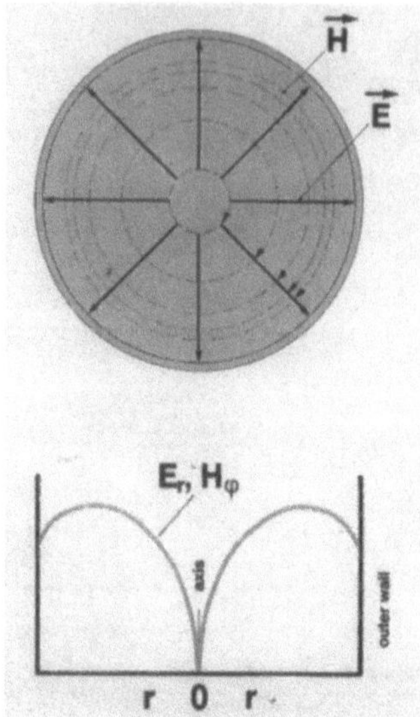

Fig.2 Distribution of field vectors and amplitude across the plasma source of the E_{01} waveguide mode

Contact Hole Etching of PSG

Small diameter contact holes having vertical walls cause problems in the subsequent metallization step. The combination of an isotropic and a following anisotropic etch step producing a countersunk etch profile helps to improve step coverage of the metallization, thus yielding low-resistance and reliable metal contacts. With the ECR source both types of etching can be produced using the same CF_4/O_2 gas mixture.

In the plasma, CF_4 decomposes into CF_x and F radicals. The fluorine radicals perform the chemical etching process, while CF_x has a tendency to form polymers which deposit on surfaces and reduce etching rates. Adding O_2 leads to the formation of transient compounds of CF_x and O_2, thereby preventing the formation of polymer films on the surfaces and increasing fluorine concentration in the gas phase as well. Both mechanisms, resulting from O_2 additions, lead to increases in isotropic etching rates [3,4].

Fig.3 shows an SEM picture of an isotropically etched profile in phosphosilicate glass (PSG). The gas for etching was a CF_4/O_2 mixture having a 40% O_2 concentration. The virgin, underlying-substrate surface and the undamaged photoresist indicate that particle energies were low. This fact is confirmed by the nearly identical etch rates in both horizontal and vertical directions.

When 13.56 MHz rf power is fed to the wafer support, purely anisotropic etching of PSG results. Fig.4 shows the profile obtained using a CF_4/O_2 gas mixture having a 12.5% O_2 concentration. For anisotropic etching, as with isotropic etching, the underlying substrate remains clean and the photoresist shows no evidence of having suffered from thermal effects.

Fig.3 Isotropic PSG etch

Fig.4 Anisotropic PSG etch

Anisotropic Etching of PolySi

One of the most difficult etch processes using conventional RIE methods is the delineation of polySi gate electrodes, because an anisotropic etch with minimum linewidth loss and high selectivity to SiO_2 is difficult to achieve. Also the high ion energies and dc self-bias voltage characteristic of parallel-plate etchers operated at low pressure may cause breakdown of the thin gate oxide [5].

The low ion energies in ECR plasmas make it possible to anisotropically etch polySi using pure chlorine. Fig.5 shows an SEM of an etched polySi track 1 µm wide, with about 200 nm polySi atop 25 nm SiO_2. The photoresist flaked off when the wafer was diced, so the result of etching can be clearly seen along the fracture. With a dc self-bias voltage as low as 30 V an anisotropic profile was obtained with no damage to the gate oxide. Etching results were independent of structural dimensions, as

Fig.5 Anisotropic etching of 1 µm wide polySi tracks using pure Cl_2 with no additives.

shown by the two SEMs in fig.6. For both 0.7 µm and 2.0 µm line widths, etching is anisotropic and the underlying substrate is free of residues. For etch rates of 55 nm/min on a 200 mm diameter wafer, the etching uniformity was ± 9 percent over the whole wafer.

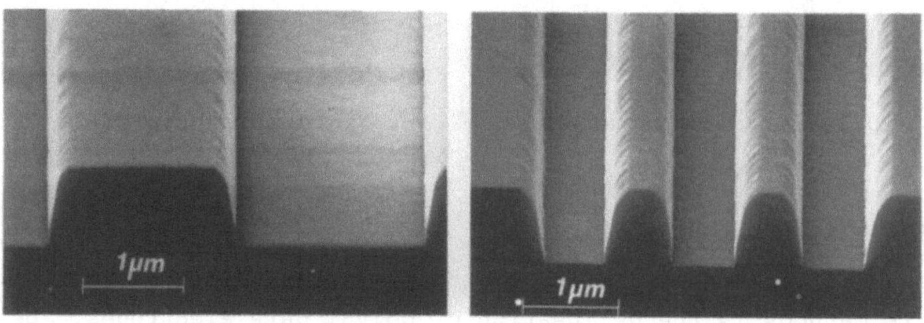

Fig.6 PolySi etch results for linewidths of 2.0 and 0.7 µm

Etching of III-V Compound Semiconductors

Fig.7 shows an SEM of a 7 µm deep trench etched into a GaAs/ $Al_{0.5}Ga_{0.5}As$ heterostructure. Using pure chlorine at a pressure of 10^{-3} mbar resulted in an etch rate of 0.7 µm/min, nearly vertical sidewalls and excellent surface morphology.

InP may be etched with a gas mixture of methane and hydrogen, the etch products being phosphine and volatile organoindium compounds [6]. The sample shown in fig.8 has been etched with a gas mixture of 10% CH_4 in H_2, giving slightly tapered etch profiles and smooth horizontal surfaces.

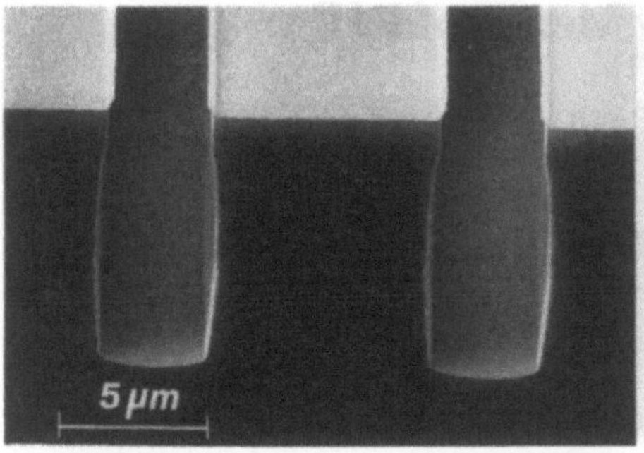

Fig.7 GaAs trench etched with chlorine chemistry

Fig.8 InP etched with a gas mixture of 10% CH_4 in H_2

References

1. Meinke, H.; Gundlach, F.W.: Taschenbuch der Hochfrequenztechnik. Berlin, Springer-Verlag 1986

2. Matsuoka, U.; Ono, K.: Magnetic Field Gradient Effects on Ion Energy for Electron Cyclotron Resonance Microwave Plasma Stream. J. Vac. Sci. Technol. A6(1) (1988) 25

3. Coburn, J.; Winters, H.: Plasma Etching: A Discussion of Mechanisms. J. Vac. Sci. Tachnol. 16(2) (1979) 391

4. Lee, Y.; Chen, M.: Silicon Doping Effects in Reactive Plasma Etching. J. Vac. Sci. Technol. B4(2) (1986) 468

5. Watanabe, T.; Yoshida, Y.: Dielectric Breakdown of Gate Insulator Due to Reactive Ion Etching. Solid State Technol. 27 (1984) 263-266

6. Hayes, T.R.; Dreisbach, M.A.; Thomas, P.M.; Dautremont-Smith, W.C.; Heimbrook, L.A.: Reactive Ion Etching of InP Using CH_4/H_2 Mixtures: Mechanisms of Etching and Anisotropy. J. Vac. Sci. Technol. B7(5) (1989) 1130-1140

Production of Resist Masks by Means of RIE

G. HELLER, K. RICHTER, E. KÄNDLER, A. GÜNTZEL

Dresden University of Technology
Department of Information Technology
Chair of Technology of Electronic Components

Summary

Resist masks for the submicrometer range are structured by the use of plasmas. The optimum etching parameters of a single-wafer diode etch reactor for achieving maximum resolution and step heights are searched for by means of statistical experimental design. There were used test structures on Si-wafers of 5", which had been coated with different photoresists and silylated after projection exposure. O_2 and CF_4O_2-mixtures were used as etching gases. Process monitoring was done with the aid of a laser reflectometer to determine the etch rate and of an emission spectrometer to detect the end point of etching.

Introduction

The structuring of resist masks with an O_2-plasma requires selective resistance to etching of the mask areas and the fast, anisotropic removal of the mask openings. A suitable solution variant for the problem is provided by the desire technique stated by B. Roland and F. Coopmans (SPIE 1986). It is based on the photo-stimulated interface silylation by Si-organic compounds, such as HMDS. The exposed and silylated areas are converted into SiOx etching barriers in O_2-plasma, the Si-free areas, however, being etched off.

Experiments and Results

The investigations into etching were carried out under the following conditions:

Use of a single-wafer diode etch reactor having an electrode diameter of 130 mm and a variable electrode spacing, fixed frequency = 13.56 MHz, setting of the high-frequency power over the bias voltage (100-450 V), process vacuum = 1 - 9 Pa.

The results were obtained by two process steps:

- SiOx barrier formation in connection with isotropic removal and very low etch rates
- Anisotropic, fast etching of the unsilylated areas by means of O_2-RIE

The safety of the residue-free etching is increased, if a suitable CF_4/O_2-mixture is used for the barrier formation, since Si-containing adsorption layers in the unexposed region are removed. The process designated as flash-etching is connected with the loss of a part of the SiOx-layer in the silylated region and has to be optimized.

Trends of the influence of bias voltage, flow rate and composition of the etching gas on the thickness (step height) and resolution of the resist mask were determined by statistical experimental design. For this method of optimization the following influencing quantities were specified:

- CF_4-flow, O_2-flow, bias voltage and flash-etching time for the 1st step at constancy of the etching parameters of the 2nd step.
- O_2-flow, bias voltage and overetching time after the end-point indication at constant processing conditions which result from the optimization of the 1st step.

Meaningful limit values from preliminary experiments were used for establishing an experimental design of fractional factorials 2^{4-1} for flash etching. The most favourable result obtained by it provided the starting point for a second experimental design with accordingly changed limit values, which was connected with the first experimental design. The regression analysis was performed with a quadratic model and yielded a deviation for ca. 6 % of the inserted values. For graphical illustration, the resolution was represented in dependence on two influencing quantities each. (Fig. 1)

If the kind of resist and the silylation conditions change, the limit values of the flash-etching time have to be newly established. A good method of prediction was derived from an observed correlation between the flash-etching time and the swelling of the resist in the silylated region. The resist development with O_2-RIE was optimized on the basis of an experimental design of full factorials 2^3 with the objective quantities of resolution, step height and etch rate. For the end-point indication an emission spectrometer proved to be better than the laser reflectometer, which gave a constant signal already when resist residues were still present.

The quadratic model used for the evaluation showed a very good agreement with the experimental results.

The following statements were possible:

- O_2-flow rate and overetching time do not have any essential influence on resolution.
- On the other hand, the etch rate and the step height are considerably determined by the O_2-flow and the bias voltage. Maximum step heights are at a limit value of the bias voltage; if it is exceeded, the step heights are considerably lowered.
- The flow rate is important for the minimization of the etch rates. At an etching result of the same value it is possible to reduce the etching times by a maximum of 50 % by means of high flow rates.

Etch rates were measured by in-situ measurements of the removal of resist with a laser reflectometer. The removal of layer results from:

$$d = \frac{\lambda \cdot m}{2n \cos \beta}$$

λ lasing wavelength
n refractive index
β angle of incidence (< 90 °C)
m number of periods

selectivity was determined with the etch rates of the unsilylated and silylated resist.

flash etching: S = 1.4

O_2-RIE (without flash-etching, with O_2-barrier formation): S=16.4

flash + O_2-RIE: in dependence on the flash-etching time

S = 3 to 10

By the use of the optimized parameters resolutions of 0.65 µm were safely achieved at an exposure of λ = 436 nm and NA = 0.3. (Figs. 2, 3)

The results were a considerable contribution to the further improvement of the required resist quality and of the preceding preparation of the silylated layers on industrial equipment. A survey of that is given by the figures. (Figs. 4,5) .

At present, the technique is used to assess the application of other kinds of irradiation.

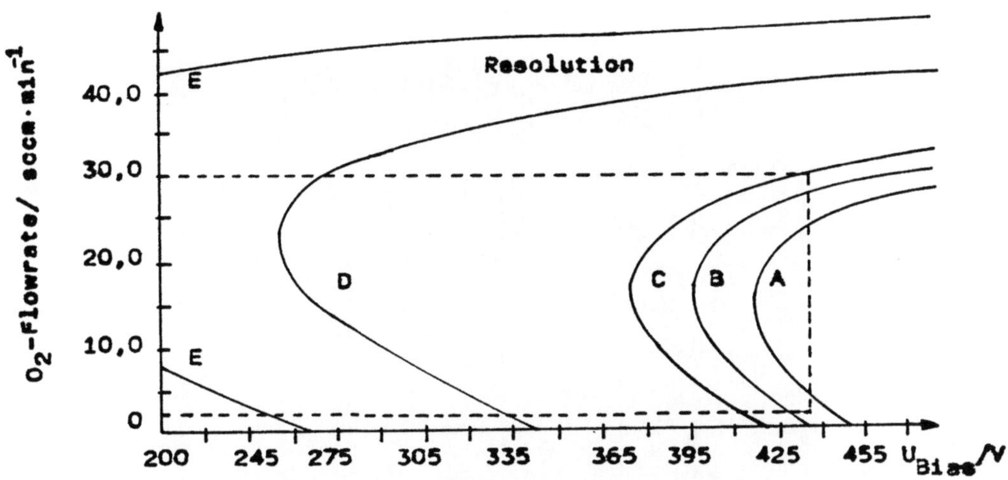

Fig. 1: Depiction of levellines for optimizing the flash-etching

Resolution: A= 0,6 µm D=2,0 µm CF_4=6,3 sccm/min
B= 0,8 µm E=3,0 µm
C= 1,0 µm

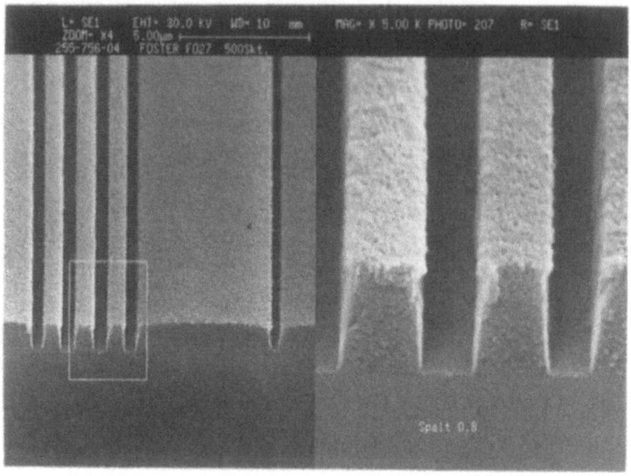

Fig. 2: REM-photo of a finished resist mask (resolution 0,8um)

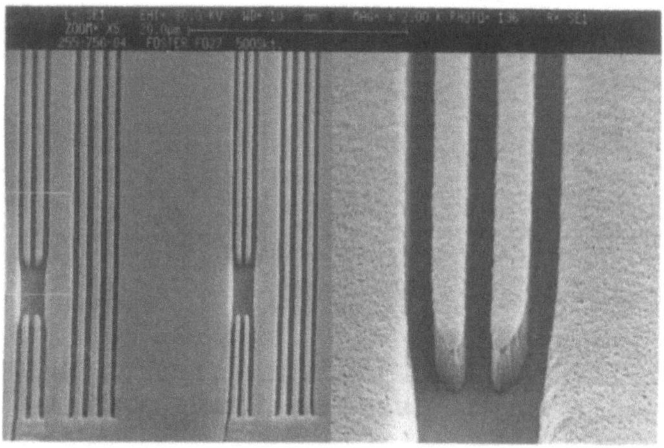

Fig. 3: REM-photo of a finished resist mask
(resolution 0,5um and 0,6um)

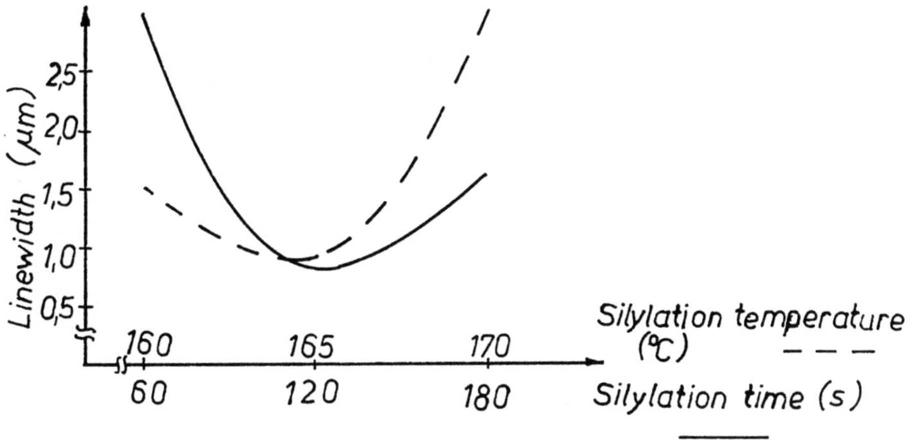

Fig. 4

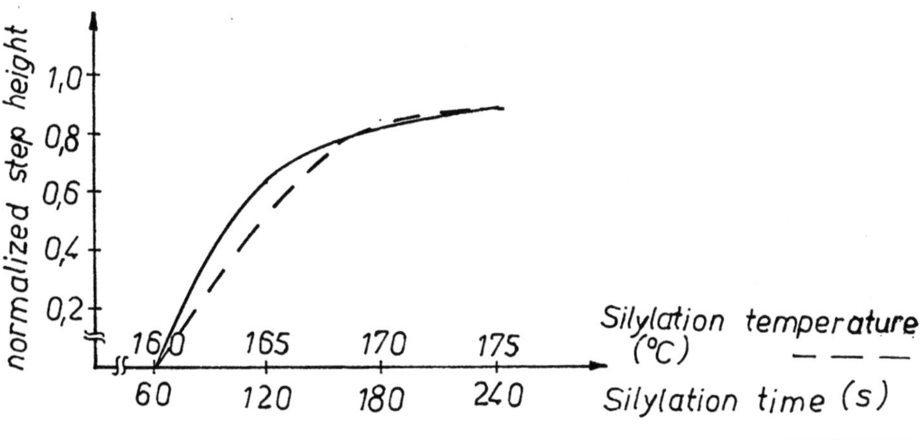

Fig. 5

Fig. 4,5: Linewith and step height as a function of silylation temperature and silylation time

Contact Methods

A new Explanation for the Degradation of Gold-Aluminium Bonds

J.F.Haag
Fraunhofer–Institut für Festkörpertechnologie
Paul–Gerhardt Allee 42
8000–München 60

INTRODUCTION

The formation of intermetallic phases in the system gold-aluminum is well known in semiconductor industries. A secondary phenomenon of the phase growth, the formation of 'Kirkendall-voids ' leads to the failure of the contact. This has been well examined on goldwire bonded to the IC aluminum metallization [1] - [6].
By bonding with aluminum wire a monometallic contact on the chip can be achieved. Even though, the problem will arise on the substrate. In the case of an hybrid, however, the gold pastes are of lower purity than the materials used in semiconductors. Impurity atoms can change the diffusion behavior pretty much, so the degradation may be changed altogether.

In this work we have examined the ageing behaviour of aluminum wire bonds (Al–1%Si wire, 25μm diameter) on five different gold thick film inks by three different manufacturers. It emerges that the theory of void formation doesn't hold true in this case. A new mechanism, the internal oxidation of the gold-aluminum intermetallics is proposed for this system.

EXPERIMENT

Preparation: the printed and fired substrates were separated with a diamond dice and subjected to the following cleaning steps prior to bonding:

- ultrasonic cleaning with acetone
- rinsing with fully deionized water
- ultrasonic cleaning with 2–propanol
- rinsing with fully deionized water

The purity of all chemicals was better than 99.8%. In order to obtain consistent results, all bonds were made by the same operator at the same wire bonder. In the whole study wire of the same specification by the same supplier was used.

For temperature storage, the substrates were put in a preheated furnace. For storage under protective gas, the furnace was rinsed with nitrogen for at least 12 hours. Afterwards the temperature was reached in about five minutes.

The substrates used for measurement in vacuum were degassed at a pressure of - $2{,}6 \cdot 10^{-6}$ Pa for at least 12 hours. To eliminate temperature deviations, measurements in air were repeated in this device, too.

Characterization: The following measurement techniques were used for testing the bonds:

Pull–Tests were carried out on a semi–automatic pull tester with a hook of 20 gf.

The *bond resistance*, which is very low for new bonds, is not easy to measure. In this work two different measurement techniques were used. Either a number of bond can be connected in series, or the resistance of one contact can be monitored 'in situ' with a four point method (Fig. 1)

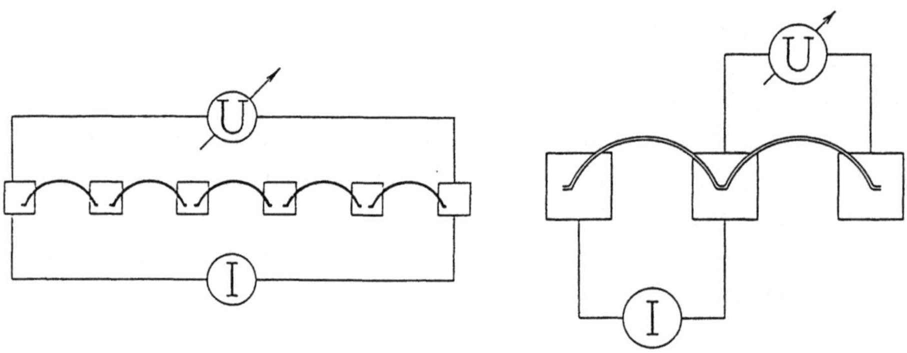

Fig. 1: Measurement of the bond resistance: a) Bond series b) Kelvin method

RESULTS

As it has been shown elsewhere [7] there is no correlation between the resistance of the bond and its pull strength. So it is not sufficient to measure only one of these. This is illustrated in Fig. 2. The pull strength (on the abscissa) first drops sharply, while the bond resistance begins to rise later on.

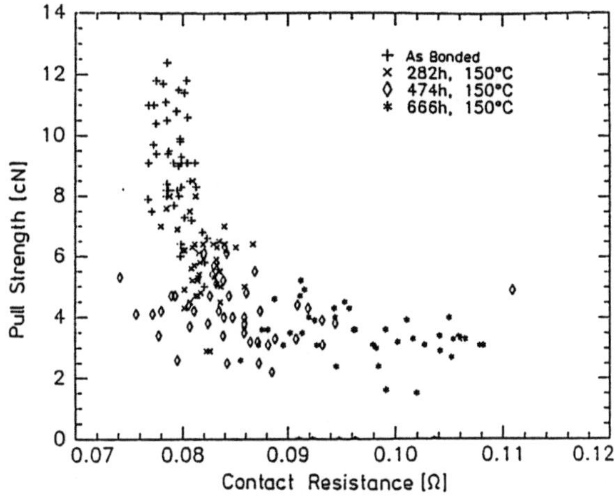

Fig. 2: Pull strength vs. resistance for different ageing time at 150° C

The degradation of the pull strength for different temperatures is illustrated in fig. 3. The pull strength drops quickly (at higher temperatures) to a value of about 2.5 cN, but stabilizes on this level for times up to 300h at 300° C.

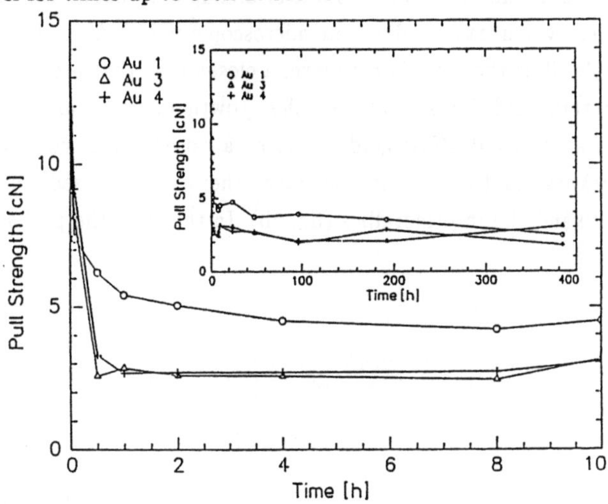

Fig. 3: Pull strength of bonds, aged at 300° C

This value is still acceptable according to MIL STD 883C. So the contacts do not fail mechanically. The contact resistance, however, rises to very high values (fig 4). Even if we don't observe a mechanical failure of the contact, it is electrically open.

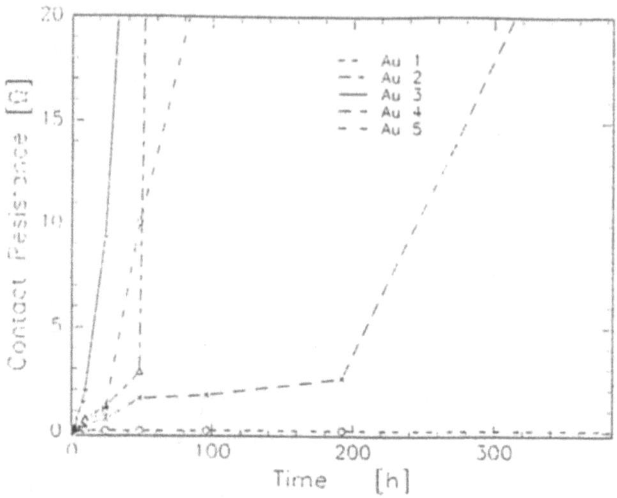

Fig. 4: Contact resistance of bonds, aged at 300°C

Such a behaviour has been explained with the formation of Kirkendall-Voids. Differences in the diffusion constant of a couple of metals can lead to the formation of vacancies, which can condense in microscopic voids. According to K.Okumara [7], on thick gold films these voids may form a closed line, that leads to zero pull strength (fig. 5). On thin gold films, however, the growth of voids may be stopped, because the intermetallics is cut off its gold supply by a closed line of voids to the gold. However as the thickness of the gold is restricted, there is no closed line of voids between the substrate and the intermetallics (fig. 5b). In this case the pull strength doesn't drop to zero.

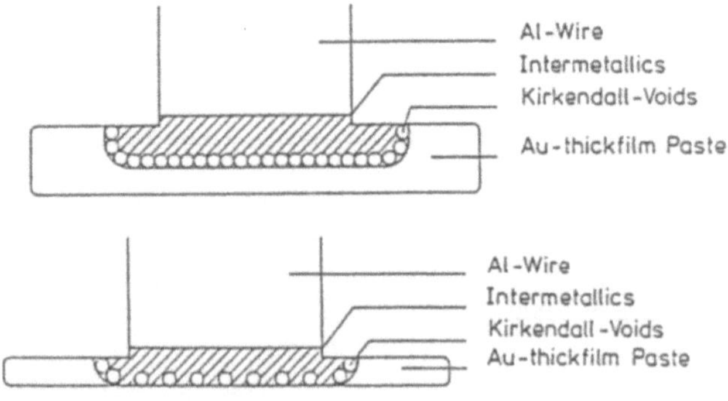

Fig. 5: a) Kirkendall-Voids may form a closed line to disconnect the bond. b) On thin films the growth of the voids is stopped.

In the present study thick film inks of 7 to 15 μm fired thickness are examined. In cross-sections voids could not be observed. For these reasons oxidation of the intermetallics is proposed as a new ageing mechanism.

This new theory can be proved by different methods:

Ageing under protective gas or in vacuum: The oxidation can be suppressed if oxygen is kept from the contacts. This can be done by ageing in protective gas or in vacuum. The results of storage in nitrogen are shown in table 1. The resistance of bonds aged in nitrogen is much lower than that of bonds aged in air.

Ink Nr	Au 1 Air N_2	Au2 Air N_2	Au 3 Air N_2	Au 4 Air N_2	Au 5 Air N_2
300°C, 9h	0,05 0,05	88 0,23	3,06 1,48	1,26 0,14	0,71 0,29
300°C, 24h	0,05 0,05	12,7 0,91	5,00 2,45	0,93 0,30	---
400°C, 2h	0,10 0,06	6,90 5,69	9,24 5,84	0,93 1,82	4,15 0,51
400°C, 16h	0,17 0,09	8,7M 973	219k 1,2k	1,4M 1,0k	300k 78,85

Table 1 Comparison between tests in air and under protective gas

Even better results are obtained by ageing in vacuum. Fig 6. shows a comparison between the resistance increase in vacuum and in air for three different gold pastes. In vacuum storage (300°C) the resistance remains at the level of new bonds, whereas the resistance of bonds aged in air reaches very high values.

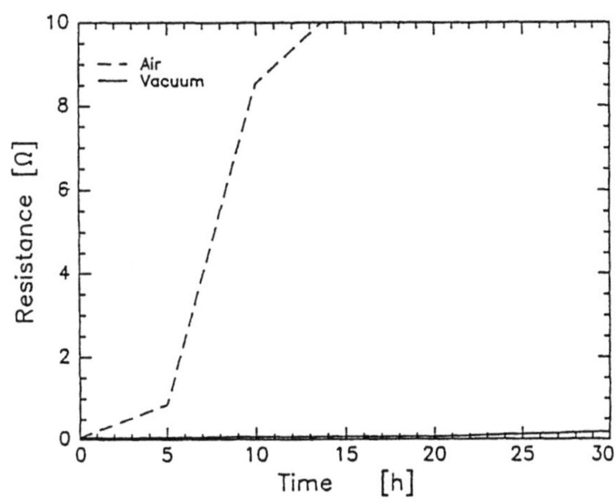

Fig 6: Comparison between storage in air and in vacuum.

Analytical Methods: Energy dispersive X–ray analysis shows an oxygen peak in the intermetallics, whereas no oxygen can be found in the wire or in the thick–film inks (Fig. 7).

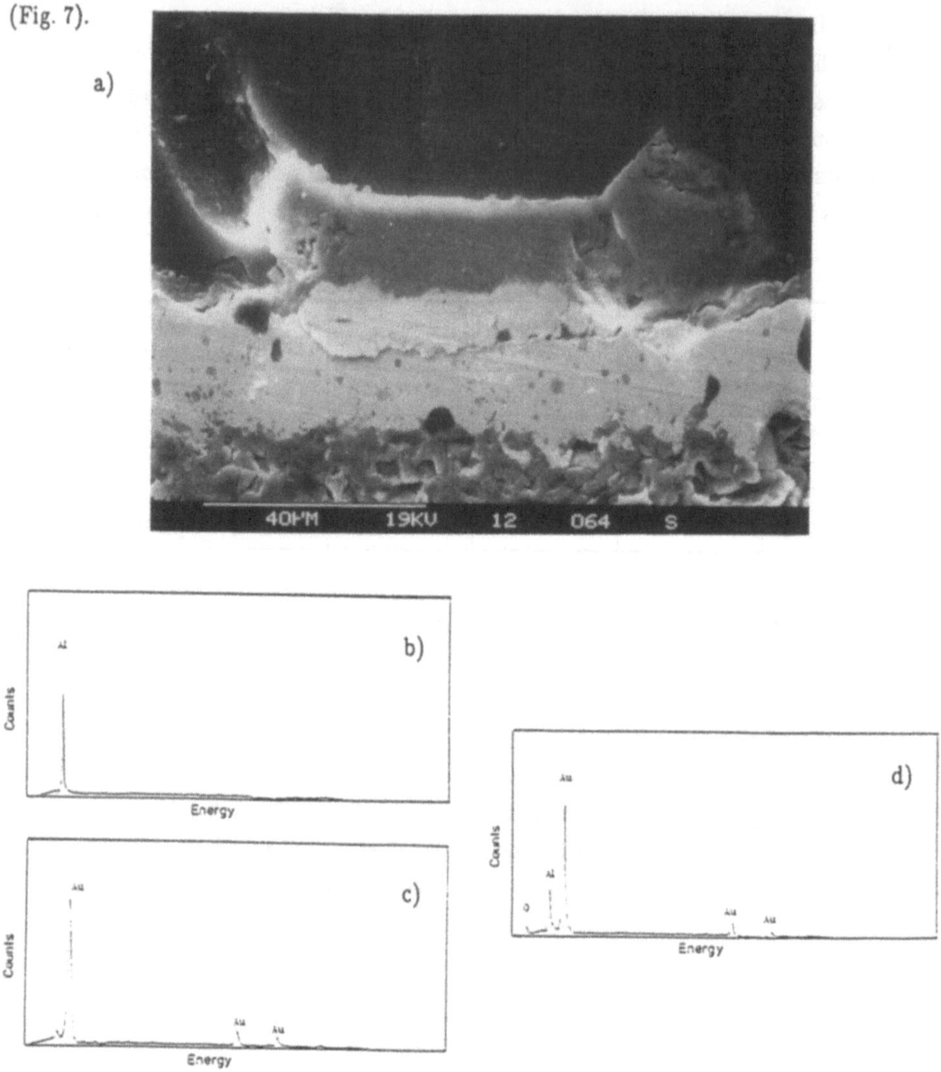

Fig. 7: a) SEM Picture of a bond cross-section and EDX-Spectra of b) wire, c) conductor and d) intermetallics

Dielectric Breakdown: With these results, the well known 'healing effect' can also be explained. It has been reported in the literature [8] that the resistance of aged bonds dropped, as the probe voltage increased. In the new theory, this is a simple dielectric breakdown: As the probe voltage increases, the dielectric strength of the oxide layer is exceeded, the dielectric breakdown occurs and the resistance of the contact is lowered irreversibly. This is shown in fig. 7. Following the breakdown occurs at 4.4 V, the resistance is so low that the current limit of the voltage source causes the flat portion of the characteristics.

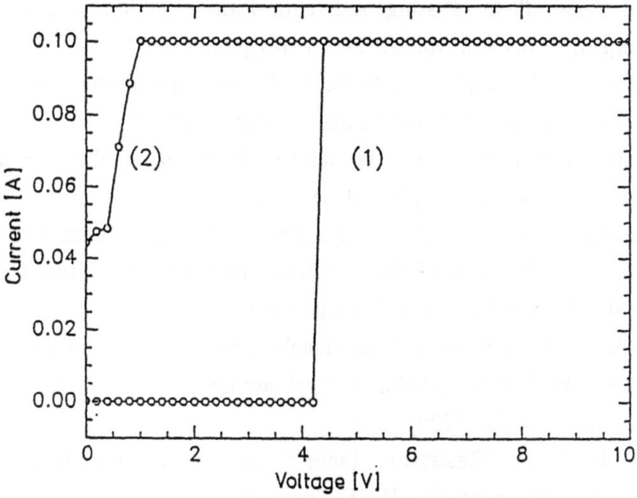

Fig. 8: The 'healing effect' can be explained as a dielectric breakdown.

CONCLUSIONS

The degradation of the pull strength and the increase of the electrical resistance are not correlated. So two different mechanisms are required to explain the degradation of the pull strength and the resistance, respectively. In this work the resistance increase of aluminum wirebonds on thick film gold pastes has been explained by a new mechanism. Oxidation of the intermetallic phases causes an insulating layer to be built up in the interface. This theory has been proved by ageing in nitrogen and in vacuum, as well as by surface analytical methods. With this theory the 'healing effect' can be explained as dielectric breakdown.

[1] I.A.Blech, H.Sello 'Some New Aspects of Gold–Aluminum Bonds', Jour. Electrochem. Soc. **13** (10) 1052 (1966)

[2] G.K.Chen 'On the Physics of Purple Plague Formation and the Observation of Purple Plague in Ultrasonically-joined Gold-Aluminum Bonds', IEEE Trans. Parts, Materials and Packaging **3** (4), 149 (1967)

[3] E.Philofsky 'Intermetallic formation in Gold–Aluminum Systems', Sol. State El. **13**, 1391 (1970)

[4] G.V.Clatterbaugh, J.A.Weiner, H.K.Charles 'Gold-Aluminum Intermetallics Ball Bond Shear Testing and Thin Film Reaction Couples', IEEE Trans. Comp. Hybrids and Manuf. Techn. **7**(4), 349 (1984)

[5] S.Weber, W.Kohl, A.Bischoff 'Metallographische Untersuchung der Intermetallischen Phasenbildung im System Au-Al unter Berücksichtigung der Randbedingungen der Halbleiterverbindungstechnik', Sonderbände der praktischen Metallographie **20**, 381 (1981)

[6] H.Devaney, P.H.Eisenberg, G.H.Ebel 'Gold–Aluminum Intermetallics: Key Parameters-Reactions-Effects & Reliability Impact' International Reliability Physics Symposium, New Orleans, 1990

[7] J.F.Haag 'Zuverlässigkeitsuntersuchungen von Aluminium Drahtbonds auf Gold-Dickschichtleiterbahnen' Verbindungstechnik in der Elektronik, DVS Berichte 129, 131 (1990)

[8] D.W.Bushmire 'Resistance Increase in Gold-Aluminum Interconnects with Time and Temperature' IEEE Trans. Parts, Materials and Packaging **13** (2), 152 (1977)

Simulation of the Inner Lead Bonding Process

A. EBERLE

Technical University of Berlin Research Center for "Microperipheric Technologies"

Summary

This investigation was undertaken to predict the thermomechanical behaviour of an inner lead tape automated bonded (TAB) structure during thermocompression bonding. Using finite-element modelling and plasticity theory, the stresses, displacements and strains were determined at the bonding interface as well as in the passivation layer system. The material parameters of the bilinear material stress-strain diagram for plated gold were obtained by fitting the results of a sensitivity analysis to experimental data achieved from a compression test.

Introduction

Tape-automatic-bonding (TAB) is used to make electrical and mechanical connections to high lead-count, large-area VLSI circuits or chips. Attractions of TAB include high bonding strength, small assembly thickness and ease of testing inner lead bonded devices prior to attachment of the chip to a package or substrate. The TAB process often begins with preparing the chip bumps before the dice are separated. Chips are presented to the inner lead bonder and all bumped pads are bonded simultaneously to the tape by thermocompression bonding. Fig. 1 shows the schematic procedure and a typical TAB structure. The thermocompression bonding process relies on the formation of a solid state bond between two surfaces by bringing the interface to an elevated temperature and deforming the members by the application of bonding pressure throughout a characteristic dwell time as shown in Fig. 2. The purpose of this work is to relate the external parameters, such as pressure, bump geometry and material properties, to the conditions that are required to attain a good bonding formation, and to avoid cracking of any underlying layers. Therefore, finite-element analysis, including elasto-plastic material behaviour is used and experimental results obtained by Dries [1] and Stapleton [2] are consulted.

Characterization of thermocompression bonding

The representation of lead attachment modelled by Dries [1] and Stapleton [2] was somewhat similar to TAB. They bonded 0.01 x 0.062 x 0.375 inch gold leads to gold plated onto rigid substrates. Deformation along the bond interface was measured experimentally by using silver markers deposited into grooves etched along the surfaces. This gave data of interfacial sliding and of extension of surfaces at the interface. We used finite-element simulation to correlate these results and to compare it with an analytical solution from slip line theory [3]. Fig. 3a shows that the analytical solution for the marker displacement is poor compared to that of an FEM analysis which takes the work hardening effect of the material and better boundary conditions into account. Fig. 3b outlines more results from the FEM analysis. The calculations of stress and shear strain support the observations of a central dead-zone. Despite high compressive stress, bonding is difficult in the dead-zone because of low shear stress, little surface elongation and a lack of plastic shear deformations. Within the outer portions of the interface, higher plastic shear strain and surface elongation, accompanied by shear stress result in good bond formation.

Simulation of the TAB bonding process

The bonding process of a TAB structure is more complex. The load is time-dependent, the geometry is rather complicated and the structure consists of materials which exhibit a big difference in behaviour.

Thermal simulation

The bonding process requires a temperature of up to 500 ^{o}C at the thermode, and the resulting thermal stresses cannot be neglected. The two-dimensional simulation is performed using a 2-D plane strain finite-element mesh (Fig. 4). A thermal transient analysis is followed by a nonlinear thermal stress analysis involving the pressure history. The thermal load of the structure, when the thermode has reached maximum temperature (after 1.0 sec), is presented in Fig. 5. The SiO_2 passivation layer acts like a thermal barrier. The difference in temperature within the layer is about 130 ^{o}C. The temperature gradient within the bump is about 35 ^{o}C.

Material data

The analysis was performed using temperature-dependent, elastic material data for all brittle materials and temperature-dependent, nonlinear material parameters for all ductile materials. Since there was no data available characterizing plated gold at about 500 °C, a sensitivity study was performed and the results were fitted to the data obtained by a compression test. By using a bilinear approach to the stress-strain diagram, it was seen that the plastic modulus had to be reduced by a factor of ten compared with data from literature (Fig. 6).

External parameter variation

As mentioned earlier, the external parameters such as bonding pressure, bump geometry and hardness were varied and their influence on layer stress and on parameters affecting good bond conditions, e.g., surface elongation, shear stress and strain, was studied.

Influence on layer stress

The distribution of equivalent von Mises stress in the gold and passivation layer, as evaluated by the FEM model is typically as shown in Fig. 7. The stress, depending on the pressure and temperature load 1.0 second after the thermode makes contact to the lead, is shown in Fig. 7 a. Within regions near the center of the bump, there is little stress and only a slight increase in stress due to higher pressure. The highest stress gradient arises directly underneath the bump's edge (relative dist. from center = 1.0), whereas the maximum stress, as well as the offset caused by higher pressure, lie outside the perimeter of the bump. Care must be taken when pressure load is greater than 100 N/mm^2, because a non planar thermode surface could cause cracking of the underlying layer if von Mises stress exceeds about 900 N/mm^2. This can easily occur if only a few bumps are in contact. Reducing the pressure from 50 to 25 N/mm^2 lets the maximum stress decrease 12 %. But, as will be shown later, this will not bond. Variation of the bump geometry was performed by changing the bump's heighth-to-width ratio. A 100x25 μm bump (h/w = 0.25) was first shortened to 15 μm giving a h/w = 0.15. This was of no influence on the distribution of layer stress. The bump model having a width of 50 μm and heighth of 25 μm (h/w = 0.5), had a maximum stress level reduced by 36 % (Fig. 7 b). Increased bump hardness showed a small constant stress offset along the interface (Fig. 7c).

Influence on bonding interface

It must be pointed out, that the stress and strain values of the node lying directly on the bump edge can not be calculated exactly, because this is a singularity. As mentioned above, a reduction of bonding pressure below 30 N/mm^2 will result in non reliable bonds (as long as other parameters are not changed), because a resulting surface elongation of under 1 % is too small (Fig. 8 a), and there is almost no shear strain and stress in the bond interface. On the other hand, a high pressure (e.g., 100 N/mm^2) induces large surface elongation, especially in the middle region, but there will remain a central dead zone because of zero shear in the center line. A good compromise seems to be a pressure of 50 to 60 N/mm^2, for the distribution of surface elongation is rather uniform and shear conditions are moderate as well as the load for the underlying structure. Experiments, performed at the Research Center, showed that the pressure value to cause reliable bonds at 500 °C lies between 50 and 70 N/mm^2 [4]. Reducing the bump heighth from 25 µm to 15 µm gives a non-uniform distribution of surface elongation along the interface, resulting in a larger dead-zone (Fig. 8b). The same effect occurs on a lower strain and stress level when the heighth of a 50 x 25 µm bump (h/w = 0.50) is reduced to 15 µm (h/w = 0.25). This means, that the heighth-to-width ratio controls the strain distribution in the interface. Therefore, a 100 µm bump should not be shorter than 25 µm, a 50 µm bump should not measure less than 20 µm in heighth. As can be seen in Fig. 8 c, a hard bump material also leads to a strong unevenness in strain. The hardness of the bump material should not exceed a value of 70 to 80 Vickers.

Conclusions

Considerations in terms of normal stress, surface extension, and interfacial shear is a first step towards characterization of thermocompression bonding. Results from experimental observations and from a finite-element analysis showed good agreement. The following observations can be supported:

1. There is a central dead-zone where bonding is difficult, despite high normal stress.

2. Extension of two surfaces in contact is sufficient for bond formation after some time at temperature (to decrease material parameters).

3. Application of an interfacial shear stress supports bonding.

Investigations of the influence of external parameters on conditions for good bonding showed:

1. Bonding pressure should lie between 50 and 70 N/mm^2.

2. There is an optimal heighth-to-width ratio for every investigated bump width.

3. Bump hardness should not exceed a value of 70 to 80 Vickers.

References

1. Dries, L.: "Deformation and Structural Changes Occurring at the Interface During Gold-Gold Thermocompression Bonding", Masters Thesis, Dept. of Metallurgy and Materials Science, Lehigh University, Bethlehem, Pa., 1973.

2. Stapleton, R. P.: "Surface Elongation and Interfacial Sliding During Gold-Gold Thermocompression Bonding", Masters Thesis, Dept. of Metallurgy and Materials Science, Lehigh University, Bethlehem, Pa., 1974

3. Ahmed, N. & Svitak, J. J.: "Characterization of Gold-Gold Thermocompression Bonding", IEEE, 25th Electronic Components Conference 1975, May, 12-14.

4. 4. Zwischenbericht zum Forschungs- und Entwicklungsvorhaben "Untersuchung der TAB-Technologien fuer die Kontaktierung von VLSI-Schaltkreisen", Forschungsschwerpunkt "Technologien der Mikroperipherik", TU Berlin, 1988

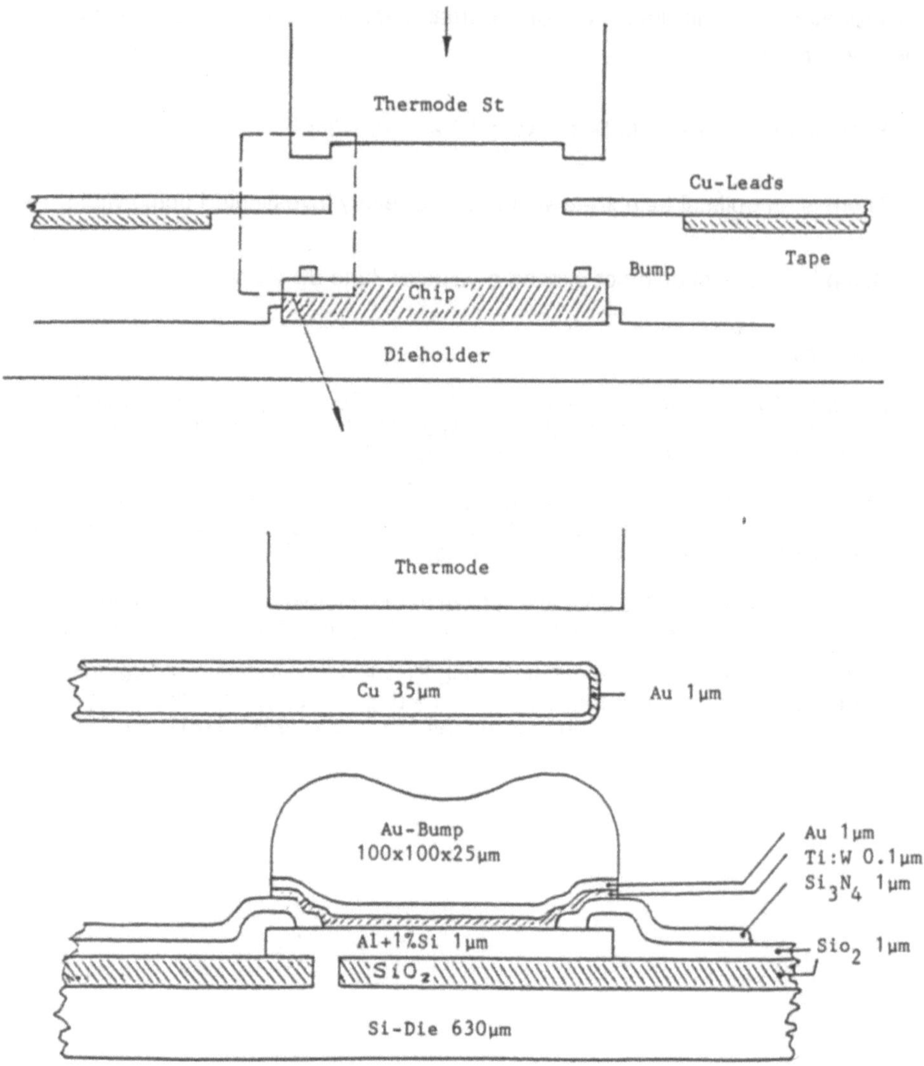

Fig. 1. Schematic drawing of inner lead attachment by thermocomression bonding for a typical TAB structure

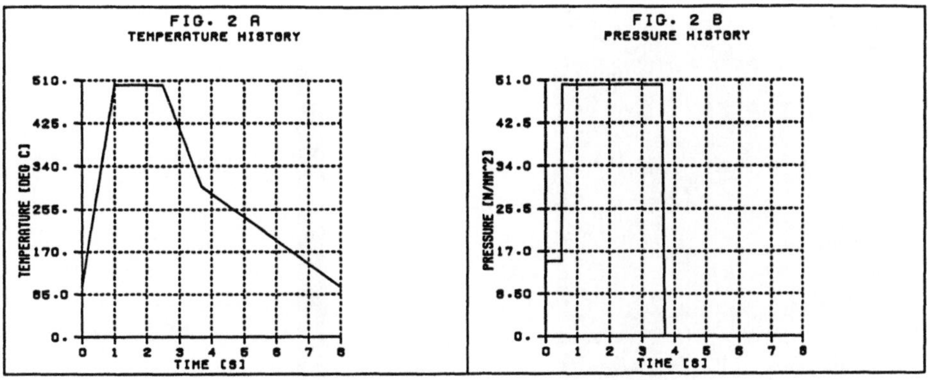

Fig. 2. Load history for thermocompression bonding, a) temperature and b) bonding pressure

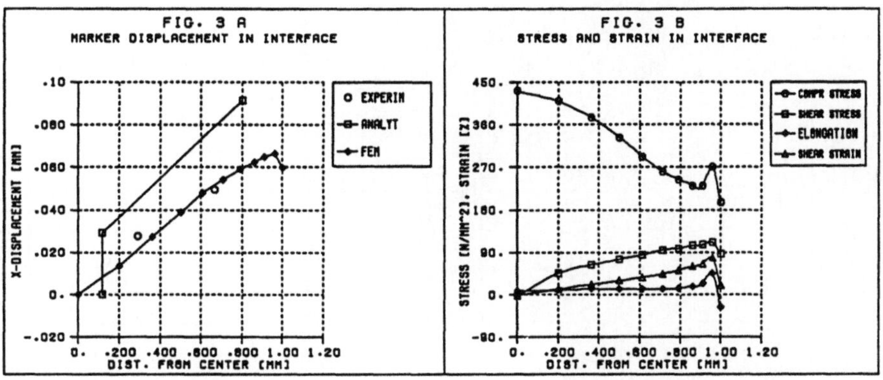

Fig. 3. a) Comparison of marker displacements evaluation with experimental results b) stresses and strains in a bonding interface configuration [2]

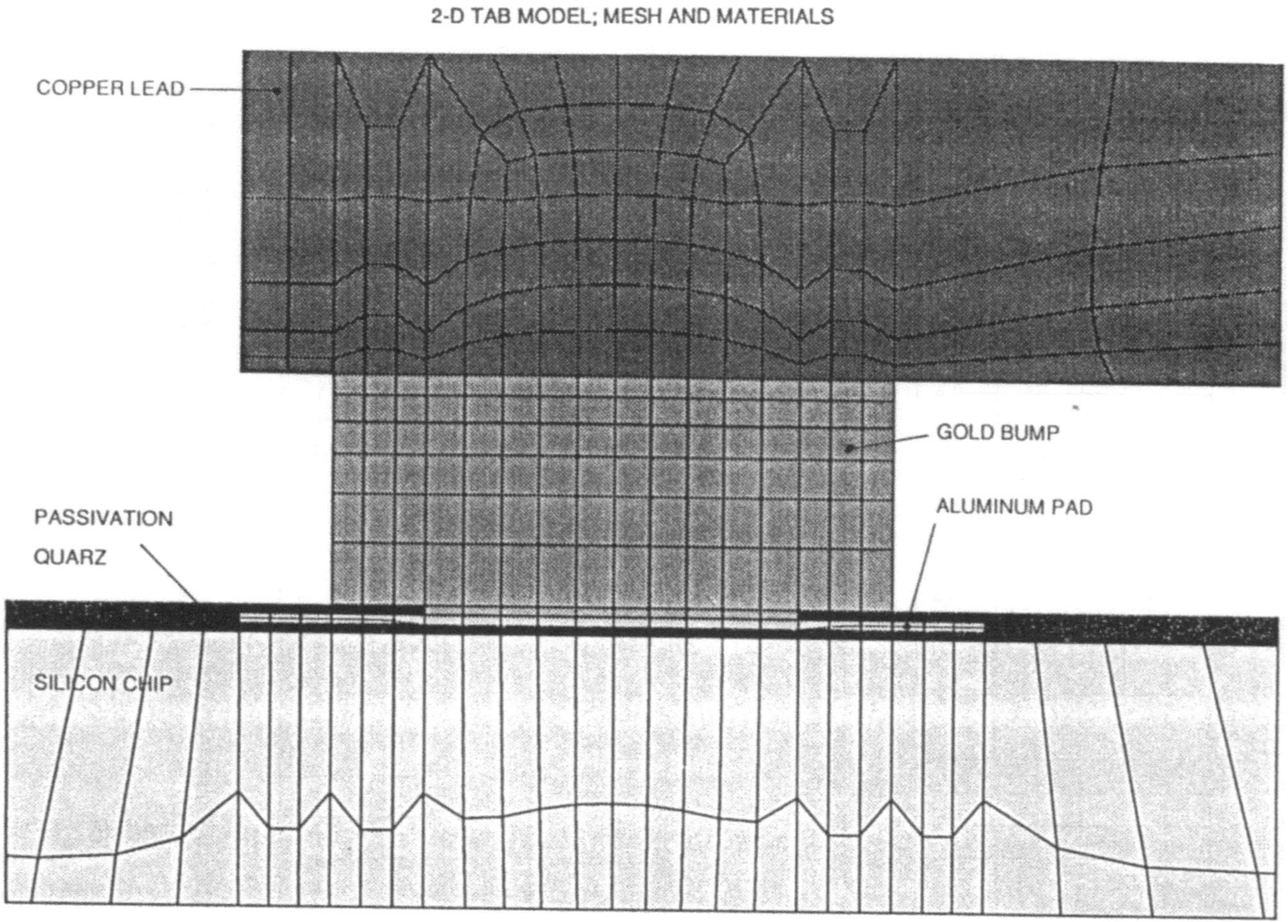

Fig. 4. 2-D finite-element mesh of the structure's upper part

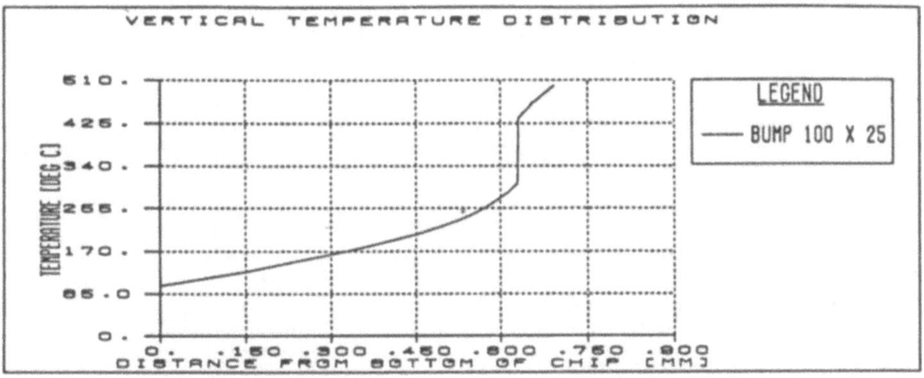

Fig. 5. Temperature distribution in a TAB structure after temperature of thermode has reached 500°C

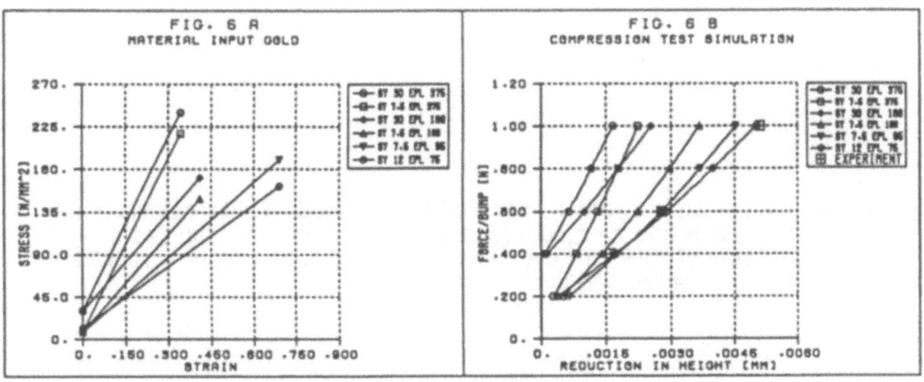

Fig. 6. a) Material input for sensitivity analysis of gold (SY=yield stress, EPL=plastic modulus) b) force-deflection diagram of sensitivity analysis compared with data from a compression test

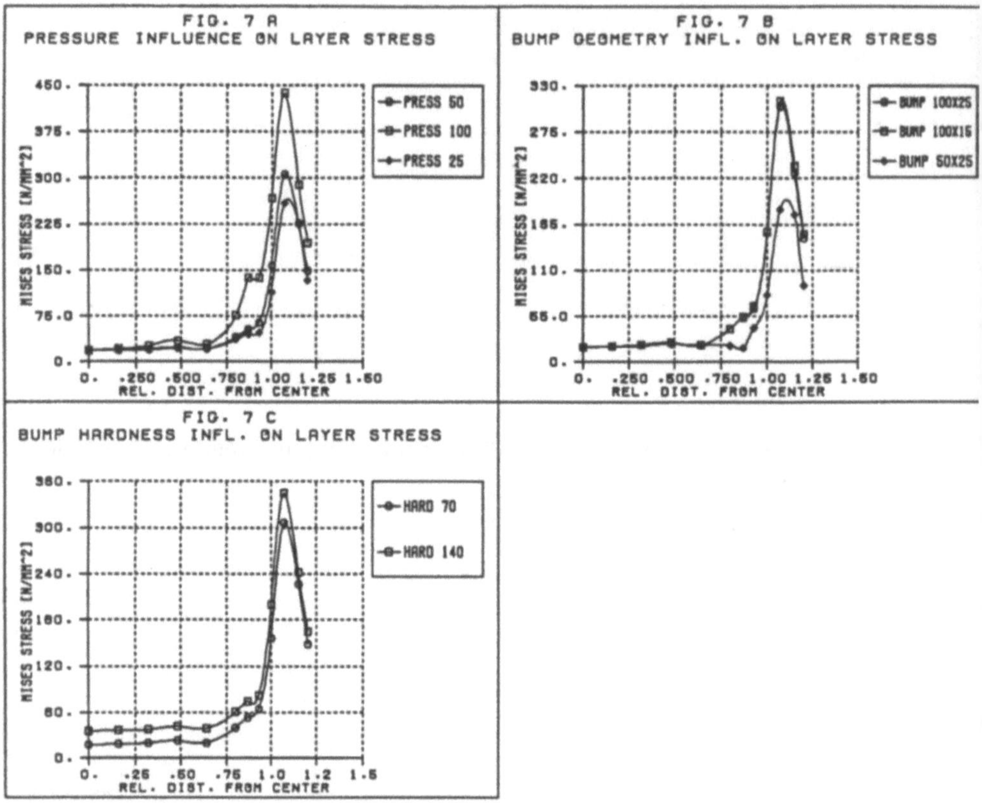

Fig. 7. Sensitivity of layer stress 1.0 second after thermode makes contact on: a) bonding pressure for a 100x25µm bump geometry b) bump geometry at a pressure of 50N/mm^2 and c) bump material hardness for a 100x25µm bump at a load of 50N/mm^2

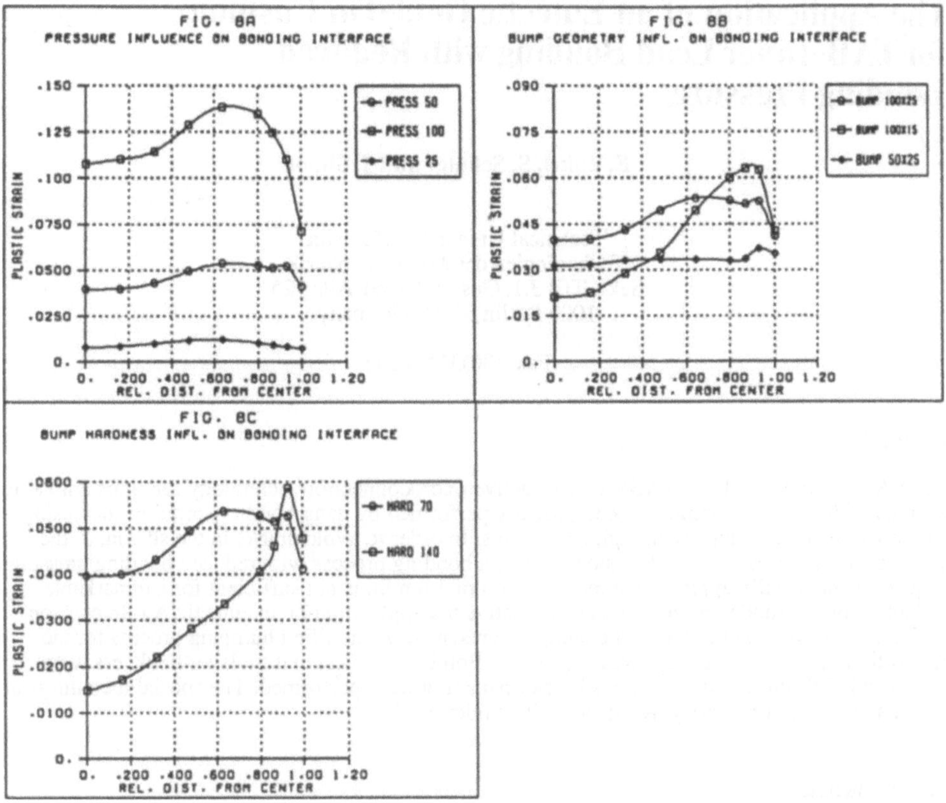

Fig. 8. Sensitivity of bond interface surface elongation 1.0 second after thermode makes contact on: a) bonding pressure for a 100x25µm bump geometry b) bump geometry at a pressure of 50N/mm^2 and c) bump material hardness for a 100x25µm bump at a load of 50N/mm^2

The Application of an Eutectic Gold-Tin Cushion for TAB-Inner Lead Bonding with Reduced Bonding Pressure

E. Zakel, S. Schuler and J. Simon

Technical University of Berlin
Technologien der Mikroperipherik
Sekr. TIB 3.1, Gustav Meyer Allee 25
1000 Berlin 65, W-Germany

Fax: (30)31472835

Abstract

Tape Automated Bonding (TAB) is an attractive microconnection technology for chips with a high number of I/O's. The bonding process, usually performed by gang bonding, requires increasing thermode forces for chips with high lead counts. In order to avoid cracks in the silicon, in the passivation layers and in the diffusion barriers a bonding process with reduced bonding pressure is required. The usually applied tape metallization of 1 um Sn is not sufficient for a remarkable reduction of thermode pressure. As an alternative the application of an eutectic Au-Sn cushion which is deposited on the top of the bumps is presented. A modified bumping process for the deposition of the solder bumps was developed. Bond parameters and tin layer thickness were optimized. Reliability investigations by thermal aging were performed. The special metallurgical aspects of the system were investigated with a microprobe.

1. Introduction

Due to the requirements of a reduced pitch TAB has an increasing importance for chips with high lead counts. This microconnection technology requires a tape which is generally gang bonded to a bumped chip. The thermode force increases proportional to the number of bondpads and can cause damages of the bonded chips. Thus gang bonding of chips with high lead counts has to meet the requirements concerning the alignment and planarity of the thermode and height tolerances of bumps and leads /1/.

1.1. The Bonding Process

Height differences between the thermode, the bumps of the chip and the leads of the tape are compensated during the bonding process. This compensation occurs by plastic deformation for thermocompression bonding and by a liquid cushion in the case of soldering. The height tolerances are caused by :

- The angular thermode misalignement resulting in a nonuniform distribution of bonding pressure on the same chip.
- Thermode inplanarity caused by nonuniform thermal expansion resulting in a curvature.
- Nonuniformity in bump height distribution on the same chip
- Nonuniformity of lead height and surface roughness of 4-7 um.

High bonding pressures can cause cracks in the passivation layer at the edge of the bumps and cracks in the Ti:W layer which acts as a diffusion barrier between the aluminum pad and the gold bump /5/. In order to avoid silicon cratering during bonding the hardness of the electrodeposited gold bumps must be reduced in a supplementary tempering step.

The usually applied tape metallization of 1 um tin is not sufficient for the compensation of height tolerances and for bonding with very low bonding forces. As an alternative the application of an eutectic Au-Sn cushion which is deposited on the top of the gold bumps has been investigated /3/. The solder bumps can be bonded to a gold plated tape with a higher reliability concerning whisker growth and bondability after extended storage.

1.2. Gold-Tin Metallurgy

Compared to tin-lead soldering systems, eutectic 80% Au 20%Sn solders show a higher thermal stability and a high resistance to thermal cycling /8,9/. The formation of the brittle intermetallic compounds $AuSn_2$ and $AuSn_4$ can be excluded by the choice of adequate bond parameters which allow the formation of the gold rich Au-Sn 80/20 eutectic alloy /4,7/. The excellent wetability of the liquid eutectic alloy allows bonding without flux application /3/. Only few publication describing the formed alloys and intermetallic compounds after eutectic soldering of TAB inner leads exist /4/. This is due to the difficulties in obtaining quantitative EDS or WDS analysis of small contact geometries. The semiquantitative analysis of T.S. Liu /4/ indicate the formation of a gold rich alloy with less than 30 wt.% Sn.

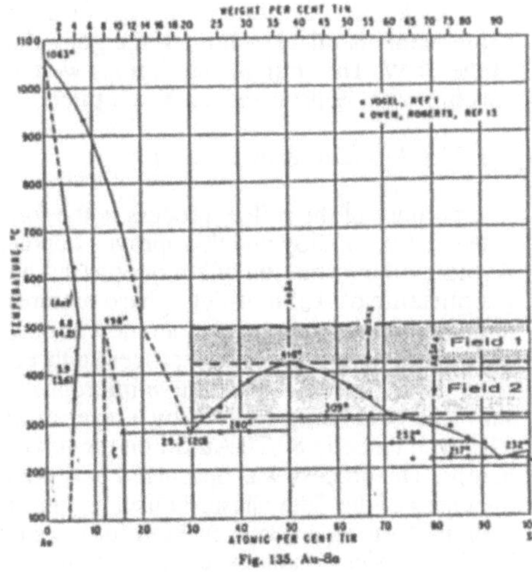

Fig. 1: Au-Sn phase diagram /7/ with 2 fields for isothermal eutectic alloy 80/20 formation

For the formation of the eutectic 80/20 Au-Sn alloy for TAB-contacts two possibilities are pointed out in the Fig.1. For temperatures higher than 419 °C (Field 1) the alloy is obtained by diffusion in the liquid phase. For temperatures between 309 °C and 419 °C (Field 2) the isothermal process has to pass concentration areas with both solid and liquid phases. Optimal thermode temperatures required for eutectic soldering to a tin plated tape are between 425-475 °C indicating that liquid phase diffusion is preferable for the relatively short bonding cycle of 0.5- 1 second /5/. Previous investigations /3/ show that the formation of the eutectic alloy by the reflow of the Au-Sn solder bumps is possible. The eutectic alloy with a liquidus temperature of T_m = 280 °C is limited by two intermetallic phases with higher liquidus temperatures /7/. The phases are AuSn (T_m=418 °C) and the Zeta-Phase (T_m= 498 °C) which contains 10 wt% Sn. Thus the tempering process has to be optimized in order to avoid the formation of either of these intermetallic compounds, which affect bondability of the solder bumps. In order to obtain the eutectic composition the isothermal process must be stopped before the formation of the Zeta phase.

1.3. Reliability Considerations

Pore formation of eutectic soldered TAB-contacts due to a Kirkendall-effect after thermal aging has been reported /2,5/. It was shown that the accumulation of tin in the bonded interface has an accelerating effect on the Kirkendall effect in this ternary system. This effect is due to a preferred diffusion of copper into the gold-tin regions and results in the formation of pores and ternary Cu-Au-Sn intermetallic phases /5/. Recent investigations show that the appearance of Kirkendall effect is depending on the type of Au-Sn phase formed during the bonding process /11/. The formation of an extended layer of eutectic 80/20 Au-Sn results in an intensive Kirkendall effect with pore formation and bump-lead delaminations after thermal aging /9,11/. The investigations show that the formation of the gold rich Zeta phase during the bonding process improves reliability of TAB-contacts. No Pore formation is observed after thermal aging, the Zeta phase acting like a diffusion barrier between copper and the eutectic layer.

2. Experimental Results

2.1. The Bumping Process for Solder Bumps

The solder bumps were deposited with the bumping process for straight wall bumps developed at the Technical University/10/. Fig. 2 shows the process scheme. Critical process steps for this modified bumping process are the photoresist compatibility with the tin electroplating solution and the chemical stability of the deposited tin in the subsequent chemical etching of the gold plating base and of the Ti:W. The straight wall bumps were deposited on a testchip with 49 pads and a 200 um pitch.

2.2. The Au-Sn Reflow Process

The purpose of the reflow process is the formation of an eutectic Au-Sn alloy which improves bondability. Avoiding the formation of a monophased high melting intermetallic of AuSn or Zeta-phase requires a narrow temperature-time program in an inert atmosphere. Fig. 3 shows that during the reflow process a three-layer structure is formed. Quantitative microprobe analysis show the gold bump covered by a layer of Zeta phase and a layer of eutectic 80/20 Au-Sn on the top of the solder bumps. The process was optimized in order to reduce the thickness of the Zeta phase. Thus a higher thickness of the liquid Au-Sn eutectic cushion for the subsequent bonding process is achieved. Fig. 4 shows a solder bump before and after the reflow process.

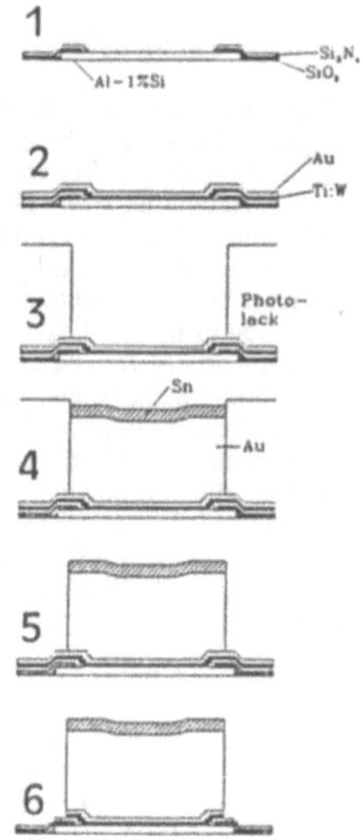

Fig. 2: Solder bump process scheme
1: Pad structure
2: Sputtering process
3: Photoresist process
4: Electroplating Au and Sn
5: Resist removal
6: Selective etching of sputtered metallization

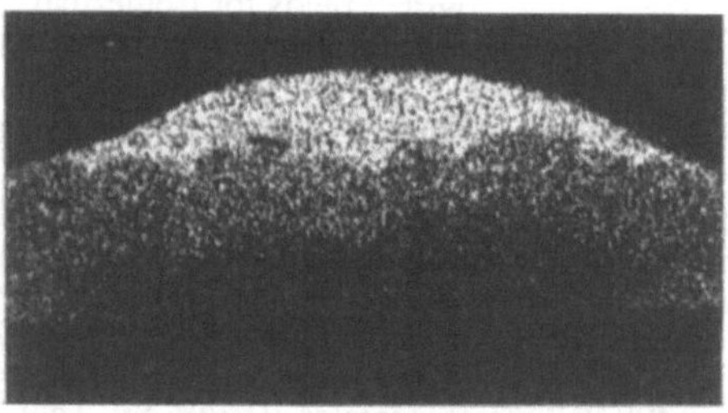

Fig. 3: Sn distibution of a reflowed bump with 3 layers: Gold, Zeta phase and 80/20 eutectic alloy

Fig. 4: Au-Sn solder bump (a) before reflow and (b) after reflow

2.3. The Bonding Process

Bonding was performed with both reflowed and not reflowed solder bumps. A three layer tape with RA-copper and a 1 um thick Au-metallization was used.Good bonding results were obtained with a tin thickness of 3.5 to 5 um on the solder bumps. A lower thickness of the tin layer reqiures increased bonding pressures. On the other hand a higher tin layer thickness can cause solder bridges between the bumps during the bonding process. Bonding of both reflowed and not reflowed solder bumps was possible without the application of flux. But different bonding parameters were required for the two types of solder bumps. Bonding temperature could be drastically reduced for the reflowed solder bumps.

Tab. 1: Variation of Bond Parameters

Solder Bump Type	Thermode Tempertature [°C] 350	400	450	500	525
No Reflow			X	X	X
AuSn Reflow	X	X	X		

The pull tests of these bond parameters show comparable results with 100 % lead fractures and slow variations of average pull strength between 56 cN and 59 cN (Tab. 2).

Tab. 2: Results from Pull-Test

Solder Bump Type	Thermode Temperature [°C] 350	400	450	500	525	
No Reflow			56.4	59.3	59.0	F*
			(4.3)	(4.8)	(2.6)	s*
AuSn Reflow	56.0	58.0	58.0			F
	(3.6)	(4.2)	(5.1)			s

*F = Mean Pull Strength [cN]
*s = Standard deviation [cN]

a
b

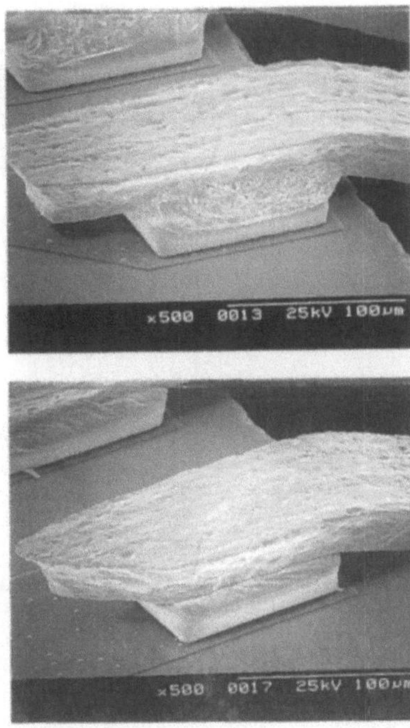

Fig. 5: Bonded solder bump (a) without reflow and (b) with reflow

Fig. 5 shows the inner lead bonds of solder bumps with and without reflow. Metallographic cross sections of the bonded chips reveal differences in the wetting behaviour of the liquid Au-Sn layer as a function of bond temperature. An increased wetability of the leads with increasing bond temperature is shown in Fig. 6. The liquid Au-Sn solder forms a meniscus on the sidewalls of the leads during the bonding process. An optimal wetability is obtained at a thermode temperature of 525 °C for the solder bumps without reflow and at 450 °C for the reflowed solder bumps. Some of the reflowed solder bumps show pores in the bonded interface after bonding at 450 °C. The optimal bond parameters with regard to optimal wetability and no pore in the bonded interface are 525 °C for the solder bumps without reflow and 400 °C for the reflowed solder bumps. The thermode force could be reduced from 30 cN/pad applied for soldering gold bumps to a tin plated tape to 10 cN/pad for the solder bumps.

2.4. Metallurgical Investigations

The cross sections in Fig.6 show differences in the bond interface metallurgy for solder bumps with and without reflow. The bond interface of the reflowed bumps consists mainly of Zeta phase with isolated spots of eutectic alloy with a total thickness of 7-10 um. The thickness of the Au-Sn alloy for the bumps without reflow is typically 1-2 um. Semiquantitative analysis of this thin layer indicate the formation of a gold rich alloy with a tin concentration less than 15 wt.%. During the reflow process a Zeta phase layer of a thickness of 3-5 um depending on the temper parameters is formed. Due to its high liquidus temperature of 495 ^{o}C the Zeta phase remains solid during the bonding process at 350-450 ^{o}C. The liquid 80/20 Au-Sn layer on the top of the reflowed bumps is mainly pushed outside of the interface forming a meniscus on both sides of the lead. The eutectic solder remainig in the interface reacts with the 1 um thick gold metallization of the lead forming a second layer of Zeta phase. This explains why the bonded interface consists mainly of Zeta phase. Isolated spots of eutectic Au-Sn are surrounded by Zeta Phase having no direct contact with the copper leads as shown in Fig. 7. This is an important fact for the thermal aging behaviour of the inner lead bonds.

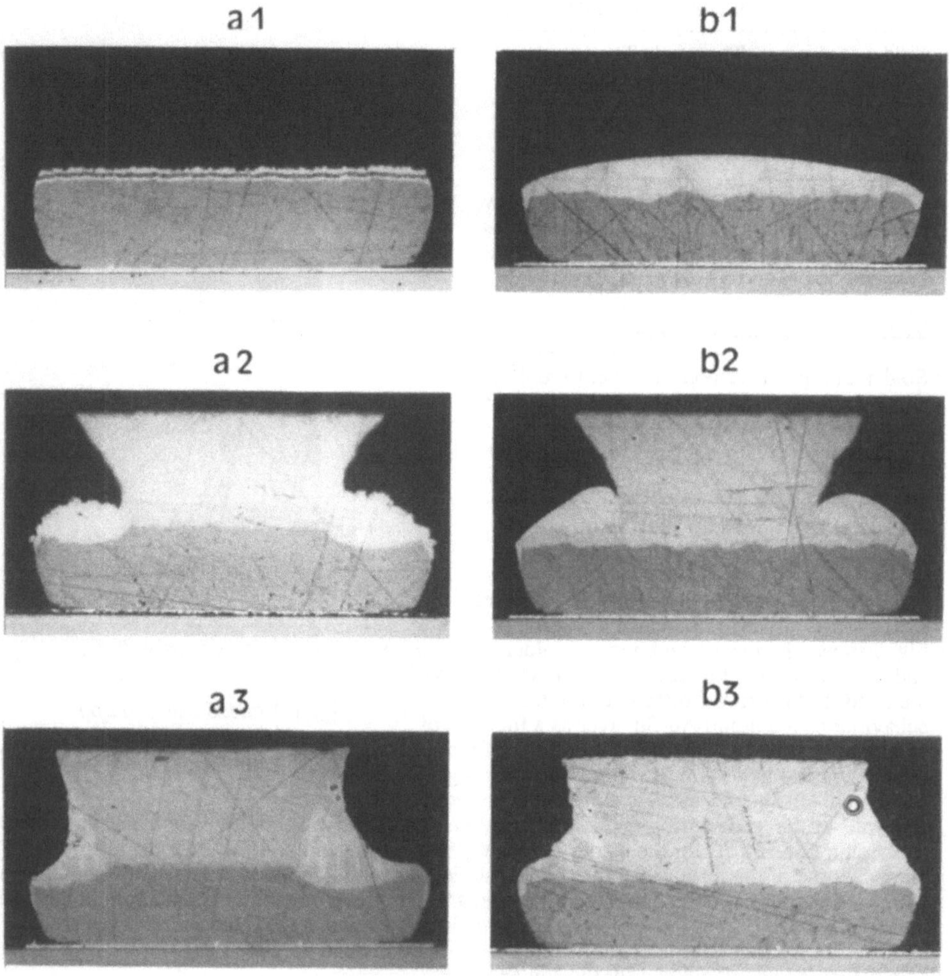

Fig. 6: Metallographic cross sections of bonded solder bumps

(a) Without Reflow
a1: Solder Bump
a2: Bonded at 450 ^{o}C
a3: Bonded at 525 ^{o}C

(b) With reflow
b1: Reflowed solder bump
b2: Bonded at 350 ^{o}C
b3: Bonded at 450 ^{o}C

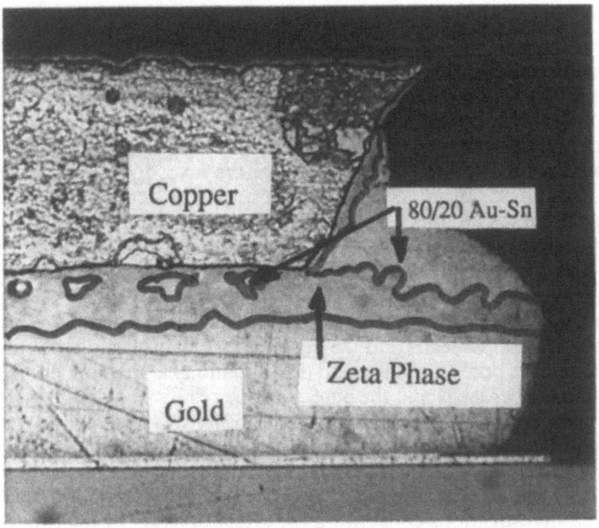

Fig. 7: Metallographic cross section of a reflowed solder bump after bonding

2.5. Reliability Investigations

Bonded chips of both reflowed and not reflowed solder bumps were thermally aged at 125 °C, 150 °C and 175 °C for up to 1000 hours. In pull test only lead fractures were observed in the initial state and after aging. A slow degradation of the contacts during thermal aging is shown in Fig. 8. Metallographic cross sections of the aged chips showed no pore formation in the bonded interfaces for both types of solder bumps.

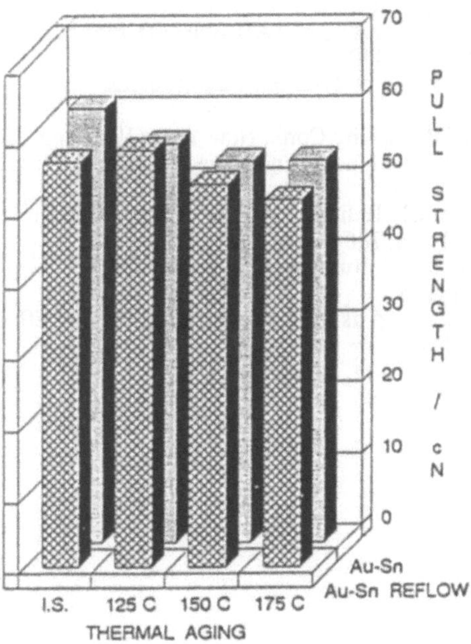

Fig. 8: Results from pull test for solder bumps without reflow (Au-Sn) and with reflow (Au-Sn Reflow) in the initial state (I.S.) and after thermal aging for 1000 hours at 125 °C, 150 °C and 175 °C

3. Conclusions

A new bumping process for TAB-solder bumps based on gold-tin was developed. A reflow process for the preformation of the 80/20 Au-Sn eutectic was optimized. Bonding of the two types of solder bumps with and without reflow was possible without flux application and with reduced bonding pressure. The bonded solder bumps show an excellent reliability with low degradation of pull forces after thermal aging. No Kirkendall effect was observed for the investigated metallurgy. This is due to the formation of the Zeta phase during the bonding process. The Zeta phase layer between the copper leads and the gold bumps inhibits Cu-Au interdiffusion and the Kirkendall effect. Bonding to a gold plated tape reduces the possibility of a direct contact between copper and the 80/20 Au-Sn eutectic. The Au-Sn alloy formed shows a concentration similiar to the Zeta-phase. The results support previous investigations which show that the appearance of the Kirkendall effect in the ternary Au-Sn-Cu system is accelerated by the formation of the 80/20 Au-Sn eutectic. A direct contact between the copper leads and the eutectic results in a unidirectional copper diffusion into the eutectic zone with pore formation and delaminations at the bonded interface. The Zeta phase between the copper leads and the eutectic zone results in a suppression of copper diffusion and hereby no pore formation. This gives the possibility to avoid the accumulation of the Au-Sn eutectic in the interface by the choice of adequate bonding parameters. A prebonding reflow process for the solder bumps offers the possibility to reduce bonding temperatures by 100 ^{0}C. On the other hand no critical tempering process is necessary for bonding solder bumps without reflow. Further investigations concerning the effect of aging of the two types of solder bumps on bondability will help to decide which type is better suited for specific applications.

4. References

/1/ P. Hedemalm and L.G. Liljestrand: Quality and Reliability of TAB Inner Lead Bonds. Proc. ITAB, San Jose (1990) pp. 88

/2/ L.G. Liljestrand: Bond Strength of Inner and Outer Leads on TAB Devices. Hybrid Circuits, 10 (1986) pp. 42

/3/ E. Zakel, J.Simon and S. Schuler: Neue Bumpingverfahren für die TAB-Technologie. Verbindungstechnik in der Elektronik, Heft 2 (1990)

/4/ T.S. Liu: Aspects of Gold-Tin Bump-Lead Interconnection Metallurgy. National ISHM-Symposium, Baltimore (1977), pp.120

/5/ E. Zakel and H. Reichl: Investigations of Failure Mechanisms of TAB-Bonded Chips during Thermal Aging. Proc. ECTC, Las Vegas (1990), pp. 450

/6/ G. Azdasht and E. Zakel and H. Reichl: Laser Soldering of TAB Inner Lead Bonds. Proc. Microsystems Conference, Berlin 1990

/7/ M. Hansen: Costitution of Binary Alloys.Genium Publishing Corp., New York (1985), pp. 232

/8/ G. Schlamp: Temperaturwechselbeständige Weichlotverbindungen für die Halbleitertechnik. Metall 39 (1985), Nr. 9, pp.810

/9/ G. S. Matijasevic, C.Y. Wang and C.C. Lee: Extremely Reliable Bonding of Large Silicon Dice Using Gold-Tin Alloy. Proc. ECTC (1990), pp. 786

/10/ O. Ehrmann, G. Engelmann, J. Simon and H. Reichl: A Bumping Technology for Reduced Pitch. Proc. ITAB, San Jose (1990), pp. 41

/11/ E. Zakel and H. Reichl: Au-Sn Bonding Metallurgy and its Influence on the Kirkendall Effect in the Ternary Cu-Sn-Au System. To be published.

Bonding Techniques on Microsystem in Cryogenics and Microelectronic Engineering

St. Howitz; D. Gerber

Technical University of Dresden

1. Summary

A number of cryogenics, infrared detectors and superconduction components working at low temperatures. The materials of detector circuits, evaluation circuits and substrats shouldn't mounted by the help of fluxes. An other restriction is to bond the materials without stress, even if they have a different thermal expansion coeffizient. One possibility to solve this problems is soldering without flux. The following experiments where realized on cryogenics, whose range of working temperatures is 300 ... 77 K. Indium, tin and bismut where applicated, because they are ductile in the whole range of working temperatures.

We describe alternative reflow soldering techniques without fluxing, which satisfy technical and scientific applications. The discussed electrostatically bonding methods, at first applicated for presoldering of glass surfaces, are also suitable in packaging of Si to Si micromechanic components.

2. Reflow soldering techniques

Problem

Base of the developed reflow soldering was a sputtered multilayer structure as shown in Fig. 1. We evaporated a solderable layer, 10 ... 15 μm thick, at the silver layer. Contamination of solder layer generates undesirable wetstop layers, for instance In_2O_3. This contamination results in closed solder oxid layers, at open air after a few hundered milliseconds, and a few minutes at vacuum ($p = 10^{-3}$ Pa) /1/. By help of nondestructive X-ray microscopy and destructive microanalytics at soldered bonds, we have verified a relationship between beginning wetstop layer and defects in solder joints /2/.

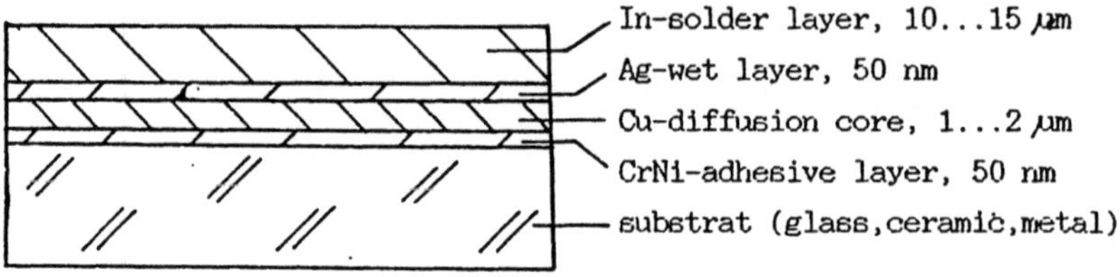

Fig. 1 Cross-section of our thin film metal plating

Reflow soldering with polymer layers

The intention to coate the solder layer with a thin film of polymer was employed to prevent formation of solder-oxid-wetstop-layer. Polymer layers with defined thickness, pinhole density, chemical and mechanical properties will be able to provide a high degree of reproducibility for bonding. The suitable thickness-proportion solder/polymer is about 50. Polymer thickness minimum for perfect reflow soldering is 80 nm /3/. Kampfrath described the properties of glow-discharged polymer layer more in detail /4/. Inspections by X-ray microscopy as shown in Fig. 2 and 3 demonstrat better conditions in solder joints we have bonded by polymer layers /5/. In AES-measurements we have shown agglomerations of the destroyed polymer in solder bulk.

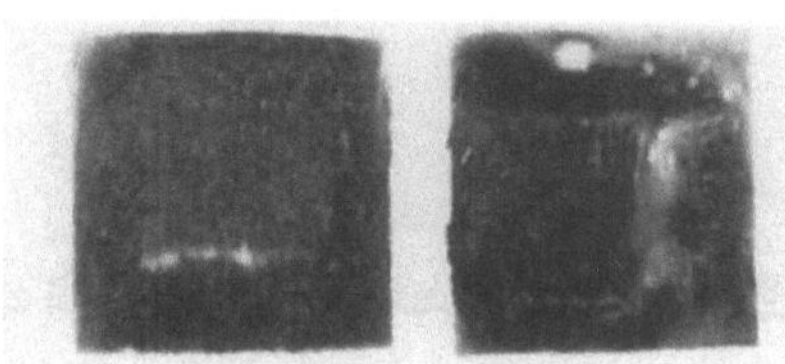

Fig. 2 Reflow solder joint without polymer

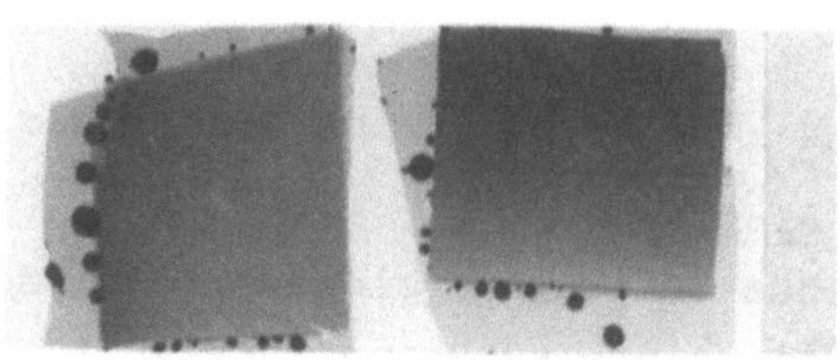

Fig. 3 Reflow solder joint with polymer

Soldering with solder ball

Soldering without flux is applicable in general for metal systems with very good solubility. We have examined soldering experiments under nitrogen with InSn48 - solder ball on the Ag-layer of the metallized substrate (Fig. 1). At first the solder ball was cleaned of flux residue after melting of the solder ball. At second the solder ball was put on the silver plated substrat, which was heated on solder melt temperature. Now the counterpart which is coated equal to the first substrat must set up in the vertical direction on molten and oxid covered solder ball. In this way the oxid boundary layer of deformed solder ball should be cracked and wetprocess starts. Optical measurements of the wet process have revealed the beginning of interdiffusion between InSn48 and Cu after a short alloy cycle of about 5 s by 50 nm Ag.

Insitu-reflow soldering in vacuum

Perfect solder joints were obtained by insitu-reflow soldering in vacuum /2/. Poor effectivity is the lack of this process. For a better effectivity we developed an alternative step by step insitu-reflow soldering with aerating at inert gas. At first step the samples must coated by solder evaporation (15 μm, $p < 10^{-3}$ Pa). At second the presoldered samples were manual mounted in the inert gas (N_2, Ar, $p < 10^{-3}$ Pa) aerated vacuum chamber, and at last the insitu-reflow soldering realized also in vacuum. Solder joints which were reflow soldered with and without aerate at inert gas are shown in X-ray microscopy images in Fig.4 and 5. We obtained solder joints without failures up to 30 minutes aerating at inert gas.

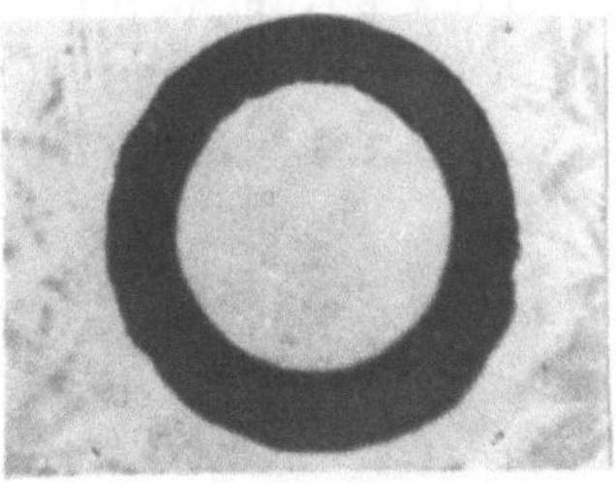

Fig. 4 Reflow solder joint by insitu soldering in vacuum /5/

Fig. 5 Reflow solder joint with aerate at N_2 or Ar /5/

3. Field assisted bonding at reflow soldering

Reliable bond strength of solderable surfaces based on intermetallic diffusion between connecting surfaces. The usual procedure is vacuum metallization. The field assisted anodic bonding /6/, a glass-metal or glass-semiconducter bonding method, was the starting point to get solderable glass surfaces without of vacuum metallization. Our experiments have shown, that anodic bonded solde layer adhere well on sodium containing glass interfaces (see Fig. 6

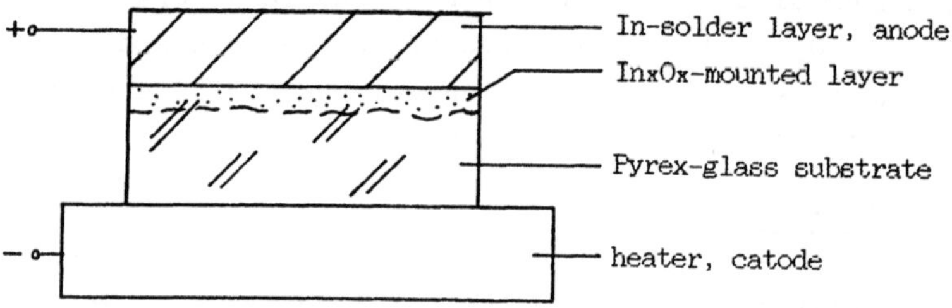

Fig. 6 Field assisted resoldering at glass interface

The evaporated solder layer runs away in durability and reliability of the intermetallic solder joint, because the process of diffusion is also continued at room temperature. This disadvantage don't exist at field assisted glass-metal bonds. Surface resistance is rapid sinking. In comparsion with pure glass regions we had measured a reduction from 10^{13} to 10^{8} Ohm cm.

Field assisted resoldered substrats which reflow soldered to glass - indium-glass area contacts we have used for testing of the bond strength. The shearing test yielded 3 ... 4,2 MPa, consequently solder breaking strength was attainable. In the same way we have bondet and tested PbSn63 solder. In this case the shearing stress yieldet 15 MPa. The fracture didn't turn out into the solder bulk but in the glass-solder interface. Only 30 % of solder shearing stress was attained.

At last a short summary of several applications:

- Improvement of bond strength at vacuum metallization coated glass for instance solder layer, mirror layer, or hybrid circuit metallizations.
- In glass interface mounted transparent and electric conductive layers are obtainable, suchlike buried layers suitable for instance in liquid crystal displays.

4. Si-to-Si selective field assisted bonding

Problem

Frye represented a Si-to-Si direct bonding /7/ by the help of high temperature and electric field. Fryes method suitable for the bonding of micromechanic components at 1150 °C and 20 V. /8/ deposited low melting point glass on Si-wafer surface and bonded at room temperature with an applied voltage of about 50 V. One of the problems in mounting of micromechanic components is the selective wafer bonding process. A look at a micromechanical valve as shown in Fig. 7 /9/ reveals the requirement of selective wafer bonding in the region of moveable valve membrane. /10/ described the use of oppositional electric fields for this reason.

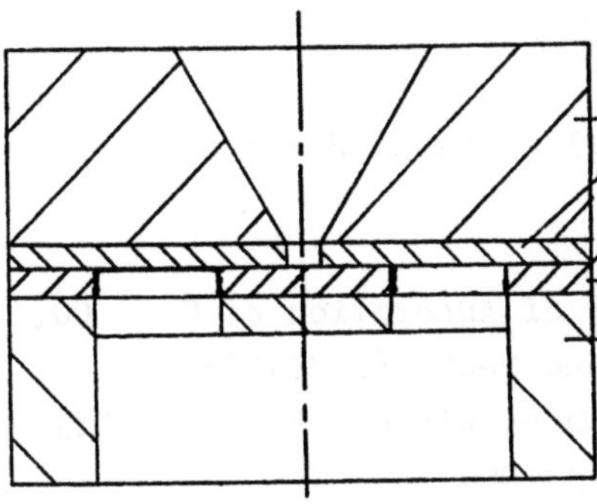

Fig. 7 Silicon to silicon bonded microvalve

An alternative selective Si-SiO_2-Si bonding technique

In analogy to chapter 3 we modified Pyrex glass surfaces by the help of Al-interdiffusion. The Al-layer was evaporated on glass surface before the diffusion. The Al-interdiffusion was realized as like as field assisted diffusion at 300 V and 400 °C as shown in Fig. 6. After Al-diffusion the Al-plating was etched completely. Our SIMS-measurements on Al-interdiffusion in glass interfaces offers Al-metal atoms up to a depth of 20 ... 50 nm. In the region of the transition both the glass surfaces with and without diffused Al exixsts a small step about 50 ... 100 nm. The idea of our alternative selective bonding process consists in use of this buried metal oxid (for instance Al_2O_3) interfaces in glass.

Wherever the Si-Si-structure shouldn't be bond, the metal oxid must be buried. Actual we are testing different methodes for buriing metal atoms or metal oxids in the glas interface, for instance low energetic metal implantation, diffusion and field assisted diffusion. The two images in Fig.8 and Fig.9 show broken regions of selective field assisted bonded Si-Pyrex bond. The bond was realized at 450 °C and with an applied voltage of about 700 V.

Fig. 8 and 9 Fracture images of selective Si-Pyrex bond

5. References

/1/ Rossnagel,S.;Dylla,M.; AES-Study of Adsorption of CO_2, CO, O_2 and H_2O on Indium; J.Vacuum Sc.and Tech. (1978)558

/2/ Howitz,St.; Reflow Solder Procedures without Flux for Cryogenics; Diss.; Tech. Uni. Dresden 1990

/3/ Hoffmann,M;Howitz,St.; Soldering with Polymer Layer; Tech. Uni. Dresden 1988

/4/ Kampfrath,G.; Properties of Glow-Discharged-Polymer; Acta Polymerica 38(1987)6

/5/ Ullrich,H.; Schlaubitz,M.; X-Ray Microscopy; Tech. Uni. Dresden 1989

/6/ Wallis,G.;Pomerantz,D.; Anodic Bonding; Pat.-US 3.397.378;1968

/7/ Frye,R.C. a.o.; A Field Assisted Bonding; J.Electrchem.Soc. 133(1986)78

/8/ Shimbo,M. a.o.; A New Developed Si to Si Direct Adhesion Method; J.Electrochem.Soc.Abstracts 86-1(1986)337

/9/ Schmidt,B.;Schubert,D.;Design of Si-Si-Microvalve;AdW-ZfK 1986

/10/ Winkler,T. a.o.; Selective Anodic Bonding of Si-Si-Components; Tech.Uni.Dresden 1990

Laser Soldering of TAB Inner Lead Bonds

G. Azdasht, E. Zakel and H. Reichl
Technical University of Berlin
Technologien der Mikroperipherik
Sekr. TIB 3.1, Gustav-Meyer-Allee 25
1000 Berlin 65, W-Germany
Fax: (030)31472835

Abstract

Tape Automated Bonding (TAB) is a modern technology which meets the requirements for micronnecting VLSI-circuits. Gang bonding with a thermode is the usually applied bond technique. The limitations for gang bonding of chips with high lead counts are the increased bond forces and the induced mechanical stress. Laser soldering is an alternative for inner lead bonding chips with high lead counts and reduced pitch. Because the microjoining of surfaces occurs via thermal energy from the laser beam, no mechanical pressure is necessary for laser soldering. Due to the optical properties of the laser beam and the possibility to reduce the laser spot, soldering of small pitches is possible.
The results of TAB inner lead bonding with a pulsed Nd:YAG laser are presented. Tapes with a tin metallizations were laser soldered to bumps consisting of gold, gold-tin and nickel-gold. The pull strength of laser soldered TAB-contacts was optimized by variation of laser power and reliability investigations were performed. A comparison of thermal aging behaviour of laser soldered contacts and thermode soldered contacts is presented.

1. Introduction

Due to the possibility to reduce the pitch TAB is a suitable microconnection technology for chips with a high number of I/O's. The increasing thermode forces required for gang bonding can cause cracks in the passivation layer, in the Ti:W diffusion barrier and in the silicon. The necessity of an optimal thermode alignment and planarity require experienced operators and improved tools /1/. The process shows a low flexibility for bonding small quantities of different chips. The necessity of a specific thermode for different chips results in higher tooling costs and supplementary time for tool change and alignment. Laser soldering is an alternative to gang bonding as no mechanical pressure is applied. Due to its flexibility different chips can be bonded easily without tool change. Bonding of small pitches is possible by focusing the laser beam. The application of laser for TAB inner lead contacts /2/ and for outer lead contacts /3/ has been reported. Actually two types of lasers are used for different industrial applications: CO_2-lasers and Nd:YAG lasers. The commonly applied metals for TAB are copper, gold and tin. A high absorption of these metals at the laser wavelength is required for a laser bonding

process. These metals show a very low absorption at the CO_2-laser wavelength of 10.6 microns. Nd:YAG lasers are better suited as tin shows an absorbtion of 53.6 % at the Nd:YAG laser wavelength of 1.06 microns /4/. On the other hand copper and gold show a low absorption /5/. From the different types of Nd:YAG lasers the pulsed laser appears as best suited for TAB inner leads as pulse width and pulse energy can easily be controlled. A reliable bonding process is possible without the danger of damaging the area around the bondpads like in the case of CW-lasers.
Concerning TAB metallurgy a degradation of TAB contacts due to a Kirkendall effect in the ternary Cu-Au-Sn system with pore formation and ternary intermetallic Cu-Sn-Au compounds has been reported /6,7/. It has been shown that a tin accumulation in the interface between the copper leads and the gold bumps accelerates the pore formation causing delaminations of the bonded interfaces /7/. No Kirkendall effect was observed for thermocompression bonded Cu-Au contacts. Recent investigations show that the composition of the Au-Sn alloy formed during the bonding process affects the Kirkendall effect /11/. The formation of the 80/20 Au-Sn eutectic accelerates the Kirkendall effect, whereas the gold rich Zeta phase with 10 % Sn inhibits copper diffusion. Gang bonding offers the possibility to avoid the accumulation of tin in the interface. For a tape with reduced roughness the molten gold-tin alloy is pushed outsite of the interface by the bonding pressure /7/. Laser soldering does not permit the exclusion of tin from the bonded interface as no bonding pressure is applied. Spletter et al. /2/ found 16 % bump-lead delaminations after temperature cycling and high temperature storage for 1000 hours at 150 °C. Compared to the lead breaks these lead lifts show 50% lower pull forces. This might indicate an influence of the Kirkendall effect as a degradation mechanism for the laser soldered contacts. Thus the object of investigations was to examine different tape and bump metallizations with regard to laser bondability, to optimize the bond process with regard to pull strength and reliability after thermal aging and to study the metallurgy of the contacts.

2. Experimental results

The laser soldering process was performed with a pulsed Nd:YAG laser of SPEKTRUM Ltd, Berlin. The experimental setup for laser soldering is shown in Fig.1. After the alignment of tape and chip, which is made with an x,y,Θ table the leads are pressed towards the chip by a vacuum chuck. The chip is held by a separate vacuum chuck. This mechanical contact between tape and bump is necessary in order to obtain an optimal thermal coupling of the laser energy. Laser bonding experiments were performed with differnt tape types and different bump metallizations shown in Tab. 1.

Tab. 1: Laser soldering results

	Tape Metallization Type	Tape Metallization Thickness	Copper Type	Copper Roughness	Bump Metallization	Bump Thickness
1	ED-Sn	2 um	***RA	4 um	Au	25 um
2	*ED-Sn	2 um	RA	4 um	Ni/Au	25/1 um
3	EL-Sn	.7 um	RA	4 um	Au	25 um
4	**EL-Sn	.7 um	ED	7 um	Au/Sn	25/3 um

*ED=Electro-deposited
**EL=Electroless deposited
***RA=Rolled Annealed

A tape with a 7 um surface roughness could only be successfully bonded to a solder bump of gold covered with 3 um tin. Laser soldering was also possible to electroless deposited Nickel bumps. Both bumping processes were developed at the Technical University /8,9/. For all of these tape-bump combinations average pull forces higher than 40 cN were measured. Further investigations were performed with a test chip with 49 I/O's and 100x100 um gold bumps. A three layer RA-copper tape with 2 um electrodeposited tin was applied. As thermal coupling between lead and bump is induced via the melting tin, a higher tin thickness is regarded as better suited for laser soldering.

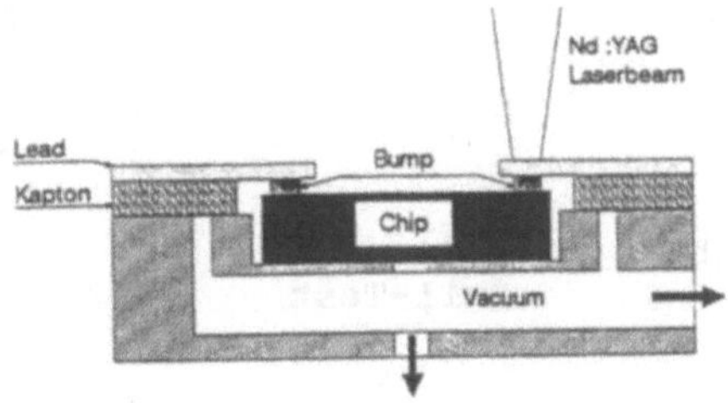

Fig. 1: Experimental setup for laser soldering

2.1. Bond Process Optimization

The bond process was investigated by increasing laser power and measuring the pull force of the contacts. Fig. 2 shows SEM-images of laser soldered contacts indicating that the lead is melting with increasing power. Increasing pull forces are measured with increasing laser power (Fig. 3). Two failure modes were observed in pull test : lead breaks and lead lifts. Bump lifts appear with higher laser powers than indicated in Fig. 4. The percentage and average pull force of each failure mode as a function of laser power is shown in Tab 2. With increasing laser power an increase in pull force and in the percentage of lead breaks is measured. An optimum with 45 cN and 100% lead breaks was achieved. Metallographic cross sections show differences in the contact metallurgy depending on laser power. In the low power range the contact between bump and lead occurs via a Au-Sn alloy with a small

a

b

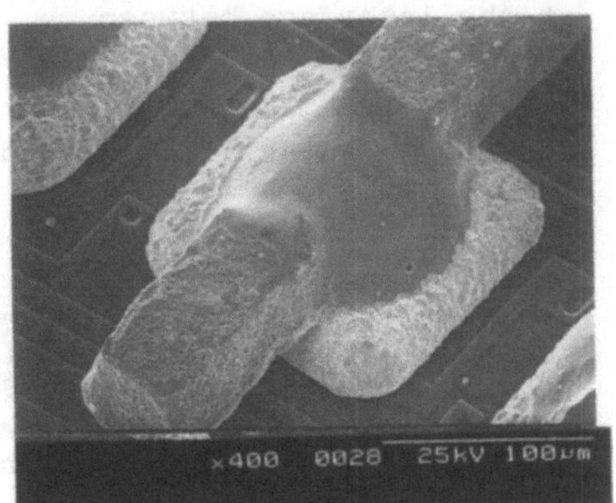

Fig. 2: Laser soldered ILB contacts at different laser powers
a: 9 relat. Units
b: 20 relat. Units

Tab. 2: Results from Pull-Test

Power [relat.Units]	Aging treatment	Total $\bar{F}$*	Total s**	Lead lift %	Lead lift $\bar{F}$*	Lead lift s**	Lead break %	Lead break $\bar{F}$*	Lead break s**
9	I.S***	25.8	4	100	25.8	4	0	–	–
	1000 h, 150 °C	0	–	100	0	–	0	–	–
13	I.S***	37.4	7	44	33	8	56	40	4
	1000 h, 150 °C	17	8	63	14.4	8	27	27	8
20	I.S***	45.5	4	0	–	–	100	45.5	4
	1000 h, 150 °C	29	6	0	–	–	100	29	6

* F = Mean pull strength [cN]
** s = Std.deviation [cN]
***I.S = Initial state

concentration of copper. Quantitative microprobe analysis show concentrations near the 80/20 Au-Sn eutectic with 62-74 wt% Au, 18-21 wt% Sn and 5-10 wt. % Cu. Because no bonding pressure is applied, the Au-Sn eutectic accumulates in the bonded interface as shown in Fig. 4. In pull test these contacts failed only by bump lift. Fig.4 shows differences in the delaminated interface in the initial state and after thermal aging. In the initial state the fracture occurs in the formed Au-Sn alloy. After thermal aging the pull force decreases from 25 cN to 0 cN. Pore formation is the cause for the observed delaminations at the interface between copper and the Au-Sn eutectic alloy. Microprobe analysis show the formation of a ternary Cu-Au-Sn alloy due to copper diffusion into the eutectic Au-Sn zones.
An increase in laser power causes the melting of larger areas of the copper lead. For the power level 20 the lead is totally molten forming a Cu-Sn alloy. Fig. 5 shows the element distribution of such a lead. No accumulation of tin in the interface between bump and lead is observed. This is due to the Cu-Sn alloy formation which results in a slow reduction of liquidus temperature from 1083 °C to 900 °C for 20 wt. % Sn /10/. As shown in Fig. 5 the bump is not affected by the high temperatures causing the melting of the lead. These contacts show high pull forces of 45 cN in the initial state. After thermal aging a degradation to 29 cN were measured. The Kirkendall effect can be excluded as a cause of degradation because no bump lifts and no pore formation was observed. The SEM investigations of lead breaks after thermal aging indicate

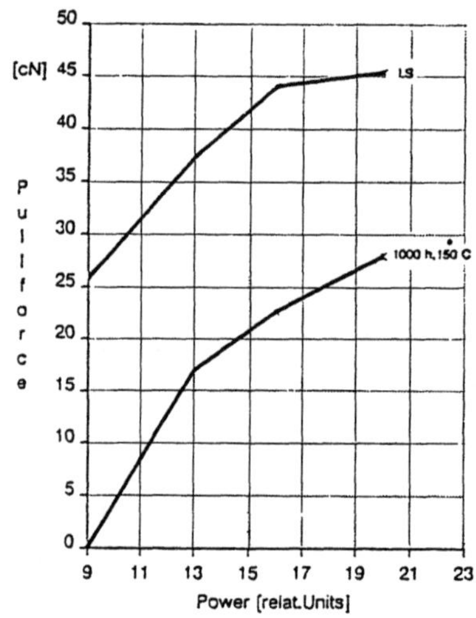

Fig. 3: Pull forces as function of laser power in the initial state (I.S.) and after 1000 hours at 150°

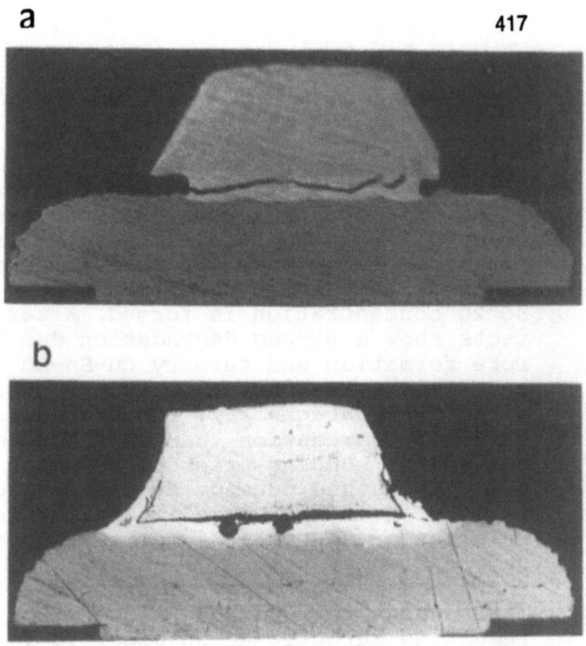

Fig. 4: Cross sections of ILB contacts in the low power range (9 relat. Units) with differences in bump lift fracture mode

a: in the initial state and

b: after thermal aging 1000 hours, 150 °C

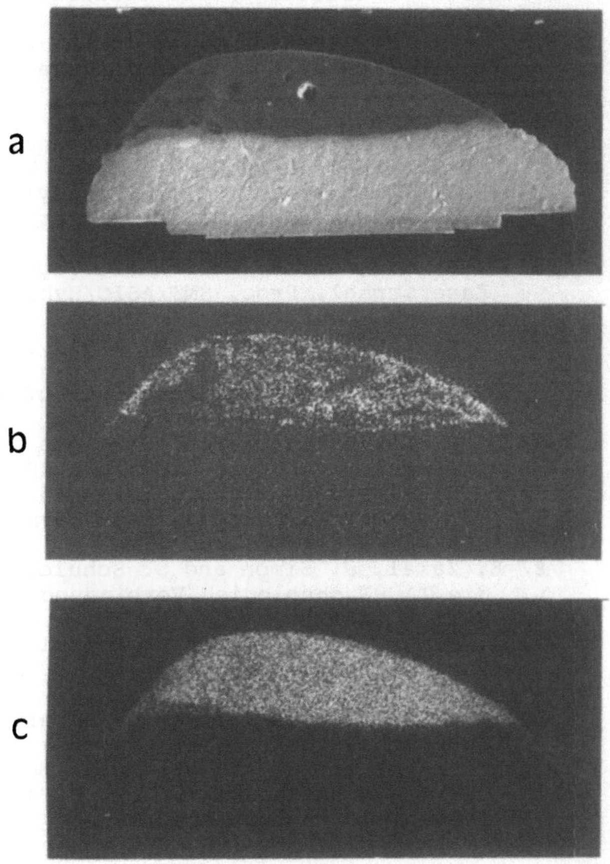

Fig. 5: SE-image and Cu-Sn distribution of a ILB contact in the high power range (20 relat. Units)

a: SE-image

b: Sn distribution

c: Cu-distribution

a brittle fracture which might be caused by the formation of brittle intermetallics.

3. Conclusions

A pulsed Nd-YAG laser was applied for contacting TAB inner lead bonds. Tin plated tapes were successfully bonded to bumps of gold, gold-tin an to electoless deposited nickel-gold. The effect of laser power on pull force and fracture mode was investigated. In the low power range a Au-Sn alloy near the eutectic 80/20 concentration is formed. After thermal aging these contacts show a strong degradation due to the Kirkendall effect. Pore formation and ternary Cu-Sn-Au intermetallics were observed. This coresponds to previous publications on thermode soldered TAB contacts /6/. Laser soldered contacts show a much stronger degradation than gang bonded contacts. This is probably due to the bonding pressure by which the formed eutectic Au-Sn alloy is pushed outside of the contact during the gang bonding process. As no pressure is applied for laser sodering an accumulation of the eutectic alloy is possible causing a stronger degradation.
Increased laser energy results in a total melting of the copper leads which form a Cu-Sn alloy with the tape metallization. As a result no eutectic Au-Sn alloy is formed in the interface, thus eliminating the cause for the Kirkendall effect. In this case the degradation measured after thermal aging might be due to the formation of brittle intermetallics.
As a result it could be shown that avoiding the accumulation of a eutectic 80/20 Au-Sn alloy in the bonded interface increases the reliability of the laser soldered contacts.

4. References

/1/ P. Hedemalm and L. G. Liljestrand: Quality and Reliability of TAB Inner Lead Bonds. Proc. ITAB, San Jose (1990) pp.88

/2/ P. J. Spletter and R. T. Crowley: A Laser Based System for Tape Automated Bonding to Integrated Circuits. Proc. ECTC, Las Vegas (1990), pp. 757

/3/ D. Knödler and R. Belschmer: Tape-Automated-Bonding mittels Laserstrahl. Proc. SMT/ASIC/Hybrid Conf., Nürnberg (1990), pp. 349

/4/ R. E. Lindquist and A. W. Ewald , Phys. Rev., 135, A191, 1964

/5/ D. E. Gray. American Institute of Physics Handbook, 3rd. Edition, McGraw-Hill, New York, 1972

/6/ L. G. Liljestrand: Bond Strength of Inner and Outer Leads on TAB Devices. Hybrid Circuits, 10 (1986), pp.42

/7/ E. Zakel and H. Reichl: Investigations of Failure Mechanisms of TAB-Bonded Chips during Thermal Aging. Proc. ECTC, Las Vegas (1990), pp. 450

/8/ E. Zakel, J. Simon and S. Schuler: Neue Bumpingverfahren für die TAB-Technologie. Verbindungstechnik in der Elektronik, Heft 2 (1990)

/9/ J. Simon, E. Zakel and H. Reichl: Electroless Deposition of Bumps for TAB-Technology. Proc. ECTC, Las Vegas (1990), pp.412

/10/ M. Hansen: Constitution of Binary Alloys. Genium Publishing Corp., New York (1985)

/11/ E. Zakel and H. Reichl: Au-Sn Bonding Metallurgy and its Influence on the Kirkendall Effect in the Ternary Cu-Sn-Au System. To be published.

Packaging

Solderable Bumps for TAB and Flip-Chip Bonding

K. DRESCHER

Technical University Dresden

Abstract

Bumps are essential parts of assembling processes in tape automated bonding (TAB) and in flip-chip bonding (FCB). Bump technology is able to solve several tasks in microsystem technologies, too. The main aims to be accomplished are a high adherence to the substrate, a very low electrical resistance, and an excellent long-term stability. Moreover we considered some more conditions, especially

- high aspect ratios
- minimum bump width and distance (pitch)
- no use of noble metals
- very broad range of melting points of the solder materials
- high uniformity of deposited bumps on large wafers
- preferential application of dry processing steps.

Bumps for TAB consist of a hard copper core covered with an solder cap. Its thickness was optimized regarding to the soldering process. The copper core is needed for avoiding shorts between leads and the chip at its edge. Bumps for FCB consist of pure solder. Their composition, dimensions, and shape are determined by the properties of the parts to be joined as well as by the mechanical strain caused by the different thermal expansion of chip and substrate. The investigated solders PbSn, In, InSn, InBi, and InBiSn have melting points in the range from 60 °C to 295 °C.

Depending on the demanded bumps height extremely thick masks must be applied. Good results were obtained using high viscous resists (AZ-type, single or multiple coating) as well as several types of dry film resists. Light sensitive polyimide can be used , too.

All kinds of bump materials have been deposited by galvanic methods. The solder material was deposited either as the final alloy or as a sandwich layer consisting of its components. We had to overcome some difficulties to etch the basic metallization which consists of an adhesion layer, a diffusion barrier and a low resistivity layer for galvanic current distribution. Finally suitable shape and surface conditions on the bump top must be formed by a reflow process.

A lot of research was spent for lift-off processing. All interesting materials have been evaporated or sputtered on the masked wafer. The main problems to be solved are a well defined shape of pattern edge and a sufficient thermal stability of mask material. Results of modelling of edge coating (for different source characteristics) and experimental results were in good aggreement. 45 μm thick $Bi_{90}In_{10}$ layers on a 20 μm thick mask could be lifted with good success.

A typical result of our research is an array of some thousends of bumps with 20 μm diameter, 40 μm pitch, and a spherical shape.

Electroless Bumping

J. Simon and E. Zakel

Technische Universität Berlin
Technologien der Mikroperipherik
Gustav-Meyer-Allee 25, D-1000 Berlin 65

Abstract

Electroless nickel bumps have been deposited selectively on the aluminum bond pads of an integrated circuit from an electroless plating solution. The bumps are mushroom shaped because no resist mask was used. A modified zincate pretreatment was used to reduce the dissolution of the aluminum. This is an important process step due to the low thickness of the aluminum bondpad. The hardness of electroless nickel is very high (about 500 HV) and cannot be reduced by tempering. Therefore inner lead bonding was performed by soldering. The bumped devices were successfully gang bonded to a tin plated tape.

Electroless Bumping

TAB technology requires bumped dies. Electroless plating was investigated for bumping instead of the commonly used electroplating process /1, 2/. This new method offers two main advantages. The bumping costs are reduced considerably because no sputtered plating base und no lithography is necessary. Beyond that Single Chip Bumping (SCB) becomes possible because the new process is independent of substrat size /3/. Electroless plating refers to the autocatalytic or chemical reduction of metal ions plated to a base substrate from an aqueous solution. On contrast to electrodeposition deposits are very uniform because no external current is required and the deposition rate is not influenced by a current distribution in the plating equipment. The number of metals to be plated successfully from electroless plating solutions is quite limited compared to electrodeposition /4/. As gold or copper are most frequently used as bump metal, electroless plating solutions for these metals would be of great interest.

Deposition rates for autocatalytic gold are very low (1-2 μm/h) and the plating solutions are very critical to stable operation. Although electroless copper is used in printed board technology, the direct application to the aluminum bond pad causes a rapid dissolution of the aluminum metallization due to the high alkalinity of electroless copper electrolytes. For this reason electroless nickel was investigated for bumping despite of its hardness. Electroless nickel deposition on integrated circuits is already reported in literature /5, 6/.

An electroless nickel bumping process must consist of the following process steps. Because of the high affinity of the aluminum to oxygen the bond pads are covered with a native oxide layer. This layer has to be removed to give a low contact resistance and a good adhesion. For this purpose the zincate treatment is described in literature /4/. The oxide layer is dissolved in the zincate solution and replaced by a thin film of zinc which prevents the aluminum from renewed oxidation until the next process step. This zinc layer is replaced by nickel from the autocatalytic nickel plating solution by an electrochemical displacement reaction. Autocatalytic nickel deposition is initiated at this nickel deposit.

Results

An integrated circuit from Siemens with 52 I/O's and a bond pad size of 110 * 110 μm^2 was used in these experiments. The results have already been reported in more detail in /7/. A modified zincate solution was applied for the pretreatment to reduce the dissolution of the aluminum. The aluminum is attacked very rapidly by standard compositions. A maximum etch rate of 0.3 μm/min has been determined for used solution. This is of great importance because of the low thickness of the aluminum in semiconductor metallizations. A double zincate treatment was performed to improve the spatial distribution of the zinc grains (fig. 1). A homogeneous zinc film is important for subsequent nickel deposition. In the case of an irregular zinc deposit the zinc film is protected by an anodic dissolu-

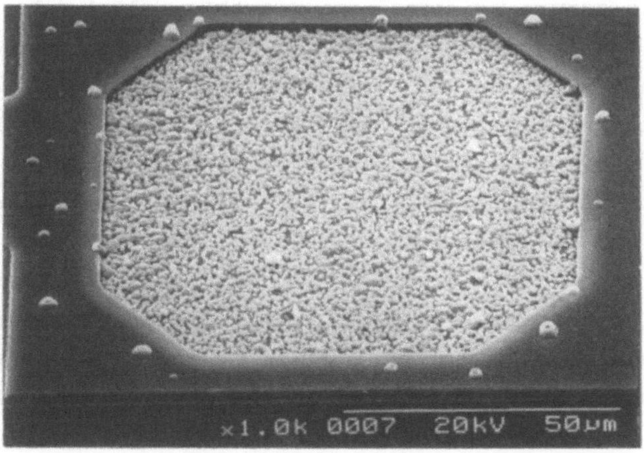

Fig. 1: Spatial distribution of the zinc grains on the aluminum bond pad after a double zincate treatment. The grains outside the bondpad are hillocks in the aluminum.

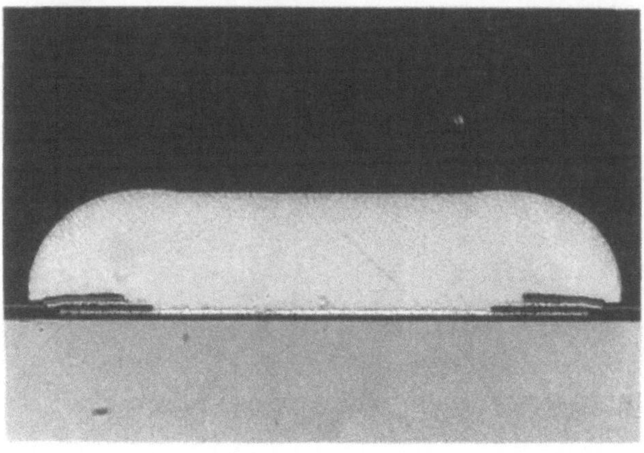

Fig. 2: Metallographic cross section of an electroless nickel bump with an intact interface

tion of the aluminum due to its position in the electrochemical series. This anodic corrosion leads to a porous interface. After the deposition of the zinc layer the samples were directly immersed into the electroless nickel plating solution. A commercial hypophosphite based electroless nickel electrolyte was used for bumping. The deposits from this electrolyte contain about 10% phosphorous. The pH-value was adjusted at 4.5 and the bath was operated at 90 °C. A plating rate of up to 25 µm/h can be achieved. The plating conditions were adjusted to assure uniform bump growth on all bondpads independent of the internal circuitry. With improper conditions nickel deposition may be prevented on certain I/O's. Because no photoresist was used the bumps were mushroom shaped. Fig. 2 shows a metallographic cross section of a nickel bump with an intact interface. To prevent the nickel from oxidation and to improve the solderability the bumps were finally coated with a thin layer of gold by immersion plating. The uniformity of the electroless nickel bumps is shown in fig. 3. The adhesion of the nickel bumps to the aluminum was measured by a shear test. Depending on the pretreatment shear forces greater 200 cN can be achieved.

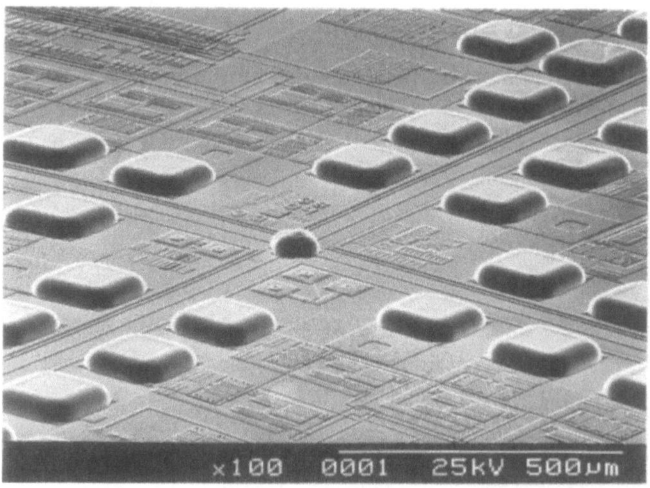

Fig. 3: Electroless nickel bumps

Inner lead bonding was performed by soldering because of the high hardness of the electroless nickel bumps (about 550 mHV200). A tape with a layer of 7-9 μm tin selectively plated to the inner leads was used. Because of the uniformity of the bumps the thickness of the tin layer mainly has to compensate the nonflatness of the thermode and to awoid a damage of the bond pad during bonding operation. The inner lead bonding was performed by gang bonding with a thermode at a temperature of 350 °C and a bond pressure of 20 g/pad using flux. An inner lead bonded chip is shown in fig. 4. Destructive pull tests were performed to test the bond strength. A mean pull strength of 33.4 g was achieved. The failure modes were lead and bump lift. The results were comparable to conventional copper bumped devices of the same type.

Fig. 4: Electroless nickel bumps gang bonded by soldering

Summary

An integrated circuit has been bumped by electroless means. The bumped dies were gang bonded to a tape whose inner leads were selectively plated with a thick layer of tin. Electroless bumping offers the great advantage to reduce the costs of the bumping process because no sputtering equipment and no lithography is required. This may be very important if only small quantities of bumped dies are required, because the process is independeent of substrat size. Electroless bumping may become an alternative to conventional wafer bumping by electrodeposition at least for certain applications.

References

/1/ T. S. Liu, W. R. Rodriges de Miranda and P. R. Zipperlin, Solid State Technology, March 1980, pp 71-76

/2/ G. Engelmann, O. Ehrmann, R. Leutenbauer, J. Simon und H. Reichl, GME-Fachbericht "Mikroelektronik", VDE-Verlag, Berlin 1989, pp 287-292

/3/ J. Simon and H. Reichl,
Proc. 2nd Int. TAB Symposium, San Jose 1990, pp 49-53

/4/ Lowenheim, Modern Electroplating,
John Wiley & Sons, 1974

/5/ K. Wong, K. Chi and A. Rangappan,
Plating and Surface Finishing, July 1988, pp 70-76

/6/ K. Yamakawa, M. Inaba and N. Iwase,
ISHM'89 Proceedings - Baltimore, Maryland, pp 620-626

/7/ J. Simon, E. Zakel and H. Reichl,
Proc. 40th Electronic Components & Technology
Conference 1990, pp 412-417

Hybridization of Cooled Mosaic Sensors by Indium-Bumps

M.Plötner, G.Blasek, G.Sadowski, S.Rzepka

Dresden University of Technology
Dept. Information Technology
Div. Technology of Microelectronics

Abstract

Flip chip technology is familar for hybridization of mosaic sensors. In the case of cooled applications solder bumps made of In or In-alloys are preferred. The high density of pixels necessary for radiation detectors results in sharp claimings on patterning sufficiently high bumps that can be accomplished by electrodeposition. Solder alloys necessary for sensors susceptible to higher temperatures are plated sandwich-like. Special requirements on joining technology, too arise from materials, geometries and working conditions. Both cold pressure welding and reflow soldering are practicable with In-bumps. Technological variants of hybridization techniques are discussed, results are demonstrated on phantom devices.

Introduction

Mosaic sensors are favoured for image performance of radiation. Such multielement detectors are to be connected with processing electronics. For many practical sensing properties a monolithic approach sensor with analyzing electronics is unfavourable or if not even impossible. Thus hybrid modules of different materials are built up. With mosaic contact pattern the electrical connections between sensor and circuit are most frequently done by solder bumps and flip chip bonding.

For some sensing properties, i.e. infrared radiation by semiconductors, the detectors have to be cooled down to 77 K. In the latter or in the case of higher thermal expansion differences (for instance with pyroelectric thermal sensors) which has to be adjusted In-solders are used for bumps because of their high ductility down to deep temperatures /1-3/. Since the pixel density determines the lateral resolution very small bump spacings compared to conventional flip chip bonding have to be obtained.

Requirements on solder bumps are established on the example of a liquid nitrogen cooled mosaic sensor. For its fabrication criteria with bump pitches down to 40 µm are deduced. By the electrochemical solder deposition general problems of In-bump technology are explained and results based on a phantom device with 32x32 contacts and 80 µm pitch are given.

<u>Requirements on the bumps and fabrication techniques</u>

Fig.1 illustrates a hybrid sensor device with bump contacts. The essential stress on the bumps is their deformation during application temperature changes due to different thermal expansion of the chips. Stresses arise which are to be compared to the ultimate stress of the bump material, adhesion of the bump socket and permissible load of the components.

Pure Indium is most suited for cooled applications especially for joints with stress sensitive sensors. When joining by reflow soldering low melting In-alloys like InBi, InSn or InBiSn are to be used at temperature sensitive detector materials. Because of their lower ductility larger bump heights are required /4/. Moreover, Sn-containing alloys are critical with respect to formation of intermetallics.

Contrary requirements on bump geometry concerning their technological realization arise: large solder socket areas for high adhesion, great heights for good plasticity and reliability. On principle the sensor pitch has to be observed and moreover heat conduction and electrical contact are to be considered.

In the case of CdHgTe arrays with 128x128 elements and 50 µm pitch assembled to Si by In-solders bump heigths of 10...20 µm according to alloy in joined state (i.e. about 20...30 µm original height) at roughly 25 µm base dimensions obey simple model considerations /5/.

Because of more complicated handling of sensors the exclusive waferscale bump fabrication on the Si-chips is fixed and in this way only simple vacuum and electrochemical methods are further considered. In both cases the solder is deposited through photoresist masks. Patterning of vacuum deposited metal films is done by lift-off process.

Fig. 2 shows bump geometries after deposition and after wet back. In general at small vias the resist thickness has to be greater than the bump height requested by the lift-off process. By cracking at hidden edges this relation may be improved.

However, it remains the conic growth of the metal in the via yielding again lower bump heights after levelling. Therefore resist layers capable of lifting-off and having at least 20 /um thickness have to be patterned with 25 /um wide vias to resolve the bump geometries required. In the case of electroplating much more solder volume can be concentrated on a given post area by the mushroom-like overgrowth. Indeed, also by this method thicker resists are of advantage especially at small pitches, but useful results are obtained already with thin resists. Owing to more simple patterning of the contact pitches desired electrochemical solder bump fabrication was preferred.

Solder bump fabrication by electroplating

The electrochemical deposition of In and In-alloys into vias with centre to centre spacings down to 40 /um was investigated /6/. Indium is usually plated from fluoroborate electrolyte. Due to intensive generation of Hydrogen it is necessary to improve adhesion behaviour of the resist. To fabricate bumps from solder alloys the sequential plating of the components was preferred. Deposition of InSn as sandwich from fluoroborate bath is very: Sn on In is very difficult, plating of In on Sn is not enough reproducible because of high polarization. For Bi-In sandwich deposition Bi is first plated from perchlorate bath containing DTPA as complexing agent and in the following In is deposited as mentioned. The low throwing power of the Bi-electrolyte leads to greater differences in the volume deposited compared to In-plating alone. The results of plating array structures with 40...80 /um pitches demonstrate this method useful for flip chip preparation (see fig. 3,4).

An essential problem arises from increasing deviations in bump height and if present in composition by decreasing via dimensions since the throwing power of both electrolytes has not yet been high enough (see fig. 5). In addition to the basic variation a distinct elevation of bumps at the chip edges occurs according to the special array read-out chip topology of mosaic contacts and free margin.

After electroplating and resist removal the under bump metallization (Al-NiCr-Cu) is etched selectively.

The mushroom-like grown solder is wet back to contactable bumps by reflowing. While this step can possibly omit by vacuum deposition it is indispensable for electroplating because of annoying reaction products of solder and etch which have to be reduced or burst away. For sandwich structures a homogeneous alloying is obtained only in this way. Owing to this the reflow fluid contains salicylic acid for activation.

Joining

The Si-chip with plated bumps and the sensor chips with plane metallizations (phantoms with contact pads for probing) are joined by flip chip technology /7/.

Highest demands are made on positioning to receive mechanically steady stable contacts. Because of the small pitches position tolerances in joined state below 2 μm have to be observed. Consequently, procedures without any active position recognition (adjusting frame or stop) are not applicable. At conventional methods with semitransparent mirrors or beam-splitter type prisms problems arise which are difficult to govern in their calibration. The use of ir-microscopes restricts the joining sequence. Therefore a direct visual adjusting with side view was choosen. The chips can be positioned by looking to the corner bumps from x- and y-direction with about 2 μm lateral tolerance, the distortion being below 3', the unbalance below 5'.

It was aimed to avoid fluxes since their entire removal out of the hybrid device cannot be guaranteed. In- and Ag-metallizations were used for contact layers on sensor phantoms. The main advantage of pressure welding is the applicability of pure In for temperature sensitive detector materials, too; the higher requirements on positioning and larger mechanical bond stresses are disadvantageous. Contrary to this in soldering the claimings for equipment are lower due to the self-alignment capability and so does the mechanical load.

The strain rate at pressure welding does not significantly influence the mechanical strength and reliability of the contacts. Temperature and pressure may be exchanged against each other in an certain range. The joining time has to be long enough to permit contact forming diffusion (60 s at 300 K).

In reflow soldering high rates of heating up cause breaking of

oxide surface layers on the bumps and so wetting of the metallizations even without a flux. In this way mechanically stable contacts arise with Ag-metallizations already in normal atmosphere, with In-layers only in vacuum (fig. 6).
The soldered devices reached a distinct higher contact yield (99% with In-Ag) after joining in normal atmosphere than the welded one (95% with In-In). The reason for this should be the differences in bump height within the array.
The resistance of the contact was about 3 for both techniques.

Summary

Structured electroplating may be usefully applied to the solder bump fabrication for flip chip interconnection with very small pitches. Vacuum deposition with lift-off patterning results in essentially higher requirements on masking layers for fine pitched bumps.
Bumps from pure In with heights according to joints and 80 μm pitches were fabricated. Smaller bump spacings (~ 40 μm) are also possible, but regarding to their joining reliability problems with the height tolerances become more and more critical. In-Bi-alloys may be plated sandwich-like. After electroplating the formation of bumps in a reducing atmosphere is necessary.
Pressure welding as well as reflow soldering are applicable to join with In-bumps in which soldering revealed higher contact yield. Cold pressure welding puts higher requirements on positioning and bump height tolerances, but on principle fluxless working is given which has to be forced by the procedure in soldering and does remain critical.

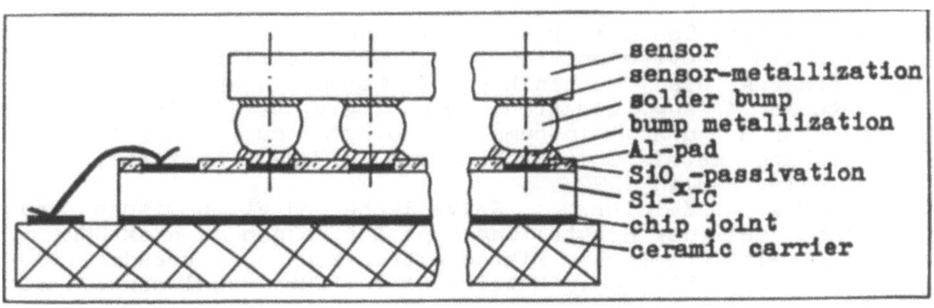

Fig.1: Hybrid sensing device assembled by flip chip

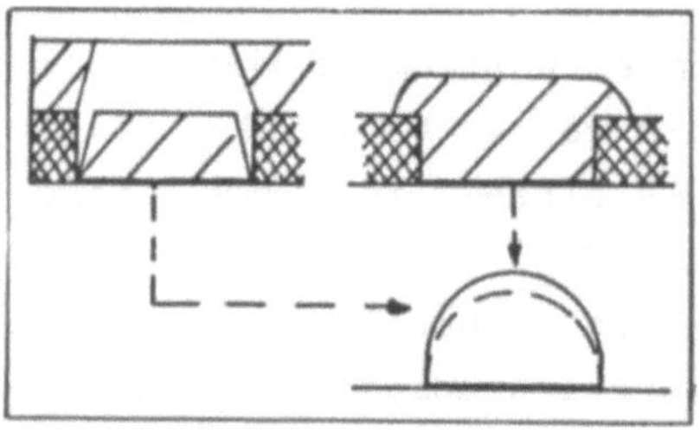

Fig.2: Bump geometry at vacuum and at electrochemical plating

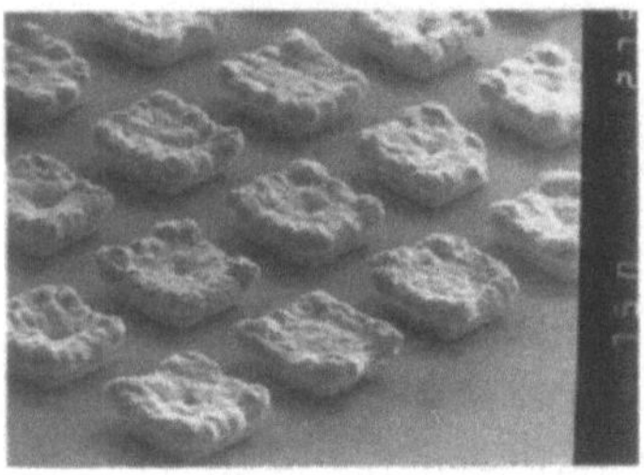

Fig.3: Solder bumps after plating (Bi+In, 80 µm pitch)

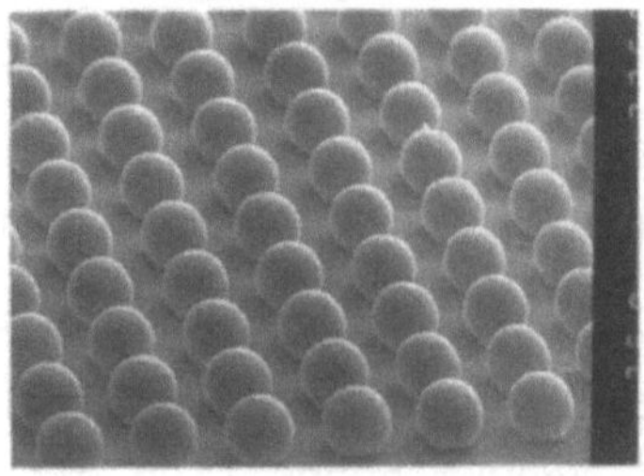

Fig.4: Solder bumps after wet back (InBi, 40 µm pitch)

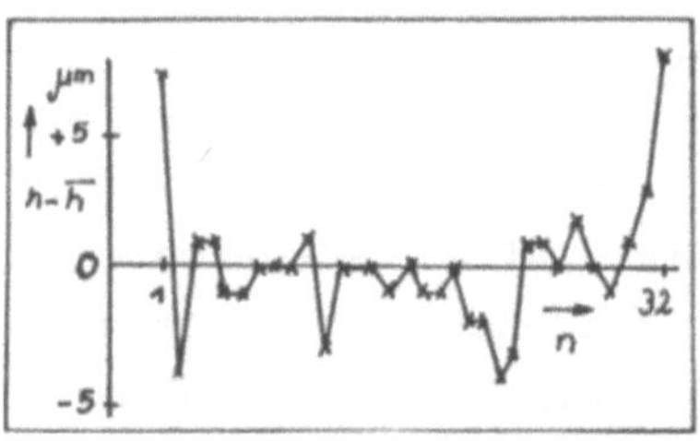

Fig.5: Deviation in bump heights within an array (In, 80 µm pitch)

Fig.6: Joined bumps (Si-chip revealed, In, soldered, 80 µm pitch)

References

/1/ Rode, J.P.: Hybrid HgCdTe Arrays, SPIE 443(1983), 120

/2/ Zanio, K.R., Bean, R.C.: HgCdTe photovoltaic detectors on Si substrates, SPIE 930(1988), 44

/3/ Watton, R.: Technology and performance for linear and two dimensional pyroelectric arrays, SPIE 510(1984), 139

/4/ Plötner, M. u.a.: Mechanik von In-Loten bis zu tiefen Temperaturen, 4.Tagg. Elektroniktechnologie, Berlin 1990

/5/ Plötner, M. u.a.: In-bump Technologie zur Hybridisierung gekühlter matrixförmiger Sensoren, 4.Tagg. Elektroniktechnologie, Berlin 1990

/6/ Sadowski, G., Zeidler, D.: Galvanische und ätzchemische Probleme bei der Herstellung von lötfähigen Kontakthügeln, Tagg. Wiss. Fortschr. Mikroelektronikmontage, Dresden 1990

/7/ Rzepka, S.: Flip-Chip-Matrixmontage für Bauelemente mit höchsten Anschlußzahlen, Wiss. Zt. TU Dresden, 1990

Development of a Fine Pitch Bumping Process

G. Engelmann, O. Ehrmann, J. Simon and H. Reichl

Technologien der Mikroperipherik
Technische Universität Berlin

Summary

To meet the future requirements of connecting high lead count, high speed integrated circuits the TAB Technology has a large potential. A high pad number on the chip requires reduced tolerances in the bump height and a reduction in pitch size, which can only be achieved by straight wall bumps /1,2/.
A process for the fabrication of such straight wall bumps is described. A thin film metallization system, photolithography of thick resist layers and electrochemical deposition of Au-bumps have been developed and adapted to fine pitch design.

Metallization System

In the first step of the bumping process conductive thin film layers are uniformly deposited over the wafer surface. In our process a barrier layer of 250 nm TiW is followed by a plating base of 200 nm Au. A diffusion barrier is necessary to prevent the formation of brittle intermetallic compounds. Small pad sizes require a good adhesion and low contact resistance of the metallization system, as well as a good thermal stability of the barrier.

The Leybold Z 660 rotating platter sputtering system, in which the film layers have been deposited has an input/output load lock and reaches a base pressure of about $1x10^{-7}$ mbar. Three targets and one heater in the vacuum chamber allow the deposition of the two layers without breaking the vacuum.

After removing the native oxide from the aluminum pads by sputter etching the TiW barrier layer is d.c. sputtered from a 200 mm magnetron cathode (10 weight % Ti). As TiW is a high stress material, the stress has to be minimized to optimize the properties of the barrier layer. Bow and warp of the coated wafers have been measured with a contactless geometry gauge. Measuring before and after the film deposition the warp induced by the film stress has been determined in dependence of the sputtering gas pressure (fig.1). The sample sputtered at the intermediate gas pressure shows no sputter induced warp.

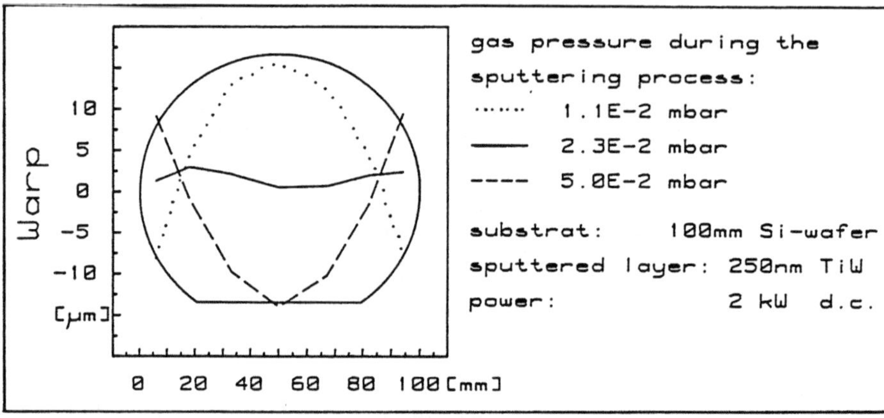

Fig.1. Film stress induced warp of wafers coated with Ti:W

Fig.2. Ti:W barrier layer of bumped pads after heat treatment at 400°C for 60 minutes, the gold bump is removed by etching
a (left) : Ti:W sputtered in 95% Ar with 5% N_2
b (right): Ti:W sutterd in pure Argon

Past studies /3,4/ have shown, that the barrier properties of TiW can be increased by sputtering in a mixture of argon and nitrogen. To analyse the barrier properties, bumped samples have been tempered at 400°C for 60 minutes in a nitrogen ambient. After the heat treatment the gold bumps were removed by wet etching. Fig. 2a shows a pad with a stable TiW layer, which was sputtered in 95% Ar with 5% N_2. A barrier layer deposited in pure argon is altered by the heat treatment and thus the etchant destroyes the whole pad (fig. 2b) .

The contact resistance of the 100 x 100 μm^2 bumps was measured less than 1mΩ/bump. This value is by far low enough to allow substantial reduction of the pad surface.

<u>The Photolithographic Structuring of the Bumps</u>

In order to take advantage of the bond pad downsizing for an equivalent reduction of the bump pitch dimensions, the horizontal growth of the gold must be limited within the entire bump height, forming in this way straight wall bumps. This is achieved using resist layers which are at least as thick as the bumps must be high (about 25 μm). For the same reason the edges of the resist structures have to be approximately perpendicular. Such structures can be realized using positive photoresist of high transparency and high viscosity. The use of this photoresist is an alternative to photosensitive polyimid /5/. In the standard process layers up to about a thickness of 40 μm are realized by double spin coating. To obtain steep structure edges a very good contact between the mask and the photoresist surface is essential. The edge bead removal and the vacuum printing are promoting the good contact.

For preserving the structures realized in the printing and development during the postbake the baking processes have to be carefully tuned up. As the prebake determines the photosensitivity of the resist its temperature should not exceed 130°C. To exclude the flowing and shrinking of the structures the postbake temperature should at least be 15°C below that of the prebake. Successful optimizing of all parameters makes a dramatic reduction of the bases of the resist openings possible. In figure 3 the linear size of the quadratic structures is 100 μm, 50 μm, 37,5 μm and 10 μm respectively. This scaling

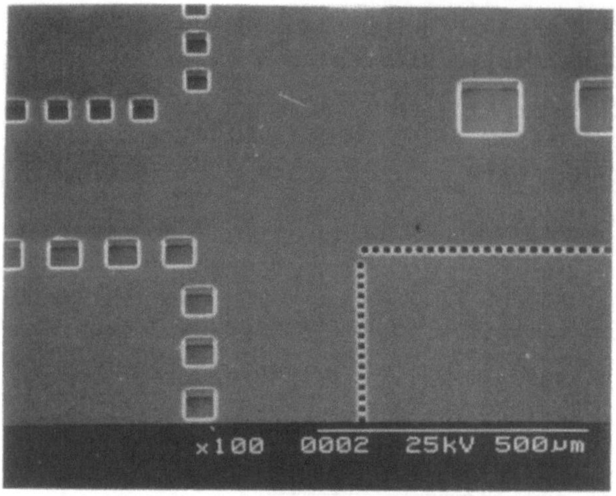

Fig.3. Detail of a Wafer with 100 x 100 μm^2, 50 x 50 μm^2, 37.5 x 37,5 μm^2 and 10 x 10 μm^2 bump openings in an approx. 30 μm thick photoresist

Fig.4. Profile of a photoresist structure with steep edges: height of the resist layer: 30 μm, pitch: 20 μm

down corresponds to an opening's pitch of 20 μm, a ten times larger I/O density or a reduction of the bump area to one hundredth of the initial size. The resist profile of such a structure is shown in fig. 4 (edge slope: 87° ±1°; aspect ratio: 3). Latest investigations show that there is even more potential in the used resist system: either the resolution of structures in the 6 μm range is attainable (for a layer thickness of 30 μm) or resist layers of at least 60 μm can be structured.

<u>Bump Plating</u>

Gold bumps were formed by electrochemical deposition from a non-cyanide plating solution. They should have high ductility and low hardness, so that a damage of the bond by cracking of the passivation or Si-cratering by plastic deformation of the bumps is avoided during innerlead bonding. The bumps have very smooth side walls and the gold has a very fine grain structure (fig. 5). As the resist and the electrolyte are compatible underplating is not observed and the resist structure is perfectly reproduced in every respect. In this way microbumps with a size of 10 x 10 μm^2 and a height of 25 μm could be produced (fig. 6). The hardness of 140 mHV25 is to high for bonding without tempering. By tempering at 400°C for 60 min the hardness decreases to 60 mHV 25.

Fig.5. Straight wall bump (25 x 100 x 100 μm^3) from a sulfite plating solution reproduces exactly the resist opening.

Uniformity of bump height over the whole wafer was achieved by appropriate adjustment of electrolyte flow and current density. In case of the used plating solution the slope of the bumps is very sensitive to the current distribution. A uniformity of the bump height of ±1 μm can be achieved within a diameter of 9 cm on a 4"-wafer for an average bump height of 25 μm.

Fig.6. A row of microbumps 10 x 10 μm^2, height: 25 μm
pitch : 20 μm

Conclusion

It has been demonstrated, that microbumps with a size of 10 x 10 μm^2 and a pitch of 20 μm can be structured, and therefore bumping is not the limiting process for the downscaling of TAB - interconnections.

References

1. Liu, T.S.; Rodriges de Miranda, W.R.; Zipperlin, P.R.: A Rewiew of Wafer Bumping for Tape Automated Bonding. Solid State Tech. (Mar. 1980) 71 .

2. Courant, P.; Dehaine, G.; Kurzweil, K.: High Density High Lead Count TAB. ISHM Procidings (1988) 117 .

3: Nowicki, R.S.; Harris, J.M.; Nicolet, M.-A.; Mitchell, I.V.: Studies of the Ti-W/Au Metallization on Alumimum. Thin Solid Films Vol.53 (1978) 195 .

4. Baker, J.E.; Blattner, R.J.; Nadel, S.; Evans, C.A.; Nowicki, R.S.: Thermal Annealing Study of Au/Ti-W Metallization on Silicon. Thin Solid Films Vol.69 (1980) 53.

5. Milosevic, I.; Perret, A.; Losert, E.; Schlenkrich, P.: Polyimid Enables High Lead Count TAB. Semiconductor International, Vol.10 (1988) 122 .

Tape Automated Bonding for High-Density Electronic Packaging

Gerard J. Plite

Martin Marietta Corporation
P.O. Box 555837, MP 1704
Orlando, Florida 32855-5837 U.S.A.

Summary

This paper summarizes Martin Marietta's experience with tape automated bonding (TAB) for high-density, high-reliability industrial and military electronic applications. The subjects addressed include tradeoff studies for various design concepts and materials; process development alternatives; and process characterization studies. Also included are results from finite-element modeling studies of TAB interconnections, along with environmental tests results. Finite-element stress modeling was done to evaluate the effect of temperature cycling on miniature TAB beam interconnections. Several beam configurations were studied, and environmental testing was done to evaluate the model predictions.

Introduction

Advances in electronic packaging technology, particularly multichip packaging, are increasing the density both of chips and their input-output (I/O) interconnections, taxing traditional packaging techniques such as wirebonding beyond their limits. Anticipating this problem, Martin Marietta began in 1985 an internally funded research and development program devoted to an advanced packaging technology called tape automated bonding (TAB). This process, in which chips are suspended on flexible tape on which is deposited conductive lines forming a chip-to-substrate interconnect, has a number of advantages over wire bonding. The primary advantage is high yield, which is essential to multichip packaging. Using probe points on the tape, a large number of chips can be fully tested, characterized, and burned in before bonding to a substrate. Other advantages include 1) significantly stronger bonds compared to wirebonding and 2) improved electrical performance due to lower inductance.

Martin Marietta has made considerable progress in developing TAB. Figure 1 shows a multichip package we have developed that contains 12 TAB-bonded processor chips. Each chip contains 122 bond pads, and there are nearly 3000 thermocompression bonds in the entire assembly.

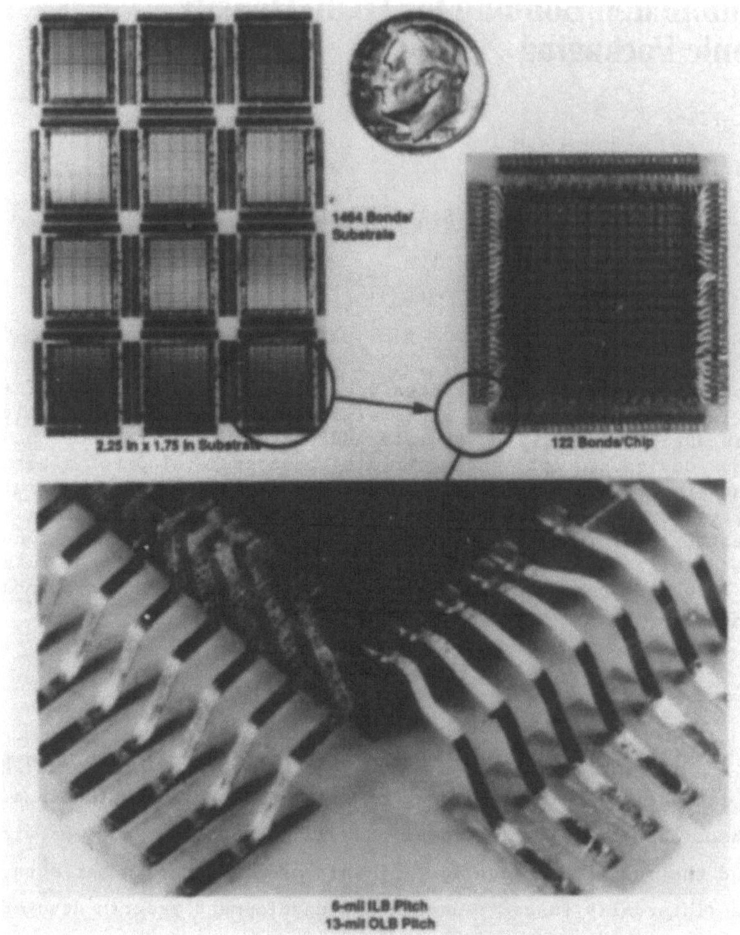

Fig. 1. Gullwing TAB Multichip Module

Martin Marietta considers multichip packaging to be a near-term solution for programs that require high density and improved performance. In the long term, we believe that chip-on-board (COB) packaging will be the next logical step in the evolution of electronic packaging for the 1990s. COB results in considerable size and weight savings and improved performance for high-speed processors. TAB will be a key to the success of both of these packaging technologies.

Trade Studies

The following section summarizes some of the results of TAB trade studies that Martin Marietta has conducted. Figure 2 illustrates three configura-

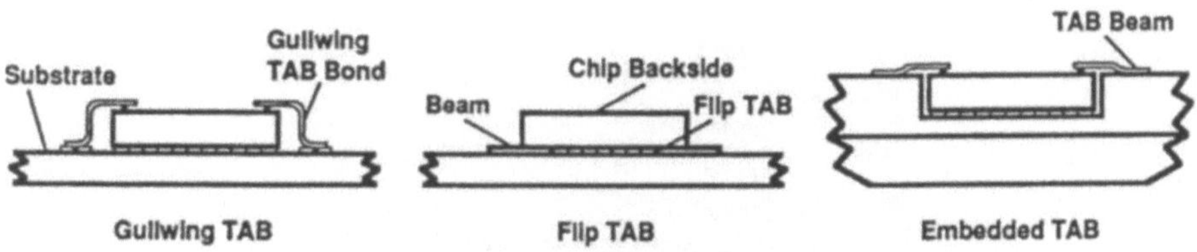

Fig. 2. Baseline TAB Interconnect Concepts

tions which have been built and tested on related IR&D studies. The first configuration, which we consider to be the most practical solution, is the gullwing lead form interconnect. This method provides a flexible interconnect link that can absorb stresses due to differences in the thermal coefficients of expansion of the lead and substrate materials. Test samples have been subjected to over 1000 thermal cycles (-65°C to +150°C) with no failures.

The second approach shown in Figure 2, flip TAB, has some advantages that make it easier to produce. By eliminating lead forming, step devices are easier to place with less lead damage. This approach also requires less space. The primary disadvantage is that it is difficult to make contact to the chip's back side. Many CMOS devices require either a ground or voltage connection to the back side of the chip.

The third approach that we have studied is called embedded TAB. This approach involves recessing the chip into a pocket in the substrate, eliminating many of the drawbacks of flip TAB. However, this approach significantly increases the complexity and cost of the substrate, as well as limiting the circuit routing area.

As part of our studies, we have also investigated the effect of both supported and unsupported three-layer tape to improve the assembly process. Figure 3 illustrates a design concept that incorporates a thin 0.020-inch-wide Kapton support ring to stabilize the beam locations and improve alignment during inner-lead and outer-lead bonding. This approach has proved to be much easier to produce than unsupported tape. Martin Marietta has also developed considerable expertise in lead forming and trimming, a critical step in the manufacturing process. A precision die set for trimming and forming high-lead-count processor chips was found to work extremely well.

Fig. 3. TAB with Support Ring

Based on the experience gained from these trade studies, we have selected the miniature gullwing TAB interconnect approach for use on future programs.

Materials and Process Considerations. Gold-to-gold thermocompression bonding has been selected as our as primary approach; however, other materials are also being investigated to improve the capability for rework and repair. The interconnect beams are gold-plated and thermocompression-bonded to gold-bumped chip pads. The chip-bumping process consists of thin-film deposition of titanium-tungsten-gold to standard aluminum chip-bonding pads, followed by electroplating 1-mil-tall gold bumps on the silicon wafer.

Environmental Stress Considerations. Stress due to differences in thermal expansion between the copper lead frame of most TAB interconnects and the mounting substrate must be considered in the detail designed for any new chip. We have performed finite-element stress modeling of several beam shapes to determine the ability of the TAB leads and bonds to survive thermal cycling. Figure 4 illustrates examples of three TAB configuration models we analyzed. The first model consisted of a straight beam interconnection of the chip to the package. The second model involved a beam that included a fanout angle. The first and second models were similar in that the inner and outer lead bonds were on the same level. The third model incorporated a formed lead configuration to bond a chip to a substrate. Each model was generated using a NASTRAN finite element program. Thermal mechanical stress for this analysis was considered to be introduced from temperature cycling between -55°C and +20°C for one set of cases, and from cycling between +20°C and +125°C for a second set. Table 1 summarizes the

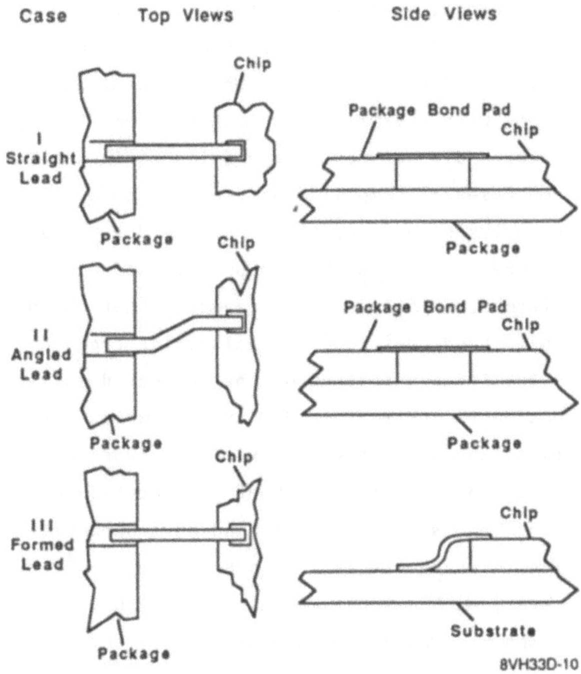

Fig. 4. TAB Stress Models

Table 1. Stress Levels for TAB Interconnections

Model Case	Loading Condition	
	+25 to -55°C	+25 to +125°C
	Maximum Resulting Stress (psi)	
Straight Leads	17,600	22,000
Angled Leads	28,200	35,200
Formed Leads	2,120	2,650

8VH33D-IA

stress levels for all three models of both loading conditions. Stress levels for the straight and angled models were relatively high (17,000 to 35,000 psi) in comparison to the formed lead configuration (2650 psi). The results in Table 1 suggest that stress levels for unformed TAB leads are approaching the tensile-strength limits of the lead materials (20,000 to 30,000 psi). Also, the formed lead configuration appears to be far superior because the overall stress levels are much less.

To verify these conclusions, we performed thermal shock testing using a 122-terminal I/O TAB interconnect test chip that contained a continuity chain to monitor failures during testing. Thermal cycling from -55°C to +125°C was done, along with thermal shock testing from -65°C to +150°C. Results from these tests did not reveal any failures in the TAB beams or bonds; however, some cases of bump lifting from the chip were found. Subsequent failure analysis revealed that this was due to wafer processing problems. Thermal cycle testing of both supported and unsupported gullwing interconnect lead form configurations has revealed that the unsupported configuration can withstand over 1000 thermal cycles from -65°C to +125°C with no failures. Preliminary testing of the supported gullwing configurations has revealed some failures after 300 cycles, which may have been caused by the thermal expansion of the polyimide support ring. Studies are continuing to investigate the reliability of this configuration. Preliminary results from both the stress analysis and environmental testing have suggest that the TAB interconnect is reliable for the configurations we have studied. However, each new design must be carefully reviewed to ensure that the design configuration does not impose any high stress levels on the bonds or wafer bumps.

Summary

TAB technology is critical to the success of future high-density packaging applications, both hermetically sealed multichip packages and future chip-on-board. Based on our trade studies of various design configurations and materials, we have selected gold-plated tape and gold wafer bumping as our baseline approach. Alternatives such as tin-plated tape and selective plating are being investigated in ongoing studies to determine if these materials are easier to manufacture and repair. Results from lead stress evaluations and thermal cycle testing have shown the TAB gullwing interconnect to be the most reliable configuration.

Acknowledgments

I would like to acknowledge the efforts of the following Martin Marietta employees who contributed material for this paper: Steve Malone and Frank Staudinger, for their work in the development and testing of the TAB samples for this project; and Ted Anderson, for his stress analysis.

Adhesiveless Copper-Polyimide Laminates for TAB's

Viktor Fronz Renker GmbH & Co. KG, Freiburg, West-Germany

Summary

For the TAB-technology usually laminates with adhesives (3-layer laminates) are used. The drawbacks caused by the adhesives are well known. Adhesiveless copper-polyimide laminates (2-layer laminates) could prevent these disadvantages.

Using the sputter technology we are able to produce 2-layer laminates. With a special pretreatment it is also possible to reach a good adhesion of the copper film on the polyimide film.

The peel strength will be discussed for 2-layer laminates. The advantages and disadvantages are pointed out for such laminates, compared with 3-layer laminates. The different physical conditions comparing the peel strength of laminates with and without adhesives has to be taken in account.

Introduction

At the present time 3-layer flexible laminates are mostly used to produce flexible PCB's and also TAB tapes. 3-layer flexible laminates are made from a plastic film (mostly polyimide) and a copper foil with an adhesive between. The adhesive layer (about 25 µm) is necessary to perform a proper adhesion between the copper layer and the polyimide layer. (Fig. 1)

For further processes as TAB tapes or using it as PCB the adhesion layer can be a disadvantage because of certain poor properties which the adhesive has e.g. failure during heat cycling or by processing 2-metall-TAB tapes. (2C TAB tapes)

To make fine pattern a thick copper film from about 35 µm can be another handicap because of the underetching during the production process of TAB tapes.

This disadvantages could be avoided or at least minimized by using
1) an adhesiveless flexible laminate and
2) a thin copper film to minimize the underetching and therefore be able to produce finer pattern lines.

Certainly a specially adapted adhesion between the copper and the polyimide has to be assured. Besides that it is probably necessary to use a semi-additive technology to produce patterns with a certain copper thickness.

We developed a 2-layer flexible laminate which mostly meets the IPC-specification. It has all the advantages I mentioned before like no adhesive and a copper layer of about 1 µm. (Fig. 2)

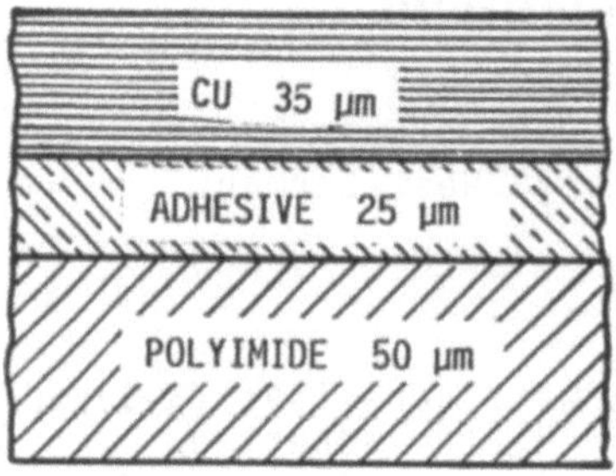

Fig. 1: 3-layer laminate

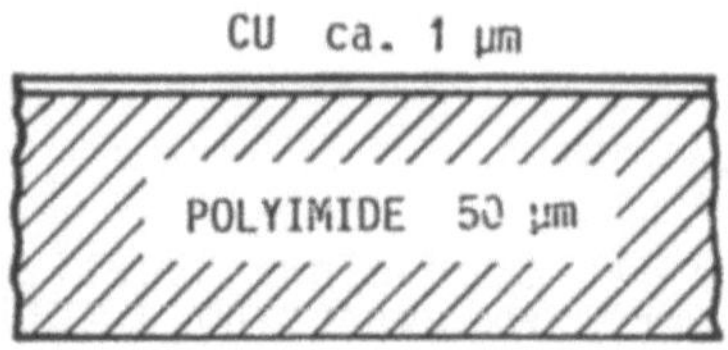

Fig. 2: 2-layer laminate

Production methods for 2-layer flexible laminates

It is well known that there are existing different ways to produce 2-layer flexible laminates. Generally all the methods can be used for metallizing plastics.

That methods are:

1. Cast-on method
   A thin copper foil is coated with a kind of polyamic acid (PMDA/ODA) and then imidized.

2. Wet chemical method
   Plating of polyimide in a wet solution. More steps are necessary for the plating process.

3. Dry chemical method
   That are all methods where the activation process is done by a dry method; e.g. screen printing of an acitvation layer and following reaction by heat. Or coating the film with a solution and activation with UV light.

4. Ion beam method
   Pretreatment or even coating by ion beam.

5. Physical method
   Using PVD or CVD methods.

Most methods are using a kind of adhesion or nucleation layer to establish a proper and sufficient adhesion. The adhesion layer is mostly chromium, palladium or even tungsten. Only the cast-on method is working so far without an adhesion layer.

We at Renker used a PVD method with a plasma pretreatment of the polyimide film. Besides, that we don't use an adhesion or nucleation layer like are mostly used to adhere the copper to the polyimide. After the plasma pretreatment we sputter copper on the polyimide. The sputtering technology has the advantage that the

atoms are bombarding the polyimide film and therefore the adhesion from a sputtered metal is usually better than for example a thermal evaporized film.
Taking everything into account we are able with the sputter technology and a prior pretreatment of the polyimide film to attain without an adhesion or nucleation layer a suitable adhesion of the copper onto the polyimide film. We are able to meet the IPC specification concerning adhesion.

The basic mechanism of sputtering

Sputtering is a coating process which involves the transport of almost any material from a source, called a target, to a substrate of almost any other material. The ejection of the source material is accomplished by the bombardment of the surface of the target with gas ions accelerated by a high voltage. Particles of atomic dimension from the target are ejected as a result of momentum transfer between incident ions and the target. The target-ejected particles traverse the vacuum chamber and are subsequently deposited on a substrate as a thin film.

When the sputtering target (referred to as the cathode) is negatively charged, electrons are ejected into the gas (generally argon) in the vicinity of the target. Under the influence of an applied voltage the electron accelerates toward the positive anode gaining energy from the electric field. As the electron travels toward the positive electrode, it may collide with a gas molecule, giving up part of its energy, leaving behind an ion and an extra free electron. The cumulative effect of this phenomenon is a self-sustaining glow discharge. The area where the ionization is taking place actually heats up and glows.

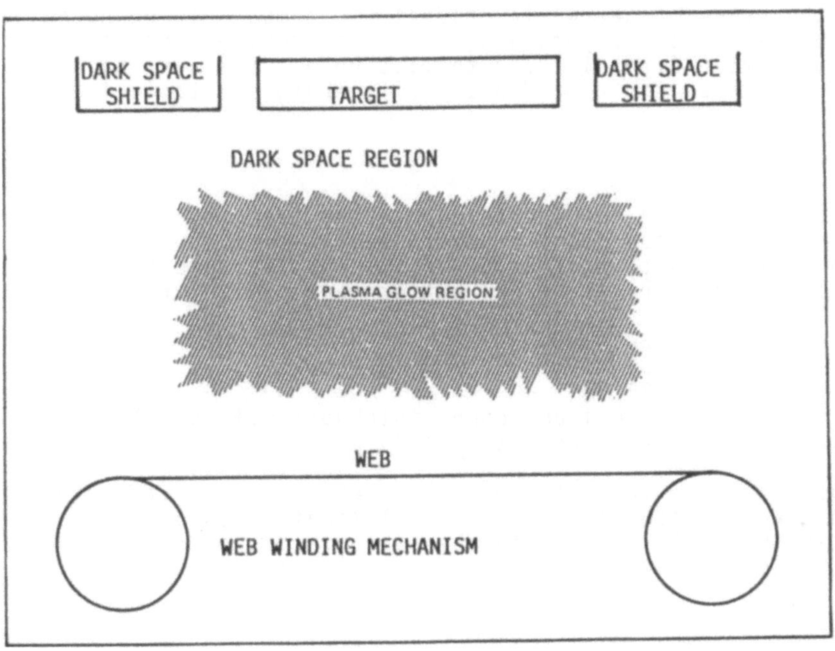

Fig. 3: Sputter mechanism

The area close to the target will not have this glow, and is thus referred to as the dark space. The further the electrons travel, the more chance they have to strike and ionize the argon molecules. Therefore, little ionization takes place close to the target's "front door". To further capitalize on this phenomenon, a ring (called a dark space shield) is placed around the target to make sure that the absolute minimum of target material is wasted by being sent out to the sides and back, but is directed instead down toward the substrate. (Fig. 3)

Comparison of 3-layer and 2-layer laminates

For sure there are advantages and disadvantages of both laminates. Important is how the laminates will get used. To produce TAB tapes using 3-layer laminates it is possible to punch the complete pattern of the polyimide film and adhere the copper foil to the punched polyimide film.

2C 3-layer TAB tapes are almost not possible to produce. With 2-layer laminates it is possible to make 2C TAB tapes. But the pattern in the polyimide film has to be etched. That means etchable polyimide film has to be used.

Furthermore with the 1 µm thick copper layer it is possible to produce very fine pattern lines. Pitches of 100 µm are easily possible. But to do so the semi-additive-technology has to be used. That means plating to a certain copper thickness is necessary. Using thick copper layers (ca. 35 µm = 1 oz.) fine pattern are not easy to produce with the subtractive technology. Underetching will occur. The yield for the production of fine pattern lines using thick copper layers can drop drastically. Besides that, the general production costs because of the tools can increase the price of the whole product. That must not be necessary using thinner copper layers and using the etching technology of the polyimide film. I tried to list up all the advantages and disadvantages for 2- and 3-layer laminates in table 1.

Adhesion

One important factor for laminates for TAB tapes is for sure the adhesion of the copper to the polyimide. To test the adhesion we use the peel strength test according to the IPC specification. 3 mm wide stripes are etched as pattern and peeled of at a angle of 90°.

peel length:	50 mm
peel speed:	50 mm/min
copper thickness:	35 µm
	The 1 µm copper thickness will be plated before the peel test to 35 µm.

To compare 3-layer laminates with 2-layer laminates it is important to take the adhesive on 3-layer laminate into account. Fig. 4 and 5 should illustrate the characteristics of the peel strength test with 3-layer and 2-layer laminates.

The adhesion strength is actually the force which is needed to seperate two surfaces with a specific area. The peel strength according to the IPC specification is a technical figure and uses only the width of the stripes which is peeled off. It does not take any care about the length which is automatically formed during the peeling process.

By using an adhesive it is for sure that by peeling the copper stripe off the adhesive will form a kind of elastic bands. Therefore the area of the seperated surfaces will increase. (Fig. 4) Because of that physical characteristic the peel strength can with a elastic adhesive automatically be increased.

Not so with a 2-layer laminate. Important is here the thickness and ductility of the copper foil which will form the radius of the copper stripes. This radius is an important factor for the peel strength test. But immediately after the polyimide surface and the copper surface are seperated no force will exist. (Fig. 5)

That is also the reason that on 3-layer laminates the peel strength is usually higher than on 2-layer laminates. On 2-layer laminates occurs at a certain peel strength a cohesive failure in the polyimide film. If that cohesive failure is reached no further adhesion is available between the copper and the polyimide. Not so on 3-layer laminates. Here usually occurs an adhesive failure because of the physical properties I pointed out.

Table 2 shows a few adhesion strength values we are able to reach so far. At about 1.7 (sometimes 1.5) N/mm the above cohesive failure is usually present.
Sample 200-5 is one of the first samples done in a production sputter coater. Here we hope to be able to reach similar values like we did with the lab coater.

Two side sputtering is also possible to produce 2-C 2-layer laminates. Adhesion values at the moment are similar to the No. 200-5.

As table 2 shows the different sides of the polyimide film give different adhesion strength. This is well known and the reason is in the production of the polyimide film. We used for all the tests Kapton 200 H, by Du Pont U.S.A.; film thickness 50 µm.

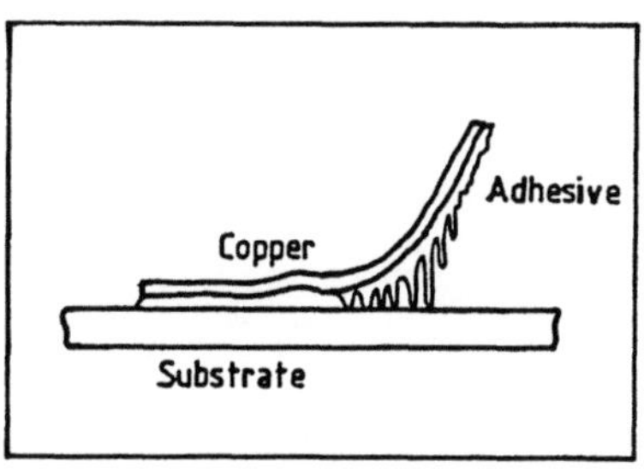

Fig. 4: 3-layer laminate

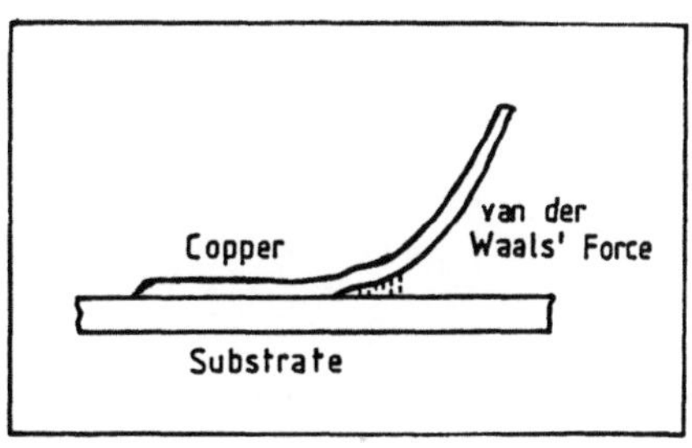

Fig. 5: 2-layer laminate

For using 2-layer laminates for TAB tapes the heat cycling and humidity cycling are also an important factor. No degradation of the adhesive is with 2-layer laminate possible. The adhesion strength drops but a certain adhesion strength still exists.

Conclusion

It has been shown how far advanced the development for 2-layer laminates is. It is possible to produce 2-layer laminates with copper on both sides without an adhesion layer and a good adhesion. The production facilities are available.

To compare 2-layer laminates with 3-layer laminates the physical characteristics has to be taken into account. The comparison is not simple to make.

To produce TAB tapes 2-layer laminates have certain advantages against 3-layer laminates. It is possible to establish fine pattern lines much easier and economical than with 3-layer laminates. The tool costs are lower because everything can be done by the photolitography process.

Certainly certain processes have to be changed but that is just a matter of time and in the long run for sure it will be easier and will be more cost effective.

Besides that, 2C 2-layer TAB tapes are much easier to produce. Maybe in the future when the technology will be completely established 2-layer flexible laminates will be so commen like today 3-layer flexible laminates in the PCB business.

Acknowledgement

The author wishes to thank to J. Fix for the technical assistance and is grateful to I. Fehrenbach for preparing the manuscript.

References

Polymer Interface and Adhesion ed. Souheng Wu, Marcel Dekker Inc. New York 1982.

Table 1:

Properties	**2-layer laminate**	**3-layer laminate**
Cu-thickness	1 µm	17 / 35 µm
adhesive	no	25 µm
PI-thickness	25 / 50 µm	25 / 50 µm
flexibility	higher, no adhesive	high
ductility	adjustable (plating)	fixed
outgasing	no	adhesive
burning	no	adhesive
fine pattern	prefered	possible
pitch	50	160
etching time	short	long
final Cu-thickness	adjustable	fixed
adhesive free	yes	no
(Cu & PI)	only PI	PI and adhesive
2ε-TAB tapes	possible	difficult
PI pattern	etching	punching
cost for test tapes	low	high (tool costs)
changes in tape pattern	easy	difficult (tools)
semi-additive technology	prefered	-
subtractive technology	-	prefered
adhesion	IPC reachable	IPC reachable

Table 2:

Adhesion strength values

Sample No.	Polyimide side	Adhesion in N/mm RT	200° C 1 h	260° C 30 sec	150° C 1000 h	85° C; 85% r.h. 1000 h	Boiling test
159 - 14	Air	1.70	1.34	1.47	0.62	0.62	
159 - 17	Air	1.86	1.76	1.71	0.63	0.48	0.52
199 - 12	Drum	1.25	1.37	1.26			
199 - 57	Drum	1.23	1.37	1.27			
201 - 10	Air	1.72	1.39	1.71			
201 - 11	Air	1.86	1.69	1.77			
200 - 5	Air	1.82	0.60	1.77			

Optical Assembly and Interconnection Technologies

Application of Micromachining and Micromechanics to Components for Optical Communication

Bernd Hillerich

Daimler-Benz AG, Research Center Ulm, Wilh.-Runge-Str.11, D-7900 Ulm, West-Germany

Summary

The state of the art of micromachining technologies and micromechanical structures used in components or subsystems for optical communication systems is reviewed. Numerous examples show that in optical communication both micromachining and micromechanics are applied with considerable success.

I.Introduction

As it is usually understood, micromachining comprises the application of methods and materials used in semiconductor technology for fabricating components and subsystems, e.g., structuring of wafers by photolithography, deposition, etching, sputtering, etc. For some time, indeed, components for optical communication have been made applying these technologies. Integrated Optics would not have reached its present state without adopting the vast knowledge of semiconductor (i.e. silicon) wafer processing. But in optics also specific methods are used: Microcleaving, localized melting by microflames, molding, etc.
Micromechanical structures are mainly used for aligning optical fibres and waveguides to each other or to other components. Here specific technologies are applied, too, e.g. precision embossing of alignments grooves.
In this paper the present state of micromachining and microstructuring in components for optical communication is discussed, omitting the technologies for optoelectronic emitters and detectors or integrated optics, which have been often reviewed elsewhere. Furthermore examples of combining several technologies in a single device are given.

II.Micromachining methods derived from semiconductor processing

1.Alignment grooves in silicon

Anisotropic etching in silicon [1] is a key technology of micromechanics. By this method very precise V-shaped grooves can be formed in <100> silicon to be used as alignment means. Chips with V-groove arrays were utilized to connect ribbons of up to 12 fibres [2]. The accuracy of fibre alignment depends on the

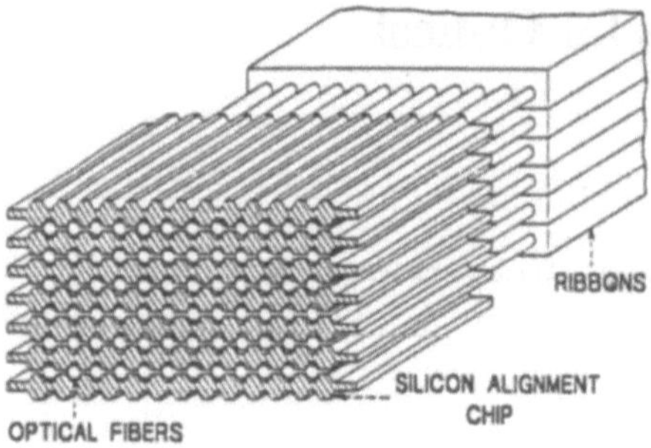

Fig.1: Stacked multifibre connector

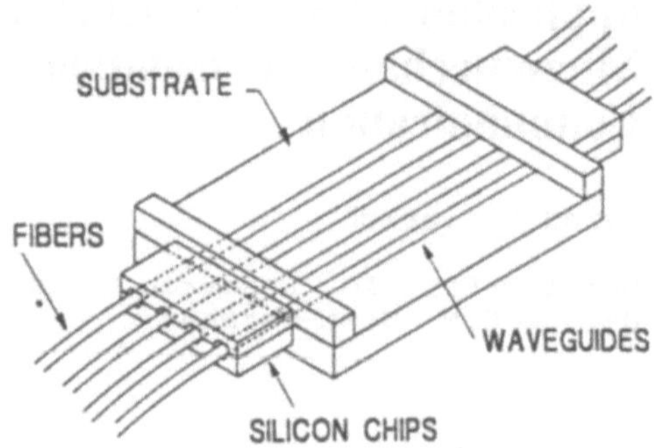

Fig.2: Butt-coupling of Si chip fibre array to planar waveguides

constancy of groove dimensions (0.2 to 1μm) and of fibre geometry (diameter, core concentricity). By etching grooves on both sides of Si-wafers, stacking of chips and fibres is possible. In this way multifibre connectors with 12x12 fibres (see Fig.1) have been realized, which, however, have to be factory-assembled . Though wafer thickness variations are a critical point in stacked arrays, low coupling losses of 0.3dB$\pm$0.2dB with multimode fibres (core diam.:50μm) were reported [3].
The method of positioning fibres in silicon-V-grooves is widely used for aligning fibre arrays to LEDs [4], to photodiode ar-rays [5], and for coupling fibres to integrated optical (IO) waveguide devices [7]. With the 'butt-coupling' technique as shown in Fig.2, the polished fibre-silicon array is (actively) aligned to the waveguides by monitoring the coupling loss and fixed by epoxy or UV-curing adhesive. However, this alignment (3 to 6 degrees of freedom !) is time-consuming and cumbersome. With a self-aligning 'flip-chip' approach [8] (Fig.3) the active alignment is reduced to one transversal adjustment. The upper silicon chip has an unetched part that overlaps with the top surface of the waveguide substrate, and a region with etched V-grooves to hold the fibres. By this chip design the fibre cores are positioned at the correct distance below the substrate surface to allow optimum coupling. For a 12-single-mode fibre array an average excess coupling loss (due to resi-dual misalignment) of 0.9dB was reported, whereas the inherent coupling loss (due to mode-mismatch etc.) amounted to 1.5dB.

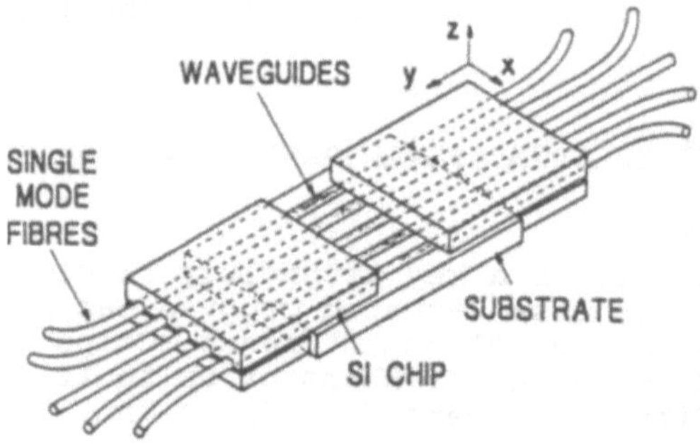

Fig.3: Self-alignment method for fibre-waveguide coupling [8]

Fig.4: Self-aligned flat fibre-photodiode coupling scheme

Another self-aligned arrangment for coupling fibres to photodiodes (or surface emitters) [9] is shown in Fig.4. After the photodiode has been positioned on the Si chip by visual inspection (using etched alignment mark), the fibre is pushed into the V-groove. Light emerging from the fibre (reflected upward by a gold-coated facet at the v-groove end) enters the photodiode without coupling loss.

2.Etched diffraction gratings

Blazed diffraction gratings can be made be etching wafers, the orientation of which deviates from <111> by the blaze angle (typically 4° to 15°) ([10], see Fig.5). For Au-coated gratings with groove periods between 2.5 and 5μm diffraction efficiencies of 80 to 90% are obtained, which competes very well with the best conventional gratings made by replication of ruled master gratings. But the silicon gratings are produced by a batch process, suitable for cheap mass production of small area gratings with high yield.
Utilizing these gratings wavelength multiplexers [11] and demultiplexers [10] with up to 10 channels - based on the principle of grating spectrometer - have been developed with insertion losses down to 1dB. By bending thin (0.1mm) grating chips, even concave (i.e.focussing) gratings were made for planar demultiplexers[12].

3.Other techniques

The feasibility to produce IO-devices on silicon substrates (e.g. by flame hydrolysis deposition of SiO_2 and reactive ion etching [13]) makes the excellent microstructuring properties of silicon available to integrated optics. An example is shown in Fig.6.

Fig.5: Etched Si-grating

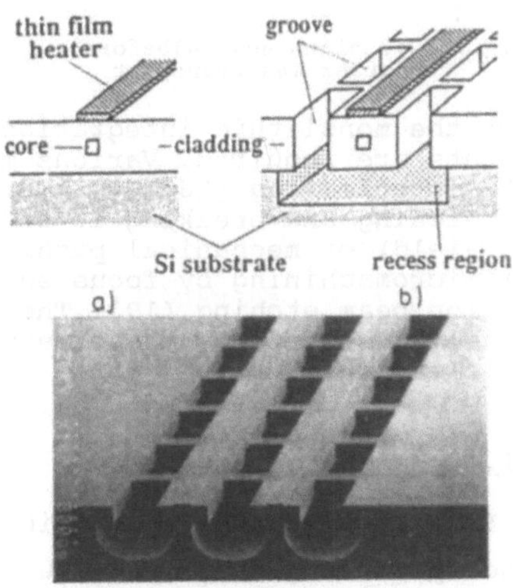

Fig.6: Structure of thermo-optic phase shifter. a)conventional, b)bridge-suspended [12], c)SEM-photo of b)

Here the Si substrate below a thermo-optic phase shifter waveguide has been hollowed out by dry isotropic etching. By this means the electrical driving power (for the thin film heater) is reduced by 1/10 due to the smaller thermal conductance. Diffractive lenslet arrays were fabricated by thin film evaporation and photolithographic techniques [14] (8-level SiO_2 film stack). Similar devices were made on GaP [15], but the steps of the lenses were smoothed out by mass transport (see Fig.7). The focussing of these lenses was nearly perfect, i.e., diffraction limited. In the same paper [15] the hybrid integration of GaAs/AlGaAs laser arrays on Si substrates was reported. To increase the otherwise thermally limited output power by a factor of 2, the Si substrate was equipped with water-cooled microchannel heatsinks made by anisotropic etching.

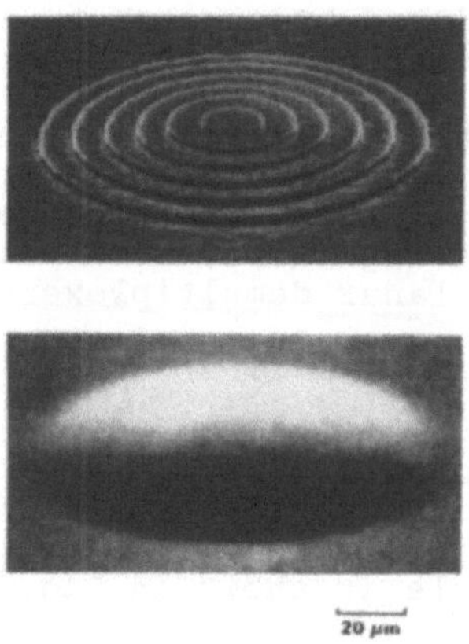

Fig.7: GaP microlens, a)before, b)after masstransport

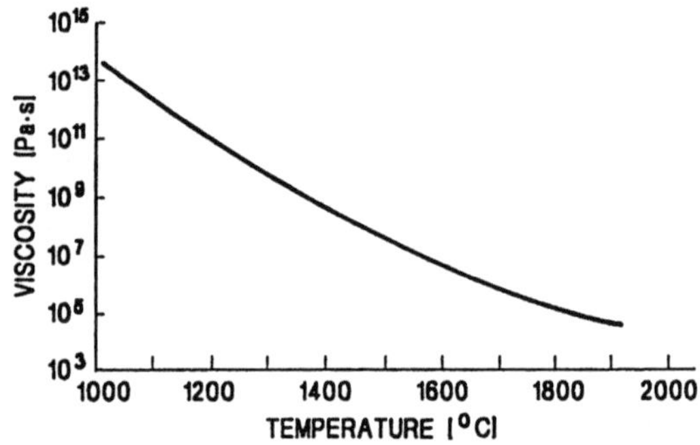

Fig.8: Viscosity of SiO_2 glass vs.temperature

For the monolithic integration of laser diodes integrated mirror facets are required. Various methods have been applied:

a) microcleaving [16,17], i.e. forming a cantilever by selective etching and breaking it off using ultrasonic excitation (bad yield) or mechanical pushing down (no batch process).
b) micromachining by focussed ion beam milling [18] or reactive ion beam etching [19]. Though the mirror surfaces are not allways perfect, these methods are more promising than microcleaving.

II. Specific methods

1.Localized melting and fusing

The specific glassy behaviour of fused silica - the base material of optical communication fibres - e.g. the smooth transition from the solid to the soft state over a large temperature region (see Fig.8), is successfully exploited for a number of applications:

a) The best method to splice (i.e.permanently connect) optical silica fibres, in terms of low coupling loss (< 0.1dB), low reflection (<-60dB), small size, ageing behaviour, and environmental stability, is by fusing the butted fibre ends. Using localized heating up to ~2000°C by a small electric arc (crossection ~1mm) the fibre ends are softened and pushed together [20]. This technique is the standard process for installing optical fibre cables and is also applied for multifibre ribbons.

b) Taper lenses, i.e. tapered fibre ends with a microlens on their tips [21], are widely used in optical transmitter units to couple the radiation of laser diodes into fibres. The taper is formed by heating and drawing the fibre in an electric arc. The taper tip is subsequently fused, which results in a nearly hemispherical lens (see Fig.9). Typical coupling efficiencies are between 25% and 50%, depending on the radiation characteristics of the laser [22].

c) Fused biconic taper couplers [23] are made by holding two or more single-mode fibres in contact with each other, heating and fusing them by a microflame, and stretching. The coupling ratio to the output fibres is governed by interference of the two lowest order modes in the fused region (see Fig.10). Besides directional couplers other functional devices as star couplers, power splitters [27], wavelength multiplexers [24], coupler ribbons [25], and polarization splitters [27] are fabricated. Inherent advantages of these devices are low insertion loss (0.1 to 1dB), low reflection, no fibre attachment problems, small size, excellent ageing and environmental stability.

These devices are not fabricated in a batch process, but within seconds (fused connections and taper lenses) or minutes (taper couplers) with high reproducibility and yield, and hence economic.

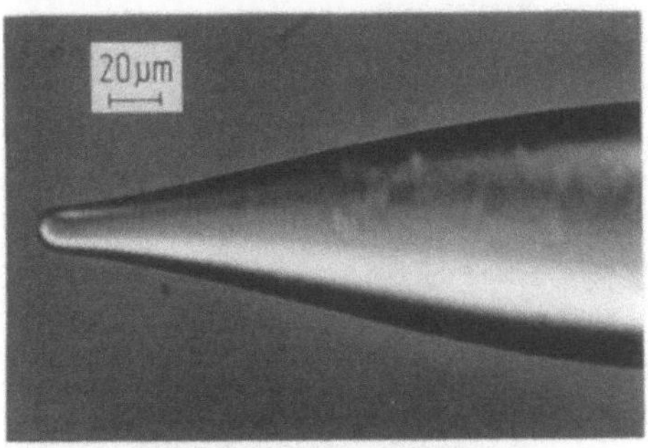

Fig.9: Microphoto of taper lens on a single-mode fibre

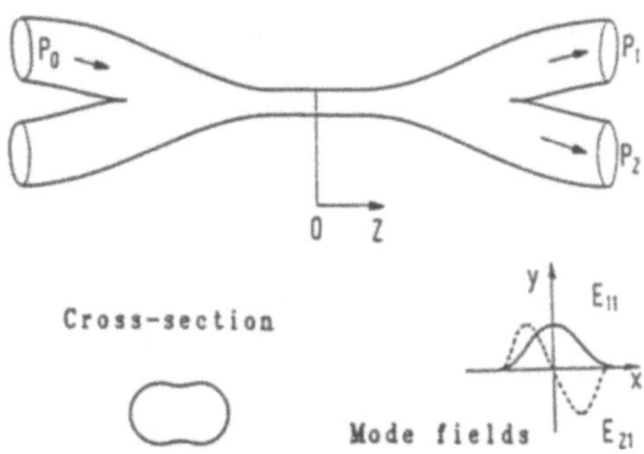

Fig.10: Principle of fused biconic taper coupler

2.Molding

By injection molding using polycarbonate, alignment V-grooves for fibre array connectors were fabricated [28]. Due to surprisingly small tolerances (better than 1µm), coupling losses for 12-multimode-fibre ribbons were as low as 0.2dB ±0.2dB. Similar good results and excellent environmental characteristics were reported for transfer-molded V-groove mass splices made of glass-filled phenolic resin [29] and for single-mode fibre splices [30].

Compared to silicon groove arrays, plastic devices are less brittle and allow more variation in shape. Some people prefer molded V-groove splices even to fused splices for reasons of handling and equipment cost. Injection molding was also used to make an active bidirectional coupler [31], incorporating beam splitter, fiber alignment groove, focussing mirror, and alignment ribs for light emitter and photodetector packaged in TO-housings (see Fig.11).

3. Further techniques to provide alignment

Alignment grooves were also embossed into metal (steel, copper etc.) by pressing an embossing die, composed of high-precision prismatic matrices made of hardened steel, into blank substrates. The reproducibility of groove dimensions is in the order of 1 to 6μm, depending on groove depth and material. Fig. 12a shows a circular substrate (diam.:5mm, INVAR-steel) with embossed alignment grooves for 3 fibres. It was used for aligning the fibres in a wavelength multiplexer (see Fig.12b) with excellent environmental performance [32].
A modified flip-chip solder bonding technique was used to assemble IO-devices with fibre arrays and electrical interfaces [33]. The self-alignment is due to surface tension of wet solder bumps as sketched in Fig.13. The accuracy of alignment is better than 2μm. Excellent environmental performance was reported. Probably this technique will have a strong impact on future work on the interfacing of IO-devices.

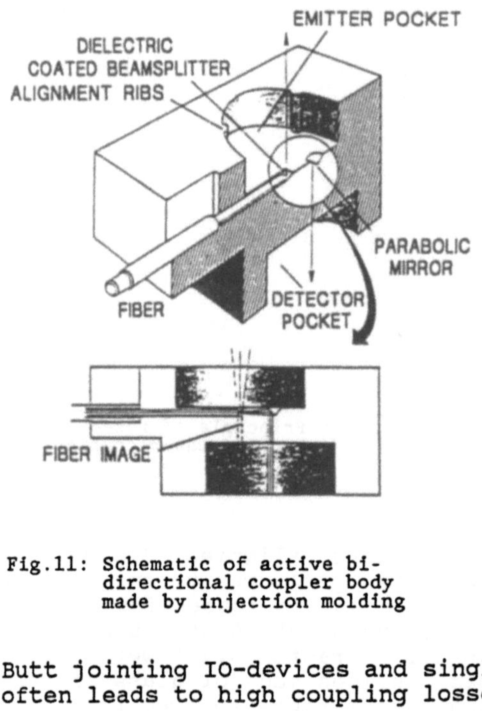

Fig.11: Schematic of active bidirectional coupler body made by injection molding

a)

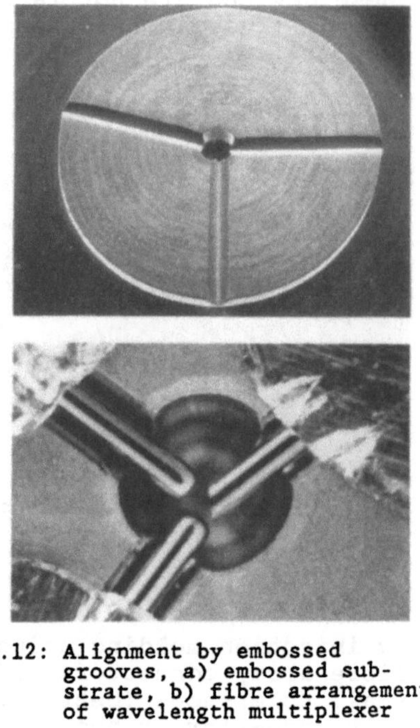

b)

Fig.12: Alignment by embossed grooves, a) embossed substrate, b) fibre arrangement of wavelength multiplexer

Butt jointing IO-devices and single-mode fibres (see Ch.II.1) often leads to high coupling losses, because the mode field radius of IO waveguides is usually smaller than that of fibres (typical values: 1 to 2μm vs. 5μm). If taper lenses are used for mode matching (Ch.II.2), the distance between taper lens and IO waveguide must be 20μm $\pm$5μm, which could cause a very complicated alignment. By a clever combination of V-grooves and pyramidal throughholes in Si chips [34], as shown in Fig.14, the otherwise cumbersome adjustment procedure is reduced to 3 degrees of freedom only.

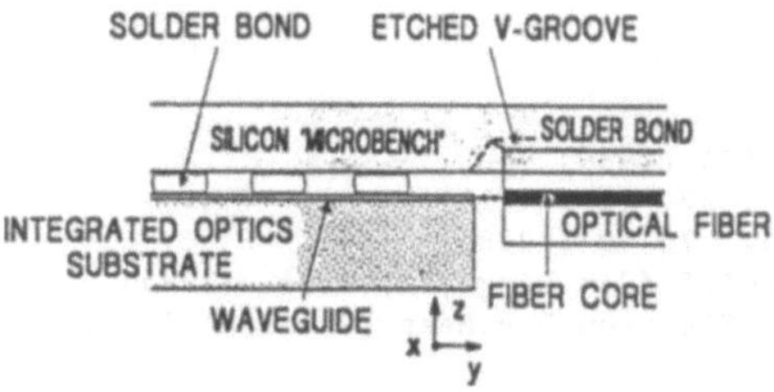

Fig.13: Self-alignment of optical fibres to an IO device by flip-chip solder bonding [33]

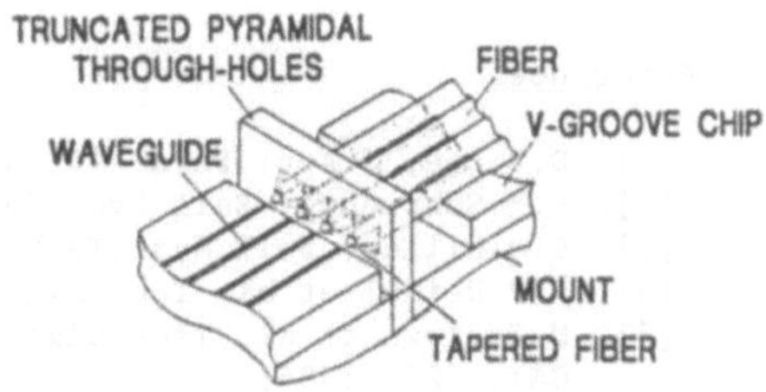

Fig.14: Alignment of taper-lensed fibre array to an IO-device by pyramidal througholes and V-grooves in silicon [34]

<u>III.Examples of applicating several techniques</u>

Fig.15 shows a transmitter/receiver module for a duplex single-mode fibre link [35]. It comprises laser diode, monitor photodiode for the laser, receiver photodiode, coupling optics for laser to single-mode fibre coupling and fibre to photodiode coupling, and wavelength selective filters. The lens system (low cost ball lens and GRIN-rod lens) and the fibres are self-aligned in V-grooves that are embossed into the steel carrier block.

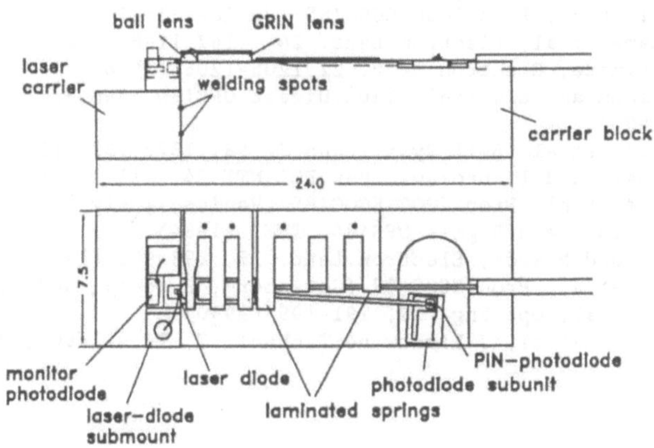

Fig.15: Schematic view of microoptic transmitter/receiver

They are clamped by laminated springs, which are welded on the carrier block by laser spot welding. Only the laser carrier is actively aligned and fixed with an accuracy of $\pm 0.2\mu m$ by laser welding. The prealigned photodiode subunit (see Fig.4) is attached, in that the receiver fibre is inserted into the V-groove. In spite of its simple assembly and low-cost design, the optical performance of this device is virtually competitive with separated transmitter and receiver units.

References

[1] D.B.Lee, J.Appl.Phys. 40, 4569-4574 (1969)
[2] C.M.Schroeder, Bell Syst.Techn.J. 57, 91-97 (1976)
[3] C.M.Miller, Bell Syst.Techn.J. 57, 75-90 (1976)
[4] P.P.Deimel et al, Tech.Digest OFC'85 (San Diego,CA),TUC4 (1985)
[5] M.G.Brown et al, Tech.Digest Int.Electron.Dev.Meet.(San Francisco, CA), 727 (1984)
[6] J.D.Crow, Tech.Digest IOOC'89 (Kobe, Japan), 21D1-5 (1989)
[7] E.J.Murphy and T.C.Rice, J.Lightwave Technol. LT-1, 479-482 (1982)
[8] E.J.Murphy and T.C.Rice, IEEE J.Quantum Electron. QE-22, 928-932 (1986)
[9] B.Hillerich and A.Geyer, Electron. Lett. 24, 918-919 (1988)
[10] Y.Fuji et al, IEEE J.Quantum Electron. QE-16, 165-169 (1980)
[11] B.Hillerich et al, Telecommunications, July 1985, 73-78
[12] Y.Fuji and J.Minowa, Appl.Opt. 22, 974-978 (1983)
[13] A.Sugita et al, Tech.Digest IOOC'89 (Kobe, Japan), 18D1-4 (1989)
[14] J.Jahns and S.J.Walker,Appl.Opt. 29, 931-936 (1990)
[15] Z.L.Liau et al, Tech.Digest IOOC'89 (Kobe, Japan), 18B1-1 (1989)
[16] H.Blauvelt et al, Appl.Phys.Lett. 40, 289-291 (1982)
[17] H.Nobuhara et al, Electro.Lett. 21, 718-719 (1985)
[18] C.R.Harriott et al, Appl.Phys.Lett. 48, 1704-1707 (1986)
[19] L.A.Coldren et al, Appl.Phys.Lett. 37, 681 (1980)
[20] I.Hatakeyama and H.Tsuchiya, Appl.Opt. 17, 1959-1964 (1978)
[21] H.Kuwahara et al, Appl.Opt. 19, 2578-2583 (1980)
[22] B.Hillerich, Opt.Commun. 68, 183-186 (1988)
[23] B.S.Kawasaki et al, Opt.Lett. 6, 321-328 (1981)
[24] K.O.Hill et al, Proc.IOOC-ECOC'85 (Venice, Italy), 567-571 (1985)
[25] I.Yokohama et al, Electron.Lett. 24, 1147-1148 (1988)
[26] D.B.Mortimore, Electron.Lett. 22,1205-1206 (1986)
[27] M.Eisenmann and E.Weidel, Tech.Digest OFC'90 (San Francisco, CA), TUG7 (1990)
[28] A.H.Cherin et al, Bell Syst.Techn.J. 58, 1825-1838 (1979)
[29] T.Yoshizawa and T.Kurokawa, Rev.ECL NTT 34, 111-117 (1986)
[30] G.LeNoane et al, Proc.IOOC-ECOC'85 (Venice, Italy), 633-636 (1985)
[31] H.O.Roberts, Tech.Digest OFC'84, TUN3 (1984)
[32] E.Weidel and M.Rode, Electron.Lett. 18, 898-900 (1982)
[33] M.J.Wale et al, Proc.ECOC'89 (Gotenborg, Sweden), 368-371 (1989)
[34] H.Inoue et al, Opt.Eng. 29, 191-199 (1990)
[35] B.Hillerich et al, J.Lightwave Technol. 7, 1654-1660 (1989)

Thermal Tunable Minimized Semiconductor Laser-Module

E. Großmann, U. Hilbk, K. Peters
Heinrich-Hertz-Institut für Nachrichtentechnik Berlin GmbH,
Einsteinufer 37, 1000 Berlin 10, Fed. Rep. of Germany

Summary

Key elements for optical coherent distribution systems are wavelength tunable semiconductor lasers. At present devices for electronic tuning are not commercially available. For that the wavelength dependence on temperature is used. Consequently, for achieving an acceptable tuning time a small heat capacity of the laser module is necessary. Further system requirements on the laser module are longterm stability and high optical power coupled into the monomode fiber. The measured characteristics of the developed laser module are shown.

Introduction

Within the scope of a research project a TV distribution system utilizing coherent optical multichannel techniques (Fig.1) is under development at the Heinrich-Hertz-Institut Berlin.

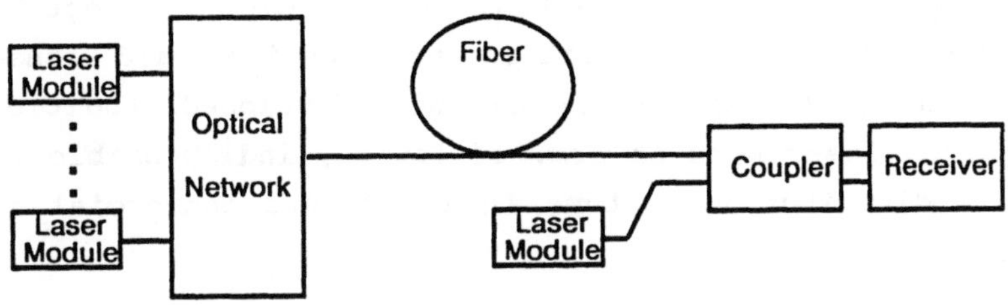

Fig.1. Principal set up of a coherent optical system

Coherent multichannel systems provide the optimal usage of the capacity of monomode fibers. In such systems modulated signals of several lasers, densely spaced in wavelengths, are combined by an optical network and fed through monomode fibers to the subscribers. Each subscriber is able to separate every transmitted channel by means of a wavelength tunable local laser. The desired signal is regained by coherent superposition of the light of the local laser and of the light of all channels in a way that the wavelengths of the selected channel and the local laser have a constant wavelength difference (Intermediate Frequency). To obtain a good system performance all lasers should have limited spectral linewidths (in our system less than 40 MHz).

Advantages of coherent multichannel systems are:

- Large span of margin loss and/or large number of subscribers can be realized
- Large number of channels with densely spaced frequency division multiplexing can be achieved
- Each desired channel can be selected at random by tuning wavelength or frequency of the local laser respectively.

The key element for coherent distribution systems is a wavelength tunable monomode semiconductor laser module with narrow spectral linewidth. Different types of semiconductor lasers have been developed suitable for coherent systems [1] with a tuning range up to 8 nm. However at present these electrical tunable lasers are not commercially available.

Therefore in the before mentioned coherent distribution system only the wavelength dependence on temperature of monomode semiconductor laser provides a sufficient tuning range. For that commercially available DFB lasers are applied as local lasers within the system. To achieve an acceptable tuning time small heat capacities are required and consequently a minimized laser module has been developed. Additionally minimized laser modules are, of course, indispensable for coherent TV distribution systems in any future commercial application.

<u>Set up</u>

The aim of the set up is the construction of a module with high output power, fast tunability and stable operation. The coherent properties of semiconductor DFB lasers are strongly influenced by reflected light entering the laser resonator. Therefore an optical miniature isolator with an isolation > 70 dB is used to protect the laser from light reflected somewhere in the fiber network. To achieve low loss inside the isolator the laser light is collimated by a micro lens and than directed through the optical isolator to a second micro lens which focuses the parallel incoming beam into the fiber (See Fig.2). To reduce reflections within the laser module end faces of both lenses and the input window of the isolator have antireflection coatings with a reflection factor of less than 10^{-3}. Moreover the rear surface of the collimating lens and the entrance window of the isolator are tilted with angles of 6° and 3°, respectively.

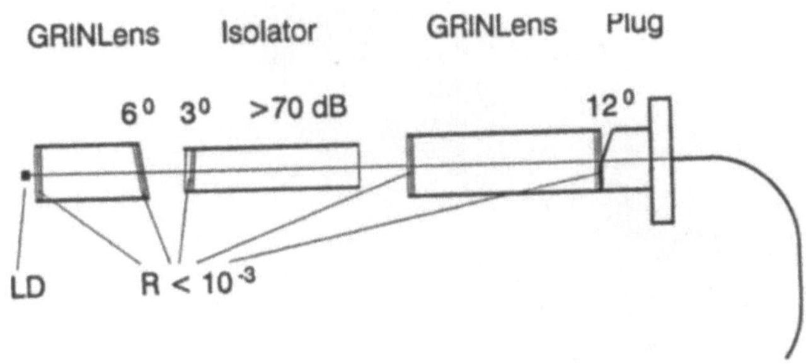

Fig.2. Schematic set up of a laser fiber coupling with two micro lenses and a miniature optical isolator

To ensure an easy mechanical alignment the lenses and the isolator are placed on two small optical benches with one V-groove. For optical coupling the laser diode is positioned in a distance of 200 μm and off axis in front of the first gradient index (GRIN) lens which has a numerical aperture of 0.6. The laser and the first lens are placed on the separate first part of the optical bench with the volume of 6*6*3 mm^3 and temperature controlled by aid of a thermal sensor and a Peltier thermo-cooler with a heat flux of Q=1.6 W. The temperature sensor is fixed as close to the laser diode as possible. The isolator and the focusing gradient index lens with a numerical aperture of 0.37 are located on the second optical bench. The dimensions of the module (see Photo) are limited by the dimensions of the optical components. The light is directly coupled into a slant polished optical plug providing the coupling possibility into different types of fibers such as monomode or polarization maintaining fibers. To ensure the proper alignment between the optical axis of the polarization maintaining fiber and the polarization of the laser radiation a cheap commercial available rotary fixture is used. Beside this flexibility the coupling direct into a plug spares an additional connector for the link to the optical fiber network. Furthermore this slant plug avoids an Fabry-Perot resonator between the fiber end at the coupling and an additional end face somewhere in the network.

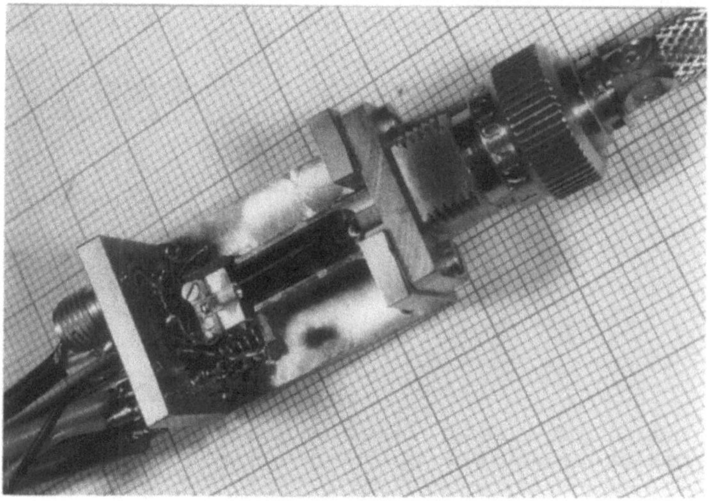

Photo: Minimized laser module

The module is assembled by positioning the optical components with piezoelectric micropositioners. After fixing the components a metallic cap is put over the system to shield radio frequencies.

Results and Discussion

Measurements of the mounted laser module on operating wavelength and optical power as a function of temperature are shown in Fig.3 with the bias current as parameter. A temperature adjustment of 10 C causes a change in wavelength of 0.92nm (Fig.3a) corresponding to a tuning range sufficient for more than 10 channels. At present the channel spacing of our system amounts to 0.08 nm. According to Fig.3b the same temperature adjustment results in an unwanted change of the optical output power of about 0.2 mW, only imposing a small penalty to the system performance. Due to the linear wavelength dependence on temperature it is sufficient to show temperature dependencies instead of wavelength dependencies on time. Fig.4 shows responses to pulses of optimal Peltier currents for heating and cooling. Increasing the temperature of the laser diode is much faster than decreasing it due to the loss of heat of the Peltier thermocooler. For the described module a temperature adjustment of 10 °C is gained in 2.5 seconds and 3.5 seconds for heating and cooling, respectively. The aim of development is to assemble a laser module with fast tunability as well as longterm stability of wavelength and spectral linewidth, as mentioned above.

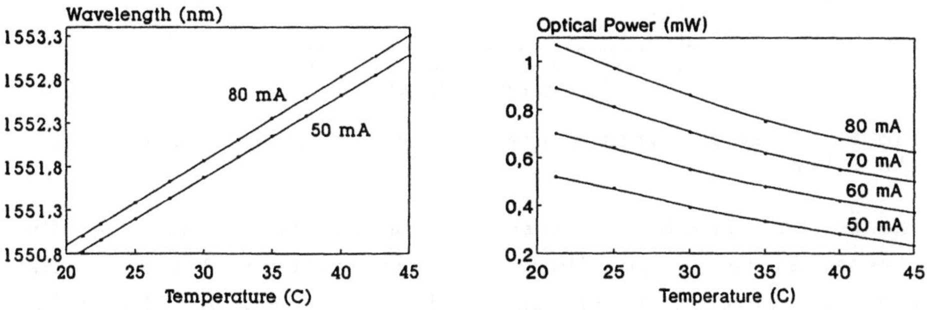

Fig.3. Wavelength (a) and optical power within the fiber (b) as a function of temperature for different laser bias currents

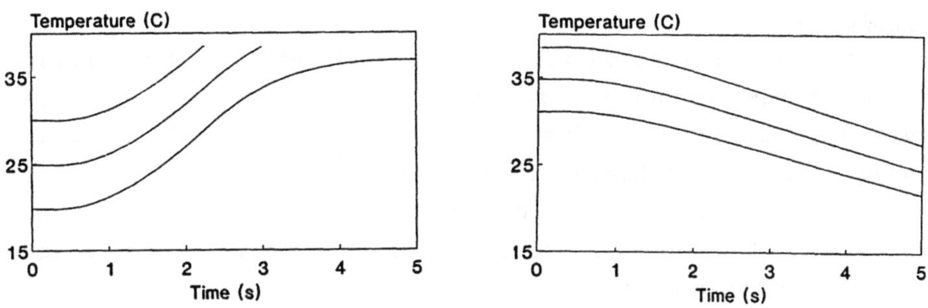

Fig.4. Temperature dependencies for heating (a) and cooling (b), the duration of the current pulse amounts to 2s and 4.5s, respectively

For a time period of 16 hours measurements on the wavelength have been performed. Changes are within the resolution of the wavelength meter (2 pm).

As mentioned previously the superposition within the resonator of backreflections from external reflectors with amplified light has a disturbing influence on the laser diode characteristics. Dependent on the phase difference between both fields the laser radiation broadens or narrows. Since the phase of reflected light is dependent on the distance of the reflector mechanical vibrations and drifts causes strong

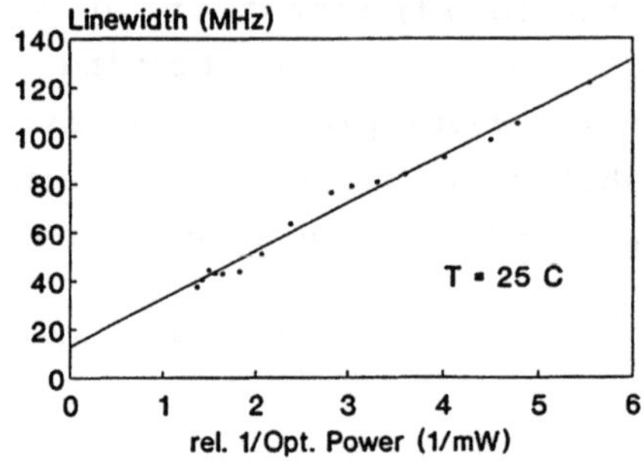

Fig.5. Linewidth versus the inverse of optical power within the fiber

change in linewidth. Therefore reflections should be suppressed as far as possible. The measured linewidth as a function of inverse power is depicted in Fig.5. The slight deviation from the expected linear behavior is caused by residual reflections from the optical isolator.

Conclusion

For the application within an experimental TV distribution system a semiconductor laser module using the laser temperature for wavelength adjustment has been developed. A tuning range of 1 nm is achieved in an acceptable time. The penalties due to power loss and residual reflections of the isolator are tolerable. The longterm stability of the module is convenient for its system application.

Acknowledgment

This work was supported by the Bundesministerium für Forschung und Technologie of the Federal Republic of Germany, contract TK 440. The authors alone are responsible for the content.

References

1. Koch, T.L.; Koren, U.:Semiconductor for Coherent Optical Fiber Communications. J. Lightwave Technol. Vol. 8 (1990) 274-293.

Micro Collimated Light Source Using Micro Fresnel Lens

Hiroshi Goto, Yoshinori Ito, Hiroshi Sekii, Shiro Ogata, and Koichi Imanaka

OMRON Corporation, Central R&D Laboratory.

Shimokaiinji, Nagaokakyo-city, Kyoto 617, Japan.

ABSTRACT

We have developed super micro collimated light source (MCL) without attaching any other external collimating lens. The MCL is assembled with a light source such as a laser diode (LD) or a light emitting diode (LED) and a micro collimating lens in the smallest commercially available laser package. In order to obtain a compact MCL and narrow and high quality collimated beam, a short focal length micro Fresnel lens (MFL) has been utilized for the first time as the collimating lens. The radiated beam from the MCL using LD light source has higher than 3 mW intensity at 50 mA driving current, narrow enough as ϕ1.3mm beam diameter with nearly Gaussian intensity profile, and lower than $\lambda/14$ (RMS value) wavefront aberration measured at 50cm distance. The beam from the MCL using LED has 0.5 mW intensity at 50 mA driving current and about ϕ1.6 mm beam diameter as narrow as in the case of LD light source. The MCL has also high reliability against temperature variation and humidity, because all of the elements are assembled in the small package filled with dry N_2 gas.

1. INTRODUCTION

Micro optical devices or hybrid integrated optical devices become very important in the development of optical application systems such as optical communication systems, optical information processing systems, and optical sensing systems to achieve the compactness and high performance of these systems. Laser diodes (LDs) or light emitting diodes (LEDs) are widely used as the light source in these systems because of small size, low driving current, and the capability of direct high speed modulation. However, the output beam emitting from these light sources have wide far-field radiation angle. Therefore, these light sources have been used with external lens such as a collimating lens or a focusing lens, resulting in the difficulty to compact collimated or focused light source.

To overcome such problem, we have developed super compact micro collimated light source (MCL)(1), which is not needed to adapt any other external lens, by using a micro Fresnel lens (MFL)(2)(3) with an LD or an LED in the same package, in order to achieve the compactness of light sources.

In this paper, we describe the construction and the characteristics of MFL and MCL.

2.MICRO FRESNEL LENS

A micro Fresnel lens (MFL) which is constructed with stepped setback of many divided annular zones as shown in Fig.1. Because the zone periods of the MFL outer portion are close to a optical wavelength,the MFL works on the theory of diffraction. Therefore an incident plane light wave along the optical axis is converted into a converging wave by the MFL. The optical properties including focal length and focal spot size are essencially the same as those of refraction type lenses, such as ordinary convex lenses and graded-index lenses.

The MFL fabrication process is shown in Fig.2. First, a master MFL was fabricated on a glass substrate by using an electron-beam lithography technique. An ITO layer was deposited on the glass substrate to prevent charging-up during an electron-beam writing, and electron-beam resist was spin-coated on the substrate. The concentric patterns of the MFL were written by computer-controlled vector scanning of the electron-beam. The blazed profile was formed by a suitably chosen electron-dose distribution, because the etching rate of the resist depends on the electron-dose during development. Next,a stamper was prepared by a conventional Nickel electro-forming method. The cross-sectional pattern of the stamper is the just inverse of that of the master. After potting the monomer of an UV-curable resin between the stamper and a base glass substrate with anti-reflection (AR) coating, photo-polymerization (2-P) process which apply UV-light to the resin resulted in replica MFLs. The cross-sectional profile of the replica MFL is same as that of the master MFL. Fig.3 shows a microphotograph of the MFL which has 780nm design wavelength, 0.25 numerical aperture, ϕ0.5mm lens diameter, and 1mm focal length. Fig.4 shows a cross-sectional scanning-electron-microscope (SEM) photograph of the MFL. An almost ideal blazed relief structure is obtained.

The RMS value of wavefront aberration of the MFL fabricated by these process is lower than 0.05λ and the first order diffraction efficiency of the MFL is 50%. By using the glass substrate as the base plate and 2-P process to fabricate replicated MFL, the MFL has several advantages such as high reliability, against temperature, humidity, low wavefront aberration,and easy fabrication.

Therefore, the MFL could be widely applicable to various micro optic devices.

8. MICRO COLLIMATED LIGHT SOURCE

The construction of the MCL is shown in Fig.5. An LD dice or an edge emitting LED dice, the MFL and a photodiode for monitoring the output power of laser diode dice are assembled in the same package. In order to get a collimated beam from the MCL, the MFL is fixed on the optical axis of diverging beam emitted from the LD or LED dice, and on the position at the distance of the MFL's focal length from the emitting point in the dice. The double quantum well laser diode[4] and edge emitting LED emitting at 780nm, fabricated by molecular beam epitaxy are applied in the MCL.

Assembling process of the MCL is as follows. First, the LD or LED dice and the photodiode were bonded on the laser stem 5.6mm in diameter. Next,a thin glass substrate with the MFL was fixed on the top of stem block with UV-curable glue. The MFL has a structual merit of easy assembling because of its thin, plane, and small shape. Therefore it is easy to fix the MFL on the stem with the glue. During UV-light exposure, the MFL was kept against the position shift caused by volume contraction of the glue. In order to keep the wavefront aberration of collimated output beam within $\lambda/14$, location tolerance between the MFL and the laser diode dice is controlled within several microns. After assembly, all of elements were sealed with a can with glass window filled with dry nitrogen gas.

The MCL has following several advantage because of its small size. There are (1) the can protects all of elements from humidity, dust, and chemicals,(2) the small heat capacity of MCL makes it easy to control MCL body temperature accurately,and (3) because of a light weight, the MCL has a resistance against physical shock such as vibrations.

Consequently the MCL has the great merits of high reliability.

4.OPTICAL CHARACTERISTICS

The collimated beam intensity distribution measured at 50 cm distance of MCL using an LD dice (MCLD) as a light source is shown in Fig.7 (a). The beam diameter of 1.3 mm at $1/e^2$ of maximum intensity and nealy Gaussian intensity distribution without any fringe were obtained. The beam intensity distribution of MCL using an LED (MCLED) is shown in Fig.7 (b) .The beam diameter of 1.6 mm and nearly Gaussian intensity distribution as same as in the case of MCLD were obtained.

The collimated power output-injection current (L-I) characteristics of MCLD and MCLED measured at 50 cm distance are shown in Fig.8 (a) and (b) respectively. The output beam from MCLD has higher than 3 mW intensity at 50 mA driving current. The coupling efficiency between radiated light from an LD dice and collimated light output is about 30 percent. The beam from MCLED has 0.5 mW intensity at 50 mA driving current. The coupling efficiency is about 30 percent same as that of MCLD.

Fig.9 shows the collimated laser beam wavefront pattern analyzed by a ZYGO WA8100 system. The wavefront aberration of 0.04λ (RMS value), which is sufficiently lower than the Marechal criterion value, is obtained. Therefore the present collimated beam can be focused into a diffraction-limited spot size.

Finaly, we evaluate temperature dependence of optical characteristics of the MCLD. Fig.10 shows the wavefront aberration value as a function of ambient temperature. The wavefront aberration value increasing with increase in temperature, but the MCLD has enough small wavefront aberration value of 0.06 even at 80 . Also the beam intensity distribution of MCLD and MCLED have not been changed with increasing temperature between 0 ℃ and 60℃. Therefore, the present MCL can be operated stably even at high temperature over 60 ℃.

5.SUMMARY

We newly developed a super compact micro collimated light source using a micro Fresnel lens as a collimating lens. The MCL could be widely applied in various optical systems because of its high optical performance, high reliability, and compactness.

Furthermore the electron-beam lithography and 2-P process to fabricate the MFL is very useful to another micro optical devices such as optical waveguides or micro lens.

6.REFERENCES

(1) S.Ogata, H.Sekii, T.Maeda, H.Goto, T.Yamashita, and K.Imanaka,"Micro Collimated Laser Diode with Low Wavefront Aberration, "IEEE Photon. Technol.Lett., vol.1, no.11, pp.354-355, 1989.

(2) T.Fujita, H.Nishihara, and J.Koyama, "Blazed gratings and Fresnel lenses fabricated by electron-beam lithography, "Opt.Lett., vol.7, no.12, pp.578-580,1982.

(3) M.Tanigami, S.Ogata, S.Aoyama, T.Yamashita, and K.Imanaka, "Low wavefront aberration and high temperature stability molded micro Fresnel lens, "IEEE Photon.Technol.Lett., vol.1, no.11, pp.384-385, 1989.

(4) H.Imamoto, F.Sato, K.Imanaka, and M.Shimura, "GaAs double Quantum well laser diode with short-period (AlGaAs)m (GaAs)n superlattice barriers, "Appl.Phys.Lett., vol.54, no.15, pp.1388-1390, 1989.

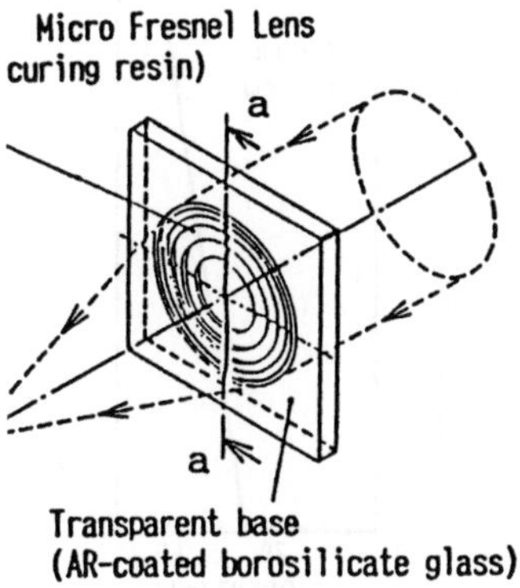

Fig.1 Outline of an MFL

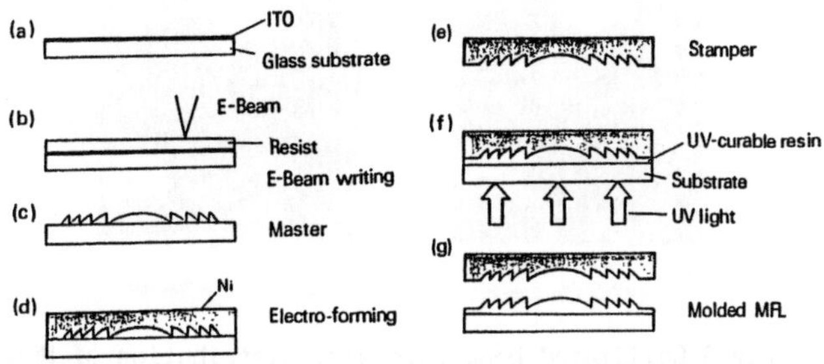

Fig.2 Fabrication process of a micro Fresnel lens

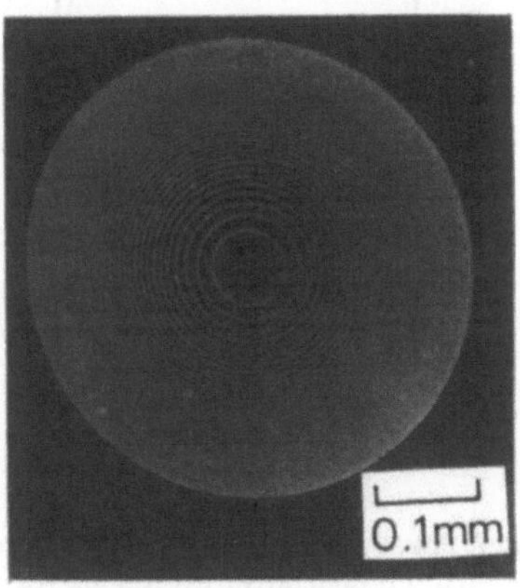

ig.3 Micro photograph of an MFL

Fig.4 Cross-sectional SEM photograph of an MFL

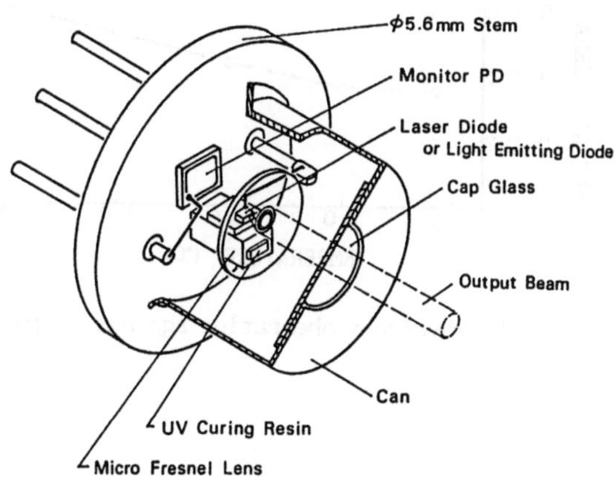

Fig.5 Construction of an MCL

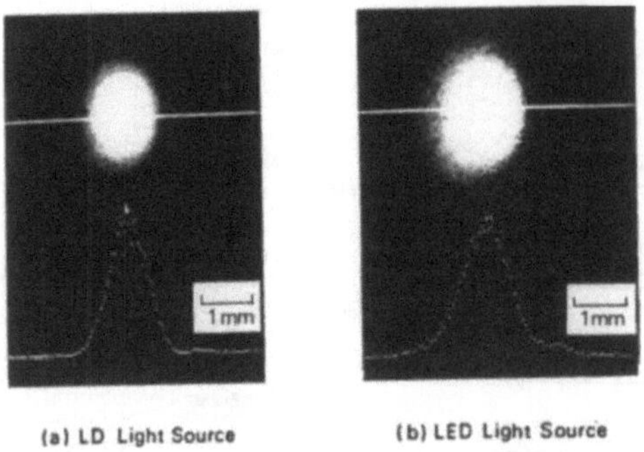

Fig.6 Collimated beam intensity distribution of MCL (a) using an LD and (b) using an LED

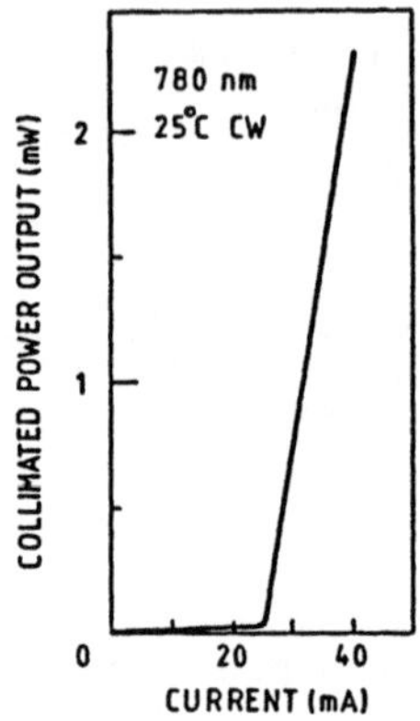

(a) MCLD

Fig.8 Collimated beam wavefront pattern of an MCLD

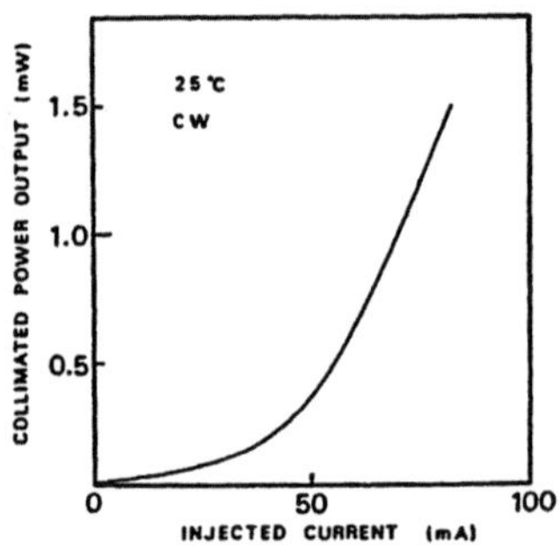

(b) MCLED

Fig.7 Power output-injection curren (L-I) characteristics of (a) M and (b) MCLED

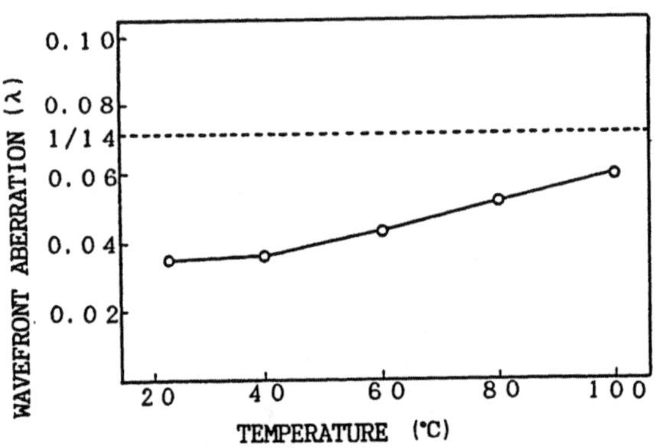

Fig.9 The wavefront aberration against temperature

Monomode Fused Optical System Elements

S. LOCHMANN; J. M. LABS; A. THIEL; K. D. LANG

Sektion Elektronik
Humboldt-Universität zu Berlin

Summary

Rapid advances in fibre optics and the resulting multiplicity of applications have increased the demand for diverse passive optical system elements. Fusion techniques play a major role in the fabrication of components with desired parameters. The paper presents a survey of the fabrication and specific properties of system elements such as coupling devices, WDM components and fibre optic filters.

1. Introduction

Despite intensive research in the field of integrated optics, numerous fibre optic elements are fabricated as discrete components to provide optimum properties that cannot yet be attained using integrated, e. g. planar, technology.
The development of such system elements is specific for the respective ranges of operation wavelengths. The examples discussed below mainly refer to the first optical window (850 nm). The reason for this is that, on the one hand, there is a wide variety of applications in this range, particularly in sensorics, and on the other, once the technology is mastered, transition to shorter wavelengths is relatively easy to perform because of the much larger extension of the field.
Apart from a few special fabrication techniques (twisting-etching, double core fibres, D-fibres etc.), abrasion and fusion have become popular as fundamental and complex techniques for the fabrication of diverse system components. A survey of actual monomode fused components is given in Table 1.
To prepare such elements, a technological work station is required permitting manipulation of optical fibres in the micrometer range (0.1 µm in some cases), variation of energy sources with respect to temperature profile and position, in-situ

analyses of various parameters (insertion losses, extinction, channel isolation etc.) and including computerized real time control. A block diagram of the setup is shown in Fig. 1.

2. Coupling Configurations of Coaxial Structure

2.1 Attenuators

Attenuators, which, amongst others, are required for level matching in fibre optic or measuring systems, can be fabricated in a simple way by controlled tapering of optical fibres, so that higher order modes are excited that can be filtered out e. g. by acrylate coating. Fig. 2 shows a typical attenuation curve for such a tapering process. Fixed attenuators are made by heating a closely limited fibre region (e. g. <2 mm) and using mostly abrupt taper geometry (Fig. 2a). Variable taper attenuators are made using more homogeneous, sometimes adiabatic tapers (Fig. 2b) taking advantage of the emission of energy portions of the fundamental mode caused by the defined curvature of the taper waist.

2.2 Wavelength Selective Filters

Wavelength selective filters, too, can be made to have various bandwidths by defined tapering of depressed cladding fibres. The principle of operation is light energy being coupled within the taper regions from the core into the fundamental mode of the coaxial optical fibre and vice versa. This manifests itself in an oscillating attenuation curve during the tapering process (Fig. 2c). The properties of coaxial optical fibres and hence their coupling behaviour depend upon the width and the lowering of the refractive index of the inner cladding region deposited in the fabrication process. To obtain high channel isolation it is necessary to employ low-loss adiabatic tapers and yet establish defined coupling between core and coaxial modes. If this cannot be provided there will occur a relatively small rise in attenuation that has to be attributed to energy transfer between the HE_{11} and HE_{12} modes of the core. A typical spectral response of a relatively broadband taper filter (3 coupling processes) is shown in Fig. 3.
Coaxial couplers operate in a similar way as the wavelength taper filters described above. To obtain the very narrow bandwidth (e. g. <0.2 nm) required for instance for external fre-

quency stabilisation of laser diodes, however, they are fabricated as optical fibres with defined coaxial structure /1/.

3. Coupling Configurations with Parallel Fibres

3.1 X-Couplers

There are numerous firms and institutions internationally that are capable of producing X-couplers. Assuming optimum optical fibre structures are given, the variation of excess losses can be considered as a measure of the reproducibility of the process technology (Fig. 4). Also the spectral behaviour, which can be influenced by the fusion geometry and is usually stated at the 3 dB point, will vary within a typical range (Fig. 5). For many applications, however, specific coupler parameters are required. This concerns mainly polarization and wavelength properties. On the basis of standard coupler technology it is possible to fabricate wavelength selective couplers by increasing the pulling and thus coupling lengths to obtain almost any bandwidth or channel separation desired. According to the typical channel separation the components can be subdivided into

- broadband WDM distinguishing between the optical windows
- narrowband WDM with channel separation ca. 20-50 nm and
- high definition WDM with channel separation <ca. 2 nm.

Fig. 6 shows a typical variation of channel separation with the number of 3 dB point passes correlating with coupling length. This can also be seen from the spectral response plotted in Figs. 7 and 8a. For this purpose the logarithmic coupling ratio is calculated which through

$$10 * \log K_1 = 10 * \log P_4/P_3 = a_{13} - a_{14} \qquad (1)$$

is related to insertion losses that are easily accessible to measurement.

For multi-channel applications the WDM components described above can be concatenated.

In contrast with wavelength selective couplers there are various applications where 3 dB signal splitting is required at two defined operation wavelengths or within a certain range of wavelengths. In the former case, so-called dual wavelength couplers are used, and broadband couplers in the latter. Fig. 8b shows a typical spectral response of a 3 dB coupler for 633 nm and 850 nm, the fabrication technology of which differs only little from that of WDM components.

In broadband couplers, use is made of the dependence of energy transfer on the phase constants of the two optical fibres, and the maximum coupling power is limited, if possible, to 50 per cent. This can be achieved technologically by

- use of different optical fibres (in diameter, profile)
- local abrasion of an optical fibre
- local etching of an optical fibre
- pretapering of an optical fibre.

Of these techniques, pretapering can best be integrated in the fabrication process. Fig. 9 shows calculated and measured spectral response of a broadband coupler in comparison with a standard coupler.

Beside spectral response, the polarization of fused couplers can be influenced in defined ways.

In polarization independent couplers it is assumed that there are no irregularities in the fused region to disturb light propagation and that the difference in coupling behaviour of the HE_{12} or EH_{12} modes resulting from the fusion geometry is minimal. The latter influence occurs only with longer coupling lengths correlating with some 20 or more 3 dB point passes and can be neglected.

Irregularities in the coupling range, however, cause an immediate energy transfer between the respective polarization states and at the same time excite lossy higher order modes. For practical purposes, to achieve polarization independence, one can postulate low excess losses (<0.2 dB). When the state of launched polarization changes, such components exhibit minimum tolerance of the splitting ratio (<1%), while higher excess losses will result in stochastic polarization behaviour because of mode coupling (Fig. 10).

Polarization maintaining couplers are generally made using polarization preserving optical fibres, which, to obtain high

extinction (>20 dB), have to be aligned in parallel with accuracy <1°.
Polarization splitters can be fabricated in the same way and differ only in that they have the arrangement of one of the two parallel fibres shifted by 90°.

3.2 Star Couplers

The fabrication of monomode star couplers is much more difficult in comparison with that of multimode components because of their higher sensitivity in signal splitting. For this reason multichannel star couplers for telecommunication are made exclusively by concatenating standard couplers.
Besides, it has become possible to fabricate monolithic star couplers with 3 channels and, according to /2/, even with 4 channels, with excess losses <0.2 dB and defined splitting ratio, symmetric or asymmetric, depending on the application. To position the parallel fibres exactly, use is made of low-melting Vycor glass, which has a lower refractive index than silica glass. The fibres are tapered together with the collapsing capillary tube under standard conditions. Major advantages of such devices are compactness, low excess losses and, in 3x3 star couplers with suitable dimensions, the capability of being applied as X-couplers with monitor output.

3.3 Y-Couplers

Y-couplers can play a major role in applications where crosstalk attenuation >60 dB is required. Fused couplers would basically be predestined to serve the purpose, yet in practice, to achieve not only good but above all stable crosstalk attenuation, considerable efforts have to be made to provide non-reflecting termination, for instance, of the unused port of an X-coupler. In Y-couplers, this problem does not arise because of the specific geometry of the fusion region where the closely adjacent parallel fibre cores are directly fused with the core of a single fibre. Drawbacks are enhanced excess losses of ca. 1.5 dB /3/ and more complicated process technology. Thus the parallel fibres first have to be thinned by etching to a diameter <20 µm, then fused, broken and spliced in a defined manner with equally etched original fibres.

Beside good crosstalk attenuation, Y-couplers exhibit a mostly desirable broadband wavelength response over ca. 300 nm.

3.4 Ring Resonators

For various applications, in particular optical frequency multiplexing, fibre optic ring resonators are of interest because of their superior properties. Direct coupled or cross coupled ring resonators are generally used (Fig. 11). While filters of great ring length are relatively easy to construct using available coupling elements, there are technological limits to a reduction of the ring length required to obtain a larger free spectral range. On the one hand, fused couplers have been reported having a component length of only 20 mm, which can well be integrated but give rise to relatively high excess losses ($>$ ca. 1 dB). On the other hand, low excess losses are associated with greater component length ($>$ ca. 50 mm). This, however, involves minimum ring lengths of ca. 160 mm corresponding to a free spectral range of ca. 1.2 GHz.
To enhance the free spectral range other techniques are required. One example is adiabatic tapering of an optical fibre to a waist diameter $<10\ \mu m$ and subsequent fixation with ultraviolet-hardening adhesive to form a ring where the coupler is realised only by mechanical contact of the thin-tapered fibre. With such ring resonators, which were first described in /4/, it was possible to obtain ring diameters of only 2 mm and thus a free spectral range of 30 GHz at relatively low insertion losses (1.2 dB).

4. Summary

The multiplicity of applications involving passive monomode system components requires specially adapted techniques permitting the fabrication of a wide variety of components with highest-quality properties measurable by the component parameters. The many-sided possibilities offered by fusion techniques are described in the context of a classification of passive optical fibre components most of which have been realised.

References

1. Böck, G.: Wellenlängenselektive koaxiale Lichtwellenleiter-Richtkoppler. Siemens-F/E-Ber. 15 (1986) 1, 32-39.
2. Mortimore, D.B.: Monolithic 4x4 Single-Mode Fused Coupler. Electr. Lett. 25 (1989) 10, 682-683.
3. Minelly, J.D.; Hussey, C.D.: Single-Mode Fibre Y-Junction Beam-Splitter. Electr. Lett. 23 (1987) 20, 1087-1088.
4. Caspar, C.; Bachus, E.-J.: Fibre Optic Micro-Ring Resonator with 2 mm Diameter. Electr. Lett. 25 (1989)

Table 1. Survey of monomode fused system components

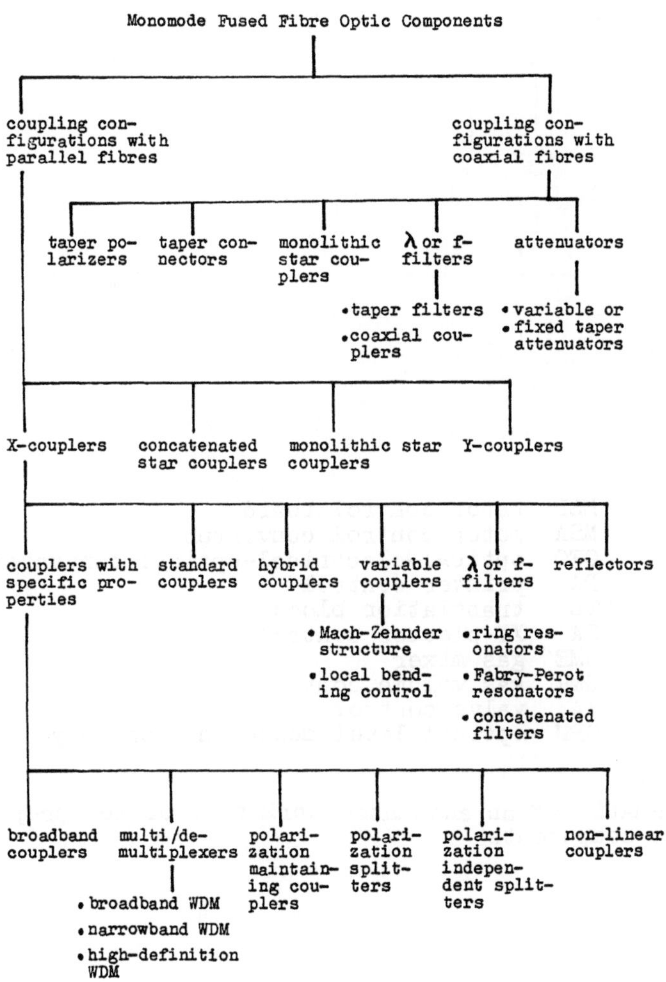

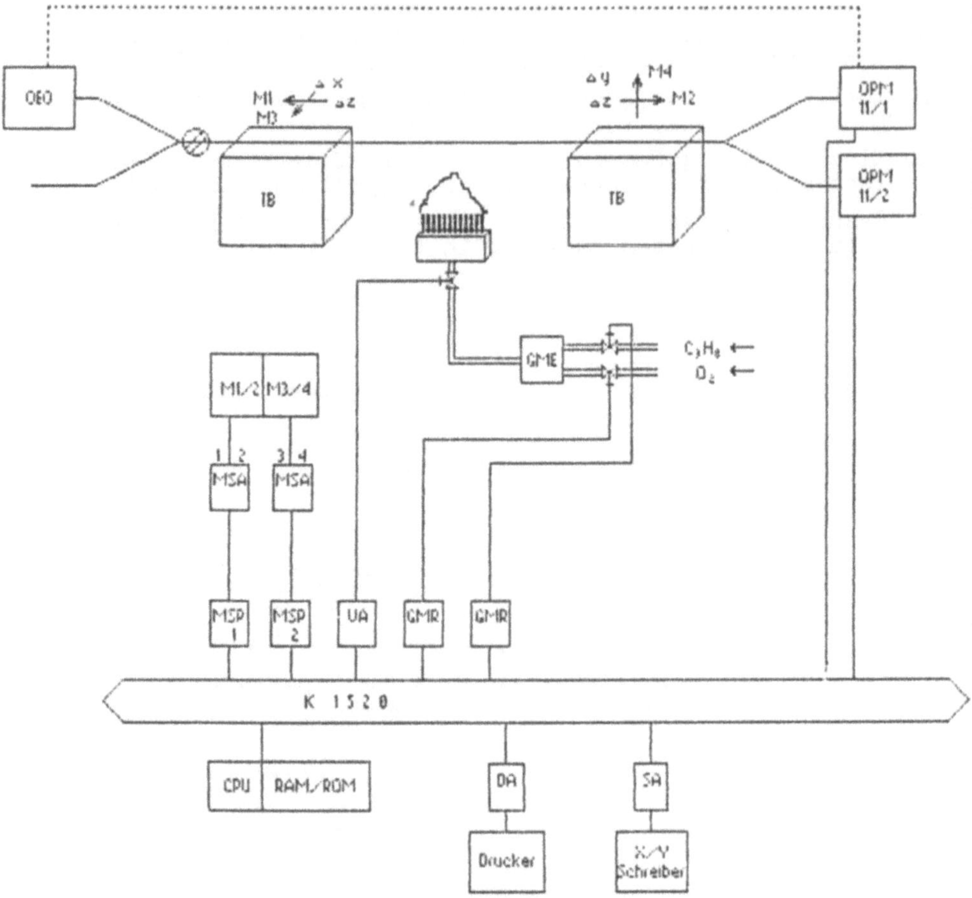

MSP motor control board
MSA motor control converter
OEO optical-electrical-optical converter
DA printer control
TB translation block
SA XY-plotter control
GME gas mixer
GMR gas volume control
VA valve control
OPM optical level measuring assembly

Fig. 1. Schematic of an automated work station for preparing fused system components

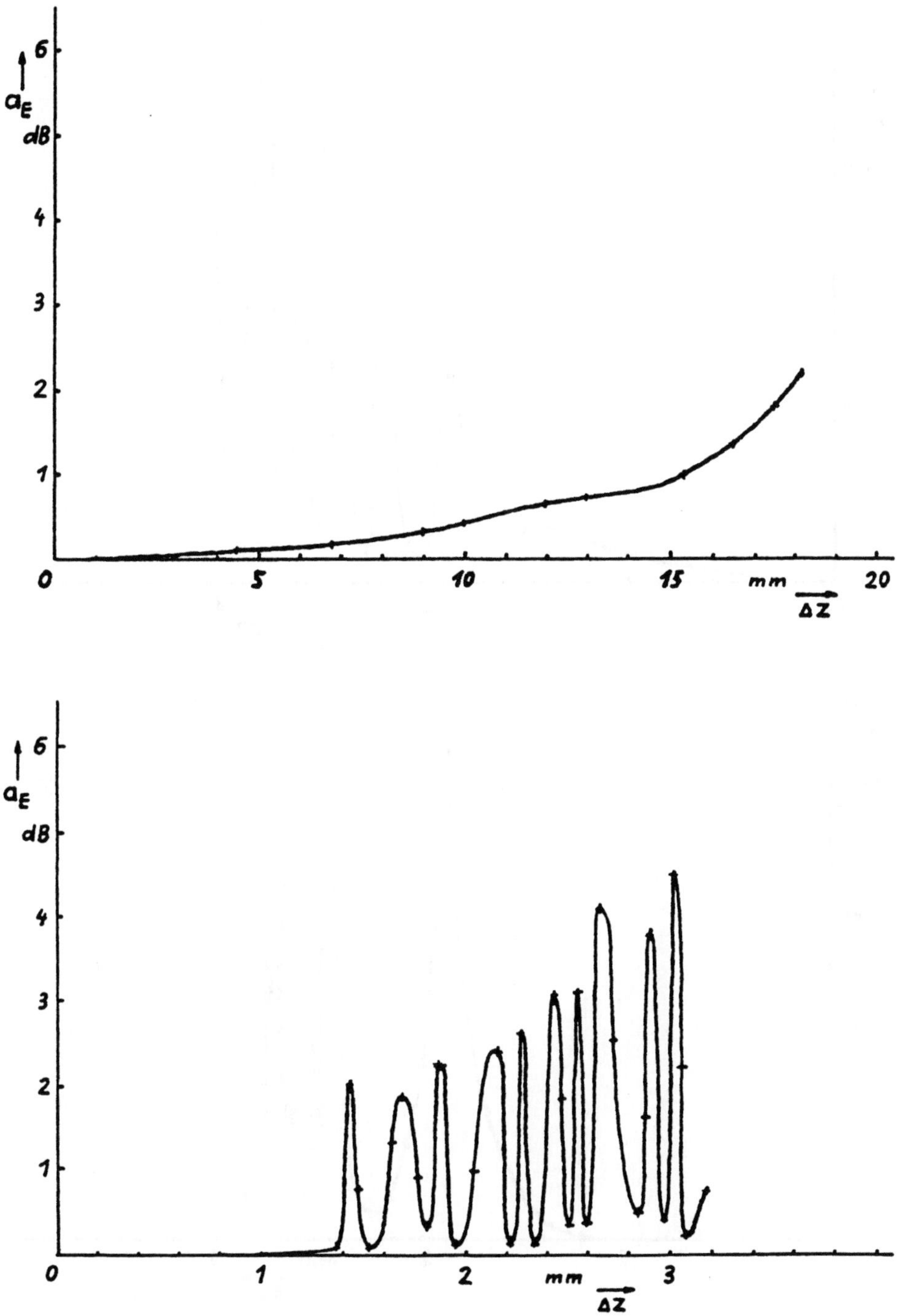

Fig. 2. Typical attenuation curve in tapering with different widths of the viscous zone

a - ln = 1.5 mm; matched cladding fibre
b - ln = 8 mm; matched cladding fibre

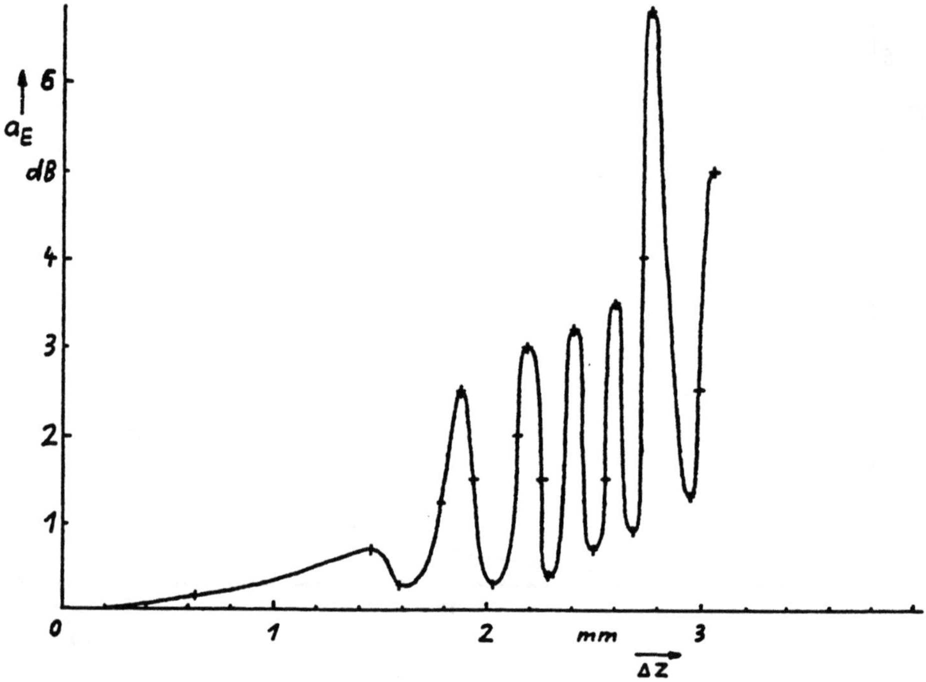

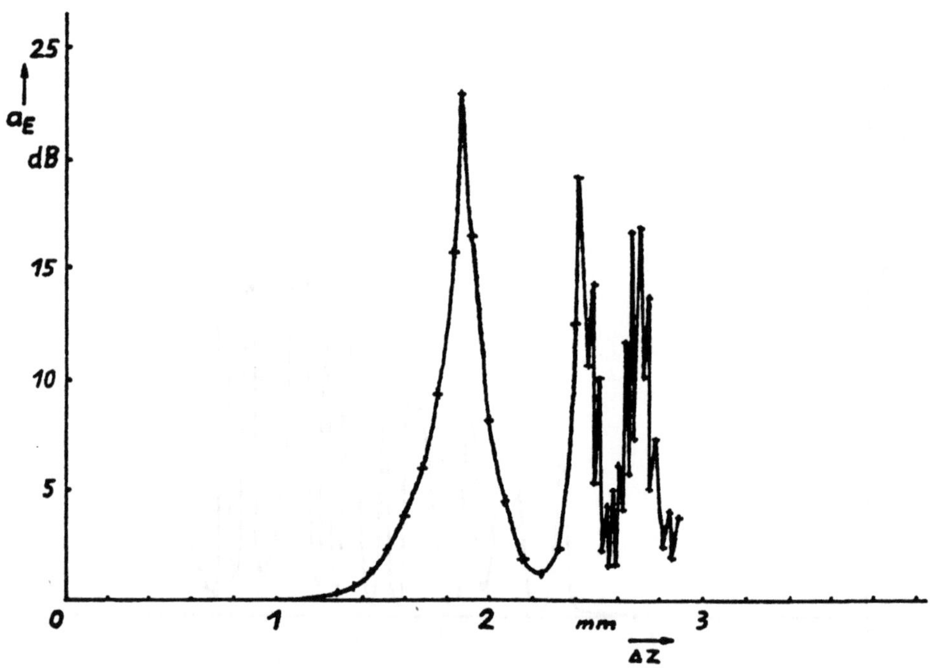

c - ln = 1.5 mm; depressed cladding fibre
d - ln = 8 mm; depressed cladding fibre

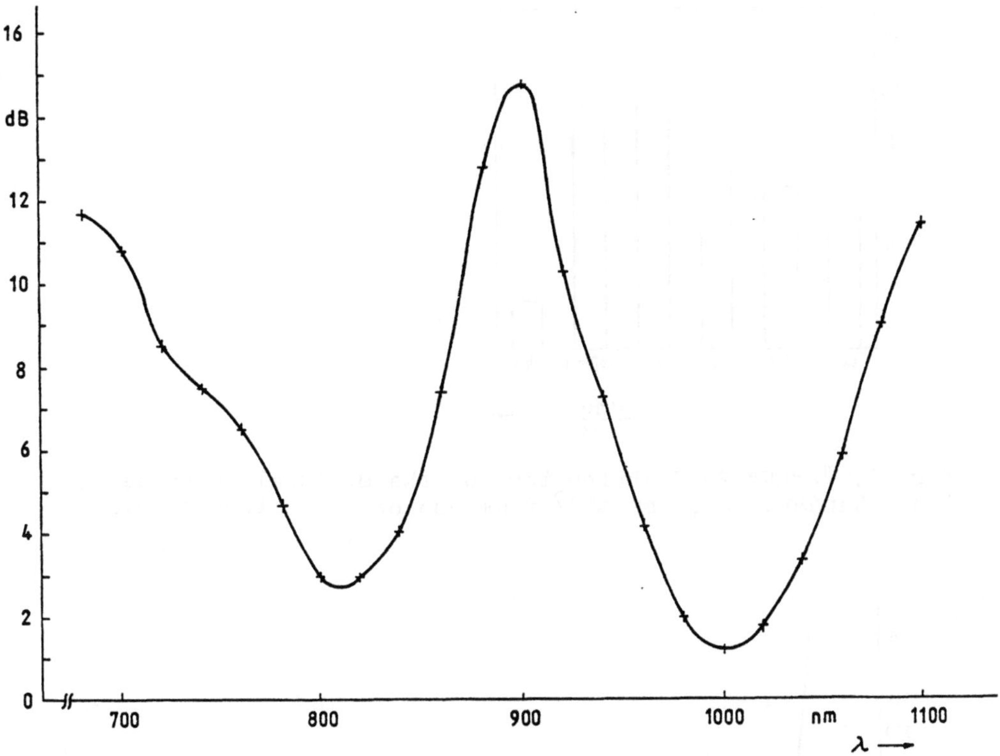

Fig. 3. Spectral response of a taper filter with 3 coupling processes

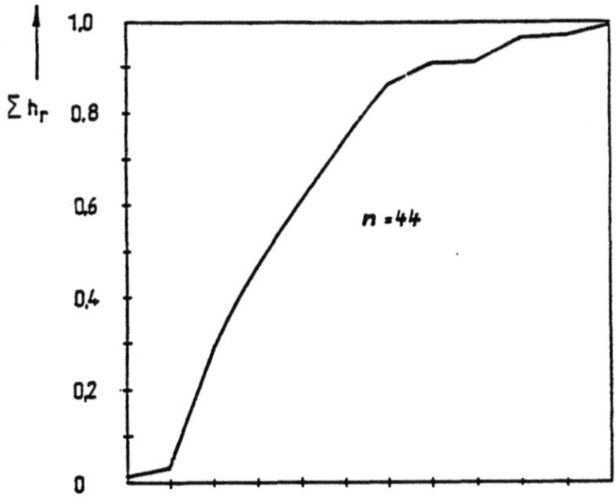

Fig. 4. Total excess losses of couplers made from one optical fibre charge; λ_c = 790 nm

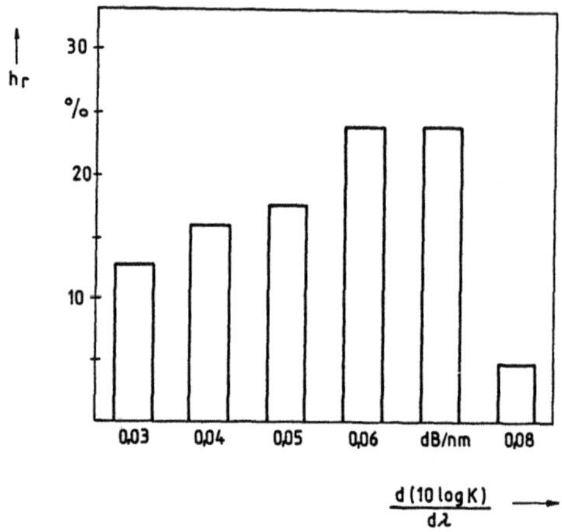

Fig. 5. Frequency distribution for the spectral response of 3 dB standard couplers made from one optical fibre charge

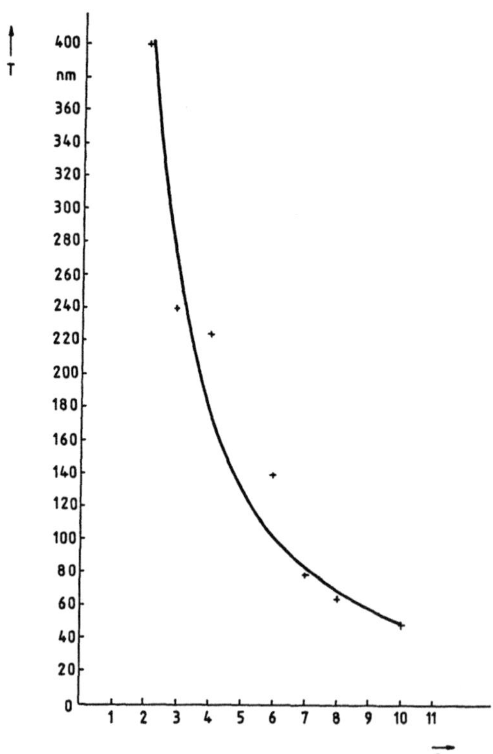

Fig. 6. Typical variation of channel separation T with the number n of 3 dB point passes

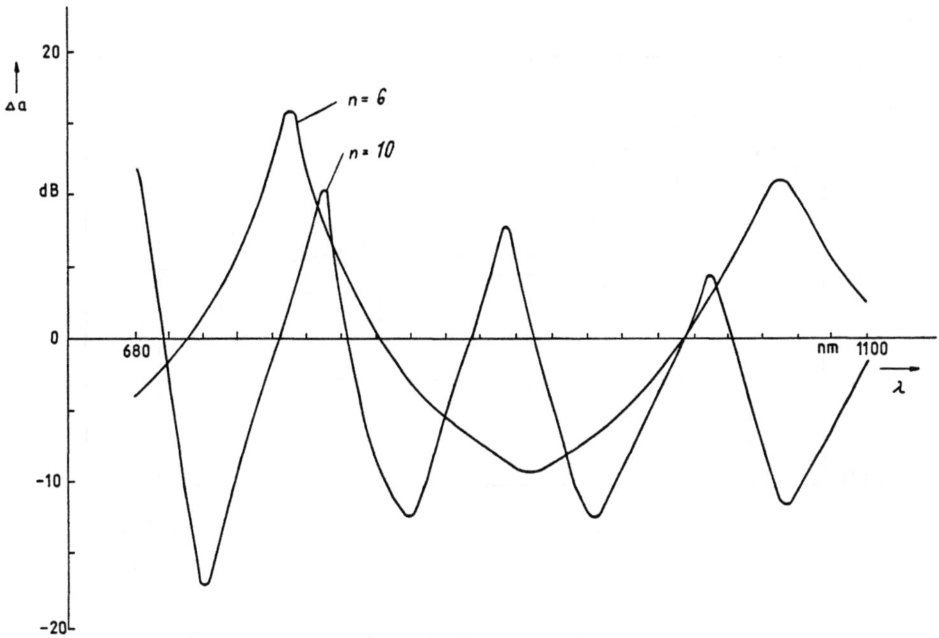

Fig. 7. Spectral response of WDM components with 3 or 10 3 dB point passes, respectively

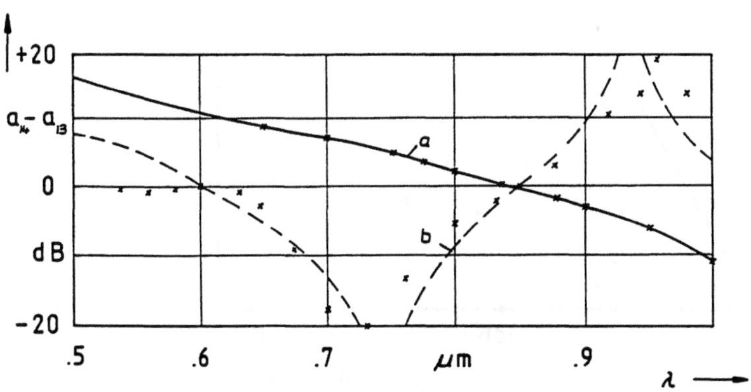

Fig. 8. Calculated and measured spectral response of a standard coupler (a) and a dual wavelength coupler for 633 nm and 850 nm (b)

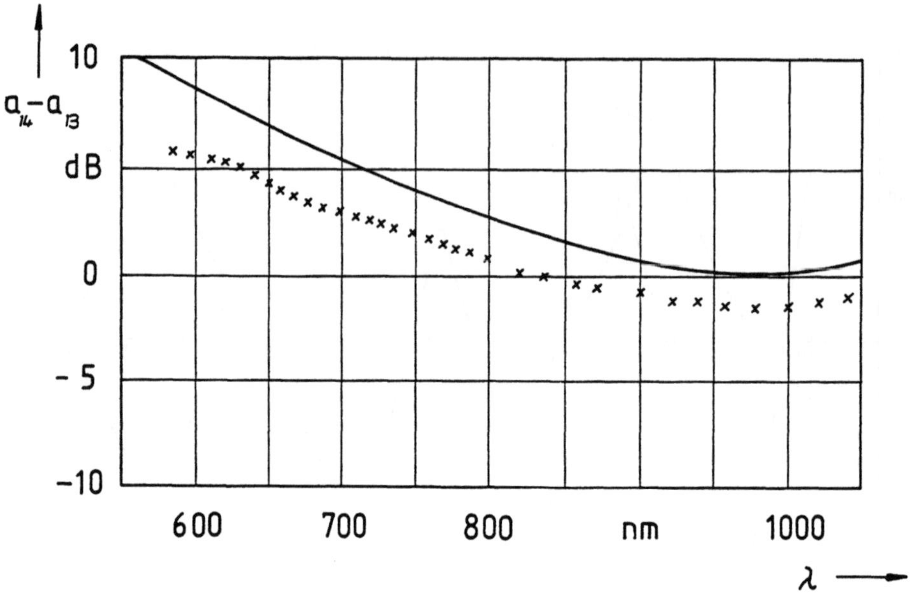

Fig. 9. Calculated and measured spectral response of a broadband coupler

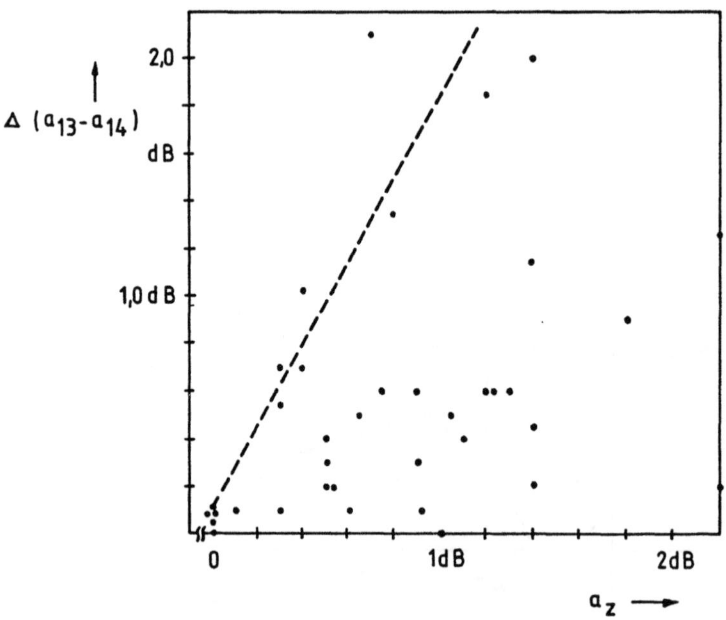

Fig. 10. Tolerance of the logarithmic splitting ratio plotted against excess losses with 360° change of the launched linear polarization state

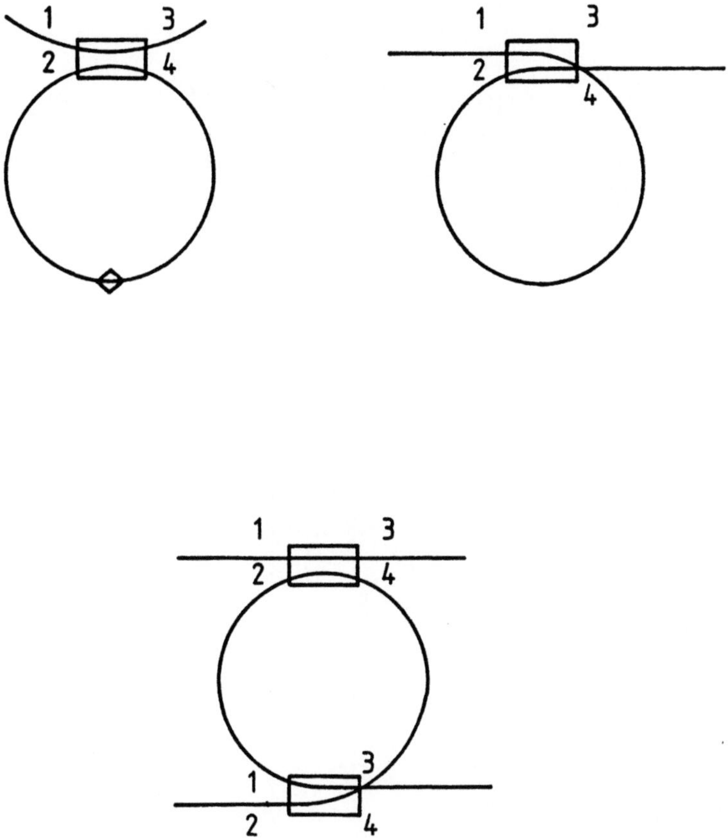

Fig. 11. Structure of fibre optic ring resonators

Sensor and Actuator Technologies

Sensor Packaging

M. Esashi

Department of Precision Engineering
Tohoku University
Sendai, JAPAN

Summary

Packaging methods for mechanical sensors and chemical sensors are described. High reliability, small size and small packaging cost are required for many sensor applications. Electrical and mechanical instabilities which originate from surface ion and packaging stress respectively are reduced. Batch-process packaging which makes it possible to assemble before dicing to each chips is developed and applied for sensors as integrated capacitive pressure sensor. Bonding methods and their applications are also described.

Introduction

High reliability is most important for sensors. Semiconductor is a functional material and has been used for sensors,but it may responds to many physical or chemical quantities . Sensors must be sensitive to only one quantity. Though optical or magnetic sensors can be used inside their carriers,some mechanical sensors as pressure sensors or chemical sensors have to be exposed and they suffer many interferences. This is the motivation for this sensor packaging study and silicon micromachining has been applied.

Small size and low packaging cost are required in many applications. For biomedical instrumentation,micro sensors have to be mounted at the tip of a catheter tube. Sensor cost should be small for disposable use.

One of key technologies for the sensor packaging is bonding. Bonding methods suitable for each applications have been developed.

Micro sensors for catheter

Catheter tip micro sensors are suitable to monitor physical or chemical quantities continuously in vivo,i.e.in blood vessel. Small size chemical sensors for the purpose have been developed using Ion Sensitive Field Effect Transistor (ISFET). The structure and the principle of the ISFET are shown in Fig.1. Practical reliability for long term use was given by the microma chined probe type structure shown in Fig.1. The whole layers of insulators including a chemically inert silicon nitride layer. The active gate region of the FET was located at the tip of the probe,and the n type drain diffusion layer between the gate and bonding pads was buried under the p^+ layer to reduce instabilities due to a parasitic channel. Moreover the surface was rounded,because a smooth surface is mechanically tough. This sensor has been used as pH and CO_2 sensors clinically.

Pressure sensors for catheter have been developed. A sidewall type which is suitable for multi point pressure sensing is shown in Fig.2. The problem due to the surface charge which modulates the resistance of the piezoresistors was solved by burying piezoresistors as shown in Fig.2. Additionally,the problem due to the packaging stress which causes thermal disturbances was solved by using athick supporting rim of the sensor chip and further by mounting on a stainless steel support with elastic material.

Fig.3(a) shows the structure of a catheter-tip piezoresistive pressure sensor. The difficulty of this kind of sensor is mainly the lead wire attachment. To reduce the assembly work and to ensure the reliability of the micro-soldering of lead wires, solder pools surrounded by polysilicon layer of 50um thickness

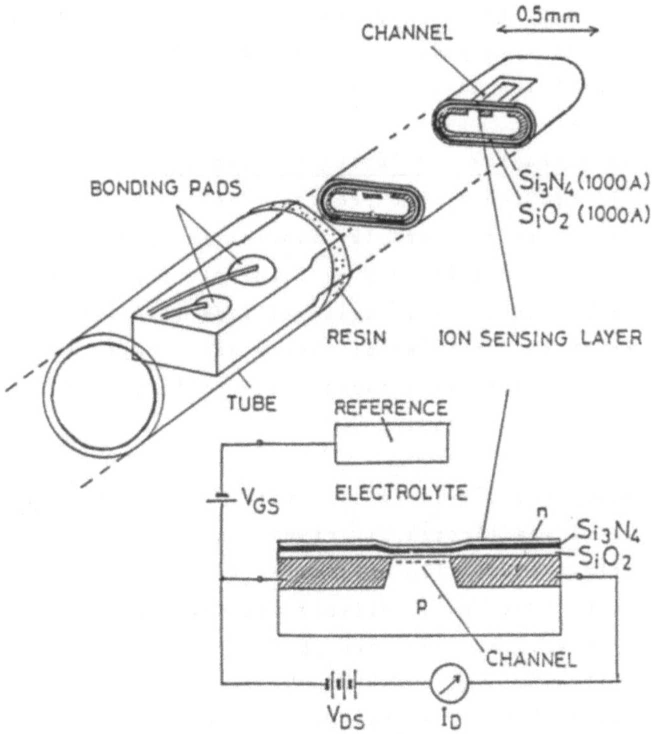

Fig.1 Structure and principle of the ISFET.

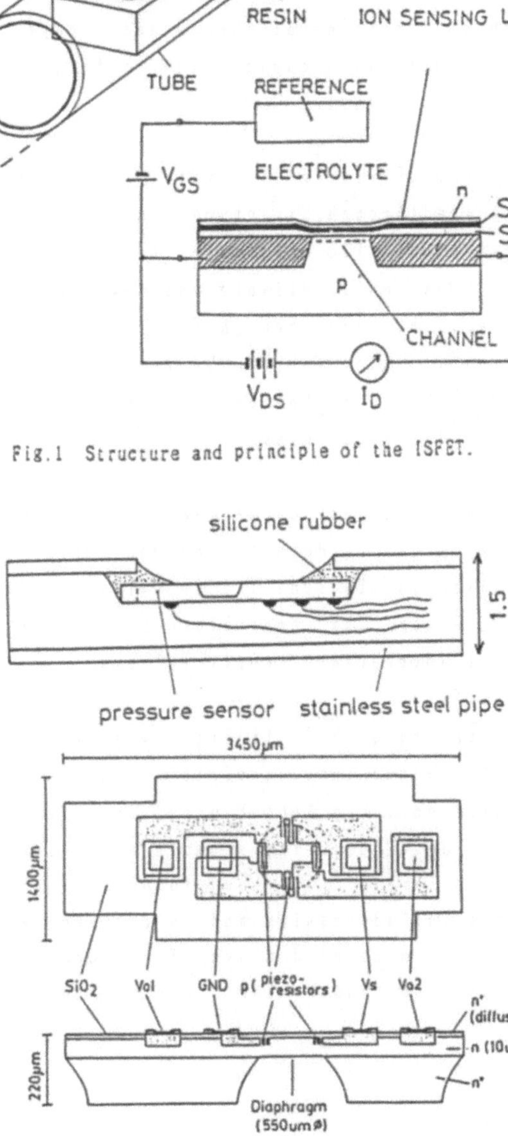

Fig.2 Pressure sensors for catheter

were formed on the sensor chip. The fabrication process is shown in Fig.3(b). Selective oxidation process called LOCOS was used to open contact holes at the bottom of the pool.

Batch process packaging and integrated sensors.

Integrated silicon capacitive pressure sensors have been fabricated. The structure and the fabrication process are shown in Fig.4 and Fig.5 respectively. A Pyrex glass plate and a silicon wafer are anodically bonded to form internal cavity for reference pressure. A sensing diaphragm and a capacitance detection circuit which is a capacitance to frequency converter are integrated in each sensor.

The feedthrough from the internal cavity is made through the glass hole. This feedthrough structure can not only seal the cavity hermetically but also make wafer level batch process packaging which gives small,reliable and low-cost sensors. Narrow holes have to be made through the glass. Electrochemical discharge drilling was used for the boring. Fig.6 is the schematic diagram of this method and it takes few seconds to make a hole in a 300μm thick glass.

The pressure-output frequency curve measured is shown in Fig.7. The output frequency can be detected by monitoring the consuming current in the power supply line,i.e.two wires detection is capable for this sensor. The temperature shift of the output can be canceled out at appropriate supply voltage.

The feedthrough structure shown in Fig.4 was applied for other sensors. A tactile imager that consists of an 8 X 8 piezoresistive silicon force sensor array on a flexible polyimide film was developed. The structure of each force sensor chip is illustrated in Fig.8. It has piezoresistors for force sensing and diodes for addressing and is packaged with the feedthrough structure mentioned above. The micro flow sensor shown in Fig.9 uses the feedthrough structure too.

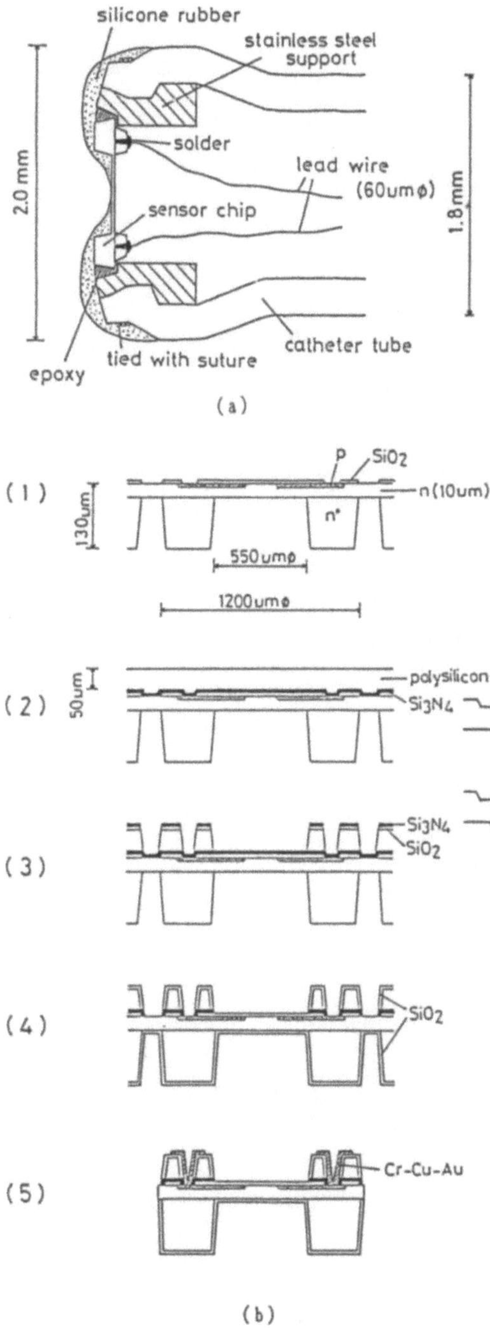

Fig. 3 Catheter tip micro pressure sensor

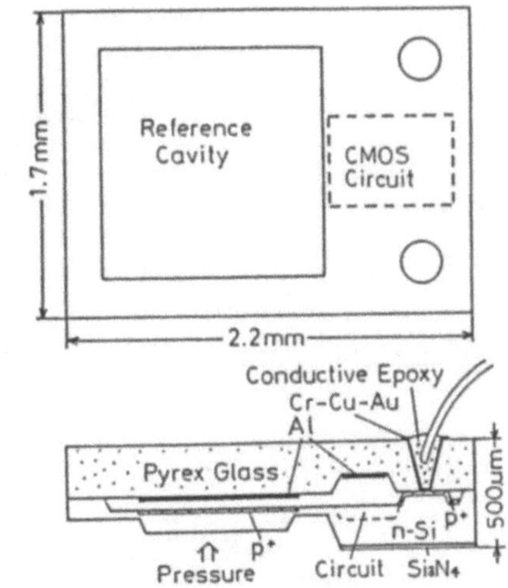

Fig.4 Structure of the integrated capacitive pressure sensor.

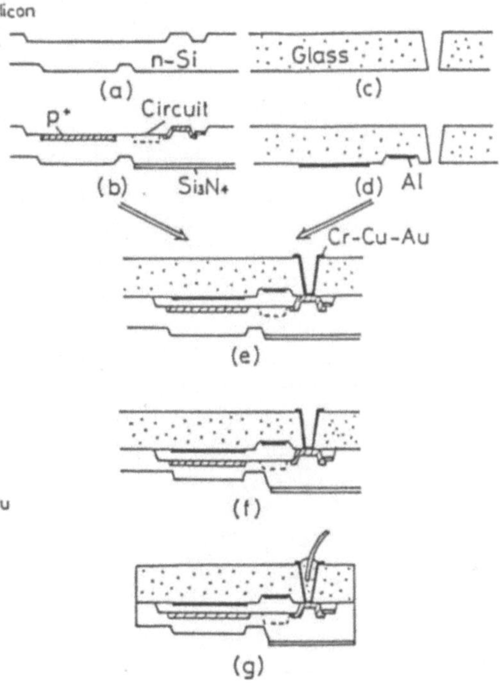

Fig.5 Fabrication process of the pressure sensor.

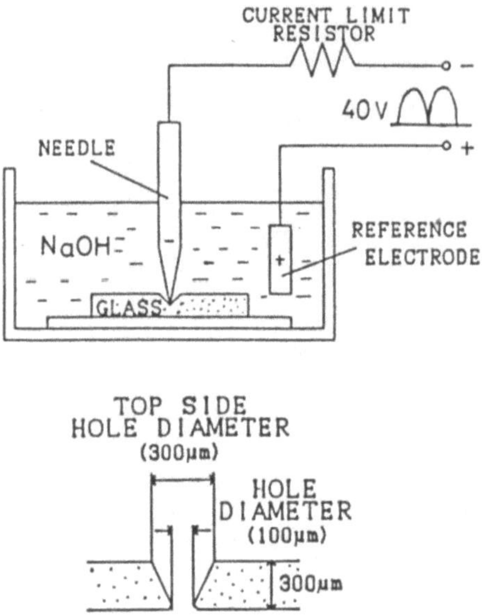

Fig.6 Electrochemical discharge drilling

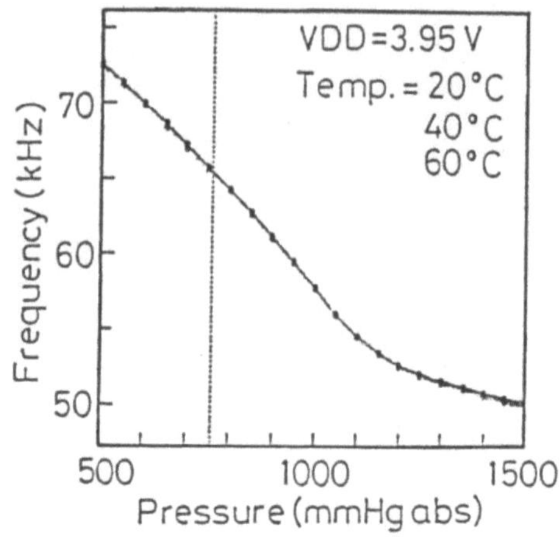

Fig.7 The relation between output frequency and applied pressure.

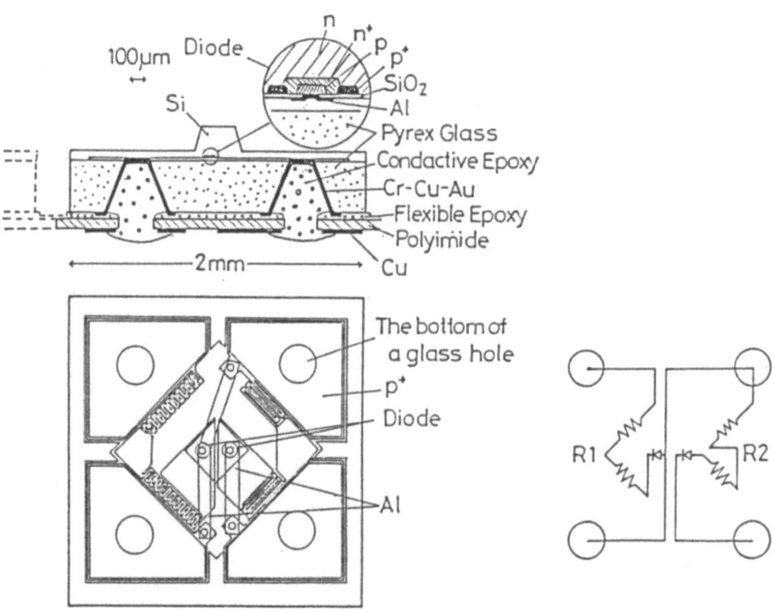

Fig.8 A tactile imager

Bonding

Pyrex glass to silicon anodic bonding as illustrated in Fig.4 has been popular for micromachined devices. However, another bonding methods are required for some purposes and examples are shown in Fig.10. Fig.10(a) is a capacitive pressure sensor which was fabricated with high temperature silicon to silicon direct bonding which solves the problem of thermal stress. Fig.10(b) is a micro pump fabricated on a silicon wafer. Room temperature anodic bonding with intermediate sputtered low melting point glass was used. The method can be used for materials with different thermal expansion coefficient. Fig.10(c) is a micro flow cell which has chemical sensors in it. Dry film was used to bond silicon and glass at about 150° C.

References

1. Matsuo,T; Esashi,M: Methods of ISFET fabrication. Sensors and Actuators. 1 (1981) 77-96
2. M.Esashi et.al.: Fabrication of catheter-tip and sidewall miniature pressure sensors. IEEE Trans. on ED. ED-29 (1982) 57-63
3. Esashi,M; Matsumoto,Y; Shoji,S: Absolute pressure sensors by air-tight electrical feedthrough structure. Sensors and Actuators. A21-A23 (1990) 1048-1052
4. Shoji,S; Nisase,T; Esashi,M; Matsuo,T: Fabrication of an implantable capacitive type pressure sensor. Transducers'87 (1987) 305-308
5. Esashi,M; Nakano,A; Shoji,S; Hebiguchi,H: Low-temperature silicon-to-silicon anodic bonding with intermediate low melting point glass. Sensors and Actuators. A21-A23 (1990) 931-934
6. Esashi,M; Shoji,S; Nakano,A: Normally closed microvalve and micropump fabricated on a silicon wafer. Sensors and Actuators. 20 (1989) 163-169

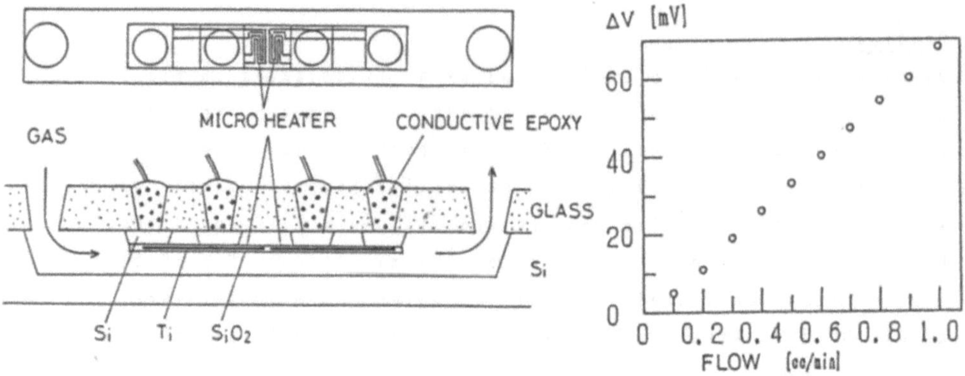

Fig. 9 Micro flow sensor

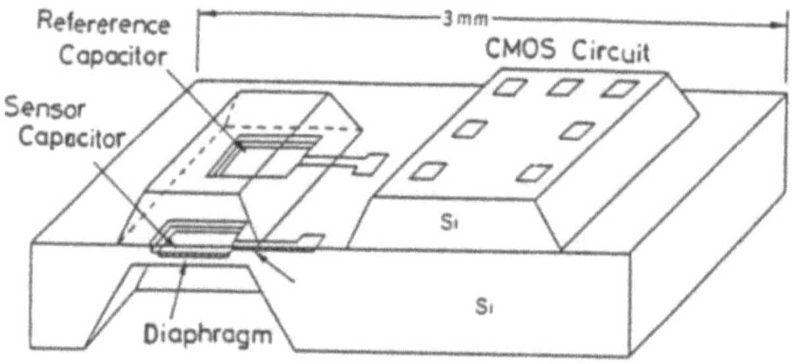

(a) capacitive pressure sensor fabricated with silicon to silicon direct bonding.

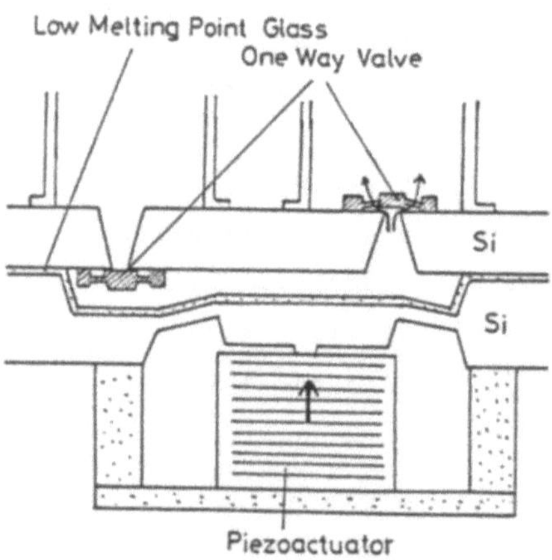

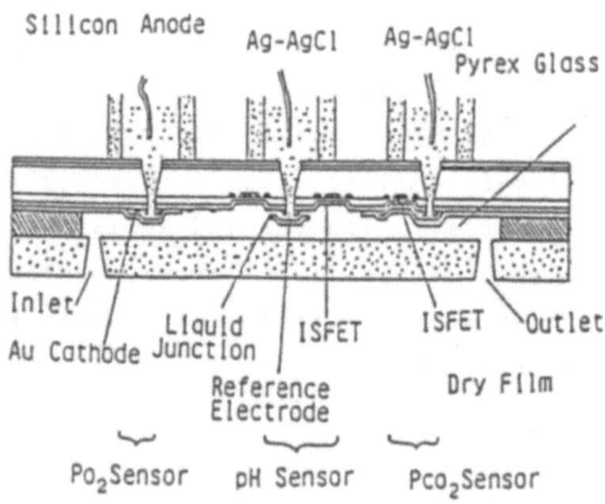

(b) micro pump

Room temperature anodic bonding.

(c) micro flow cell. Dry film was used to bond silicon and glass.

Fig. 10 Bonding

A Method for Thermal Insulation of Microstructures

I. STOEV , N. BOGDANOVA

Institute of Solid State Physics,
Bulgarian Academy of Sciences, Sofia, Bulgaria
Higher Institute for Electrical and Mechanical
Engineering, Sofia, Bulgaria

Summary

The possibilities for utilizing the anisotropic etching for fabrication of suspended microstructures, situated at 45^0 over against (110) edge, have been discussed. The obtained structures can be with or without SiO_2 sublayer. They can be used for flow sensors as well as for other sensor structures.

Introduction

For the formation of some kinds of sensors it is necessary to realize a thermal insulation between the sensitive element and the wafer as well as between the sensitive element and the other structures on the substrate. This technological problem is closely linked with the flow sensor practical implementation.

Usually the temperature differences, for example between two resistors, are to be determined, the one of which is heated or cooled. Their responce, sensitivity and selectivity are defined to a large extent by the thermal insulation of these structures from the substrate.

The silicon technology and deep anisotropic etching have led to the creation of a new generation of sensors with threedimensional configuration. Thin diaphragms or cantilever formation have became possible. The flow sensor disposition on thin silicon or oxide diaphragm increases considerably their response. The oxide cantilever or bridge disposed up on a groove in silicon wafer practically insulates the sensitive element from the substrate. Such kind of diesign allows for measurements to be conducted, registrating only the temperature changements, brought by the gas under investigation.

Technological and design problems

The subject of investigation of this paper are the possibilities which the anisotropic etching offers for forming of suspended microstructures on silicon substrate.

Different methods for obtaining of suspended microstructures, utilizing anisotropic etching of the backside of the wafer have been known. These methods are complicated and unreliable, because they are related with double side photolithography and complex technological cycle. Considerably more simple is the method, utilizing p^+- stopping layer. It permits opening of V-groove from the face side under the active element. In this case an epitaxial layer is necessary.

The utilizing of SiO_2 sublayer leads to a substantiel simplification of the technological order. Depending on the oxide bridge orientation towards (110) at wafers with (100) orientation two cases are possible. The first one is connected with standard trimmimg at anisotropic etching in parallel or perpendicularly as over against the (110) edge. To obtain an oxide bridge, vertical to V-groove it is necessary to penetrate deeply under the oxide into the volume of the silicon wafer. This is achieved by forming a deep cavity [1]. The second one approach is completely opposite and utilizes structure orientation at 45° over against the (110) edge. In this case etcher attacks besides standard crystalographic edge (111) and such created by the presence of convexed edges, for example edge (211). Due to that, the velocity of etching of the silicon under the oxide bridge is considerably higher. In this way an air cavity is created [2]. A characteristic feature of this oxide bridge is that it is situated diametrically to the square aperture in oxide mask. Underetching of the two ends of the bridge is faciliated by the mutual intersection of two planes (111) and the correspondent stopping of the etching without creation of big stress under the oxide.

Our experiment is related to the investigation of the possibilities for development of this approach with a view to delivering the sensor element from the oxide bridge. This method leads to additional decrease of the resistor thermal inertion. Figure 1 shows the basic idea of the developed structure. A trimming at 45° against (110) edge is used. But sensor element and oxide under it intercept the rectangular groove. Their two ends make 45° and 135° angles. Due to that, during the etching the etcher attacks the silicon laying under the resistor with a different speed. The highest underetching speed is obtained under the middle part of the bridge, where the V-groove depth is the biggest and the silicon etching is accomplished along curves shown in the figure by dotted line. As a result of silicon etching out mechanical stress is engendered in the oxide, leading to oxide bridge destruction and resistor suspension [3].

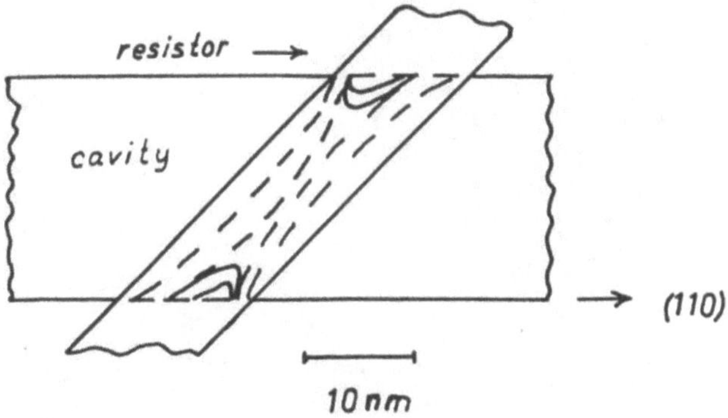

Fig. 1 Schematic presentation of the Si etching process under sensor structure

Results

The obtained results for the sensor layer structures from Al, Ta and Ti deposited by RF sputtering show that the mechanical stability of the resistor is influenced by the oxide sublayer thickness. The last is expressed in uniform bending in the middle of the resistor after the oxide layer detachment. In not a single experiments occured neither resistor breakage, nor resistor value changement. An anisotropic etching for V-groove formation under the oxide layer was carried out by standard microelectronic etcher like EDP, KOH, N_2H_4. From the mechanical stability point of view the last solution showed the best results. It is most lixely due to the higher speed of etching of the convexed angles.

Figure 2 shows Ta sensor element, situated at 45° over against rectangular anisotropic etched grooves. The carrying oxide thickness is 300 nm. With its detachment in the middle of the sensor structure is occured peculiar bending. The resistor resistivity before and after etching is 750 Ω . An experiment was carried out by small oxide bridges with different thickness. As a result it was found out that the characteristic bending is obtaind at oxide thickness up to 800 nm.

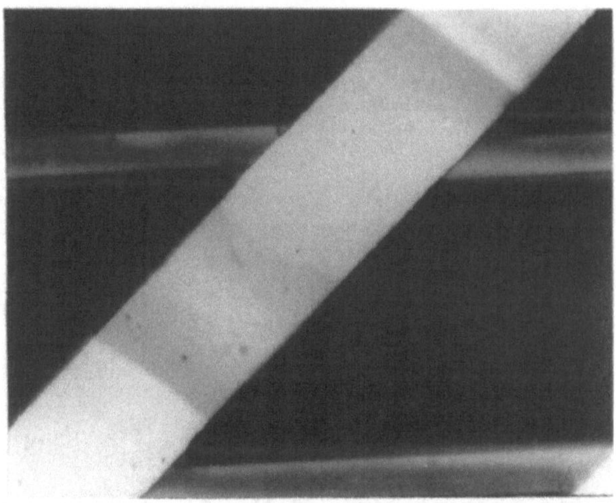

Fig. 2 Microphotography of Ta resistor without SiO_2 sublayer

Figure 3 shows the general view of a Ta sensor element with oxide bridge (thickness 1000 nm). Bending, typical of previous structure was not observed. In this case the oxide has got the necessary mechanical strength.

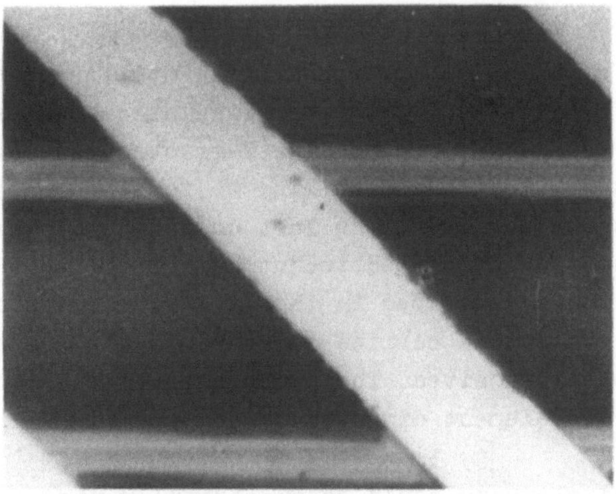

Fig. 3 Microphotography of Ta resistor with SiO_2 sublayer

The thermal inertness of the structures was investigated. Ta resistor before and after anisotropic etching was measured using Wheatstone bridge. In terms of constant air flow the bridge was debalanced up to 35 mV. Then the flow was turned off and the time for reaching equilibrium was measured. Figure 4 shows the measurement results of sensor structures before etching (curve 1), of sensor structures with oxide bridge (curve 2) and of suspended sensor structures (curve 3).

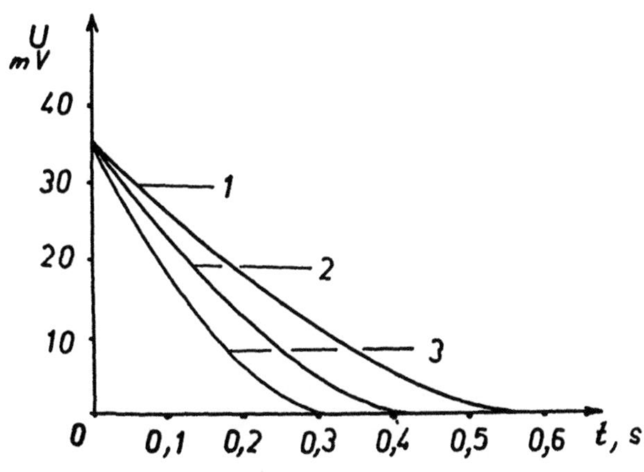

Fig. 4 Thermal inertness of Ta resistor

Conclusions

The results show that it is possible to utilize anisotropic etching for fabrication of thermal insulated bridges from metal layers, going over rectangular V-grooves. An advantage of these structures is the possibility to form sensor structures with and without sublayer. The last ones have diminshed thermal inertness. The described technological cycle is simplified compared to those described in the literature. A single time anisotropic etching is carried out through the resistor which is already been deposited. There are not restrictions concerning the sensor element dimention.

The described elements could be successfully used for fabrication of flow sensors and other sensor structures.

References

1. Bergveld, P.: Sensors and actuators, Twente. Sensors and Actuators 17 (1987) 3 - 26

2. Parameswaran, M.; Baltes, H. P.; Ristic, Lj.; Dhaded, A. C.; Robinson, A. M.: A new approach for the fabrication of micromechanical structures. Sensors and Actuators 19 (1989) 289 - 307

3. Stoev, I.: A method for fabrication of thermal insulated structures on silicon. Bulgarian Patent 89254 (1989)

Machining of Metal Foils for Use in Microcompact Heat Exchangers

Craig R. Friedrich and Robert O. Warrington

Department of Mechanical and Industrial Engineering
Louisiana Tech University
Ruston, Louisiana 71272 USA

Summary
The stress distribution in a thin foil subject to micromachining is modeled to predict the allowable depth of cut with a diamond turning tool. From these results a fixturing system is devised and microflow channels are cut into 100 micrometer thick aluminum foil using a conventional computer numerically controlled lathe. The cutting method and results and the results from the numerical model are presented. The application for use in microcompact heat exchangers is also presented.

Introduction

The evolution of micromechanical systems is requiring many components to be miniaturized. Among these components are heat exchangers for use in microelectronics cooling and aerospace applications. A microcompact heat exchanger, using diamond-cut flow channels, has been developed by Bier, et al [1,2] at Kernforschungszentrum Karlsruhe, West Germany. The heat exchanger uses aluminum or copper foils which have had rectangular flow channels cut into the surface using a machine equipped with air bearings. The group at Louisiana Tech University is currently using a CNC lathe equipped with mechanical roller bearings to do preliminary testing and design. The purpose of this work is to study the cutting and fixturing environment so that the heat exchanger design may be structuraly and thermally optimized for a particular application, when manufactured on an air-spindle machine.

Stress Distribution Model

The purpose of the stress distribution model is not to predict actual stresses at the cutting interface, but rather to predict the relative stress level at the base of the foil for proper fixturing. A segment of foil being cut is modeled as shown in Figure 1. The model is especially useful for pre-

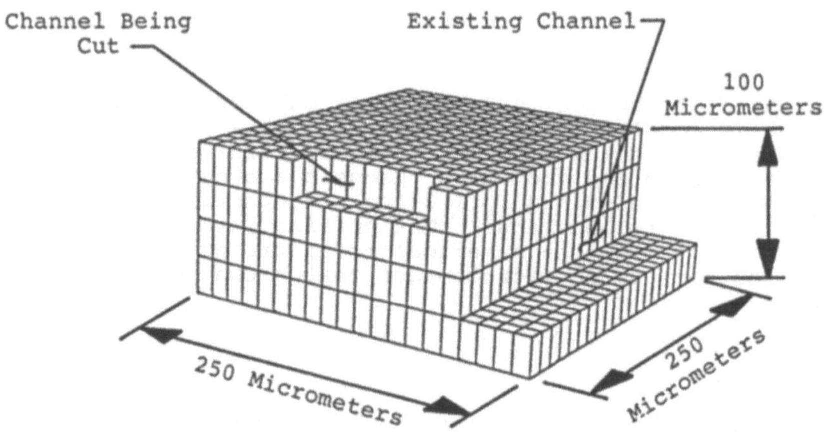

Fig. 1. Finite element model of segment of foil being cut

dicting the relative stress level in an adjacent flow channel which has already been cut. The channels are modeled as being square although the channels actually being cut are currently trapezoidal in shape. The trapezoidal shape of the diamond is to help prevent premature failure of the tool due to excessive vibration from the mechanical bearings. The material used in the model is 2024-T4 aluminum with a yield strength of 172 MPa and an ultimate tensile strength of 303 MPa. The magnitude of the cutting forces for this material are taken from the study by Strenkowski and Carroll [3]. It is reasonable to assume that for a given cutting force vector direction and a constant shear plane angle of the material, the magnitude of the stress at a point in the material should vary with the material strength. This simplifying assumption is used to determine the locus of points in the material which are subject to a certain fraction of the yield strength, for example. A cross-section of the model is shown in Figure 2. This result shows

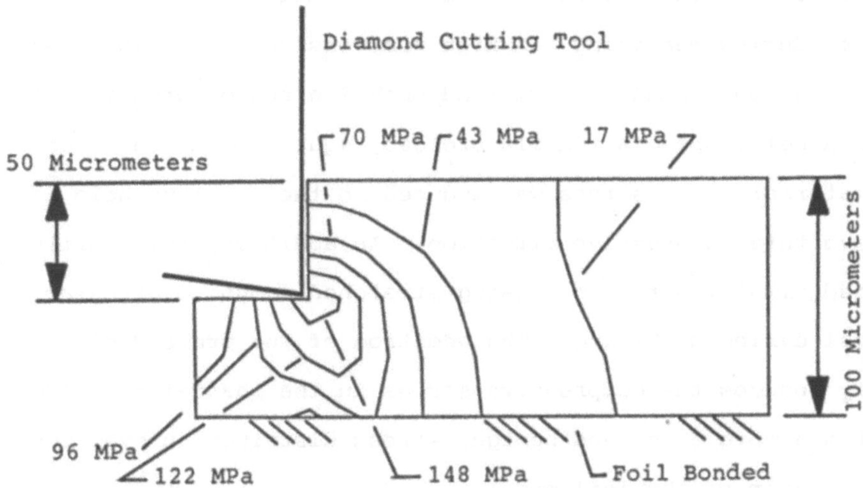

Fig. 2. Foil cross section showing maximum shear stress distribution for a 50 micrometer deep plunge cut

that for a single 50 micrometer deep plunge cut, the distortion energy theory predicts that some yielding of the foil will take place near the base. This result coincides with observation of the bottom of the foil after actual machining of a 50 micrometer deep channel with a single cut. Machining has also shown that smoother channels are produced when the depth of cut is less than 2.5 micrometers. This result is especially critical when machining soft aluminum, such as 1145, which is nearly pure.

The stress distribution model is also used to predict the behavior of an adjacent, previously cut, channel when a new channel is being cut. As shown in Figure 1, a new 25 micrometer deep channel is being cut adjacent to an existing 75 micrometer deep channel. An investigation of the tensile stress in the cutting direction, in the vicinity of the cut channel, revealed that a relatively high compressive stress was present at the channel when the foil was held onto a fixed surface during machining. A high compressive stress in a very thin area could lead to localized buckling and deformation of the channel shape. A tensile preload, equal to one-third of the material yield stress was induced in the model to help relieve this compressive situation. In addition, the tensile preload simulates the foil being stretched around a circular mandrel during machining. The addition of the preload significantly reduces the compressive stress in the channel and also creates a more even, and benign, stress distribution over the bottom layer of the foil model.

Fixturing Method

A fixturing method was built as a result of the cutting model. The fixture consists of a five-inch diameter 6061-T6 aluminum mandrel which is four inches wide. The mandrel has a milled slot which accepts a mechanism for clamping the ends of the foil together. This assembly is then placed in the CNC lathe and the channels are cut. This arrangement is shown in Figure 3. This is similar to the arrangement used by the group at Karlsruhe, but was developed as a result of the cutting model. One shortfall of this fixturing method is the inability to accurately measure the tensile preload in the foil. Inspection of the channels from a number of machining tests has lead to a subjective "feel" for the correct amount

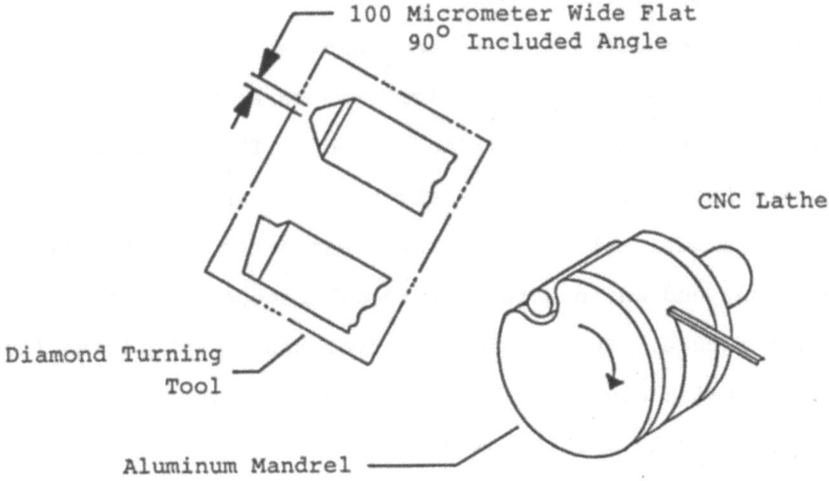

Fig. 3. Fixturing arrangement for foil machining

of preload. The milled slot and the clamping mechanism do not create enough imbalance to cause a vibration level above the ambient vibration of the lathe bearings. This conclusion has been reached by an extensive study of the vibration and damping characteristics of the lathe using a Fast-Fourier-Transform analyzer. The spindle is run at a speed which produces a small amplitude of vibration.

Future Work

The foil strips are cut into square coupons so that they may be stacked and bonded or clamped to produce a crossflow heat exchanger. The present design will be extended to produce a three-fluid heat exchanger by the addition of holes which are perpendicular to the foil surface. These holes will be on-the-order of tens-of-micrometers in diameter and will be produced on a machine similar to one developed by Masaki, et al [4]. The crossflow and three-fluid heat exchanger designs are currently being optimized for structural strength, heat transfer efficiency, and minimum weight. The Micromanufacturing Laboratory is currently being equipped with air spindles, positional stages with 50 nanometer incremental resolution, and an interferometric microscope is planned for the near future. In addition, a novel thermal stabilization scheme is being developed which uses evaporative cooling to maintain thermal equilibrium during machining.

References

1. Schubert, K.; Bier, W.; Linder, G.; Seidel, D., "Herstellung und Test von kompakten Mikrowarmeubertragern," Chem.-Ing.-Tech. 61 (1989) Nr.2, pp.172-173.

2. Bier, W., et al, "Manufacturing and Testing of Compact Micro Heat Exchangers with High Volumetric Heat Transfer Coefficients," for presentation at American Society of Mechanical Engineers Winter Annual Meeting, November 1990.

3. Strenkowski, J. and Carroll, J., "A Finite Element Model of Orthogonal Metal Cutting," ASME Transactions-Journal of Engineering for Industry, Vol. 107, November 1985, pp.349-354.

4. Masaki, T., et al, "Micro Electro-Discharge Machining and Its Applications," Proceedings IEEE Micro Electro Mechanical Systems Workshop, February 1990, p.21.

Novel Stress Free Assembly Technique for Micromechanical Devices

H.L. Offereins, H. Sandmaier
Fraunhofer– Institut für Festkörpertechnologie
Paul–Gerhardt–Allee 42
8000 München 60 (F.R.G)

Abstract

A novel technique for the stress free assembly of micromechanical devices is presented. In contrast to conventional assembly methods, the micromechanical device is connected to the housing via an intermediate part made of silicon with a bellow structure. This concept guarantees a mechanical decoupling of the chip and allows an efficient fabrication by full wafer processes.

Introduction

Over many years particular attention has been paid to the packaging of semiconductor sensors and micromechanical devices, due to the important role they play on the accuracy of the entire device [1]. Complex constructions for the assembly of e.g. mechanical sensors mounted in single chip processes led to the fact, that the price of a sensor was primarily determined by the assembly technique. The packaging of a sensor can be fabricated far more efficiently, if measures for a mechanical decoupling are already taken in the silicon element.

Fig. 1 shows a cross section of a conventional semiconductor pressure sensor assembly setup [2,3,4]. The sensor chip is connected to the housing via a thermal conformed intermediate made of silicon or pyrex glass. The mounting of these devices is made by gluing, soldering, alloying, anodic bonding or silicon fusion bonding. In order to reduce the stresses caused by the packaging, the intermediate is connected by a tube to the housing. One disadvantage of this concept is the demand that all the devices used exhibit a conformity in thermal behaviour.

An analysis of this conventional sensor assembly shows following sources of stresses in the diaphragm of the sensor chip caused by the packaging:

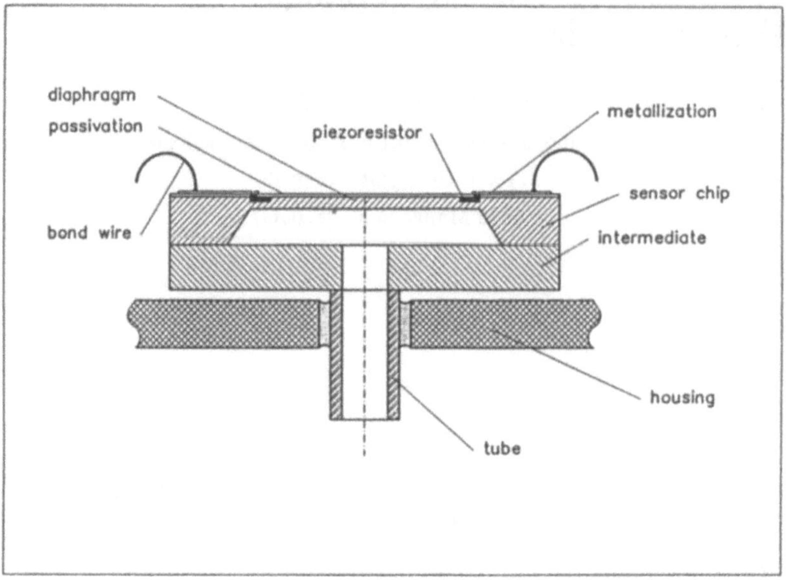

Fig. 1: Conventional sensor assembly setup for a piezoresistive pressure sensor

Different thermal expansion coefficients for housing and intermediate materials lead to stresses in the contact area of the intermediate to the tube. When different materials are used for sensor chip and intermediate, this effect also causes stresses in the contact area of the intermediate to the sensor chip. FEM simulations on this setup showed that these stresses have strong influence on the stress distribution in the area of the piezoresistors.

Stresses in the diaphragm, which are induced by the packaging deteriorate the characteristics of the sensor. They primarily influence the long term drift, the offset as well as its temperature dependance, the temperature cycle hysteresis, and the transient response to an abrupt temperature change [1,5]. The development of micromechanical sensors with high accuracy requires an almost completely stress free assembly technique.

Principle

A complete mechanical decoupling is possible, if an assembly setup as shown in Fig. 2 with a structured intermediate is used. By a doublesided anisotropic etching process the intermediate is formed to a bellow structure as shown in Figs. 2 and 3. In Fig. 4, an intermediate with an anisotropically etched bellow structures is shown.

Through the bellow structure, strains caused by relative displacements of the sensorchip to the housing are concentrated in the easily deformable intermediate. Thus, stresses induced by the packaging are relaxed in the intermediate and have no influence on the diapragm of the sensor chip. The complete decoupling can be seen in Figs. 3 and 4 as the results of calculations with a finite element program.

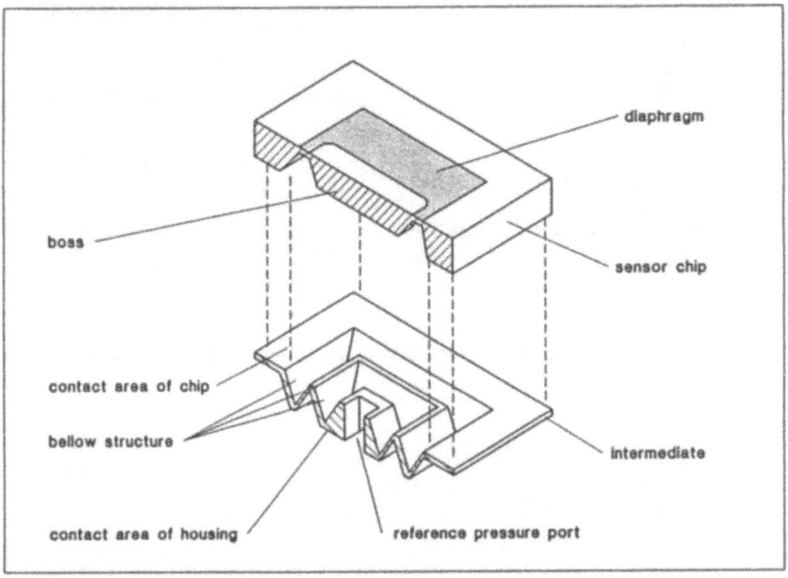

Fig. 2: Sensor assembly setup with structured intermediate

For obtaining a mechanical model for the calculation, the setup in Fig. 1 was sectioned in the contact area of the tube to the intermediate. In the finite element model, the mechanical influence of the housing due to different thermal expansion coefficients was simulated by displacements which act on the intermediate in its contact area to the tube. The amount of the displacements was calculated with a temperature change of 450K, as it appears during an anodic bonding process. The inner volume of the sensor was subjected to normal pressure.

The effect of the bellow structure is demonstrated in Figs. 3 and 4: the displacements caused by the packaging at the inner ring of the intermediate are relaxed by the deformations of the bellow structure. The stresses in the diaphragm are exclusively caused by the pressure in the interior of the setup. In comparison to the setup with a nonstructured intermediate, the influence of the packaging on the stresses in the diaphragm can be reduced up to 99%.

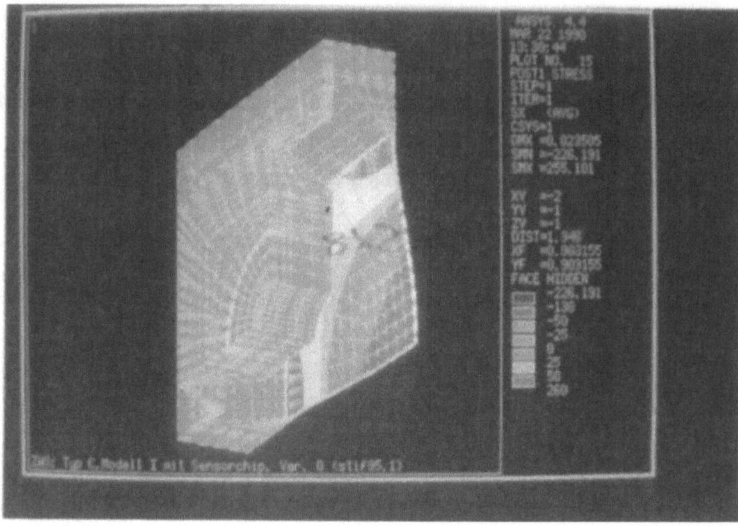

Fig. 3: Stresses in a pressure sensor with an anisotropically etched intermediate

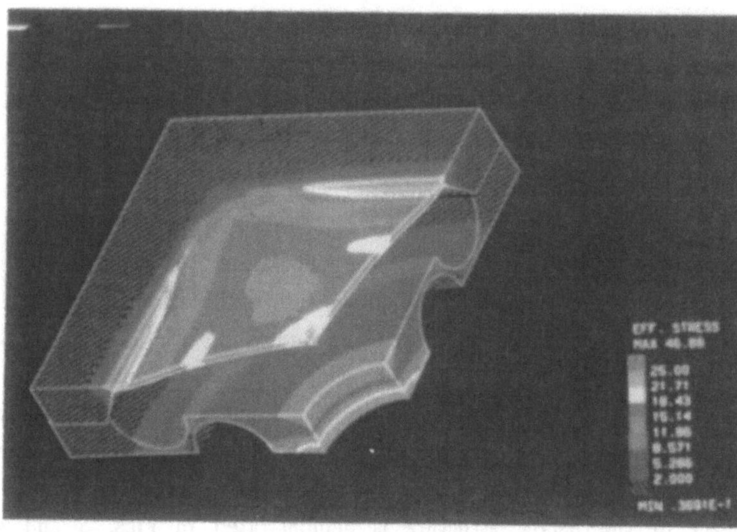

Fig. 4: Stresses in a pressure sensor with isotropically etched intermediate

In case of full wafer bonding, the wafer including the sensor chips and the wafer with the intermediates may have different or contrary bendings. In order to induce no stresses in the diaphragms during this process, the bending of the wafer including the sensor chips should not be influenced by the bonding process. Structuring the intermediate with a thinned contact area to the chip, it has far less resistance to deformation than the bulk of the chip. Thus, all the deformations necessary for the mounting are done by the intermediate.

The geometry of the intermediates can be varried within the limits of the etching technology. In the case of anisotropic etching, different etching depths for the thinned contact area to the chip and each of the closed V–grooves are possible. The number of closed V–grooves are restricted by the dimensions of the sensor chip.

The disign optimization for the best geometry and dimensions of the intermediate was carried out with the help of a finite element program. The variation of the geometry was restricted to the dimensions of a 500mbar pressure sensor and a wafer thickness of 380μm. The minimal thickness of the thinned areas in the intermediate was decided as the double diaphragm thickness.

The simplest form of a bellow structure is a thin contact area to the chip. Properly designed, this form already reduces the stresses caused by the packaging in comparison to the variant with a nonstructured intermediate up to 80%. The advantages of this design are its small need of space and the simple, one sided etching process.

The intermediate variant with a thin contact area to the chip and one closed V–groove in the upper side shows no substantial improvement compared to the simply thinned contact area. The reduction of stresses caused by the packaging is about 83%. This variant requires increased spatial requirements and a special compensation structure preventing the underetching of the convex corners in a bent V–groove [6].

The best mechanical decoupling was achieved with a thin contact area to the chip and double sided V–grooves (Fig. 2). With this variant, the stresses caused by the packaging are reduced up to 95%. By using a 525μm silicon wafer for the sensor chips, this value can rise up to 99%. This gives the most suitable design for low pressure sensors, where the stresses caused by the packaging have the strongest influence on the sensor output signal and spatial requirements play a secondary role.

Conclusions

The novel stress free assembly technique has the following advantages:

- Reduction of fluctuations in the characteristic of the devices.
- Improvement of the long term drift and temperature dependance of micromecanical devices.
- The mounting of the micromechanical devices to the intermediates

with bellow structures can be carried out effectively in a full wafer process.

– As a result of the mechanical decoupling, a simple and cheap assembly in a housing is possible.

– The simulation and the design optimization can be carried out without considering the packaging.

The assembly technique presented can be used for almost all micromechanical sensors. There is no restriction to the sensor principle, as piezoresistive, capacitive or resonative structures. It also shows advantages for actuators and micromechanical systems.

Acknowledgements

The authors wish to thank B. Folkmer for his valuable support in the finite element simulations.

References

1. M. Esashi, H. Komatsu, T. Matsuo, M. Takahashi, T. Takishima, K. Imabayshi, H. Ozawa, Fabrication of Catheder–Tip and Sidewall Miniature Pressure Sensors, IEEE Trans. on Electron Devices, Vol. ED–29, No. 1, pp. 57–63, 1982
2. P. Mihm, Aufbau und Gehäusung von Silizium Drucksensoren, AMA Seminar, pp. 285–295, 1989
3. G. Pollak–Diener, Verbindungsverfahren für Mikrosensoren und Mikroaktuatoren, AMA Seminar, pp. 269–284, 1989
4. F. Breimesser, M. Poppinger, A. Schwaiger, Piezoresistive Pressure Sensor with Silicon Diaphragm, Siemens Forsch. und Entwickl.– Ber., Bd. 10, Nr. 2, pp. 72–77, 1981
5. H.L. Offereins, H. Sandmaier, Stressfreie Chipmontage, me, Bd. 4, Heft 1, Beilage Mikroperipherik I, VDE– Verlag GmbH, Berlin, 1990
6. H.L. Offereins, H. Sandmaier, K. Kühl, G. Meyer, Methods for the Fabrication of Convex Corners in Anisotropic Etching of (100)– Silicon using Aqueous KOH, Eurosensors IV, Karlsruhe, October 1–3, 1990

The LIGA Process for Microsystems

W. Ehrfeld

Steag Aktiengesellschaft, Bereich Mikrotechnik,
Griesbachstraße 10, D-7500 Karlsruhe 21

Summary

Threedimensional microstructures can be fabricated from metals, metal alloys, plastics and ceramic materials with submicron tolerances by means of the so-called LIGA process which is based on a combination of lithography, electroforming and molding processes. Since a large variety of materials can be processed without any restrictions in the cross-sectional shape of the microstructures, a wide range of applications exists. Present development work deals with sensors for automotive technology, optical components for satellite experiments, optical demultiplexers and fiber-chip coupling elements for optical communications, micronozzles and microvalves for fluid technology, microfiltration devices, detachable electrical micro-connectors, ultrasonic sensors for medical diagnosis, and various further microstructure products. In view of the versatility of the LIGA process, it is to be expected that it will play a major role in realizing future innovative microsystems.

Introduction

When regarding microelectronic circuits as the most prominent microstructure products, it may be clearly stated that microfabrication is of fundamental importance for all fields of modern technology. However, it should be also emphasized that the overall potential of microfabrication is not only restricted to microelectronics. Besides the standard planar VLSI fabrication processes, new methods for generating threedimensional microstructures have been developed which allow to manufacture advanced micromechanical and microoptical components. Thus, a broad technological basis exists for producing microsystems which may comprise electronic, mechanical, optical, chemical and other functions.

A particularly promising method for producing threedimensional microstructures is based on a combination of deep-etch lithography and subsequent high-precision replication processes /1/. This method is known under the acronym LIGA which characterizes its essential fabrication steps comprising lithography, electroforming and molding processes (in German: Lithographie, Galvanoformung, Abformung). It has been developed at the Karlsruhe Nuclear Research Center and will be commercialized by German industrial firms. In May 1990, the companies Hoesch AG, Hüls AG, Rheinmetall GmbH, Steag AG and VEW AG signed a contract with the object of founding a joint enterprise for the industrial implementation of the LIGA method.

The LIGA method is superior by the fact that it combines a wide variety of materials with flexible design possibilities. Plastics, metals and ceramics as well as combinations of such materials can be utilized and there are no restrictions in designing the cross-sectional shape of high aspect ratio microstructures. Accordingly, a promising potential exists for realizing advanced microsystems by means of the LIGA method.

Principle of the LIGA Method

The LIGA microfabrication method can be characterized by the fact that, in the first step, a microstructure with a high depth to width aspect ratio is generated by means of deep-etch lithography. This microstructure serves as a threedimensional template for subsequent replication steps which usually comprise microelectroforming and micromolding processes.

Preferably, deep-etch lithography is carried out by means of highly collimated X-rays generated by a synchrotron radiation source which allows to transfer a quasi-twodimensional pattern of an X-ray mask into a resist layer with a thickness of several hundred micrometers (Fig. 1). The lateral dimensions achievable in this way are in the micrometer and the tolerances in the submicrometer range so that an aspect ratio exceeding a value of 100 can be realized. For special applica-

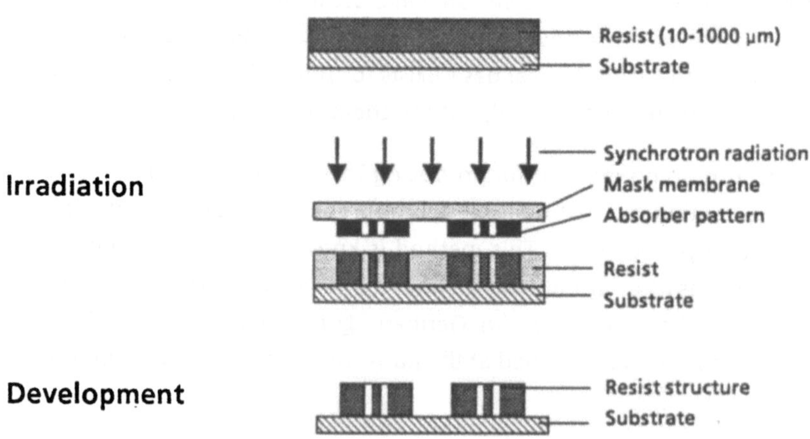

Fig. 1. Generation of a high aspect ratio resist relief by deep-etch X-ray lithography.

tions which do not require extremely high aspect ratios, other lithography methods can also be applied, e. g. deep UV, high energy electron beam or reactive ion etching processes.

If an optically transparent X-ray resist like poly(methylmethacrylate) is used in deep-etch lithography, extremely precise rib waveguide structures or other microoptical devices can be directly generated. By means of three-layer resists consisting of PMMA and suitable fluorinated polymers, slab waveguides with a step profile of the index of refraction have been realized.

For producing metallic microstructures, the resist structure is generated on an electrically conductive substrate and metal is deposited from a suitable electrolyte (Fig. 2). In this way, a complementary metallic microstructure is obtained

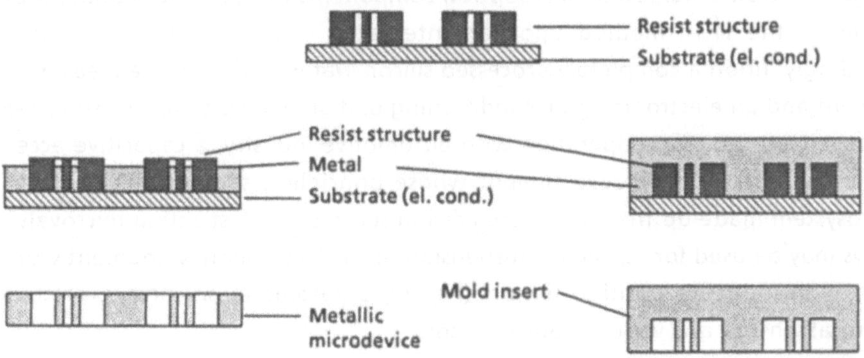

Fig. 2. Generation of a metallic microstructure and a mold insert, respectively.

which may be the final product or a mold insert for a subsequent replication step. Usually, the mold insert is made by continuing metal deposition until a thick layer is grown above the resist relief so that a stable metallic microbody is obtained.

In the next step, the mold insert is used for producing a large number of microstructures from plastics by injection molding with thermoplastic materials or by reaction injection molding with casting resins. Such a micromolding process of high aspect ratio microstructures can be realized only since the surface roughness of the sidewalls of the mold insert generated by deep-etch X-ray lithography and electroforming is extremely low. Usually, the mold insert is evacuated in the micromolding process to avoid gas inclusion in the mold cavities.

There are various ways to apply a plating base onto the bottom of the plastic relief, so that templates for a second electroforming process can be generated for a mass fabrication of metallic microstructures. In recent experiments, microstructures from ceramic materials (PZT, ZrO_2) were fabricated by means of slip casting processes using plastic microstructures as disposable molds which are depolymerized during firing of the ceramic microstructures.

Microstructure Products

Although the LIGA method has been realized on a laboratory scale only, there is a large number of interesting concepts for microstructure products. In order to prove the technical feasibility, prototype specimens have been fabricated for some special applications /2/. The main fields of application will be shortly described in the following.

Sensors: Micromechanical or microoptical components can be directly built up by means of the LIGA method upon an integrated electronic circuit or, correspondingly, upon a completely processed silicon wafer. In this way, a measuring element and an electronic signal conditioning unit are joined to form a so-called smart microsensor. In cooperation with automotive industry, a capacitive acceleration sensor is under development whose principle is shown in Figure 3. A microsystem made up from such acceleration sensors and fast acting microvalve arrays may be used for automatic suspension control. Further development work deals with inductive position sensors (Fig. 4), ultrasonic sensor arrays, electrochemical sensors, and various optical sensors.

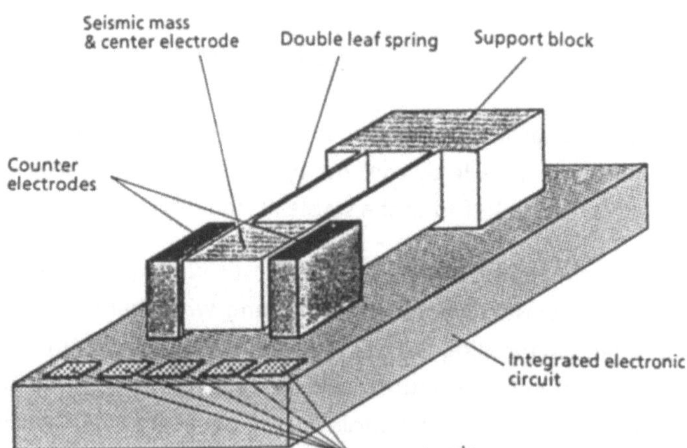

Fig. 3. Principle of a capacitive acceleration sensor with integrated signal conditioning.

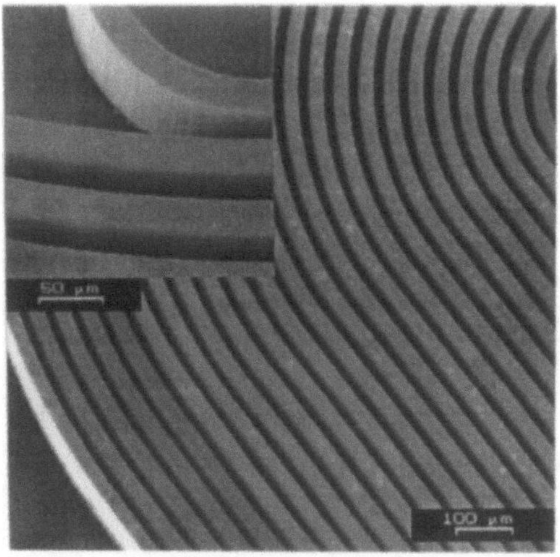

Fig. 4. Scanning electron micrograph of a copper coil of an inductive position sensor (conductor width = 20 µm, conductor height = 100 µm)

Microoptics and Integrated Optics: The present development work deals in particular with the fabrication of resonant meshes, simultaneous planar spectrometers, optical couplers, and various microoptical components. A resonant mesh fabricated by the LIGA process shall be used as an infrared band pass filter in the ISOPHOT satellite experiment (Fig. 5). A prototype of a simultaneous spectro-

Fig. 5. Resonant mesh infrared band pass filter with a crossed dipole pattern (dipole length = 18.5 µm, dipole width = 3 µm).

meter whose principle is shown in Figure 6 has been proven successfully. It shall be applied in photometric analysis of mixtures as well as in optical communication technology for wavelength division demultiplexing.

Electrical and Optical Interconnection: In many applications, the technical realization of complex microsystems will require high performance interconnecting elements which may be fabricated successfully by means of the LIGA method. Scanning electron micrographs of electrical multipin microconnectors are shown in Figure 7. They have 130 poles per cm and can be equipped with precise guiding

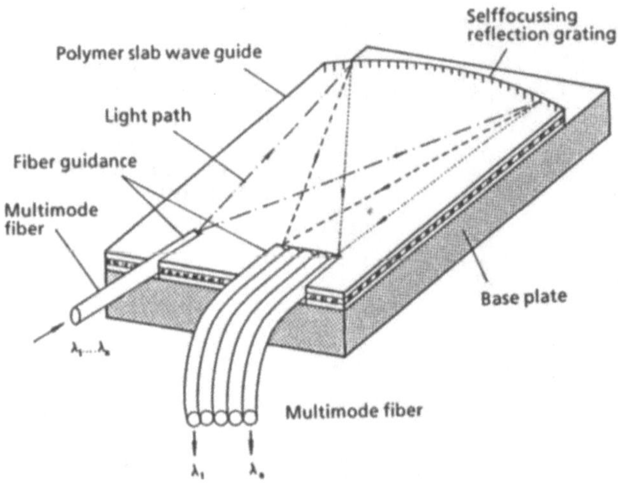

Fig. 6. Schematic representation of a simultaneous micro-spectrometer.

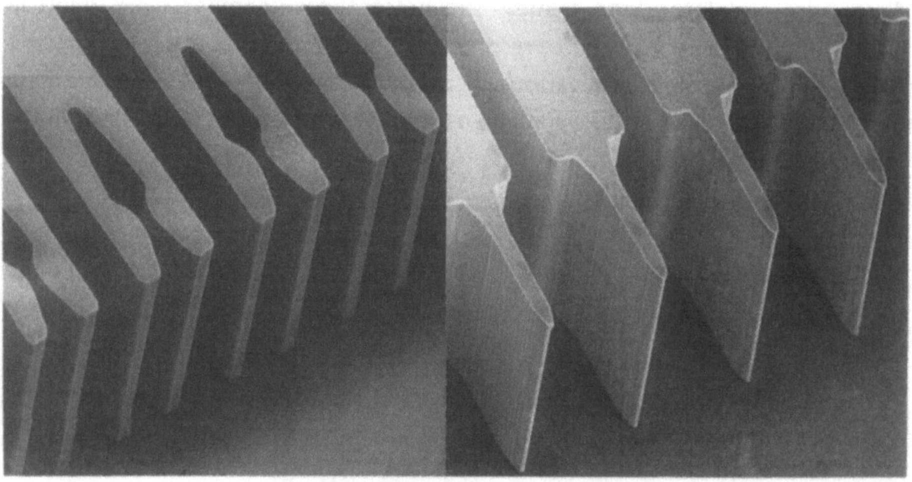

Fig. 7. Scanning electron micrographs of an electrical microconnector (center spacing = 80 μm, structure height = 180 μm).

structures which simply allow a multiple nondestructive insertion and separation. Elements for coupling of monomode fibers with integrated optical chips are under development. They utilize fixed metallic guiding structures in combination with leaf springs for a precise alignment of the optical fibers relative to the optical chip (Fig. 8). Equal or matched materials can be used for the optical chip and the substrate so that problems due to different coefficients of thermal expansion can be eliminated.

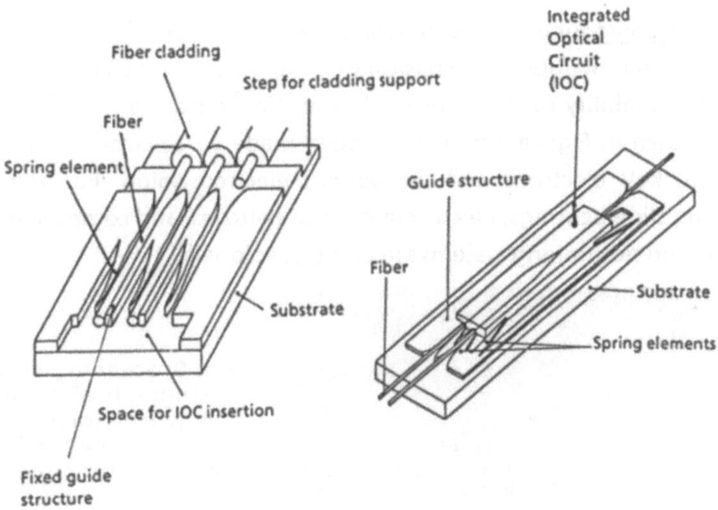

Fig. 8. Schematic representation of fiber-chip coupling elements.

Fluid Technology: The first development work on the LIGA method aimed at the fabrication of micron-sized nozzles for uranium isotope separation. Current work deals with microfiltration membranes, spinneret plates for synthetic fiber production, microvalves, nozzles for fuel injection, ink jet nozzles, microchannels and mixing elements. In contrast to standard microfiltration membranes, identical pore sizes in connection with an extremely high porosity are achievable. In spinneret plates, much smaller capillaries can be realized than by spark erosion. Microvalves will be equipped with integrated microactuators using piezoelectric or electrostatic forces which allow to achieve an extremely short response time.

Bio- and Medical Technology: The present development work deals with components for advanced endoscope and cardiac catheter systems, sensors for measuring gas composition in respirators, components for X-ray and ultrasonic devices, and microfiltration modules for separating cells from biological fluids. In the case of endoscopes and catheters, the traumatic effects of invasive instruments shall be reduced by miniaturization and, in addition, more complex systems shall be realized. The gas sensor development aims at cost reduction and shorter response time. In the case of X-ray and ultrasonic devices which consist of arrays of X-ray sensitive or piezoelectric microstructures, a higher resolution is achievable than with standard components. Microfiltration modules can be fabricated with exactly uniform pores and shear stresses which may destroy the membranes of living cells can be reduced by means of special flow configurations.

Micromechanical Components: Most microdevices mentioned above in connection with special fields of application basically belong to micromechanics. In order to illustrate the capability of the LIGA method in this respect, a metallic microgearwheel is shown in Figure 9 which has been made by electron beam lithography with 50 keV electrons and subsequent electroforming. Furthermore, spring couplings, slide bearings, electromechanical switches, and components for microactuators and positioning systems are being developed.

Fig. 9. Scanning electron micrograph of a metallic gearwheel (diameter = 20µm, thickness = 2.7 µm module = 0.3 µm).

Conclusion

The variety of microstructure products which have been discussed in this paper may be regarded as a promising basis for the commercialization of the LIGA process and its application in realizing microsystems. On the way from demonstration on a laboratory scale to industrial fabrication, however, many problems have to be solved which comprise development of design rules, materials research, production equipment and test procedures as well as market studies concerning the future demand for microstructure components and microsystems.

References

1. Ehrfeld, W.; Becker, E.W.: Das LIGA-Verfahren zur Herstellung von Mikrostrukturkörpern mit großem Aspektverhältnis und großer Strukturhöhe. KfK-Nachrichten 19 (1987) 167-169.

2. Ehrfeld, W.; Bley, P.; Götz, F.; Hagmann, P.; Maner, A.; Mohr, J.; Moser, H.O.; Münchmeyer, D.; Schelb, W.; Schmidt, D.; Becker, E.W.: Fabrication of Microstructures Using the LIGA Process. Micro Robots and Teleoperators Workshop, IEEE Catalog Number 87TH0204-8 (1987).

Movable Microstructures Manufactured by the LIGA Process as Basic Elements for Microsystems

J. MOHR, C. BURBAUM, P. BLEY, W. MENZ, U. WALLRABE

Kernforschungszentrum Karlsruhe, Institut für Mikrostrukturtechnik,
Postfach 3640, D-7500 Karlsruhe 1, West Germany

Summary

By the LIGA process microstructures with structural heights of several hundred micrometers and dimensions down to one micrometer can be produced in submicron accuracy. Using a special sacrifical layer technique it is possible to fabricate movable as well as fixed microstructures in one process step. These microstructures are used as microsensors or microactuators. As first examples capacitive acceleration sensors as well as microgears with a shaft bearing used as microturbines have been produced by this extended LIGA technique.

1. Introduction

The essential components of microsystems are beside the electronic circuitry sensors and actuators made up by micromechanical and microoptical elements /1/.Up to now the main technology for the fabrication of micromechanical elements is the well-known silicon technology /2/. By photolithography and anisotropic etching microstructures like cantilevers, membranes or bridge structures are produced to fabricate sensor for pressure, position or acceleration measurements for example /3/. As it is evident the geometries of the structures fabricated by anisotropic etching of silicon are restricted to the crystallographic orientation of the silicon. A process for the production of microstructures with any geometry like gears or other movable microstructures was published by Howe /4/ and Muller /5/. It is based on the structurization of low-pressure chemical vapor deposited polysilicon by photolithography and etch techniques. In this process the heights of the structures are limited to about ten microns.

An alternative method for the fabrication of free shaped three dimensional microstructures with great structural heights is based on deep-etch synchrotron lithography and high precision replication processes /6/. This process known under the acronym of "LIGA" (in German: Lithografie, Galvanoformung, Abformung) has been developed at the Karlsruhe Nuclear Research Center and will be commercialized by the Steag company in cooperation with other German industrial companies. The LIGA process allows the fabrication of microstructures of any lateral shape with structural heights of more than one hundred micrometers, lateral dimensions down to one micrometer and submicron accuracy over the total height of the structure /7/. Moreover, different materials like plastics, metals and ceramics or a combination of these materials can be used. The range of application of the LIGA technique covers various technological disciplines like micromechanics, integrated optics /8/, sensor and actuator technology /9/ as well as chemical, medical and biological engineering /10/. Applications have been realized by fabrication of microstructures in LIGA technology either fixed on a base plate or as self-supporting foil. To obtain movable microstructures with the LIGA-process a sacrificial layer technique has been developed which is described in this paper.

2. Sacrificial Layer Technique

In the first step of the LIGA process a thick resist layer is irradiated through a mask by synchrotron radiation. By developing the exposed resist layer a form is produced which can be filled by means of electroforming with a metal. After stripping the unexposed resist a metallic microstructure is obtained which is anchored to the substrate. If parts of the microstructure shall be movable they have to be realized on a sacrificial layer which can be etched after electroforming. If, in addition, the microstructure have any electrical function an insulating layer is required.Therefore the substrate used in the extended LIGA process must be insulated and covered with a structurized conductive layer. The details of the fabrication method of movable microstructures with electrical functions are evident from the process sequence shown in Fig. 1.

In the first step thin metallic layers are sputtered onto an insulating substrate (e.g. silicon wafer or ceramics) which ensure on one hand good adhesion to the substrate, and act on the other hand as a reliable basis for the following electro-deposited structures of the LIGA process (a). These metallic layers are patterned by conventional photolithography and wet etching (b) in order to ensure electrical insulation between the different components of the microstructure. On the substrate prepared in this way a titanium layer of a few microns is sputtered. Structurization is carried out by photolithography adjusted to the first layer and wet etching, too. This layer acts as a sacrificial layer. In our case titanium is used because of good adhesion to the resist used in deep-etch synchrotron radiation lithography /11/. Additionally Ti has a high selectivity against the metals mostly used to produce metallic microstructures by the LIGA process such as Au, Cu or Ni /12/. However, also other materials are suitable as sacrificial layers, for example metallized plastic layers.

This substrate prepared in the mentioned manner is then used in the typical LIGA processing. A resist of several hundred micrometers in height, e.g. PMMA, which is polymerized directly onto the substrate, is exposed to synchrotron radiation through a x-ray mask precisely adjusted to the structurized sacrificial layer (d). By this adjustment the movable parts of the microstructures are positioned precisely on top of the sacrificial layer whereas the fixed parts are placed on the metallized areas of the substrate. The irradiated resist is subsequently removed (e) and a complementary structure is build up by electrodeposition of a metal (f). After stripping the resist the titanium sacrificial layer is etched selectively using hydrofluoric acid against the metals above and underneath. After this etching process parts of the microstructure remain firmly attached to the substrate whereas the other parts are free to move (g).

If the microstructures produced by this sacrificial layer technique do not have any electrical function, a metallic substrate can be used instead of the double-layered substrate. The sacrifical layer is structured directly on top of the substrate followed by the LIGA process described above.

3. Examples of Movable Microstructures

Thus a large variety of microstructures can be produced with this extended Liga technique. Both sensors and actuators can be realized. In the following sections, two examples are presented, a capacitive acceleration sensor and a microgear with a shaft bearing.

3.1 Capacitive Acceleration Sensor

A schematic representation of the acceleration sensor manufactured by the extended LIGA technique is shown in Fig. 2. Between two stationary electrodes firmly attached to a Si-wafer as a substrate a freely movable seismic mass is suspended to a cantilever. The small gap between the stationary electrodes and the movable seismic mass varies under acceleration and the resulting change in capacity can be measured. Fig. 3 shows a SEM micrograph of the seismic mass fixed to the cantilever which can move between the two fixed counter electrodes. The width of the cantilever is about ten micrometers, the gap between the seismic mass and the counter electrodes is only three micrometers in width and the total height of the structure is about 100 µm. The sensor is designed to detect signals with a sensitivity of 20 %/g; the resonance frequency is calculated to be 500 Hz. The motion of the seismic mass has been demonstrated by magnetic fields as well as by electrical fields. This proofs that the sacrifical layer underneath the large areas of the seismic mass can be etched completely. Therefore the geometries of the microstructures are not restricted by any limitations of the etch process.

3.2 Microgears With a Shaft Bearing

A further example of movable microstructures are microgears with a shaft bearing. The gears are located on top of the sacrificial layer whereas the shaft is placed directly onto the substrate (s. Fig. 1). After removing the sacrificial layer the gears are free to rotate on the anchored bearing. Fig. 4 is a SEM micrograph of a microgear made of copper before etching the sacrificial layer. The sacrificial layer can still be seen on top of the substrate in front of the gear. In this example the diameter of the gear is 200 µm, the gap between the bearing and the gear is about ten micrometers. Due to the fact that the shaft and the gear are produced in one mask step any alignment procedure of the finished components is unnecessary. Therefore even smaller dimensions are possible without any problem. Fig. 5 shows the same gear after a running time of about 30 minutes. In this case the gear was driven by an air flow. The shaft and the extensions of the rotor blades in particular show deformations due to the soft copper. Using harder metals like nickel the wear could be expected to be reduced drastically. In this case life time of the gears might be increased considerably. In the first experiment the gears were prevented from sliding off the shaft by cover plate from glass. In order to hold the gears onto the shaft it is possible to cover the shaft by another retaining layer which is also patterned by photolithography and electroforming prior to the stripping of the x-ray resist. The gears shown here represent the preliminary stage of microturbines, microtransmissions or electrically driven micromotors.

4. Conclusions

The two examples are presented to show the feasibility of the extended LIGA process. Numerous applications are under development especially in the field of sensors and actuators. The LIGA technique has a number of unique features which in combintion with the Si-microtechnique could open up a wide field of new applications. The two most important advantages are firstly the free lateral structuring and consequently the optimization of the design, without the limits of any crystallographic anisotropy and secondly the variety of materials which can be used, fascilitating the application of physical effects and structural strength to fabricate microactuators.

Future work will concentrate on the production of movable microstructures using electromechanical, magnetic or thermomechanical effects to obtain microactuators by the LIGA process. A technique to produce graduated structures is under development. This technique will allow the fabrication of gear trains with different gears on one axle. In order to achieve an inexpensive mass fabrication adjusted moulding technique on substrates with a structurized sacrificial layers is under test.

References

/1/ A. Heuberger (ed.): Mikromechanik, Springer Verlag, Berlin (1989)

/2/ K.E. Petersen: Silicon as a mechanical material, Proc. IEEE 70 (1982) 420

/3/ W. Benecke, in Mikromechanik, edited by A. Heuberger, Springer Verlag, Berlin (1989) 343

/4/ T.T. Howe, in Polysilicon Films and Interfaces, edited by C.Y. Wang, C.C. Thompson and K.-N. Ting, Material Research Society, Pittsburgh (1988), Vol.106, 213

/5/ R.S. Muller, in 7th IEE Japan Sensor Symposium, IEE Japan, Tokyo, (1988) 193

/6/ E.W. Becker, W. Ehrfeld, P. Hagmann, A. Maner, D. Münchmeyer: Fabrication of microstructures with high aspect ratios and great structural heights by synchrotron radiation lithography, galvanoforming, and plastic moulding (LIGA process), Microelectronic Engeneering 4 (1986) 35

/7/ W. Ehrfeld, H.J. Baving, D. Beets, P. Bley, F. Götz, J. Mohr, D. Münchmeyer, W. Schelb: Progress in Deep-Etch Synchrotron Radiation Lithography, J. Vac. Sci. Technol. B6 (1988) 178

/8/ B. Anderer, W. Ehrfeld, J. Mohr:Grundlagen für die röntgentiefenlithographische Herstellung eines planaren Wellenlängen-Demultiplexers mit selbstfokussierendem Reflexionsbeugungsgitter, KFK report, Kernforschungszentrum Karlsruhe (1990)

/9/ W. Ehrfeld, R. Einhaus, D. Münchmeyer, H. Strathmann: Microfabrication of Membranes with extreme Porosity and Uniform Pore Size, J. Of Membrane Science, 36, (1988) 67

/10/ J. Mohr, W. Ehrfeld, D. Münchmeyer, A. Stutz: Resist technology for Deep-Etch Synchrotron Radiation Lithography, Makromol. Chem., Macromol. Symp. 24 (1989) 231

/11/ A. Maner, W. Ehrfeld: Electroforming Techniques in the LIGA Process for the Production of Microdevices, Proceedings INTERFINISH (1988) 577

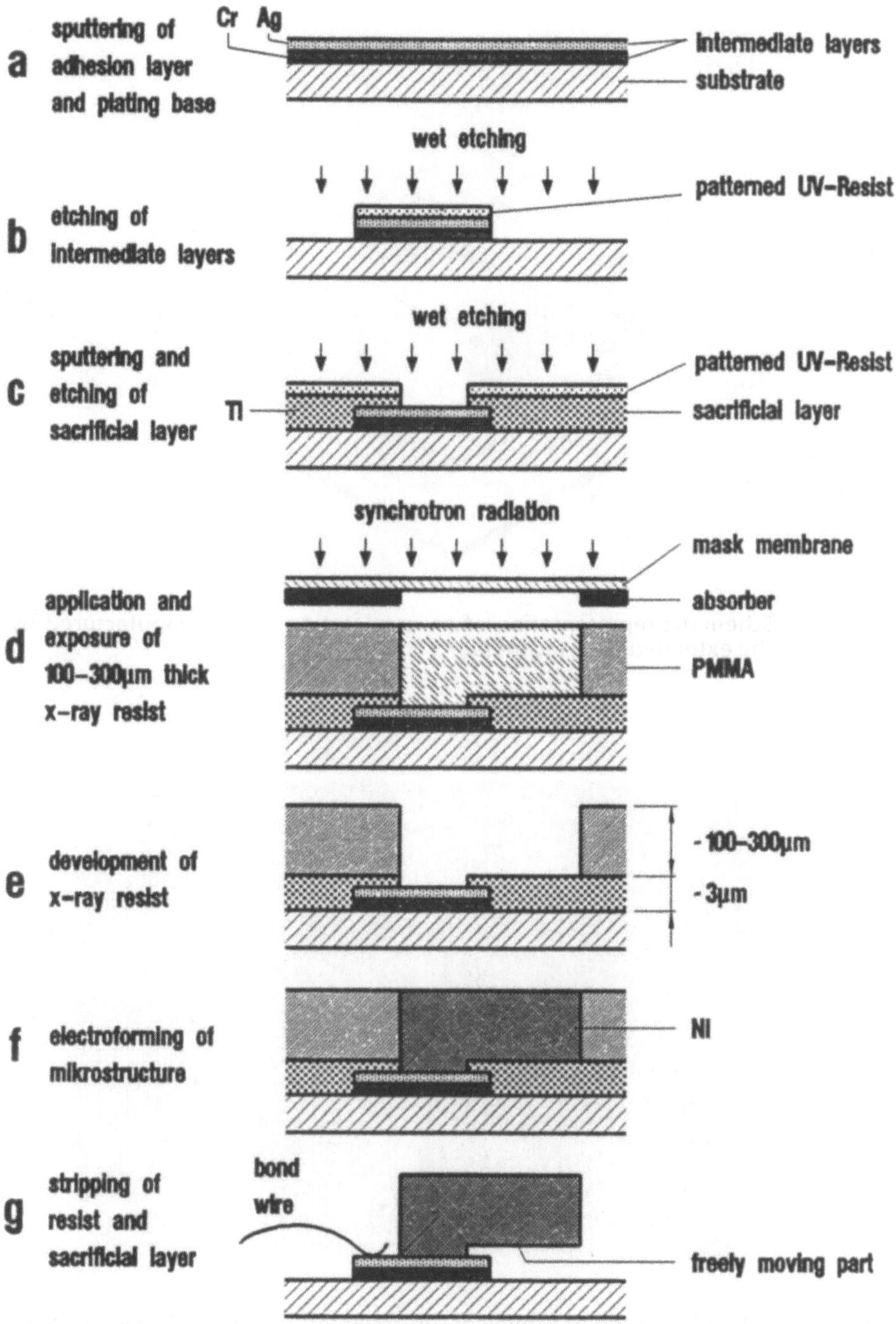

Fig. 1: Schematic representation of the fabrication steps of movable microstructures with electrical functions by the LIGA process.

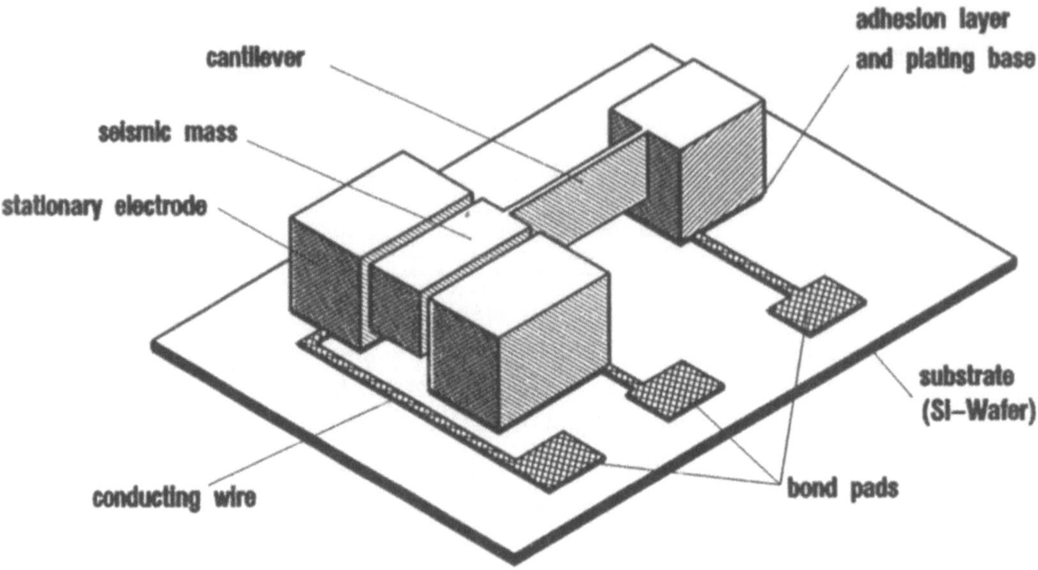

Fig. 2: Schematic representation of an acceleration sensor manufactured by the extended LIGA technique

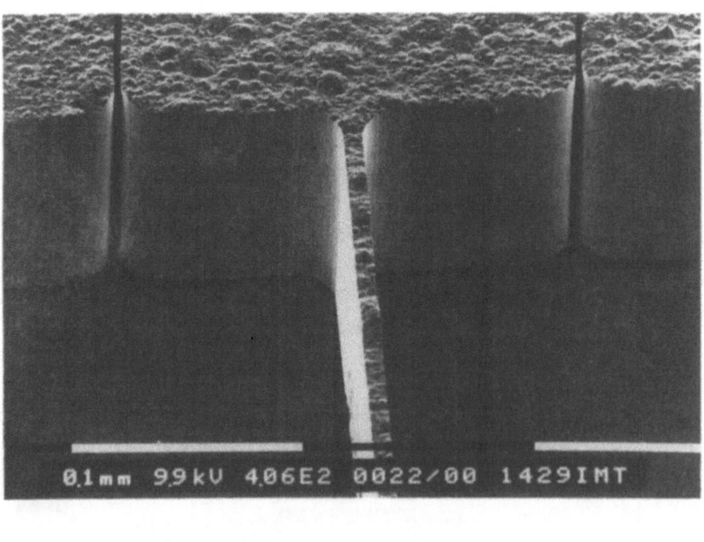

Fig. 3: SEM micrograph of a detail of the acceleration sensor. The seismic mass fixed to the cantilever can move between the two fixed counter electrodes.

Fig. 4: SEM micrograph of a microgear with a shaft bearing manufactured by the extended LIGA technique. The gear is already fixed on the sacrifical layer.

Fig. 5: SEM micrograph of a microgear made of copper after rotation of about 30 minutes.

Ion Beam Techniques for Micro Electro Mechanical Systems

Hiroki KUWANO

NTT Applied Electronics Laboratories,
Nippon Telegraph and Telephone Corporation
Musashino, Tokyo 180, JAPAN

Abstract

Ion beam techniques are discussed for Micro Electro Mechanical device fabrication. An ion beam with a energy range of several volts to several Megavolts is widely applied for semiconductor device fabrication. Generally, ion beam energy, direction, and density are highly controllable. Energetic ions produce a number of effects: sputtering, deposition, ion beam induced chemical reaction, and ion implantation. Ion interactions with solids are briefly summarized. A broad ion beam application to a magnetic recording head are shown. A focused ion beam and a fast atom beam have particular usages for surface modification, three dimensional processing, and device repairing. Other applications demonstrated and suggested are discussed.

1.Introduction

In micro electro mechanical device fabrication particular problems exist such as structural, behavioral, and tribological problems. In particular, the thin film and micro dimension mechanical properties of the device are sensitive to surface morphology, internal film stress, and the interface between the film and the substrate.

An ion beam, which is generated in a plasma source separated from the substrates, is widely applied for the semiconductor device fabrication due to its high controllability. Ion beam processing has possibilities to resolve the problems of micro electro mechanical device fabrication. Ion beam techniques are useful for precise patterning and three dimensional etching, and film formation, since an energetic ion direction, its energy, and species can be highly controlled by applying an electric or magnetic field. Many other surface modification techniques are also useful such as ion implantation, ion beam induced-chemical reactions, crystallization, and amorphousization.

Regarding ion interactions with solids, there are vast amounts of literature because of the widespread use of ion im plantation for semiconductor device fabrication [1-6]. In this article, ion interactions with solids will be briefly summarized. A broad ion beam application to a magnetic recording head will be shown. In addition, a focused ion beam technique and a fast atom beam technique will be discussed for micro electro mechanics device applications.

2.Interactions of Ions with a Solid Surface

The interactions of an ion beam with a solid surface or depositing film surface are strongly dependent on the incident kinetic energy of the beam [7]. Figure 1 shows typical energy ranges for a few selected methods. There are three principal ranges for the ion beam technique. In the 0.1 to 100 eV ion energy range, ion beam deposition is performed. In the energy range of several 10 eV to several keV, sputtering for deposition and etching is performed. Ion implantation is used as a doping technique in the semiconductor field and as a surface modification technique to change mechanical and tribological characteristics of device.

As shown in Fig. 2, a number of collisions occur among the lattice atoms as an energetic ion slows down and comes to rest in a substrate. From these collisions, sufficient energy may be transferred from the ion to displace an atom from its lattice site. Sometimes, a displaced atom is removed from the substrate. The removal of the atom is called sputtering.

The displaced atom can in turn displace another atom, and so on, creating a cascade of atomic collisions. This could leads to a distribution of vacancies, interstitial atoms, amorphous regions, and other types of defects.

$$Sputtering\ Yield = \Lambda F_d(E_0) \qquad (1)$$

$$= \Lambda \alpha N S_n(E_0) \qquad (2)$$

where Λ contains all the material properties such as surface binding energies and $F_d(E_0)$ is the density of deposited energy at the surface and depends on the type, energy, and direction of the incident ion and target parameters, α is a correction factor, N is the atomic density of target atoms, $S_n(E)$ is the nuclear stopping cross section, and

$$NS_n(E) = (dE/dx)_n \qquad (3)$$

the nuclear energy loss.

Derivation of Λ involves describing the number of recoil atoms that can jump over the surface barrier and escape from the solid. The details of the derivation for the linear cascade regime are given by Sigmund. The result is

$$\Lambda = 0.042/NU_0 \qquad (4)$$

where U_0 is the binding energy. Since the sputtering yield is related to the binding energy of atoms in the material being etched, it is possible to vary the binding energy's value by introducing reactive gases. These gases can react with the etched surface and vary the binding energy, and consequently the etch rate. For example, activated chlorine or fluorine-contained species react with a Si material to form loosely bound or even volatile compounds, and thereby increase the etch rate.

When an energetic ion penetrates a substrate it undergoes a series of collisions with atoms and electrons in the target. In these collisions the incident particle loses energy at a rate of dE/dx, depending on the energy and mass of the ions as well as on the substrate material. In Fig. 3, the total path length R is determined by the rate of energy loss along the path of the ion,

$$R = \int_{E_0}^{0} \frac{1}{dE/dx} dE = \int_{E_0}^{0} ((dE/dx)_n + (dE/dx)_e)^{-1} dE \qquad (5)$$

where E_0 is the incident energy of the ion as it penetrates the substrate, the subscripts n and e denote nuclear and electronic collisions, respectively.

Nuclear collisions are responsible for the production of lattice disorder by the displacement of atoms from their lattice positions. Electronic collisions involve much smaller energy loss per collision, negligible deflection of ion energy, and negligible lattice disorder. Nuclear collisions dominate at low energies while electronic collisions dominate at high energies.

Since mechanical and tribological properties can be easily improved without the use of deposition or etching processes, the ion implantation technique seems to stand out for micro elctro mechanical device fabrication.

3.Applications of Broad Ion Beam

An ion beam technique is applied to a high density magnetic recording head. A magnetic recording head is a kind of micro electro mechanical device. Figure 4 shows a thin film magnetic recording head [8] for a high density magnetic storage device.

There are three important ion beam procedures for the fabrication: thin pole piece magnetic film formation using a ion beam technique, planarization, and Cu coil patterning. Recently, it became clear that pole piece material requires a higher saturation magnetization Bs to develop thin film heads for high-density magnetic recording systems. CoZrRe amorphous film seems to be most attractive material as its Bs is 1.4 T. Use of an ion beam sputtering technique enables simultaneous zero-magnetostriction and high-permeability of the CoZrRe amorphous film as shown in Fig.5 [9].

An effective planarization technique is required for fabri cating 3-dimensional micro electro mechanical devices. A planarization technique which is not subjected to material restrictions is necessary. In Fig.6, a broad ion beam planarization process including film deposition is shown [10]. In the figure, film deposition and etching simultaneously occur.

The Cu thin film patterning was achieved by nitrogen ion beam etching and a Ta film mask as shown in Fig. 7 [11]. The etching ratio of Cu to Ta is as high as 20. This indicates the fine-pattern generation ability of Cu films.

In Fig.8, a cross-section photograph of CoZrRe amorphous film heads with a Bs of 1.2 T, developed by the ion beam techniques described above is shown [9]. The head has a double-layer 31-turn Cu coil, which is almost completely planarized.

4.Applications of Focused Ion Beam

The focused ion beam application has been grown as a result of the invention of liquid metal ion sources and by the utilization of focusing columns with mass separation capability [12- 21]. The ability to sputter and to also induce deposition by causing breakdown of an adsorbed film has produced an immediate application of focused ion beams to micro electro mechanical device fabrication procedures such as direct etching, direct deposition, direct implantation, and pattern repairing without mask.

Recently, a proposal was reported for device transplantation using a focused ion beam [22]. There are two candidates for fixing a transplantation device to the substrate: sputtered-atom redeposition and focused ion beam-induced deposition. Transplantation of stainless strip to the desired place on the silicon substrate was examined as shown in Fig. 9.

5.Applications of Fast Atom Beam

Fast atom beam techniques for etching, sputter deposition, implantation, and analysis are easily applied to insulators and composite materials because charging of the specimen surface is not a serious problem in the process. These techniques are applied for film formations [23-25], secondary ion mass spectrometry [26], and etching [27-30].

A fast atom is defined as an energetic uncharged atom having an energy significantly above the thermal energy, namely above 10 eV. To obtain a fast atom beam, energetic ions emitted from the ion source must be neutralized. The neutralization of ions is most effectively achieved in during charge transfer.

Fast atom beam techniques are applied to SiO_2 etching and GaAs etching. In Fig. 10, a schematic GaAs/AlGaAs U-shaped LD structure and a SEM photograph of the LD is shown [30]. The totally reflecting mirrors were made by reactive fast atom beam etching with Cl_2 gas. The reflecting mirror loss is about 0.95 dB per reflection.

In Fig. 11, 0.2-μm-wide and 1.0-μm-deep fine patterns were obtained by CF_4 reactive fast atom beam etching onto SiO_2 sub strate [28]. The notches of the Cr mask pattern edge were transferred to the SiO_2 pattern edge as shown in the figure.

In Fig. 12, the concept of bi-layer friction-reducing coatings [24,25], which could be applied to micro electro mechanical systems, is to provide a hard bottom layer coating as well as a solid lubricant top layer coating. The hard coating protects the substrate from mechanical damage. Hard coatings are achieved with a dual fast atom beam system, which is shown in Fig. 13. One of the fast atom source is used to sputter a target, while the other is used to bombard the substrate during film growth. The substrate was bombarded principally to improve the crystalline structure and the adhesion between the substrate and films.

BN film deposited on silicon substrates using BN target , which obtained with fast atom bombardment of their growth surface was partly identified as a poly-crystalline cubic BN including amorphous [25]. B, TiB_2 and B_4C coatings on silicon substrates using each compound target, which also obtained with fast atom bombardment of their growth surface at room temperature, showed single or poly-crystalline structures. On the other hand, coatings obtained without fast atom bombardment at room temperature showed amorphous structures [24,25].

Solid lubricant coatings are formed by an Mo-S complex target [23]. The bi-layer solid lubricant, especially BN bottom layer film, had 1/5 of MoS_2 solid film lubricant friction coefficient, less than 0.01 and had high durability as shown in Fig. 14 [24,25].

6.Summary

Ion beam techniques have been discussed for micro electro mechanical device fabrication. Some distinctive applications using broad ion beam techniques, focused ion beam techniques, and fast atom beam techniques have been described. Although an ion beam technique can not fully be applied to micro electro mechanical device fabrication, it is a useful and practical technique because of its flexibility and controllability.

Acknowledgments

The author is grateful to the many colleagues, Tago, F.Shimokawa, K.Yanagisawa, T.Ohkubo, C.Nishimura, and Y.Nagai who have shown experimental results and answered queries.

References

[1]Mayer,J.W., Eriksson,L., and Davies,J.A., Ion Implantation in Semiconductors, Academic Press, New York, 1970.
[2]Dearnaley,G., Freeman,J.H., Nelson,R.S., and Stephen,J., Ion Implantation, North-Holland, Amsterdam, 1973.
[3]Brice,D.K., Ion Implantation Range and Energy Deposition Dis- tributions, Vol.1, Plenum Press, New York, 1975.
[4]Williams,J.S., and Poate,M., eds., Ion Implantation and Beam Processing, Academic Press, New York, 1984.
[5]Ziegler,J.F., ed., Ion Implantation Science and Technology, Academic Press, Orlando, Fla., 1984.
[6]Feldman,L.C. and Mayer,J.W., Fundamentals of Surface and Thin Film Analysis, North-Holland, Amsterdam, 1986.
[7]Takagi,T., J. Vac. Sci. Technol., A2(1984), 382.
[8]Mikazuki,T., Kishigami,J., and Koshimoto,Y., IEEE TRANSACTIONS ON MAGNETICS, Vol.MAG-21(1985), 1539.
[9]Nishimura, Yanagisawa, K., Tago, A., and Kishigami,J., IEEE TRANSACTIONS ON MAGNETICS, Vol. MAG-25(1989), 2728.
[10]Tago,A., Yanagisawa,K., Yachi, T., and Mino,M., Technical Report of Jpn. Electric Engineering Society, MAG-90-45(1990), 103.
[11]Yanagisawa,K. and Tago, A., Digest of Papers 1989 2nd Micro Process Conference, Tokyo, 174.
[12]Melngailis, J., J. Vac. Sci. Technol. B5(1987), 469.
[13]Seliger,R.L., Ward,J.W., Wang,V., and Kubena,R.L., J. Appl. Phys. Lett., Vol.34(1979), 310.
[14]Kubena, R.L., Lee, J.Y.M., Jullens, R.A., and Stevens, E.H., IEEE Trans. Electron Devices Vol.31(1984), 1186.
[15]Kuwano,H., Yoshida,K., and Yamazaki,S., Jpn. J. Appl. Phys., Vol.19(1980), L615.
[16]Kuwano,H., Takaoka,H., and Ozawa,A., J. Vac. Sci. Technol., B3(1985), 1357.
[17]Morimoto, H., Sasaki, Y., Watanabe, Y., and Kato, T., J.Appl. Phys. Vol.57(1985), 159.
[18]Komuro, M., Hiroshima, H., Tanoue, H., and Kanayama, T., J. Vac. Sci. Technol. B1(1983), 985.
[19]Ochiai,Y., Gano,K., and Namba, S., J. Vac. Sci. Technol., B1(1985), 67.
[20]Gamo,K., Takakura,N., Samoto,N., Shimizu,R. and Namba, S., Jpn. J. Appl. Phys., Vol.23(1984), L293.
[21]Komuro,M., Atoda,A., and Kawakatu,H., J. Electrochem. Soc. Vol.126(1979), 483.
[22]Ohnishi, T., Kawanami, Y., and Ishitani, T., Proposal for Device Transplantation using a Focused Ion Beam, Jpn. J. A.P., Vol.29(1990), L188.
[23]Kuwano,H. and Nagai,K., J.Vac.Sci.Technol.,A3(1985), 1809.
[24]Kuwano,H. and Nagai,K., J.Vac.Sci.Technol.,A4(1986), 2993.
[25]Kuwano,H., in Synthesis and Properties of Boron Nitride, edited by Pouch, J.J. and Alterovitz, S.A., Trans Tech Publica- tions, Switzerland, 1990.
[26]Nagai,K. and Kuwano,H., in Ion Mass Spectrometry, SIMS4, edited by Benninghoven, A., Okano,J., Shimizu,R., and Werner, H.W., Springer, Berlin, p448. 1984.
[27]Revell,P.J. and Evans, A.C., Thin Solid Films, Vol.86(1981), 117.
[28]Kuwano.H. and Shimokawa, F., J. Vac. Sci. Technol. B6(1988), 1565.
[29]Shimokawa,F., Tanaka,H., Uenishi,Y., and Sawada,R., J. Appl. Phys. Vol.66(1989), 2613.
[30]Shimokawa,F., Tanaka,H., Sawada,R., and Hara,S., Appl. Phys. Lett. Vol.56(1990), 1617.

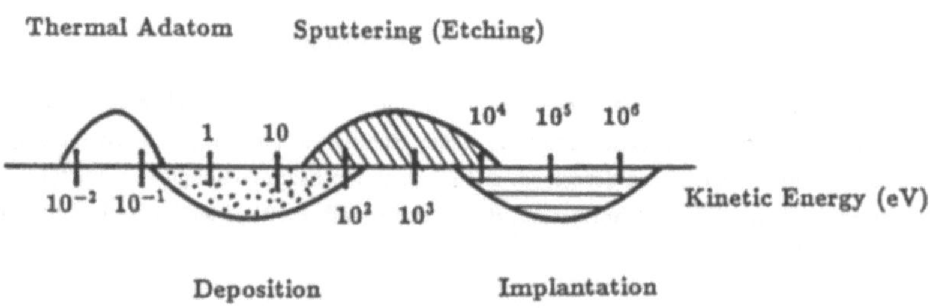

Fig. 1 Range of kinetic energy corresponding to various engineering applications.

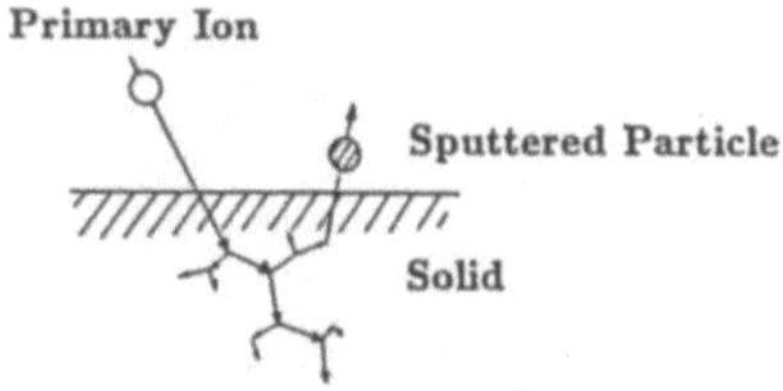

Fig. 2 Schematic of ion-solid interactions and the sputtering process.

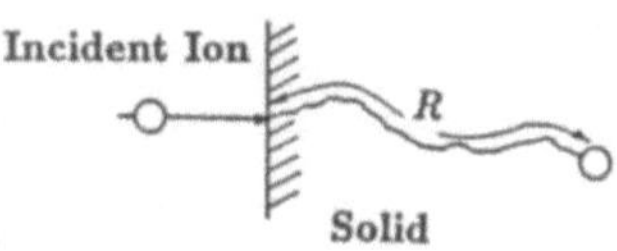

Fig. 3 Penetration of an ion incident on a substrate with total path length R.

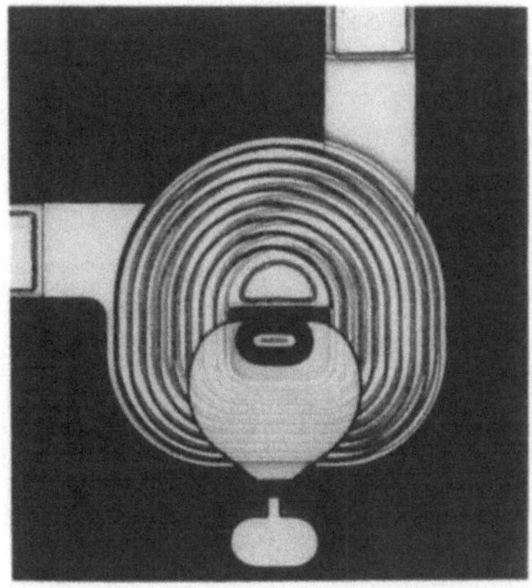

Fig. 4 A photograph of a thin film magnetic recording.

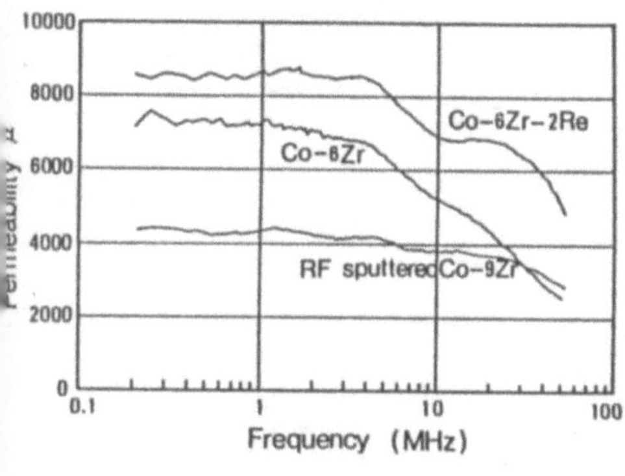

Fig. 5 Permeability dependence on frequencies.

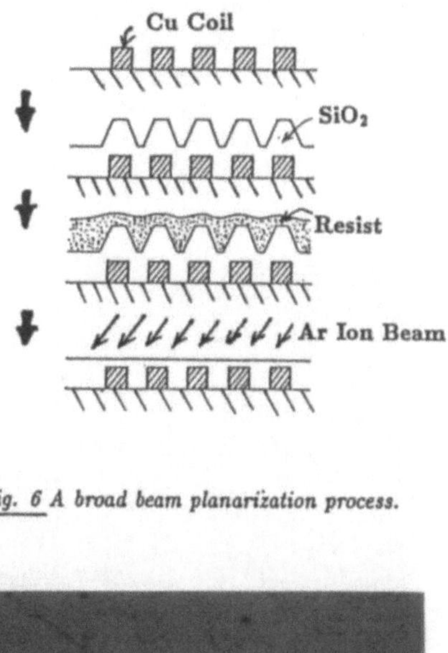

Fig. 6 A broad beam planarization process.

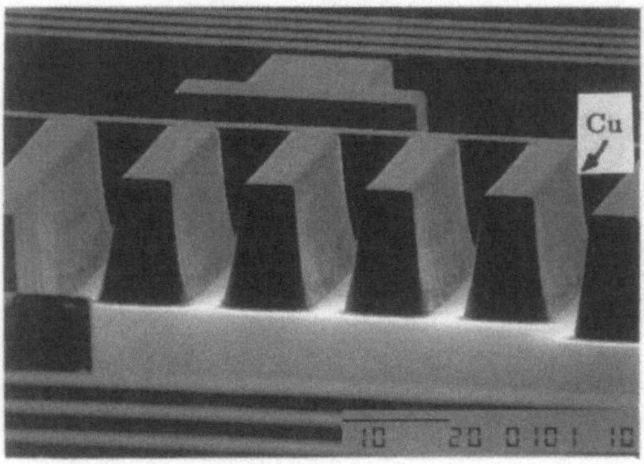

Fig. 7 A scanning electron microscopy photograph of Cu thin film patterning with nitrogen ion beam etching.

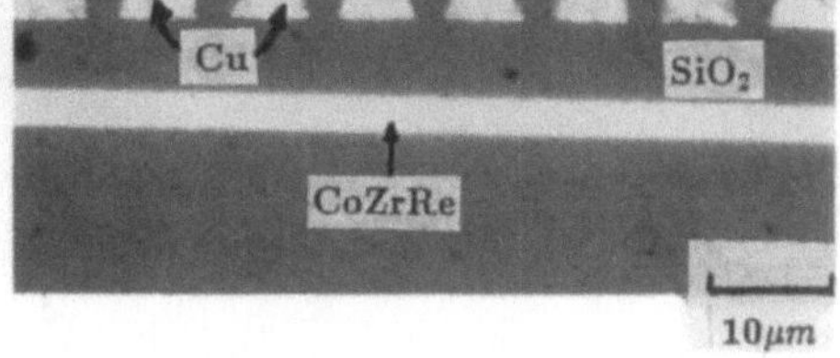

Fig. 8 A cross-section photograph of CoZrRe amorphous film head.

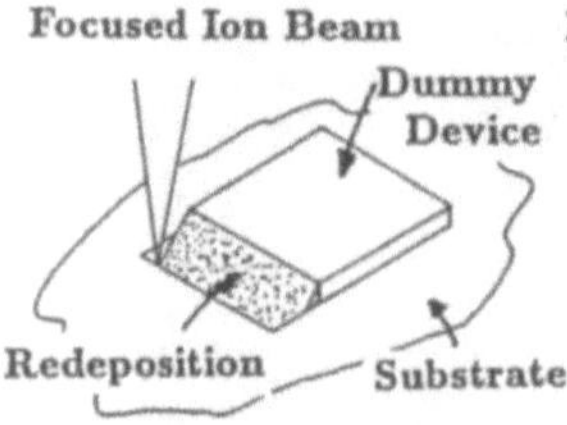

(a)Sputtered-atom redeposition.

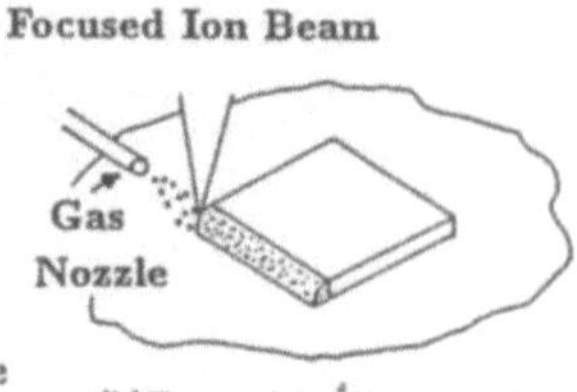

(b)Focused ion beam-induced deposition.

Fig. 9 Transplantation to a desired place on a substrate with a focused ion beam.

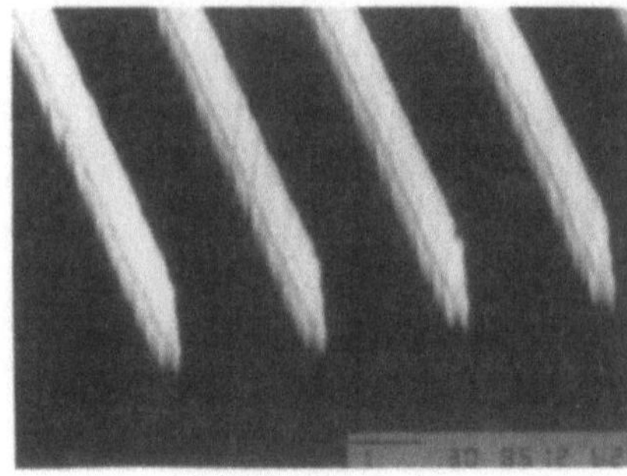

Fig. 11 A scanning electron microscopy photogra[ph] of 0.2-μm-wide fine patterns onto SiO_2 substrat[e] with CF_4 reactive fast atom beam etching.

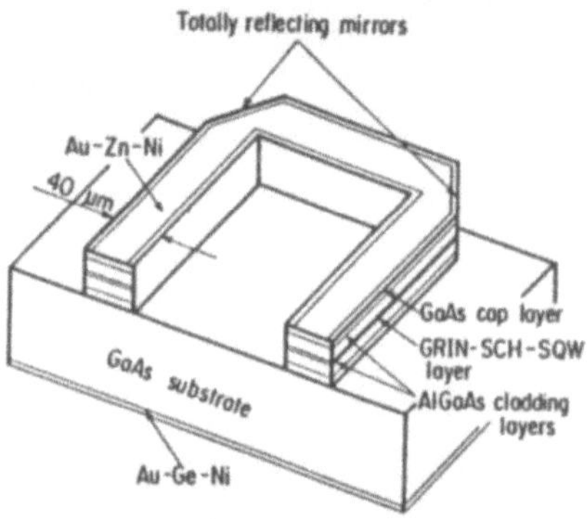

(a) Schematic of GaAs/AlGaAs U-shaped LD structure.

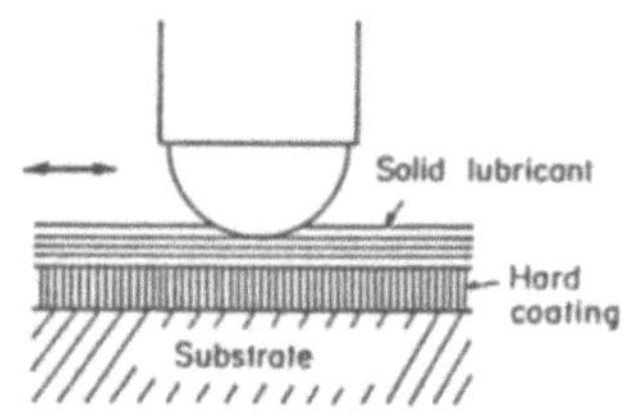

Fig. 12 A concept of bi-layer friction-reducing coatin[g]

(b) A scanning electron microscopy photograph of the LD with Cl_2 reactive fast atom beam etching.

Fig. 10 GaAs/AlGaAs U-shaped LD.

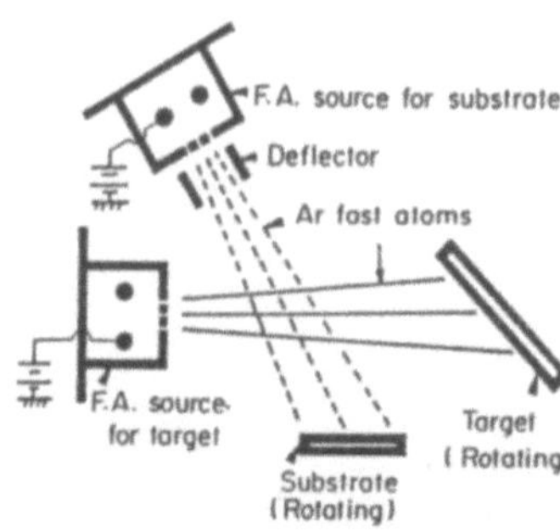

Fig. 13 Schematic of dual fast atom beam system.

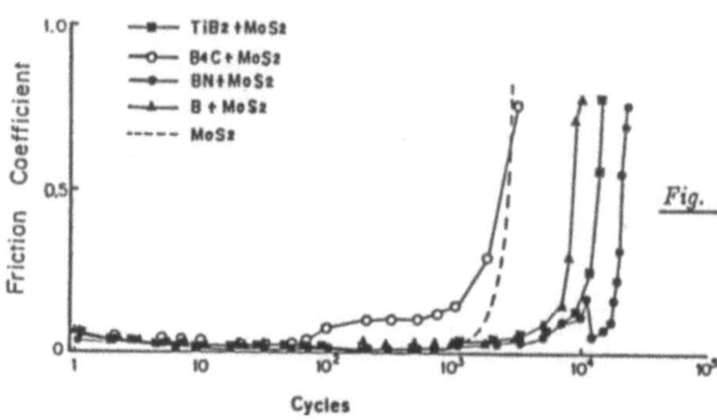

Fig. 14 Friction coeficients of bi-layer coatings.

A Novel Mask Compensation Pattern for Etched Microstructures with Several Convex Corners

G. GERLACH

Technische Universität Dresden
Sektion Informationstechnik

Summary

Convex corner cutting under the etch mask of convex corner structures anisotropically etched in (100) single-crystal silicon arises in the fabrication process of three-dimensional microstructures. This paper presents a novel design of complex corner compensation mask with small geometrical dimensions for microstructures with high packing density. These complex patterns possess defined convex corners and slits. During etching the undercut process begins at the convex corners of the etch mask and reaches after a defined time the mask openings in the compensating patterns forming new convex corners. In this way the etch front of the undercut process will be reflected for one or more times. Experimental results of an example for making V-groove-crossings show satisfactory corner compensation.

Convex corner cutting under the etch mask of mesa structures anisotropically etched in (100)single-crystal silicon arises in the fabrication process of three-dimensional microstructures [1]. In a lot of purposes this undercut process leads to an unwanted rounding of the convex corner. Two methods are usually used to correct this effect [2]: (i) the addition of compensating patterns to the mask corners and (ii) the tilting of the mask side with respect to the wafer flat by a small angle. The first method is yet the more important and most used way for the fabrication of three-dimensional micromechanical structures.

Nowadays several kinds of geometrical forms of such compensation corners are known, for example square, rectangular, triangular, circular, or trapezoid shapes, and combinations of squares and rectangles. Their design and size depends (i) on the wanted etch depth in the non-masked areas, (ii) on the available space between neighbouring masked areas, and (iii) on the undercut etch rate of the anisotropical etch solution. For high packing densities of three-dimensional microstructures with convex corners on the silicon wafer only a very small area is available for realizing corner compensation patterns. In this case the well-known compensation shapes often cannot be used (Fig.1).

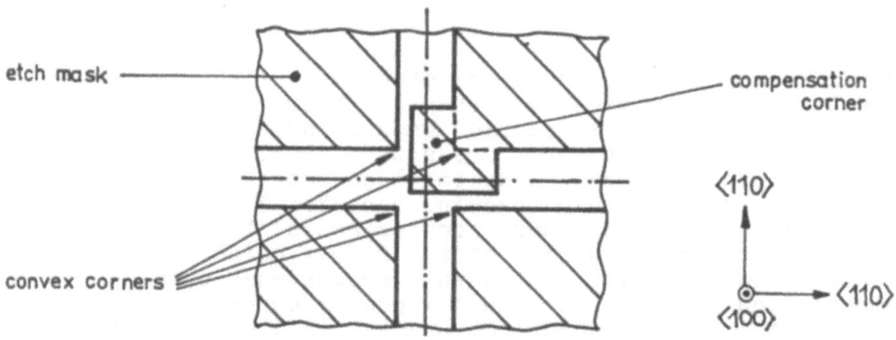

Fig.1. Non-compensable convex corner cutting with square compensation corners

Complex Compensation Patterns

If two or, such as shown in Fig.1, four convex corners of an etch mask are bordering each other in a small distance, the corner undercutting can be compensated by joined compensation mask parts of the several convex corners. Such an complex compensation pattern is shown in Fig.2. In the region of the connecting bridges between the mask parts only concave corners appear preventing the undercutting. Convex corner cutting begins at the outermost convex points of the compensation pattern. If the undercut process reaches the narrow slits in the mask new convex corners will be formed at previous concave corners leading to a "reflection" of the undercut front. The undercut length l results from the undercut etch rate R_u and the etch time t:

$$l = l_1 + l_2 = R_u \cdot t \qquad (1)$$

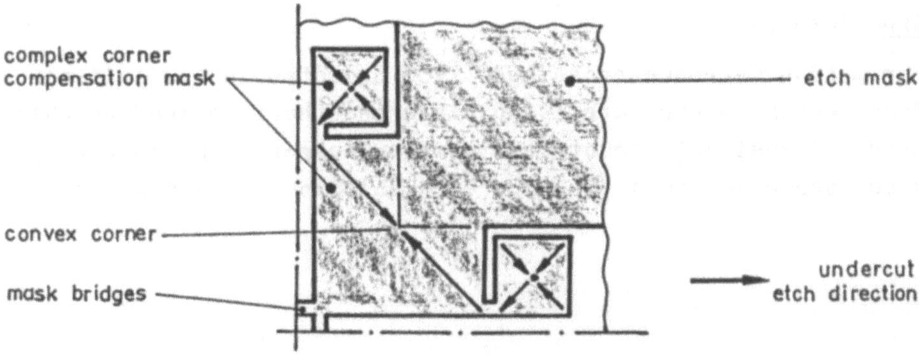

Fig.2. Principle of complex compensation patterns [3]

If there hasn't an etch stop mechanism an effect in the depth of the silicon wafer l is given by

$$l = \frac{R_u}{R_d} \cdot d \qquad (2)$$

where R_d is the depth etch rate and d the etched depth. For a given undercut compensation length l the width w of the compensation mask will be decreased by the number of "reflections" of the undercut etch rate at convex corners developing during the etch process (Fig.3).

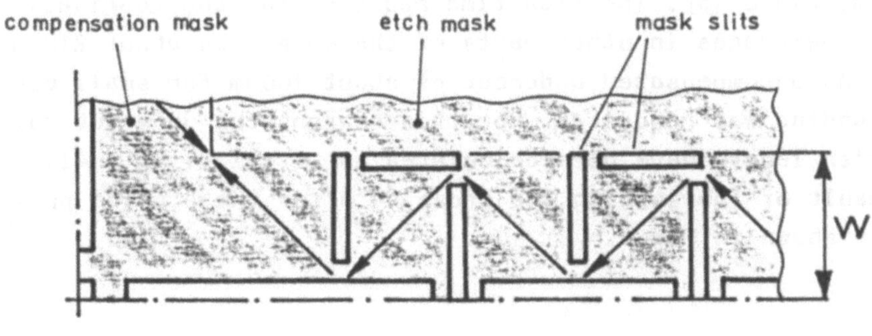

Fig.3. Complex compensation pattern with a lot of undercut etch front reflections

Experimentals

For making V-groove-crossings ⟨100⟩ oriented in (100)silicon four convex corners are lying close together. To produce this three-dimensional structure a complex compensation pattern was used, shown in Fig.4 [4]. The etch solution was a mixture of

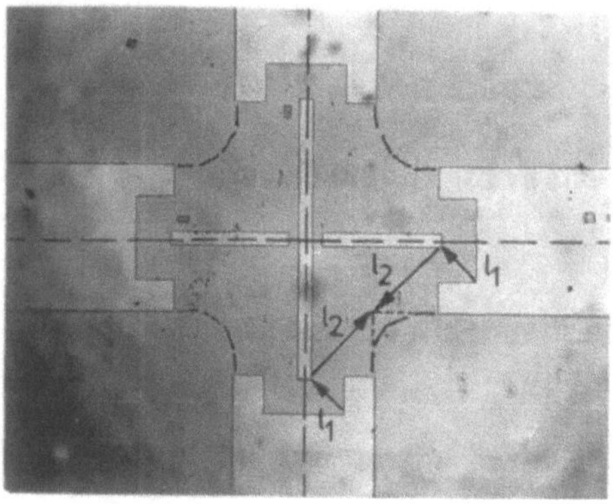

Fig.4. Complex compensation pattern for producing of V-groove-crossings

30 weight percent potassium hydroxide in deionized water at 80°C. The undercut etch rate R_u for such an etch solution is given by Fig.5 [5]. The etch time required for the fabrication of thin membranes in other parts of the wafer was about 215 minutes. An uncompensated undercut of about 100µm for small corner rounding was requested. For this reasons the undercut compensation length have been calculated as $l = l_1 + l_2 = 410$ µm. The result of fabricating the crossing after the etching process is shown in Fig.6.

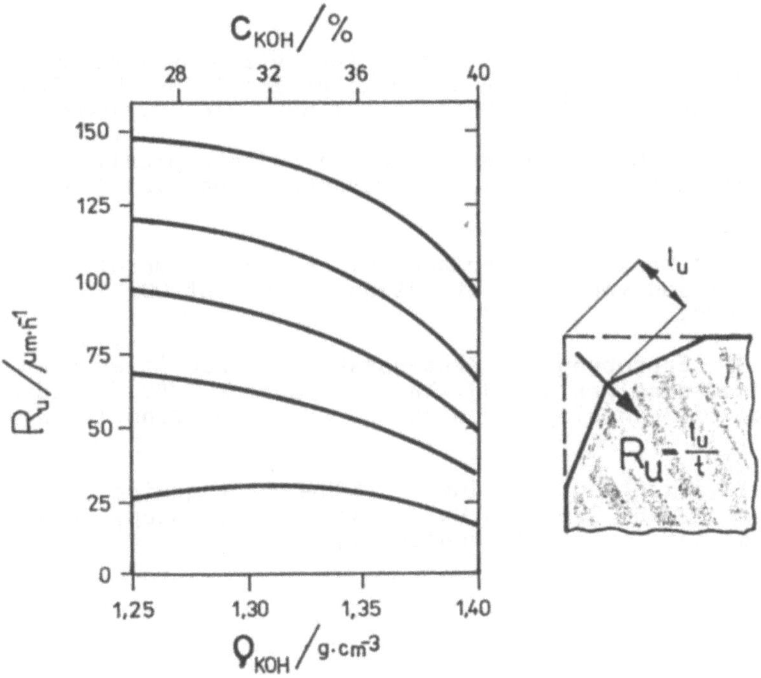

Fig.5. Undercut etch rate R_u of binary etch solutions of potassium hydroxide

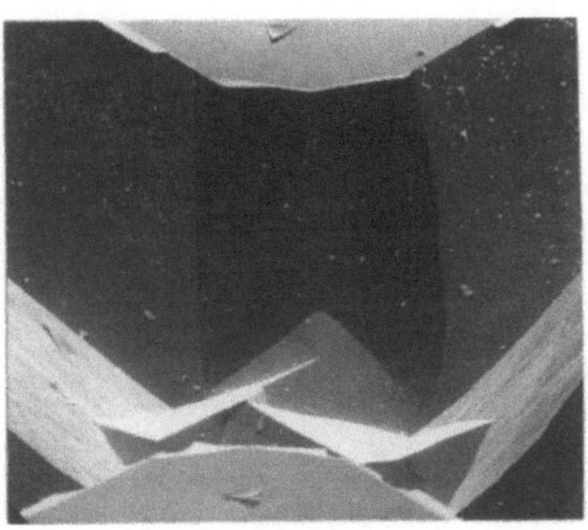

Fig.6. REM micrograph of the produced V-groove-crossing from Fig.4 (enlargement 200x)

References

1. Kragness, R.C.; Waggener, H.A.: Verfahren zum Herstellen formgebender Einätzungen an einer monokristallinen Halbleiterscheibe. German patent DE AS 1.621.532, Dec 19, 1967.

2. Abu-Zeid, M.M.: Corner undercutting in anisotropically etched isolation contours. J. Electrochem. Soc. 131 (1984) 2138-2142.

3. Jenschke, W.; Kühne, S.; Gerlach, G.: Kompensationsätzmaske zur Kompensation von Eckenunterätzungen. Patent DD 276.366, Febr 21, 1990.

4. Gerlach, G.: Erprobung der Kantengestaltung zum Ätzen des Miniaturdruckwandlers IBW 88. In: Res. Report 5/89, Dresden: University of Technology, Department of Information Technology 1989.

5. Gerlach,G.: Ätzen von Silizium mit KOH-H_2O-Ätzgemischen. Res. Report 12/85, Dresden: University of Technology, Department of Information Technology 1985.

Wafer Bonding and Etchback Techniques for Silicon-on-Isulator Systems

C. Harendt, W. Appel, H.-G. Graf, B. Höfflinger
and E. Penteker

Institut für Mikroelektronik Stuttgart, Allmandring 30a,
D-7000 Stuttgart 80, phone (0711) 685 5928

Summary

Thermal bonding of oxidized wafers is used to produce dielectrically insulated silicon for device fabrication. The two step bonding process is described and experiments with different wafer surfaces including prestructured and nitride coated wafers are presented. Thinning techniques for Silicon-on-Insulator (SOI) fabrication are compared regarding film uniformity and material quality. SOI films in the thickness of 1-10 microns have been obtained with an etchback process using a p^+-etchstop layer.

Introduction

Dielectrically insulated silicon is a desirable starting material for several device appplications (power and smart power devices, sensors, high temperature applications). As a result of complex fabrication methods of the substrates, the use of silicon-on-insulator (SOI) has been restricted thus far to special devices used mainly for space applications. Recently, the wafer bonding technique gained increasing interest because of its flexibility, simplicity, and material quality [1,2,3]. Additionally, the bonding of two wafers offers possibilities for complex structures required in micromechanical devices or sensors. This paper describes the bonding process for plain and prestructured wafers and compares different thinning techniques for SOI production.

Wafer Bonding

Wafer bonding proceeds via two steps : The pairing of the precleaned hydrophilic surfaces of two mirror polished wafers at room temperature and the subsequent annealing at temperatures above 800°C. During the room temperature step the final position of the wafers is fixed by attractive interations between the hydrophilic surfaces. Particle contamination or surface damage causes disbonded areas (voids), which remain nearly unaffected during high temperature treatment. Direct process control of the room temperature bonding by infrared transmission [4] allows the detection of these voids, which are indicated by interference rings. Fig. 1 shows such a bonding sequence; the remaining void is caused by a scratch on a wafer surface. Bonding experiments with 100mm wafers covered with native oxide or thermal oxides (20 - 2000nm) have shown that homogeneous bonds can be obtained using particle free wafers carefully cleaned with standard RCA treatment. Good bonding results have also been obtained with LPCVD-nitride coated wafers, while plasma oxide surfaces require an extra polishing step before bonding.

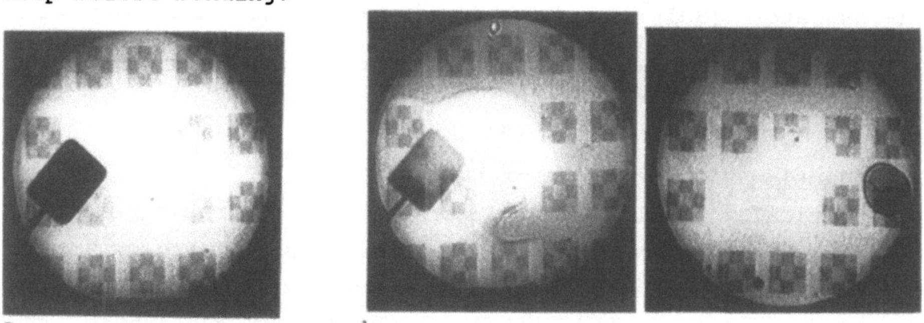

a b c

Fig 1. Infrared transmission images of a prestructured wafer during room temperature bonding. a) t=0, b) t=2 sec, c) t=3 sec

Prestructured wafers with etched grooves give bonding results similar to those of plain wafers, providing the bonding surface is smooth. Fig. 2 shows a scanning electron micrograph of a cleaved edge of the wafer from Fig. 1 after thermal annealing. Inspection of the bonding interface with the infrared transmission and scanning electron microscopy showed no indication of extra voids adjacent to the grooves. Bond strength measurements with the crack propagation method [2,4] gave values identical to those achieved with unstructured wafers.

Thermal treatment in a nitrogen or oxygen atmosphere for 30 min. at 800°C was found to increase the interfacial adhesion sufficiently for further processing. Higher annealing temperatures (1000 - 1200°C) result in tight bonding of the wafers with bond strengths in the range of the SiO_2-bond [1,4].

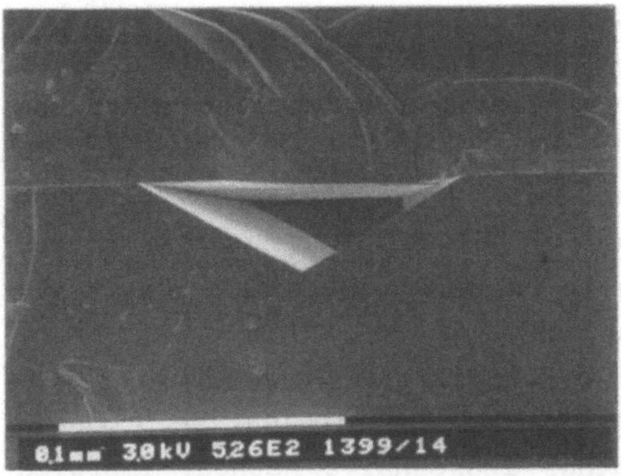

Fig.2 Scanning electron micrograph of a cleaved edge of a prestructured wafer pair annealed for 2 h at 1000°C.

Temperature shock cycles (-65 - +150°C), a technique used for IC reliability test according to MIL standard 883C, was chosen to qualify the bond. Infrared images of bonded wafer pairs with voids (Fig. 3) indicate no additional disbonded areas caused by the stress after 10 cycles for wafers annealed at 800°C and 1000°C. In contrast, thin SOI films (1 and 2 um) that were annealed at 800°C for 30 min prior to the etchback procedure, showed indications of local disbonding of the films after 10-20 stress cycles. Mechanical stress caused by bonding nonuniformities or small size voids, which are not observable with the IR technique, are responsible for this effect. A thermal annealing step, e.g. a short oxidation step (15 min, 900°C) after etchback, increased the bond strength and homogeneity sufficiently to solve the problem. No disbonded spots could be observed after several cycles of temperature stress on those samples.

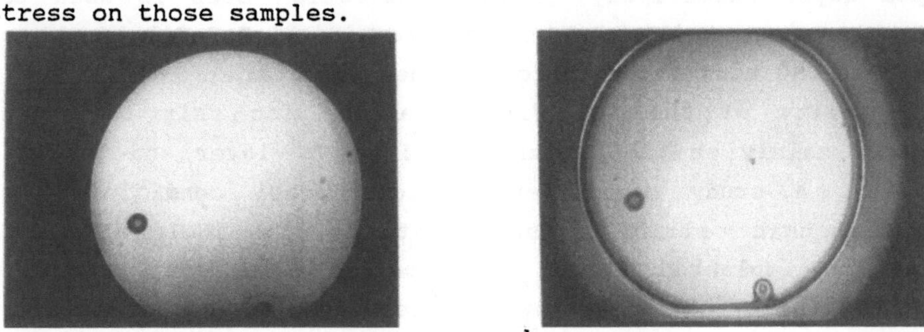

a b

Fig 3. Infrared transmission images of a bonded wafer a) after annealing (30min, 800°C), b)after 10 temperature cycles.

Thinning Techniques for Silicon-on-Insulator

The simplest method for SOI fabrication is to thin the upper wafer mechanically by grinding, lapping, and polishing to the desired thickness. Precise SOI-thickness control, however, is impossible with this technique. With standard polishing equipment using a chemo-mechanical polish, thickness variations of >3um for 100mm wafers are to be expected, which limits the achievable SOI-film thicknesses to 10-20 um. For many applications such as power or bipolar devices this is the required thickness range of SOI material, which can be produced in high quality with this rather inexpensive method. Improved thickness control can be obtained using the selective polishing technique with oxide spacers as polish-stops [5].

The Etchback Process

Thin SOI-films have been fabricated with a selective etchback technique using a p+-etchstop layer [2,4]. Before bonding a high boron doped layer with a peak concentration $>5 \cdot 10^{19} cm^{-3}$ covered with a low doped epi layer of the desired thickness and a thin oxide layer were produced, as shown in Fig. 4. In order to prevent a spreading of the boron profile, all process temperatures, including epitaxial growth and bond anneal, are kept below 900°C. After bonding and annealing, the upper wafer is mechanically thinned to about 100um. A preferential alkaline etch such as KOH/isopropanol, which stops at implanted p+-layers with a selectivity of 150:1 [6], is applied to remove the substrate wafer until the etchstop is reached. A second etch ($HF/HNO_3/CH_3COOH$, ratio: 1/3/8) removes the high boron doped layer selectively. A final oxidation step is used to extract boron tailing from the remaining SOI-film and to increase the bond strength for further processing.

The quality of the SOI film obtained with this technique depends mainly on the properties of the p^+-layer and the epi quality. A study of different etchstop/epi conditions [6] revealed that reduced pressure epitaxy with $SiHCl_3$ at 850°C grown on implanted boron etchstops shows defect densities $<1000 cm^{-2}$ if the wafers were treated with rapid thermal annealing for activation of the boron. After overetching, implanted etchstop layers tend to show etchpits, that are probably caused by local defect clustering, while p^+-layers obtained by diffusion lead to a better SOI homogeneity. Table 1 summarizes some of the data. The final thickness of the SOI-film after optimization of the etchants depends on the thickness homogeneity of the epi-layer and the etchstop. It varies between 70nm and 150nm, giving nonuniformities of 3% to 8% for 5 and 1 um SOI films, respectively.

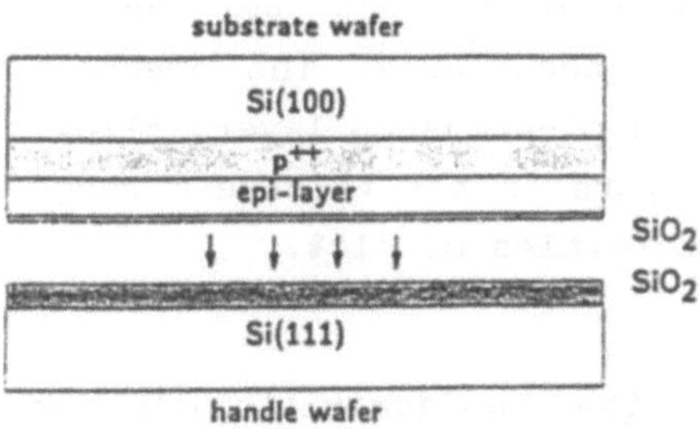

Fig 4. Schematic view of the wafers prepared for etchback: substrate wafer with p^+-layer, epitaxial layer and thin oxide, handle wafer with oxide.

Fig 5. Scanning electron micrograph of a 2 um thick SOI-layer. The 500 nm oxide was etched with HF to indicate the structure.

sample	mean thickness after final etch (nm)	standard deviation (nm)	
p^+implant 1um epi	887	71	(8%)
p^+implant 2um epi	1906	117	(7%)
p^+implant 5um epi	5155	129	(3%)
p^+diffused 1um epi	978	65	(7%)

Table 1 Comparison of different SOI film thicknesses obtained with optical thickness measurements of 25 points per wafer.

Conclusions

Homogeneous bonded wafers have been obtained for different interfacial oxide thicknesses using the described bonding process. Bonding experiments with wafers coated with other materials (LPCVD-nitride, plasma deposited oxide) showed the flexibility of the technique. Transmisson infrared imaging allows direct control of the bond quality and homogeneity. Bonding of structured wafers offers a wide range of applications for sensors, micromechanical devices or three dimensional systems.

Bulk quality SOI films with thicknesses >20 um can be fabricated with mechanical thinning techniques. The etchback procedure presented, when used with a p^+-etchstop layer, allows the production of full wafer SOI layers in the thickness range of 1 - 10 um with thickness nonuniformities of <10%.

Acknowledgements

This work has been supported by the Bundesministerium für Forschung und Technologie (BMFT). The authors alone are responsible for the context. The assistance of Prof.C.E.Hunt, Davis, CA, on preparation of diffused etchstop layers is gratefully acknowledged.

References

1. J.B.Lasky, "Wafer bonding for silicon-on-insulator technologies", Appl.Phys.Lett.48 (1986) 78-80
2. W.P.Maszara, G.Goetz, A.Caviglia and J.B.McKitterick, "Bonding of silicon wafers for silicon-on-insulator", J.Appl.Phys. 64(1988)4943-4951
3. H.Gotou, A.Sekiyama, T.Seki, S.Nagai, N.Suzuki, M.Hayasaka, Y.Matsukawa, Y.Kobayashi, S.Enomoto and K.Imaoka, "A 256 kBIT SOI-FULL-CMOS-SRAM", IEEE Proc. IEDM 89, 1989, 912-913
4. C.Harendt, H.-G.Graf, E.Penteker and B.Höfflinger, "Wafer Bonding: Investigation and in-situ observation of the bond process", Sensor & Actuators A 23, 1990, 927-930,
5. R.P.Zingg and B.Höfflinger, "Novel fabrication technique for dual-gate MOS transistors", Microelectronic Eng. 10 (1990),115-126
6. C.Harendt, C.E.Hunt, W.Appel, H.G.Graf, B.Höfflinger and E.Penteker, "Silicon on Insulator Material by Wafer Bonding", submitted to J.Electronic Mat.

Basic Components and Applications

Sensors for Physical Parameters

Silicon Microsystems as an Industry

James W. Knutti

ICSENSORS, Inc.
1701 McCarthy Blvd., Milpitas, CA. 95035
(408) 432-1800

ABSTRACT

Silicon microstructure technology provides a basis for manufacturing complete microsystems or microsubsystems on a chip. Examples of significant new microsystems currently undergoing commercial development include optical signal processing modules, safe and arming subsystems and micromechanical logic, integrated pressure regulators, medical devices, and chemical and gas handling and analysis systems. Certainly the concept of a chemistry laboratory on a chip is feasible using these techniques.

INTRODUCTION

Two ingredients are needed to ensure successful commercial development of advanced silicon microsystems. The first is a collection of high performance sensors, actuators and mechanical structures that are easily modified and applied to each requirement. Second, the complimentary technology in each field of application must be adaptable to take full advantage of the strengths of the silicon microstructures. For example, an integrated Fabry-Perot optical filter depends upon the suspended plate structure and capacitive position sensing used previously in silicon accelerometers combined with an electrostatic actuator to set plate spacing and hence control the tuning of the device. Traditional optical coating technology and optical signal processing technology have been adapted for use in the silicon part to achieve a high performance, tunable filter at a very low cost. Similar coupling of a silicon pressure sensor and integrated valve structure results in a miniature pressure regulator. Integration of similar silicon structures along with techniques from the field of chemical analysis results in a chemical analysis microsystem, such as the miniature gas chromatograph. The development of each of these applications has followed this pattern.

This paper will present design, performance and economic considerations, and results obtained while developing several different integrated silicon micro-systems including the applications outlined above. Technology partnerships have been important factors in each case. The requirements on the microstructure and the role of the system technology will be described.

SILICON TECHNOLOGY

Silicon is an ideal base material for integrated microsystems. In addition to its compatibility with many complimentary technologies, four basic sets of silicon capabilities can be combined to realize a complete microsystem.

1. Mechanical Microstructures
2. Integral Signal Processing
3. Sensing Mechanisms
4. Actuating Mechanisms

As a mechanical material, silicon is noted for its strength and stability. Of equal importance is the manufacturing technology that allows low cost batch fabrication of precisely dimensioned parts. Modeling and finite element analysis are used to design complex structures, and technologies such as electrochemical etch stopping provide accurate dimensions. Fusion bonding and other wafer lamination technologies are used to combine multiple wafers into compound structures containing sensor, actuator and integral control circuitry. Active circuits can be built on the same silicon wafer, as demonstrated in several sensor designs

A number of sensing principles are used in silicon that are applicable to integrated microstructures. Common approaches are piezoresistive, capacitive and field sensing. Actuation ranges from electrostatic and electromagnetic through piezoelectric, thermal phase change and differential coefficient of expansion.

MARKET CONSIDERATIONS

A primary advantage of silicon as a base material for microsystems is its inherent manufacturability. Well developed processes provide:

1. Tight dimensional control
2. Device to device uniformity
3. Batch processing and high volume capability
4 Low manufacturing and assembly cost
5. Very small size
6. Low power consumption
7. High speed

Photolithographic processing, electrochemical etch stops and accurate anisotropic etching result in very tight dimensional control and device uniformity. This is critical in such applications as self-testable accelerometers and integrated optical filters, where very accurate plate spacing is required. Inherently low manufacturing costs open up many new high volume markets for these subsystems, such as automotive air-bags and active suspensions, or fiber optic demultiplexors. Very small size is also significant in areas ranging from safe and arming modules through hand-held consumer and medical devices. Similarly, low power consumption reduces battery requirements in portable consumer and medical products and industrial devices as well as reducing the need for heat dissipation. High speed allows closed loop control in such applications as gas analysis, medical devices, optical processing and munition triggering systems.

CONCLUSION

Silicon microsystems provide an effective solution to many present and future market needs. These are markets where traditional technologies were not feasible on the basis of cost, performance, size, chemical compatibility or the ability to manufacture. The silicon microsystem will be a component part in many future products in much the same way that the microprocessor has provided the basis of many of the products manufactured today. Similarly, successful development of microstructure markets will depend upon a close coupling between microsystem component manufacturers and end use manufacturers in each market area. The basis of this success will be a thorough understanding of the capabilities and limitations of the technology and a realistic matching with market needs where silicon microsystems truly provide an edge.

Micromachined Silicon Devices Integrated with Standard IC Processes

W. Riethmüller

Fraunhofer Institut für Mikrostrukturtechnik
Dillenburgerstr. 53
D - 1000 Berlin 33

Abstract

The technology of silicon micromechanics for the fabrication of sensors and actuators offers the unique possibility of creating microsystems by monolithic integration of such micromachined devices with integrated circuits for signal processing. In this paper we will discuss the advantages and disadvantages of such types of smart sensors and actuators. The main technological problems regarding monolithic integration of micromechanical devices in IC processes are shown. Most of them are related to the wet anisotropic selective etching of silicon. Also the advantages and disadvantages of developing a completely new process or using standard IC processes are presented. A process based on a standard IC process is briefly described and preliminary results are presented.

Introduction

The interest in microelectronic compatible sensors and actuators is growing rapidly. The main driving force for developing these types of sensors comes from the automotive industry. But also for other consumer products, for process and factory automation, for medical applications and for the growing demands of environmental control, new advanced transducers need to be developed.

Most of them are based on silicon using highly developed IC process technology and special micromachining steps. The use of silicon led directly to a concept of smart transducers with electronic and mechanical subcomponents combined to form a monolithic device /1,2/. The disadvantage of these smart transducers is the complex production process resulting in a lower yield and a higher price/device. There are also some restrictions in the spectrum of possible materials and in the design of the transducer element.

The advantages of monolithically integrated transducers have been discussed extensively in the past and there is no doubt that new qualities in the field of signal conditioning and processing can be achieved /3-8/. The general trend towards miniaturized systems requires technological concepts for the realization of monolithic multifunctional microsystems.

Concepts for monolithic integration of transducers and ICs

There are in principle four methods of monolithic integration of ICs and micro-mechanical elements :

a) Use the existing micromachining process and add steps to realize electronic elements.
b) Develop a completely new process
c) Use a modified standard IC process and add steps for realization of the mechanical device.
d) Use a standard IC process and add new steps for the realization of the mechanical element.

A comparison of these 4 methods can be summarized as follows:

A: The advantage of this concept is that the process could be optimized for the transducer, but the electronic elements would lose performance.

B: Initially the most interesting approach because of the possibility to find the optimum properties for electronic and mechanical elements. However, this concept will only be economical if high volumes are required and/or the devices offer outstanding behaviour. For most systems, this concept will not satisfy these conditions.

C: This concept can be a good solution if the modifications of the IC process are rather small with marginal influence on electronic parameters. For example, a Standard Buried Collector (SBC) process on (100) substrates and a swapped buried layer process have been developed /9/. Additional steps to realize the silicon membrane for pressure sensors via anisotropic etching and electrochemical etchstop are added. The change in parameters are compensated by design rule modifications.

D: Using this procedure, existing integrated circuits can be combined with mechanical devices. All circuits can be used without redesign, and process characterisation and qualification have only to be done from the point where the transducer specific steps are added. This method, based on an SBC process, is presented in references /10/ and /11/.

The Wet Anisotropic Selective Silicon Etch Process

The availability of anisotropic and selective wet etchants for silicon has made this the most widely used material for microtransducers. Anisotropic wet etching is not a standard process step in silicon IC technology. Therefore, most of the problems occurring during the realization of smart transducers are related to this process step. The efforts in solving these problems could be divided into three parts: etch stop technique, anisotropic etch and passivation.

A: Etch stop technique

Commonly used etch stop techniques are based on p^+-silicon or electrochemical effects. The advantage of the p^+- etchstop is the intrinsic etchstop. The disadvantage is the high boron concentration around 10^{20} cm^{-3} in the etch stop layer. This high concentration induces high mechanical stress and a high defect density in the layer. If the layer is realized with an epitaxial process, these problems can be solved by codoping with germanium /12/, which at least allows the establishing of well-defined and controlled intrinsic stress in the layer. Furthermore, these p^+-layers are very sensitive to the high temperature steps within IC-processes. During these process steps the boron concentration of the p^+-layer is lowered and the stress problem arises due to the different diffusion coefficents of boron and germanium and the resulting inhomogeneous doping profile. For SBC processes, the critical process steps are the buried layer diffusion and the p- diffusion for transistor isolation. For CMOS processes it is mainly the well diffusion. We found that using SBC processes with 10 µm thick n-doped epitaxial layers causes the peak concentration of the p^+-layer to fall below 5×10^{19} cm^{-3}, which is too low for a well-defined etch stop. For CMOS processes with typical 5 µm deep wells, the boron concentration peak is around 6×10^{19} cm^{-3} and, therefore, at the lowest level for an etchstop. Also, the leakage currents of the devices are then outside of the valid ranges. In both cases, we started with 4 µm thick epitaxial layers with a 1.5×10^{20} cm^{-3} boron concentration.

For the use of p^+-etchstop layers within IC processes, an additional layer for the electronic devices must be added on top of the p^+-layer. The doping concentration (n or p) in this layer must be around 10^{16} cm^{-3} for SBC processes and around 10^{14} cm^{-3} for CMOS processes. We have developed a 3-step epitaxial process to achieve these layers and to avoid autodoping effects during the epitaxial growth. Figure 1 shows the doping profile of such an n-p-p^+-structure on a p-substrate realized with this process and measured with the spreading resistance method /13/.

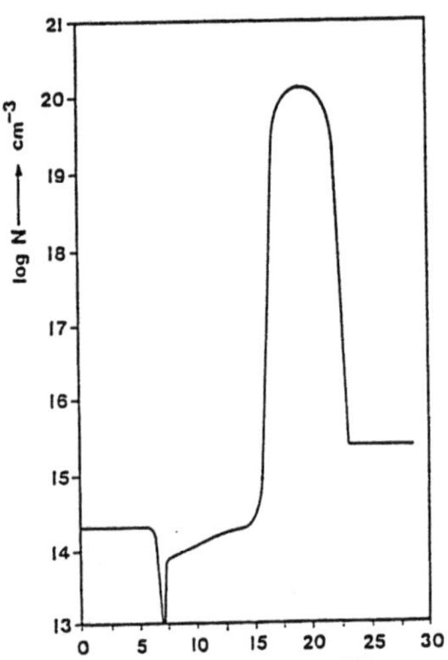

Fig. 1: Spreading resistance measurement curve of an epitaxially grown p^+-p-n layer structure on a p-substrate with a special 3-step process.

Consequently, we summarize that p^+-etchstop could only be used with NMOS, PMOS or SBC processes without a buried layer.

The electrochemical etchstop on reverse-biased pn-junctions is the best option for the combination with IC processes because there are no limitations in the choice of doping levels for substrate and epitaxial layer. The disadvantage of this etch stop technique is that it is not an intrinsic etchstop.That means that the etch stop fails if the electrical contact fails.

A very promising etchstop technique can be used for substrates with buried isolation layers (SiN, SiO). Up to now, this technique has not been well-investigated and differing results have been published /12, 14, 15/. In the future, however, this technique will be further developed because of the interest in SOI and SIMOX techniques for ICs.

B: Anisotropic Etch

Table 1 shows the main characteristics of various anisotropic etchants with respect to their use within IC processes. Results published in literature as well as in our own work are presented.

A comparison of the different etchants can be summarized as follows:

EDP is a suitable etchant from the technological point of view. On the other hand, it is hard to handle, especially under cleanroom conditions. Hydrazine also exhibits very good etching behaviour and is compatible with IC facilities, but it must be handled very carefully. KOH delivers the best etch results, but attacks aluminum and cannot be used within an IC facility. Contamination of the substrate and the equipment with alkaline ions has to be controlled carefully. Ammonium hydroxide remains the best etchant with respect to IC process compatibility and handling, but the surface roughness must be lowered.

Etchant	Etchrate	Anisotropy (etchrate (100)/(111))	Selectivity to p^+	Surface roughness with pn-etchstopp	Selectivity to thermally grown SiO_2	Selectivity to LPCVD-Si_3N_4	Selectivity to Al	Clean room compatible	IC-process compatible	Handling a.s.o.
EPW (EDP)	low 0.3 - 1.0 µm/min	low 1:30	high 1:450	bad >1 µm	good <10 nm/h	very good <1 nm/h	very good <1 nm/h	no	yes	not easy to use, special refuse
KOH	high 0.5 - 2.0 µm/min	high 1:100	low 1:50	very good <15 nm	bad >50 nm/h	very good <1 nm/h	rapidly etched	yes	no	easy to use and to dispose
KOH:IPA	0.5 - 1.0 µm/min	like KOH	high 1:350	bad pyramids	like KOH	like KOH	like KOH	yes	no	like KOH
N_2H_4	1.0 - 3.0 µm/min	low 1:15	very high >1:1000	good in a small process window	good <10 nm/h	very good <1 nm/h	very good <1 nm/h	yes	yes	toxic, explosive, special refuse
NH_4OH:Si	0.2 - 1.0 µm/min	low 1:30	very high >1:1000	bad pyramids	very good <1 nm/h	very good <1 nm/h	very good <1 nm/h	yes	yes	concentration loss during etching, easy to dispose

Table 1: Important characteristics of commonly used etchants with respect to IC-processes

C: Passivation

The data in Table 1 demonstrate that LPCVD-Si_3N_4 is an excellent passivation layer in all etchants discussed earlier, and thermally grown SiO_2 in most of them, but they could not be used as an end passivation layer for ICs. If a wafer is etched in one of these etchants after the IC process is completed, the problem arises that topographic steps up to 2 µm high within the IC must be protected. In addition, if KOH is used, the standard aluminium metallisation must be passivated carefully. For both of these problems, the normal passivation layers of ICs such as PECVD-SiN or SiO are not sufficient due to their high etchrates. Therefore, additional layers have to be deposited, and must be removed after anisotropic etching . The number of useful passivation materials is small because most of the organic layers, resists, etc. will not withstand the entire etch process.

<u>A process for integrated micromachined transducers</u>

Based on the results and strategic considerations described earlier, we have developed a process for smart micromachined transducers.

The process is based on an industrial 3 μm - polysilicon gate - CMOS process with Al/Si - metallisation and a double well technique with n - epitaxial layer on 100 mm diameter wafers. No IC process modifications were made except that the substrate was changed from an n^+ to a p-type. Four to six mask levels were added, mainly at the end of the process, in order to realize the micromechanical element and for the passivation layers during anisotropic Si-etching. Si-etching is performed with KOH and an electrochemical etch stop. Two different test chips have been fabricated, each consisting of 4 different piezoresistive accelerometers and an operational amplifier. Measurements of the OpAmp show that all electrical parameters of the devices fall within the range for the conventional standard CMOS process. Figure 2 shows part of one of the testchips with a piezoresitive Wheatstone bridge and the OpAmp.

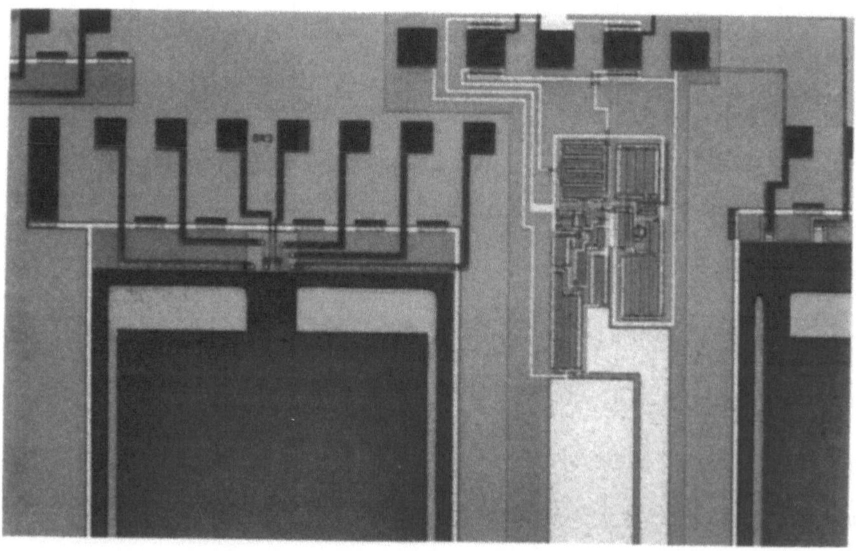

Fig. 2: Part of a testchip with various piezoresistive accelerometers and an operational amplifier. Fabricated with a 3 μm CMOS process.

Conclusion

It has been shown that standard IC processes with no or with only marginal modifications can be used for the fabrication of smart transducers with monolithically integrated micromechanical elements. These processes are well established and characterised, and represent a good starting point for the realization of complex multifunctional microsystems. The main problems during processing are related to the deep anisotropic etching of silicon. Research on new anisotropic etchants such as NH_4OH will be very helpful in minimizing the technological expense. Nevertheless, a process for integrated sensors such as accelerometers has been developed. This process is based on a 3 μm industrial standard CMOS process and silicon etching with KOH and electrochemical etchstop. Preliminary test devices have been sucessfully processed and the electronic devices show no parameter change compared to devices fabricated with the conventional process.

References

/1/ S. Middelhoek, A.C. Hoogerwerf, Smart Sensors: When And Where?, Sensors and Actuators 8 (1985), 39 - 48

/2/ J. M. Giachino, Smart Sensors Sensors and Actuators 10 (1986), 239 - 248

/3/ S. Middelhoek, P.J.French, J.H. Huijsing, W.J. Lian, Sensors With Digital Or Frequency Output, Sensors and Actuators 15 (1988), 119 - 133

/4/ C. Steinbrüchel, The Mechanical Field Effect Transistor: A New Force Sensor, J. Vac. Sci. Technol. A7 (3), May/June 1989, 847 - 849

/5/ B. Puers, L. Reynaert, W. Snoeys, W.M.C. Sansen, A New Uniaxial Accelerometer in Silicon Based on the Piezojunction Effect, IEEE - TED Vol. 35, No. 6, June 1988, 764 - 770

/6/ W. Benecke, L. Csepregi, A. Heuberger, K. Kühl, H. Seidel, A Frequency- selective Piezoresitive Silicon Vibration Sensor, Transducers '85, Philadelphia June 1985, p. 105

/7/ M.A. Schmidt, R.T. Howe, S.D. Senturia, J.H. Harltonidis, Design and Calibration of a Microfabricated Floating-element Shear-stress Sensor, IEEE - TED Vol. ED - 35, No. 6, 1988, 750 - 757

/8/ K. Petersen, A. Shartel, N. Raley, Micromechanical Accelerometer Integrated with MOS Detection Circuitry, IEEE - TED Vol. ED - 29, No. 1, 1982, 23 - 26

/9/ S. Ansermet, D. Otter, R.W. Craddock, J.L. Dancaster, Cooperative Development of a Piezoresitive Pressure Sensor with Integrated Signal Conditioning for Automotive and Industrial Applications, Sensors and Actuators A21 - A23 (1990), 79 - 83

/10/ M. Nakamara, K. Murakami, H. Nojiri, T. Tominaga, Novel Electrochemical Micromachining and its Application for Semiconductor Acceleration Sensor IC, Transducers' 87, 112 - 115

/11/ S. Yamashita, K. Shimaoka, H. Funabashi, S. Sugiyama, I. Igorashi, A Fully-integrated Pressure Sensor, Technical Digest of the 8th Sensor Symposium, 1989, 13-16

/12/ L. Csepregi, K. Kühl, R. Nießl, H. Seidel, Technologie dünngeätzter Siliziumfolien im Hinblick auf monolithisch integrierbare Sensoren, BMFT - Forschungsbericht T84 - 209, 1984

/13/ E. Krullmann, FhG - IMT Berlin, private communication, not published

/14/ V. Gueorgiev, L. Popova, I. Petrov, I. Stoev, Anisotropic Etchants Inhibiting Properties of Silicon nitride Compound Layers Produced by Ion Implantation, Phys. State Solidus Appl. Res. 94 (1986), p. 856

/15/ B. Diem, R. Truche, S. Violet-Bosson, G. Delapiere, "SIMOX" (Separation by Ion Implantation of Oxygen): A Technology for High-temperature Silicon Sensors, Sensors and Actuators, A21-A23 (1990), 1003-1006

Design of an Integrated Humidity Sensor with Digital Output in Double Metal CMOS Technology

Thomas Boltshauser, Edoardo Charbon and Henry Baltes

Physical Electronics Laboratory, Institute of Quantum Electronics
Swiss Federal Institute of Technology Zurich
ETH-Hoenggerberg, CH-8093 Zurich, Switzerland

Summary

We present an integrated humidity sensor system designed for the 2μm, 5V, double-metal CMOS technology *CMN20a* of VLSI Technology provided through the multiproject wafer service of the Swiss Center of Electronics and Microtechnology CSEM. The design includes a humidity sensitive capacitor, a reference element, signal processing circuits, and digital control and conversion circuits, all integrated on the same chip. The sensor specific analog circuitry has been designed using the analog design expert system IDAC developed by CSEM. Test results on a previous humidity sensitive capacitor in CMOS technology are reported.

Introduction

A variety of moisture sensors based on humidity-sensitive electrical properties of materials and devices have been investigated hitherto [1]. Various shapes of electrodes and different dielectric materials (electrolytes, organic polymers, porous ceramics) have been employed. The diffusion of moisture in polyimide films and their aging has been examined with particular care [2]. Availability and excellence of CMOS technology triggered the development of integrated capacitive humidity sensors with e.g. direct voltage [3] or frequency output [4]. Such integrated sensors offer the crucial advantages of small overall size due to minimal external circuitry, reduction of parasitic effects, low power consumption, and inexpensive batch fabrication. Here we describe an attempt to include analog-to-digital conversion on the sensor chip in order to ease modification of the digital output signal and interfacing with a computer. We have used the expert system IDAC for automatic generation of the necessary analog circuitry.

Transducer Principle

The humidity sensitive capacitor consists of an interdigitated lateral structure and ground plate screen formed by the two metal layers of the 2μm, 5V, CMOS process *CMN20a* of VLSI Technology Inc. shown in Fig.2. The ambient air modulates the impedance between the finger-shaped electrodes. This results in a capacitance C proportional to the relative humidity [4]. For ease of integrability and design we have decided to realize the following two capacitance measurement methods.

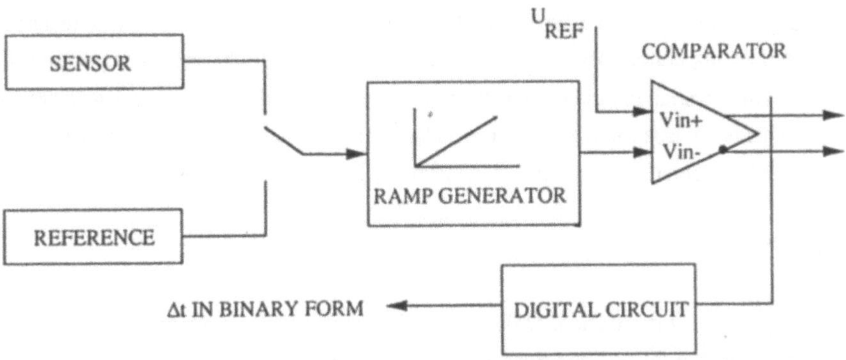

Fig.1. Block diagram of the analog part

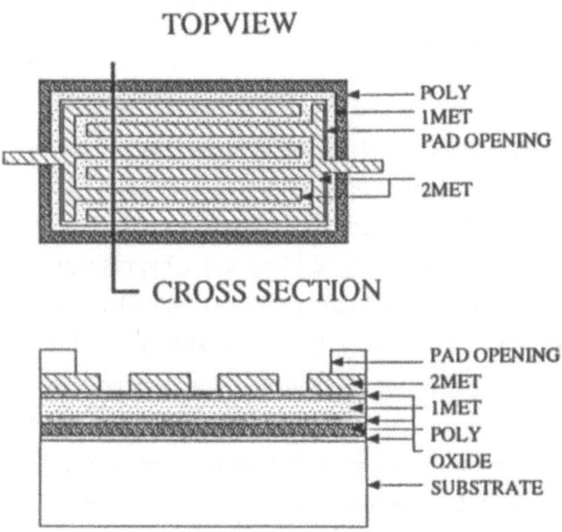

Fig.2. Lateral capacitor design

Fig.3. Photomicrograph of Sensor and Reference (R)

(i) *Conversion of capacitance in a period of time.* This *time-mode* has 3 basic phases: preparation, charging and discharging. A voltage ramp is generated and the time Δt required for the rising voltage to reach the reference voltage U_{ref} is measured:

$$C\frac{\partial U(t)}{\partial t} = i_c \;\rightarrow\; U(t) = \frac{1}{C}\int_0^t i_c\,dt \tag{1}$$

$$U(t) = \frac{1}{C} i_c t + U_0 \;\rightarrow\; \Delta t = \frac{CU_{ref}}{i_c} \qquad \text{with } U_0 = 0 \text{ if C is discharged at } t = 0. \tag{2}$$

Here, Δt depends only on C and U_{ref} because i_C is constant.

(ii) *Conversion of capacitance into a frequency* [4]. In this *frequency-mode*, a capacitor is first charged with a constant current generating a voltage ramp. When the voltage reaches the value U_{ref}, the comparator goes high and excites a counter that discharges the capacitor during the time Δt. An artificial triangle wave is thus generated whose *charge and discharge times are the same*. This is not true for analog charge and discharge systems. Every ramp increments an additional 12 bit counter, which allows to measure the frequency of the wave.

Analog Part

For separate characterization, analog and digital parts have been integrated each with their own contact pads, but the pad positions are adjacent for ease of interconnection. Fig.1. shows the block diagram of the analog circuit. Its purpose is the generation of analog waveforms such as ramps whose rise time Δt depends on the electrical characteristics of the sensor. A comparator discriminates the signals coming from the ramp generator.

Sensor S and Reference R. Fig.2. shows a design and Fig.3. a photomicrograph of the interdigitated lateral capacitor. Two metal-2 electrodes consisting of 24 fingers each (3 μm x 350 μm) are located above the polysilicon and metal-1 layers (both connected with pads) . The minimal distances permitted by CMN20a were used for maximizing the capacitance. The finger structure is depassivated by designing a pad opening, which may be filled with a hygroscopic, planarizing layer such as polyimide [2]. An identical, but not passivated capacitor called reference R has been integrated as well. Covered with a sufficiently thick passivation layer, R is expected to ignore the ambient moisture. Reference and sensor are evaluated simultaneously for compensation of temperature effects etc.

Ramp generator. We distinguish four phases: During phase 1 both capacitors, S and R, are completely discharged by closing S1 and S2. Closing also S3 and S4 stabilizes the circuit and prevents charge injections [5]. Phase 2 prepares the circuit for charging S and R. S4 (S3) is opened in order to couple the ramp generating circuit to the chosen element S (R). During phase 3, S1 (S2) is opened for starting the charge transient of S (R). For minimizing charge injection S1 and S2 are provided with a dummy transistor. Phase 4 starts when the capacitor charging process is completed and the voltage reaches U_{ref}. At this point Sl (S2) and S3 (S4) are closed again in order to discharge the capacitor and to permit another cycle. Because of the independence of each circuit, either S or R can be measured separately. In the *frequency mode* it is possible to measure the frequency of a periodic triangular wave based on the ramp generated by charging S or R. This is done by counting a series of ramp cycles. After conclusion of a ramp the counted time Δt is stored and an equal interval of time is allowed for complete discharge of S (R) and circuit stabilization, after which a new cycle starts.

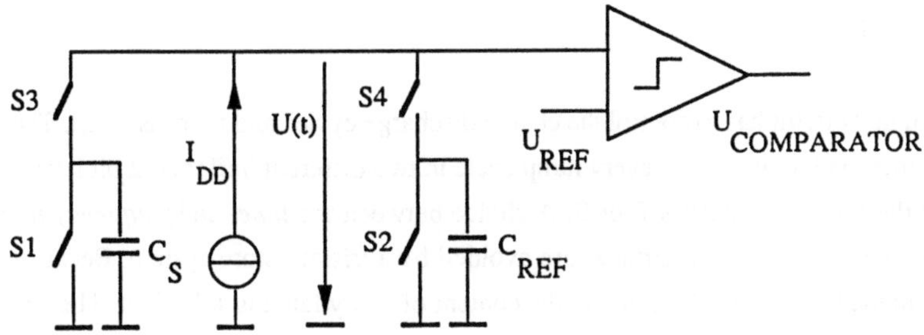

Fig.4. Ramp generator

Comparator. The most complex component of our analog circuit is the comparator adapted from the IDAC library [6] (see Fig.5.), designed to compare an input voltage with U_{ref}. The most important comparator specifications are the offset voltage V_{off} and the decision time t_{dec} ,the comparator's response time. Our simulation yields $V_{off} = 8.0mV$ and $t_{dec} = 70ns$ slightly below the specifications. After simulation, a physical layout of the comparator was generated by ILAC [6]. The layout has a standardized I/O configuration for its integration in VTI compatible designs. Maximal tolerated supply voltage is 5.5V, maximal power dissipation 100mW.

Design rule check. The analog part has been checked by VTIdrc (design rule checker) stating more than 400 design rule violations due to our design of pad openings and contact structures not allowed by the standard design rules, though compatible with the fabrication process. A *window* drc, however, run for the entire analog part except for the sensor did not announce any errors.

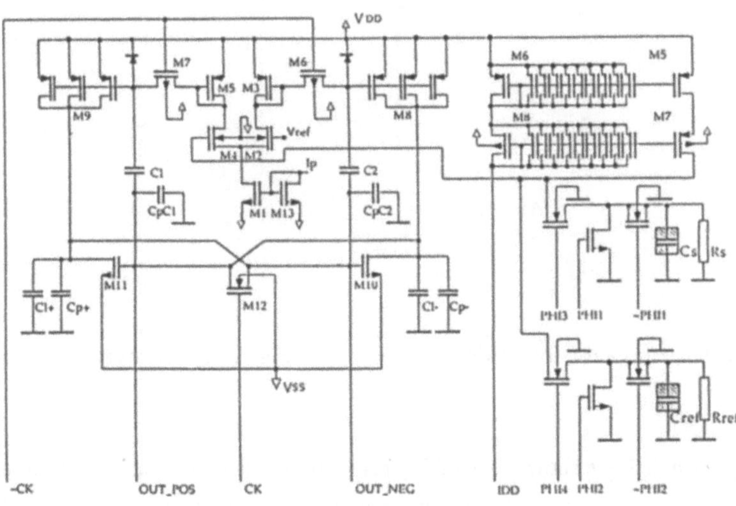

Fig.5. Analog circuit

Digital part

The digital circuit has to control the charge/discharge cycles and to measure Δt. The circuit stores the binary value of Δt after every ramp cycle in two different buffers (12 bit latch) corresponding to the selected capacitor S or R. A choice between the *time-* and *frequency-mode* can be made. Large input-output interfaces are avoided by a tristate-store system. Because of this concept a unique bus (12 bit) for reading the content of every latch is sufficient. The correct logical behaviour and reliability of the entire digital circuit has been checked by logical simulations. The whole spectrum of transitions has been investigated and the timing has been analyzed carefully for a frequency of 10 MHz. VTIdrc (design rule checker) and VTINetCompare have been run. Neither detected any error, although NetCompare was run under worst conditions of allowed parasitic effects. Therefore VTI accepted the entire design for fabrication and denied their warranty only for the analog part of the chip.

Characterization and Test

The device is presently under test. We verify the correct logical functions of the time- and frequency-mode with the logic-tester IMS XL60 at the Integrated Systems Laboratory of ETH Zurich (Prof. W. Fichtner). The analog part is tested separately, in particular the frequency response as a function of an external capacitance and the limits of its linear range.

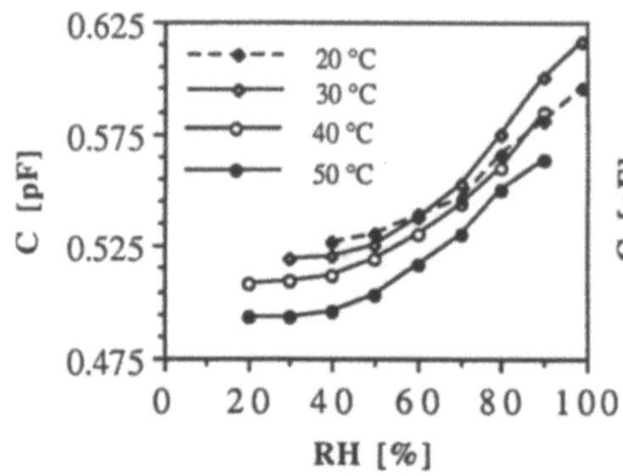

Fig.6. Capacitance response for 10 kHz

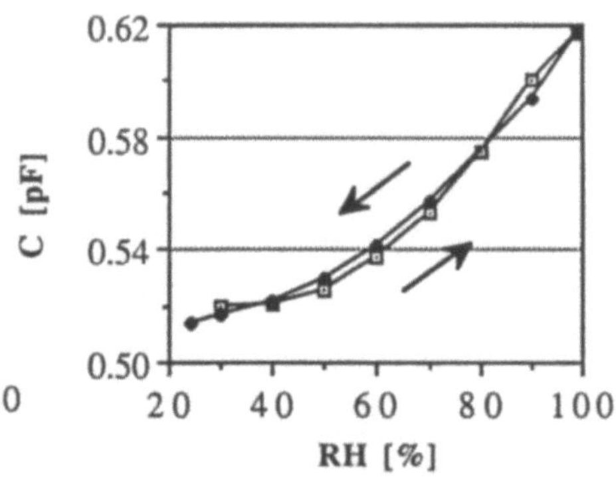

Fig.7. Hysteresis response for 30 °C

In Figs. 6. and 7. we show the humidity response of a lateral capacitor structure similar to that of Fig.2., but with a seven times smaller area (about 130 μm²) fabricated in SACMOS 3μm technology by Faselec AG, Zurich. Between 20% and 90% relative humidity (RH), we found a 15% change in capacitance at a temperature of 30 °C at 10 kHz. The capacitance values have been measured with an accuracy of 0.05% using a HP Precision LCR Meter. The sensor's response time is below 10 s.

Conclusion and Outlook

We reported the operating principle, design, and simulation of an integrated humidity sensor system fabricated in standard 2μm CMOS technology by VLSI Technology Inc. We presented test data on the humidity response of CMOS lateral capacitors used as sensing element. The software packages VTItools, IDAC, ILAC, SPICE and others have been used for design and simulation of our system. Experiments with hygroscopic polyimides serving as dielectric material to fill and planarize the interdigitated capacitor structures and the 3-D modelling of such sensors are in progress [7].

Acknowledgement

It is our pleasure to thank Dr. N. Felber and Mr. P. Amsler of the Integrated Systems Laboratory of ETH Zurich, for their assistance in analog and digital testing.

References

1. N. Yamazoe and Y. Shimizu, Humidity sensors: principles and applications, *Sensors and Actuators 10* (1987), 379 - 398.
2. M. C. Buncick and D. D. Denton, Effects of aging on polyimide: a study of bulk and interface chemistry, *1990, IEEE Solid State Sensor and Actuator Workshop, Hilton Head island, SC, USA, Technical Digest*, in press.
3. S. V. Silverthorne, C. W. Watson and R. D. Baxter, Characterization of a humidity sensor that incorporates a CMOS capacitance measuring circuit, *Sensors and Actuators 19* (1989) 371 - 383
4. H. P. Baltes, E. Charbon, M. Parameswaran, and A. M. Robinson, Humidity sensitive oscillator fabricated in double poly CMOS technology, *Sensors and Actuators B1* (1990) 441 - 445.
5. P. E. Allen, D. R. Holberg, *CMOS analog circuit design*, Holt, Rinehart and Winston, New York, 1987.
6. IDAC 3.1 and ILAC 1.2c Manual, CSEM (Centre Suisse d'Electronique et de Microtechnique S. A., Neuchâtel), Switzerland, March 1989
7. L. Chandran, T. Boltshauser, H. Baltes, Electrical Permittivity and Surface Conductivity of Some Humidity Sensitive Materials, Third International Meeting on Chemical Sensors, September 24-26, 1990, Cleveland, Ohio, USA.

CMOS Compatible Capacitive Pressure Sensor with Read-Out Electronics

M. Kandler, J. Eichholz, Y. Manoli, W. Mokwa

Fraunhofer Institute of Microelectronic Circuits and Systems
Finkenstr. 61, D-4100 Duisburg 1, F.R.G.

Abstract

The fabrication of capacitive pressure sensors using silicon integrated circuit processing is described. The pressure sensors were fabricated using planar etch processing techniques. This results in very small sensors having membrane diameters between 50 μm and 150 μm. The pressure dependence was studied up to 5 bars. Concepts for on chip signal conditioning by using the switch-capacitor method are given.

1. Introduction

Sensing of pressure in various fields like hydraulics, pneumatics or in the medical field is of great importance. The use of silicon technology for the production of such sensors is suitable for mass production thus resulting in low prices. Even for a rough environment a high sensor realiability and robustness is desirable. These requirements can be met by combining silicon sensors with on-chip electronics to a "smart sensor".

2. Sensor Fabrication

In contrast to a commercially available silicon pressure sensor, where the fabrication of the pressure sensitive membrane needs a wafer backside etching process, the pressure sensor presented here was fabricated by using planar etch processing /1/. This technique is compatible to standard CMOS processing. The fabrication starts with the deposition of silicon oxide which then is patterned and etched to leave an island. This oxide island serves as a sacrificial layer and defines the membrane region. Next a layer of polycristalline silicon is deposited on the wafer which forms the pressure sensitive membrane. The membrane is achieved by laterally etching of the sacrificial oxide layer via etch channels. These channels are sealed by depositing silicon dioxide on the membrane. The membrane and the silicon substrate are forming the plates of a capacitor (Fig. 1).

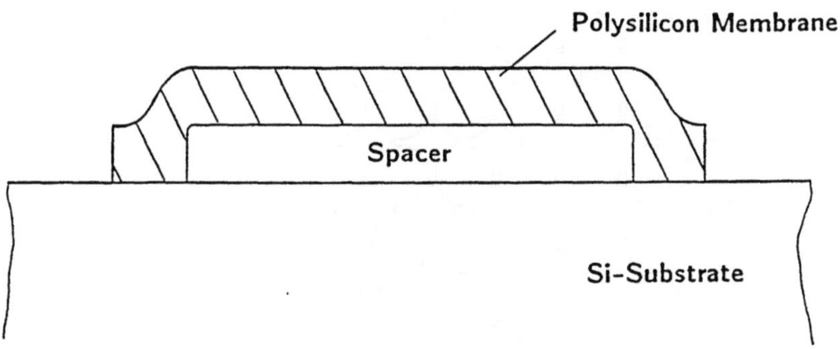

Fig. 1: Polysilicon pressure sensor

When a pressure is applied to the membrane it bends down, resulting in an increase of capacitance. Because of the vacuum sealing the fabrication results into an absolute capacitive pressure transducer. Sensitivity and pressure range can be determined by membrane diameter, membrane thickness and spacer height. Membrane diameters up to 150 μm were obtained without showing any initial deflection due to built in strain, having thicknesses between 1 μm and 2 μm. For our prototypes we used spacer thicknesses between 0,5 μm and 1 μm. The capacitance of the membrane itself is very low. For example a 100 μm in diameter membrane has a capacitance value of about 0,017 pF without pressure. Therefore to get larger signals we designed sensor arrays with up to 81 individual pressure sensors which are switched in parallel (Fig. 2). This results into capacitance values between 1 to 2 pf.

The pressure sensor is highly overpressure resistant since the lateral dimensions of the device are much larger than the vertical dimensions. The membrane cannot deflect larger than the height of the cavity. For example a 70 μm in diameter membrane, designed for a pressure range of 6 bars could withstand an overpressure of more than 100 bars without damage.

Fig. 2: Pressure sensor array (membrane Ø 70 μm)

3. Experimental Results

The pressure sensors were tested at a pressure of 1 to 6 bars. The capacitance values were measured with a HP4275A meter. As a pressure reference we used a MKS Baratron type 122 A. Fig. 3 shows the change

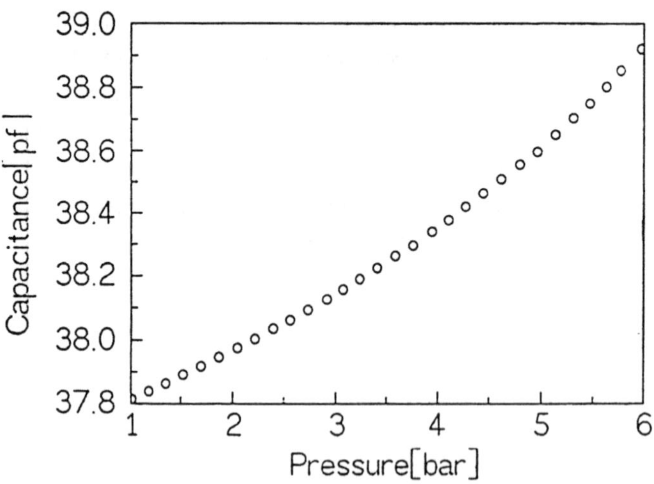

Fig. 3: Change of capacitance as a function of pressure (membrane Ø 70 μm)

of capacitance as a function of pressure between 1 and 6 bars. There is a non linear increse in capacitance due to the reciprocal distance dependence of the capacitance. The parasiteric capacitance is caused by the capacitance of the polysilicon to the silicon substrate in the anchoring regions and therefore is not pressure dependent.
The stability of the sensors was studied for 44 days (Fig. 4). The change of capacitance during this time are within the experimental limits of the measurement equipment.

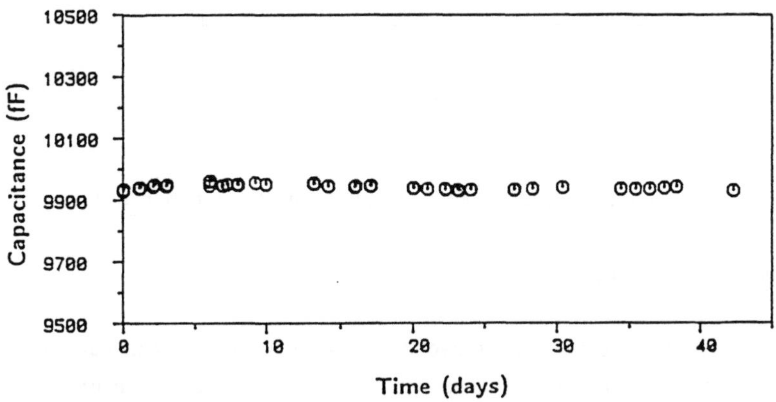

Fig. 4: Long term measurement at a pressure of 1 Bar.

We also studied the temperature dependence between 20°C and 70°C (Fig. 5). A linear increase of capacitance was found. A pressure independent reference capacitor which was formed by the same fabrication stepts but without etching away the sacrificial layer shows the same temperature dependence. This indicates that the temperature dependence is mainly due to the temperature dependence of the parasiteric capacitance. By simply taking the difference of both capacitance values, the temperature dependence is strongly supressed.

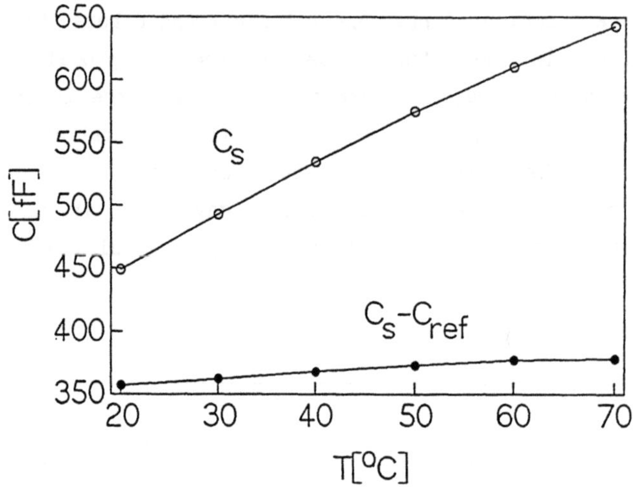

Fig. 5: Temperature dependence of a sensor array (Ø 70 μm membrane).

4. Sensor Read-out Electronics

Having CMOS compatible capacitive sensors the well known switch-capacitor method for on-chip signal conditioning can be used. The base circuitry is shown in Fig. 6. The output voltage is a function of the quotient of the capacitance values.

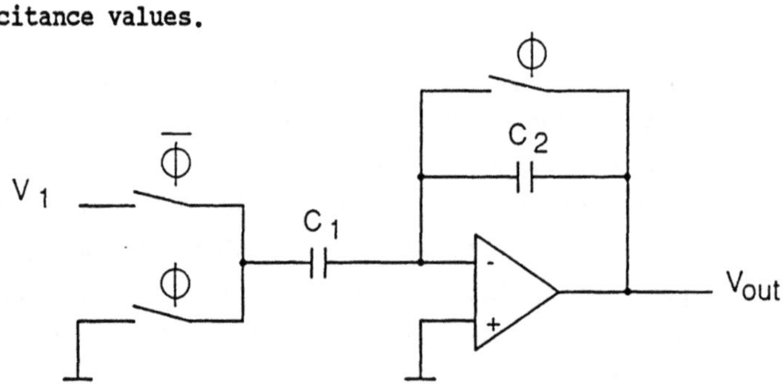

$$V_{out} = -V_1 \frac{C_1}{C_2}$$

Fig. 6: Base circuitry for a switch-capacitor read-out electronic

By replacing either C_1 or C_2 by the pressure sensor the output voltage becomes proportional or inversly proportional to the sensor capacitance.

The SC-circuit shown in Fig. 7 realizes a function where the output voltage is porportional to the reciprocal of the difference of the two feedback capacitors C_2-C_3.
This function is appropriate for a linearization of the sensor signal, when C_2 is substituted by the sensor capacitance and C_3 is substituted by the reference capacitance.

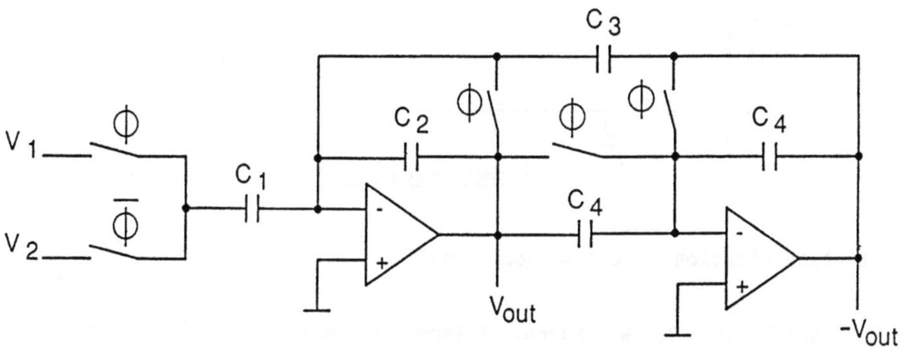

$$V_{out} = (V_1 - V_2)\frac{C_1}{C_2 - C_3}$$

Fig. 7: Switch-capacitor circuitry, where the output signal is proportional to the reciprocal of the difference of the feedback capacitances C_2 - C_3.

Fig. 8 shows how this circuit linearizes the sensor signal.
Taking the difference of the sensor capacitance and the reference capacitance also increases the relative change of the output signal, compared with the relative change of the sensor capacitance itself.

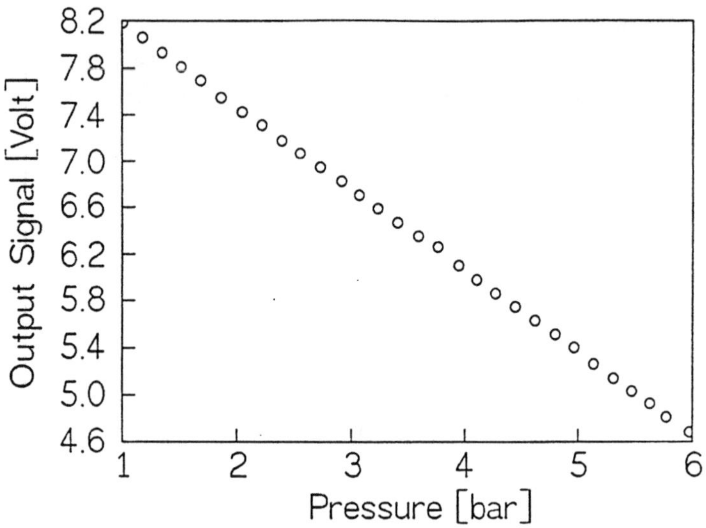

Fig. 8: Linearization of the sensor signal

/1/ H. Guckel and D. W. Burns, "Planar Processed Polysilicon Sealed Carities for Pressure Transducers Arrays", IEDM 1984, 223-225

Piezoresistive Pressure Sensors Representing the 2nd Generation Avoid the Physical Limits Based on Conventional Designs

H. SANDMAIER

Fraunhofer-Institut für Festkörpertechnologie,
8000 München 60, Federal Republic of Germany

Summary

A novel design of piezoresistive, low pressure sensors is presented. This design exhibits a number of advantages compared to the conventional. The main objective of this development was realizing a sensor with high sensitivity, high pressure overload, and low nonlinearity. This paper describes the theory for the design of the piezoresistive low pressure sensor. Through the application of a square diaphragm bossed in the center, a piezoresistive low pressure sensor for the pressure range of ±10 kPa could be realized, exhibiting excellent sensitivity and low nonlinearity of 35 mV/V F.S.O. and 0,05 % respectively.

Introduction

There is a large demand for low cost sensors due to the steadily increasing application of microprocessors in metrology, automation, and control system engineering. Conventional sensors can not meet the requirements of many applications for cost effectiveness and for high accuracy simultaneously. Further development of technology demands sensors of small size and of simple construction. The application of the semiconductor technology makes the fabrication of micromechanical sensors possible which fulfil both requirements.

At present micromechanical pressure sensors are most highly developed. They consist of a thin etched diaphragm deforming under the influence of pressure. The deformation will be converted into a resistance change with the help of piezoresistors. Connecting four resistors to a Wheatstone bridge, the change of resistance will be transferred into a voltage change, yielding the output signal. In the low pressure range the conventional designs of pressure sensors come to their limits, due to the increasingly nonlinear pressure dependence of the output signal of the sensor. The demand for the realization of sensors with a linear output characteristic lead to the development of bossed diaphragms [1 - 5].

1. Optimization of the sensor structure

A simplified representation of a piezoresistive semiconductor pressure sensor is shown in Fig. 1. The input pressure will be transformed to physical quantities according to the depicted blocks, which in the end lead to the output quantity ΔV.

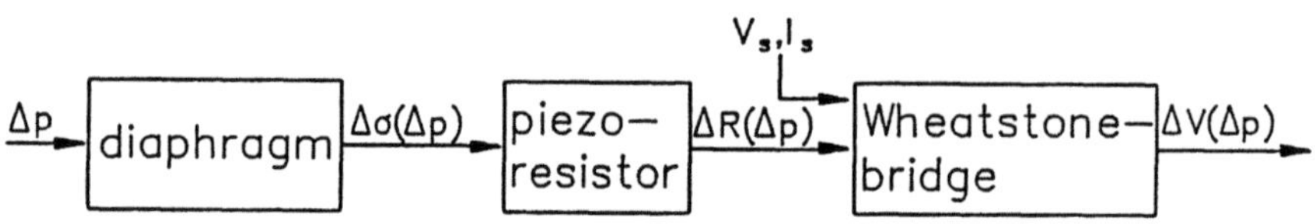

Fig. 1: Block circuit diagramm of a piezoresistive pressure sensor
Δp: pressure change, Δσ: change of mechanical stress, ΔR: resistance change, ΔV: voltage change; V_S, I_S: supply voltage and supply current of the Wheatstone bridge

The characteristics of the output signal of the sensor and, in particular, its nonlinearity can be influenced by all blocks. For this reason, the realization of high precision pressure sensors requires the exact knowledge of the transfer function of each block shown in Fig. 1.

1.1 Diaphragmm

For an exact determination of the output signal of the sensor, nonlinear dependences have to be taken into account. Analytical calculations of the transfer characteristic of the first block in Fig. 1 yield, that for circular and rectangular diaphragms the increase of the nonlinearity in the low pressure range is caused by the large deflection of the diaphragm. These deflection is the consequence of the fact that high sensitivities of the sensor in the low pressure range require very thin diaphragms. Such diaphragms expand and blow up like a balloon resulting in a nonlinear behaviour. Analysis with the help of nonlinear theories lead to the recognition that for circular and for rectangular diaphragms [6, 7], a reduction of this nonlinear effect through a change of the size of the diaphragm is not possible.

For this reason the development of low pressure sensors with linear output characteristic requires a way different form that of the conventional diaphragm design [3]. In order to prevent a blow up, the center of the diaphragm has to be bossed. In the past, circular ring diaphragms have been employed in order to reduce this effect [8, 9]. However, results of stimulations have shown that this form makes only small improvements possible. Nevertheless, the principle of bossing can also be transferred to other forms of diaphragms. Simulations with the help of the finite element method and extensions of the existing analytical theory [6] lead to a new design employing a bossed square diaphragm. Such diaphragms exhibit significant advantages over circular ring diaphragms.

1.2 Piezoresistor

The realisation of piezoresistive pressure sensors with linear output signal requires an exact knowledge of the transfer characteristic of the piezoresistive effect. The nonlinear dependence of this effect was determined experimentally. For this purpose a special set-up has been developed which, through a controlled and well-defined deflection of a cantilever with piezoresistors on top, makes the measurement of the effect possible. Both longitudinally and transversally arranged p-type resistors have been studied. Although the transverse effect exhibits a larger nonlinearity in the sensor design actually realised, four transversally loaded resistors have been applied. The reason for this is the measured quadratic stress dependence of the resistance change which in the case of the sensor design shown in Fig. 2 and in the case of current supplied Wheatstone bridge renders a complete compensation of the nonlinearity feasible [6].

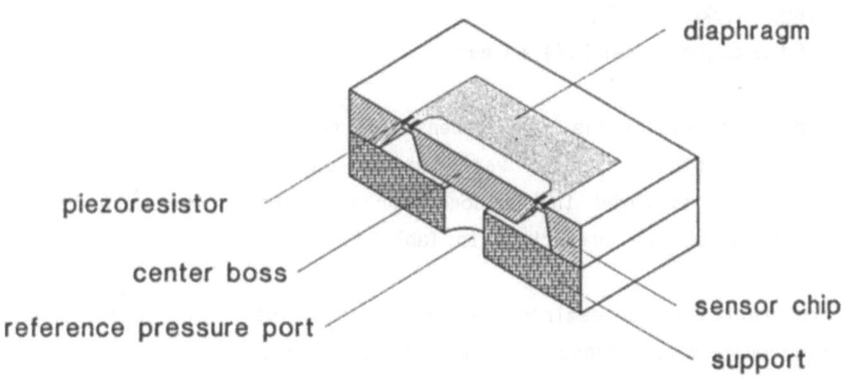

Fig. 2: Illustration of the bossed diaphragm with resistor arrangement

For the process used (see Table 1) the dependence of the transverse piezoresistive effect ($\Delta R/R = f(\sigma)$, where σ = stress) can be described by the following equation:

$$\frac{\Delta R}{R} = -4{,}87 \cdot 10^{-10}\sigma + 32 \cdot 10^{-20}\sigma^2 \qquad [\sigma \text{ in } N/m^2] \qquad (1)$$

1.3 Wheatstone bridge

The resistance changes of the four transverse resistors are converted into a voltage change employing a Wheatstone bridge. There is no additional nonlinearity resulting from the use of the Wheatstone bridge for the typical resistance changes of a few percent. The use of this circuit offers the possibility of compensating both the nonlinearity as well as the temperature dependence of the first two blocks depicted in Fig. 1. In order to optimize the design of a precision pressure sensor, the Wheatstone bridge has to be taken into account.

2. Fabrication process and experimental results

Based on the design shown in Fig. 2, a pressure sensor for a nominal pressure range of ±10 kPa has been realized. The theoretically determined sensitivity and nonlinearity of the sensor are 35 mV/V F.S.O. (Full Scale Output) and 0,05 % respectively, employing the following geometrical dimensions:
diaphragm size : 3.5 x 3.5 mm^2
diaphragm thickness: 20 μm
size of the center boss: 2.84 x 1 mm^2

With the exception of the three-dimensional structuring, the technological realization of the sensor was carried out by means of processes conventionally employed in semiconductor technology. The most important processing steps are summarized in Table 1.

The theoretically determined sensor characteristic was experimentally verified on pressure sensors fabricated in the above manner. The measured nonlinearity of the sensors output signal amounted to < 0.05 %. In spite of the high sensitivity of 35 mV/V F.S.O., a pressure overload of a factor of 10 could be realized by means of this design and the technology employed. The pressure overload can be increased drastically employing overpressure protections on both sides of the bossed diaphragm.

- Starting material	: {100} n-type silicon
- Oxidation (wet)	: 1000°C, 100 min. (0,5 μm)
- Photoresist process	: piezoresistors
- Implantation	: boron, 50 keV, $7 \cdot 10^{14}$ cm^{-2}
- Photoresist process	: removal of field oxide in the diaphragm area
- Oxidation (dry)	: 1000°C, 95 min., (0,1 μm)
- Annealing	: 1000°C, 65 min., N_2
- Photoresist process	: contact areas
- Metallisation	: 1 μm Al
- Photoresist process	: Al structuring
- Sintering	: 450°C, 15 min., 10% H_2/90% O_2
- Plasmanitride deposition	: backside (o,6 μm)
- Photoresist process	: etchwindow definition and opening
- Anisotropic etching	: 33% KOH, 100°C, 105 min.

Table 1: Technological processing of the sensor

Fig. 3: Front and backview of the pressure sensor chip developed

3. Advantages of the new design

Compared to the conventional design of a piezoresistive pressure sensor, the sensor discussed in this paper offers significant advantages as a consequence of the application of a bossed square diaphragm with resistors subject to transverse load. The following provides a summary of these benefits:

- high sensitivity and minimal spatial requirements of the diaphragm
- small nonlinearity in the low pressure range
- high pressure overload
- minimization of the influences of fluctuations in the fabrication process and of the assembly stresses
- compensation of the alignment errors of the resistors
- improvement of the long-term stability and temperature dependence because
  - *the metallization can be placed outside the diaphragm since the contact areas of the resistors are already on the bulk material,
  - *in the area of the diaphragm and in particular across the resistors, the field oxide can be substituted by a high quality passivation.

Acknowledgments

The author would like to thank K. Kühl and I. Schulz for the execution of the technological work.

References

1. H. Sandmaier et al., Piezoresistive Low-pressure Sensor with High Sensitivity and Accuracy, Transducers '89, Montreux, Vol. 2, pp. 142 - 145 (1989)
2. L. Christel, et al., Single-cristal Silicon Pressure Sensors with 500 x Overpressure Protection, Transducers '89, Montreux, Vol. 2, pp. 84 - 88 (1989)
3. M.-H. Bao, et al., Micromachined Beam-diaphragm Structures Improve Performances of Pressure Transducer, Transducers '89, Montreux, Vol. 2, pp. 137 - 141 (1989)
4. J. Mallon et al., Low Pressure Sensors using bossed Diaphragms and Etch-stopping for Precision Control, Transducers '89, Montreux, Vol. 2, pp. 89 - 95 (1989)
5. H. Reimann, New Mechanical Structures to Achieve Low-Pressure Silicon Sensors and Accelerometers, Sensor 88, Nürnberg, pp. 265 - 280 (1988)
6. H. Sandmaier, Untersuchung des nichtlinearen Verhaltens piezoresistiver Niederdrucksensoren auf der Basis von Silizium, Dissertation, TU München (1988)
7. K. Suzuki et al., Nonlinear Analysis of a CMOS Integrated Silicon Pressure Sensor, IEEE Trans. on Electron Dev., Vol. ED-34, No. 6, pp. 1360-1366 (1987)
8. H. Sandmaier et al., Reduzierung der Nichtlinearität von Siliziumdurcksensoren für kleine Drücke, NTG-Fachbericht 93, Sensoren-Technologie und Anwendung, Bad Nauheim, S. 159-164 (1986)
9. M. Shimazoe, et al., A Special Silicon Diaphragm Pressure Sensor with High Output and High Accuracy, Sensors and Actuators, Vol. 2, No. 3, pp. 275-282 (1982)

High Performance Miniaturized Piezoresistive Pressure Transducers

ROBERT M. WHITTIER

Endevco Corporation
a Subsidiary of Allied Signal Aerospace Corporation

Summary
This paper describes recent advances for several silicon diaphragm piezoresistive pressure transducers. It also reviews the advantages of stress-concentrated micromachined diaphragms, in particular the two island design with transverse strain gages. Recent diaphragm designs have provided significantly smaller output errors over wide temperature ranges, such as -55°C to +125°C, and have added on-chip zero adjustment. This eliminates zero temperature compensation in many applications. Packaging approaches are also described which isolate the piezoresistors from the pressure media, but still allow the pressure diaphragm to be placed close to the pressure boundary to provide high frequency response.

Stress-Concentrated Micromachined Diaphragm

Silicon diaphragm piezoresistive pressure sensors usually employ junction isolated resistive elements diffused into flat edge supported diaphragms. Resistors are located at positions of high positive and negative stress, and an active four arm wheatstone bridge is usually used. Performance can however be enhanced by shaping the diaphragm to concentrate the stresses at the specific locations of the piezoresistors. Concentrating stress in gaged areas, while maintaining low stress levels elsewhere, gives greater sensitivity for the same resonant frequency than is possible for a flat diaphragm. This concentration is achieved for the subject sensors by forming two islands on what would normally be a flat diaphragm. These islands must be close together, so that when deflected, the stress is high in the narrow grooves. Figure 1 shows a cross-section of such a diaphragm with the letters A, B and C pointing to the groove bottoms.

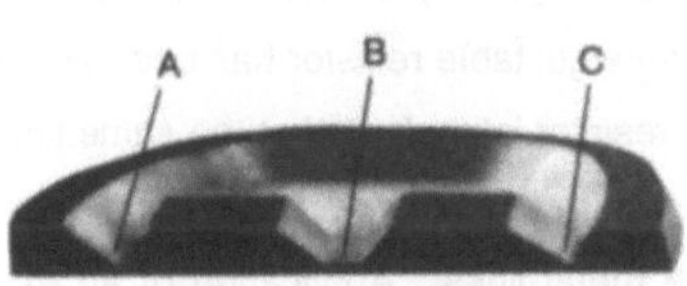

FIGURE 1: Photomicrograph of diaphragm section through notches and islands (O.D. 1.25mm).

The micromachining of these diaphragm shapes is accomplished by etching in caustics such as potassium hydroxide (KOH). This anisotropically etches the single crystal silicon providing excellent dimensional control. One requirement is to obtain excellent dimensional control for the groove bottoms. In general this is a function of the photolithographic and etching process control.

Structural material for these microsensors is N-Type (100) silicon starting with wafers 125-300 microns thick. Piezoresistors are P-Type [110] orientation with the gages extending along the length of the groove bottoms. With the application of pressure the gages are subjected to a stress crosswise to their length. Resistance change is a result of the transverse strain coefficient, and because the non-linearity of a positive and negative stressed transverse gage are in opposite directions there is a self-correction in the wheatstone bridge circuit (1). Resistance changes of as much as 3% can be achieved while still providing low wheatstone bridge non-linearities. The end result is high sensitivity, typically 200 mV FS with about 0.1% non-linearity.

Survival to overrange pressures is also enhanced by this diaphragm design. In part this is due to the fact that less area is at high stress which reduces the probability that a surface or internal fault will be at high stress.

Performance Advances

Design and process improvements have been incorporated into the micromachined diaphragm to enhance its performance. Each of these borrow upon technologies previously developed for integrated circuit fabrication.

Ion implantation of the piezoresistive circuit replaces the thermal diffusion process. By using this process, better repeatability is achieved from unit-to-unit on a wafer and between wafer-to-wafer, and batch-to-batch. As a result, improved specifications are practical for performance over wide temperature ranges.

Providing zero electrical output from the wheatstone bridge is often a requirement, however this is not feasible because of photolithographic and fabrication process variables. For many transducers output at zero pressure is adjusted by adding resistors across, or in series with, a piezoresistor within the bridge. The result is that the bridge may be balanced at about +25°C, but it varies with temperature. To improve upon this, an adjustable resistor has been added to the bridge circuit on the silicon diaphragm chip. This trim resistor is implanted at the same time as the bridge resistors; and as a result it has the same temperature coefficient of resistance. It is not one element but is a ladder network connected by fusible metal links. Application of an external voltage at special contact points opens the desired links to balance the bridge.

The performance of a typical sensor is shown by Figure 2. This data was measured by placing a single OTC resistor in series with the voltage input to the bridge circuit. Resistance value for such a resistor is a function of the microsensor basic design and is normally not individually trimmed.

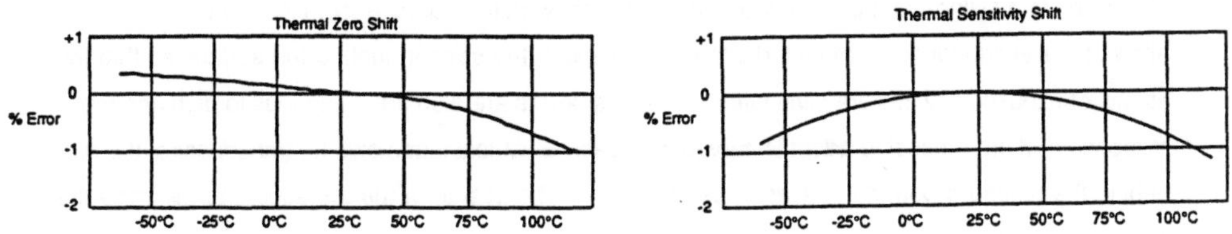

FIGURE 2: Zero and Sensitivity Change vs. Temperature, Reference 25°C.

Another feature of this new diaphragm system is that a thin metallic coating is applied to the diaphragm. This is done by sputtering metal onto the silicon wafers before the wafers are separated to provide the individual sensor chips. By doing this the light sensitivity of the pressure sensor is significantly reduced.

Microsensor Designs

The complete silicon microsensor is fabricated using two micromachined silicon wafers. One is used to form the thin diaphragms and electrical circuits. The other wafer is used to form additional structural support and the vacuum reference cavity, or pressure reference port. These two wafers are glass bonded together to form a sandwich and then are sawed apart to form the individual microsensors. Wafer bonding is accomplished at about 500°C in special fixtures which compress the two wafers together. If the sensor has an absolute pressure reference the bonding is done at vacuum; this provides a high quality and well sealed reference.

Although the supporting wafer is less complex from a semiconductor standpoint, it plays a critical role in providing the required sensor performance. It adds stiffness to the edge support of the diaphragm, adequate structure for bonding the microsensor to a transducer casing, and a means to isolate mounting stresses from the diaphragm. Thermal expansion differences and stresses from inputs such as those induced by external mounting forces must all be reduced to provide optimum performance.

Several physical designs have been created based on these approaches, with the variations between these designs being a result of requirements for physical size and shape, pressure range, pressure media, and electrical connection.

<u>Gage Pressure Sensor, Probe Configuration</u>

The probe is a cylindrical transducer of small diameter with the microsensor close to one end. This allows the transducer to be mounted into structures with the sensor flush to the surface so that the effects of acoustic cavities are eliminated. Small diameters are also advantageous to high frequency response and are easiest to adapt to curved surfaces. Electrical leads exit the transducer at the rear behind the pressure boundary. Care must be taken in the design of the transducer to minimize the output errors which may result from mounting the transducer.

To provide compatibility with many different types of pressure media this transducer has been designed with gold contacts and wires. (Most sensors use aluminum contact systems which have minimal corrosion resistance). The piezoresistive circuit is exposed to the pressure in the reference cavity with only the wiring exposed to the front side. The reference is either the absolute reference, gage reference, or differential side. Aluminum nitride has been chosen as the support for the microsensor because of its high stiffness and fairly good expansion coefficient match to silicon. Nail head terminals are bonded into this part to transmit the signals from the front to the back. Figure 3 shows this packaging scheme. Detail configuration of the microsensor is shown by Figure 4.

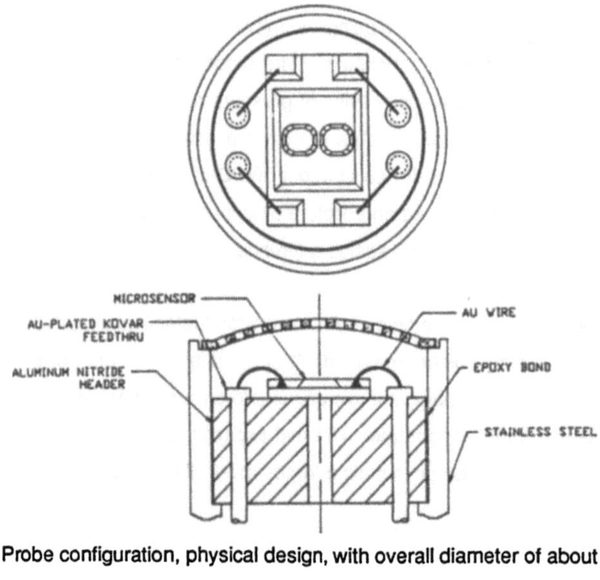

FIGURE 3: Probe configuration, physical design, with overall diameter of about 2.5mm.

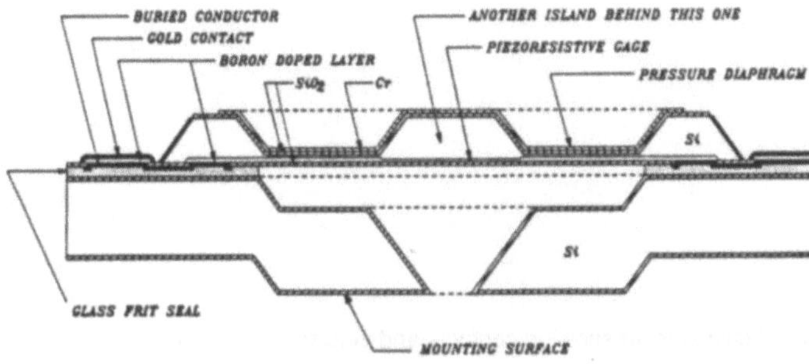

FIGURE 4: Cross section of microsensor showing piezoresistors exposed to one side and contacts exposed to other side.

Absolute Pressure Sensor, Flat Configuration

Certain applications require pressure transducers which can be bonded to a structure without altering the structure or significantly changing the dimensions of the surface. Dynamic pressure measurements on an aerodynamic body in a wind tunnel or flight test are typical of these. What is really desired is that the pressure transducer has no thickness. In addition to being exposed to pressure the transducer can also be exposed to at least two other significant dynamic inputs, which if not considered in the transducer design, can cause error. One of these inputs is dynamic strain and the other is sudden changes of light, or thermal radiant energy.

A transducer which is extremely thin will be more vulnerable to strain inputs than one which is thick. It must include a high degree of internal strain isolation between the structure and the diaphragm. If it also occupies considerable area it would also be more susceptible to bending inputs. For this application a microsensor has been designed which occupies minimal area, 1.65 mm by 1.17 mm, and minimal thickness, 0.30 mm. It is bonded to a .025mm thick polyimide circuit interconnect and is surrounded by a metal casing. The diaphragm is protected from particulate damage by a screen; this screen and the thin film metal on the diaphragm reduce the thermal radiant inputs.

Figure 5 pictures the mounting side of this microsensor; the principal mounting area is a small 1mm^2, which assists with the strain isolation. A cross section of the microsensor in Figure 6 shows the two island style diaphragm, absolute reference cavity, and electrical contact system. Electrical connection to the sensor from the external circuit traces is made by .05mm diameter aluminum wires from underneath the top portion of the sensor. This minimizes the overall thickness for the transducer package. A thickness of 0.76 mm is quite practical and includes the metal case.

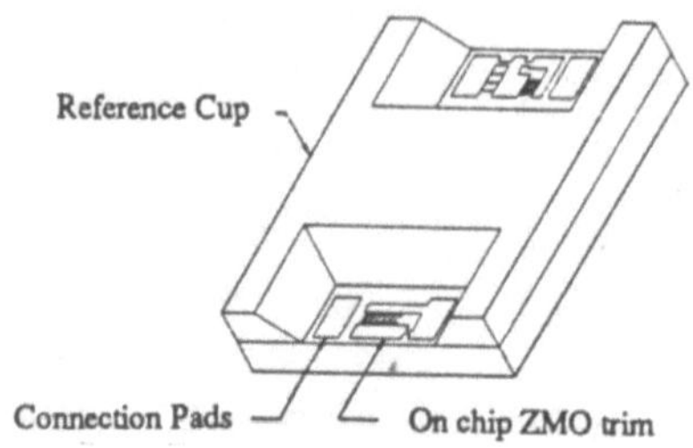

FIGURE 5: Bottom of microsensor showing contacts and zero trim resistors.

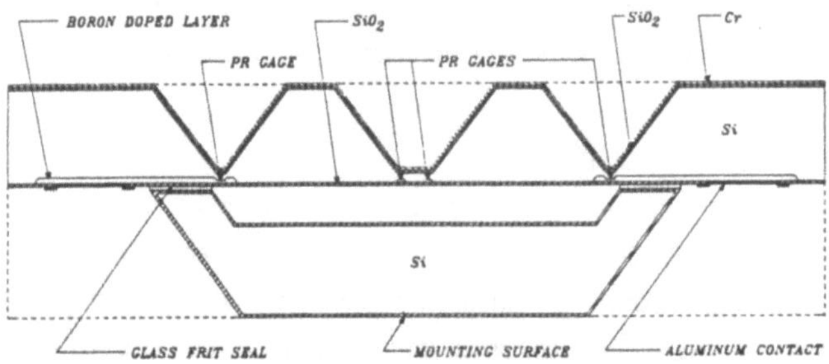

FIGURE 6: Cross section, piezoresistors in absolute pressure reference cavity.

Conclusions

The techniques and design approaches described have resulted in pressure transducers having very good performance qualities. Errors over wide temperature ranges are significantly less than from the earlier designs. The probe style has been fabricated with ranges down to about 14kPa (2 psig) and the flat style to ranges of 103kPa (15 psia). Typical performance characteristics are shown in Table 1.

Characteristic	Result
Full Scale Output	200 mV
Output at Zero Pressure, 25°C	2% FSO
Non-Linearity, BSL	0.1%
Pressure Hysteresis	.05%
Resonant Frequency, 15 psia Range	180kHz
Compensated Temperature Range	-18°C to +93°C
Thermal Zero Shift	.006%/°C
Thermal Sensitivity Shift	.01%/°C
Maximum Temperature Range	-54°C to +121°C

TABLE 1: Typical Performance Results

References: [1] L.B. Wilner, "A Diffused Silicon Pressure Transducer with Stress Concentrated at the Transverse Gages"; ISA 23rd International Instrumentation Symposium, Las Vegas, NV, May 1977

Fully Integrated Silicon Pressure Sensor with On-Chip CMOS Evaluation Circuit and On-Chip Trimming

H.-J. Kress, F. Bantien, J. Marek, and M. Willmann

Robert Bosch GmbH, Automotive Equipment Division K8

Tübinger Straße 123, 7410 Reutlingen,

Federal Republic of Germany

Abstract

A silicon pressure sensor with an integrated signal conditioning circuit has been developed for automotive applications. The sensor is fabricated with a standard 3μm CMOS process. The membrane is formed by anisotropic etching in (100)-oriented silicon using an electrochemical etch stop. A complete signal conditioning circuit is added to amplify the bridge output signal. Furthermore, the sensor offset, the sensitivity, and the temperature effects on offset and sensitivity values are adjusted individually. Compensation is done for a temperature range from -40 to +125°C. On-chip trimming is provided for the adjustment of the correction values.

Introduction

Silicon piezoresistive pressure sensors show significant advantages compared to e.g. metal-film pressure sensors. Large piezoresistive coefficients, good elastic properties, and small size of the silicon sensors are the most important ones.[1] Consequently, there has been a wide dissemination of silicon pressure sensors in many application fields such as automotive equipment, process control, and medical systems. *Integrated* pressure sensors, however, are not very common in industrial applications, although they offer further advantages in size and cost reduction.

Nevertheless, some integrated sensor systems have been reported recently.[2-4] The sensor elements and signal conditioning circuits have been combined by using different fabrication processes and following different compensation concepts. The reported sensors require process modifications for the circuit fabrication. Either the surface impurity concentration had to be changed,[2,3] or laser trimmable thin film resistors have been introduced for compensation purposes.[4]

The sensor described in this paper is based on a standard CMOS process without any significant modifications for the circuit processing. Merely the starting material differs from that of a standard CMOS process. In the following paragraphs, details of the integrated sensor fabrication and the circuit design will be presented and finally some results will be shown.

Technology

The fabrication of integrated pressure sensors requires the combination of micromachining techniques and IC-processing steps. The micromachining technique is based on an anisotropic etching process with an electrochemical etch stop. The IC-process used is a conventional CMOS-process.

Since the membrane thickness determines the pressure range, thickness control plays an important role for the micromachining process. The simplest solution, namely using time determined etch stop techniques, leads to undesirable high tolerances in thickness variation. On the other hand, buried p^+-layers are acting as effective etch stop interfaces. However, a heavily boron doped region leads to local mechanical strain. Moreover, the out-diffusion of boron ions during the IC process will cause several problems. According to the literature, the electrochemical etch stop technique has been proved to be the best solution for an exact thickness control and thus it has been consequently used.[1]

The starting material for the integrated sensor fabrication is p-doped, (100)-oriented silicon. Prior to the CMOS process, an n-type epitaxial layer is grown on the substrate material. The IC process begins with the definition of the p-well regions. In addition, the four piezoresistors are diffused simultanously. The utilization of the low doped p-well for the bridge resistors guarantees an extremely high k-value which is independent of process parameter variations. Higher doped p^+-regions connect the piezoresistors with aluminium lines leading to the circuit input.

The IC fabrication on the front side is carried out and completed by the deposition of the passivation layer and the following definition of the passivation windows. The PE-nitride which is used as passivation layer on the frontside is deposited on the backside of the wafer additionally. There it acts as an etch mask for protection during the anisotropic etching process in KOH. A special wafer carrier is provided to protect the frontside during the etching process.

Finally, the pressure sensor wafer is attached to a corning #7740 (Pyrex) glass wafer using anodic bonding in a vacuum chamber to enclose a reference vacuum for measurement of absolute pressure.

Mechanical Design

The piezoresistive pressure sensors are developed for the pressure range from 50 kPa to 1000 kPa. The particular measurement range is designed by adjusting the thickness of the sensor membrane, i.e. the thickness of the epitaxial layer. Thus, various sensors are processed with only one set of lithography masks.

Sensor membranes with sharply defined membrane edges are fabricated by electrochemical etching. The position of a membrane edge with respect to the front side structures, e.g. the piezoresistors is dominantly effected by the front-to-backside alignment and is controlled accurately. The relative position of the piezoresistors, however, is of considerable interest for the sensitivity of the sensor element. A test structure has therefore been developed prior to the design of the integrated sensors.

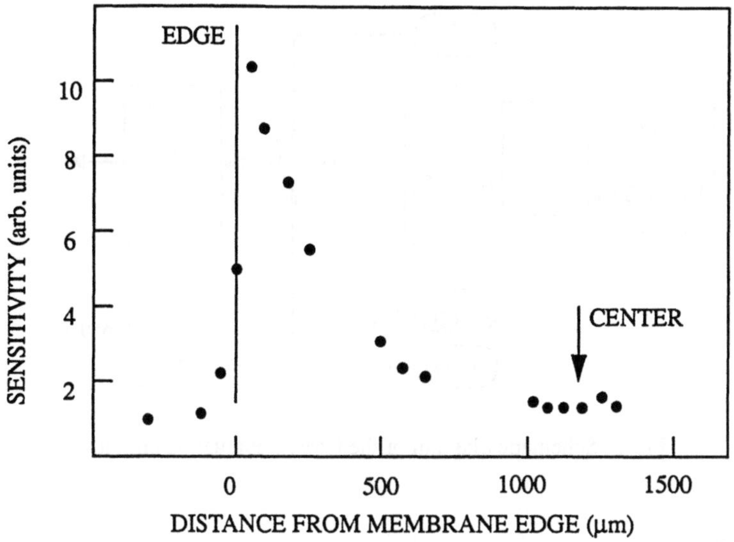

Fig. 1: Sensitivity in respect to the resistor position on the membrane

The test structure contains a large number of piezoresistors which are arranged along the centerline of a quadratic sensor membrane in different distances from the membrane edge. The resistors as well as the membrane edges are oriented along the <110>-directions on a (100)-oriented silicon wafer. The membrane is deflected by a differential pressure and the piezoresistive signal, i.e. the change of the resistors is recorded.

Figure 1 shows the relative changes of the resistivity in respect to the position of the resistor on the membrane. In addition, the position of the membrane edge and the center of the membrane, as determined from transmission microscopy, is indicated. It is obvious, that the maximum of sensitivity is located near the membrane edge. The maximum is found to be located about 30 to 40 µm within the membrane. With increasing distance from the edge, the sensitivity decreases rapidly. However, the decrease is not as strong as the decrease towards the membrane edge.

It is remarkable, that the piezoresistive changes in the center of the membrane are negligible, despite the fact that the strain has a local maximum at this position. Thus the longitudinal and transverse piezoeffects compensate each other.

The pressure sensor consists of four piezoresistors located near the edges of the membrane. While two of them are stressed longitudinally, the other two resistors are stressed transversely. According to the test design, the resistors are placed near the points of maximum sensitivity.

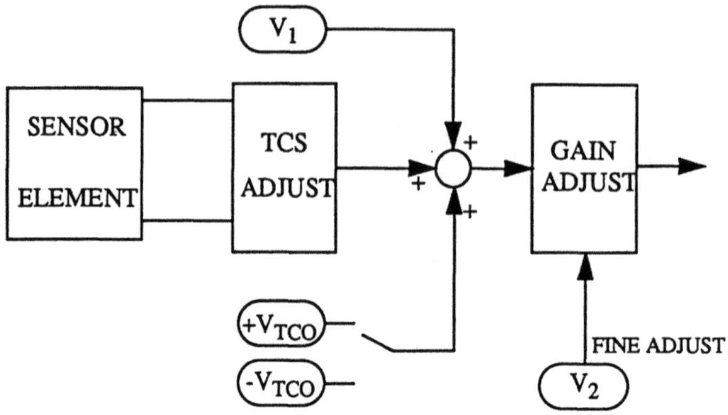

Fig. 2: Schematic diagram of the CMOS evaluation circuit

Circuit Design

The CMOS evaluation circuit is designed to eliminate process related statistical fluctuations of the signal offset and its temperature coefficient (TCO) as well as the fluctuations of the sensitivity. Moreover, the negative temperature coefficient of the sensitivity (TCS) which is caused by the temperature dependent piezoresistive coefficient π_{44} is compensated. Additionally, the sensitivity of the pressure sensor has to be improved in order to allow direct connection to an external A/D converter. A schematic diagram of the chosen circuit configuration is shown in Fig. 2. This configuration allows to compensate the above mentioned terms *independently.*

The first stage consists of a standard instrumentation amplifier which is used to compensate the negative temperature coefficient of sensitivity. The output of the first stage is directly connected to a summing amplifier. A voltage V_{TCO} which is proportional to the absolute temperature is added in order to balance the temperature dependence of the bridge offset value. The sign of the compensation voltage can be changed for positive and negative TCOs, respectively. In addition, an offset coarse adjustment is provided in this stage in order to restrict the output voltage to a usable range. The following output stage is used to adjust the final amplification and to guarantee sufficient driving capability. Moreover, a voltage V_2 is added to give the final offset fine adjustment.

The circuit configuration of the first stage (Fig. 3a) allows to compensate the negative TCS over a wide temperature range. The compensation is achieved by a temperature dependent amplification A(T):

$$A(T) = 1 + \frac{2R_2(1+\alpha_2\Delta T)}{R_1(1+\alpha_1\Delta T)} \quad . \tag{1}$$

Suitable resistor values R_1 and R_2 with the corresponding temperature coefficients α_1 and α_2 are used to achieve the desired positive temperature coefficient of amplification. The tuning of the amplification temperature coefficient is done by trimming the resistance value of R_1. Thus we are able to follow the standard IC process without any additional lithography levels for the circuit fabrication and without changing any process parameters.

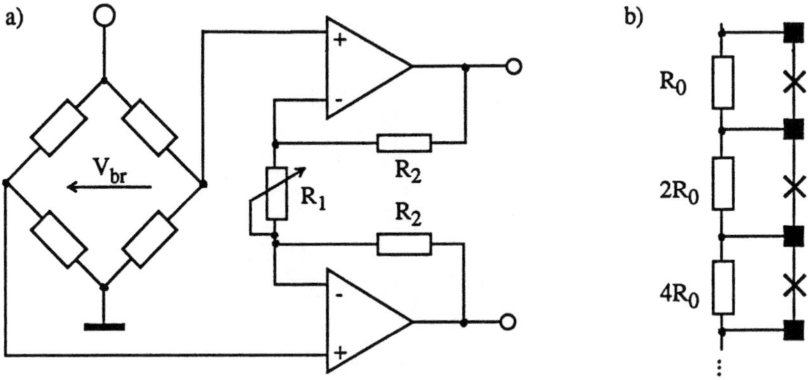

Fig.3: a) Schematic circuit diagram of the first signal conditioning circuit stage
b) Trimming resistor diagram

The trimming of all adjustable resistors used in the circuit is done by metal links. Binary weighted resistors are initially shortened via metal shunts as it is shown in Fig. 3b. The desired resistors are then "switched on" by applying a voltage to the neighbouring pads which burns off the metal link.

Results

Simulations have been carried out to determine the variation of amplification with respect to the temperature. The results show sufficient linearity of the amplification variation over the whole temperature range for automotive applications. The nonlinearity error is smaller than 1% of the full-scale output FSO in the temperature range from 0 to +125°C. Over the full temperature range from -40°C to 125 °C the nonlinearity error is smaller than the desired value of 2 %FSO.

Figure 4 shows a top view photograph of the fully integrated pressure sensor. The chip size of the sensor is 3.9 * 3.6 mm^2. The membrane with the four piezoresistors is clearly visible in the center of the chip. The evaluation circuit is arranged around the sensor diaphragm. The circuit has been designed for a wide range of sensor characteristic variations. A great number of metal links and accompanying passivation windows is therefore visible.

The characterization of an uncompensated 100 kPa sensor exhibits a sensitivity of about 40 mV/(V*bar). The nonlinearity error of the sensor is less than 0.1 %FSO. The temperature coefficients are TCO = -0.05 mV/(V*K) and TCS = -0.20 %FSO/K, respectively.

With increasing membrane thickness the actual measurement range increases and the absolute value of sensitivity is found to be lower. The full-scale output, however, is further improved. The FSO of a discrete portion of the 350 kPa sensor is of the order of 300 mV with an excitation voltage of 5V.

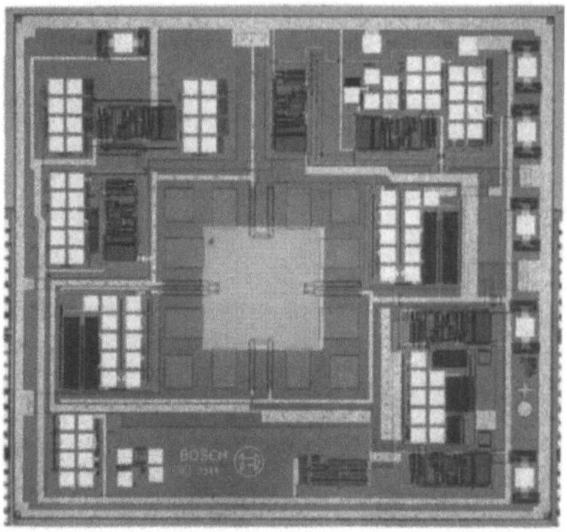

Fig. 4: Photograph of the integrated sensor chip

Conclusion

A pressure sensor with an integrated CMOS evaluation circuit and on-chip trimming is fabricated. The sensor chip includes compensation functions for offset, sensitivity, and the associated temperature coefficients. A conventional CMOS process without any significant process modifications concerning the circuit fabrication has been used. The micromachining process is based on an anisotropic etching process in combination with an electrochemical etch stop. The sensor exhibits an excellent sensitivity combined with extremely small linearity error.

Acknowledgement

We like to thank G. Findler and H. Baumann for their excellent support in the sensor fabrication as well as G. Schelling and U. Schneider for their contributions to packaging and characterization. Helpful encouragement concerning the circuit development was provided by H.-D. Schwarz.

References

[1] K. E. Petersen, "Silicon as a Mechanical Material," *Proc. IEEE*, vol. 70, pp. 420-457, May 1982

[2] T. Ishihara, K. Suzuki, S. Suwazono, M. Hirata, and H. Tanigawa, "CMOS Integrated Silicon Pressure Sensor," *IEEE J. Solid-State Circuits*, vol. SC-22, pp. 151-156, April 1987

[3] H. Tanigawa, T. Ishihara, M. Hirata, and K. Suzuki, "MOS Integrated Silicon Pressure Sensor," *IEEE Trans. Electron Devices*, vol. ED-32, pp. 1191-1195, July 1985

[4] S. Yamashita, K. Shimaoka, H. Funabashi, S. Sugiyama, and I. Igarashi, "A Fully Integrated Pressure Sensor," in *Technical Digest of the 8th Sensor Symposium*, pp. 13-16, 1989

Optimization of Bipolar Magneto-Transistors

H. Kaneko, H. Muro and P.J. French
Central Engineering Laboratories, Nissan Motor Company Ltd.,
1 Natsushima-cho, Yokosuka 237, Japan.

Summary

The magneto-transistor has the potential of high sensitivity, including 3-dimensional applications, while maintaining a compact and simple structure. It does, however, have problems with linearity, offset and temperature coefficient of sensitivity. A range of magnetic sensor structures including NMOS devices, Hall plates and both npn and pnp bipolar transistors are compared. This paper concentrates on design considerations for lateral pnp devices and, in addition to the sensitivity, examines the above mentioned problems for these structures in order to optimise sensor characteristics.

Introduction

When a carrier is subjected to a magnetic field perpendicular to the flow, it is deflected in the third mutually orthogonal direction. This is the basic operation of both Hall plates and magneto-transistors. Hall plate devices are widely available and yield a sensitivity of ~7.2%/T [1]. Magneto-transistors, on the other hand, have the potential of a much higher sensitivity.

Magneto-transistors can either use the deflection of majority or minority carriers, depending on where in the transistor a deflection is detected. These devices can also be applied to measuring 3 dimensions of magnetic fields. A frequently used device for measuring fields perpendicular to the plane of the chip is the lateral pnp transistor, whereas the vertical npn transistor is often applied to fields within the plane of the chip. Great improvements in the device sensitivity can be achieved by using the field assisted technique [2,3] and by careful consideration of the sensor design.

The sensitivity, S, of a device is given in terms of the magnetic induction field, B, by:

$$S = \{\Delta V_H/(V_B B)\} x 100 [\%] \quad 1$$

for voltage output devices, where V_B is the applied voltage and ΔV_H the output voltage. For current output devices the sensitivity equation becomes:

$$S = (1/B)\{(I_{c1}-I_{c2})/(I_{c1}+I_{c2})\}\times 100\ [\%] \qquad 2$$

where I_{c1} and I_{c2} are the two collector currents. For current output devices a further definition is used which describes the current collection efficiency, α, and is:

$$\alpha = \{I_{c1}+I_{c2}\}/I_e \qquad 3$$

where Ie is the emitter current.

Comparison of Magneto-Transistor Structures

Magneto-transistors can be fabricated by either MOS or bipolar technology. MOS magneto-transistors can use either; a) side contacts (similar to a Hall plate) with voltage output or, b) a split drain with a differential current output. Sensitivities of 6.4%/T and 4.8%/T have been reported for types a and b, respectively [4,5]. MOS devices cannot, however,be easily applied to lateral magnetic fields.

Bipolar magneto-transistors detect a magnetic field by an offset of collector current, as defined in equation 2. The main types of magneto-transistors are as follows:

1) Dual collector lateral PNP
2) Dual collector field assisted lateral PNP
3) Dual collector lateral NPN
4) Dual (multi) collector vertical NPN

where types 1,3 and 4 usually measure magnetic fields parallel to the chip surface and type 2, fields perpendicular to the chip. Sensitivities of 240%/T and 56%/T have been achieved by using structures 2 [3] and 3 [6], respectively.

A variety of vertical npn transistor structures have been applied to lateral field measurements [7] yielding sensitivities of ~4%/T. A multi collector vertical npn transistor structure has further been applied to 3-D magnetic field measurements [8]. An alternative technique is to use a lateral pnp transistor. This uses the vertical motion of carriers below the emitter. Sensitivities of 130%/T [9] and 400%/T [10] have been achieved by this method. A modification to this technique is to use a "Suppressed Sidewall Injection Magneto-Transistor" (SSIMT) [11]. Using this type of device, sensitivities of as high as 3000%/T have been achieved. However, the output is highly sensitive to biasing conditions and a more stable bias point yields an output of ~200%/T. Among these, field assisted lateral pnp magneto-transistors were fabricated and evaluated because of their high and stable sensitivity and simple structure.

Field Assisted Lateral pnp Magneto-Transistor

The field assisted technique is where an additional electric field is set up across the device to enhance the hole motion towards the collector. The basic layout and its cross-sectional view of a field assisted pnp transistor is shown in Fig. 1.

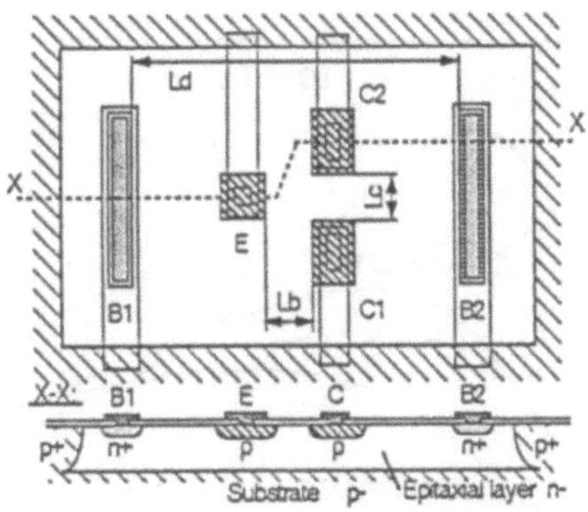

Figure 1. Layout of a field assisted lateral pnp transistor.

In addition to processing parameters, which effect the carrier mobility, and thus sensitivity, an important factor in controlling sensitivity is the separation of the two collectors, Lc. Other design considerations are the base length, Lb, and the distance between the base contacts Ld. Test devices were fabricated by a standard bipolar process using a 12um n-type, 2Ω cm epilayer on p-type substrate.

Table 1. Measured sensitivity, offset and current collection efficiency for various samples at room temperature, for Ie=10μA, Vb=10V, and the base contact distance, Ld=228μm.

	Standard Lb=34μm Lc=32μm	Short Base Lb=18μm Lc=32μm	Long base Lb=18μm Lc=32μm	Wide collector Lb=34μm Lc=64μm
S (%/T)	223~228	233~236	233~243	261~369
Offset(G)	-361~-64	-247~160	-642~332	-2052~54
α (%)	9.0	9.4	7.0	0.1

The sensitivity, offset and current collection efficiency were measured for various samples and these results are listed in table 1. The sensitivity was proportional to the applied bias voltage, Vb, and increased with collector separation, Lc, while it did not depend on Lb. The base contact spacing, Ld, did not affect sensitivity indicating a non-uniform field distribution in the base region.

The effect of both Lb and Lc on the current collection efficiency is shown in Fig. 2.

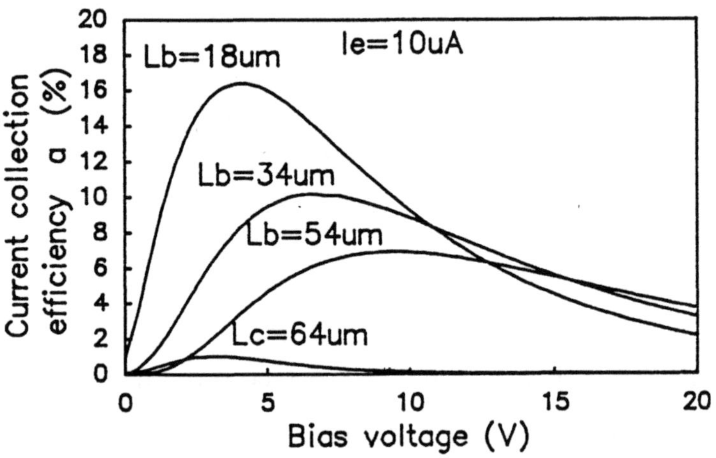

Figure 2. Effect of base bias on current collection efficiency.

The peak value of α decreases and the corresponding bias voltage increases with increasing Lb due to increasing recombination. For low Vb, the drift velocity toward the collector of the injected holes increases with Vb causing an increase in α. A further increase in Vb, however, results in a higher proportion of the injected holes being collected by the other base contact and not the collector, causing a decrease in α.

The temperature characteristics of sensitivity and electron drift mobility, which are normalized to room temperature, are shown in Fig. 3. The temperature effect on sensitivity and mobility can be approximated by $T^{-2.5}$ and $T^{-2.8}$, respectively, yielding a numerical expression for the temperature dependence of $S \propto \mu_n \cdot T^{-0.5}$, as shown in Fig.3.

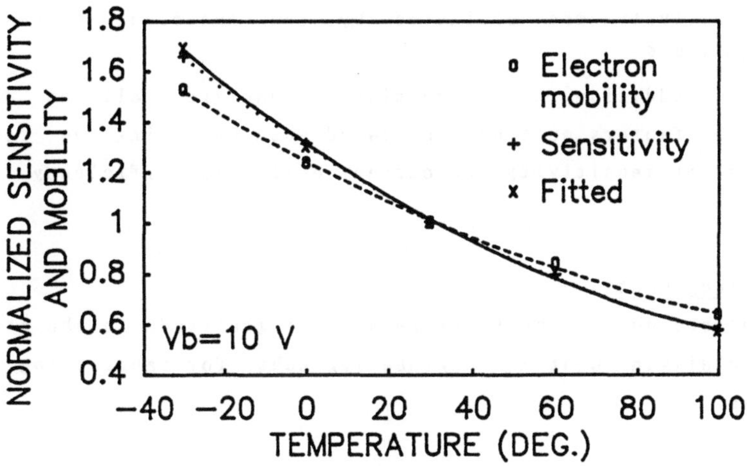

Figure 3. Temperature dependence of sensitivity and electron drift mobility (dashed line..measured electron mobility, dotted line..measured sensitivity and solid line..fitted curve-$S \propto \mu n \cdot T^{-0.5}$).

Design Considerations

A higher sensitivity can be obtained with larger collector separation, Lc, and a higher bias voltage, Vb, at the sacrifice of poor α. In addition, a larger Lc cause rather large offset variations. It is therefore neccesary to balance these factors when designing magneto-transistors. The causes of offset are believed to be; 1) Misalignment, 2) non-uniformity in the film and, 3) mechanical stress. Methods of reducing offset are presently being investigated.

The temperature compensation can be easily achieved by biasing the base with current source with a temperature dependence of $T^{2.5}$.

Conclusions

The field assisted lateral pnp magneto-transistors with various configurations have been evaluated. The sensitivity has been shown to increase with collector separation and applied bias voltage, but is independent of base length. The current collection efficiency, on the other hand, was found to be strongly dependent upon all these factors. The temperature dependence of sensitivity was estimated by the numerical expression $S \propto \mu_n \cdot T^{-0.5}$.

In addition to processing parameters, all the above mentioned factors must be considered when balancing the requirements for sensitivity and current collection efficiency.

Acknowledgements

The authors would like to express their gratitude to the staff of the Nissan Central Engineering Labs. for fabricating the devices.

References

1) Randhawa, G.S.: Monolithic Integrated Hall Devices in Silicon Circuits, Microelectron. J. 12 (1981) 24-29.

2) Davies, L.W. and Wells, M.S.:Magneto-transistor Incorporated in an Integrated Circuit. Proc. IREE (Australia) (1971) 235-238

3) Halbo, L. and Haraldsen, J.: The Magnetic Field Sensitive Transistor - A New Sensor for Crankshaft Angle Position, SAE 800122 (1980) 79-82.

4) Hirata, M. and Suzuki, S.:Integrated Magnetic sensor. Proc. 1st Sensor Symposium, Japan (1981) 305-310.

5) Misra, D., Viswanthan, T.R. and Heasell, E.L. A Novel High Gain MOS Magnetic Field Sensor, Sens. & Act. 9 (1986) 213-221.

6) Vinal, A.W. and Masnari, N.A.: Magnetic Transistor Behavior Explained by Modulation of Emitter Injection, not Carrier, Deflection, IEEE Electron Dev. Lett. EDL-3 (1982) 203-205.

7) Zieren, V.: A New Silicon Micro-Transducer for the Measurement of the Magnitude and Direction of a Magnetic Field Vector. Proc. IEDM (1980) 669-672.

8) Kordic, S.: Integrated 3-D Magnetic Sensor Based on an npn Transistor, IEEE Electron Dev. Lett. EDL-7 (1986) 196-198.

9) Roumenin, CH. S., and Kostov, P.T.: Optimised Emitter-Injection Modulation Magneto-transistor, Sens. & Act. 6 (1984) 19-33

10) Popovic, P.S. and Baltes, H.P.: Dual Collector Magneto-transistor Optimised with respect to Injection Modulation, Sens. & Act. 4 (1983) 155-163.

11) Ristic, Lj., Smy, T., Baltes, H.P. and Filanovsky, I.: Suppressed Sidewall Injection Magneto-transistor in CMOS Technology, Proc.Transducers '87, Tokyo, (1987) 543-546.

Silicon RF Power Sensor from DC to Microwave

P. Kopystynski, E. Obermeier
Institut für Festkörpertechnologie der Fraunhofer-Gesellschaft, München (now Technical University Berlin).

H. Delfs, W. Hohenester, A. Löser
Rohde & Schwarz GmbH u. Co. KG, München

Summary

Precision measurement of RF power is important in RF and microwave instrumentation. Mostly diode detectors are used, but the result depends on the detector circuit and the RF waveform. Power measurements independent of waveform require a thermal sensor. The power is absorbed in a resistive termination, and the temperature rise is measured. The paper describes layout, fabrication, and results of a monolithic silicon power sensor. This work was supported by the BMFT.

Introduction

Our most important criteria for the development of a microelectronic power sensor were the following:

- Sensitivity should be as high as possible.
- Burn-out power should be as high as possible.
- Sensor should be suitable up to 18 GHz or higher with low reflectivity.
- Sensor should be suitable down to 0 Hz, to facilitate DC-calibration.
- Sensor should be drift-free.
- Sensor should be as linear as possible.
- Time constant should be very short for fast measurement.

The first two criteria are contradictory. High thermal resistance of the absorber leads to high temperature rise at a given power, i.e. the burn-out temperature will be reached with relatively small power. A tradeoff is required. The possible ratio of maximum admittable and minimum detectable power determines the dynamic range, which is mainly determined by the temperature stability of the materials and the noise of the temperature sensor. A dynamic range of 10^5 ... 10^6 seems possible.

A monolithic silicon power sensor with similar criteria was developed by Hewlett-Packard [1], but it is not suitable down to DC, and the concept of realization differs considerably.

Preliminary Approaches [2]

The first approach to a high sensitivity power sensor used a Kapton sheet as a carrier. On one side of this sheet was a Nichrome resistor of .4 mm x .4 mm as a power absorber, on the other side a meander of pure nickel with a TCR of 3000 ppm/K as a temperature sensor. The thermal resistance of the cell was 3000 K/W (Fig. 1).

The very high thermal resistance turned out to be impractical, because a power of 30 mW led to remaining changes in the cell, and burn-out ocurred at 80 mW. The Kapton sheet bent under thermal load due to thermal expansion, which affected the nickel sensor resistor in an unreproducible way.

The Kapton sheet was replaced by borosilicate glass sheets of 80 µm or 150 µm thickness. The thermal resistance dropped to 2000 K/W and the burn-out power doubled. At the same time, the time constant of the cell was increased from 130 to 300 msec.

It was planned, to use two cells in a bridge configuration, and a closed loop would drive one cell as a reference with DC power, so that the bridge of the sensor resistors was at balance. In this case the DC power in the reference cell is equal to the RF power in the measurement cell. The effective time constant of such a system is much shorter, than the time constant of the cells themselves, because the bridge is already in balance when the final temperature is not yet reached. Differences in the thermal time constant of both cells, however, result in measurement error at short measurement times.

The glass cell avoided the problems of the Kapton cell, but it turned out, that the nickel thermal sensors never became stable enough and suffered remaining changes of value and/or TCR at less than 100 mW. Therefore we looked for temperature sensors, which do not use thin films and rely on more stable bulk effects. Two types of silicon temperature sensors were tested in preliminary tests by cementing the silicon sensor element on the backside of a glass cell. The first silicon sensor was of the spreading resistance type in n-type silicon. It turned out to be rather noisy. The second was a thin strip of high resistivity silicon with two ohmic contacts, used as a thermocouple. The results were favorable even in a very simple arrangement. Therefore this principle was chosen for further development work.

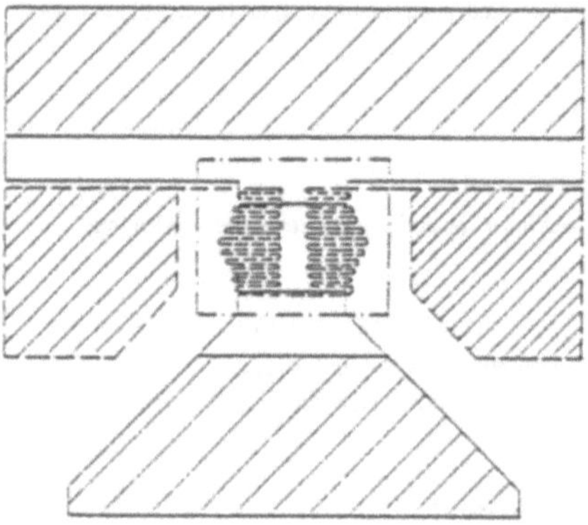

Fig. 1: Layout of Kapton Cell

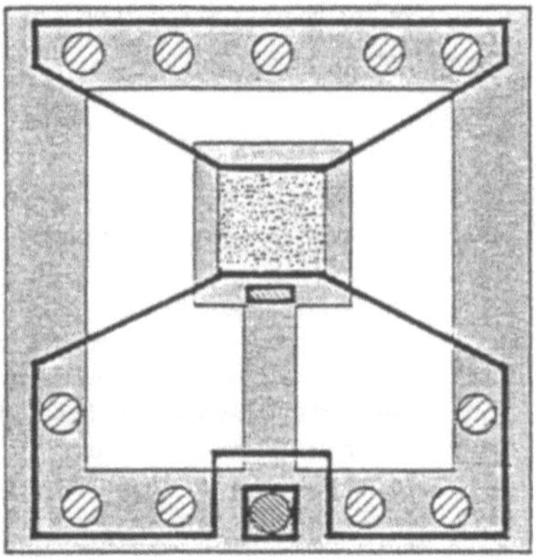

Fig. 2: First Proposal for Silicon Cell

Layout of a monolithic silicon cell

The first layout proposal is shown in Fig. 2. A thin film absorber resistor of 200 µm x 200 µm is placed on an oxidized silicon chip. To give this absorber the optimum thermal resistance, the chip was to be anisotropically etched through from the backside to the oxide layer, except at a small silicon island, carrying the resistor, which was connected with the silicon frame by a narrow bridge of silicon. The width of this bridge was to determine the thermal resistance. Two ohmic contacts at both sides of this silicon bridge would form a thermocouple, which would span about 85% of the temperature rise between the middle of the absorber and the frame. With a 10 000 µm² cross section of the bridge the thermal resistance would be about 400 K/W, the burn-out power was estimated to be 1 W and the time constant was 10 msec.

The rather bizarre shape of the cell was chosen, because it did not seem possible, to realize the cell on a thin silicon membrane, because such a thin sheet of remaining silicon normally requires an etch stop in any form; it could be a p-n-junction or a highly doped layer [3,4]. Both possibilities could not be used here, because the Seebeck coefficient of a silicon thermocouple decreases strongly with increasing impurity concentration (Fig 3.),[5]. To realize a high sensitivity thermocouple, the silicon material has to have high resistivity.

Analyzing the manufacturing process it turned out, that in order to realize the convex corners of the structure, complicated and difficult to design etch delay compensation structures would be needed. The thin oxide or oxide/nitride layer, which would be left free over the frame after etching, would make the chip extremely fragile. The danger of thermal warping could not be excluded. So it was considered less difficult, to realize a whole area thin silicon membrane without an etch stop layer. To get the required thermal resistance, the remaining silicon sheet was to be 8 µm thick. It ist not easy, to get such a thin layer without an etch stop, but it is facilitated by the fact, that it is already optically transparent. The remaining thickness can therefore be controlled by optical transmission measurement. The layout of this cell is shown in Fig. 4.

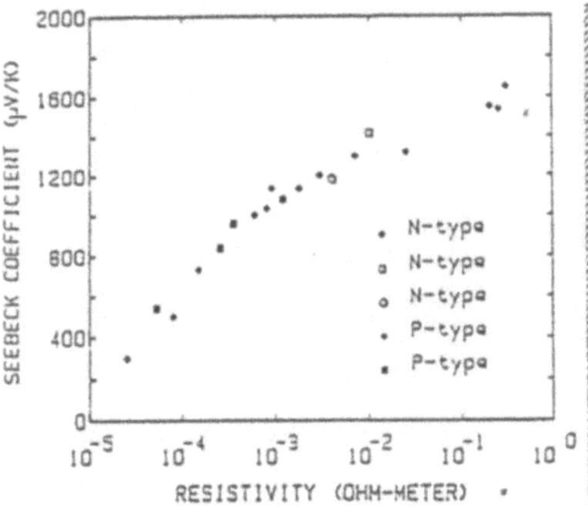

Fig. 3: Seebeck Coefficient over Silicon Resistivity

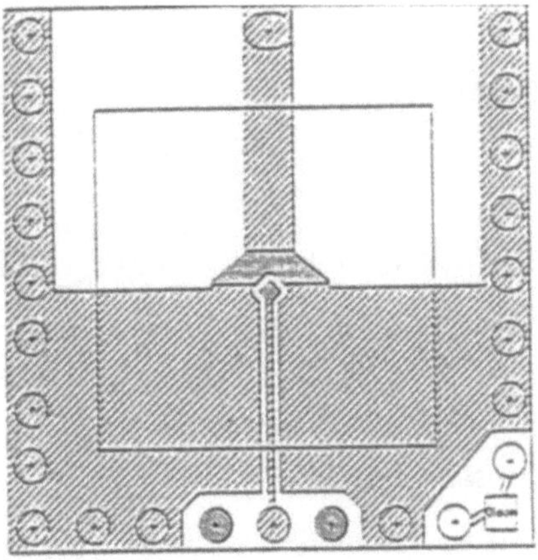

Fig. 4: Layout of Silicon Membrane Cell

The time constant of the cell was calculated to .4 msec, more than one order of magnitude less, than with the first design. Initially, a rectangular resistor was planned for the new design. But with the thin silicon membrane the thermal flow is radial rather then linear, as in the first layout. Therefore the largest part of the temperature drop occurs immediately at the absorber. A calculation of the temperature distribution over the chip showed, that even if the first junction of the thermocouple was placed immediately at the resistor, the thermocouple would span less than 55 % of the temperature difference between the middle of the resistor and the silicon frame. Therefore the absorber resistor was split in a trapezoidal form, and the thermocouple junction was placed in the middle. So the thermocouple efficiency could be raised to over 75 %.

High frequency requirements

To be used to very high frequencies up to 27 GHz the sensor cell has to fulfill additional requirements. For that purpose the power absorbing resistor must have not only the ohmic value of 50 Ω +-.5%, but the transmission line leading to the resistor itself is to be designed with an impedance of 50 Ω from the connector to the very small resistive element itself. To understand that, we must see the construction of the whole coaxial adapter of the sensor, which is shown in Fig. 5.

On the high frequency side we have a 3.5 mm coaxial line. An aluminum oxide substrate of .38 mm thickness is mounted within the outer conductor, whereas the inner conductor ends in a stripline on the substrate. This stripline is tapered down from a width of 1.86 mm to .25 mm. In order to maintain the impedance of 50 Ω, ground planes on either side of the stripline, connected with the outer tube, are added on the substrate metallization. Together with the stripline they form a coplanar line of the required impedance. The inner conductor ends on the substrate in a mounting pad for one bump of the sensor chip. Behind this pad the ground planes from both sides merge, and the ground connection of the chip is realized with low inductivity using 16 bumps.

On the chip, the inner conductor continues as a stripline, which has 50 Ω impedance together with the ground plane on the substrate. The distance to the substrate is to be maintained precisely, and a bump height of 31 +-1 µm is required. The discontinuities at the bump and the resistor terminals are compensated by additional conductor structures.

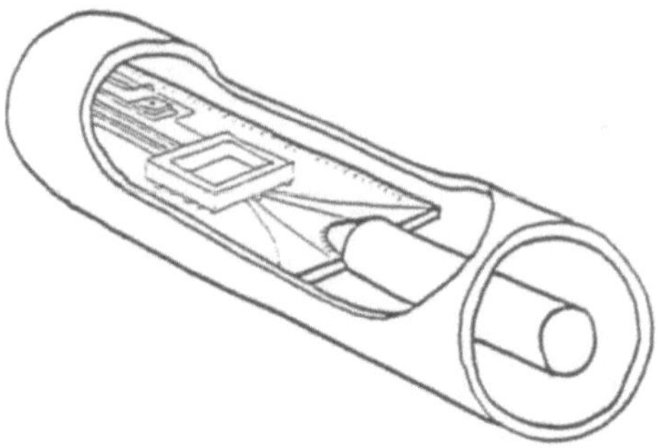

Fig. 5: Sensor Construction

Fig. 6: Sensor Chip Mounted on Substrate

As we pointed out above, the resistivity of the silicon must not be less than .1 Ωcm, because the Seebeck coefficient of the thermocouple would suffer. Calculations show, that the high frequency requirements on the resistivity of the starting material are still more stringent and require a material of more than 50 Ωcm. With lower resistivity a fraction of the incoming RF power would be capacitively coupled into the bulk silicon under the oxide/nitride layer and that would cause an unwanted frequency dependent reflectivity of the power sensor. Such a high value of the silicon resistivity, however, cannot be tolerated for the thermocouple, because the resistance of the thermocouple is to be less than 10 kΩ, but with 50 Ωcm silicon the thermocouple resistance would rise to 55 kΩ. Therefore the region under the ground plane of the chip, where a lower resistivity can be tolerated without high frequency problems, and where the thermocouple is located, is selectively implanted with a low dose, to reach a sufficiently low value of the thermocouple resistance and still maintain a sufficiently high value of the Seebeck coefficient. The dependence of the Seebeck coefficient on the implant dose was determined on a test chip with a thin silicon membrane and different implant doses.

Technology and process

The following process (simplified) was designed, to fabricate the sensor chip:

- Starting material p-silicon 200 Ωcm
- Wet oxidize, etch, boron-implant thermocouple region.
- Phosphor-implant n region of reference diode.
- Phosphor-implant n+ contact of diode.
- Boron-implant p+ contacts of diode and thermocouple.
- Anneal, oxidize, deposit 1500 A LPCVD nitride.
- Open contact windows
- Sputter NiCr, Ni, Au.
- Etch metallization, etch Au-Ni over resistor.
- Deposit on both sides of wafer 4000 A PECVD-nitride.
- Etch contact areas and backside areas for membrane etch.
- Electroplate Bumps
- Anisotropic etch backside to form 8 µm silicon membrane.
- Laser trim resistor to 50 Ω.
- Cut wafer into chips.

Particularly critical process steps are the bumping process, the membrane etch, and the laser trim.

The bumping was subcontracted to Microelectronic Marin. The close absolute tolerance of the bump height is unusual.

The ansisotropic etch takes about 24 hours in KOH at 60°C. Faster etching at higher temperature turned out to be impractical, because the etch rate was too high, to attain the final membrane thickness with sufficient accuracy, and because the etch rate was not uniform enough at higher temperatures, to yield an uniform membrane without an etch stop. The membrane thickness was optically determined by evaluating the transmission spectrum. Interference between the front and back side reflection results in transmission maxima and minima, which give a precise measure for the membrane thickness.

The laser trim is somewhat critical too, because the laser energy must be carefully optimized and controlled, in order to trim the resistor without damaging the unterlying passivation and the silicon membrane.

Fig 6. shows a SEM-picture of the mounted chip and Fig 7 a picture of the chip itself. The finished chips turned out to be surprisingly rugged. They could be handled with a vacuum pick-up without any difficulty. A critical point in the soldering process of the chips to the ceramic substrate was to maintain the height of the bumps with sufficient accuracy. For that purpose a fotostructurable resist was applied to the substrate, which left the contact areas free and served as a plating resist and later as solder resist. A layer of pure tin was plated on the contact areas. Together with the gold layer of the substrate and the bumps it formed the solder connection of the bumps. By carefully controlling the temperature over time, the amount of gold dissolved into the tin could be controlled, and a reproducible distance between chip an substrate was reached.

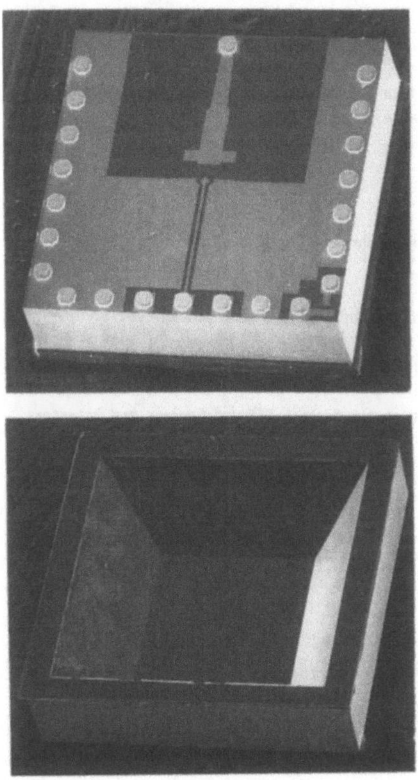

Fig.7: SEM-Pictures of Sensor Chip

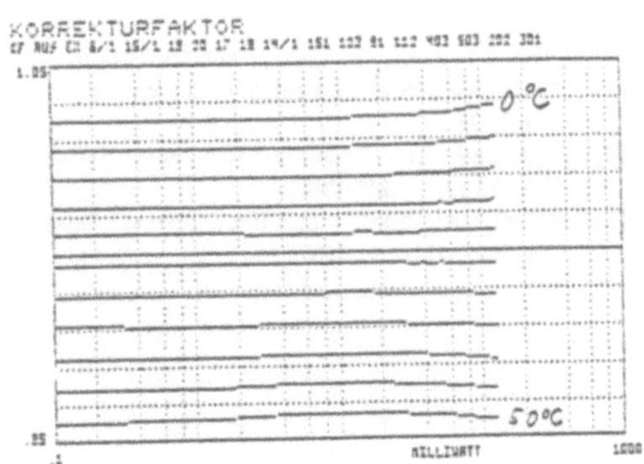

Fig. 8: Correction Factor for Power Sensor.

Results

About 20 sensors were used to characterize the final results. The sensitivity is 250 μV/mW, differs from sensor to sensor within a few percent and is well in accordance with the initial calculations. The deviation from linear response is 2% at 100 mW. The reason is the nonlinearity of the thermocouple characteristic and the temperature dependence of the heat conductivity of silicon.

At low power levels the DC response turns out to be polarity sensitive. The difference at 10^{-4} W ist about 1 %. The reason is the Peltier-effect at the resistor terminals, which are not symmetrical with regard to the thermocouple contact. DC-calibration is possible, however, by taking the median value.

Sensor sensitivity depends also on temperature with a temperature coefficient of .16 %/K. To compensate for this dependence, the temperature of the sensor must be known. Initially the diode on the chip was planned as a temperature reference. It turned out, however, to be plagued by electrical insulation problems. Therefore a separate diode chip was mounted on the ceramic substrate near the sensor chip.

From the temperature dependence and the nonlinearity of the sensors results a correction factor, which is shown for a typical sensor in Fig. 8. This correction factor is individually measured for each sensor and stored in an EPROM, which is put into the same probe housing with the sensor. There is no electrical connection between the sensor and the EPROM within this housing; the correction is performed in the processor of the corresponding instrument. The combination of sensor and correcting EPROM within one housing, however, facilitates changing of the probes without recalibration of the instrument.

The high frequency behaviour of the sensor is excellent: The VSWR at 27 GHz is less than 1.15. That means, that more than 99% of the high frequency power is absorbed in the probe. The frequency dependence of the sensitivity shows a 5% drop at 27 GHz. Each probe head will be calibrated individually by comparing with a reference normal.

The maximum nominal power of the sensor is 100 mW. At more than 500 mW remaining changes may occur. The dynamic range between maximum admittable and minimum detectable power is 10^6.

Literature

[1] J.C. Lamy, "Microelectronics Enhances Thermocouple Power Measurements", HP Journal, Vol. 26, No.1, 16-18, 1974

[2] H. Delfs, F. Dosch, A. Löser, BMFT-FB-T 84-161

[3] H.A. Waggener, "Electrochemically Controlled Thinning of Silicon", Bell System Tech.J. Vol.50, 1970, 473-475

[4] "Electrochemically Thinned N/N+ Epitaxial Silicon - Method and Applications", J. El.chem.Soc. Vol. 188, No.7, 1971, 1240-1246

[5] A.W. Van Heerwaarden, P.M. Sarro, "Thermal Sensors Based on the Seebeck Effect" Sensors and Actuators, Vol.10, 321-346, 1986

Self-testable Accelerometer Microsystem

Stephen C. Terry, Diederik W. de Bruin, Henry V. Allen

IC SENSORS
Milpitas, California 95035
USA

Abstract

Self-testability in accelerometers has application in measuring their sensitivity and frequency response during fabrication, without the need for applying mechanical vibration, and in verifying their operation and calibrating the sensitivity over temperature in the field. The self-test force in devices that have been reported to date has only been able to deflect the seismic mass to about 15 % of maximum range. The most effective way to increase the electrostatic force is to narrow the gap between the electrode on the mass and the over-force stops on the silicon top cap. However, this affects the operation of the sensor, both during regular use as well as in self-test mode. The device becomes more overdamped and the output signal may exhibit more non-linearity. These effects can be minimized by a careful design of the structure. Overall, minimizing the electrode gap in order to maximize the force is only recommended when the self-test feature is used for qualitative measurements where a large full scale deflection is necessary. For quantitative measurements, a spacing which produces a 10 % deflection seems a good compromise between reasonable electrostatic force and accuracy.

Introduction

Self-testable sensors incorporate built-in actuators which allow testing of the device during assembly or during operation. The concept of a self-testable accelerometer has previously been presented.[1,2] However, as these devices come into production, there is a need to address important issues related to optimizing the electrostatic force which is at the heart of the actuation mechanism. This paper discusses some performance issues associated with increased self-test force.

The self-testable accelerometer is a modification of a piezoresistive, doubly-supported cantilever structure with top and bottom caps as first reported in 1988[3]. In order to incorporate the self-test function, the over-force stops on the caps which limit travel have been enlarged to form electrodes. The cross-section of the device is depicted in figure 1. A metal electrode is deposited on the seismic mass and connected across the flexures to an outside bond pad. The top cap is electrically connected to the substrate and serves as the counter electrode to the one on the proof mass. A unique feature of this improved structure[4] is that no metal electrode is required on the cap.

In opperation, the mass is deflected toward the top cap when a voltage is applied between the electrode and the top cap and substrate. To avoid damage by electrostatic discharge, the electrode is connected to the substrate through a reversed-biased diode. Since p-type piezoresistors are used in this device to sense deflection of the proof mass, the substrate must be biased to the most positive voltage on the chip. Consequently, the electrode voltage must be kept negative with respect to the substrate.

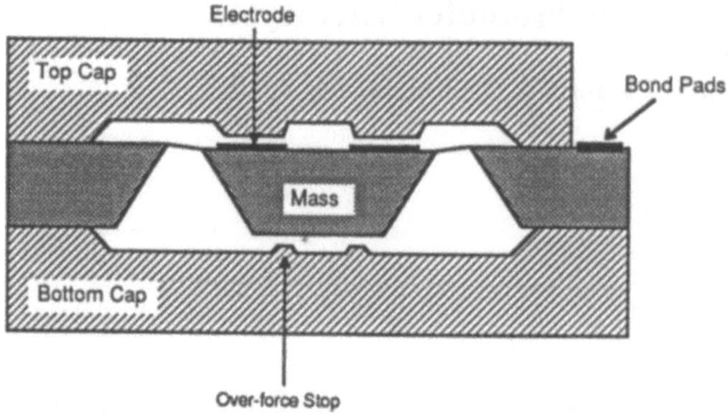

Fig. 1. Cross section of the self-testable accelerometer. Electrostatic force is generated between the electrode and the corresponding overforce stops on the top cap.

Optimizing the electrostatic force

In many self-test applications, it is desirable to electrostatically deflect the proof mass a distance which corresponds to the rated full scale acceleration of the device. This is significantly greater than the deflections achieved with the first self-testable accelerometers, thus necessitating design changes to increase the self-test force. It will be shown that modifying the structure to increase the self-test force is not without trade-offs.

The electrostatic force (in g's) is a function of the applied voltage V, the electrode area A and the gap spacing between the electrodes:

$$F_{el} = \frac{e_0 A V^2}{19.6\, m\, (x_0 + x)^2} \tag{1}$$

where x_0 is the gap at zero deflection and x is the deflection of the mass from its rest position. Referring to Figure 1, $x > 0$ for displacements of the mass away from the top cap and towards the bottom cap.

Since the electrostatic force is proportional to the electrode area, it can be increased by utilizing as much of the proof mass surface as possible for the electrode. This is limited only by the area required for the metal interconnections between sensing resistors. Increasing the applied electrode voltage has more effect because the force increases with the square of the voltage. The maximum voltage that can be applied is, however, limited by the breakdown voltage of the ESD protection diode and the dielectric layer under the electrode.

The electrostatic force is also inversely proportional to the square of the gap between the electrodes on the mass and the cap. Thus, if the gap is decreased, a larger electrostatic force is generated at the same applied electrode voltage. However, this reduction in spacing between the seismic mass and the over-force stops limits the maximum travel of the mass, so the gap must not be chosen smaller than the deflection of the mass at the maximum range. Though the reduction of the gap is an attractive approach to increasing the self-test force, there are important second-order effects.

Regenerative feedback of the electrostatic force

The restoring force of the silicon springs is proportional to the deflection of the mass, while the electrostatic force is inversely proportional to the square of the gap between the electrodes. At some applied electrode voltage, the restoring force is no longer able to counter the electrostatic force, and the mass will move until it rests against the over-force stops. This theoretically occurs when the mass has traveled $^1/_3$ of the gap spacing, if the restoring force of the springs is linear over displacement. Measurements indicate that the real value is closer to half the gap spacing due to the increase of the spring constant with increasing deflection.

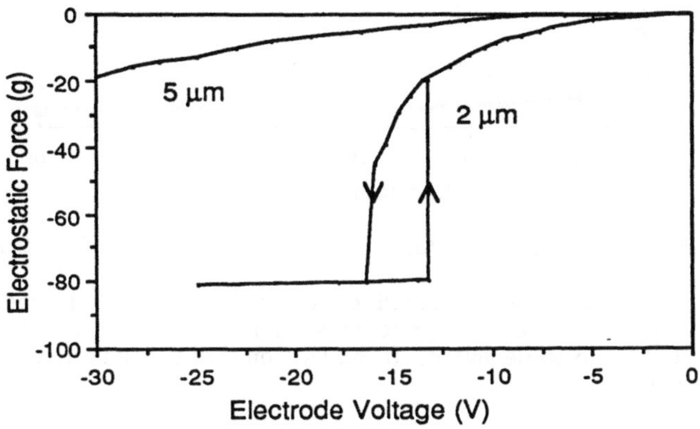

Fig. 2. Electrostatic force vs electrode voltage for two values of the initial gap spacing

After the mass has been deflected against the stops, the output of the accelerometer will be equal to its maximum value, and verification of full travel is thus possible. At that point, the electrode gap is determined only by the thickness of the dielectric layer covering the over-force stops.The 'holding' force is consequently very large, and the mass will not release until the voltage is reduced to the point where the restoring force once again exceeds the electrostatic force. The resulting hysteresis in the force versus voltage curve in shown in Figure 2 for a device with a 2-μm electrode gap.

Non-linearity of the electrostatic force

A deviation from an ideal parabola in the force versus voltage curve exists because the $(x_o + x)^{-2}$ term in (1) cannot be simplified to $1/x_o^2$. This may effect the accuracy of the self-test function. The self-test function will generally be implemented by applying a voltage pulse to the electrode and measuring the change in output of the accelerometer. If the deflection of the seismic mass is small compared to the gap, the output of the sensor will be the sum of the signals due to the electrostatic force and the externally applied 'mechanical' g-force, both of which contribute to the deflection x. Ideally, with an external g-force present, the output change upon applying the self-test voltage is proportional to the electrostatic force. For devices with narrower gaps, this is not necessarily true, because the gap can no longer be considered constant. If the deflection is not small compared to the initial gap spacing, the mechanical acceleration will modulate the electrostatic force through the $(x_o + x)^{-2}$ term in (1). This adds a non-linear component to the response of the device to a mechanical acceleration. Since the gap is narrowed by increasing voltage, the non-linearity will also increase with applied voltage. Figure 3 shows the observed total harmonic distortion at the output of the sensor when subjected to a sinusoidal vibration with an amplitude of 5 gRMS. The non-linearity increases rapidly with increasing electrode voltage in devices with a 2 μm gap, while it remains small in devices with 5 μm gap.

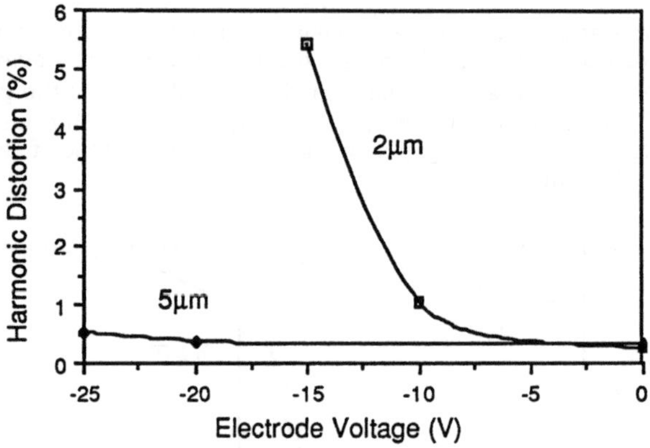

Fig. 3 Total harmonic distortion for a 1 g signal as a function of electrode voltage.

The sensitivity to mechanical accelerations also increases with increasing electrode voltage. Figure 4 depicts the sensitivity to a mechanical excitation with 1 g amplitude for different values of the electrode voltage. It can be seen that the sensitivity is modulated by the applied voltage.

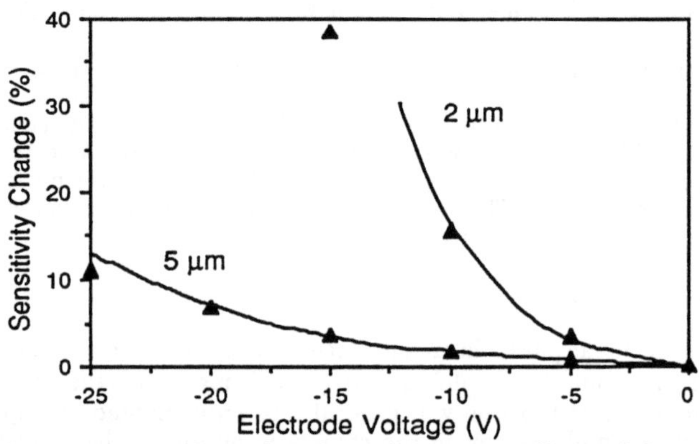

Fig. 4 Sensitivity increase vs electrode voltage for a 1 g small signal for two gap spacings.

The lines in figure 4 represent a simulation of this effect using a model with a linear spring constant, while the dots are measured values. The simple model is in good agreement with the measured data.

Increased Damping

In the earlier versions of the accelerometer, cavities were etched in the top and bottom caps to obtain a critically damped system using squeeze-film damping. The damping ratio is inversely proportional to the cube of the depth of the cavities and is further determined by the stiffness of the silicon support of the seismic mass. The desired gap to achieve critical damping in a sensor like the one being discussed is about 20 μm. Because the gap between the electrodes is smaller than this to achieve high electrostatic force, the sensor will be overdamped.

The damping can be improved by etching air-handling channels in the electrode. This way, the electrode looks like a plurality of stops rather than a single mesa. The damping is then set by a contributions of the electrode areas, where the gap is small, and of the air handling channels. The object in the design is to place multiple channels in the structure while maximizing the electrode area. The damping ratio of a device is also increased with increasing electrode voltage because the gap becomes progressively more narrow. This is particularly apparent for devices with a 2 μm initial gaps as shown below.

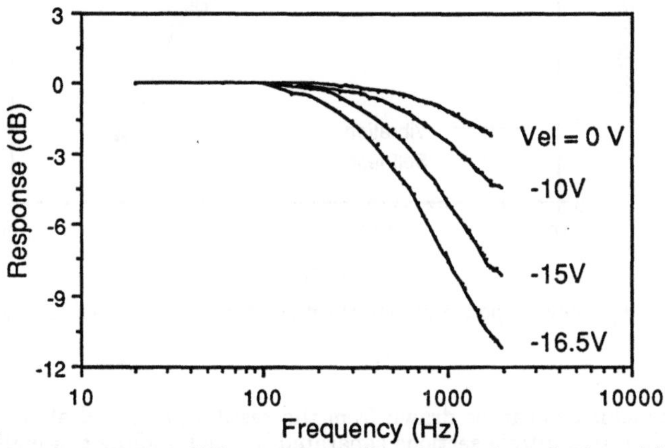

Fig. 5. Frequency response at different electrode voltages for a device with x_0=2 μm.

Self-test during manufacturing

Besides verification of performance and calibration during operation, the self-test feature can also be use to screen a device during manufacturing. By applying a voltage at the electrode, the functionality of the sensor can be determined and a rough value for the sensitivity can be computed. In addition, it is possible to utilize sinusoidal control voltages of several frequencies and thus measure the frequency response without the need for mechanical vibration. If the electrode voltage is:

$$V_{el} = B + C \sin(wt) \tag{2}$$

then the resulting electrostatic force follows from substituting V_{el} in (1):

$$F_{el} = \frac{e_0 A a}{19.6 m (x_0 + x)^2} \tag{3}$$

where $a = B^2 + 0.5 C^2 + 2BC \sin(wt) - 0.5 C^2 \cos(2wt)$

If B=0, the output of the accelerometer contains only the second harmonic. The frequency response can be determined by measuring the output signals at the double frequency for several values of w. However, because of the diode used to protect the electrode from ESD, the sine wave cannot be centered around zero, and an offset B must be added to the electrode voltage. In that case, the response contains signals with both the fundamental and second harmonic frequency and it is necessary to use a frequency-discriminating AC voltmeter or spectrum analyzer to determine the frequency response. Figure 6 shows a comparison of the measured frequency responce from an accelerometer obtained by applying a mechanical acceleration and a sinusoidal self-test voltage. For the purposes of characterization during manufacture, the agreement is quite good.

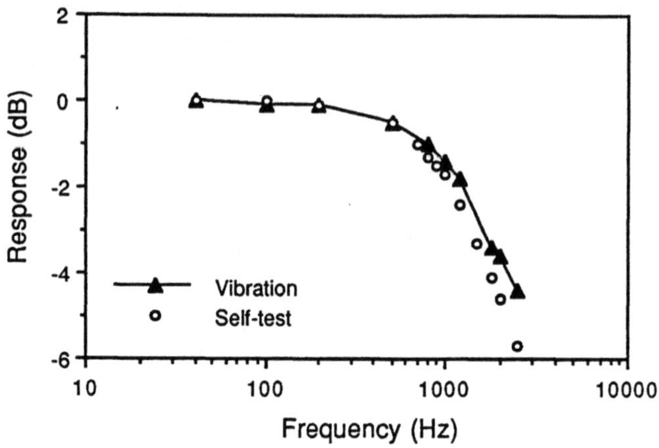

Fig. 6. Frequency response measured mechanically and using the self-test.

Conclusion

A number of conclusions can be drawn from the results presented above. First, breaking up the electrode into multiple sections separated by air handling channels is necessary to minimize the effect of viscous air damping between the plates. Second, when a self-test voltage is applied while mechanical acceleration is being measured, the effective sensitivity increases with increasing electrode voltage as a result of modulation of the electrostatic force. Third, the self-test can only controllably deflect the mass to about half the gap, after which the mass is pulled against the stops when the restoring force of the flexures can no longer counter the electrostatic force. Finally, only when full-scale electrostatic deflection is required at an electrode voltage below 20 V does the initial gap need to be reduced substantially.

Acknowledgement

The authors would like to thank Allan Crabill, who led the team that processed the devices, for his contributions.

References

1. Henry V. Allen, Stephen C. Terry, Diederik W. de Bruin, "Self-Testable Accelerometer Systems", Proceedings of IEEE Micro Electro Mechanical Systems, IEEE Catalog No. 89THO249-3, February 1989, pg 113-115.

2. Henry V. Allen, Stephen C. Terry, Diederik W. de Bruin, "Accelerometer Systems with Built-in Testing", Abstracts of Transducers '89, Montreux, Switzerland, June 1989, pg 148-149. (to be published in Sensors and Actuators)

3. Stephen C. Terry, "A Miniature Silicon Accelerometer with Built-in Damping", Technical Digest of IEEE Solid-State Sensor and Actuator Workshop, IEEE Catalog No. 88THO215-4, June 1988, pg 114-116.

4. D. W. deBruin, H. V. Allen, S.C. Terry, "Second-order Effects in Self-testable Accelerometers", Technical Digest of IEEE Solid-State Sensor and Actuator Workshop, IEEE Catalog No. 90CH2783-9 June 1988, pg 149-152.

Micromechanical Capacitive Acceleration Sensor with Force Compensation

H. F. Schlaak[+], F. Arndt[+], A. Steckenborn[+],
H. J. Gevatter[++], L. Kiesewetter[++], H. Grethen[+++]

[+] Siemens AG, Produktion und Logistik, Berlin
[++] Technische Universität Berlin,
Institut für Feinwerktechnik und Biomedizinische Technik
[+++] Siemens AG, Werk für Textendgeräte, Berlin

Summary

A micromechanical acceleration sensor consisting of a differential capacitor has been developed. The seismic mass is etched in silicon forming simultaneously the middle electrode and is suspended on 8 cantilever beams consisting of highly boron doped silicon. The mass is embedded in a glass-silicon-glass-sandwich between the counter electrodes which are placed in 5 µm deep etched grooves in the glass substrates. The complete fabrication including the anodic bonding of glass and silicon is processed on 4"-wafers. The sensor design with smooth cantilever beams (10 µm thickness, 1.3 mm length) provides a high sensitivity. The sensor element operates in an electronic force compensation circuit with electrostatic reset to keep the seismic mass in a virtual idle position. The circuit provides an error compensation. The measurement resolution has been performed within µg-range. The fabrication tolerances of all important dimensions have been evaluated to 1 %. They correspond to the results of sensor characteristics measurements.

Introduction

Acceleration is an important measurement quantity for a lot of electronic systems with two main application fields: the measurement of vibration or motion (e.g. platform stabilisation) and the inertial guidance or navigation to determine the absolute position (e.g. scanning devices for text or graphics). Due to the increasing field of micromachining, miniaturized accelerometers with extremely small seismic masses deflected by acceleration can be fabricated by three dimensional etching techniques. Two functional principles have been performed: the piezoresistive /1/ and the capacitive accelerometer /2,3/. Piezoresistive accelerometers are already commercially available. They are suited for low sensitivity and high bandwidth. However, they show a rather high temperature sensitivity and drift while the usable temperature range is restricted. The capacitive scheme exhibits high precision with low bandwidth and low measurement range. The sensor can be used in a force compensation circuit where the seismic mass is kept in a virtual idle position by electrostatic force. Thus a high dynamic range with linear signal output can be performed, even though the sensor characteristics itself is nonlinear /9/.

In this paper we present a capacitive micromechanical accelerometer with fully symmetric suspension working in a novel force balancing circuit with error compensation. The system is suitable for µg-resolution.

Sensor Design

The basic structure of the sensor element is shown in Fig. 1 a. The quadratic seismic mass is suspended by 8 diagonal spring beams in full symmetry. Since the seismic mass is fixed by spring pairs on top and bottom, twisting and bending of the mass is nearly completely suppressed. The probe mass serves as common middle electrode of a differential capacitator. The displacement of the probe mass caused by acceleration is measured by the capacitance change between the fixed electrode of each side and the movable one in the middle. Fig.1b shows the sensor cross section with the seismic mass and the suspension made of a three-layer-silicon-chip. The spring beams and the movable electrodes on each side of the mass consist of p^{++}-silicon epitaxial layers. The silicon chip is mounted between two glass plates carrying the fixed counter electrodes which are placed on the bottom of grooves etched into the glass plates.

For a high resolution sensor a design with rather large electrodes and very smooth springs has been derived /4/. The main characteristics of the sensor element are listed in Table 1 using the density ρ_{Si} = 2,33 g/cm^3 and the Young modulus for <100>-silicon along the cantilever axis E_{Si} = 1.3 10^5 N/mm^2/5/.

Table 1: Main characteristics of sensor chip

seismic mass:	area	A	3 x 3	mm^2
	thickness	t_m	520	µm
	mass	m	11	mg
capacitator:	electrode gap	h_o	5	µm
	capacitator	C_o	15.9	pF
spring suspension:	effective length	l	1.3	mm
	beam width	w	160	µm
	beam thickness	h	10	µm
	total spring constant	c_t	0.08	N/mm
	resonance frequency	f_o	400	Hz
sensitivity		S_c	4.5	pF/g

The spring constant has been estimated by the model of a beam suspended on both ends /6/

$$c_x = E_{Si} \cdot w \cdot (h/l)^3, \quad (1)$$

with the total spring constant $c_t = 8\ c_x$ and the resonance frequency $f_o = 1/2\pi \cdot \sqrt{c_t/m}$. The sensor sensivity is expressed by the capacitance change due to acceleration:

$$S_C = \frac{dC}{da} = \frac{m \cdot C_o}{c_t \cdot h_o} = \frac{\varepsilon_o \cdot \rho_{Si}}{8 \cdot E_{Si}} \cdot \frac{t_m}{w} \cdot \left(\frac{A}{h_o}\right)^2 \cdot \left(\frac{l}{h}\right)^3 \qquad (2)$$

and is dominated by the electrode gap h_o and the relative spring length l/h.

Fabrication Technology

To fabricate the silicon chips 10 µm thick p^{++}-boron-doped silicon layers (10^{20} cm^{-3}) are grown on both sides of double sided polished silicon wafers. The p^{++}-layers are patterned with the cantilever structure by RIE-etching using an Al-mask. The seismic masses and the contact holes (see d in Fig. 1b) are formed by anisotropic wet etching with EDP (see SEM photographs in Fig. 2). The cantilever suspension at the seismic mass has a triangular shape (Fig. 2b) to achieve on one hand a solid spring attachment, on the other hand a corner compensation for the anisotropic etching process /7/. Thus edge roundings usually occurring at convex corners have been avoided.

For the fabrication of the electrode chips 4"-Pyrex-glass-wafers (Corning 7740) are used because the thermal expansion is matched rather well to silicon, which is important for the anodic bonding process. The electrode grooves are etched using a Ti/Au-mask in a buffered hydrofluoric acid optimized for low surface roughness. The electrodes consist of a 540 nm thick evaporated Ti/Ni/Au layer system with a 200 nm SiO_2-layer for electrical isolation. The 3-wafer-sandwich is joined by anodic bonding in a single run and finally cut by a dicing saw.

The fabrication tolerances have been evaluated during the wafer processing where the most important dimensions, the electrode groove depth and the cantilever width of several chips on different wafers have been measured. The results show an accuracy of about 1 %, while the deviations on each wafer are less than 0.5 %.

Measurement Electronics

The electronic force compensation system is shown in Fig. 3 /8/. The difference current $I = \omega_o \cdot W_o \cdot \Delta C$ depends on the carrier (frequency ω_o, amplitude W_o) and the capacitance change ΔC caused by acceleration. The PI controller provides the output signal U_x proportional to the acceleration a. The measurement resolution can be estimated by the amplifier input noise current: assuming $\Delta I_{min} = 6$ pA, $W_o = 2$ V, carrier frequency $f_o = 50$ kHz delivers $\Delta C = 0.01$ fF equivalent to $\Delta a = 2$ µg (with $S_c = 4.5$ pF/g).

Unsymmetries of the electrode gap h_o and the electrode area A due to fabrication tolerances cause an offset and a quadratic nonlinear error of the output signal U_x. The electronic system in Fig. 3 provides an error compensation by tuning the drivers where ΔU_O compensates the offset and the amplification difference of U_x eliminates the nonlinearity /9/.

Experimental Results

The static sensor system characteristics has been measured in the earth gravity field by a precision turntable. The difference between the measured output signal U_x and the calculated linear regression line is shown in Fig. 4. The quadratic error measures 1.5 mg with an operating range of 0.5 g corresponding to 0.3 % relative error. The dynamic behaviour has been studied with closed and open loop systems. The corner frequencies f_c are about 100 Hz and 0.8 Hz respectively (Fig. 5 a,b). With decreased air damping by evacuation the bandwidth can be enlarged to f_c = 77 Hz at 2 mbar (Fig. 5 c).
In conclusion an acceleration sensor system with µg-resolution and quadratic error compensation has been performed which is capable for position sensing.

Acknowledgement

The authors wish to thank W. Funck, M. Groszkowski, W. Gruhl, G. Jantke, M. Klug and H. Kretschmer for their expert help in processing of the sensor elements, H. Gaus and Y. Yu for the sensor characterization.

References

/1/ **S. Terry:** A Miniature Silicon Accelerometer with Built-in Damping. Technical Digest of Solid-State Sensor and Actuators Workshop, Hilton Head Island, USA, June 6-9, 1988, pp. 114-116.

/2/ **F. Rudolf et al.:** Precision Accelerometers with µg Resolution. Sensors and Actuators, A21-A23 (1990), pp. 297-303.

/3/ **H. Seidel et al.:** Capacitive Silicon Accelerometer with Highly Symmetrical Design. Sensors and Actuators, A21-A23 (1990), pp. 312-315.

/4/ **H. Gaus :** Mikromechanisch hergestellte Beschleunigungssensoren. Seminarband "Planartechnik für Drucker und Scanner", Timmend. Strand, Germany, Sep. 28-Oct.1, 1988.

/5/ **S. Johannson:** Micromechanical Properties of Silicon. Uppsala University, Dissertation, 1988.

/6/ **F. Sass, Ch. Bouché, A. Leitner:** Dubbel-Taschenbuch für Maschinenbau. Springer-Verlag, Berlin, Heidelberg, New York, 1974, p. 426.

/7/ **B. Puers; W. Sansen:** Compensation Structures for Convex Corner Micromachining in Silicon. Sensors and Actuators, A21-A23 (1990), pp. 1036-1041.

/8/ **Y.Yu :** Elektronische Signalaufbereitung für kraftkompensierte Sensoren. Seminarband "Planartechnik für Drucker und Scanner", Timmend. Strand, Germany, Sep.28-Oct.1,1988.

/9/ **H.-J. Gevatter; H. Grethen:** Kennlinie eines Beschleunigungssensors mit elektrostatischer Kraftkompensation. Technisches Messen, tm 56 (1989) 2, pp. 93-98.

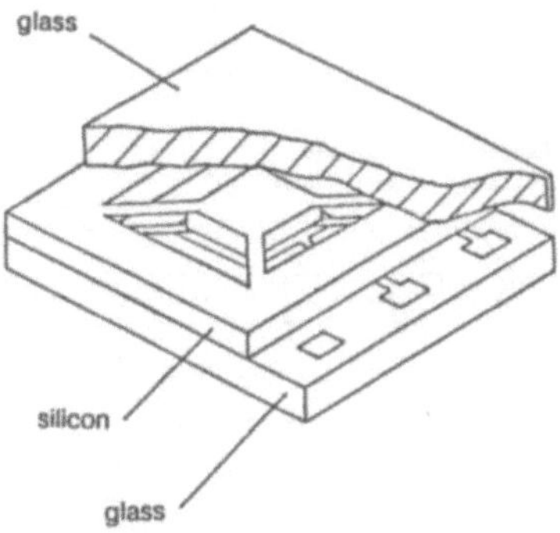

Fig. 1 a: Schematic view of sensor element

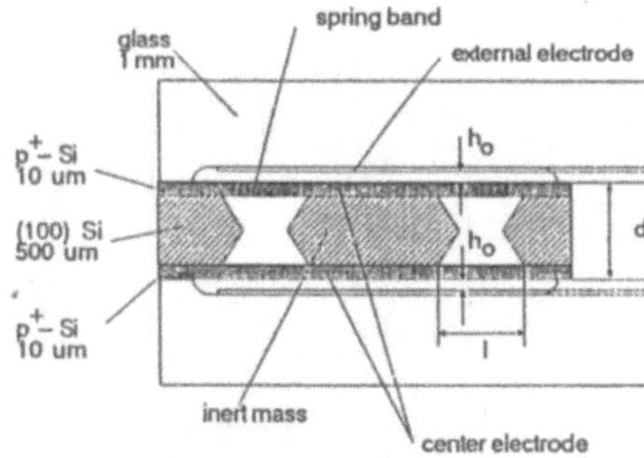

Fig. 1 b: Cross section of sensor

Fig. 2 a:
Scanning electron micrographs of the silicon seismic mass with spring suspension.

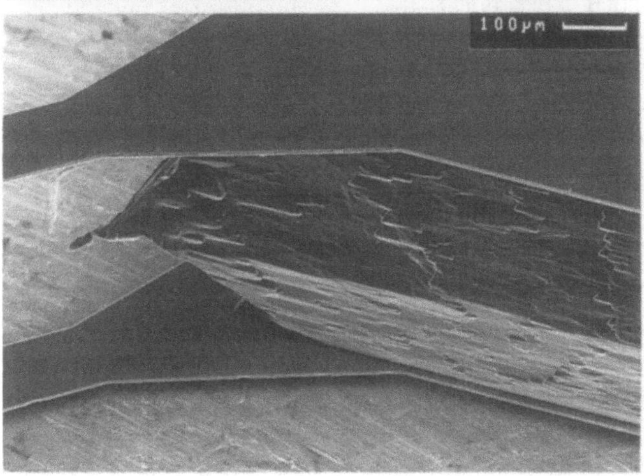

Fig. 2 b:
Detail of underetched cantilever suspension at the seismic mass

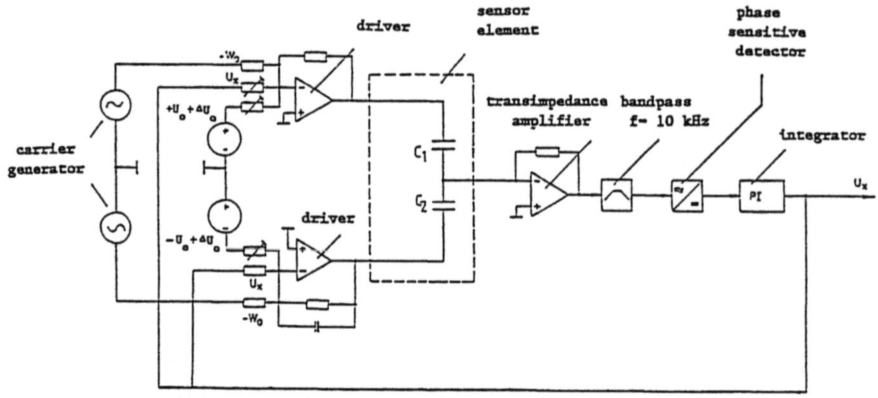

Fig. 3 : Closed loop force balancing circuit with error compensation ΔU_o

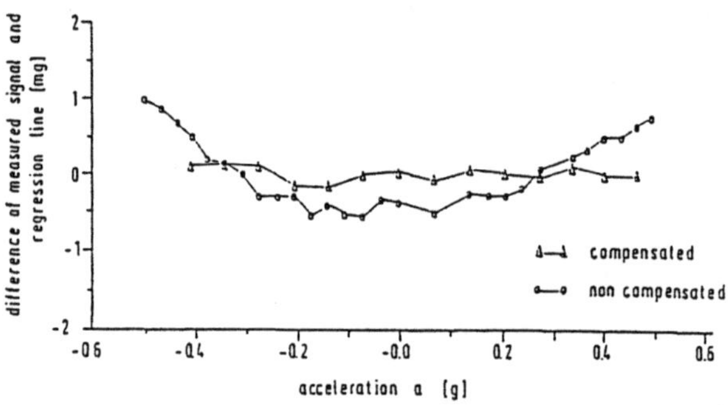

Fig. 4:
Difference of measured signal and linear regression;
quadratic nonlinearity 1.5 mg for 0.5 g range;
Compensation by balancing the driver amplification factors for U_x.

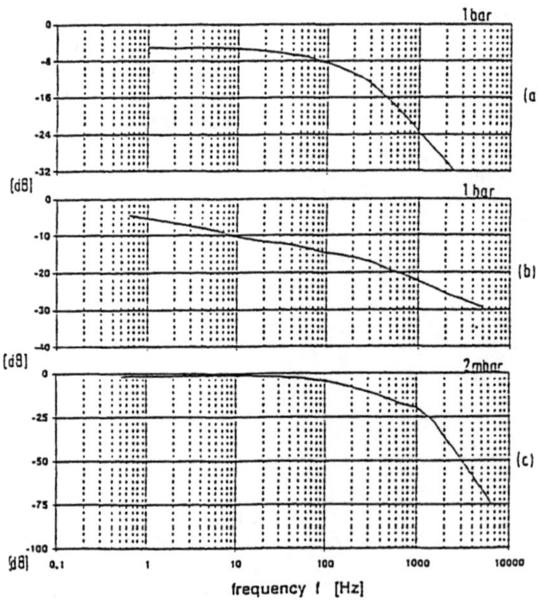

Fig. 5:
Dynamic frequency response with :
(a) closed loop
(b) open circuit
(c) open circuit with reduced pressure

Capacitive Accelerometer Made by Silicon Micromechanics

U. E. Gerlach-Meyer

Electronic Systems Department
Battelle-Institut e.V.
Am Römerhof 35, D-6000 Frankfurt am Main 90

Summary

A capacitive accelerometer has been fabricated from Si 110 wafers using the silicon micromachining technique. The sensor basically consist out of an elastically suspended plate between two fixed capacitor plates.

The central plate is equipped with several slits which reduces the air damping and enhances the dynamical behavior. A sensor with a sufficient number of slits detects accelerations up to 1000 Hz with a signal deviation of less than 10%, while a sensor with no slits reaches only about 50 Hz. The experimental data has been fitted with the theoretical curve, and the damping and resonance frequency is determined for a different number of slits.

1. Introduction

All accelerometers known today are using a spring mass principle. An elastically attached mass follows the fixed part of such an accelerometer with a delay when an accelerations acting on the device. Thus the detection results from a distance measurement and is connected with the motion of the movable part of the sensor and its dynamic behavior.
Using miniaturised devises the air damping has an important impact and the detected acceleration values decrease strongly with the frequency of the stimulating force.

2. Sensor design and preparation

A capacitive accelerometer has been prepared from Si 110 wafers using the silicon micromachining technology. The sensor basically consist out of an elastically suspended plate between two fixed plates forming a differential capacitor. The central plate is attached by means of two elastic beams in a cantilever design. The fixed parts are prepared by vapor-deposition of alumina on pyrex substrates, which are attached to the silicon by anodic bonding. The two parts of the differential capacitor are electrically connected in a bridge circuit. The capacity change due to acceleration is detected as an unbalancing of the bridge.

The capacitor spacing is several µm and the plate area is a few mm^2 with a mass m of some mg. The dimension of the elastic beams compromises between the sensitivity and measuring range or natural frequency, respectively [1].

3. Dynamic properties

Under periodic acceleration we consider the spring mass system of the sensor to be a forced oscillator with the equation of motion [2]

$$d^2s/dt^2 + 2\varrho\, ds/dt + \omega_o^2 s = F(t) \qquad (1)$$
$$\omega_o^2 = k/m$$

where ω_o indicates the natural frequency of the undamped system, k the spring constant, m the mass, and ϱ the damping constant and s the motion of the sensor; t is the time and F the exiting force.
The amplitude A of the sensor depends strongly on the frequency to be

$$A(\omega) = F/\left((\omega_o^2 - \omega^2)^2 + 4\varrho\omega^2\right)^{1/2} \qquad (2)$$

and a phase shift φ of

$$\tan\varphi = 2\varrho\omega/(\omega_o^2 - \omega^2) \qquad (3)$$

of the outcoming signal in relation to the exiting force with $\varphi = 90°$ for $\omega = \omega_o$. $A(\omega)$ depends strongly on the frequency showing a pronounced peak at $\omega = \omega_r$ as long as ϱ is smaller than ω_o. For $\varrho > \omega_o$ the amplitude decreases strongly with frequency.

For a given ω_o we can minimize the frequency dependence when the sensor works near its aperiodic limit, i.e. when the damping ϱ over ω_o is near unity. The amplitude versus frequency has no longer a resonance peak but decreases with frequency.

At a step like excitation and $\omega_o = \varrho$ the sensor reaches its end value comparatively fast, with no oscillations. The amplitude decrease is about 10% for $\varrho = 0.3\omega_o$. For $\omega_o^2 = 2\varrho^2$ the same decrease is reached not before $\varrho = 0.7\omega_o$, but with a small overshooting of less than 1%. That means for a given resonance frequency, the damping should be adjusted to reach the nearly aperiodic limit of $\omega_o^2 = 2\varrho^2$ in order to obtain an optimal frequency behavior of the sensor.

4. Air damping

Our sensor as described above is strongly over critically damped. The signal decreases strongly with frequency and in case of a step like excitation the signal reaches its final value only slowly.

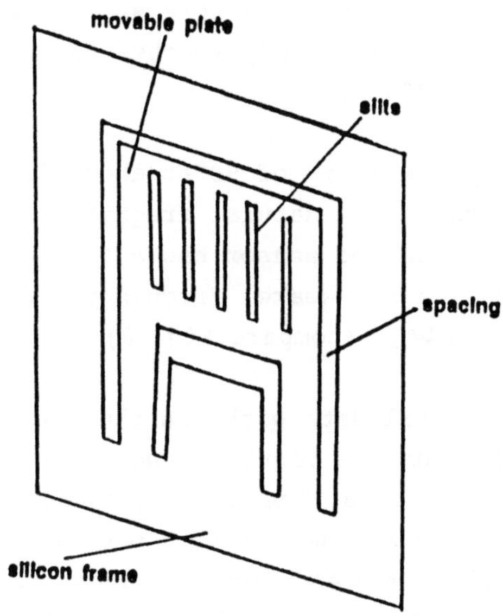

Fig.1: Schematic view of the middle plate of the sensor with surrounding Si frame

Damping is caused by the air which has to stream inside the small slit between the capacitor plates, when the middle plate is moved due to an acceleration. The air has to stream over a long distance from the center of the plates to the edges where it can escape to the other side of the differential capacitor.

One can reduce the damping effect and enhance the dynamical behavior by introducing several slits into the central plate, which allows the air to escape to the other side of the differential capacitor after a shorter flow distance. We have etched five slits into the middle plate of the sensor (compare Fig. 1).

5. Experimental results

With such a slitted sensor we came very close to the nearly aperiodic behavior of $2\rho^2 = \omega_0^2$. Up to 1200 Hz or $.52\,\omega_0$ the amplitude A decreases less than 10% of its $A(\omega = 0)$ value. The impact of the slits is demonstrated by comparing the frequency behavior of a sensor with 5 slits with that of a sensor with only a single slit on the outer side of the central plate (lower curve Fig. 2). The signal of this sensor decreases strongly with frequency and already reaches the 10% limit at a frequency of less than 200 Hz.

Both sensors have a phase shift of 90°at about 2300 Hz, which has been determined to be the natural frequency $f_0 = \omega_0/2\pi$. As expected the higher damped sensor has a stronger phase shift at low frequencies and weaker frequency dependence of the phase shift around ω_0, (compare Fig. 3).

We fitted the experimental data with the theoretical curve of equation (2) and (3) (compare Figure 2 and 3) and obtained $f_0 = 2300$ Hz for both sensors. The damping depends strongly on the number of slits. For the sensor with one slit ρ/ω_0 was determined to be $\rho/\omega_0 = 3$ and for the sensor with five slits we obtained $\rho/\omega_0 = 0.8$, which is nearly the optimal value of $\rho/\omega_0 = 1/\sqrt{2}$.

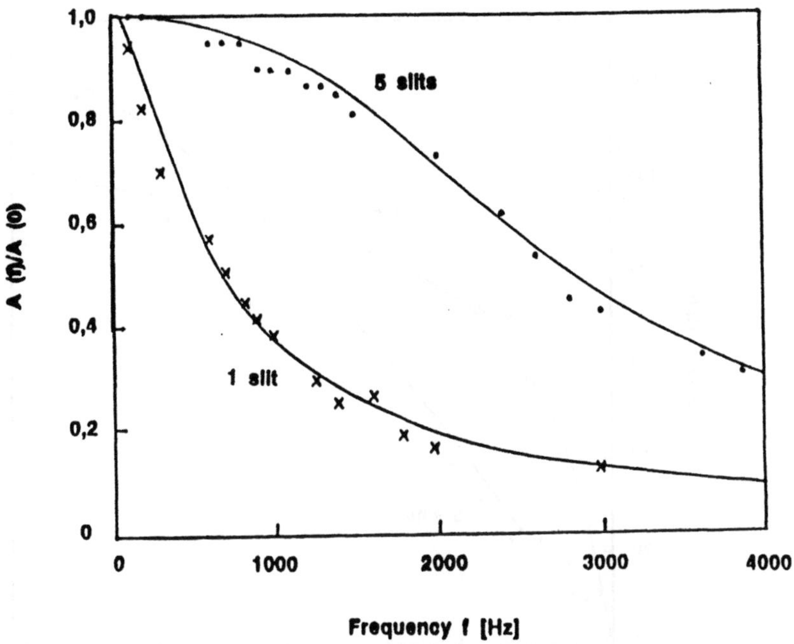

Fig. 2 Standardizised amplitude (sensor signal) versus frequency. The experimental results are given by the points for a sensor with 5 slits and crosses for a sensor with 1 slit. The curves show the theoretical fits with $\varrho / \omega_0 = 0.8$ (upper curve) and $\varrho / \omega_0 = 3$ lower curve.

6. Conclusion

The damping of an acceleration sensor with given resonance frequency can be adjusted by reducing the flight path of the air by several slits inside the central plate. Assuming the air damping roughly proportional to the flow path the distance of the slits can be estimated to reach an optimal damped sensor. The etching of the slits can be easily included in the fabrication process without additional work steps.

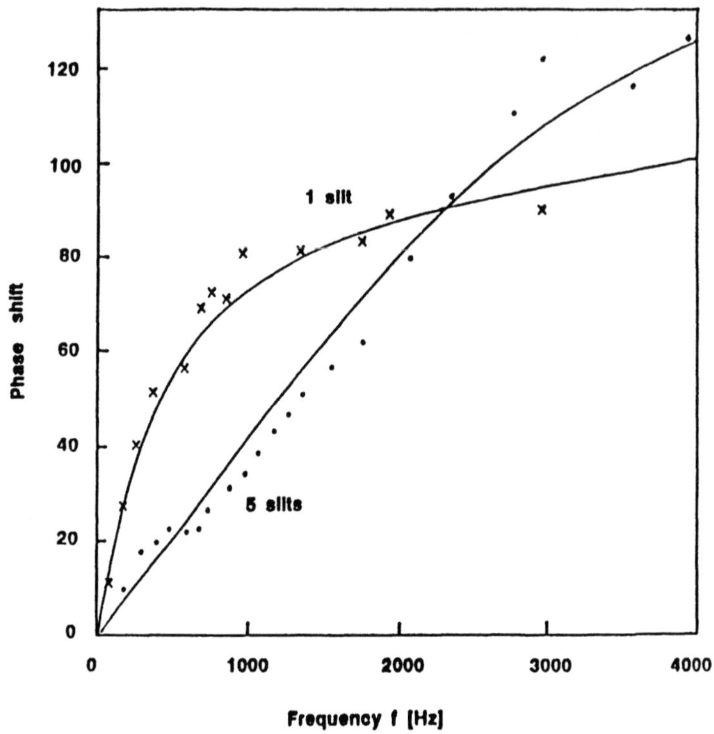

Fig. 3 Phase shift of the sensor signal versus frequency of the exiting signal. The experimental results are given by the points for a sensor with 5 slits and the crosses for a sensor with 1 slit. The curves shows the fits with $\varrho / \omega_0 = 0.8$ (5 slits) and $\varrho / \omega_0 = 3$ (1 slit). The natural frequency is determined to be $f_0 = 2300$ Hz.

Literature

/1/ H. Seidel and L. Csepregi: Design optimation for cantilever types accelerometers, Sensor and Actuators 6 (1984), 81-92

/2/ F. Sauter: Differentialgleichungen der Physik, Sammlung Göschen 1070, Berlin 1966.

Micromechanical Acceleration Sensor

U.Breng*, T.Geßner*, P.Lorenz**, M.Rauch**, W.Leyffer*,
M.Dittmann**, B.Küttner**, G.Schwenzer*, M.Wetzel**, N.Zichner*

Technische Universität Chemnitz
Sektion Physik-Elektronische Bauelemente *
Sektion Informationstechnik **
PSF 964, Chemnitz, 9010, GDR

Abstract

A miniature silicon acceleration sensor has been developed, consisting of a capacitive accelerometer and a electronic system. A seismic mass suspended by thin bars is used as sensing element. Signal conversion is achieved by capacitance measurement to fixed electrodes. The accelerometer consists of a silicon sensor element and a glass plate with contact pads. A microelectronical processing technology is used for the fabrication of the anisotropic etching of monocrystaline silicon and simultaneously lithography on the top and on the backside of the wafers. An outline of the fabrication process is given. The sensor application take place as threshold accelerometer, that means the generation of an electrical output signal from a adjustable acceleration value.

Introduction

Sensors and actuators are used in a variety number of areas and applications. Acceleration sensors generally consist of a spring-mass-system. A cantilever-type accelerometer with monocrystalline piezoresistors [1] and capacitive accellerometers with a seismic mass suspended by torsion parts as sensing element are known [2].

The silicon micromechanical acceleration sensor described in this paper is based on the capacitive accelerometer and a electronic system. The seismic mass is suspended by springs located on a frame. For this purpose anisotropic etching and specially developed lithography techniques are used. The fabrication process is simple and compatible to standard IC production steps. This acceleration sensor may be applied to damage sensors in special machines, for instance in washing robots.

Design And Structure

The capacitive accelerometer consists of three parts: the spring part with a seismic mass, the glass basis plate (PYREX) together with a metallized Al-electrode system and the cover to realize a mechanical limitation for the seismic mass. Fig.1 shows a shematic reperesentation of this accelerometer. The cover and spring parts were fabricated out of silicon material. The seismic mass is suspended by angular springs with a thickness df located on a frame. The capacitive mode of operation is realized by a capacitor between the Al-electrode system on the fixed basis plate and the metallized bottom of the seismic mass. The distance x_{rel} to the basis part defines the zero point in the capacity elongation characteristic.

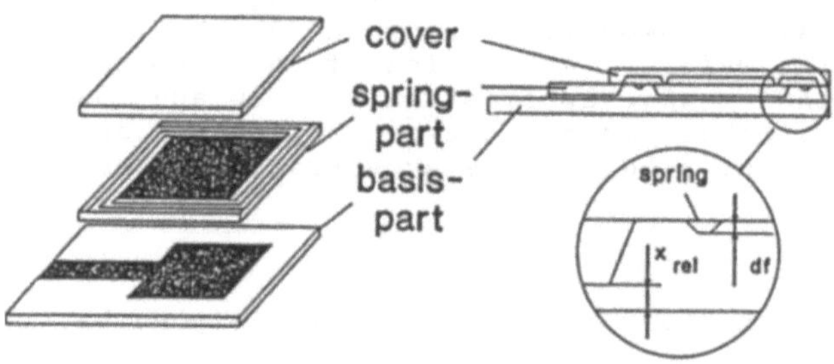

Fig.1. Sensor design

The distance x of the capacitor is changing as a function of acceleration and therefore effects a capacity acceleration dependence. The capacity change is analysed by a integrated electronic system. The integreated circuit is based on two oscillators, which one oscillator being a reference system and the second oscillator is connected to the capacitve accelerometer. The frequency comparison is performed by a comparator. In the case of threshold accelerometer the comparator generats a digital signal, which shows that a fixed acceleration value is reached. The packaging of the sensor elements and the accelerometer with the integrated circuit is done in the following way:

- anodic bonding between basis and spring parts
- adhesion of the cover to the spring part
- one or two packages for the sensor element and the integrated circuit respectively.

Sensor Fabrication Process

Fig.2 shows some technology steps for the fabrication of spring parts. 3-inch-Si-wafers were covered with 1000nm SiO_2 by wet oxidation. Then 150nm Si_3N_4 were deposited by a LP-CVD-process (Fig. 4a). The simultaneous lithography on the top and on the the backside of the wafers was developed by means of mask aligner from Carl Zeiß JENA / Elektromat Dresden. A special marker field was transferred to the backside of the wafer by a mirror-mask- system. This transfer took place simultaneously with the first mask level (spring structure) and therefore an adjustment between the top and the backside of the wafers was possible. Then the exposure of the backside level took place by using these adjusting marks. The alignment accuracy of the developed process was ±5μm. After plasmachemical etching the Si_3N_4 and following wet etching the SiO_2 on the backside of the wafers the spring structure was carried out by plasmachemical etching the Si_3N_4 on top of the wafers (Fig.2b).

The anisotropic silicon etching was performed with KOH (30%) at 80°C by controlling the etching time [3...5]. Therefore the measurement of the wafer thickness was necessary. Fig.2c shows the first etching process to realize the spring line region. The etching depth was determined by the spring thickness df and the value x_{rel} (Fig.1).

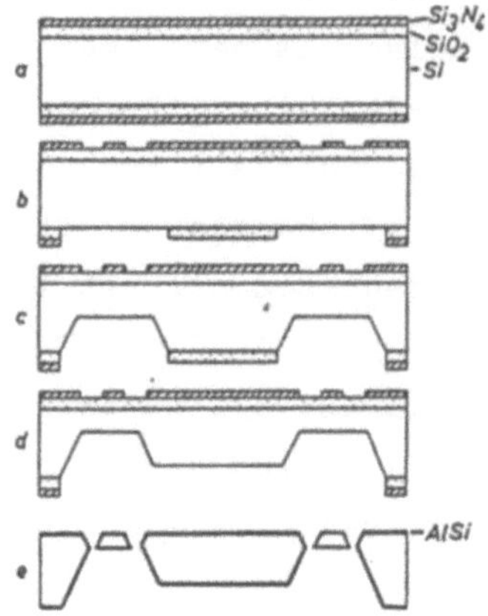

Fig.2. Fabrication procedure
a) initialstate b) after pattern steps c) after first silicon etching d) after second silicon etching e) finish structure with AlSi-metallization

In the case that df < x_{rel} a protective resist layer on the top of the wafers was necessary during the following SiO_2 wet etching process on the backside of the wafers. Following to that a silicon etching process in the spring line and the seismic mass regions was performed (Fig.2d). After SiO_2 etching on the top of the wafers the last silicon etching was realized. After silicon oxide and silicon nitride etching 1μm thick Al-Si layers were deposited on the top and on the backside of the wafers by sputtering technique.
In the case that df > x_{rel} the protective layer was necessary after the first silicon etching process on the backside of the wafers and the value x_{rel} could be realised during the third etching process.
The values of x_{rel} and df in this fabrication process are approximatly ±5μm.

The fabrication of the cover parts was realized by using an anisotropical silicon etching process only on the top side of the wafers.
The metallic electrode system on the glass basis part was realized by using an Al-Si layer system. The advantages of PYREX-glass as the material used for the basis plate are a decrease of the background basic capacity and facilities during the sensor packaging by using the anodic bonding process [6].

Application

The spring thickness df and the seismic mass distance to the basis part determine the sensor function. Fig.3 shows calculations of spring force and acceleration values as a function of the seismic mass elongation x for three spring thicknesses df by using the finite element method [7]. The maximum possible elongation x_{max} is the value x_{rel}. The figure indicates that

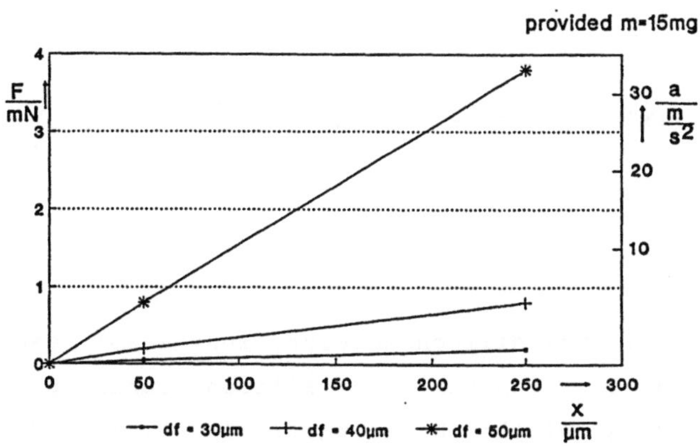

Fig 3. Theoretic F(x)-dependence (by using of finite element method

the spring thickness df can be very effectively employed for sensor applications by various acceleration values.
Fig.4 shows the measured results by testing a acceleration

sensor of df=50μm and x_{rel}=50μm (chip size: 10x12x0.8 mm) without basic part. There is a principal agreement between the calculated dependence in Fig.3 for df=50μm and the measured curve in Fig.3 is existing. Differences are based on variations of df during the fabrication process and measurement faults. The capacity value of the sensor without elongation is 10pF.

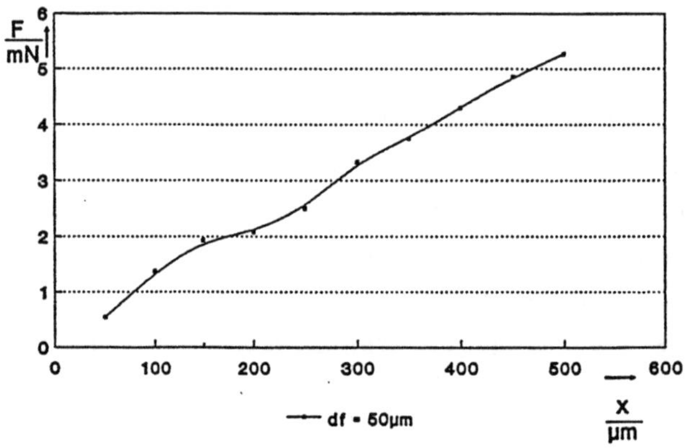

Fig.4. Realized F(x)-dependence with df=50μm

Conclusions

The developed miniature silicon acceleration sensor can be used as threshold accelerometer. The silicon planar process for the fabrication of the three parts is described. On application with acceleration values between 0.1·g and 100·g is possible. Future investigations should concern reliability tests and developments of the fabrication process, that means for instance the use of highly boron doped silicon as etching stop during the silicon etching process.

References

[1] Seidel, H.; Csepregi, L.: Sensors & Actuators, 6 (1984) 81-92

[2] Endolf, F. et al.: Silicon microaccelerometer. Transducers (1987), 112

[3] Heuberger, A. (ed.): Mikromechanik, Springer 1989

[4] Kloeck, B.: Study of Electrochemical Etch-Stop for High Precision Thickniss Control of Silicon Membranes. IEEE Trans. ED, 36, 4 (1989) 663-669

[5] Leyffer, W.: Untersuchungen zur Strukturierung einkristalline Siliciums, Tagungsberichte 7. Kolloquium Gerätetechnik Karl-Marx-Stadt (1989(

[6] Pollack-Diener, G.: Montagetechniken für Silicium-Sensoren, Technisches Messen, tm 56, 11 (1989) 422-427

[7] Bathe, K. (ed.): Finite-Elemente-Methoden, Springer 1986

CMOS Readout Electronics for Capacitive Acceleration Sensors

Rolf Buckhorst[*], *Bedrich J. Hosticka*[*] *and*
Helmut Seidel[**]
[*] *Fraunhofer Institute of Microelectronic Circuits and Systems,*
D-4100 Duisburg, Federal Republic of Germany
[**] *Messerschmitt-Bölkow-Blohm GmbH,*
D-8000 Munich, Federal Republic of Germany

Abstract:

In this contribution we will present a readout electronics for a capacitive acceleration sensor, both fabricated in silicon. The sensor is formed by an arrangement of a sensing capacitor and a reference capacitor. The capacitance of the sensing capacitor is inversely proportional to the active acceleration of the sensing capacitor. The readout electronics uses a switched-capacitor amplifier for determining the ratio of the reference capacitance and the sensing capacitance. The output signal of the system, which consists of the readout electronics and the acceleration sensor, is linearly proportional to the active acceleration. Since the output signal is symmetrical with respect to the ground, it has the same sign dependence as the measured acceleration. Using two external adjustable voltages the electronics makes it possible to trimm the sensitivity and the offset. The smallest detectable change of capacitance is about 7 fF. All components which are needed to operate the switched-capacitor amplifier (clock generator, etc.) are integrated in the readout electronics. The chip was fabricated in a 2 μm silicon-gate n-well CMOS technology. It contains not only the readout electronics but it is used as carrier for the acceleration sensor described above.

Sensor Description

The sensor is formed by a double capacitor using one moveable and two fixed plate electrodes. The moveable electrode represents a seismic mass. The position of the seismic mass between the plate electrodes depends on the active acceleration of the sensor. The moveable electrode is made of an etched silicon slab which is connected to the surrounding silicon frame by eight thin elastic silicon bars. These bars are located at the corners in diagonal directions always at the top and at the bottom (see Fig. 1), thus forming a highly symmetrical arrangement. Both backplate electrodes are located in etched glas trenches. The glass is fixed to both sides of the silicon chip by anodic bonding. A photolitographically structured chrome-gold layer is used as electrode material. The symmetric mounting results in a high axis selectivity with low cross sensitivity. In addition, this construction method offers advantages when compared with other capacitive sensor concepts, (e.g. plates with bars only at the top or bottom or cantilever beam sensors), because the deflection of the plates is parallel to the chip plane. This yields a highly linear sensor. The acceleration-sensitive plate has the same

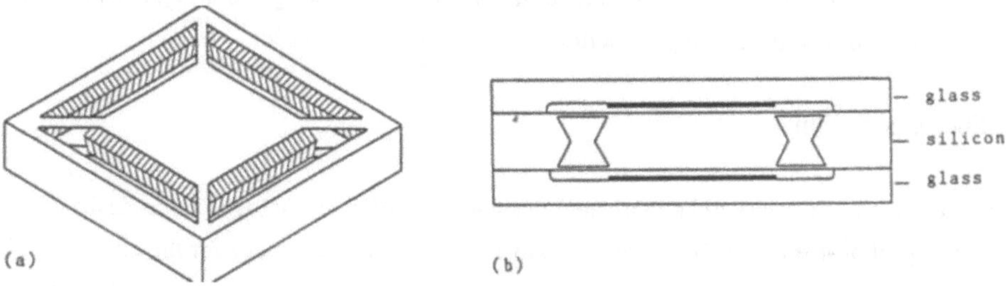

Figure 1: Schematic of the capacitive accelerometer: a) top view b) cross section of the seismic mass

thickness as the wafer, while the elastic bars have only a thickness of several micrometers. The typical electrode gap of five micrometers produces a sensor capacitance between 2 and 10 pF for sensor areas between 6.5 mm * 3.5 mm and 8.7 mm * 3.5 mm. Measurements of the sensor capacitance show a sensitivity of 1pF/g and a resolution in the 1 mg-range.

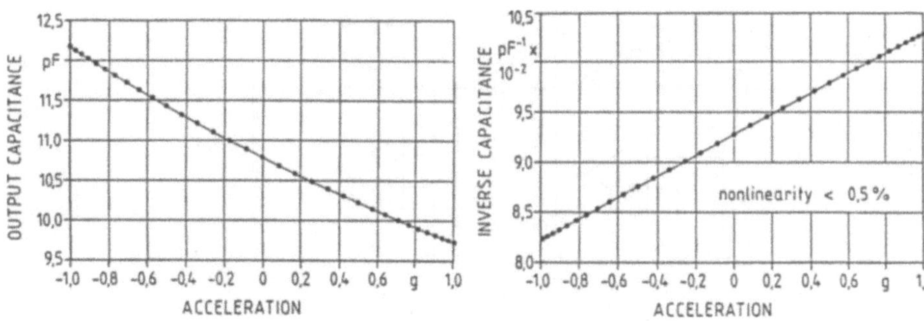

Figure 2: Measurement of the capacitance dependence on the active acceleration

The sensor capacitance can be described by the usual capacitor equation:

$$C = \frac{\varepsilon_0 A}{d} \qquad (1)$$

where ε_0 is the permittivity, A is the capacitor area, and d is the electrode gap. This has been corroborated by measurements, as depicted in Fig. 2, which confirm that the acceleration a is proportional to d. The inverse capacitance shows a linearity error smaller than 0.5%. The acceleration-sensitive capacitor is located close to a reference capacitor, which has been constructed in a similar manner. However, this capacitor exhibits no sensitivity to acceleration and can thus be used for temperature compensation. In the following the arrangement of the reference capacitor and the sensing capacitor will be referred to as sensor.

Amplifier Description

As the readout amplifier was to be fabricated in standard CMOS technology, it was designed in switched-capacitor technique (SC). The components which are necesssary for the operation of the SC-amplifier, like oscillator, clock generator, and clock drivers, have been integrated on the same chip but will not be discussed here. As described above, the sensor consists of the reference capacitor C_1 and sensing capacitor C_2 (see Fig. 3) while all on-chip capacitors have fixed values. The switches are operated using nonoverlapping digital clock signals. The resulting charge transfer yields an output voltage that corresponds to the amplified variation of the sensing

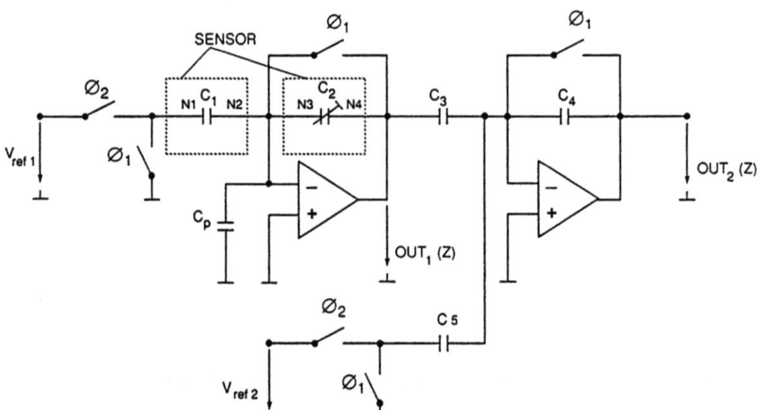

Figure 3: Schematic of the amplifier cascade

capacitance C_2. The output signal $Out_2(z)$ has typically a reset phase and an active phase. During the active phase the output signal is proportional to the active acceleration and in the reset phase $Out_2(z)$ is 0V. In addition, there is a signal that indicates the active phase by a high level. This signal is generated by the clock generator and is available at a second output of the circuit. It can be used as a strobe signal for an A/D-converter to digitize the output signal $Out_2(z)$. The SC-amplifier shown in Fig. 3 consists of two operational amplifiers, six transmission gates, and three capacitors. The amplifier of the front stage is an operational transconductance amplifier (OTA) /2/ with a measured unity gain bandwidth of 5.7 MHz, a phase margin of 60°, and an open loop gain of 71 dB. The second stage employs an OTA followed by a class-B output buffer /3/, /4/. This amplifier has the capability to drive a load of 10 kΩ parallel to 50 pF. It has a unity gain bandwidth of 4.8 MHz, a phase margin of 60°, and an open loop gain of 75 dB. A measurement of the amplifier cascade shows a linearity better than ± 0.19%. The maximal resolution of input capacitance amounts to 7fF. The usable bandwidth is 10 kHz, while the clock frequency is 100 kHz. The circuit dissipates 65 mW at ± 5 V power supplies.

Amplifier Performance

Both sensor capacitors, C_1 and C_2, exhibit substantial stray capacitances to the substrate due to long leads (about 12 mm). This parasitic capacitance is designated as C_p in Fig. 3. C_p is of the same order as C_1 and C_2 and was about 1 pF in our case. For this reason it is important that the amplifier concept is insensitive to stray capacitances /5/. The output voltage function of the first amplifier stage is:

$$Out_1(z) = -\frac{C_1}{C_2}\,\frac{A_0}{A_0 + 1 + \dfrac{(C_1 + C_p)}{C_2}}\,V_{ref1}(z) \tag{2}$$

Taking into account the open loop gain A_0=71dB and considering $C_1 = C_2 = 2$ pF we obtain a constant gain error less than 0.1%. This means that the influence of C_p is negligible. The output voltage $Out_2(z)$ is described by the following equation:

$$Out_2(z) = V_{ref1}(z)\,\frac{C_1}{C_2}\,\frac{C_3}{C_4} - V_{ref2}(z)\,\frac{C_5}{C_4} + V_{Offset} + V_{CFT} \tag{3}$$

The voltage V_{CFT} is generated by the clock-feedthrough of the transmission gates. The design of the amplifier eliminates the dependence of this voltage on the signal and, hence, we can treat V_{CFT} as a constant offset voltage. The voltage V_{Offset} is the offset voltage of the output amplifier. If we set the ratios C_3/C_4 and C_5/C_3 to unity and consider Eq. 1 defining the sensing and the reference capacitance, we obtain $Out_2(z)$ as a function of A_1 and A_2 (the areas of capacitor plates), d_1 (the fixed electrode gap of the reference capacitor C_1) and d_2 (the electrode gap of the acceleration-dependent capacitor C_2):

$$Out_2(z) = V_{ref1}(z) \frac{A_1}{A_2 d_1} d_2 - V_{ref2}(z) + V_{offset} + V_{CFT} \tag{4}$$

As the description of the sensor has shown, d_2 is proportional to the active acceleration. This means that the output voltage $Out_2(z)$ of the amplifier is proportional to the active acceleration a and a linear dependence between output voltage $Out_2(z)$ and acceleration a has been established. The sensitivity of the sensor system is determined by the factor:

$$V_{ref1}(z) \frac{A_1}{A_2 d_1} \tag{5}$$

Eq. 5 suggests that the sensitivity can be adjusted by the voltage $V_{ref1}(z)$ independently of the geometrical size of the capacitors. Thus it is possible to set the sensitivity to a nominal value. This allows a proper scale adjustment to ensure operation within a desired range inside the operating limits of the amplifier. If there is no active acceleration on the sensor capacitor, the voltage $Out_2(z)$ should be equal to zero. At that time the electrode gaps of the reference capacitor d_1 and the sensor capacitor d_2 are ideally equal. For this case $Out_2(z)$ should be adjusted to 0V to compensate the voltages V_{Offset} and V_{CFT}. If there is no active acceleration and d_1 is not equal to d_2, this error can be corrected by $V_{ref2}(z)$, too. After the adjustment of the zero the output voltage $Out_2(z)$ shows the same sign dependence as the active acceleration. Accordingly, the calibration of the sensor is possible by trimming the voltages $V_{ref1}(z)$ and $V_{ref2}(z)$ to make the output signal $Out_2(z)$ independent of manufacturing tolerances.

Layout and Mounting

The SC-amplifier was integrated together with all peripheral circuits necessary for its operation in a 2 µm silicon n-well CMOS technology. The size of the chip is 4.9 mm * 9.6 mm. This large die area is determined mainly by the dimensions of the sensor, because the chip serves as carrier for the sensor. The sensor is glued on the chip and the electrical connections are bonded with aluminium wires (see Fig 4). The integrated SC-amplifier, the oscillator, the clock drivers, and the clock generator require a die area of 2.8 mm^2 which is about ten times smaller than the area needed for the sensor (i.e. 30.4 mm^2). To operate the system composed out of sensor and readout electronics, only one piezoelectric crystal and two external capacitors are required.

Figure 4: The photo shows the sensor mounted on the silicon substrate containing the readout electronics

Summary

In this contribution a readout electronics for a capacitive accelerometer, both fabricated in silicon, was presented. The chip containing the readout electronics also provides all peripheral circuits necessary for the operation and serves as carrier for the accelerometer.

References

/1/ H. Seidel, "Capacitive silicon accelerometer with highly symmetrical design", Sensors and Actuators, A21-A23, pp 312-315, 1990

/2/ B. J. Hosticka, "CMOS operation amplifiers" in Design of MOS VLSI Circuits for Telecomunications (ED.: Y.Tsividis, P. Antognetti), Prentice Hall, N. J., 1985

/3/ W. C. Black, Jr., "High-speed CMOS A/D conversion techniques", Ph. D. Dissertation, University of California, Berkeley, 1980

/4/ W. C. Black, Jr., D. J. Allstot and R. A. Reed, "A high performance low power CMOS channel filter", IEEE J. Solid-State Circuits, vol. SC-15, pp. 929-938, Dec. 1980

/5/ U. Kleine, "Integrierte Schalter-Kondensator-Filter in CMOS-Technologie", Ph. D. Dissertation, University of Dortmund, Dortmund 1985

Gas and Ionic Sensors

Development of Basic Structures for Integrated Chemical Sensors for SO_2 and NH_3

J. Lin, S. Möller, E. Obermeier

TU Berlin, Technologien der Mikroperipherik
Berlin (W), W-Germany

Abstract

Sensor components have been developed which are suitable for monolithic integration in a system consisting of a sensor and signal processing circuitry. The basic elements used are interdigital capacitors (IDC) which permit measurement of both conductivity and capacitance changes of the sensitive material. Planar IDC structures of different electrode widths were coated with an organically modified silicate film by means of a spin-coating process. The dielectric properties of this film change in relation to the SO_2 and NH_3 concentrations to be detected. The measurements show that the response limit lies below the MAK-value (MAK-value SO_2: 2 vpm; NH_3: 50 vpm). The sensors are characterized by high repeatability and reversibility, short response time, low cross sensitivity towards other gases and towards humidity as well as good long-term stability.

Introduction

The basic structure of the sensors described consists of an interdigital capacitor (IDC) on an insulated substrate such as quartz or oxidized silicon. On the substrate, two interacting comb-type metallic electrodes produced from an NiCr/Au system form the interdigital capacitor which is covered by the sensitive material consisting of an organically modified silicate (Fig. 1). The IDC structure represents a basic structure which can be used for examining a large number of chemo-sensitive materials, the conductivity or dielectric constant of which change in relation to the concentration of the test substance.

Simulation

To provide support for the development of the IDC design, simulations were carried out to determine the capacitance value of various structures. The method of calculation is based on an analytical procedure with respect to the superimposition of element capacitances making use of the method of conformal transformation [1]. In this procedure, the electrode width (or the distance between electrodes) and the electrode height as well as the effective capacitor surface area were varied. In the case of all IDC structures, a ratio between the electrode width and distance between electrodes of 1 : 1 was selected. Figure 2 shows simulated and measured capacitance values as a function of the electrode width.

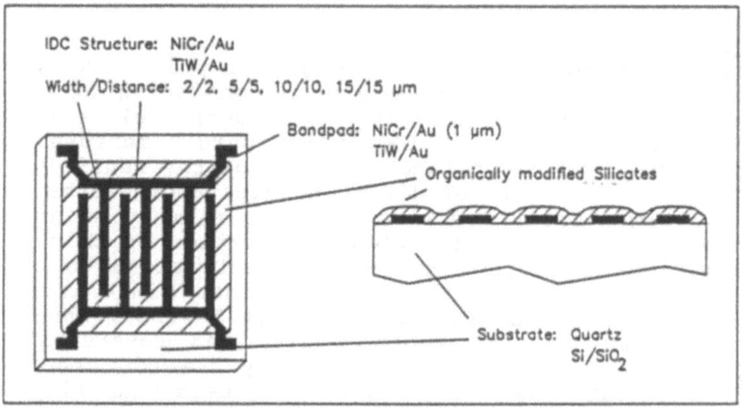

Fig. 1: Design principle of an IDC gas sensor with organically modified silicate layer

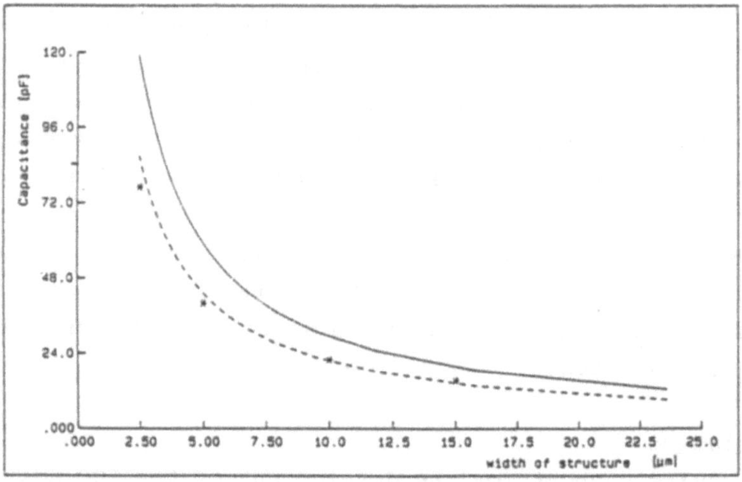

Fig. 2: Calculated and measured capacitance values of interdigital capacitors in relation to electrode width (electrode width : distance between electrodes = 1 : 1)

Experimental

Planar IDC structures with an electrode height of up to 200 nm and a widths of 2.5, 5, 10 und 15 μm were produced photolithographically (electrode width : distance between electrodes = 1 : 1). For metallization, sputtered films of NiCr (as an adhesive layer) and gold were employed.

As gas-sensitive materials, three different films were optimized on the basis of organically modified silicates [2]. Two films (NND/PTMS and NNDE/PTMS) are suitable for SO_2 detection, this applying for concentrations in the MAK range (MAK-value for SO_2 = 2 vpm) and for concentrations up to 6000 vpm. The mechanism of gas adsorption is based on the adduct formation of tertiary amines and SO_2. For measurement of NH_3 concentrations, a Cu-containing silicate film was developed utilizing the adsorption of NH_3 on Cu. The IDC structures were coated with the silicate solutions by means of a spin-on process followed by hardening.

Results

The sensors developed were tested under a calibration gas atmosphere. For this purpose a computer controlled gas test station was established. The sensors to be tested are installed in a test chamber and receive the required calibration gas concentration which is regulated by a mass flow controller.

In order to determine the lower detection limit of the sensors, the sensors coated with NND/PTMS were measured in the range of the MAK-value up to 10 vpm. Figure 3 shows the absolute change in capacitance of such a sensor in relation to the SO_2 concentration in the range from 0 to 10 vpm.

The reversibility and the cross sensitivity to other gases such as NH_3, CO, CO_2 and CH_4 were likewise examined. Figure 4 shows the time curve of the relative capacitance changes of a sensor coated with NNDE/PTMS with changing exposure to various SO_2 concentrations and NH_3 (up to 1 %). The examination revealed no cross sensitivity to NH_3.

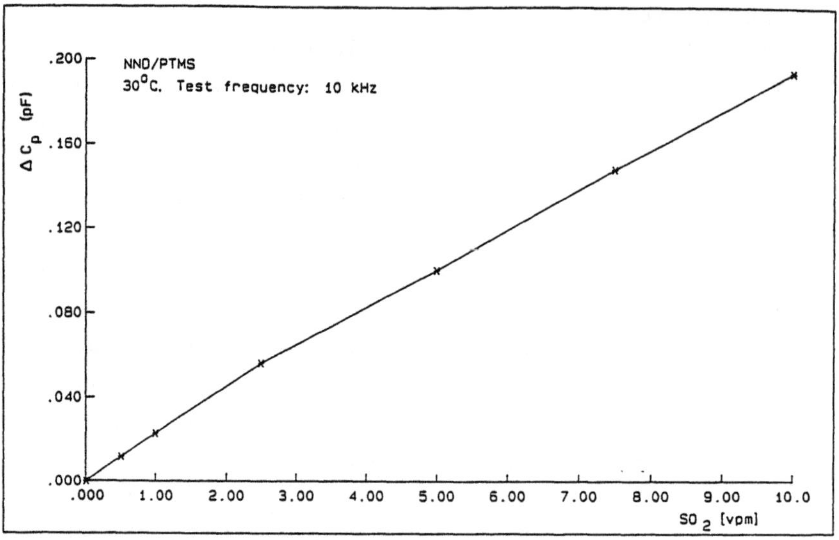

Fig. 3: Absolute changes in capacitance of an IDC coated with NND/PTMS as a function of the SO_2 concentration

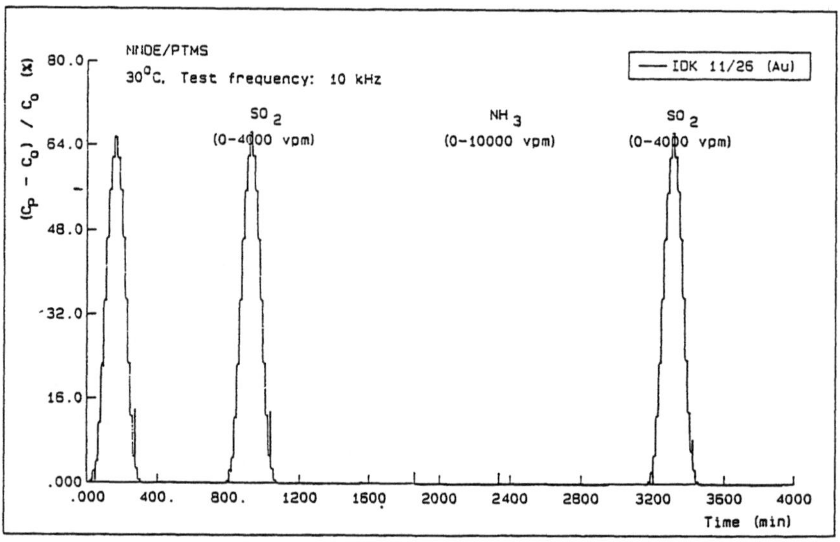

Fig. 4: Cross sensitivity vs. NH_3 concentration of IDCs coated with NNDE/PTMS

The transient response to an abrupt jump in calibration gas concentration is made clear in Figure 5.

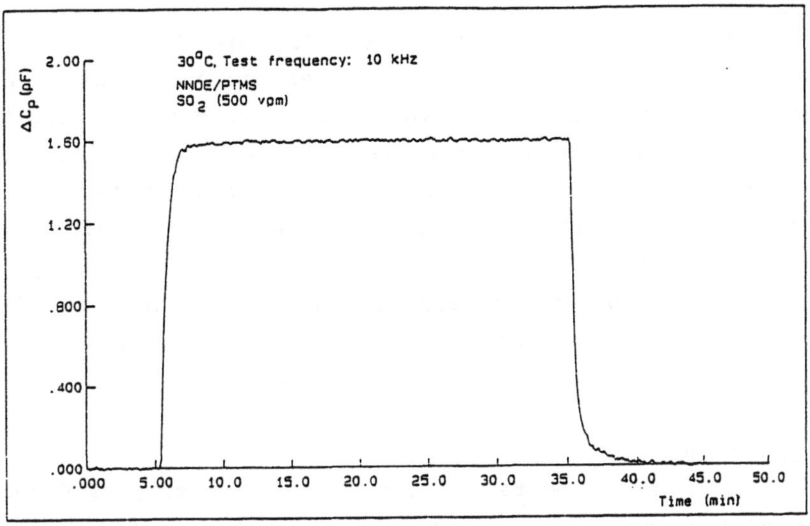

Fig. 5 Change in capacitance for abrupt exposure to SO_2 (500 vpm) vs. time for NNDE/PTMS

Figure 6 shows the absolute change in capacitance of a sensor coated with 143C1 in relation to the NH_3 concentration over a range from 0 - 500 vpm. From the shape of the curve it will be seen that the sensor is suitable for concentration measurements in the MAK-value range (MAK-value of NH_3: 50 vpm) and is even sensitive to lower NH_3 concentrations.

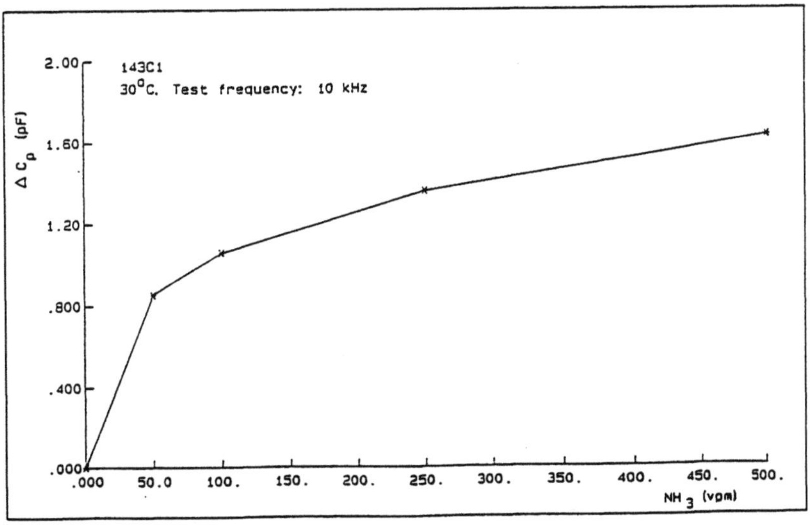

Fig. 6: Absolute changes in capacitance of an IDC coated with 143C1 in relation to NH_3 concentration

Discussion

From the results presented for SO_2 and NH_3 gas sensors it becomes obvious that IDC structures are well suited as basic elements for chemo-sensitive sensors. It can be added that due to the low surface area required in view of the electrode width of only a few μm, these structures can be integrated on one chip together with a signal processing circuitry so that changes in capacitance can be converted into a voltage signal. The SO_2 sensors developed are characterized by good reversibility, low cross sensitivity towards gases such as NH_3, CO, CO_2 and CH_4 and towards humidity.

Acknowledgement

This work was supported by the Federal Ministry of Research and Technology and carried out in cooperation with the Fraunhofer Institute for Silicate Research, Würzburg, and the University of the Bundeswehr, Institute of Measurement and Automation Technology, Munich.

References

[1] R. K. Hoffmann, Integrierte Mikrowellenschaltungen, Springer-Verlag, Berlin Heidelberg New York Tokyo 1983

[2] F. Hutter; K. H. Haas; H. Schmidt, Proc. of 2nd Int. Meeting on Chemical Sensors, Bordeaux, 1986, pp. 443-446

Technology and Application of Suspended Gate Field Effect Transistors for Gas Detection

H. LORENZ, M. PESCHKE, H. RIESS and I. EISELE

Institut für Physik, Fakultät für Elektrotechnik
Universität der Bundeswehr München, D–8014 Neubiberg, FRG

Abstract

A new technology for the fabrication of a suspended gate field effect transistor (SGFET) has been developed. With the help of a modern lift–off process sequence it is possible to use physical as well as electrochemical thin film deposition methods for the incorporation of the gas sensitive layer. Sputtering the active layer before the gate metallization allows a mass fabrication which is completely in accordance with semiconductor production. With different gas detecting materials such as SnO_2 and Pt a SGFET device is well suited for pattern recognition with sensor arrays. Due to the small effective sensor area of 30 μm x 20 μm the element can easily be integrated in micro systems.

Introduction

The basic principles for the use of SGFETs as gas sensors are well known [1]. The airgap between the insulator and the gate structure allows the gas to flow underneath and adsorb at the chemically active layer which is located directly below the gate metallization. The chemisorbed or physisorbed gaseous species changes the semi–conductor–gate work function difference ϕ_{ms}, which causes a measurable shift in the threshold voltage of the FET. The application of such a device has been described previously for Pd/NH_3, SnO_x/NH_3, Pd/H_2 and SnO_x/H_2 systems [2]. The sensitive layer has been deposited until now exclusively by electrochemical means. Recently a new method for physical deposition has been developed [3]. It allows to use general thin film techniques such as sputtering, CVD or plasma deposition. This offers the advantage of depositing materials of great interest such as non–soluble metal oxides. Moreover all chips on a wafer can be processed at the same time. The present paper describes the lift–off technique for physical deposition of the sensitive layer and presents results with SnO_2 and Pt modified SGFETs.

Fabrication Process

The schematic diagram for the fabrication of a SGFET with a physical deposited active layer is shown in Fig. 1. Except for the gate metallization the SGFET corresponds to a conventional n–channel MOSFET. It is constructed on a 50 Ωcm p–type (100) silicon substrate according to the well established procedure with n^+ doped source and drain regions. The gate lenght L amounts to 20 μm and the gate width W to 30 μm. The insulator consists of a double layer with 100 nm thermally grown SiO_2 and 100 nm LPCVD deposited Si_3N_4.

In contrast to the previous fabrication of the suspended gate structure, a lift–off technique is now introduced for patterning. Using a negative resist (AZ 5214, Hoechst) an overcut profile is achievable, which is well suited for lift–off. Next, the spacer material (sputtered SiO2 of 200 nm thickness) between the insulator and the suspended gate is deposited. At this point the sensitive layer SnO_2 (100 nm) is reactively sputtered in an oxygen/argon plasma with 5×10^{-3} mbar partial pressure for oxygen and argon. Both, the spacer and the active layer are patterned in one step by lift–off.

It is obvious that for the deposition of the sensitive layer, general thin film techniques, like sputtering, reactive sputtering, evaporation, CVD or plasma depositon, are practicable. A second lift–off procedure for the patterning of an adhesive intermetallic Ti/W layer (50 nm) follows. In this case the active gate area is omitted from the deposition by photoresist. This protection is necessary because it has been shown that Ti contamination in the gate region causes severe stability problems. The final metallization is a layer of Pt (1000 nm for mechanical purposes), also patterned by lift–off in order to avoid ion milling or electrochemical etching. Although the active area is 30 μm x 20 μm, the suspended gate shows a very good mechanical stability, with smooth surfaces and edges. The SiO_2 spacer is removed in buffered HF, thus forming an air gap inside the gate structure.

Considerations for signal evaluation

The exposure of the sensitive layer to certain gases results in a work function change of this surface. This change causes a shift in the threshold voltage V_t due to the the workfunction difference ϕ_{ms} in the SGFET according to:

$$V_t = \phi_{ms} - \frac{Q_{it} + Q_{ox}}{C_o} + 2\,\phi_f + \frac{1}{C_o}\left(2\,q\,N_a\,\epsilon_s\,\epsilon_o\,(2\phi_f + V_{sb})\right)^{\frac{1}{2}}$$

where Q_{it} and Q_{ox} are the semiconductor–insulator interface states and insulator charge density, ϕ_f the Fermi level, N_a the acceptor concentration and V_{sb} the substrate bias. ϕ_{ms} is the work function difference between the gate and the semiconductor. C_o is

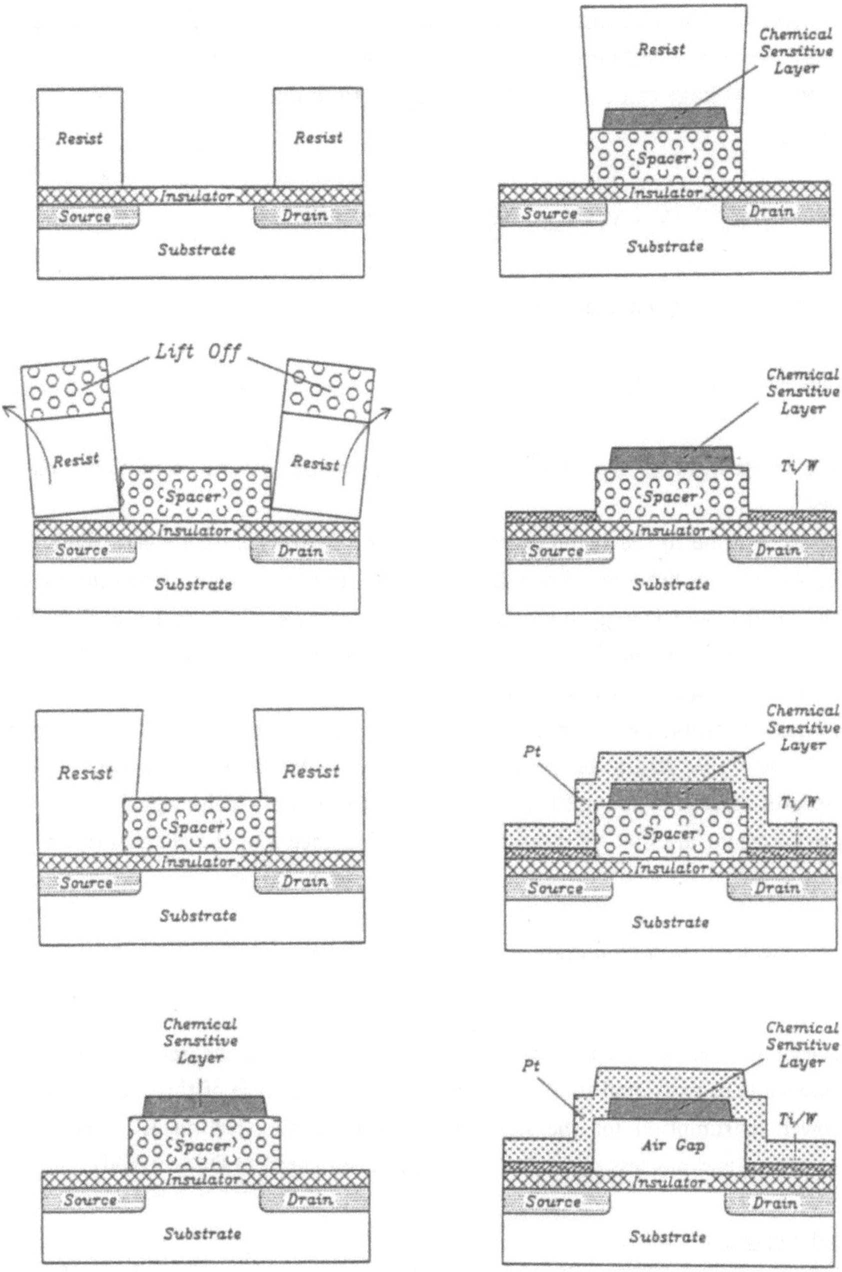

Fig. 1. Schematic plot of the SGFET fabrication. (a) Basic structure with defined resist. (b) Deposition of the spacer, patterned by lift–off. (c) Define resist for (d) the deposition of the sensitive layer. (e) Define resist for (f) the deposition of Ti/W layer. (g) Gate metallization with thick Pt. (h) After etching the spacer, the final structure is achieved.

the value of the gap and insulator capacitance in series. The I–V–characteristics of a MOSFET can be divided into the linear and the saturation region. In the linear region the threshold voltage can be derived with respect to source drain voltage V_{ds} and gate voltage V_g:

$$I_{ds} = \mu C_o \frac{W}{L} (V_g - V_t) V_{ds} - \frac{V_{ds}^2}{2}$$

For the saturation region one obtains:

$$I_{ds} = \frac{1}{2} \mu C_o \frac{W}{L} (V_g - V_t)^2$$

For the integration of the SGFET in a sensor array, several conditions must be fullfilled:

1) The access time to the signal of every sensor in the array must be very short. When the supply voltages are applied to the SGFET the transistor needs some time to give a reliable signal (conditioning). Therefore significant changes of the potential distribution within the transistor should be avoided. It was empirically established, that drain source and gate voltages cannot be turned off.

2) Another condition for an integrated system is low power consumption. From this point of view the saturation region is not very useful because for each device a drain–source current of about 100 μA is necessary to get a low noise signal.

Both requirement are met by measuring the conductance g of the channel in the linear region:

$$g = \left. \frac{\partial I_{ds}}{\partial V_{ds}} \right|_{V_g} = \mu C_o \frac{W}{L} (V_g - V_t) - V_{ds}$$

If the conductance is measured with an AC signal V_{ds} can be set to zero and the conductance is proportional to V_t and thus ϕ_{ms}. The advantage of this method is a very low power consumption for the signal evaluation. In addition an AC–method with lock–in–techniques has less distortion and noise as compared to DC–methods.

General behaviour

The behaviour of SGFETs with sputtered Pt and SnO_2 as sensitive layers has been investigated with respect to oxidizing and reducing gases, temperature and humidity. In Fig. 2 the response at T=160°C with respect to different concentrations of NH_3 and NO_2 in synthetic air is shown. For both sensors the exposure of NO_2 leads to a positive

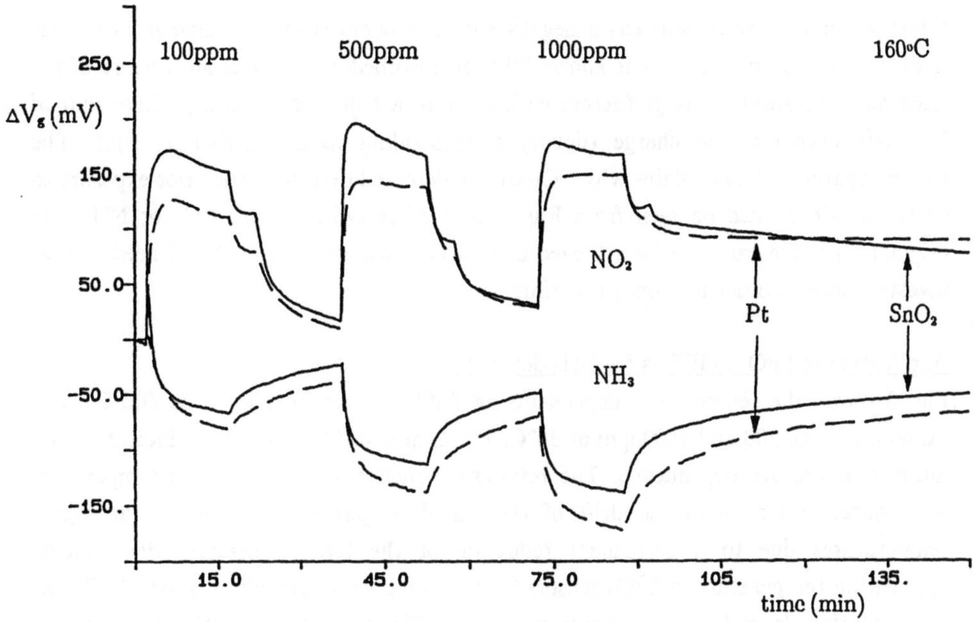

Fig. 2 Time response to NO_2 and NH_3 in synthetic air at 160°C for Pt- and SnO_2 SGFETs.

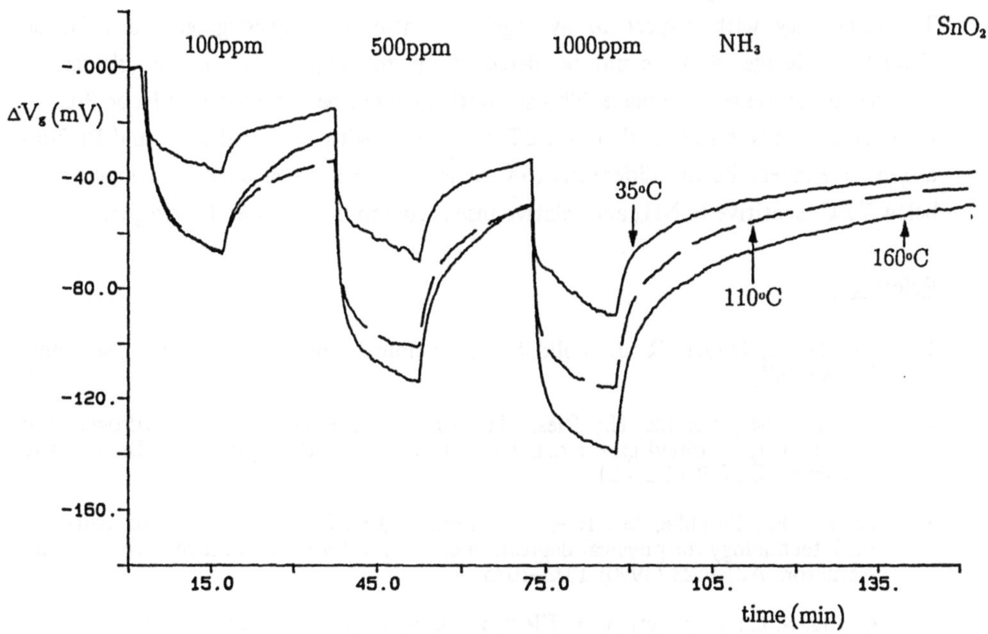

Fig. 3 Temperature dependence of SnO_2 SGFETs to NH_3 in synthetic air.

threshold voltage shift, whereas a negative shift is oberserved for ammonia and other hydrogen containing gases. It is known [4] that adsorbed NO_2 causes an increase of the fixed negative surface charge (acceptor) leading to a higher Φ_{ms}, i.e. a positive shift of V_t. NH_3 decreases the charge (donor) corresponding to a negative V_t shift. The electronegativity also explains why NO_2 can be detected despite of the prior exposure to synthetic air. It can be seen from Fig. 2 that after switching off NO_2 or NH_3, the original V_t value can only be achieved after a long recovery time. It will need further investigations in order to reduce this time.

Application of SnO_2 SGFETs for NH_3 detection

Fig. 3 shows the temperature dependence of SnO_2 sensors for NH_3. The SGFET was exposed to 100, 500 and 1000ppm at 35°C, 110°C and 160°C. The results indicate a non linear temperature dependence. The response increases with increasing temperature and concentration. Also the drift of the signal is getting more evident at higher temperatures due to a permanent reduction of the SnO_2. Therefore the prefered application temperature of SnO_2 sensors for the detection of amonia is below 70°C. The concentration dependence was measured up to 1000ppm with a detection limit around 1ppm. Below 100ppm the concentration behaviour is logarithmic, at higher values a square root behaviour was found.

It also was observed that SnO_2 sensors are relative insensitive to humidity. Even for 20% humidity the response is less than 15mV at 35°C. The rise time is within seconds.

The selectivity with respect to hydrogen has also been investigated. At 35°C no significant change of V_t could be detected up to 500ppm H_2, i.e. no detectable dissoziation processes take place. This separation cannot be achieved with Pt or Pd.

In conclusion it was shown, that SGFET gas sensors with sputtered SnO_2 and Pt films as sensitive layers, behave differently to various gases. SnO_2 sensors are at temperarures below 70°C sensitive to NH_3 and relative insensitive to humidity and hydrogen.

References

1. Janata, J.; Huber, R. J.: Solid State Chemical Sensors. New York: Academic Press (1985)

2. Peschke, M.; Lorenz, H.; Riess, H.; Eisele, I.; Recognition of hydrogen and ammonia by modified gate meallization of the suspended-gate FET. Sensors and Actuators B1 (1990) 21–24.

3. Lorenz, H.; Peschke, M.; Riess, H.; Janata, J.; Eisele, I.; New suspended-gate FET technology for physical deposition of chemically sensitive layers. Sensors and Actuators A21–A23 (1990) 1023–1026

4. Chang, S. C.: Transactions on Electron Devices, IEEE, ED, 26, (1979) 1875

Preparation of TiO_2 Films for Oxygen Sensors

D. Fischer, H. Köppen

Electronic Systems Department
Battelle-Institut e.V.
Am Römerhof 35, D-6000 Frankfurt am Main 90

Summary
Properties of TiO_2-films for oxygen sensing were investigated. Especially the influence of sensor characteristics on crystal phase (anatase-rutile) and structure of the films was studied. Preparation was made by sputtering and ICB (Ion Cluster Beam deposition). Changes in phase and structure due to variation of deposition parameters were measured by X-ray diffraction. The experiments showed a clear advantage of ICB for our purpose. The results and the influence on sensor characteristics are discussed.

1. Introduction

At temperatures exceeding about 600°C stoichiometry and therewith electrical conductivity of TiO_2 becomes dependent on surrounding oxygen partial pressure p_{O2}. The relationship $\sigma = A\, p_{O2}^{-n}$, where n depends on the diffusion mechanism in the TiO_2-lattice, is the base for use of TiO_2 as an oxygen sensor. The high temperature application requires

- good adherance to the substrate during thermal cycling up to 1000°C
- no structural change during sensor operation
- production of highly oriented rutile films, preferably with (110)-texture.

Since our laboratory has been equiped with an ICB deposition system, we tried to test its ability for making oxyde films, especially TiO_2, and compare the results to those of sputtered films.

2. TiO_2 film preparation

2.1 Sputter deposition

Films were prepared by DC magnetron sputtering in a parallel plate sputtering system V400 from Leybold. The metallic Ti target was reactively sputtered onto silica and sapphire substrates in a Ar/O_2-plasma. Substrate temperature was usually 300°C. Stoichiometric films are only obtained at higher oxygen flow when the target surface itself is fully oxidized. At this point a sharp drop in sputtering rate occurs because of the lower sputtering yield of the oxide. For variation of the film properties oxygen flow, substrate temperature and bias voltage were varied.

2.2 ICB deposition

TiO_2-films were prepared in a Mitsubishi ICB system, shown schematically in fig. 1. ICB is basically a special high vacuum evaporation process (1,2) where the evaporated flux is partially ionized in an electron shower with succeeding acceleration towards the substrate surface. By variation of ionizing current and acceleration voltage the mean kinetic energy per deposited atom can be varied within a wide range. The maximum degree of ionization measured in the substrate plane is about 10%.

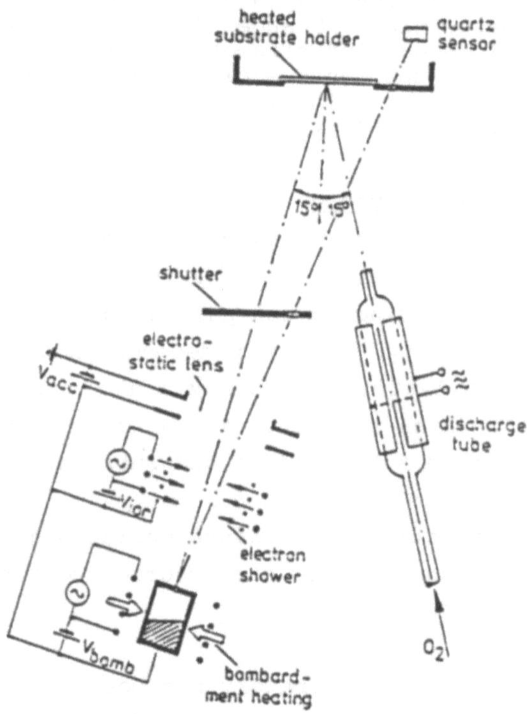

Fig. 1 Schematic setup for reactive ICB

For TiO_2-formation TiO was evaporated from a tungsten crucible in oxygen partial pressure. For variation of the film properties oxygen flow, acceleration voltage and ionization current were varied.

3. Results and Discussion

For a first test of high temperature stability, films 0.5 µm thick on quartz and sapphire substrates were heated for 20 hours in air to 1000°C. After this treatment the sputtered films became turbid whereas the ICB films remained clear. This means, that crystallite structure of the sputtered films was changed at high temperature.

TiO_2 appears in three crystal modifications, i.e., rutile, anatase and brookite, although the latter is never seen in thin films. Basicaly it is known, that at low substrate temperatures and low kinetic energy of the deposited atoms films made by PVD processes mostly consist of anatase. By increasing substrate temperature and kinetic energy more and more rutile is formed. On the other hand anatase is transferred to rutile at temperatures of 800°C which are applied for our sensor measurements.

sample no.	1	2	3	4	5	6	7	8	9	10	11
varied parameter	basic set	T_s = 400°C	substr. unheated	j_{O_2} = 20 cm^3	j_{O_2} = 10 cm^3	R = 16 $\frac{nm}{min}$	V_{acc} = 3 kV	I_{ion} = 500 mA	I_{ion} = 100 mA	I_{ion} = 0	V_{acc}=0 I_{ion}=0
anatase 101	4	6	19			4	9		4	2	18
anatase 004	6	10		4		3	3.5	3.5	4.5	6	
anatase 200			11						2		2.5
rutile 110		6		7	10			6			

Table 1 Relative peak hights from X-ray diffraction measurements for ICB TiO_2 films deposited on quartz glas substrates by variation of deposition conditions

Since we wanted to study texture influence, preferential orientation of the rutile modification has to be established already during film formation.

In order to study the influence of deposition parameters on TiO_2 film texture, ICB depositions were made by variation of the following basic set of parameters: V_{acc} = 6 kV; I_{ion} = 300 mA; $T_{substrate}$ = 300°C; jO = 50 cm^3/min; R = 6 nm/min. Table 1

shows the relative peak hights of different crystallite orientations from X-ray diffractometer measurements for films deposited on silica substrates. The parameter deviating from the basic set used for film no. 1 and its value are also indicated. It is seen, that pure rutile films with [110] texture are only obtained at low oxygen flow. The oxygen pressures in the vicinity of the substrate are not exactly known because they are measured near to chamber wall. With this approximation 10 ml/min = $2 \cdot 10^{-5}$ mbar and 50 ml/min = $8 \cdot 10^{-5}$ mbar. Since the mean free path of Ti ions in O_2 at 10^{-4} mbar is about 50 cm and the distance from source to substrate is 30 cm there is a high probability for the accelerated Ti ions to be scattered before reaching the substrate. At lower O_2 pressure more and more the full energy is transferred to the growing film surface. The degree of ionization of Ti atoms for V_{acc} = 6 kV and I_{ion} = 300 mA was measured at substrate location in vacuum to be about 5%. Therefore the mean energy per deposited atom at low oxygen pressure is about 100 eV. At higher oxygen pressures part of the Ti ions don't meet the substrates because of scattering. This reduces the mean energy and thus the rutile content as seen from Table 1.
The results for TiO_2 films deposited by ICB on sapphire substrates came out to be very different. No anatase was seen, even at V_{acc}, I_{ion} = 0. In case of high oxygen partial pressure during film formation, the most intense diffractometer peak showed rutile [301] orientation. The rutile [110] orientation as on the amorphous silica substrates did not occur. This is certainly due to the influence of the sapphire crystal lattice. Again at lower oxygen partial pressure textured films were obtained.

Sputtered films were prepared by variation of oxygen partial pressure, deposition rate and bias voltage. Fig. 2 shows diffractometer traces of 0.3 µm thick films prepared on quartz and sapphire. We were not able to get pure rutile films with our sputtering system. Even an increase of substrate temperature to 400°C gave only 60% rutile and 40% anatase. It may be possible by modification of the system to get a more moderate slope in sputtering rate around the stoichiometry point instead of a sharp drop. By stabilization of sputtering operation within this slope Schiller et al. (3) reported that pure rutile films are obtained at bias voltages and oxygen partial pressures similar to those used by us. It is further seen from fig. 2 that crystallite orientation on sapphire substrates was similar to that on quartz. This was the most surprising difference compared to ICB. It shows that the film formation process may be governed by other parameters. Especially the argon partial pressure may influence adatom mobility. X-ray fluorescence measurements showed an argon content in the films of about 1%.

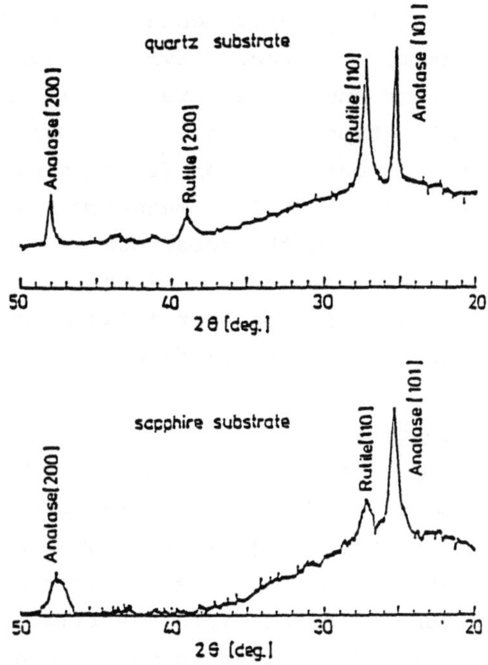

Fig. 2 X-ray diffraction patterns of sputtered TiO_2 films.

As a summary of these results we can say that ICB is more suitable than sputtering to modify morphology of TiO_2 films in a well defined manner.

Conductivity measurements of the films were carried out with a simple two electrode arrangement. The Pt electrode films were deposited by sputtering through a mask. Pt wires were connected to this film by gap welding. The sample was mounted in a small oven and heated to 800°C. The oven could be connected either to a gas mixing system for establishment of a continuous N_2/O_2-flow, or to a vacuum system. O_2 partial pressures could be varied from 1 bar to 10^{-3} bar in the mixing system and from 10^{-2} bar to 10^{-10} bar in the vacuum system.

Conductivity measurements showed, that the influence of film orientation on conductivity was only indirect. Most of the ICB-films and all sputtered films showed deviations from the normal n-type conductivity of pure TiO_2 at higher oxygen partial pressures. For films prepared on sapphire substrates it was proven by SIMS measurements that diffusion of the three valent Al from the substrate at the measurement temperature of 800°C into the lattice of the four valent Ti ist the reason for this deviation. For films on quartz no deviation occurs for pure rutile (110)-films.

4. Conclusion

TiO_2 films were prepared by reactive magnetron sputtering and by reactive ICB. Pure rutile structure with preferred crystallite orientation (texture) could only be obtained by ICB. Suitable film texture seams to minimize impurity diffusion from the substrate into the film. Impurities of lower chemical valence than Ti lead to p-type conduction in the n conducting pure TiO_2 and thus to a deviation of the simple relationship $\sigma = A\, p_{O_2}^{-n}$. A linear relationship of $\log \sigma$ vs. $\log p_{O_2}$ was achieved for ICB films of suitable texture on quartz and sapphire in the pressure range between 10^{-10} bar and 1 bar O_2.

Acknowgledgement

The authors would like to thank to Dr. K.D. Schierbaum of Tübingen University for making the SIMS measurements and to the german ministry of research and technology for financial support.

References

(1) K. Fukushima, I. Yamada, T. Tagaki, J. of Appl. Phys. 58, 4146 (1985)

(2) T. Takagi, K. Matsubara, H. Takaoka, J. App,. Phys. 51, 5419 (1980

(3) S. Schiller, G. Beister, S. Schneider, W. Sieber, Thin Sol. Films 72, 475 (1980)

Solid State Chemical Sensors in Thick Film Technique

W.F. Chu, H. Erdmann, M. Ilgenstein, and V. Leonhard

Battelle Institute e.V., Am Römerhof 35, D-6000 Frankfurt 90

Summary

Chemical sensors with solid ion-conductors have been prepared by thick film screen printing method. Solid ion conductors Nasicon ($Na_{1+x}Zr_2Si_xP_{3-x}O_{12}$) and YSZ (yttria stabilized zirconia) have been deposited onto α-alumina substrate. An all-thick film hydrogen sensor based on Nasicon electrolyte with sodium tungsten bronze (Na_xWO_3) reference electrode and porous platinum working electrode and an oxygen sensor based on zirconia electrolyte with Pd/PdO-reference electrode have been developed. Development of carbon dioxide sensor based on Nasicon or ß-alumina with sodium carbonate working electrode will be the objective of the current research work.

1. Introduction

The principles of chemical sensors with solid electrolytes are the chemical reactions taking place in the electrochemical cells and the resulting e.m.f. (electromotive force) of the cells. An electrochemical cell can be constructed that contains a solid electrolyte membrane separating two electrode compartments. There may be similar or dissimilar materials in the electrode compartments, e.g. oxygen gas at two different pressures. The e.m.f. of a cell reaction at each electrode is given by the Nernst equation:

$$E = E_o + \frac{RT}{nF} \ln \frac{[Ox]}{[Red]} \qquad (1)$$

where R is the gas constant, T the temperature, n the number of electrons exchanged, and F is the Faraday constant. [Ox] and [Red] are the concentrations of the oxidized and the reduced species, respectively. Combination of the Nernst equations for the reactions occurring at each electrode gives then the overall e.m.f. of the cell. The measured e.m.f. ist proportional to the concentration gradient of the species in the electrode compartments. If one electrode compartment contains species of known concentration, the concentration of the species in the other electrode compartment can be easily calculated.

Nasicon (sodium superionic conductor) has a Na^+ ion conductivity in the range of $10^{-4}\ \Omega^{-1}\ cm^{-1}$ at 25 °C. The chemical formula is $Na_{1+x}Zr_2Si_xP_{3-x}O_{12}$, which has a frame work built of corner-sharing ZrO_6 octahedra and $(P,Si)O_4$ tetrahedra and a three-dimensional network of tunnels in which Na^+ ions reside [1].

Zirconium dioxide (zirconia) is polymorph; it has three crystal structures. They are monocline, tetragonal and cubic. The cubic form which exhibits highest oxygen ion conductivity is stable at high temperatures (up 2372 °C to 2680 °C) [2]. Cubic zirconia can also be stabilized down to room temperature by doping with di- or trivalent oxides such as MgO, CaO, Y_2O_3 or Yb_2O_3. The increase of oxygen ion conductivity of doped zirconia is related to the increased concentration of oxygen ion vacancies in lattice. The electrical (ionic) conductivity of CaO-stabilized zirconia lies in the range of $5 \cdot 10^{-2}\ \Omega^{-1}\ cm^{-1}$ at 1000 °C.

2 Preparation of sensor materials

There are two ways of preparing Nasicon. One is the powder metallurgical method by mixing and sintering of the starting materials of Na_2CO_3, ZrO_2, SiO_2 and $NH_4H_2PO_4$. The another one is the chemical sol-gel method by hydrolysis and subsequent condensation of an alcoholic solution of $Zr(OC_3H_7)_4$ and $Si(OC_2H_5)_4$ with an aqueous solution of NaOH and $NH_4H_2PO_4$. By the chemical sol-gel method, it is possible to vary the stoichiometric sodium content x. With x = 2.0 to 2.2, the substance has the highest sodium ion conductivity. Moreover, sintering-active and homogeneous materials could be obtained by the chemical sol-gel method [3,4].

Stabilized zirconia can also be prepared by the chemical sol-gel method. Zr-alcoholate $Zr(OC_3H_7)_4$ and $Y(NO_3)_3 \cdot 3H_2O$ are dissolved in ethanol. The alcoholic solution is then hydrolyzed at 80 °C with an alcoholic aqueous solution of NH_4OH. The condensed substance could be calcined at 850 °C - 1100 °C and then sintered at about 1500 °C. Materials prepared by the powder metallurgical method using ZrO_2 and Y_2O_3 must be sintered at higher temperatures > 1700 °C.

3 Thick film technique

Screen printing technique has already been applied successfully in the electronic industries for integrated circuits. Pastes of conducting leads like Ag, Pt, or Pd are commercially available. Solid electrolytes and reference electrodes, however, are mostly oxide ceramics. Preparation of suitable pastes of solid electrolytes and metal or metal oxide electrode materials would be the key step for the development of thick film chemical sensors.

The electrical conductivity of solid electrolytes depends strongly upon the density of the material. High density can usually be achieved by sintering pressed pellets made of fine, non-agglomerated powders. For thick films printed on α-alumina substrate, however, highly sintering-active, submicron powders are not suitable duo to the strong shrinkage of the film after sintering. The sintering activity of fine powders need to be decreased by annealing at suitably high temperatures prior to the preparation of printing pastes. There exists an optimum for the density , the electrical conductivity of the material and adhesion of the film.

Screen printing pastes are prepared from powder materials. Parameters such as grain size, grain form, grain distribution, surface properties and packing density of the powder have to be optimized. In addition, materials for the paste preparation such as adhesive substance, solvent, binder, vehicle substance and surface active component are added to the starting powder materials. A suitable printing paste exhibits thixotropic properties. The viscosity of the paste decreases upon pressure application during the printing process. Toward the end of the printing process, the viscosity of the paste increases as the printing pressure is reducing. Finally, the screen printed thick films are dried and sintered at high temperatures. The sintering temperature, time and atmosphere are also important for the good quality of the thick films.

Chemical sensors consist of layers of electrolyte, electrodes, conducting leads, protecting cover and heating element on the reverse side. Attention has to be payed to the compatibility of these single layers with each other. Materials requiring highest sintering temperature should be deposited and sintered at first. Finally, layers are produced at lower sintering temperatures and, in some cases, in innert atmosphere.

4 Sensor structures

4.1 Hydrogen sensor

Hydrogen sensor based on Nasicon electrolyte with a reference electrode of sodium tungsten bronze has been prepared. Sensor chips mounted in a standard 24 pole DIL-housing have the dimension of $1 \times 2 \times 0{,}2$ cm^3 with α-alumina as substrate. On the back side of the substrate, heating element consisting of ruthenium oxide thick film was also deposited by the screen printing technique. The working temperature of the sensor was 200 °C. Layers of Nasicon, gold lead, platinum measuring electrode, silver lead for heating element, ruthenium oxide heating element, tungsten bronze reference electrode and finally, protecting glass covering layer were successively screen-printed onto the substrate and the printed substrate was sintered in air or in nitrogen gas after each printing step.

This hydrogen sensor could be installed in all areas where hydrogen gas concentration might be in the explosion range of about 6 to 67 % by volume in air. H_2 gas concentration of about 100 ppm to 100 % in air could be detected. The sensor signal has a value of about 300 mV upon changing of the measuring gas of air to a mixture of air with 4 % hydrogen gas by volume. A response time for the 90 % signal change was recorded to be 1 sec. Fig. 1 shows the principle construction of the sensor. Fig. 2 shows a typical measuring curve of time-independent potential change at different hydrogen gas concentrations.

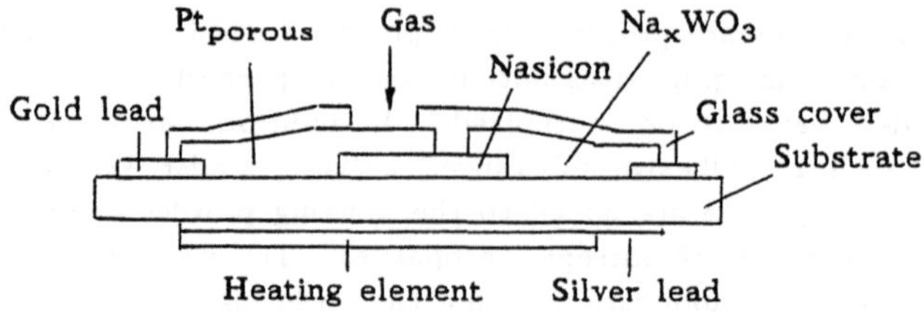

Fig. 1 principle construction of the sensor

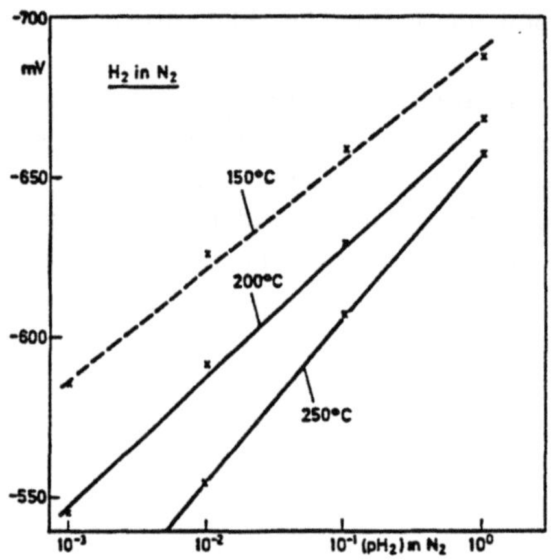

Fig. 2 Potential change at different hydrogen concentrations

4.2 Oxygen sensor

An oxygen sensor based on zirconia electrolyte with platinum measuring electrode, Pd/PdO-reference electrode and Pt-heating element on the reverse side of the α-alumina substrate has been prepared. The sensor chip has a dimension of 7 x 10 mm^2. Fig. 3 shows the principle construction of the potentiometric sensor.

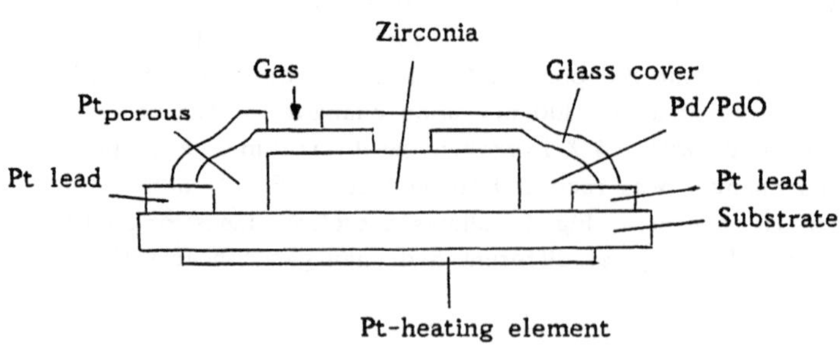

Fig. 3 Principle construction of the potentiometric sensor

In gas mixtures with oxygen concentrations > 10%, deviation from the Nernst equation (1) was observed. The reason for this deviation was the insufficient gas tightness of the electrolyte layer and of the glass cover. Oxygen gas could diffuse through the pores in the electrolyte layer and in the glass covering layer into the reference electrode compartment. This caused the deviation of oxygen partial pressure in the reference electrode compartment from the values corresponding to the Pd/PdO equilibrium.

Sensors of the limiting current mode are more advantageous. A reference electrode of constant oxygen partial pressure is not necessary. The measured current which flows through the cell at a constantly applied voltage depends, under the limiting current conditions, linearly upon the oxygen concentration, C_{O_2}, in the measuring gas [5].

$$I_{lim} = 4\ FD\ (\frac{Q}{L})\ C_{O_2} \qquad (2)$$

I_{lim} is the limiting current, F the Faraday constant, D the diffusion coefficient, Q the effective cross section and L the effective length of lthe diffusion region. The limiting current conditions could be obtained by covering the cathode with a ceramic layer with defined pore structure [6]. In this research work, the screen printed electrolyte layer which covered the Pt-cathode served itself as diffusion barrier for oxygen gas. Fig. 4 shows the schematic construction of the sensor. Experimental results shows, however, that the pores in the electrolyte layer are not sufficiently fine enough to restrain diffusion. The currents were no longer independent of the applied voltage at oxygen concentrations greater than 0,1 %.

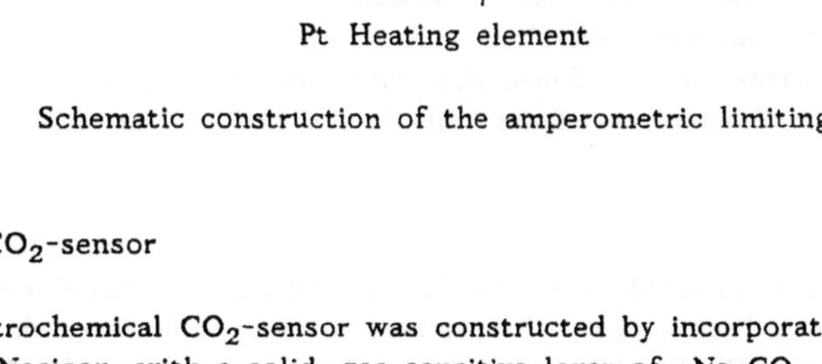

Fig. 4 Schematic construction of the amperometric limiting current sensor

4.3 CO_2-sensor

An electrochemical CO_2-sensor was constructed by incorporating a sodium conductor, Nasicon, with a solid, gas sensitive layer of Na_2CO_3 as measuring electrode [7,8]. The electrochemical cell could be described as follows:

$$Pt,\ CO_2,\ O_2\ |\ Na_2CO_3\ |\ Nasicon\ |\ O_2,\ CO_2,\ Pt \qquad (3)$$

The reactions taking place at the electrodes are:

Anode: $$Na_2CO_3 \longrightarrow 2\ Na^+ + \frac{1}{2}\ O_2 + CO_2 + 2\ e^- \qquad (4)$$

Cathode: $$2\ Na^+ + \frac{1}{2} O_2 + 2\ e^- \longrightarrow Na_2O \qquad (5)$$

Overall: $$Na_2CO_3 \longrightarrow Na_2O + CO_2 \qquad (6)$$

The resulting e.m.f. of the cell could be expressed as follows:

$$E = -(G^o_{Na_2O} + G^o_{CO_2} - G^o_{Na_2CO_3}) - \frac{RT}{2F} \ln (a_{Na_2O} \cdot P_{CO_2}) \qquad (7)$$

where G^o are free enthalpies of the species and a_{Na_2O} the activity of Na_2O in Nasicon.

Preliminary results confirmed, in accordance with works published by Maruyama et. al. [8], that the e.m.f. of the cell depends logarithmically on the CO_2 partial pressure in a CO_2/O_2 gas mixture.

References

1. H. Kohler, Dissertation, Karlsruhe 1985
2. A.R. West, Solid State Chemistry and its Applications, John Wiley & Sons Ltd., 1984
3. U. Warhus, Dissertation, Stuttgart 1986
4. K. Zucholl and H. Erdmann, Chemische Sensoren mit Festkörperionenleitern in Dickschichttechnik, GdCh Tagung Elektrochemie und Elektronik, 12-14. Okt. 1988, Basel, Switzerland
5. H. Dietz, Gas-Diffusion-Controlled Solid-Electrolyte Oxygen Sensors, Solid State Ionics 6 (1982) 175-183
6. W.F. Chu, F-J. Rohr, and R. Böhm, Elektrochemische Meßvorrichtung, DE P 3131103.2, GBM 8123075.3, EP 82107018.2 (06.08.1981)
7. G. Hötzel, Dissertation, Sutttgart 1986
8. T. Maruyama, S. Sasaki, and Y. Saito, Potentiometric Gas Sensor for Carbon Dioxide Using Solid Electrolytes, Solid State Ionics 23 (1987) 107-112

Acknowledgement

The present work has been conducted with financial support of the German Ministry of Research and Technology. The partners of this joint research project are: Max-Planck-Institut für Festkörperforschung, Stuttgart, Institut für Physikalische und Theoretische Chemie, Universität Tübingen, and Mineralogisch-Petrographisches Institut, Universität Tübingen. The excellent corporation of the partner institutes is gratefully acknowledged.

Ionic Sensor Layers on Microelectronic Structures for the Detection of Solvent Vapours

F. L. Dickert*, G. Bertlein and K. Reif
Institute of Physical and Theoretical Chemistry, Erlangen University, Egerlandstr. 3, D-8520 Erlangen, FRG

G. Mages and H. Kimmel
Siemens AG, Central Research and Development, ZFE ME AMF 3, Paul-Gossenstraße 100, D-8520 Erlangen, FRG

Summary
Several types of microelectronic devices such as interdigital structures, SAW-resonators, and ChemFETs were used for the development of chemical sensor systems for the detection of organic solvents and ammonia in the air and waste water. The chemically sensitive coatings are based on reversible equilibria in which hydrophobic ions are involved. An incorporation of solvent vapours leads to a change of the ion concentrations. A large variety of pollutants such as alcohols, ketones, esters, and even aromatic and halogenated hydrocarbons can be analyzed in this way.

Microelectronic Structures

Both passive and active electronic elements were used for the construction of sensor systems for the detection of organic solvent vapours. Gold or platinum electrodes as interdigital structures on inert materials such as quartz enable the design of resistive and capacitive sensor elements. The electrodes were manufactured with a structure width of 5µm varying from 100nm to 500nm in height depending on the thickness of the chemically sensitive coating. These structures with an approximate area of $0.25cm^2$ can easily be combined in arrays and therefore pattern recognition can be performed to gain an electronic selectivity of the sensors [1].
Another type of passive element used were surface-acoustic-wave resonators. Electrode structures of down to 2.1µm in width on piezoelectrical quartz were convenient for establishing a resonant circuit with a broad band amplifier in the feedback loop. The result of vapour absorption was directly monitored by the resonance frequency of the device which was up to 430 MHz and therefore a high sensitivity was obtained.
A n-channel depletion type MOSFET with a 30µm wide gate area was

used. After coating the gate isolator (SiO_2/Si_3N_4) with the sensitive layers, the electrical contact was realized by a perforated gold electrode. The drain source current was kept constant at 50μA and the resulting gate voltage was monitored as a function of solvent vapour concentration.

Principles of the Sensor Layers

Reversible chemical reactions which lead to ion formation were chosen to detect solvent vapours in the air. The first step is the solvent absorption which can be optimized by adapting the polarity of the layers to that of the solvent. For this reason ions with hydrophobic properties are applied and the solvent is enriched in the layers. This first step is followed by a chemical reaction. The ion concentration in the layers is changed by solvation effects stemming from the incorporated solvents. This concept can be realized by two distinct types of ions, namely carbenium ions and macrocyclic complexes.

The principle of these ideas is demonstrated in Fig.1 for carbenium ions. By interacting with an acid ingredient, thin layers of the lactones (I) yield highly coloured carbenium ions due to the cleavage of the lactone ring [2]. The absorption of organic solvent vapours leads to a diminution of the layer's acidity and therefore the back reaction occurs. This equilibrium can be followed by optical absorbance, via changes in the capacity or resistance, and preferably with SAW devices.

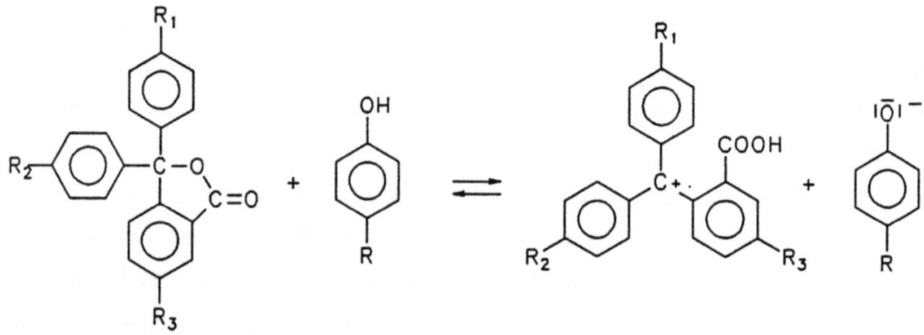

(I) (II)

Fig.1. Formation of carbenium ions from a lactone by the interaction with a phenole

Metal ions with hydrophobic properties can be obained by the complexation with macrocyclic ligands [3]. In our case the metal ions are not completely surrounded by a crown ether allowing an additional anion to coordinate at the free site. The incorporation of solvents with coordinating properties leads to a substitution process and ions are formed in this way (Fig.2). The described effect can easily be followed by a.c. conductivity monitoring.

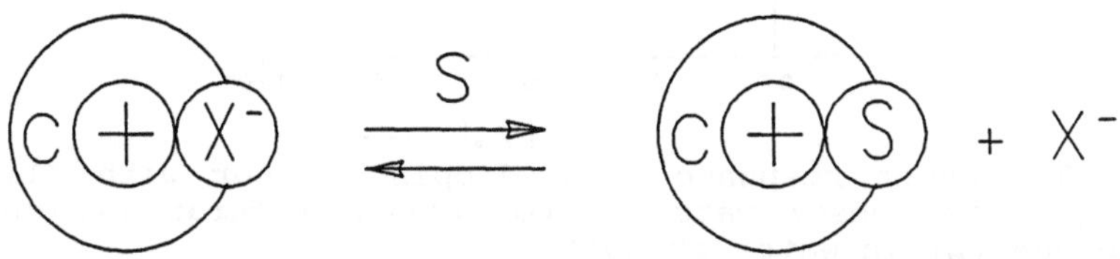

Fig.2. Dissociation reaction of a macrocyclic complex by interaction with a donor solvent molecule S, (C: crown ether)

Carbenium Ions

The reaction in Fig. 1 can be monitored more simply by dielectric measurements than by optical absorbance. The carbenium ions are formed by an excess of a phenolic ingredient in the sensor layers. The phenol forms a network of hydrogen bonds and large ions such as the triphenyl carbenium cations show structure breaking effects. An incorporation of organic solvents leads to a decrease of the ion concentration in the layers which is accompanied by the formation of hydrogen bonds and an increase in the dielectric constant. Vapour detection can be executed by means of interdigital structures due to this effect. In comparison to optical detection, pattern recognition [1] is easily performed in this way. For this purpose, an array of four interdigital structures was assembled and different types of triphenyl methane dyes were used. The sensor array and the water phase were separated by a membrane capable of being penetrated by the organic solvents [4] and in this way a dipping sensor was created. Excellent linear relationships between the capacity of the sensor element and the solvent content are observed (Fig.3). Furthermore, the effects of two solvents are additive and a linear model for data evaluation can be used.

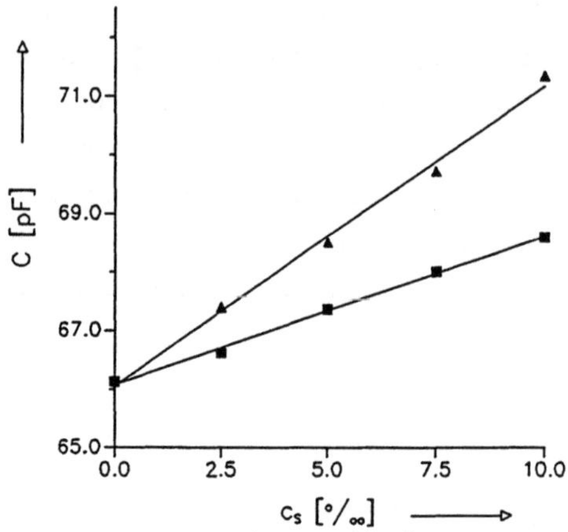

Fig.3. Sensor response of a dipping sensor with (II) (Fig.1) as sensor material to tetrahydrofuran (▲) and acetone (■) in water at 20°C

A most pronounced sensitivity can be achieved with SAW-devices. The sensor response to tetrahydrofuran is demonstrated in Fig.4.

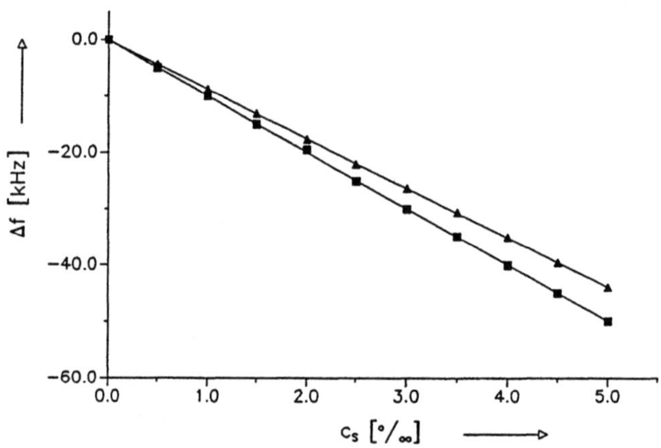

Fig.4. Frequency shift of a 300 MHz SAW-resonator coated with the carbenium ion (II) as sensor material (Fig.1) due to tetrahydrofuran in the air at variable humidities (▲: 30%, ■: 50%, 25°C)

The sensor layer thickness is only 40 nm and therefore very fast response times of only a few seconds are observed. Some ppm of organic solvents in the air are easily detectable. The results in Fig. 4 show that there is a negligible water cross sensitivity at very low concentrations of an organic pollutant in the air. An increase is observed, however, for a higher tetrahydrofuran concentration. This effect is due to the fact that tetrahydrofuran is miscible with water and, in parallel with the enrichment of this component in the sensor layer, the water absorption is favoured.

Macrocyclic Complexes

The reaction shown in Fig. 2 can be carried out by the crown ether [15]crown-5 which is a tetra or penta dentate ligand with oxygen donor atoms. In the case of octahedral metal complexes, anions additionally coordinate at the metal ions which can be substituted by the analyte according to Fig. 2. Mechanically stable sensor layers are preferentially obtained by polymeric crown ether complexes either with formaldehyde resins or by linking the complex to the backbone of a polymer.

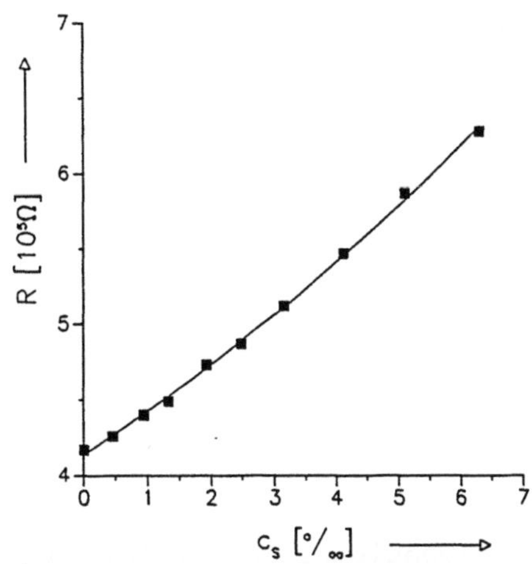

Fig.5. Sensor response of [15]crown-5 / KCl to trichloroethene

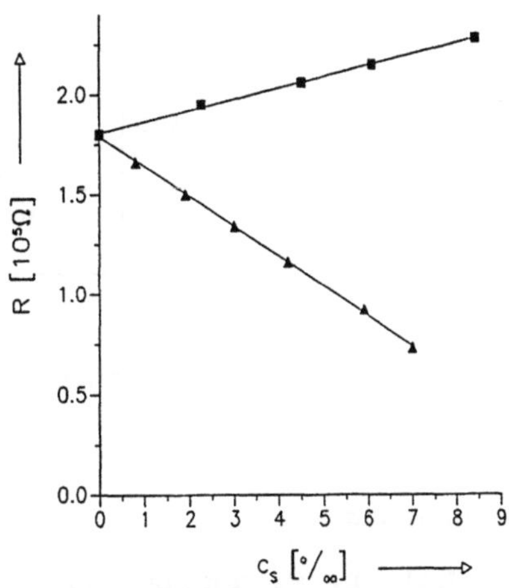

Fig.6. Sensor responses of the [15]crown-5/KCl complex with an excess of the ligand, ■: CH_2Cl_2, ▲: C_2H_5OH

The reactivity of the metal crown ether complexes can be modified by varying the ring size of the ligand, the ion radius, and the cation nature. Even inert solvent vapours are accessible in this way since there exists a competitive absorption process of solvents in the sensor layers. Incorporation of hydrophobic solvent vapours leads to a reduction of the water content in the layers (Fig. 5). The sensitivity of the sensor layers can also be modified by the addition of coordinating ingredients. Thus inert and polar solvent vapours can be differentiated by a positive and a negative slope, respectively (Fig.6). Again the use of sensor arrays allows the application of the methods of pattern recognition [1].

These crown ether complexes are also suitable as gate materials for a field effect transistor, since the shift of the ion formation equilibrium leads via a field strength effect to a change of the charge carrier concentration between drain and source (Fig.7).

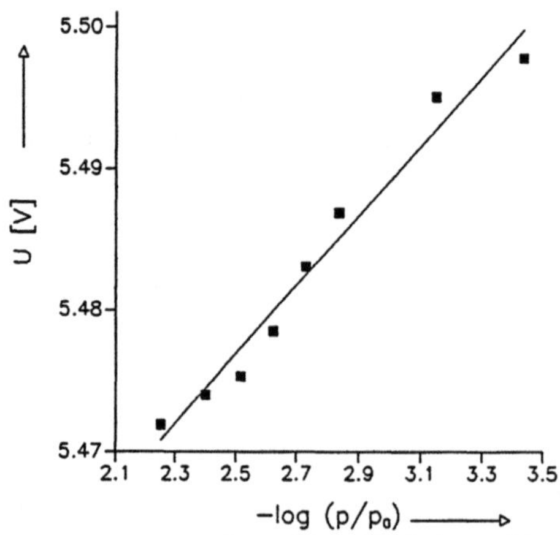

Fig.7. Sensor response of a ChemFET to ethanol

References

1. Müller, R. and Lange, E.: Multidimensional Sensor for Gas Analysis. Sensors and Actuators **9** (1986) 39-48.
2. Dickert, F. L.; Lehmann, E.; Schreiner, S.; Mages, G. and Kimmel, H.: Substituted 3,3-Diphenylphthalides as Optochemical Sensors for the Vapours of Polar Solvents. Anal. Chem. **60** (1988) 1377-1380.
3. Dickert, F. L. and Zeltner, D.: Polymer Benzo[15]crown-5 Complexes as Sensor Materials for Solvent Vapours - Aromatic, Halogenated Hydrocarbons and Polar Solvents. Angew. Chem. Adv. Mater. **101** (1989) 833-844.
4. Dickert, F. L.; Schreiner, S.; Kimmel, H. and Mages, G.: A Fiber Optic Dipping Sensor for Organic Solvents in Waste Water. Anal. Chem. **61** (1989) 2306-2309.

An Integrated ISFET-Sensorarray with a CMOS Signal-Processing Circuits

E. Müller, S. Koch, P. Woias

Chair for Integrated Circuits, Technical University,
Arcisstr. 21, 8000 München 2

Abstract

An integrated multi-sensor chip, based on a sensorarray of four ISFETs, is presented. An interface circuit adjusts a proper operating point for the ISFET and provides low output impedance. The possibility to multiplex the output signals of the four sensors reduces the number of terminal wires.
It could be shown that by means of a combination of an ISFET sensor and an on-chip signal-processing circuit a new device has been created that provides significant advantages over single-chip ISFETs.

Introduction

One of the most studied solid-state chemical transducers is the ion-sensitive field effect transistor (ISFET).
In addition to other advantages over conventional electrodes, such as low output-impedance, small dimension, low cost due to mass fabrication, the ISFET has one major benefit: It allows to integrate sensor and electronic signal processing circuit on the same chip, by using a fabrication method compatible to the requirements of a CMOS and ISFET process.
The sensorarray presented in this paper is based on ISFETs which have been developed in out group [1,2].
This sensorarray consists of four ISFETs with a silicon nitride gate sensitive for H^+-ions, an analogue signal-processing circuit and an adjacent multiplexer.

Circuit design

Each of the ISFETs is part of the input-stage of an operational amplifier used in a unity gain feedback configuration.
Apart from the well-kown advantages of those amplifier circuits, such as low output-impedance, simple thermal compensation and a linear correlation between input- and output-variable [3], this concept offers a selection of further advantages for the design of the ISFET devices.

Since the ISFETs does not have to drive off-chip loads, it is not necessary to use a large W/L ratio for their channel region. Thus a square shape was applied for the sensitive area of the FETs. This square shape offers advantages for the later application of a biologically active layer, and for the packaging of the sensorchip.
In former work [1] it had been shown that the ISFETs should be operated in their triode region. Therefore a special circuit for the input-stage of the amplifier was developed to apply a constant drain-source voltage to an ISFET and to a compensating MISFET. For that purpose a series connection of two transistors with common gates was used. Figure 1 shows the schematic of this stage.

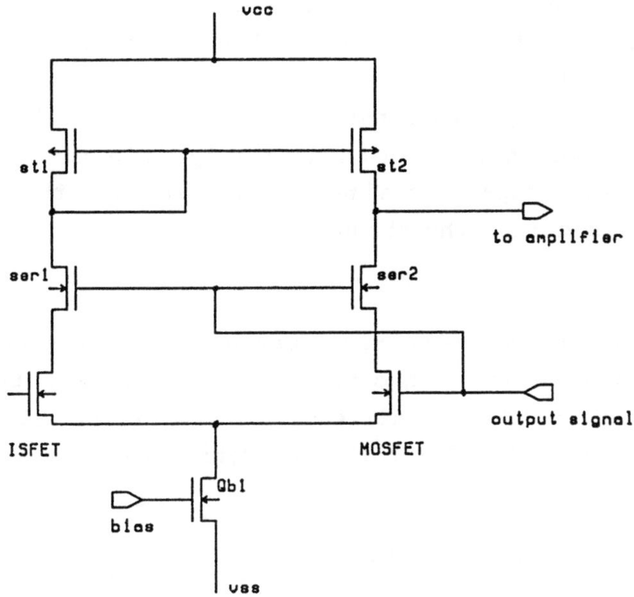

Figure 1 : amplifier input stage

To improve the durability of the passivation layer, which is mainly responsible for the limited lifetime of the sensor, diffusion regions connect source and drain of the ISFET to the interface electronic. Thus the topographical uneveness of the ISFET's surface could be reduced remarkably. Because of the applied concept of differential amplifying the higher lead resistance does not affect the transfer characteristic of the circuit.

In order to construct the sensorarray with few external connections, a four channel multiplexer was integrated on the chip. Therefore the circuit can be operated with only five terminal wires. Those are necessary for positive and negative power supply voltage, providing control signals for the multiplexer and one line to transmit the output signal.

For the design of the sensorarray great emphasis has been laid to the fact that the circuit should contain as few transistors as possible.
Thus an integrated sensorarray, produced by the Siemens AG in Munich, was built, which is hardly inferior to an individual ISFET with regard to size of the chip and the number of waste circuits during the production process.

Results

To find out the electrical parameters of the chip easily, one can operate the sensorarray with a MISFET, being the input transistor of the differential amplifier stage instead of using an ISFET. The following data sheet was measured by that way.

Offset voltage:	< 3 mV
Closed loop gain:	0.98
Open loop gain:	97
3 dB frequency:	75 KHz
Quiescent current:	1.09 mA
Temperature Drift:	0.17 mV/^{0}C
ON-resistance of analogue switch:	200 Ohms

To demonstrate proper operation of the whole chip, the voltage to voltage response characteristic of a single ISFET is shown in figure 2. The linear rage of that transfer curve could be improved by increasing the power supply voltage. The voltage limit is determined by the breakdown voltage of the passivation layer.

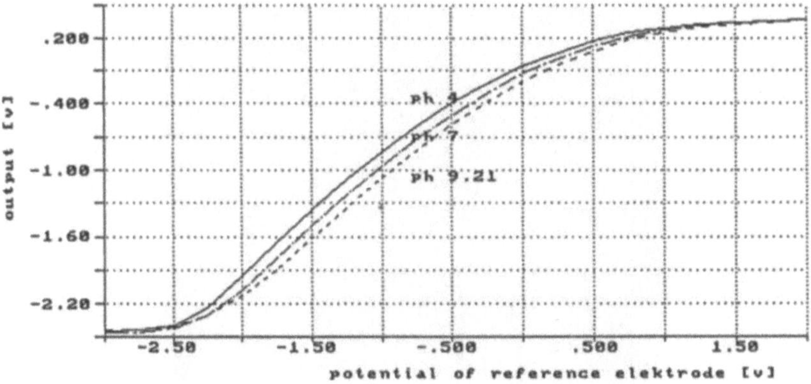

Figure 2: Output signal of one ISFET versus potential of the reference electrode.

Is was found that the pH-sensitivity of the circuit varies from chip to chip in the range of 50 up to 54 mV/pH. Figure 3 shows response of the four channels of the ISFET-array. One can recognize those four ISFETs in the upper half of figure 4. By using a titan-platinum-gold metall layer the solution contacts around the ISFETs work as a pseudo-reference elektrode.

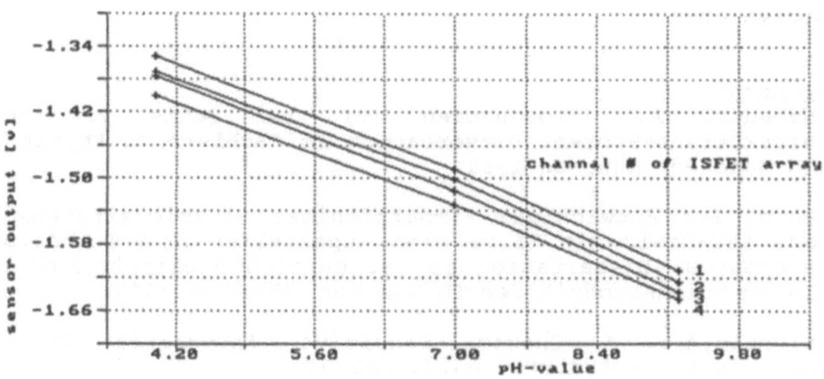

Figure 3: Response of sensorarray versus pH

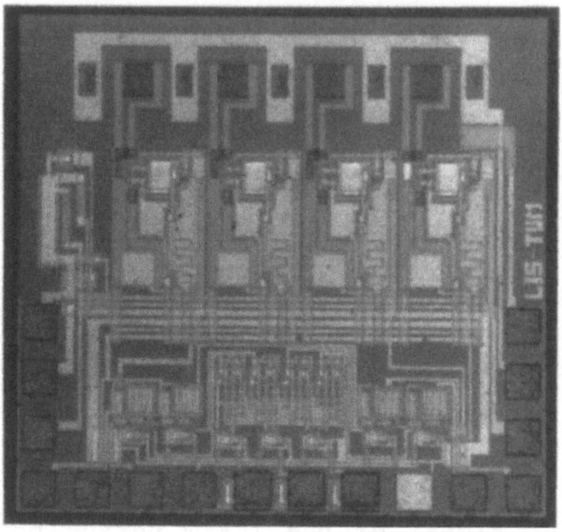

Figure 4: Photo of the integrated sensor-array

Conclusion

It was shown that the concept of developing a solid state chemical sensor together with its electronic signal processing on one chip offers a number of advantages for the design of the transducer itself.

The aim of further investigations on the sensorarray is the measurement of biological reactions, by depositing a biologically active layer (e.g. enzyme membrane) on the FET's sensitive area.

References:

1. Näbauer, A.: Biosensoren auf der Basis von Feldeffekttransistoren. Dissertation im Fachbereich Physik der TU München, München 1988.

2. Berg, P.: Entwicklung einsatzfähiger Flüssigkeitssensoren auf FET-Basis mit Hilfe eines speziellen CMOS-Technologie Prozesses. Dissertation im Fachbereich Elektrotechnik und Informationstechnik der TU München, München 1989.

3. Sibbald,A.: A Chemical-Sensitive Integrated Circuit. Sensors and Actuators, Bd.7, S23-38.

CVD-Al_2O_3-Layers: Applicability in Sensor Devices

J. RÖBER, E. ERBEN, H. CEBULLA

Technische Universität Karl-Marx-Stadt
Sektion Physik / Elektronische Bauelemente

Summary

A convenient method for measuring the ionic concentration in aqueous solutions is the use of ion sensitive field effect transistors (ISFET). The performance of these sensors and especially sensitivity and long- time- stability are mainly influenced by the membrane layer situated above the transistor gate. The aim of our investigations was to test the aluminum oxide layers deposited by CVD technique with regard to their applicability as membrane layers. This paper contains a brief description of the deposition technique and statements about the performance of so manufactured sensors.

Deposition Process

The Al_2O_3 layers were deposited by means of low pressure CVD (LPCVD) using gas mixtures of H_2, CO_2 and $AlCl_3$ /1/.
The reaction can be described as

$$2\,AlCl_3 + 3\,H_2 + 3\,CO_2 \rightleftarrows Al_2O_3 + 3\,CO + 6\,HCl \qquad (1)$$

The essential deposition reaction is the hydrolysis of $AlCl_3$:

$$2\,AlCl_3 + 3\,H_2O \rightleftarrows Al_2O_3 + 6\,HCl \qquad (2)$$

The water needed for the hydrolysis is formed in the reaction zone

$$H_2 + CO_2 \rightleftarrows H_2O + CO \qquad (3)$$

A schematic drawing of the experimental apparatus is shown in Fig. 1.

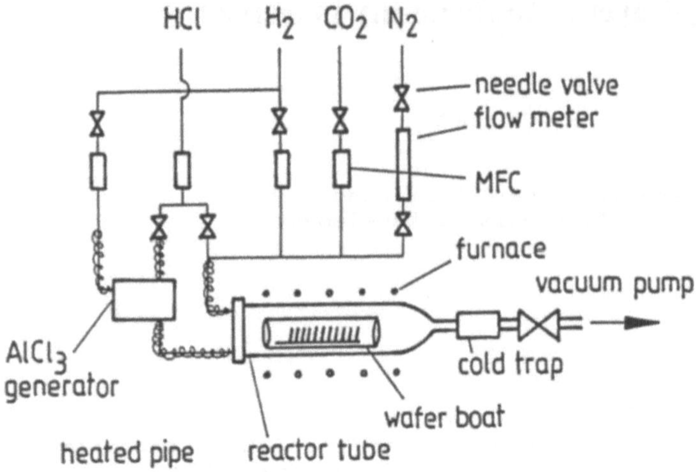

Fig. 1. Schematic drawing of the CVD apparatus

The deposition reactor is a resistive heated horizontal quartz glass tube with an inner diameter of 120 mm. The 3" Si- wafers are standing upright in a quartz wafer boat.
The reaction temperature was in the range of 1150 - 1250 K. The system pressure was approximately 300 Pa controlled by a valve in the exhaust line. The $AlCl_3$ vapour needed for the reaction was in-situ produced by reaction of HCl gas with aluminum grain at 490 K in an $AlCl_3$ generator. The flow rate of $AlCl_3$ vapour is regulable in the range of 100 - 500 mg/min while that of the inside the $AlCl_3$ generator admixed H_2 is always 200 cm^3/min. CO_2 was fed into the reactor with a flow rate of 100 - 200 cm^3/min.
The Al_2O_3 films deposited under these conditions show a polycristalline structure. The deposition rate was approximately 1 nm/min. The LPCVD process allows the deposition of films thinner than 30 nm. Such thin films deposited by LPCVD are smoother than films deposited by APCVD (air pressure CVD).

Sensors on the Basis of CVD - Al_2O_3 - Layers

Al_2O_3 films deposited by CVD were tested for various sensor applications. These films are of large interest especially for use in chemical sensor devices (ISFET, /2/).
In the usual arrangement of ISFET's the Al_2O_3 layer with a thickness of 70 nm is situated above the original SiO_2 gate material in direct contact to a solution. This means high requirements to this material with respect to its chemical stability. In comparison with other insulating materials like SiO_2 the Al_2O_3 realizes these requirements very good.
The investigation of the long-time-stability of our sensors shows, that the parameter drift of Al_2O_3 membrane layers after 800 days in measuring solution is in the range of 1 % (Fig. 2).

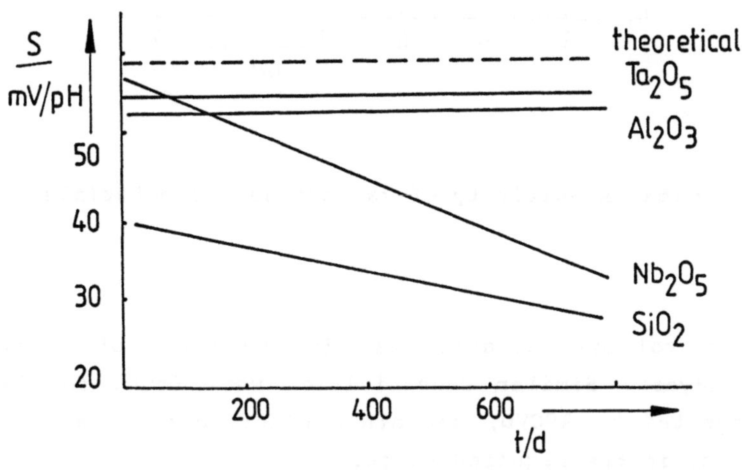

Fig. 2. Long-time-stability of various sensor materials in the measuring solution

The Al_2O_3 layers didn't resolve in the measuring solutions and it couldn't be proved diffusion of ions (especially Na^+) from the solution into the Al_2O_3 layer.

The experiment for testing the pH- sensitivity of the Al_2O_3 layers shows a behaviour near the Nernstian (Fig. 3).

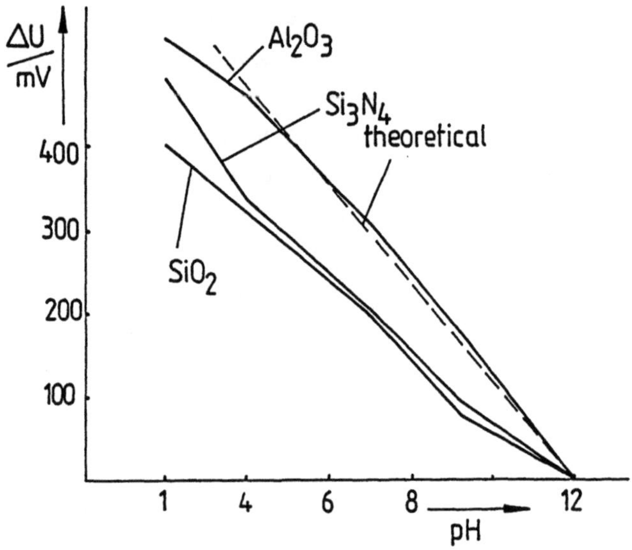

Fig. 3. Chemical sensitivity of various sensor materials

The measured voltage variation was in the range of 50 mV/pH for Al_2O_3 layers. Similar sensitivities were found for Ta_2O_5 layers deposited by APCVD, but other electrical properties of Ta_2O_5 are not in the required range.

In summary we can say that Al_2O_3 layers deposited by LPCVD satisfy the requirements for membrane layers in ISFET devices very good.

References

1. Röber, J.; Cebulla, H.; Feicke, A.; Erben, E.: Herstellungsprozeß dünner Al_2O_3- Schichten mit Hilfe des Niederdruck-CVD- Verfahrens.
   Wiss. Zeitschr. TU Karl-Marx-Stadt 31 (1989) 625

2. Leimbrock, W.: Dissertation B, TU Karl-Marx-Stadt 1990

Sensors for Biological and Medical Application

Biochemical Sensors for Microsystems

F.W. SCHELLER, R. HINTSCHE, U. WOLLENBERGER
Central Institute of Molecular Biology, Academy of Sciences of the GDR, Berlin

Summary

Biosensors are based on the combination of a chemical sensor with a biocomponent - typically enzymes or immunoglobulins - which interact specifically with the analyte. In this way a large spectrum of biological active substances gets accessible to quantitative measurements. Nowedays technologies are available to miniaturize biosensors thus affering the step wise integration into modular microsystems.

Living systems communicate with their surroundings by sensitive (chemo)receptors. In these chemoreceptor systems molecular recognition and signal transduction, including amplification, are performed by biochemical reactions. The high selectivity and adaptable sensitivity of the biological receptors have attracted the interest of analytical chemists. A breakthrough in applying immobilized enzymes in analytical chemistry was made by L.C. CLARK and C. LYONS in 1962 /1/. Starting from their membrane covered oxygen electrode, they entrapped an enzyme solution by a semipermeable membrane in front of the indicator electrode. This step resulted in the reusability of the enzyme and represents the birth of the first biosensor.

Types of Biosensor

Basically there are two types of biosensor differing in the mode of signal generation. Direct bioaffinity sensors utilize a binding event to detect substances. The binding of the incoming analyte to the complementary immobilized bioligand results in a change in the conformation of the biomolecule and/or physical changes in the immobilization medium such as changes in charge, thickness, temperature or optical parameters (colour or fluorescence). The second basic sensor type is the enzymatic/metabolic biosensor. Here recognition of the substrate by the immobilized receiver (enzymes on different levels of integration, i.e., purified enzymes, organelles, micro-organisms or tissue slices) is followed immediately by chemical conversion to the corresponding product, which is detected.

Sensor Integration

The development of biosensors may be divided into three generations charactarized by the interrelation between the functional steps - molecular analyte recognition, physicochemical signal generation and signal processing /2/. The first generation uses single or sequentially coupled enzymes for analyte conversion to generate a transducer accessible substance.

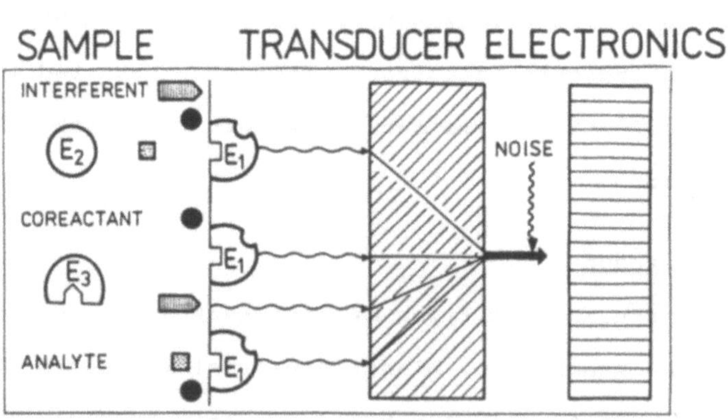

First generation

integration of auxiliary enzymes or coreactants, e.g., cofactors or mediators, in the enzyme layer, or their fixation at the enzyme protein, improves the analytical parameters and simplifies the handling, e.g., a reagentless measuring regime (second generation). This trend of increasing sensor integration includes the electronic signal processing in the sensor body, thus resulting in miniaturization and multifunctional sensing (third generation).

First generation

Amperometric enzyme sensors for the determination of about 80 different substances have been described in scientific and patent literature /2/. Analyzers based on enzyme electrodes for the determination of glucose, galactose, uric acid, choline, ethanol, lysine, lactate, sucrose, lactose, and α-amylase activity have been commercialized. (Tab. 1). Linear calibration curves of these sensors usually extent over two to three decades, with a detection limit of 1 to 10 micromolar concentrations. The most evident advantage of enzyme electrodes is their negligible enzyme consumption; this is typically in the range of a few milliunits per sample.

Up to now, glucose oxidase based electrodes clearly dominate both in research and in practical application of biosensors.

The traditional glucose oxidase (GOD) based sensors rely on anodic peroxide detection. So do the commercial laboratory analyzers of Yellow Springs (USA), Fuji Electric (Japan), ZWG (GDR), Eppendorf (FRG) and the Lithuanian Academy of Sciences (USSR) (Tab. 1). Two obstacles - the interference by electro-

Tab.1: Enzyme-Electrode Based Analyzers

Company	Model	Analyte	Measuring range (mM)	Sample throughput (h^{-1})	Coefficient of variation (%)	Stability
Yellow Springs Instruments, USA	23A	Glucose	1 - 45	40	< 2	300 samples
	23L	Lactate	0 - 15	40		
	27	Ethanol	0 - 60	20	< 2	
		Lactose				
		Galactose	0 - 55	20	2	
		Sucrose				
Academy of Sciences GDR	Glukometer	Glucose	0,5-50	60-90	1,5	>1000 samples
		Uric acid	0,1- 1,2	40	2	10 d
Fuji Electric, Japan	GLUCO 20	Glucose	0 -27	80-90	1,7	> 500 samples
		α-Amylase		30	4 - 5	
	UA-300A	Uric acid		50-60	3	
Daiichi, Japan	AutoSTAT GA-1120	Glucose	1 -40	60-120	1	
Radelkis, Hungary	OP-GL-7110S	Glucose	1,7-20	40	5 -10	240 d
USSR	ExAn	Glucose	2,5-30	20	3	
La Roche, Switzerland	LA 640	Lactate	0,5-12	20-30	< 5	40 d
Omron Tateisi, Japan	HER-100	Lactate	0 - 8,3		< 5	10 d
Seres, France	Enzymat	Glucose	0,3-22	60		
		Choline	1,0-29	60		
		L-Lysine	0,1- 2	60		
		D-Lactate	0,5-20	60		
Tacussel, France	Glucoprocesseur	Glucose	0,05-5	90	< 2	>2000 samples
Prüfgeräte-Werk Medingen, GDR (Eppendorf, FRG)	ADM 300	Glucose	1 -100	30	< 2	>2000 samples
	ECA 20 (ESAT 6660)	Glucose	0,6-60	120-130	< 1,5	10 d
		Lactate	1 - 30	120	< 2	14 d
		Uric acid	0,1- 1,2	80	< 2	10 d

chemically oxidizable sample constituents and the restriction of the measuring range by the need of cosubstrate supply - set limits to this principle. By optimizing the permeability characteristics of the membrane both problems have been minimized.

For GOD membrane electrodes the enzyme loading test reveals the transient to the internal diffusion controlled regime, i.e., complete substrate conversion, at 0.5 to 10 U/cm^2. These membranes are useful for the determination of typically 500 - 2000 samples in an interval of 5 - 30 days.

An extremely fast-responding glucose sensor is obtained when the enzyme is fixed in polyurethane directly on the surface of a plane platinum electrode produced by thin film metalization techniques /3/ . After addition of glucose, the steady state current is reached within 1 - 4 s. The sensitivity for glucose is considerably higher than that of the GOD membrane sensors. The linear region is up to 2 mM glucose. Obviously the GOD-polyurethane layer has an extremely high permeability. The high amount of glucose is probably not completely converted by the immobilized GOD. This leads to a higher sensitivity for interfering substances as compared to that for glucose.

Similar analytical parameters have been obtained for the determination of lactate, lysine, xanthine and hydrogen per oxide using the respective enzymes entrapped in polyurethane direct at the surface of thick film metal electrodes (Tab. 2)

Tab. 2: Analytical characterictics of enzyme loaded thick film electrodes

Analyte	Enzyme	Linear range, (µM)	Response time, (s)	Measuring time, (min)	Coefficient of variation, (%)	Stability, (d) 90 %	60 %	40 %
Lactate	LOD	2 - 500.	10	1	-		4	
Glucose	GOD	5 - 2000	10	1	4	>7	>50	90
Xanthine	XOD	- 5	12	1	-			2
Lysine	LyOD	0.5 - 20	12	1	2		2	3
H_2O_2	POD	20 - 600	15					

Alternatively to the entrapment of GOD the adsorption of the enzyme at monolayers of surface - active substances is used /4/. The isotherms at the air-water interface indicate that only with positively charged substances or in mixtures of stearic acid with cetyltrimethyl-ammonium bromide the negatively charged GOD penetrates the lipid layer. These stable layers were deposited by using the Langmuir-Blodgett technique on Pt-electrodes prepared by thick film metalization on ceramic substrates. These enzyme-lipid covered electrodes responded within 10 s to the addition of glucose thus indicating the maintenance of the GOD activity. A good linear correlation between the current signal and the surface pressure of layer formation, i.e., the amount of the enzyme, is found. Multilayer structures consisting of up to 10 lipid-protein layers were prepared by successive dipping of the plane electrode. The sensitivity to glucose increases by a factor of almost 2 when two layers on the electrode instead of only one are applied. A further increase of the number of layers enhances the sensitivity with saturation at 10 layers. This tendency indicates the transition from kinetic regime to mass transfer control of the sensor signal.

Second Generation

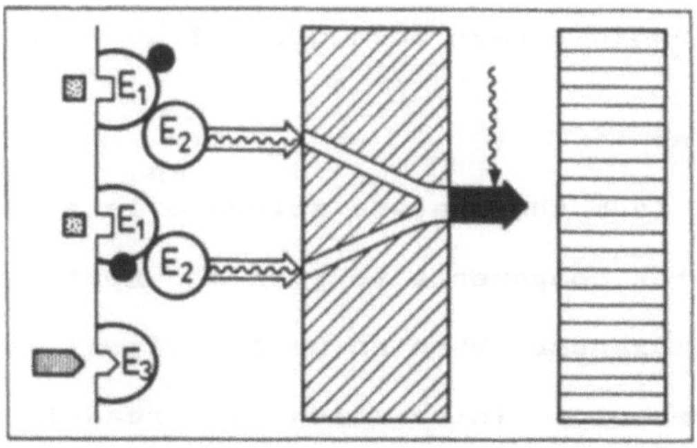

With the aim to improve the analytical performance -i.e., specificity, sensitivity, and handling - coreactants and auxiliary enzymes are integrated into the biological layer of the biosensors.

Multi-parameter sensing

Up to now, enzyme electrode-based analyzers have been one-parameter devices. However, it is highly desirable to be able to measure several different substances in parallel with one and the same analyzer, without changing the sensor part. One approach to accomplish this is the simultaneous application of several different enzyme sensors to the same sample solution. Another principle for simultaneous multi-parameter determination is the integration of several independently working enzymes in one membrane, e.g., GOD, uricase, and lactate monooxygenase for the simultaneous determination of glucose, uric acid and lactate. Alternatively, different enzymes, e.g., for glucose, urea, and lipids, are fixed at the sensitive regions of a multiple sensor chip /5/. Furthermore, several substances are accessible by only one biosensor using coupled enzyme reactions. By a hexokinase/glucose-6-phosphate dehydrogenase sequence the substrates glucose-6-

phosphate and glucose, the cofactors $NADP^+$ and ATP, and even the inhibitor fructose have been quantified /2/.

Coupled reactions

In the simplest form the analyte molecule interacts directly with the receptor components to give an appropriate output signal of the biosensor. When no useful signal is obtained in the analyte-receptor interaction a readily measurable substance can be formed in a successive enzymatic conversion. Substances disturbing either the enzymatic reaction or the indicator system can be eliminated by additional enzyme layers. Furthermore, the sensitivity of biosensors can be increased by recycling the substrate to amplify product formation /2/.

Group effect sensing

Generally, sensors using microbes for analyte recognition suffer from the multireceptor behaviour of intact cells resulting in a decreased substrate specificity. On the other hand, this ability to recognize a group of substances is exploited in sensors for complex variables, such as the sum of biodegradable compounds. This principle of determining the biological oxygen demand (BOD) has provided impetus to environmental control /6/.

<u>Third Generation</u>

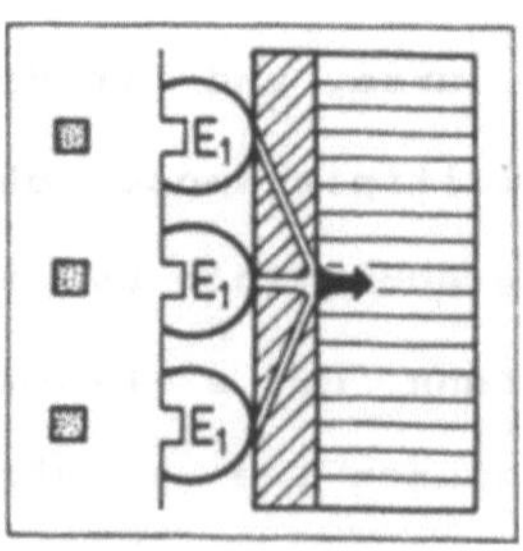

The goals of miniaturizing biosensors are the reduction of the required sample volume, the multicomponent analysis of complex chemical substances by multiple microsensors, and the cost reduction by mass production. Three basic types of chemical microsensors have been used so far in chipbiosensors

- ion sensitive field effect transistors
- gas sensitive MOS capacitors
- amperometric, potentiometric and conductometric thin film metal electrodes (TFME).

Table 3: Chip biosensors

Analyte	Enzymes	Transducer
Glucose	Glucose oxidase (GOD) (+ lactonase)	pH FET, TFME
	GOD + peroxidase (POD)	pF^- FET
Maltose	Glucoamylase + GOD + POD	pF^- FET
Lactate	Lactate oxidase (LOD)	TFME
	LOD + POD	pF^- FET
Urea	Urease	pH FET, NH_3-MOS FET
Lipids	Lipase	pH FET
Peptides	Trypsin	pH FET
ATP	ATPase	pH FET
Acetylcholine	Cholinesterase	pH FET
Creatinine	Creatininase	NH_3-MOS FET
NADH	Hydrogenase	H_2-MOS FET

Amperometric, potentiometric and conductometric base sensors are nowadays produced in large scale by using technologies of microelectronics /7/. Highly uniform and potentially cheap biosensors can be fabricated from these transducers when microelectronics-compatible technologies for the enzyme fixation are available. The need of enzyme-free reference sensors and counter electrodes on the same chip demands the structurization of the enzyme layer on the wafer /8/.

Two alternative procedures have been proposed. Nakamoto et al. /9/ masked the passive region of the wafer with a positive photoresist. The enzyme molecules are fixed in the incovered area by typical immobilization methods, e.g., a combination of aminopropyl-silanization of the surface and enzyme adhesion by glutaraldehyde. The enzyme layer has a typical thickness of 1 μm and the minimum width and line separation are both 1.5 μm.

In contrast, Shiono et al /10/ deposited the enzymes only at the sensitive region, e.g., at the gate of ISFETs, from an enzyme solution in a negative photoresist. Thus ISFET-biosensors for glucose, urea and lipids with a response time of 4 s and a sensitivity of 85 % after 2000 measurements have been developed. When this enzyme FET is in a flow-through cell, which ensures both isolation and electrical connection of the conductive fields to the measuring device, it can be used without any polymeric encapsulation or wire bonding. This technique promises to reduce the cost per sensor.

In our group the following procedures of enzyme coverage of electronic elements have been developed

- Polyurethane

  Elastic polyurethanes (PU) strongly adhering to silicon sensor chips were selected to fix the biocomponent. The dispersed enzyme in a solution of PU in an organic solvent is mechanically patterned onto the chip by simple dropping, dipping or spinncoating procedures. The porous PU film is cross-linked by reaction with polyisocyanates. In order to adjust the functional parameters as sensitivity, dynamic range and response time, the pore sizes have been varied by changing the conditions of the cross-linking reaction, e.g., concentration and viscosity of the agents, temperature and crosslinking time. Good adhesion, retention of the enzyme molecules, and low diffusional resistance for both analytes and products have been achieved by the cross-linking procedure. Using TFMEs and ISFETs, glucose and urea sensors with long term stability and short response times have been prepared.

- Plasmapolymers

Acrylic acid derivatives or aromatic amines were deposited precisely as thin films of plasmapolymers onto silicon wafers at the active areas of TFMEs or FETs using masks of microelectronic fabrication processes. Free carboxyl- or amino-groups suitable to bind GOD covalently are formed. For GOD an apparent activity up to 100 mU per/cm^2 has been found.

Heat or laser pyrolysis of aromatic hydrocarbon plasmapolymer films give porous carbon layers, which adsorb irreversibly enzymes from aqueous solutions. In this way the wafer can be loaded with enzyme after cycle I of the chip fabrication process or, if less stable enzymes are used, after completing the single sensor device in cycle II.

- Electropolymerization of pyrrole and aniline

Polymeric films are formed by the electrochemical polymerization of aromatic components which entrap effectively enzyme molecules. In this way the deposition is restricted to the active area of the metal electrode and the successive immobilization of different enzymes is possible.

Conclusions

The restricted functional stability of the biocomponent which is typically between several hours and several weeks demands that the sensor should be an exchangable part of the analyzer. Therefore a modular analyzer concept effects the present state of biotechnology. Depending on the further progress in stabilizing biomaterials the biosensor will be integrated step by step in the total microsystem. The firt step which is partially realized is the integration of electronic signal processing within the sensor body.

REFERENCES

/1/ Clark, L.C., Lyons, C., Ann. NY Acad. Sci. 102 (1962) 29-45

/2/ Scheller, F., Schubert, F., Renneberg, R., Mueller, H.-G., Jaenchen, M., Weise, H., Biosensors 1 (1985) 135-160

/3/ Hintsche, R., Dransfeld, I., Scheller, F., Pham, M.T., Hoffmann, W., Hueller, J., Moritz, W., GBF Monographs 13 (1989) 151-154

/4/ Zaitsev, S. Yu., Hanke, Th., Wollenberger, U., Zubov, V.P., Ebert, B., Scheller, F., J. Electroanal. Chem., in press

/5/ Karube, I., Sci. Technol. Jpn. 5 (1986) 35-40

/6/ Riedel, K., Renneberg, R., Kuehn, M., Scheller, F., Appl. Microbiol. Biotechnol. 28 (1988) 316-318

/7/ Bergveld, P., Biosensors 2 (1986) 15-33

/8/ van der Schoot, B. H., Bergveld, P., Biosensors 3 (1987/88) 161-186

/9/ Nakamoto, S., Kimura, J., Kuriyama, T., GBF Monographs 10 (1987) 289

/10/ Shiono, S., Hanazato, Y., Nakako, M., Maeda, M., GBF Monographs 10 (1987) 291

Miniaturised Thin-Film Biosensors

F. PITTNER, T. SCHALKHAMMER, E. MANN-BUXBAUM, G. URBAN

Inst. f. Allgemeine Biochemie d. Universität Wien und Ludwig Boltzmann Forschungsstelle f. Biochemie, Währingerstr. 38, A-1090 Vienna and Inst. f. Elektrotechnik d. Technischen Universität Wien, Gußhausstr. 25, A-1040 Vienna, Austria.

Summary:

Electrochemical biosensors based on covalently immobilized enzymes have been constructed using advanced thin film technology and employing tetrahalogene 1,2- or 1,4 benzoquinones as new immobilizing agents. For special applications, thin film electrodes were combined with permeation selective, derivatized polypyrrole layers as sensory modifying devices.

Introduction

In clinical diagnosis as well as in many other fields there is an increasing demand for miniaturized and integrated biosensors. However, various problems have to be solved, to allow the construction of highly satisfactory sensors for practical use in great numbers: The immobilized biosystem has to be stable. Also the selectivity of the electrochemical transducer and the reproducibility with respect to response and temperature dependence must not be neglected. [1 - 5]
Especially thin film technology is able to provide the high purity and reproducibility required of the electrode surface and the high spatial resolution of the electrode structure.
Our aim was to obtain devices for the quick and highly reproducible measurement of small samples. As it is not feasible to employ macroscopic techniques (e.g. membranes, gels ..) for microdevices stepwise derivatisation, starting from the metal electrode surface, turned out to be the most practical. Thin film electrochemical electrode arrangements with covalently coupled enzymes were developed in order to obtain stable, reproducible, fast and highly sensitive glucose sensors for discontinuous in vitro or continuous use in vivo.
In this paper glucose sensors will be presented based on immobilized glucose oxidase and metal thin film devices.

Materials and methods

Sodium silicate glass sheets of 0.3 mm thickness and Upilex 0.1 mm foils were used as electrode carriers and coated with thin titanium layers up to a thickness of 50-100 nm as an adhesion layer. Platinum, rhodium or palladium layers up to a thickness of 100 nm being evaporated on top of the titanium film act as electrochemical electrodes. Structuring of these thin films was performed by a lift off technique. These layers were insulated by a 3000 nm siliconnitride layer and structurized by plasma etching. The platinum surface was cleaned by etching with oxygen plasma. The Ag/AgCl reference electrode was produced by evaporating and structuring a silver film, which was subsequently chlorinated with 10 mM $FeCl_3$.

Using these procedures 2 different electrode types were constructed:

A 4 - electrode electrochemical cell (Fig. 1) with an outer diameter of 2.5 mm (possessing 2 identical working electrodes) was produced for differential measurement.

A 3 - electrode miniaturized electrochemical cell (Fig. 1) with an outer diameter of 0.2 mm was developed for in vivo testing.

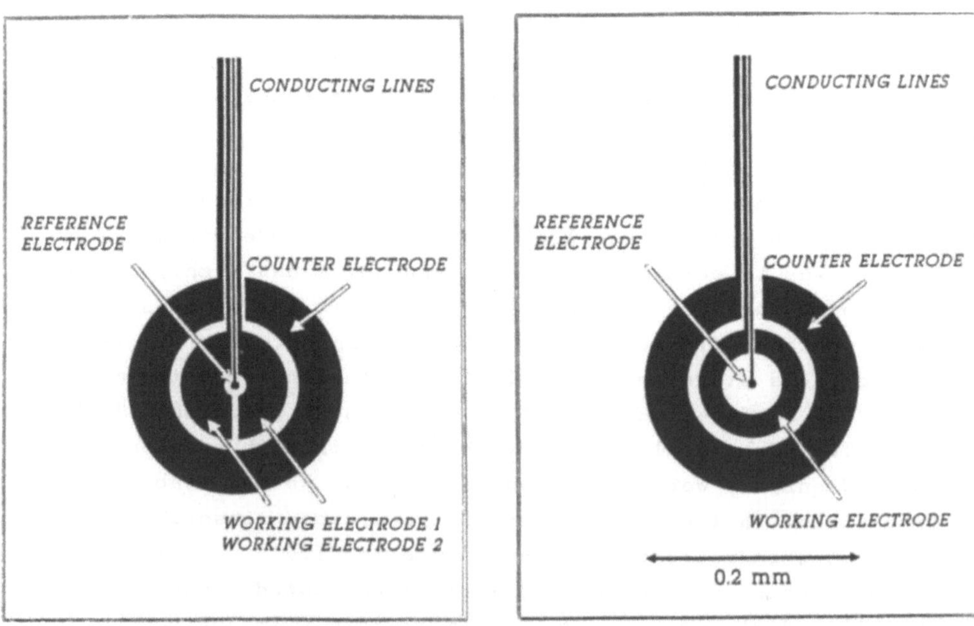

Fig. 1. Electrode types

Preparation of the biosensors using 1,2- and 1,4-arenequinones

a. Activation of the metal surfaces:
Platinum electrodes were oxidized by a coupled oxygen plasma. The oxidized electrodes were rinsed with water, followed by acetone, dried and immediately derivatized with silane.
b. Coupling with 3-amino or mercaptopropyl-triethoxysilane
The oxidized electrodes were silanized by spinning of a 1 % solution of the silane in propanol/water (95/5; v/v) on the electrodes and drying in an oven at 90°C for 3 min.
c. Reaction with o- and p- chloranil
The sililated electrodes were kept in a 1 % solution of chloranil in toluene at 40°C for 30 min. The electrodes were rinsed several times with toluene followed by acetone. The derivatized electrodes were stored dry in the dark.
d. Coupling of enzymes
To couple glucose oxidase to the activated surface, the electrodes were immersed in a solution of 5 mg/ml glucose oxidase in a 0.1 M phosphate buffer, pH 7.0 for 2 h. The electrodes were rinsed several times with 4 M saline to eliminate adsorbed protein. A polymeric top layer may be applied to prevent unwanted adsorption of most proteins and to minimize clotting effects in biological samples.

Preparation of biosensors using substituted polypyrroles

a. Cleaning of the electrode
Platinum thin film electrodes were cleaned by ultrasonication in distilled water, rinsed with acetone and dried carefully under dust free conditions.
b. Coating with polymerized substituted homopolypyrroles
The following pyrrole derivatives were used as monomers for the coating of platinum electrodes with thin polymeric layers. 1-(Ω-Carboxymethyl)-pyrrole (I), 1-(Ω-Carboxyethyl)-pyrrole (II), 1-(Ω-Carboxypropyl)-pyrrole (III), 1-(Ω-Carboxypentyl)-pyrrole (IV), 1-(Ω-Carboxydecyl)-pyrrole (V), 1-(1,3 Dicarboxypropyl)-pyrrole (VI), 1-((1-Carboxy 3-methyl)butyl)-pyrrole (VII), 2-(1-Pyrrolo)-acetylglycine (VIII), 1-Dodecylpyrrole (IX), 1-(4-Carboxybenzyl)-pyrrole (X), 1-(4-Nitrophenyl)-pyrrole (XI), 1-(4-Carboxyphenyl)pyrrole (XII), 3-Carboxymethyl-4-methyl-pyrrole (XIII), 1-o-Tosylpyrrole (XIV), 1-p-Tosylpyrrole (XV), 1-Benzenesulfonylpyrrole (XVI), 4-(3-Pyrrolo)-4-hydroxybutyric acid (XVII), 4-(3-Pyrrolo)-4-keto-

butyric acid (XVIII), 3-((Hydroxy 4-nitrophenyl)methyl) pyrrole (XIX), 3-((Keto 4-nitrophenyl)methyl)pyrrole (XX).
A 0.5 % solution of the above mentioned compounds in acetonitrile containing 2.5% of $LiClO_4$ was transferred to a glass cell and bubbled with Ar for 10 min. The electrode immersed in the deaerated solution was now cycled 10-30 times between -300 and +1600 up to +2000 mV (100 mV/s). If the substituted pyrrole was polymerizable the formation of a thin, brown polymeric layer was observed.

c. Coating with polymerized heteropolypyrrole

The electrolyte of the above mentioned composition was complemented with additional 0.05% pyrrole. The electrode was cycled up to 10 times between -300 and +1600 mV (100mV/s). A black polymer layer of variable thickness was obtained.

d. Coupling of the enzymes

The polypyrrole coated electrodes were incubated with a saturated solution of N-cyclohexyl-N'-(2-(N-methylmorpholino)-ethyl)-carbodiimide-4-toluenesulfonate for 60 min at 25°C without shaking. The electrodes were rinsed several times with water and immediately reacted with a 5 mg/ml enzyme solution in phosphate buffer pH 7.0; 0.1 M for 2 h.

Results

Covalent immobilisation with 1,2- and 1,4- arenequinones

Covalent coupling of glucose oxidase, glucuronidase, β-glucosidase and alkaline phosphatase was carried out employing an improved technique adapted from the chemistry of quinones [6]. When using immobilized amines or thiol groups on the electrode surface, the quinone reagent reacts at one halogen position, eliminating hydrohalogenic acid. Thus the halogen atom in para position (p-quinones) or ortho position (o-quinones) is highly activated by mesomeric and inductive effects. Electrode material activated in this way may be stored dry and in the dark at 4°C for at least 6 months.
To couple the enzyme the activated electrodes should be dipped into a buffer solution of the enzyme for 1-12 h. The activated halogen atom is now capable of reacting with the amino and mercapto groups of proteins, peptides and coenzymes.

Immobilisation on substituted polypyrrole layers

By comparing the properties and limits of polypyrrole-glucose-oxidase electrodes, two fundamental advantages of electrodes covered by substituted polypyrrole layers having no redoxactivity showed up:

* about 20 fold increase of response per unit area due to the porous surface
* significant permeation control by the polymeric layer for interfering electroactive substances resulting in a distinct increase in selectivity (e.g. 50 up to 200 fold suppression of ascorbic acid in glucose monitoring)

Optimal results for electrodes of the following type could only be obtained by synthesizing modified polypyrrole layers without significant conductivity and redoxactivity. The substituted polypyrrole layer acts as a highly effective barrier for interfering redoxactive compounds such as ascorbic acid, bioactive amines, sulfhydryl containing peptides and proteins, a.s.o. but having a high permeability for e.g. hydrogenperoxide.

Various N-3-substituted pyrrole monomers have been synthesized in order to prepare layers possessing these properties. Polymeric films could be obtained with pyrroles I - V, VIII - XI, XIII and XVII - XX. Homopolymers, heteropolymers and sandwiched types of these polymer layers were prepared. The various layer types led to different properties of the resulting enzyme sensors. It is necessary to introduce functional groups to obtain a modified polypyrrole layer which can be used for covalent coupling of the enzymes. Thus hydrophilic polypyrrole layers containing carboxy groups turned out to give best results.

Discussion

1. The tetrahalogene 1,2-benzoquinones react with amino and mercaptogroups substituting the halogen in ring position 4 eliminating hydrohalogenic acid. This leaves a highly activated halogen atom in ortho position (in para position for 1,4-benzoquinones) which can be used for coupling of proteins, peptides and ligands under mild aqueous conditions, employing amino- or mercaptogroups on their surface.

2. Stable enzyme monolayers with high oxidase activity and storage stability were formed.
3. 0.64 mm^2 - electrodes had a typical response of 15 nA under physiological glucose concentrations. The response time was 10 - 20 s.
4. Using polypyrrole as enzyme carrier the electrodes gave optimum results if the polypyrrole layer was redoxinactive and showed only low conductivity.
5. These polypyrrole coatings were obtained by polymerizing pyrrole with an excess of N-alkylpyrrole monomers having a carboxygroup for the coupling of glucose oxidase.
6. On these polypyrrole electrodes the electrocatalytic reaction proceeds at the metal surface. To obtain low background currents and high selectivity the modified polypyrrole layer acts as a highly effective barrier for interfering redoxactive and electrode contaminating compounds such as ascorbic acid, bioactive amines, proteins a.s.o. but has a high permeability for hydrogenperoxide.
7. 0.64 mm^2 polypyrrole coated electrodes had a typical response of 100-200 nA under physiological glucose concentrations. The response time was 20 - 30 s.
8. These electrodes were able to enhance glucose response to interfering ascorbate response ratio from 1:3 to 1:0.05. Thus the electrodes were used in bioreactors without significant decrease in activity during a fermentation period.
9. For prolonged use in body fluids and bioreactors the new polymer coated electrodes had a significantly increased selectivity and stability.

References:

1. Kimura, J. ; Ito, N.; Kuriyama, T.; Kikuchi, M.; Arai, T.; Negichi, N.; Tomita, Y. J. Electrochem. Soc. 136 (6) (1989) 1744 - 1747.

2. Abel, P.; Fischer, U.; Brunstein, E.; Ertle, R. Horm. Metab. Res. Suppl. Ser. 20 (1988) 26-29.

3. Honold, F.; Cammann, K. Horm. Metab. Res. Suppl. Ser. 20 (1988) 47 - 49.

4. Kulys, J.J. Biosensors 2 (1) (1986) 3-13.

5. Malmros, M.K.; Gulbinski, J.; Gibbs, W.B. Biosensors 3 (2) (1987/1988) 71 -87.

6. Mann-Buxbaum, E.; Pittner, F.; Schalkhammer, T.; Jachimowicz, A.; Jobst, G.; Olcaytug, F.; Urban, G. (1989/1990) Sensors and Actuators Vol. B 1, 518 - 522

Acknowledgement: This work was supported by the Fonds zur Förderung der wissenschaftl. Forschung, Project Nr. S4908-MED

Chip-Biosensors – Integration of Bioactive Components with Transducers Made in Silicon Technology

RAINER HINTSCHE

Central Institute of Molecular Biology, Academy of Sciences of the GDR, Robert-Rössle-Str. 10; DDR 1115 Berlin-Buch

Biological active systems used in biotechnology in practice show in general a limited life time ranging between minutes and a few weeks. Therefore transducers for biosensors should be low cost and simply exchangeable in order to allow a short time use. with this aim potentiometric, amperometric and impedimetric devices - ion sensitive field effect transistors (FET) and thin film noble metal electrodes (TFME) have been developed and covered with enzyme containing layers.
Using the well known pH-sensitive FET a multichannel device has been constructed. Six pH-sensitive FET gates are arranged in the nonencapsulated active arrea of a device (fig.1).

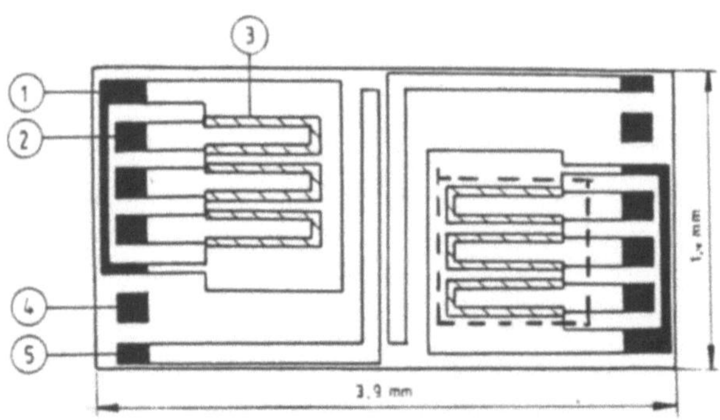

Fig.1: Scheme of the multichannel FET device (1 source, 2 drain, 3 gate, 4 connector, 5 TFME

Extending the FET application in biosensors, fluoride ion sensitive FETs have been developed and combined in a coresponding six channel device /1/. At all those channels the responses can be measured simultanously and thus the accuracy may be improved by computer aided calculation of the average of responses. Also the life time of the sensor device is prolongated by the possibility of ignoring incorrect channels. The

combination of three pH- and three pF- sensitive gates at one device results in a bifunctional transducer, which enables the construction of a difference sensor with integrated pseudoreference electrodes /2/.
Comparing miniaturized carbon-polyurethane composite electrodes /3/, and TFME on ceramics, glas and silicon wafers the TFME made in silicon technology reveal the best reproducibility, the smoothest surfaces and the lowest cost. The deposition process of the metal layers on the silicon wafers was optimized. The resulting electrodes are mechanically stable and also in water at potentials between +1.5V and -1.5V. The following typicall layer combinations have been used:

amperometric and impedimetric electrodes:

* Si / SiO_2 / Ti / Au * Si / SiO_2 / Ti / Au / Pt
* Si / SiO_2 / Si_3N_4 / Ti / Au / Pt

pH- sensitive electrodes:

* Si / SiO_2 / Ti / Sb / SbO

reference electrodes:

* Si / SiO_2 / Ti / Ag / AgCl

The chips are mounted and bonded at test strips or chip holders. A gape at the chip holder contains the flow trough channels and protects the connectors at the chip from the water solution. The simple masks are an advantage of the TFME system, which allow an inexpensive variation of different types of electrodes as well as size and number of electrodes at one chip.
The covering enzyme membranes at FETs and TFMEs must well adhere at the sensor surface and exhibit a minimal diffusion resistance for substrates and products. Furthermore the used technologies should be compatible to the microelectronic production. Two new coverage procedures have been developed. In the first /4/ the enzyme molecules are included in an elastic polyurethane film (PU), which is covered at the sensor chip by simple dropping, dipping or spinning procedures. The leakage of the macromolecules prevented by crosslinking with polyisocyanates. Main functional parameters as sensitivity, dynamic range and response time can be varied by changing the

crosslinking agents and/or the conditions of the crosslinking reaction.

The second procedure /5/ realizes partially a dry process for the coverage. Acrylic acid derivatives or aromatic amines were deposited precisely as thin films of plasmapolymers onto silicon wafers at the active areas of TFMEs or FETs using the masks of the microelectronic fabrication processes. At the surface of the formed polymer film free carboxyl- or amino-groups are identified by ESCA spectroscopy. Those functional groups can be covalently coupled with enzymes by conventional bifunctional reagents. At the constructed glucose TFMEs an appearent activity up to 100 mU glucose oxidase /cm² has been found.

Both coverage methods which are developed at the cheep TFMEs and transfered to FETs, because of the similar properties of the surfaces of the chips.

Thus glucose and urea FETs using the devices mentioned above have been constructed /1,2/. One to three amperometric TFMEs in combination with reference electrodes have used to measure glucose /3/ or lactat or peniciline. Also impedimetric TFMEs for glucose glucose (fig. 3) and potentiometric TFMEs for glucose (fig. 2), penicilline and urea have been constructed.

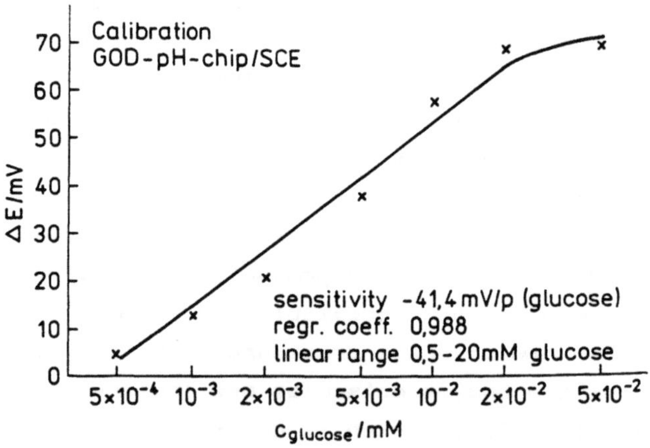

Fig.2: Calibration curve of a potententiometric glucose TFME chip (glucose oxidase in polyurethane at pH sensitive Sb-TFME - chips)

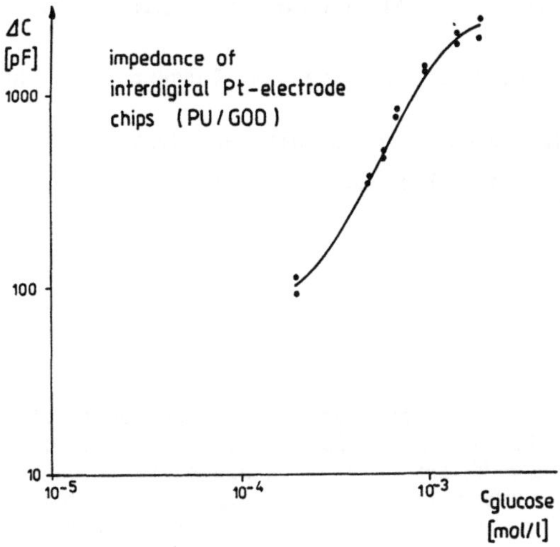

Fig.3: Calibration curve of a glucose TFME chip (impedance measurements at 1kHz; 50 um interdigital distances of electrodes)

Literature:

1. I.Dransfeld, R.Hintsche, W.Moritz, M.T.Pham, W.Hoffmann, J.Hueller, Anal. Letters 23 (1990) 437 - 450
2. R.Hintsche, I.Dransfeld, F.Scheller, M.T.Pham, W.Hoffmann, J.Hueller, W.Moritz, Biosensors 1990 in press
3. R.Hintsche, F.Scheller, studia biophysica 119 (1987) 179-182
4. R.Hintsche, G.Neumann, I.Dransfeld, G.Kampfrath, B.Hoffmann, F.Scheller, Anal. Letters 22,9(1989)2175-2190
5. G.Kampfrath, R.Hintsche, Anal. Letters 22,11(1989)2423-2431

The project is done in cooperation with Technical University Chemnitz (TFME) (Dr.G.Kampfrath, B.Hoffmann, J.Hahn), Central Institute of Nuclear Research, Dresden-Rossensdorf (FET) (M.T.Pham, W.Hoffmann, J.Hueller); Humboldt University Berlin (fluoride sensor) (W.Moritz); Central Institute of Molecular Biology, Berlin-Buch, (biosensors) (I.Dransfeld, B.Möller).

Smart Silicon System for a Multi Sensor Catheter

Y. Manoli, J. Eichholz, M. Kandler, N. Kordas, A. Langerbein, W. Mokwa

Fraunhofer-Institute of Microelectronic Circuits and Systems
Finkenstr. 61, D-4100 Duisburg 1, Fed. Rep. of Germany

Abstract

A data acquisition system for intravascular measurement of blood parameters is presented. The system consists of a set of silicon chips that can be embedded into a catheter of 2 mm in diameter. It can process the signals of a large number of sensors including on-chip signal conditioning, amplification and A/D conversion. CMOS techniques for sensor fabrication allow a monolithic realization.

Introduction

In the field of intensive care medicine continuous information about a variety of blood parameters can be of vital importance. A number of these parameters, however, not only require frequent blood sampling but their analysis is also often fairly time consuming. Having a monitoring system that samples a number of parameters can increase the efficiency of the diagnosis by enabling the doctor to do "on-line" analysis. This single monitoring system could also decrease the stress on the patient who is usually surrounded by a large number of instruments.

For the realisation of such a system several requirements have to be met. The miniaturization should be mature to a degree that allows data acquisition in such sensitive organs as veins, arteries or the heart. It should be flexible enough to allow the implementation of sensors for physical and chemical parameters as well as the necessary electronics to deal with this multitude of variables. The use of fiberoptics and spectrophotometry seems a reasonable choice for the miniaturization. This alternative becomes less feasible for a large number of parameters. Silicon on the other hand has not only reached a degree of miniaturization that enables the integration of complex electronic systems on a single chip but also possesses a number of properties that allows the detection of physical or chemical parameters. The only widely used silicon sensor in the medical field today is the pressure sensor which is applied extra-corporeally.

The "intelligent" catheter presented in this paper combines sensors for a variety of parameters with on-chip sensor readout electronics in a size enabling it to be used as a data acquisition system for intravascular measurements of blood parameters.

<u>Sensor/Electronic System</u>

The system consists of a set of silicon chips that can be embedded into a catheter of 2 mm in diameter which at the same time allows injection of medicine directly into the blood stream. The chips are bonded onto a polyamide film with gold plated copper connections. The tape is connected to the outside through a cable of five wires threated in a catheter lumen. Another lumen or even two are free for infusion purposes.

The system is designed to process signals of 16 sensors or more (Fig. 1). This includes on-chip signal conditioning and amplification of different types of silicon sensors. Using CMOS techniques for sensor fabrication,

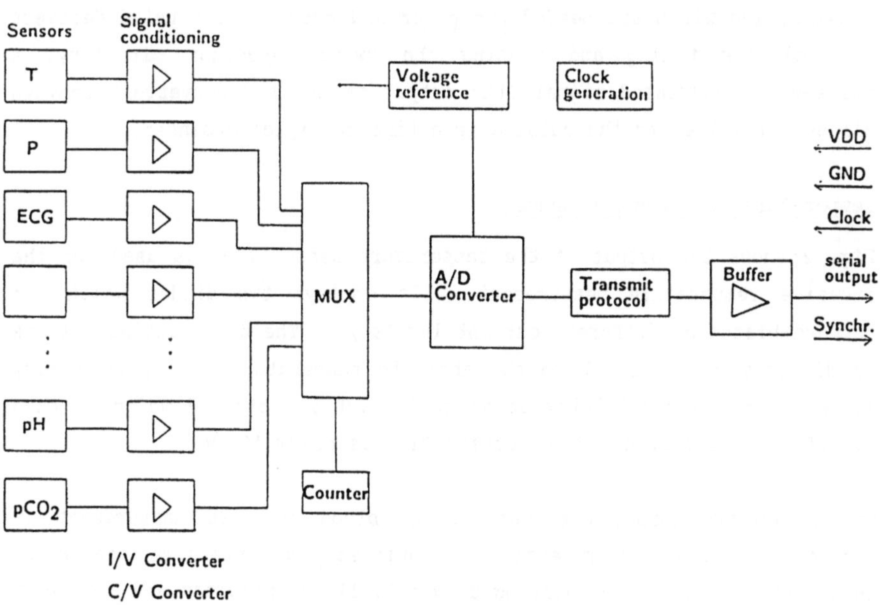

Fig. 1: Block diagram of the catheter electronics

signal processing circuitry and sensors can be brought together on a single chip. Such miniaturization is vital considering the chip-size limitation of 5 x 1 mm^2. The use of CMOS circuitry keeps the power dissipation of the system below 10 mW avoiding the problems of increased blood temperature.

The present system now in prototype production consists of a temperature and a pressure sensor. Three more sensors and associated electronics are being developed for phase two of the program. These include a blood flow sensor, ECG electrodes and amplifiers as well as a circuitry to measure blood conductance. Future sensor development will include chemical sensors for pH, pO_2, pCO_2, Na^+, Ka^+, etc.

In order to achieve electromagnetic compatibility in medical applications reliable sensors as well as robust read-out electronics for amplifying the small sensor signals are required. A direct transmission of the sensor signal should be avoided. After amplification the different sensor signals are multiplexed and fed into an analog-to-digital converter. The digital transmission minimizes noise susceptibility and reduces the necessary number of connections to the outside. Two wires are needed for power and ground and a third delivers a clock for timing and running the switch-capacitor circuitry. A pulse-width modulated signal (the output of the A/D converter) conveys all sensor values to the outside in a time-multiplexed manner.

Temperature and Pressure Sensor

Fig. 2 shows the output of the temperature sensor that is used for the invasive temperature measurement. It employs two diodes which are forward-biased at different current levels /2/. The differential voltage is directly proportional to the absolute temperature and shows a very linear slope. A sensitivity of about 0.4mV/K is obtained which is then amplified to 100 mV/K. Power consumption is below 100μW.

The capacitive sensor employed in the blood pressure measurement is fabricated using planar etching techniques. The pressure sensor is formed by a capacitor consisting of a polysilicon membrane and a silicon substrate (Fig. 3). The membrane is obtained by laterally etching of a sacrificial oxide layer, the spacer, via etch channels. These channels are sealed by depositing silicon dioxide on the membrane. The pressure sensors are very small, having diameters of up to 150µm. Sensor arrays, which deliver larger signals, can easily be designed /3/.

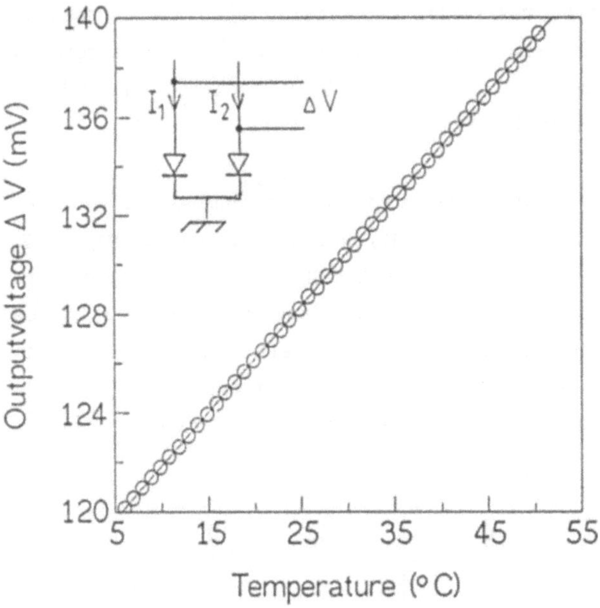

Fig. 2: Output singal of the temperature sensor

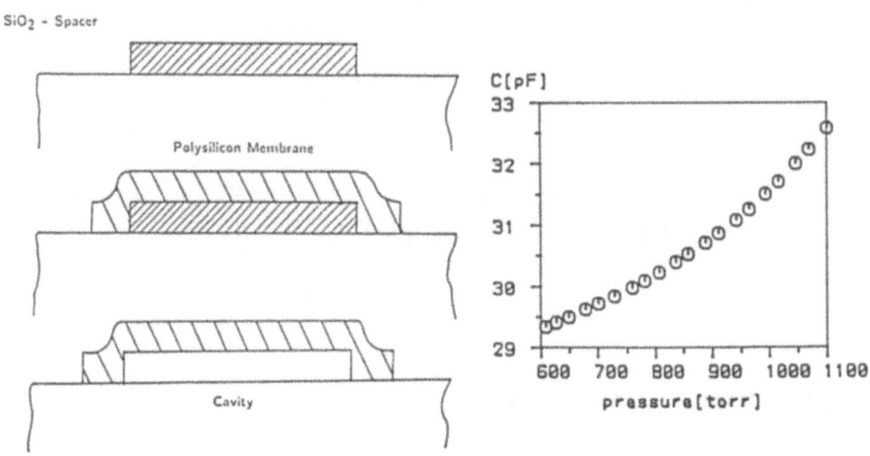

Fig. 3: Polysilicon pressure sensor and output capacitance

Readout Electronics

MOS analog integrated circuit design techniques make use of device matching properties rather than relying on absolute values of devices. Whereas absolute values of resistors, capacitors, etc. are subject to process variations in the order of 10 - 30 %, the matching errors of these elements are in the order of 0.1 - 0.5 %. In addition, the voltage and temperature coefficients do also match very well.

This property of MOS processes has led to the application of the switched capacitor (SC) circuit technique, which makes use of capacitor ratios as precise circuit elements. This technique is most appropriate in the case of the sensor working as a charge source. The charge collected on a capacitor can be nondestructively sensed due to the almost infinite input resistance of the MOS transistor gate and converted into a voltage. In Fig. 4 a typical two-stage SC readout amplifier is shown.

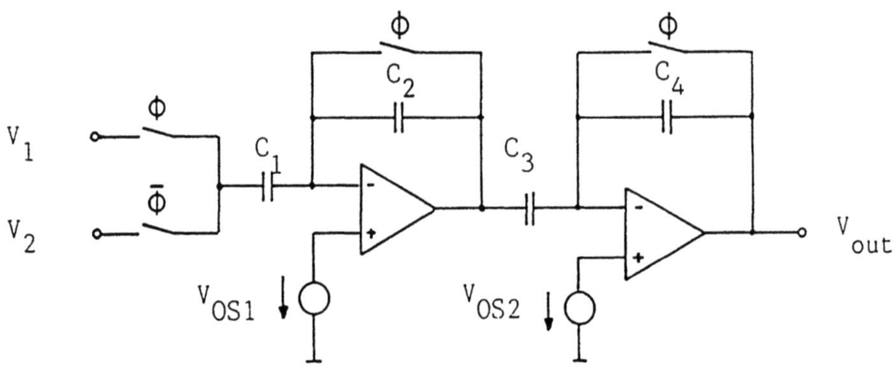

$$V_{out} = \frac{C_3}{C_4} \cdot \frac{C_1}{C_2} (V_1 - V_2) + V_{OS2}$$

Fig. 4: Two-stage SC-readout amplifier

It can be used to amplify the difference of two input voltages (V_1, V_2) as is the case in the temperature sensor. If the capacitor C_1 is exchanged for a pressure sensitive capacitive sensor and V_1, V_2 are constant reference voltages then a variation in pressure will result in a change in the output voltage. The pressure independent reference capacitor C_2 reduces the effect of offset, temperature and process variations.

To reduce the amplifier offset and 1/f-noise, special circuit techniques like correlated double-sampling (CDS) have been applied. An additional reduction of white noise is gained by implementing the second stage as an integrator.

References

/1/ Y. Manoli, B. J. Hosticka, W. Mokwa, W. Brockherde, "Silicon Microsystems", Int. Forum on ASIC and Transducer Technology, May 1990, pp. 87 - 93.

/2/ B. J. Hosticka, J. Fichtel, G. Zimmer, "Integrated Monolithic Temperature Sensor for Aquisition and Regulation", Sensors and Actuators, 6 1984, p. 191.

/3/ M. Kandler, J. Eichholz, Y. Manoli, W. Mokwa, "Smart CMOS Pressure Sensor", 22nd International Symposium on Automotive Technology and Automation, 1990, pp. 445 - 449.

/4/ U. Schöneberg, B. J. Hosticka, G. Zimmer and G. J. Maclay, "A Novel Readout Technique for Capacitive Gas Sensors", Sensors and Actuators B1, 1990, pp. 58 - 61.

A Digital Output Monolithic Temperature Sensor for Invasive Applications

N. Kordas, J. Eichholz, A. Langerbein, Y. Manoli, W. Mokwa

Fraunhofer-Institute for Microelectronic Circuits and Systems
Finkenstr. 61, D-4100 Duisburg 1

Abstract

Measurement of temperature inside the human body is of great importance in many applications. Temperature sensors having small size, good reproducibility and high resolution together with low power consumption are needed. In this paper we present a silicon temperature sensor fabricated in a standard CMOS-technology with integrated bias, amplification and A/D-conversion. The sensor is based on two forward biased pn-diodes with different areas and current densities. A temperature coefficient of 0.43 mV/K before amplification is obtained. The resolution in a temperature range from 10°C to 50°C is better than 1×10^{-3} and a reproducibility of 0.07°C is achieved. The sensor signal is amplified by using switched-capacitor circuitry and converted into a pulse width modulated signal by a single-slope converter.

Introduction

Silicon based integrated sensors offer the possibility of combining signal generation and signal processing on a single chip. There are many advantages in such a monolithic "smart sensor". Small systems with high resolution, good reproducibility and high immunity against electromagnetic interference can be produced. By using standard technology, a low cost mass production is possible.

The above advantages will lead to a further expansion of such systems in many fields in the following years. One interesting application is medical measuring, where miniaturizing allows in vivo measurements with low stress for the patient. The demands concerning power consumption, reliability and size however are very high in this field.

Our sensor allows the measurement of temperature in a range from 10°C to 50°C. Two pn-diodes with different areas are biased with different current densities, which leads to a temperature dependent voltage difference between them with a very linear slope. Amplification is done by a two-stage SC-amplifier, which also subtracts an offset voltage from the sensor signal. The amplified signal is then transformed into the digital form by a single-slope A/D-converter.

Two reference voltages are also converted to allow compensation of the temperature dependence of the A/D-converter. All signals are transmitted serially over one line, which brings the number of electrical connections up to four. In this application no external elements have been used.

Temperature sensor

Silicon based elements usually have a temperature dependence, which is undesired in most cases. This dependence however offers the possibility of temperature measurements. One suitable effect is the temperature dependence of the voltage drop over a pn-diode, which is forward biased with a constant current [1]. The characteristic of the diode is

$$I = J_s \cdot A \cdot \left[\exp\left(\frac{V_D}{V_T}\right) - 1\right] \tag{1}$$

where I is the current, J_s is the reverse current density, A the diode area, V_D the voltage drop over the diode and V_T the temperature voltage ($V_T = kT/q$, k = Boltzmann-factor, q = unity charge, T = absolute temperature). The voltage

$$V_D = \frac{kT}{q} \ln\left[\frac{I}{J_s A} + 1\right] \tag{2}$$

is a linear function of the temperature. Using only one diode has several disadvantages:

a. the bias circuitry must be temperature independent
b. the signal is distorted by parasitic resistors
c. the performance depends on technological parameters

These disadvantages can be avoided by using two diodes with different areas and different current densities (Fig. 1a). The voltage difference

$$V_{D1} - V_{D2} = \frac{k}{q} \ln\left[\frac{I_1 A_2}{I_2 A_1}\right] \cdot T \tag{3}$$

is also a linear function of T, but the logarithmic term contains only ratios. In our sensor, $I_1 = 10 \cdot I_2$ and $A_2 = 13 \cdot A_1$, which leads to a temperature coefficient of 0.43 mV/K (Fig. 1b). The diodes are fabricated in a standard CMOS-Process on a p-substrate with n-well.

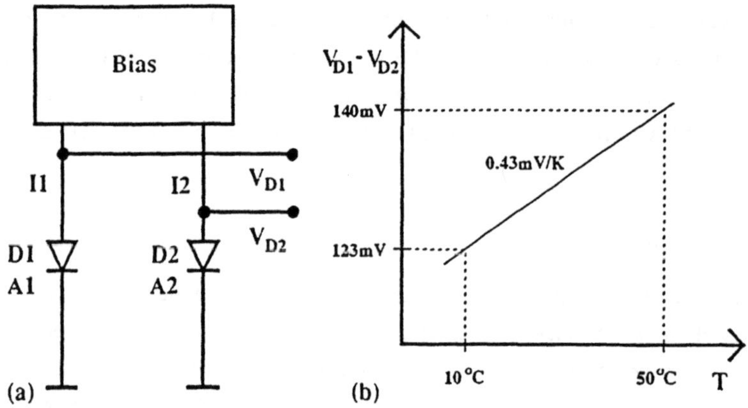

Fig. 1a: Forward-biased pn-diodes b: difference of the voltage drop

Amplifier

Amplification is done by a two stage switched-capacitor amplifier (Fig. 2). SC-circuits offer some advantages compared with resistor based amplifiers:

a. capacitors can be integrated on a much smaller area than resistors
b. there is no permanent current flow from sensor to amplifier
c. thermal noise caused by resistors is avoided

On the other hand, new problems concerning clock-feedthrough etc. must be solved.

The first stage has an amplification of C_1/C_2 and the second, which is built as an integrator, of n C_3/C_4 (n = 14) so that the total amplification is 298. To increase the resolution in the interesting range, the voltage difference at 10°C is subtracted. This is done after the first stage. The final voltage is

$$V_{out} = \left[\frac{C_1}{C_2}\left[V_{D1} - V_{D2}\right] - V_{ref2}\right] n \cdot \frac{C_3}{C_4} + V_{ref2} \qquad (4)$$

The amplified signal covers a range from 2 to 7 volts, which is the input range of the following A/D-Converter.

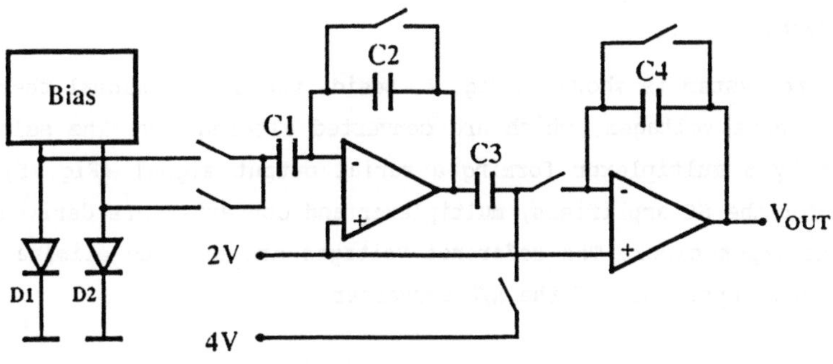

Fig. 2: Two-stage SC-amplifier

A/D-Converter

In order to provide an undistorted transmission of the signal, it is transformed into the digital form by an A/D-converter. The converter must be very small and should use no external components. A good concept for this purpose is the pulse-width modulation [2]. The principle is shown in Fig. 3a. When switch 1 is closed, the capacitor C is charged to the input voltage (sample-phase). After this, S_1 is opened and S_2 is closed, connecting a current source to the capacitor. The constant current produces a linear ramp (convert-phase). The comparator changes its output, when the voltage becomes lower than a reference voltage. Thus, a pulse width representing the analog voltage is produced (Fig. 3b). As this signal is digital, it has a high immunity against distortion.

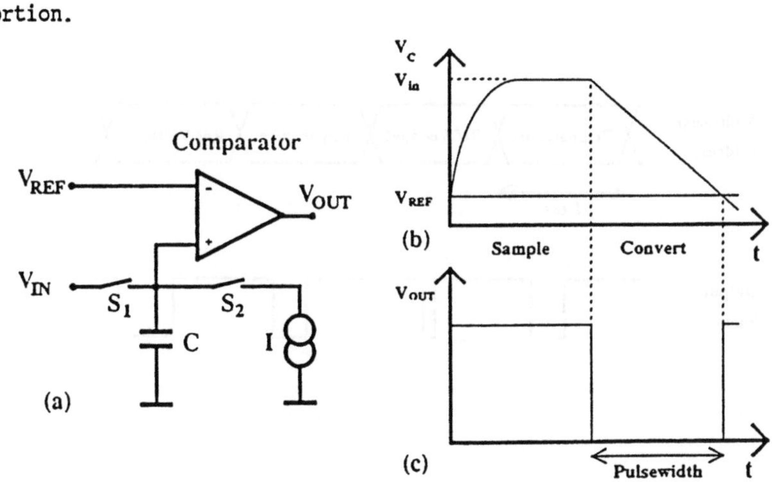

Fig. 3a: Principle of the single-slope-converter b: Voltage at the capacitor c: resulting pulsewidth

Total System

The entire system is shown in Fig. 4. Beside the sensor signal there are two reference voltages, which are converted alternately. The selection is done by a multiplexer forming a serial output signal (Fig. 5). The clocks for the SC amplifiers, multiplexer and converter are derived from a 32 KHz input clock. The reference voltages are used to eliminate the temperature dependence of the A/D-Converter.

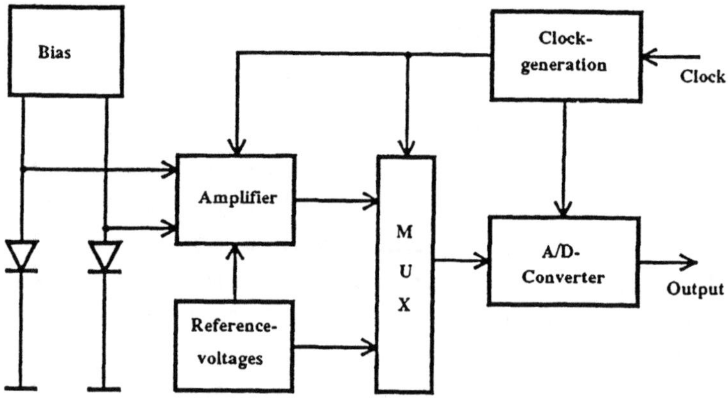

Fig. 4: Block diagramm

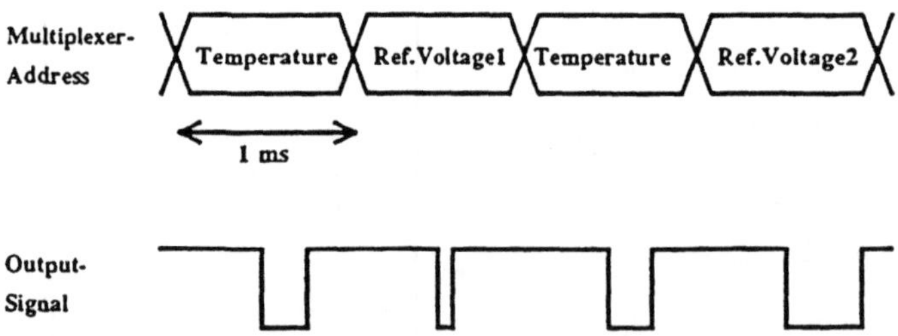

Fig. 5: Output sequence

Results

Our concept of temperature measurement with on chip signal processing has been realised within a multisensor system for a catheter based measuring of a number of physical and chemical parameters. In this example, two chips habe been used to integrate the electronic. One chip contains clock generation and the A/D-converter, the other one contains sensor, amplifiers and multiplexer.

Temperature measurement was possible in a range of 11°C to 50°C with a resolution of 0.05°C equivalent to 10 Bits. The maximum linearity error was 0.1°C and the reproducibility better then 0.07°C. As a reference sensor we used a PT-1000 resistor.

Fig. 6 shows the difference of our sensor to the PT-1000 reference after the numerical correction using the two reference voltages.

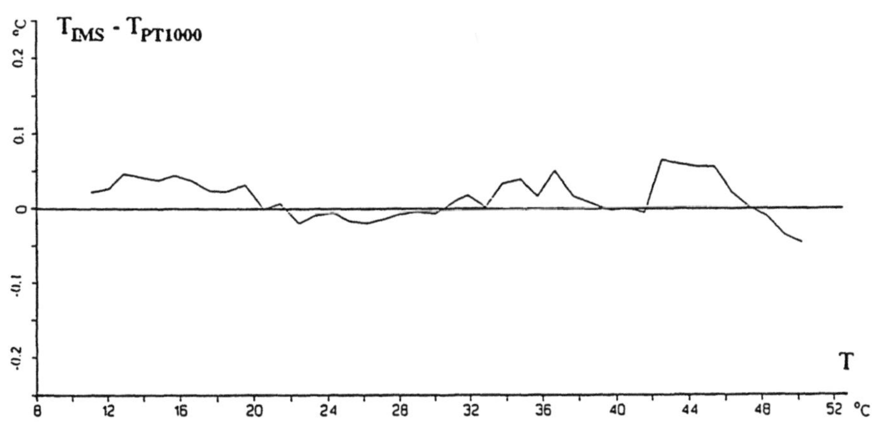

Fig. 6: Resulting Error

References

1. Hosticka, B.J.; Fichtel, J.; Zimmer, G.
   Integrated Monolithic Temperature Sensors for Acquisiton and Regulation, Sensors and Actuators, Vol. 6, No. 3, pp. 191 - 200, 1985
2. Smarandoiu, G.; Fukahori, K.; Gray, P.R.; Hodges, D.A.
   An All-MOS Analog-to-Digital Converter Using a Constant Slope Approach, IEEE Journal of Solid-State Circuits, June 1976.

Optical Sensors

New Nanomechanical Integrated Optical Devices as Modulators, Switches, and Microphones; Integrated Optical (Bio-)Chemical Sensors

W. LUKOSZ

Optics Laboratory, Swiss Federal Institute of Technology
8093 Zürich, Switzerland

Summary

A new physical principle for modulation and switching of guided modes in integrated optics (IO) is described. The changes in the effective refractive index of the guided modes required for operation of IO components are induced by varying - by a few nanometers - the width d of a gap between a dielectric plate E and the waveguide, thus changing the interaction of the evanescent field of the guided mode with this element E. The gap is filled with either a gas or a compressible medium of low refractive index, or is evacuated. IO devices fabricated in silicon technology with micromachined effective-index-shifting elements E are proposed as modulators and switches actuated by electrostatic or piezoelectric forces, and as force sensors and microphones.
As a second subject, our work on IO (bio-)chemical sensors is briefly reviewed.

Introduction

In integrated optics, the effective refractive index N is perhaps the most important quantity of a guided wave or mode. The phase velocity of the mode is given by $v_p=c/N$, where c is the velocity of light in vacuum. The operation of active integrated optical (IO) components, as, for example, phase modulators, interferometers, Bragg reflectors, and directional couplers, is based on changes ΔN in the effective indices of the guided modes. Normally, the required index changes ΔN are induced by electrooptically varying the refractive index of the waveguide material.

In this paper, I describe a different physical principle to induce the effective index changes ΔN required for IO device operation, which does not rely on any effect changing the refractive index of the waveguide material. According to the new principle, effective index changes ΔN are obtained as follows: a small dielectric plate E is positioned opposite the waveguide F, separated from it by a, for example, air filled gap C of width $d<\lambda$, where λ is the wavelength. The evanescent field of the guided mode reaches out of the waveguide through the gap C into E. This interaction causes the effective index N to depend on d. Surprisingly, nanometer variations of the gap width d are sufficient to induce in appropriately designed waveguides an effective index change of $\Delta N \approx 5 \cdot 10^{-4}$, which is a typical value required for IO device operation.

If the element E or its mounting are elastic, nanometer gap width variations Δd can be induced reversibly by a small applied force K. With, for example, electrostatically or piezoelectrically generated forces K, nanomechanical IO components could be operated for communication purposes. In a reciprocal way, variations in phase or intensity of a guided wave can be used to determine the force K. Then the IO device becomes a force sensor. The force K can be exerted by sound or ultrasound. IO microphones and hydrophones could be particularly interesting applications, as they are immune to electromagnetic interference.

Theory of Nanomechanically Induced Effective Index Changes

Figure 1 shows schematically the effective-index-shifting element E, a nonabsorbing dielectric plate of length L, refractive index n_E, and thickness d_E, positioned opposite a section F′ of a planar or strip waveguide F, and separated from it by a small gap C of width d. The medium of low refractive index n_C in the gap is air, an inert gas, a compressible solid medium, or vacuum. We consider the case in which the effective index N of the guided mode in the waveguide section F′ satisfies the condition $N>n_E$.

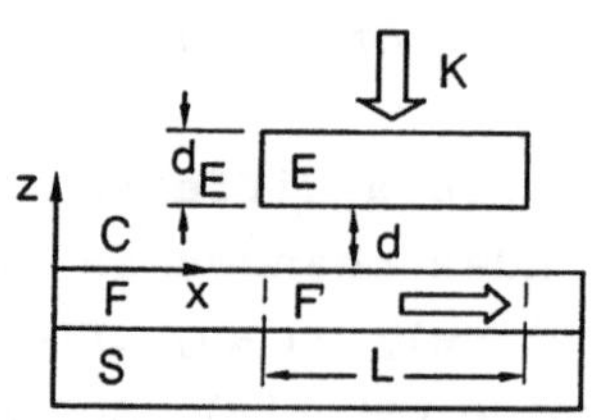

Fig.1. Schematic cross-section through waveguide and effective-index-shifting element E. F, waveguide; S, substrate; C, gap; d, variable gap width; K, force.

Then the evanescent field of the guided mode in the gap C penetrates into the element E also as an evanescent wave; no power is radiated into the element E. The waveguide can be either a planar or a strip waveguide. For a planar waveguide, the effective index N(d) can be rigorously calculated as a function of d from the mode guiding condition of the 2-layer waveguide formed by the waveguiding film F and the gap medium C, sandwiched between the two half-spaces, the substrate S and the element E.

The effective index changes ΔN induced by small gap width variations Δd are $\Delta N=(\partial N/\partial d)\Delta d$. I calculated the derivative $\partial N/\partial d$ from perturbation theory. For very small gap widths $d \ll \Delta z_E$, it can be expressed in analytical form as

$$\frac{\partial N}{\partial d} = \frac{n_F^2-N^2}{Nd_{eff}} \frac{n_C^2-n_E^2}{n_F^2-n_E^2} \left[\frac{(N/n_E)^2+(N/n_C)^2-1}{(N/n_E)^2+(N/n_F)^2-1}\right]^{\rho}, \tag{1}$$

where $\rho=0$ for TE_m modes and $\rho=1$ for TM_m modes. The effective waveguide thickness d_{eff} is defined as

$$d_{eff} = d_F+\left[(N/n_F)^2+(N/n_S)^2-1\right]^{\rho}\Delta z_S+\left[(N/n_F)^2+(N/n_E)^2-1\right]^{\rho}\Delta z_E, \tag{2}$$

where d_F is the thickness of the waveguiding film F, and

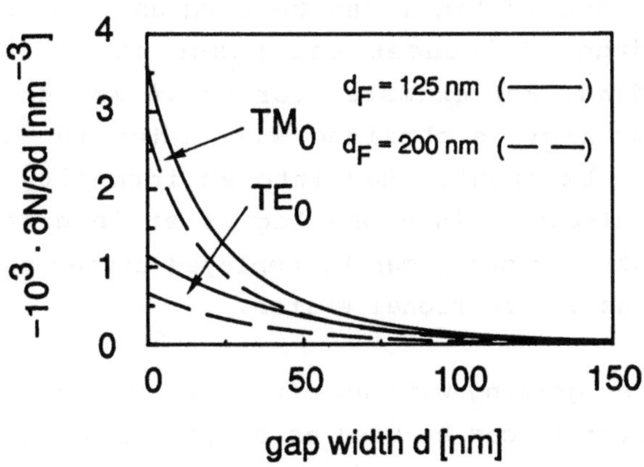

Fig.2. Calculated effective refractive index derivative $\partial N(d)/\partial d$ versus gap width d for the TE_o and TM_o modes in Si_3N_4 waveguiding films F on SiO_2 substrates S and an element E of refractive index $n_E=n_S$. Parameter is the film thickness d_F. $n_C=1$; $\lambda=632.8$ nm.

$\Delta z_j=(\lambda/2\pi)(N^2-n_j^2)^{-1/2}$ is the penetration depth of the evanescent field into the medium j. In the limit of larger gap widths $d>\Delta z_C$, I obtained

$$\frac{\partial N}{\partial d}=\frac{n_F{}^2-N^2}{Nd_{eff}}\,\frac{n_C{}^2-n_E{}^2}{n_F{}^2-n_C{}^2}\left[\frac{(N/n_E)^2+(N/n_C)^2-1}{(N/n_E)^2+(N/n_F)^2-1}\right]^{\rho}\exp(-2d/\Delta z_C), \qquad (3)$$

where d_{eff} in this case is given by Eq. (2) in which the index E has to be replaced by C. Since $\partial N/\partial d$ is inversely proportional to d_{eff}, it follows that high sensitivities $\partial N/\partial d$ can be obtained with thin waveguides. Figure 2 shows the rigorously calculated derivative $\partial N/\partial d$ versus d for Si_3N_4 waveguiding films F of refractive index $n_F=2.0$ on silicon substrates with an SiO_2 buffer layer of refractive index $n_S=1.46$. The element E was assumed to have the same refractive index as the substrate ($n_E=n_S$). At median gap widths $d_0<25$ nm, gap width variations of $\Delta d<1$ nm suffice to induce an effective index change of $\Delta N=5\cdot10^{-4}$, while Δd's of only a few nanometers are required at median gap widths of $d_o\approx100$-200 nm to induce the same index change ΔN.

Examples of IO Nanomechanical Devices

We discuss examples of IO devices with effective-index-shifting elements E. An element E opposite a waveguide section F' of length L as shown in Fig.1 can be used as a phase modulator. A gap width change Δd induces the phase shift $\Delta\Phi=2\pi(L/\lambda)\Delta N=2\pi(L/\lambda)(\partial N/\partial d)\Delta d$. For example, for L=0.6 mm and $\lambda=632.8$ nm, a phase shift of $\Delta\Phi=\pi$ is obtained with $\Delta N=5\cdot10^{-4}$. This phase modulation can be transformed into an intensity modulation by inserting the element E into one leg of an interferometer (see Fig.3). The interferometer can be employed either as an intensity modulator, or as a directional switch.

A surface relief grating with an effective-index-shifting element E positioned over it can be used as a tunable Bragg reflector. A guided mode incident perpendicularly onto the grating lines is retro-reflected, if the Bragg condition $2N\Lambda=m\lambda_m$ is satisfied, where Λ is the grating period and m=1,2,... is the diffraction order. The width $\delta\lambda_m$ of the reflected wavelength band is given by the relation $\lambda_m^2/\delta\lambda_m=2NL$, where L is the length of the Bragg reflector. With an effective index shift ΔN in the grating region, the reflected wavelengths λ_m are shifted to $\lambda_m+\Delta\lambda_m$, where $\Delta\lambda_m/\lambda_m=\Delta N/N$ and $\Delta\lambda_m/\delta\lambda_m=2(L/\lambda)\Delta N$. The Bragg reflector is switched from the reflecting to the transmitting state (i.e., $\Delta\lambda_m=\delta\lambda_m$) by an index shift of $\Delta N\geq\lambda/2L$, for example by $\Delta N\geq 5\cdot 10^{-4}$ if $L/\lambda=1\cdot 10^3$.

If a surface relief grating is used as an input coupler, its incoupling efficiency can be modulated with the element E. This follows from the incoupling condition $N=n_{air}\sin\alpha + \ell\lambda/\Lambda$, where α is the angle of incidence of the laser beam incident onto the grating and $\ell=1,2,\ldots$ is the diffraction order.

IO Fabry-Perot interferometers and ring resonators have transmission wavelengths λ_m given by the relation $NL_R=m\lambda_m$, where the integer m is the axial mode number and L_R is the length of the ring or twice the mirror separation in the Fabry-Perot resonator, respectively. The bandwidth $\delta\lambda_m$ of transmitted wavelengths is determined by the finesse F of the resonator; we have $\lambda_m/\delta\lambda_m=Fm$. With an effective-index-shifting element E of length L located in the IO resonator, the transmission wavelength can be shifted to $\lambda_m+\Delta\lambda_m$, where $\Delta\lambda_m/\lambda_m=(L/L_R)(\Delta N/N)$ and $\Delta\lambda_m/\delta\lambda_m= (FL/\lambda)\Delta N$. For a resonator with a finesse of F=10 and with $L/\lambda=1\cdot 10^3$, an effective index shift of $\Delta N\geq 1\cdot 10^{-4}$ switches the resonator from the transmitting to the reflecting state ($\Delta\lambda_m=\delta\lambda_m$).

Experiments

Experimentally, we verified the existence of the $\Delta N(d)$-effect, and we demonstrated the feasibility of IO modulators and microphones based on input grating couplers [1] and a two-beam interferometer [2]. We used SiO_2-TiO_2 waveguides with refractive index $n_F\approx 1.8$ and thicknesses $d_F\approx 140$-180 nm on glass substrates (with

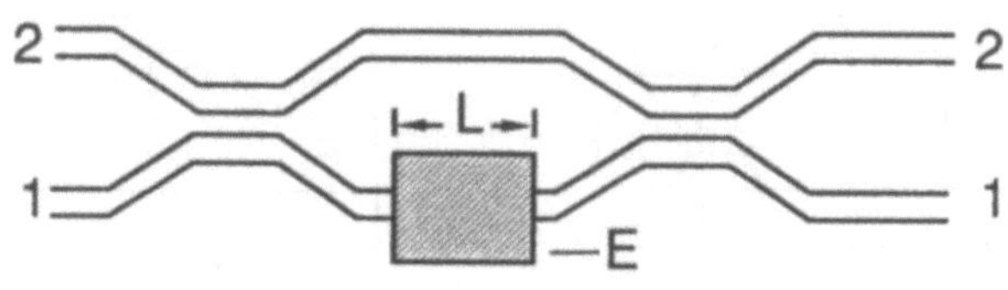

Fig.3. Schematic top view of IO Mach-Zehnder interferometer with an effective-index-shifting element E in one leg.

$n_S \approx 1.47$) at the wavelength λ=632.8 nm. The element E was not micromachined but a small glass plate.

Micromachined IO Elements

Micromachining has become a reality. By combination of standard integrated circuits technology, isotropic and anisotropic etching, and anodic glass-to-silica bonding, elements such as membranes, paddels, bridges, and cantilever beams with micrometer dimensions were fabricated. Such elements have resonance frequencies ranging from kHz to several MHz. I hope that these micromachining techniques can be combined with waveguide fabrication methods and envisage that with such techniques, waveguides and effective-index-shifting elements E can be fabricated. Thus nanomechanical devices with response times of milliseconds to submicroseconds could be realized.

Two different forms of micromachined effective-index-shifting elements E systems are schematically shown in Fig.4. In Fig.4a, the element E is a paddel hanging over a strip waveguide. In Fig.4b, the element E is a bridge spanned over a longitudinal section of the planar or strip waveguide. This bridge could be fabricated by etching a gap C of length L and width d into the material E, which outside this waveguide section forms a protective layer on the waveguide. The waveguides could, for example, be made of Si_3N_4 on silicon wafers with an SiO_2 buffer layer, both for visible and infrared wavelengths, and in doped silicon itself for infrared wavelengths. Micromachined elements E could be made of SiO_2, or of Si with an SiO_2 film on the side of the element E facing the waveguide.

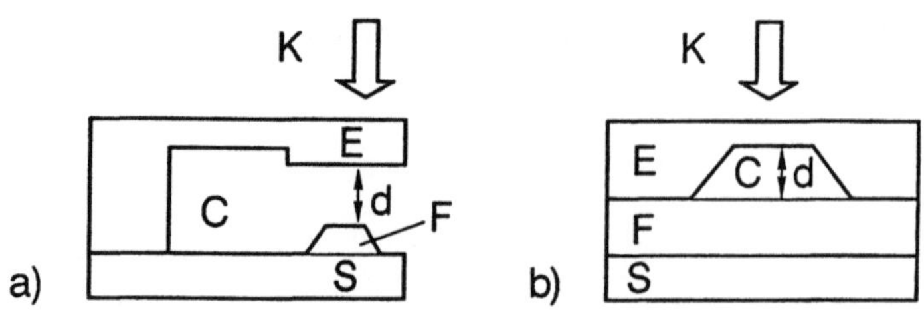

Fig.4. Schematics of two proposed forms of effective-index-shifting elements E: a) paddel, and b) bridge. F, waveguide; S, substrate; C, gap; K, force.

Let us characterize a micromachined element E by its mass M, resonance frequency ν_o, and damping constant γ. Dynamically it behaves as a damped harmonic oscillator. For critical damping, i.e. for $\gamma=\omega_o=2\pi\nu_o$, the system has a response time $\tau\approx1/2\nu_o$. Its steady state displacement Δd produced by a constant force K is given by $M\omega_o{}^2\Delta d=K$. An element E made of SiO_2 with length L=1 mm, width L_y=0.1 mm, and thickness d_E=1 μm, has a mass of M=0.25 μg. For a switch with a response time $\tau\approx1$ μs, we calculate that a force $K\approx2.5\cdot10^{-6}$ N is required to produce a gap width change of Δd=1 nm. If a response time of $\tau\approx0.1$ ms is sufficient, as in the case of a force sensor or microphone, we find assuming an element E with a larger mass of M=2.5 μg, that Δd=1 nm is obtained with a force of $K=2.5\cdot10^{-9}$ N.

The forces K required for IO modulators and switches can be exerted electrostatically. If the element E forms one of the plates of a condensor, the required force K can be generated with low voltages. An attractive feature of such IO devices is that the necessary electronic IC's could be integrated on the same silicon substrate. The displacement Δd of the element E could also be actuated piezoelectrically. Since the element E can be driven at frequencies much higher than its resonance frequency ν_o, modulation frequencies up to tens of MHz appear possible.

Conclusions and Outlook

New nanomechanical IO devices were described. Important applications may evolve if the effective-index-shifting elements can be fabricated by micromachining in silicon technology. Attractive features of the proposed nanomechanical IO devices are that no electro-optic materials are required for their fabrication and that they are integrable with IC's.

IO (Bio-)Chemical Sensors

Starting point of our work on IO sensors was the discovery that thin monomode SiO_2-TiO_2 waveguides responded sensitively to changes in relative humidity of ambient air [3]. Generally, IO sensors work as follows: changes ΔN of the effective refractive indices N of guided modes are induced by a) the adsorption of molecules on the waveguide, and b) changes in the refractive index n_C of the sample covering the waveguide. The effective index changes ΔN can be determined with input and output grating couplers from changes of the in- and out-coupling angles, respectively, and in interferometers from the phase changes of a guided mode. The sensor sensitivity is inversely proportional to the effective waveguide thickness d_{eff}. This explains the high sensitivity of the thin SiO_2-TiO_2 waveguides (with $n_F \approx 1.8$ and $d_F \approx 100$-200 nm) which we used, whereas other waveguides normally used in IO with thicknesses d_F of several micrometers have sensitivities that are orders of magnitude smaller.

(Bio-)chemical sensors have a 'chemo-responsive' coating on the waveguide surface which specifically or selectively binds an analyte. Examples are affinity sensors, in particular immunosensors, where the chemo-responsive coatings are monomolecular layers of receptor molecules binding the corresponding ligands, and antibody (Ab) molecules binding the corresponding antigens (Ag). The formation of the Ab-Ag complex is detected. We recently demonstrated the feasibility of IO immunosensing [4-7].

References

1. Lukosz, W.; Pliska, P.: Nanomechano-Grating-Coupler-Modulator and IO Microphone. 5th European Conf. on Integrated Optics: ECIO '89, Paris, April 1989, Proc. SPIE 1141 (1989) 201-207.

2. Lukosz, W.; Pliska, P.: Integrated Optical Interferometer as a Light Modulator and Microphone. EUROSENSORS IV, Karlsruhe, October 1990, (paper submitted to Sensors and Actuators).

3. Lukosz, W; Tiefenthaler, K.: Directional Switching in Planar Waveguides Effected by Adsorption - Desorption Processes. 2nd European Conference on Integrated Optics, Florence 1983, IEE Conference Publication No. 227, London 1983, pp. 152-155.

4. Lukosz, W.; Brenner, Th.; Briguet, V.; Nellen, Ph. M.; Zeller, P.: Output-Grating-Couplers on Planar Waveguides as Integrated-Optical Sensors. 5th European Conf. on Integrated Optics: ECIO '89, Paris, Proc. SPIE 1141 (1989) 192-200.

5. Nellen, Ph. M.; Lukosz, W.: Integrated Optical Input Grating Couplers as Chemical-and Immuno-Sensors. Sensors and Actuators B1 (1990) 592-596.

6. Lukosz, W.; Nellen, Ph. M.; Stamm, Ch.; Weiss, P.: Output Grating Couplers on Planar Waveguides as Integrated Optical Chemical Sensors. Sensors and Actuators B1 (1990) 585-588.

7. Lukosz, W.; Principles and Sensitivites of Integrated Optical and Surface Plasmon Sensors for Direct Immunosensing. BIOSENSORS '90, Conf. Abstracts, pp. 59-60, [see also pp. 61-62 and 245-246] (submitted to Biosensors & Bioelectronics).

High Resolution Interferometric Displacement Sensor Using Integrated Optics in Glass

D. Jestel, A. Baus, E. Voges

Universität Dortmund, Lehrstuhl für Hochfrequenztechnik,
Postfach 500500, D-4600 Dortmund - 50, FRG
Tel.: FRG (231)7553909, Fax: FRG (231) 7554631

Summary

A fibre-coupled, compact and stable integrated-optic Michelson-interferometer for 0.633 μm wavelength with ion exchanged waveguides in glass is demonstrated for optical displacement and position measurement. The integration of thermooptic modulators allows the combined detection of fast forward/backward movements with orthogonal signals, and high-resolution (1 nm) displacement detection with optical heterodyning for measuring distances in excess 10 cm.

Device Configuration

Optical interferometric sensors are highly sensitive. For an improved resolution on displacement sensing an optical heterodyning technique is applied. To overcome the limitations of movement velocity, due to the relatively low modulation frequencies in integrated-optic devices in glass, a new concept is presented. The fibre-coupled optical chip provides in a combined manner orthogonal detector signals for high-speed forward/backward movements with $\lambda/4$-accuracy and optical heterodyning by single sideband detection [1,2] for high-resolution (1 nm) displacement measurements. Undesired backreflections are avoided when using glass substrates with the same refractive index as the fibres and the collimating lens. Compared with electrooptical material there is no optical damage for short wavelength. The waveguides in this device are designed for $\lambda = 0.633$ μm.

The device configuration of this integrated-optic interferometer as shown in Fig.1 has one signal and two reference arms. The integrated-optical chip is connected to a HeNe-laser by a polarization maintaining input fibre. The incident light is divided by a Y-junction. The symmetric directional couplers (DC) act as 3 dB-beamsplitters. Both reference arms are terminated by 500 nm thick evaporated aluminum mirrors on the waveguide endfaces. The interference signals are fed to photodiodes via butt-coupled multimode fibres. A thermooptic phase modulator (PS1) within one reference arm

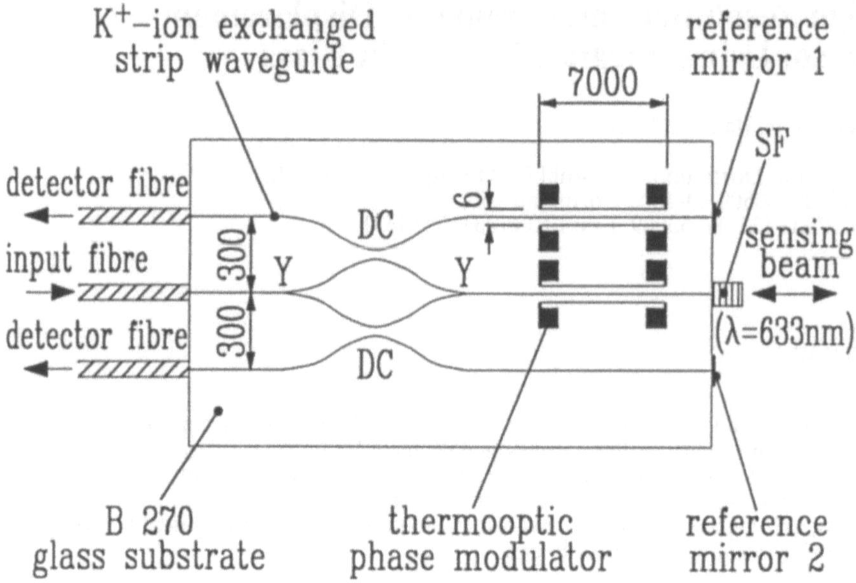

Fig.1. Integrated optical Michelson-Interferometer in glass with ion exchanged waveguides and thermooptic phase modulators. The indicated units are in μm; chip-size: 40 mm x 4 mm x 2 mm; DC: directional coupler (3 dB); Y: Y-junction; SF: Selfoc-lens

provides a phase difference of 90° between both reference arms yielding orthogonal detector signals for a fast forward/backward detection. A quarter pitch gradient index lens with an antireflective coating is attached directly to the endface of the signal waveguide in order to collimate the light of the sensing arm providing displacement measurements up to more than 10 cm distance.

Fig.2 shows the schematic of the Michelson-Interferometer configuration with two reference arms Ref.1 and Ref.2 and the sensing arm in the centre where the optical phase difference of the measurand φ gets the factor of 2. The optical phase difference is transferred to a low frequency phase difference by an optical heterodyning technique.

The sensing arm contains an additional thermooptic phase modulator PS2 for high-resolution optical heterodyning by phase modulated single sideband detection. Analog and digital driving signals at 1 kHz repetition rate are applied to the thermooptic modulator for single sideband detection. Applying this heterodyne detection modus, a displacement resolution of 1 nm is achieved. This phase modulation affects both interferences with the light in the reference arms 1 and 2 on the photodetectors PD1 and PD2 in exactly the same manner so that the orthogonality is undisturbed.

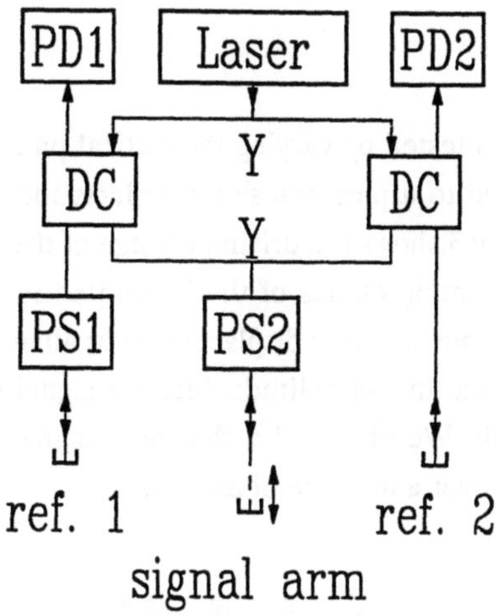

Fig.2. Block schematic of the interferometer with two reference arms 1 and 2 and the sensing arm in the centre. The thermooptical phase modulator PS2 for single sideband detection allows a measurement of the optical phasedifference φ with high accuracy. DC: directional coupler, Y: Y-junction, PS: phase shifter (PS1: static phase shift 90°, PS2: optical heterodyning), PD: photodetector

Device fabrication

The strip waveguides, designed for single mode operation at $\lambda = 0.633\ \mu m$, are fabricated by a field assisted potassium ion exchange in B270 glass (DESAG). The ion exchange process is driven by an applied electrical field of 50 V/mm at a temperature of 450°C. The process is computer controlled yielding reproducible coupling ratios of the directional couplers. The minimum centre-centre waveguide distance of the 3 dB directional couplers is 7 μm [3]. The parameters of the ion exchange process are optimized for minimum coupling losses to the polarization maintaining fibre. Fibres and collimator lens are butt coupled to the ion exchanged waveguides and fixed by UV-curing epoxy.

Efficient phasemodulators in glass could be achieved by using the thermooptical coefficients of the material. The phaseshifters (PS1 and PS2) in Fig.2 are heating electrodes fabricated by a photolithographically structurized evaporated 400 nm aluminum layer. For a small separation of 6 μm between the electrodes (with 4 μm width and 7 mm length) the drive power for π phase shift is reduced to 40 mW at 2 V driving voltage and the 3 dB modulation frequency is in excess of 10 kHz. In this case, however, the excess loss for TE-polarized light due to the metal electrodes without a buffer layer amounts to about 0.25 dB/mm.

Experimental results

The device performance is tested by varying the optical path length of the sensing arm. A mirror which is attached to a piezotranslator reflects the light back into the sensing arm. The lower trace of Fig.3 shows the driving voltage of the piezotranslator. The upper trace demonstrates the counting modus of the integrated interferometer. The signal of one of the photodiodes is shown versus displacement at 40 mm distance between optical chip and mirror. A clear and high-amplitude detector signal is achieved due to the low losses of the optical circuit. We observed a change in signal amplitude less than 10 %, when scanning the mirror over a distance of 25 mm.

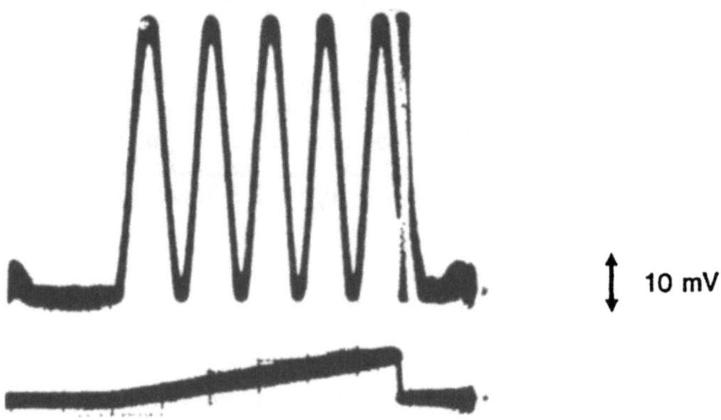

*Fig.3. Counting mode of displacement sensing.
upper trace: detector signal versus object displacement (1.5 μm)
lower trace: driving voltage of the piezotranslator*

For optical heterodyning by single sideband detection, a staircase phase modulating function is applied to one phase modulator. The phase modulation is adjusted according to the square root phase-voltage characteristic of the thermooptic phase modulators. The measured image sideband suppression is better than 30 dB with an resulting maximum absolute phase error of about 1°. Fig. 4 demonstrates the highly linear conversion of optical phase shifts $\delta\varphi = \delta d \cdot 4\pi/\lambda$ for a displacement δd into electrical phaseshifts for the heterodyne detection modus. A resolution of $\delta\varphi \approx 1°$ corresponding to $\delta d = 1$ nm is achieved. A high stability of the sensor system is achieved by the integration of several passive components and the phasemodulator on one chip. Within several minutes, after applying drive signals to the thermooptic modulator, a stable operation is achieved.

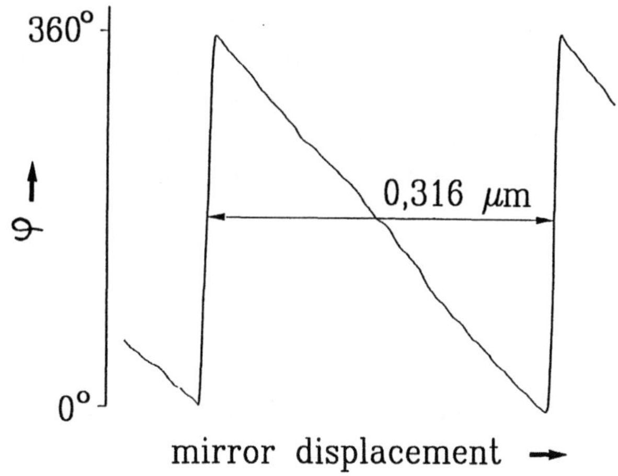

Fig.4. Measured electric phase φ_{el} versus object displacement for the heterodyne mode of operation.

Fig.5. Interferometer on top of the electronics for forward/backward counting and phasemodulation for SSB-detection. By the computerized data acquisition the measurand is displayed in user defined units or corrected for standard conditions.

Fig. 5 shows a complete sensor system with the integrated-optic interferometer setup on top of the specially developed high-speed forward/backward counter which measures fast displacements in quarter wavelength units. A display of the displacement in user defined units is given by the computerized data acquisition which reads out the counter information in order to keep the high dynamik range of the interferometric sensor system.

Conclusion

This concept of combining two detection methods with outstanding features is realized on glass with ion exchanged waveguides. The combination of orthogonal signal detection and heterodyning provides measurement of fast object movements with a resolution of 1/4 wavelength and slow object movements with a resolution below 1/100 wavelength for position detection. Applications of this integrated interferometer are highly sensitive position sensors or systems for measuring variations of optical path length. The data acquisition system keeps the high dynamic range and resolution of the interferometric sensor.

References

1. Voges, E., Ostwald, O., Schiek, B., Neyer, A.: 'Optical Phase and Amplitude Measurement by Single Sideband Homodyne Detection', IEEE Journal of Quantum Electronics, Vol. QE-18, No. 1, January 1982, 124-129

2. Eberhard, D., Voges, E.: 'Fiber Optic Interferometer with Digital Heterodyne Detection Using Lithiumniobate Devices', Proceedings of the International Conference on Optical Fiber Sensors, New Orleans, (1988), 313-316

3. Jestel, D.: 'Integrated Optical Michelson-Interferometer in Glass', Int. Congress on Optical Science and Engineering, Hamburg, 19-23 September 1988, SPIE Proc. Vol. 1014, Micro-Optics (1988), 31-34

Integrated Optical Devices in $LiNbO_3$ for Sensors

W. Konz

Fraunhofer-Institut für Physikalische Meßtechnik, Freiburg

Summary
A short description of the key components of optical devices for sensors will be given. Two examples demonstrate the importance of integrated optical circuits for sensor applications. It is reported about some components fabricated at IPM.

1. Introduction

One of the reasons for the use of integrated optics for sensors is the request to cancel the bulk optic components in optical measurement systems. A further reason is the possibility of signal processing with active integrated optical circuits. The integrated optical components have much higher speed capability and lower drive voltage compared with the bulk-optic counterparts. Another advantage is the possibility to combine several optical functions onto one chip.

2. Key components

The passiv and activ integrated optical key components used in integrated optics for sensors are listed in table 1.

Y-junction

Y-junctions are inherently symmetric devices, beeing relatively insensitive to fabrication variations. The realization of a device with a 50:50 split ratio depends only on the avoidance of defects in the junction region, so that careful attention to processing may be expected.

Polariser

With a dielectric/metal layer over an optical waveguide the TM mode of the incident light will be suppressed due to the resonant coupling with a surface plasma mode. Another possibility to realize a polariser is the proton exchange technique for lithium niobate devices. The extraordinary index of the waveguide arise with proton-exchange while the ordinary index becomes lower. Thus the extraordinary mode is guided, the ordinary mode on the other hand is radiated into the substrate.

TE/TM mode converter

Mode conversion between the orthogonal TE and TM components can be accomplished by the acoustooptic or electrooptic effect. The conversion is achieved electrooptically by utilizing an off-diagonal element of the electrooptic tensor to cause mixing between the orthogonal and normally uncoupled TE and TM modes.

TE/TM mode splitter

Mode splitting is usually performed by utilizing the birefringence of the substrate material. A directional coupler may be dimensioned in such a way, that the TE and TM mode will be seperated in the output arms.

Directional coupler

The directional coupler is based on the exchange of optical energy that occurs between closly spaced waveguides. Light of the input waveguide couples into the second and vice versa as a result of the overlap in the evanescent fields of the two guided waves. The effect can be used to build up a beamsplitter. To fabricate a beamsplitter with a 50:50 splitting ratio is rather difficult and electrical tuning will be necessary.

Phase modulator

A phase modulator consist of a monomode strip waveguide enbedded in an electro-optical material for example lithium niobate. The refractive index of the guided mode changes by an applying electrical field (Pockels effect). The changed refractive index causes also a change of the propagation constant ß. The electrode geometry, as also the drive voltage, depends on the crystal orientation and the device structure.

Frequency shifter

Frequency shifting may be realized by using a sawtooth phase modulation. Ideal serrodyne frequency shifting is performed by using a sawtooth phase modulation with a perfectly linear ramp, an infinitely fast falltime and a peak amplitude exactly equal to $2m\pi$ rad, where m is an integer. Another way is the use of an integrated optical Bragg-cell. In this case the light will be deflected by a moving grating, produced via a surface acoustic wave. The deflected light is Doppler-shifted about the frequency of the acoustic wave.

3. Examples of IO-circuits for sensors

The fiber gyro is the best known sensor in the field of optronics and also a good example for demonstrating the essential components needed for the signal processing of optical sensors. The sensing element is a fiber coil which detect the turning rate. The minimum configuration needs two symmetric Y-junction, a polariser and a phase modulator in the case of the phase modulation technique. Another way is the modulation of the frequency /1/. This may be realized with an integrated optical Bragg-cell instead of the phase modulator. The principle is shown schematically in figure 1. The modulation of the phase or frequency is needed for measuring the sensitive part of the interferometer characteristic curve to get a $\pi/2$ phase offset at small turning rates and also to get an ac-signal for better signal processing.

Another good example for demonstrating integrated optical components for sensors is an optical IC for heterodyne-detection optic which is used for high accurate measurement of velocity, displacement and position with a laser Doppler velocimeter /2/ figure 2. The following optical components are integrated on a substrate: Y-branches, frequency shifter, mirror, TE/TM mode converter and TE/TM mode splitter. The incident TE guided mode is first divided into two parts at the Y-junction. In the upper arm the light transmits across the mode splitter and is coupled into a polarization maintaining fiber. The output light from the fiber is focused after a quarter-wave plate. The reflected light having the orthogonal polarization state and excites the TM mode as signal light whose frequency is Doppler-shifted. The other part of the incident TE mode is converted to the TM mode by the mode converter so that the polarization state coincides with the signal light. The light reflected from the mirror is used as the reference light after the frequency is shifted. A phase modulator driven by a sawtooth-wave voltage is used as frequency shifter. Finally the reference and the signal lights are combined at a Y-junction for heterodyne detection.

4. Devices fabricated at IPM

Among the variety of the devices fabricated at IPM the following components are for sensor applications. About the very simple elements like Y-junctions and mirrors is not reported.

4.1 Phase modulator

The device is developped and fabricated in x- and z-cut lithium niobate for = 840 nm. Different electrode designs were needed for this two crystal directions. The electrodes of the x-cut phase modulator lies at both sides of the waveguide to use the strongest electrooptic coefficient r_{33} of the electrooptic tensor. In the case of the z-cut modulator a centered electrode lies above the waveguide seperated by a SiO_2 buffer layer (figure 3). The technical datas are listed in table 2.

4.2 Braggmodulator

We have fabricated an integrated optical Bragg-cell as a frequency shifter as shown in figure 4. In our application YX-$LiNbO_3$ (Y-cut, propagation of the light in x-direction) is used. The structure consist of two monomode waveguide tapers tilted by an angle of $2\theta_B = 1.8°$. The contour of the waveguide tapers is parabolic, yielding minimal coupling between the fundamental and the higher modes in the waveguide transition. The monomode waveguide is expanded up to 50 μm. A propagating grating is induced by a SAW (surface acoustic wave) in z-direction via an interdigital transducer (IDT) with an aperture of 0.75 mm centered to the waveguide crossing. The crosstalk between the tilted tapers is better than -32 dB without SAW interaction and with fibers being coupled to both ends of the Bragg modulator. The diffraction efficiency defined as I_D/I_{out} (I_D the deflected, I_{out} the non-deflected intensity without SAW) was 40% at an electrical power of 800 mW /3/. Further datas are listed in table 3.

Table 1: Passiv and active integrated optical components used for sensor applications

mirror
Y-junction
polariser
TE/TM mode converter
TE/TM mode splitter
directional coupler
phase modulator
frequency shifter

Table 2: Data of the phase modulator

Substrate material: $LiNbO_3$, x-cut
Operating wavelength: = 840 nm
Polarisation: TE
Drive voltage: U_{π} = 5 V
(Electrode length 15 mm,
Electrode gap 10 μm)

Table 3: Data of the Bragg modulator

Substrate material: $LiNbO_3$, y-cut

Optical data:

Operating wavelength: = 840 nm
Bragg angle: Θ_B = 0.9°
Polarisation: TM
Extinction ratio: 26 dB
Insertion loss: I_L = 10.2 dB
Optical crosstalk: <-32 dB
Deflection efficiency: I_D = 40%
(800 mW electrical power)

Transducer data:

Impedance: Z = 50 Ω
Acoustic wavelength: = 12 μm
Acoustic frequency: F_o = 287.4 MHz
Bandwidth (3 dB): F = 8 MHz
Insertion loss: I_L = 2 dB
Aperture: 0.75 mm

Literature

/1/ H. Shimizu, R.Ishikawa, K. Kaede, "Integrated optical frequency modulator for fibre-optic gyroscope using frequency modulation method", Electronics Letters 22 (1986) pp.334 -335

/2/ H. Nishihara, M. Haruna, H. Toda, "Integrated Heterodyne optics for fiber sensors", Optoelectronics

/3/ W. Konz, W. Lotze, D. Dolde, H. F. Schlaak, "An integrated optical Bragg-cell with tapered waveguides

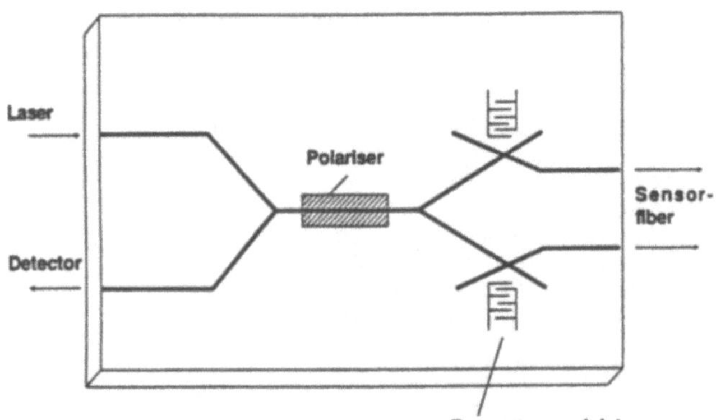

Figure 1: Integrated optical circuit (schematic) for a frequency modulated fiber gyro. In the case of a phase modulated gyro the Bragg modulators may be replaced by phase modulators.

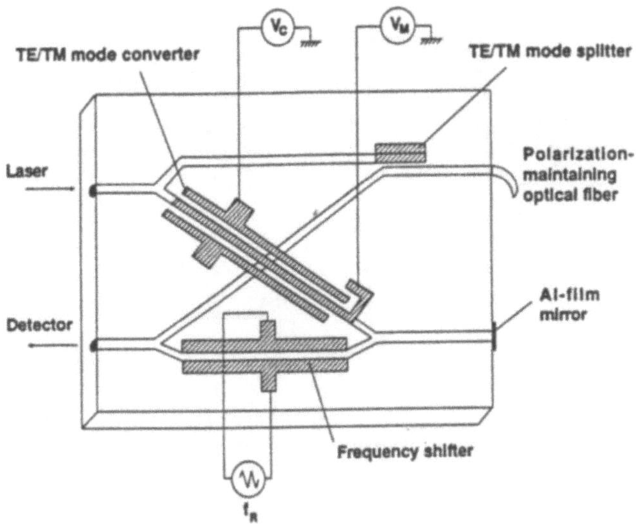

Figure 2: Configuration of the optical IC for the fiber LDV /2/

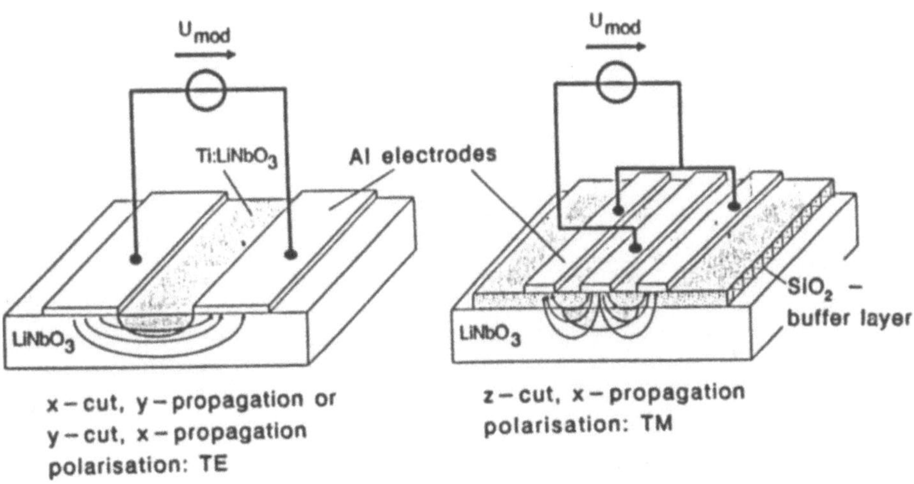

Figure 3: Electrode configurations for integrated optical phase modulators

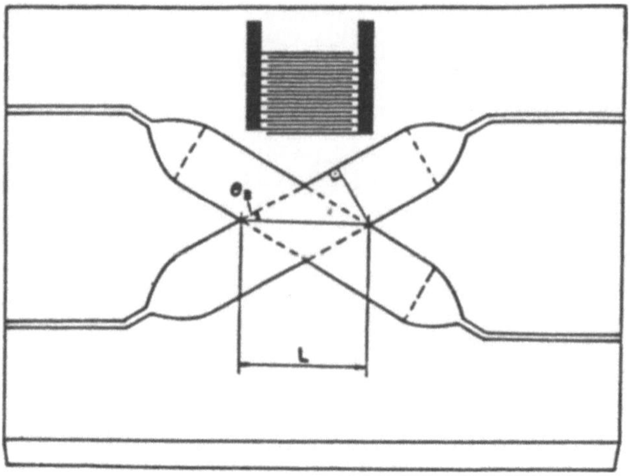

Figure 4: Integrated optical Bragg modulator

Optical Components

Optoelectronic Components and Modules for Multigigabit/s Fibre Optic Transmission

U. SPALTHOFF

SEL-Alcatel, Research Centre, 7000 Stuttgart 40, Lorenzstr. 10

Summary

Laser diodes and photodiodes for multigigabit/s transmission have been realized. The optimized performance of the semiconductor chips is exploited by packaging techniques with low parasitics suitable for microwave frequencies up to 13 GHz. Using an AlN ceramic laser submount in conjunction with YAG laser welding for fixing of the fibre a laser module with excellent high frequency performance and good long-term stability resulted. It is designed for manufacturing at relatively low cost. For sensitive detection at high bit rates pin-photodiode packages with more than 13 GHz bandwidth and avalanche photodiodes having 70 GHz gain-bandwidth-product have been realized.

Introduction

Fibre optic transmission systems are preferred for long distance transmission at high bit rates. Only a few years ago the term 'high bit rate' referred to 34, 140, or 565 Mb/s, sufficient for trunk lines building a back bone for public telephony systems and for ISDN. Currently installed trunk lines transmit up to 2.4 Gb/s per fibre.

For future B-ISDN and IBCN the bit rate on the trunk lines has to be increased. In a first step approximately 10 Gb/s, later-on 25 Gb/s will be needed. Two different approaches towards this goal seem feasible. Frequency domain multiplexing using coherent detection is attractive because it provides the possibility to use the enormous bandwidth of single mode fibres, i.e. several tens of THz without the need for electronics having the same bandwidth. Currently dominating is time domain multiplexing of digital channels combined with intensity modulation and direct detection. Using this technique record bit rates of 16 Gb/s were achieved /1/. By optical time domain multiplexing even higher bit rates are possible, e.g. 20 Gb/s /2/.

This paper deals mainly with laser diodes and photodiodes for multigigabit/s IM/DD systems, for digital- as well as analogue modulation. Apart from trunk systems many other applications are visible. Examples are video distribution, local area networks, metropolitan area networks, internal links for broadband switching systems, satellite communications, feeders for phased array radar antennas, or delay lines for satellite testing equipment.
Current high manufacturing costs of lasers and detectors suitable for telecommunication use prevent introduction of 'fibre-to-the-home' systems. Thus manufacturing volume stays low and cost savings due to high volume manufacturing cannot be achieved. Our approach to overcome this 'egg and hen' problem is to develop device structures and manufacturing methods which can be used for all applications mentioned above with only minor modifications.

Laser diode

10 Gb/s transmission requires laser diodes having rise and fall times of around 50 ps each. This has been achieved using the SIBH laser structure, characterized by a lateral current blocking layer of semi-insulating iron doped InP. This layer can be grown by metal organic vapor phase epitaxy /3/. Details of the structure and fabrication are reported elsewhere /4/. Fig. 1 shows the cross-section of a SIBH laser together with a typical light output characteristic. Compared with conventional BH and DCPBH lasers the capacitance is reduced significantly to approximately 4 pF. In conjunction with appropriate doping levels of the active heterostructure and an optimized structure of the laser cavity this enables fast sinusoidal modulation up to 13 GHz, corresponding to large signal modulation capability of approximately 8-10 GHz /5/. This is confirmed by measurement of the eye pattern. First accelerated aging experiments show stable behaviour of the lasers.

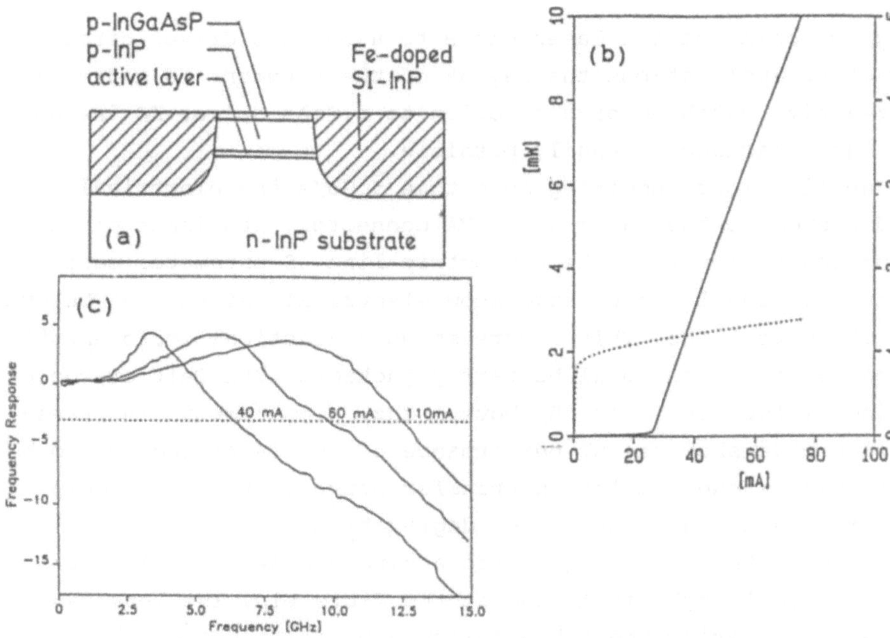

Fig. 1: Cross-section (a), P/I characteristic (b) and frequency response (c) of a SIBH laser diode

---

Laser package

The intrinsic bandwidth of the laser chip can only be utilized using microwave compatible mounting techniques. In conventional laser packages the chip is soldered on Cu, diamond or BeO submounts. The submount is soldered onto a Peltier cooler which stabilizes the laser temperature. As the Peltier cooler introduces a large capacitance, it degrades the high frequency performance above 3 GHz. Therefore we chose a package construction without Peltier cooler. The laser submount is made from AlN ceramic. AlN combines electrical insulation and good thermal conductivity. In addition its thermal expansion coefficient matches that of the laser chip, so degradation by thermal stress is avoided. These submounts can be fabricated at very low cost by conventional thin film techniques as known from alumina ceramics.

For high frequency application a microstrip line is used for connection of the laser chip to the package feedthrough. An optional resistor can be integrated on the submount for impe-

dance matching of the laser diode to a 50 Ohms driver circuit. For other applications the layout of the submount is modified. So reactive matching or connection to a GaAs driver IC inside the laser package is easily possible.
For applications requiring more than 5 Gb/s the electrical feedthrough to the laser is a SMA connector. Its inner pin is connected directly to the microstrip line. S-parameter measurements of the laser package show electrical reflections below -10 dB up to 5 GHz. This figure shows a drastical improvement compared to conventional butterfly packages with Peltier cooler having reflections >-10 dB above 1 GHz. Measurements in transmission mode show the HF performance of the laser package to be dominated by the modulation transfer curve of the laser chip itself without package-induced degradation.
Coupling of the laser light into a single mode fibre is improved by a small spherical lens at the fibre tip. The lens can be formed by tapering the fibre followed by melting the end or by grinding and polishing. The lens radius is chosen in accordance with the intended application: small lenses with radius 10 μm give high coupling efficiency but at the expense of very critical alignment tolerances. For many applications a radius of 20 or 30 μm proved to be a good compromise between coupling efficiency and alignment tolerances and resp. stability.
The fibre is glass-soldered into a stainless steel tube. After alignment of the fibre lens to the laser it is fixed into the package by YAG laser welding. Then the package is hermetically sealed by welding the tube to the wall and by a cover lid.

Pin-photodiode and avalanche photodiode

To achieve high receiver sensitivity the capacitance of the photodiode has to be minimized. We realized mesa and planar type InGaAs/InP pin photodiodes. The layers can be grown by LPE or MOVPE. To reduce the pn junction area to the light sensitive area of 70 μm diameter we chose a back-illuminated structure. The light sensitive area is adapted to the light spot diameter coming from a multimode fiber pigtail. With mesa type devices the chip capacitance is around 0.2 pF, planar type devices have a little higher capacitance of 0.3 pF due to the additional pn junction fringe. Dark currents of the devices at -5 Volts are 0.2 nA, quantum efficiencies at 1300 and 1550 nm exceed 80 %. After mounting on alumina strip lines they show flat HF response up to 13 GHz. Above 13 GHz the sensitivity decreases by the combined influence of carrier transit time, RC roll off and bond wire inductance.

The avalanche photodiode has separated absorption, grading, and multiplication layers (SAGM-structure). The layers can be grown by LPE or MOVPE. With MOVPE we incorporated up to 9 grading layers with increasing band-gap between the InGaAs absorption and the InP multiplication layer. The grading layers prevent hole trapping at the heterointerface, caused by the band-gap difference between InGaAs and InP. Best results were achieved with mesa-type devices having a diffused pn junction and an additional diffused guard ring. Once again the diodes have the back-illuminated structure. Dark currents at 0.9 V_b are in the 10 nA range , homogeneous multiplication up to M >= 25 is achieved. Chip capacitance is 0.3 pF. S-parameter measurements reveal a gain-bandwidth product of M=70 GHz in 1300 nm devices with InGaAsP absorption layer and of M=25 GHz in 1550 nm devices with InGaAs absorption layer.

Photodiode package

For pin and avalanche photodiodes the same packaging method can be applied. They are mounted on alumina ceramics on top of a laser drilled hole, which permits entrance of light from the fibre pigtail. In contrary to laser diode packages in most cases a preamplifier has to be incorporated inside the package to achieve highest sensitivity. Due to the low dark current and low capacitance of the photodetector the receiver sensitivity is essentially dependent from the preamplifier electronics.

For measurement applications and at ultra high bit rates the photodiode package is realized using an 50 Ohms matched micro strip line and an SMA connector. The package without preamplifier is used as 'optical probe head' in conjunction with oscilloscopes, spectrum analyzers and network analyzers. It enables measurement of short optical pulses and rapidly modulated optical signals. Fig. 2 shows the frequency response of this type of package to be flat up to 13 GHz, limited by the photodiode chip performance. The package introduces virtually no degradation as demonstrated by the S_{11} curve 2a). It has been tested in a 20 Gb/s receiver of a 2*10 Gb/s OTDM experiment/2/.

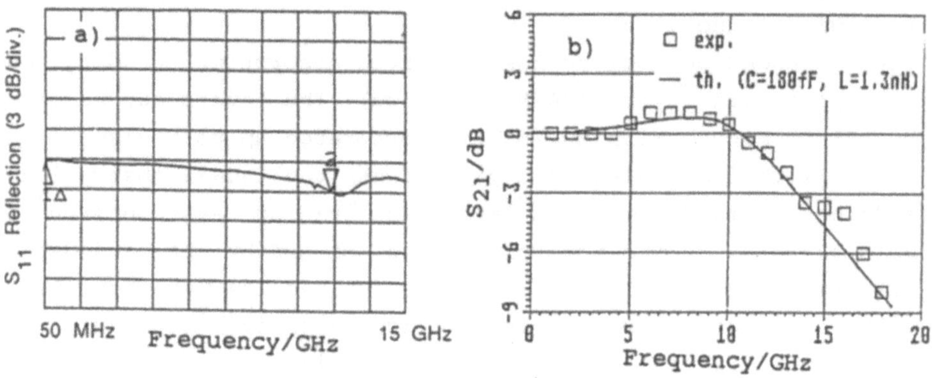

Fig. 2: Frequency response of photodiode package; a) S_{11} back reflections; b) S_{12} modulation transfer curve

Acknowledgements

Thanks are due to all collegues, who contributed to the reported results. The work described in this paper was partially financed by Deutsche Bundespost and by the European Community within RACE projects 1031 and 1057.

References

/1/ A.H. Gnauck, J.E. Bowers: 16 Gb/s Direct Modulation of an InGaAsP Laser. Electron. Lett. 23, 801 (1987)

/2/ B. Wedding, Th. Pfeiffer: 20 Gb/s optical pattern genera tion, amplification and 115 km fibre propagation using optical time division multiplexing. To be published at ECOC 1990, Amsterdam

/3/ P. Speier, G. Schemmel, W. Kuebart: MOVPE Growth and Characteristics of Fe-doped semi-insulating InP Layers. Electron. Letters 22, 1216 (1986)

/4/ R. Weinmann et al.: InGaAsP/InP Laser für sehr hohe Übertragungsgeschwindigkeiten.ITG-Fachbericht 112, p. 43; Berlin, Offenbach 1990

/5/ R.S. Tucker, J.M. Wiesenfeld, P.M. Downey, J.E. Bowers: Propagation delays and transition times in pulse-modulated semiconductor lasers. Appl. Phys. Lett. 48, 1707 (1986)

Elements for Integrated Optics Coherent Receiver Networks

H. Heidrich, D. Hoffmann, H.-P. Nolting, M. Schlak, C.M. Weinert

Heinrich-Hertz-Institut für Nachrichtentechnik Berlin GmbH., Berlin, Germany

Summary

The key elements of a coherent optical receiver network using a polarization diversity scheme are summarized. The principles of operation and design of the elements for monolithic integration of the waveguide network are sketched and first results are presented.

Introduction

A prerequisite for the economical realization of future optical communication systems is the availabiltity of monolithic integrated optics especially in the subscriber area. The potential of monomode fibre transmission can be widely exploited by the implementation of Coherent Multi Channel (CMC) broadband services in the customers access which will be allocated in the 1.55 μm wavelength window. In the coherent receiver the transmitted signal must be heterodyned with the local oscillator signal with the same state of polarisation (SOP). During transmission with single mode fibres the SOP changes and polarisation matching strategies are necessary, e.g. polarisation control or polarisation diversity reception at the receiver site [1]. In the polarisation control scheme of fig. 1a no penalty in the receiver sensitivity occurs because the SOP of the received signal is matched to the SOP of the local oscillator (LO).

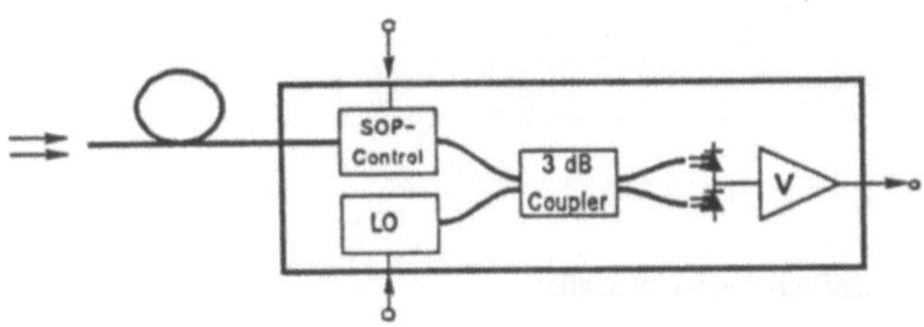

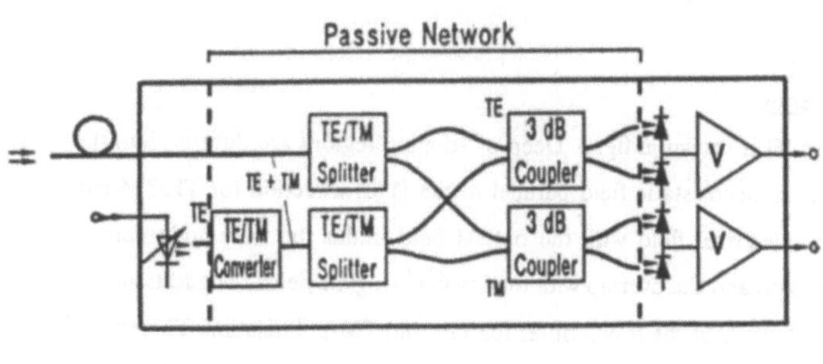

Fig. 1a: Polarisation control receiver OEIC
Fig. 1b: Polarisation diversity receiver OEIC

However, as continuous reset-free transformation must be guaranteed the SOP-control device operation is quite complex [2]. Therefore, at present, the polarisation diversity reception scheme is favoured, especially for the subscriber line. A possible architecture is shown in fig. 1b: The TE- and TM-polarisation modes of the received signal are split in a mode splitter (MS), heterodyned in 3 dB couplers and then detected separately. The TE/TM-mode converter (MC) is required to change the TE-polarized light of the laser into equal amounts of TE and TM-polarisation. In this contribution we describe the passive optical elements of a polarisation diversity receiver network. The integration with lasers and detectors has to be provided but is not discussed here.

Common Waveguide for Integration

The monolithical integration of waveguide devices with lasers requires a (001)-InP substrate and waveguides oriented along [110]. For fabrication of an integrated optical polarisation diversity network a "common waveguide" has to be defined in iterative optimizations of the separate devices. The MC requires a low background doping level ($<10^{16}$ cm^{-3}) of the device material and strong lateral confinement of the waveguide whereas for the 3 dB-couplers and especially the MS low lateral confinement would be favoured. The result of our iterative design process is the definition of a common rib waveguide (RWG) as shown in fig. 2. The waveguide layer configuration is especially designed for the fabrication of process tolerant couplers with in situ process control during dry etching [3, 4]. The fabrication process of the waveguide is described in detail in the discussion of the TE/TM mode splitters.

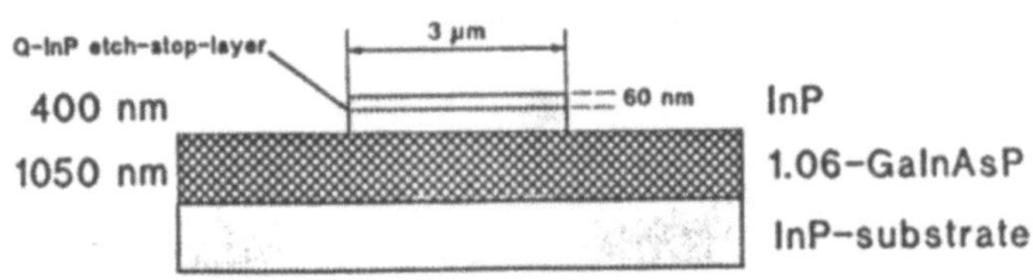

Fig. 2: Common waveguide configuration for integration

TE/TM-Mode Converter

The MC architecture [5] is shown in fig. 3. Deep p^+-doping regions are diffused into the n^--layer aside the waveguide to create an electrostatic field parallel to the $[1\underline{1}0]$ direction for TE/TM-mode conversion. The overlap of the static electrical field with the optical field under the waveguide determines the degree of TE/TM-mode conversion and the overlap with the lossy p^+-region yields the additional loss of the device.

The propagation constants β of the RWG differ for TE- and TM-polarisation. Therefore the control voltages for the electro-optically induced mode conversion have to be matched to the TE/TM-beat length Λ_0. In the ideal case the applied voltage has to change sign with a period of $\Lambda_0/2$. This is realized by fabricating electrodes on both sides of the waveguide with the control voltage applied alternativly on the left and right side (fig. 3). As for actual devices Λ_0 will not coincide with the electrode period length given by the mask pattern the use of electrodes consisting of a multiple of subelectrodes (in our case 4 subelectrodes per period) makes it possible to tune the control voltages to the actual TE/TM beat length Λ_0. Additionally, this configuration allows a reset-free continuous operation of the device as an optical compensator for arbitrary polarisation transformation ranging between pure phase-conversion and pure TE/TM-conversion [2]. Furthermore the device can be used as an optical quasiisolator in a future integrated circuit [6].

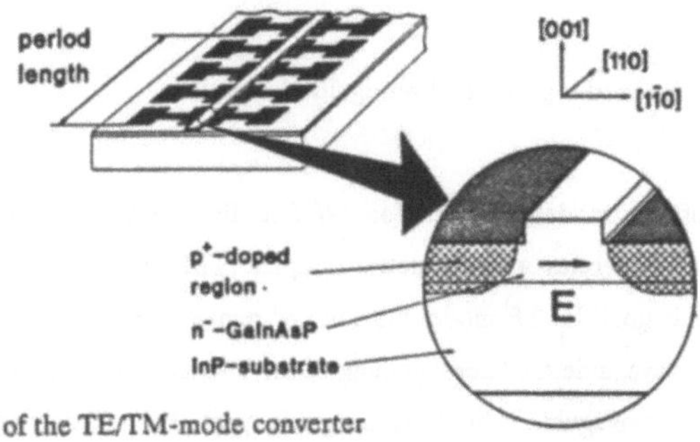

Fig. 3: Architecture of the TE/TM-mode converter

First MC devices have been fabricated. 45° mode conversion has been achieved at an effective voltage of 32 V [7]. The conversion efficiency κ was about 20% of the calculated value. The control voltage of an optimized device can be expected to be less than 10 V for 45° polarisation conversion. Using the common waveguide an electro-optical TE/TM-conversion efficiency of 1.2 °/V mm with moderate additional losses due to the overlap of the optical field with the lossy p^+-regions are expected when the p^+-regions are located close to the edge of the rib [8].

TE/TM-Mode Splitter

Mode splitters separate the TE- and TM-polarisation of the optical wave to two different output ports. This function can be verified on GaInAsP/InP by a directional coupler e.g. based on two parallel RWGs with a metal cladding on top of one of the ribs as sketched in fig. 4a. [9]. The fabrication of this structure is performed employing a selfadjusting process (c.f. fig. 4b): A sputter deposited Au-layer is used to configure the waveguide pattern in the Au-level by ion beam etching (IBE) using Ar-ions. Then the rib waveguides are generated in a reactive ion etching (RIE) process. The end of the etching process reaching the GaInAsP/InP-interface can be controlled either by measuring the phosphorus concentration of the plasma with a mass spectrometer [3] or by measuring the reflectance of the surface with an ellipsometer [4]. In a final photolithographic process one waveguide branch of the directional couplers is covered with resist for protection for the partial removal of the Au-layer by wet chemical etching. This process is compatible for the waveguide fabrication of the whole receiver circuit if it is done after the "hot" diffusion process for the MC.

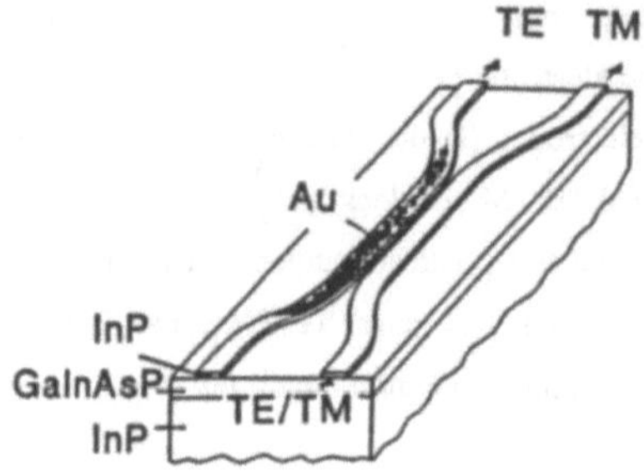

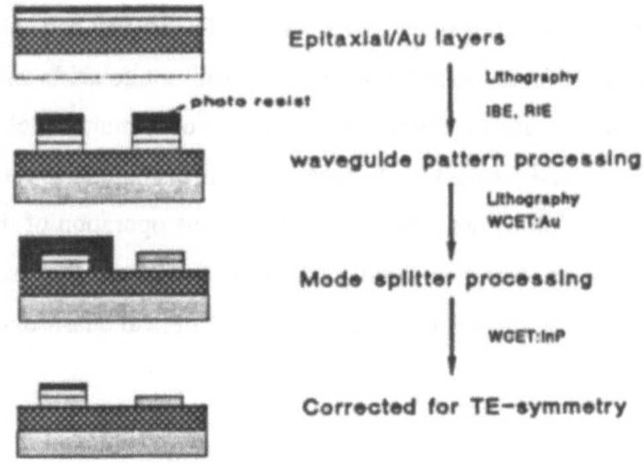

Fig. 4a: Mode splitter architecture
Fig. 4b: Selfadjusting process for mode splitter and waveguide fabrication.

The MS can act in two distinct modes of operation [9] (c.f. fig. 5): 1. For a waveguide with strong confinement the metallization generates an asymmetric coupler (TM-asymmetrical coupler type). As the influence of the metallization is small for TE-modes the coupler remains nearly symmetrical for this mode. 2. For weak confinement of the waveguide the lateral guiding of the TM-mode vanishes due to the metallization of the rib and for TM only one waveguide exists (TM-antiguide type). Therefore, independent of the device length, the TM-mode is guided to the bar state output port. TM-crosstalk and TM-excess loss due to the metallization are negligible .

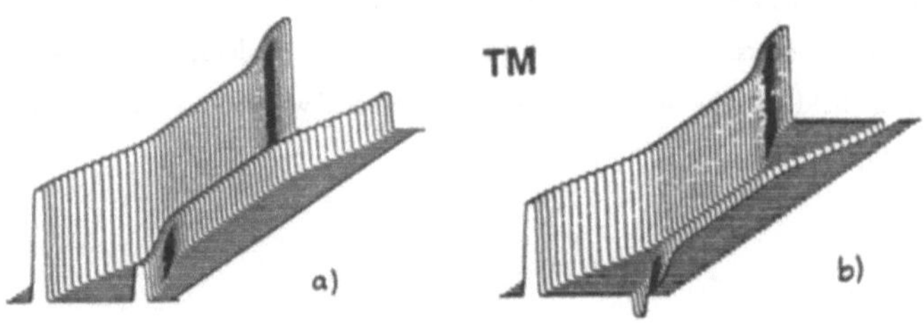

Fig. 5: Effective indices (beam propagation method calculation with complex indices) of different types of mode splitters: (a) TM-asymmetrical coupler type; (b) TM-antiguide type

To demonstrate the different operation principles the experimental crosstalk data versus device length (Device data: wavelength of $\lambda = 1.56\ \mu m$, waveguide width 3 μm, coupler gap 2 μm) are plotted in fig. 6a and 6b, respectively. The TM- and TE-crosstalk of the TM-asymmetrical coupler type oscillate, however, with different beat lengths due to the different asymmetries. In order to optimize this type of device two interdependent characteristics have to be tuned: The TE-mode asymmetry induced by the metal layer has to be eliminated and the TE- and TM-coupling lengths have to be matched. The residual losses were 0.3 dB for TE and 1.1 dB for TM.

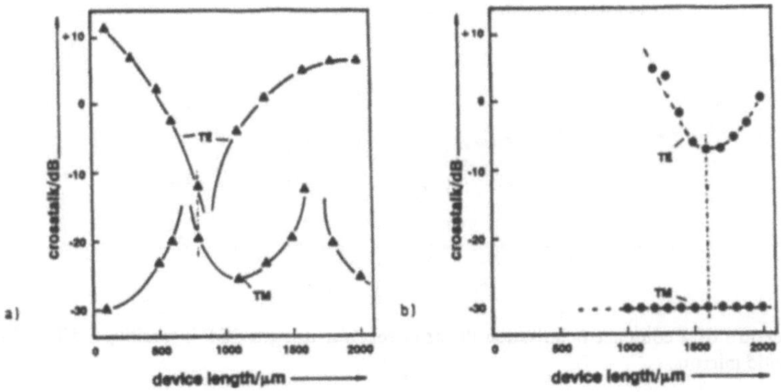

Fig. 6: TE-crosstalk ($10lg(I_=/I_X)$) and TM-crosstalk ($10lg(I_X/I_=)$) vs coupler length of the mode splitter devices
a) TM-asymmertrical coupler type, b) TM-antiguide type.
The vertical dashed-dotted lines indicate the operating points of each device.

For the TM-antiguide type an output signal for TM could only be measured in the bar state output port independent of the device lengths (TM-crosstalk <-30 dB). In the first fabricated device with the same physical dimensions of the coupler waveguides for TE only a relatively poor minimum crosstalk of -7.2 dB at a length of 1.6 mm (3 coupling lengths) has been achieved because the metal-cladding disturbed also the symmetry for TE. In a second device where the coupler asymmetry for TE is correced by changing the waveguide width of the metal-clad RWG by +5% and that of the other waveguide by -5% the TE-crosstalk could be improved to -12.2 dB. The residual losses were 0.3 dB for TE and 0.2 dB for TM [9].

3 dB-Coupler

Simulations on standard directional coupler structures verified that the coupling characteristics using RWGs with large lateral confinement differ significantly for TE and TM and that process tolerant couplers require gap widths g as small as possible [10]. However, standard photolithography limits g to about 2 µm. For realization of SOP insensitive 3 dB-couplers within the margin of ±10% we are investigating asymmetrical X-structures [11] and directional couplers with contradictionally tapered waveguide widths [12]. Such devices would allow the fabrication of a compact receiver waveguide network as sketched in fig. 7 where 45° waveguide mirrors are used in addition [13].

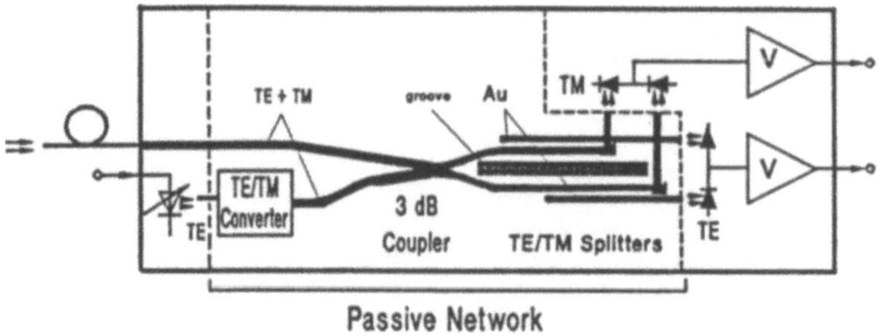

Fig. 7: Architecture of a compact polarisation diversity receiver using a SOP insensitive 3 dB-coupler and waveguide mirrors.

Acknowledgement

The work concerning the polarisation diversity receiver network was carried out within the RACE program, project "Subscriber Coherent Multichannel (CMC) System" supported by the EC, the Ministry of Research and Development of the Federal Republic of Germany, and the Senate of Berlin (West).

References

1. T. Okoshi, S. Ryn, K. Kikuchi: in Proc. IOOC/OFC '83, Tokyo, Japan, June 27-30, 1983, paper 30C3-2.
2. H.-P. Nolting, H. Heidrich, D.Hoffmann, German patent DE: P 36 31 798.5, 18.9.1986, US patent US-PS 4 832 431.
3. H. Schmid, F. Fidorra, D. Grützmacher, Inst. Phys. Conf. Ser. No. 96 (1988) 431.
4. R. Müller, in Proc. 2nd Internat. Conf. on Indium Phosphide and Related Materials, April 23-25., 1990, Denver (USA).
5. M. Schlak, H.-P. Nolting, P. Albrecht, W. Döldissen, D. Franke, U. Niggebrügge and F. Schmitt, Electron. Lett. 22 (1986) 883-884.
6. C.H. v. Helmolt, H. Heidrich, N. Schunk, J. Opt. Commun. 8 (1987) 94-97.
7. M. Schlak, C.M. Weinert, H.-P. Nolting, in Proc. 16th Europ. Conf. on Optical Commun. (ECOC), Sept. 16-20, 1990, Amsterdam (Netherlands).
8. P. Albrecht, M. Hamacher, H. Heidrich, D. Hoffmann, H.-P. Nolting, M. Schlak, C.M. Weinert, in Proc. Top. Meeting on Integrated Photonics Research, March 26.-28. 1990/ Hilton Head, Carolina (USA), Techn Dig. Series Vol. 5, (W14), 163.
9. P. Albrecht, M. Hamacher, H. Heidrich, D. Hoffmann, H.-P. Nolting, C.M. Weinert, IEEE Photon. Technol. Lett. 2 (1990) 114-115.
10. unpublished
11. W.K. Burns, F. Milton, IEEE J. Quantum Electron. QE-16 (1980) 446-454.
12. Y. Cai, T. Mizumoto, Y. Naito, J. Lightwave Technol. 8 (1990) 90-98.
13. P.Albrecht, W.Döldissen, U. Niggebrügge, H.-P. Nolting, H. Schmid, in Proc. 13th Europ. Conf. on Optical Commun. (ECOC), Sept. 13-17, 1987, Helsinki (Finland), Techn Dig. I, 239-241.

Wavelength Division Multiplexers for Optical Communication Systems

C. BORNHOLDT, F. KAPPE, H.-P. NOLTING, R. STENZEL, H. VENGHAUS, C. M. WEINERT

Heinrich-Hertz-Institut für Nachrichtentechnik GmbH
D 1000 Berlin 10 (F.R.G.)

Summary

Various wavelength division multiplexer/demultiplexer structures are discussed, which are suited for integration into opto-electronic integrated circuits. Three devices recently realized in the GaInAsP/InP material system are presented in more detail.

Introduction

Wavelength Division Multiplexing (WDM) is a powerful technique to exploit the high transmission capacitiy of optical fibres, and wavelength multiplexers/demultiplexers ('muldex') are key components of corresponding communication systems and customer access connections. According to their importance WDM devices have received considerable attention for many years and a large number of different varieties has been proposed and realized already.

Muldex structures have been made as fibre devices, on glass, on $LiNbO_3$, and on the basis of GaAs/AlGaAs or InP plus its lattice matched quaternary alloys GaInAsP or AlGaInAs. The choice of a certain material and a particular design depends on many criteria including the desired channel spacing/channel width, operation wavelengths, crosstalk, affordable tolerances, reliability and cost.

Devices made from fibres are already commercially available, and structures on glass [1] or $LiNbO_3$ [2] have also attained a high degree of maturity. Large efforts are currently spent on the development and improvement of semiconductor based components. They offer the possibility to be integrated into complete opto-electronic integrated circuits (OEIC), and by this means high reliability at low cost is expected in the long run. GaAs/AlGaAs devices are suited for operation in the 0.8 μm wavelength range, while InP and its alloys are adequate for operation at 1.3 μm or 1.55 μm, where the transmission characteristics of the optical fibre are most favourable. We will give a few examples of WDM devices recently realized in the InP material system, which is best suited for long haul and high bit rate communication. GaAs based devices are very similar as far as the basic structure and operation principles are concerned.

Most of the WDM structures realized so far fall into three distinct classes differing by the typical halfwidth of the respective filter transmission curves. Broadband devices are capable to separate 1.3 μm and 1.55 μm. Filter widths in the range of 10 nm or several 10 nm have been obtained with various interferometric structures, and muldex devices belonging to these two categories will be presented in more detail in the following. Filters with halfwidths of 1 nm or less are generally realized using Bragg gratings [3,4]. Gratings in waveguides in the III-V material system exhibit small coupling coefficients in general, and in order to achieve a sufficiently pronounced stop band a large number of grooves is required, which simultaneously leads to narrow filter characteristics.

Broadband multiplexer/demultiplexers

A Mach-Zehnder Interferometer (MZI) with appropriate design represents a broadband muldex, and corresponding examples reported in Ref. 5 exhibited crosstalk between -10 dB and -14 dB. A symmetrical directional coupler represents a broadband muldex as well, if the structural parameters are properly chosen. We realized a device in rib waveguide technology using a liquid phase epitaxy grown GaInAsP waveguiding layer (bandgap equivalent wavelength λ_g = 0.97 μm). Waveguides were structured by reactive ion etching with rib height control by in-situ ellipsometry [6]. Typical widths of the waveguides and of the gap between the guides in the coupling region are 2 - 3 μm, while the length of the coupling region itself ranges from 1 mm to 2 mm. Variation of the far-end isolation as a function of operation wavelength is shown in Fig. 1. Best values are below -23 dB, and if -15 dB is required for system applications, the device can operate within a wavelength range of ± 30 nm at 1.3 μm and ± 15 nm at 1.55 μm, respectively. Polarization dependence of the device is negligible, and the additional attenuation of the coupler is 0.3 dB.

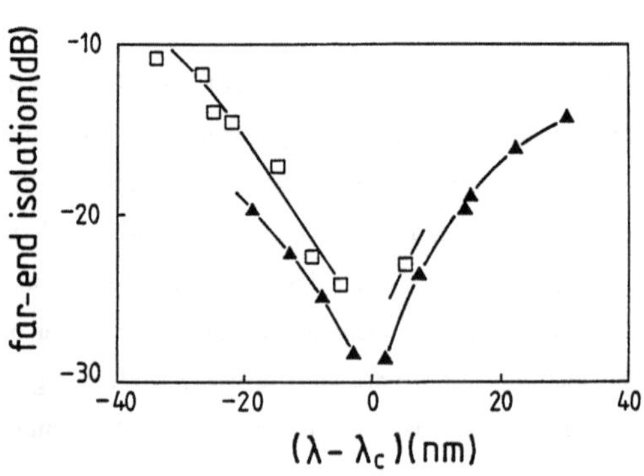

Fig. 1:

Dependence of far-end isolation on operation wavelength. Center wavelengths λ_c = 1303 nm (▲) and 1545 nm (□).

If the gap between the waveguides is reduced to zero, one gets a zero-gap coupler as reported recently for 1.3 μm / 1.55 μm channel separation [7]. The buried waveguides used led to a very short coupling region of 178 μm only. However, the crosstalk at 1.3 μm was unsatisfactory, a pronounced polarization dependence was also observed and fabrication tolerances are critical for this type of device.

Intermediate halfwidth structures

Two mode interference filters can be designed with halfwidths of the order of 10 nm [8-10]. This matches particularly well fibre optic systems with operation wavelengths either in the 1.3 μm or 1.55 μm region. Interferometers exhibit a quasi periodic sequence of transmission maxima and minima, the channel width is essentially inversely proportional to the length of the active part of the device, and actual device lengths are typically several or even many mm.

One variant of a MZI has been recently developed, which is particularly promising with respect to a short device length, although it has not yet been optimized for this purpose [11]. The device constists of three sections, the first and the last of which are 3-dB-couplers, while the central part is a strongly asymmetric directional coupler, where the asymmetry assures essentially no coupling in this region. As a consequence of the waveguide asymmetry the propagation constants corresponding to both waveguides exhibit different wavelength dependence, and this dispersion difference determines the filter characteristics. The waveguide asymmetry has been realized by different waveguide heights, and this gives rise to a difference in effective refractive index of the waveguides as shown in the left part of Fig. 2. Transmission curves of the filter measured for TE and TM polarization are also shown in Fig. 2.

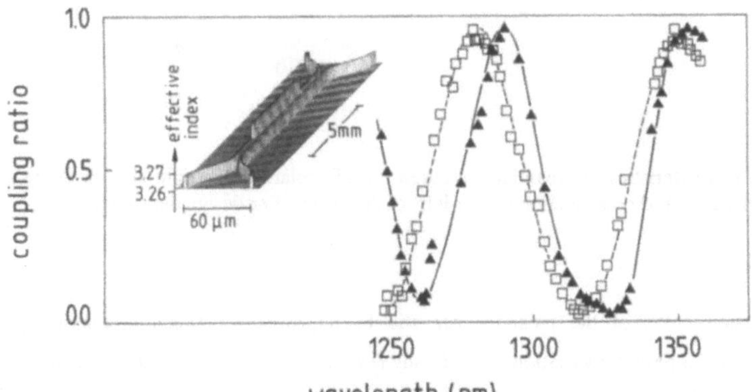

Fig. 2:

Transmission characteristics of MZI with asymmetric rib waveguides in central region, measured for TE (□) and TM (▲) polarization. Structure of device visualized by variation of effective index along waveguides (left part of figure).

Another muldex with similar halfwidth, but a sinc^2-shaped transmission curve (instead of quasi periodic) is the so-called meander, proposed theoretically and discussed in detail in Ref. 12 and realized recently. The meander is a strongly asymmetric codirectional coupler with periodic spatial variation of the coupling constant κ. A particularly simple way to realize the variation of κ is a periodic change of the distance between the two waveguides of the coupler (hence the designation 'meander', cf. insert in Fig. 3).

The fabrication of a meander with a specific center wavelength and filter halfwidth requires the proper choice of various parameters such as the composition and the thickness of the waveguiding layer, the geometry and the distance of the waveguides, the period length Λ, and the total number of periods N. An example of a meander filter curve is shown in Fig. 3. The device was fabricated with the following parameters: Bandgap equivalent wavelength of the GaInAsP layer $\lambda_g = 1.06\ \mu m$, thickness $t = 700$ nm, rib heigths 70 nm and 180 nm, waveguide width: 2 μm, separation varying between 2 μm and 3 μm, period length $\Lambda = 100\ \mu m$, number of periods $N = 13$.

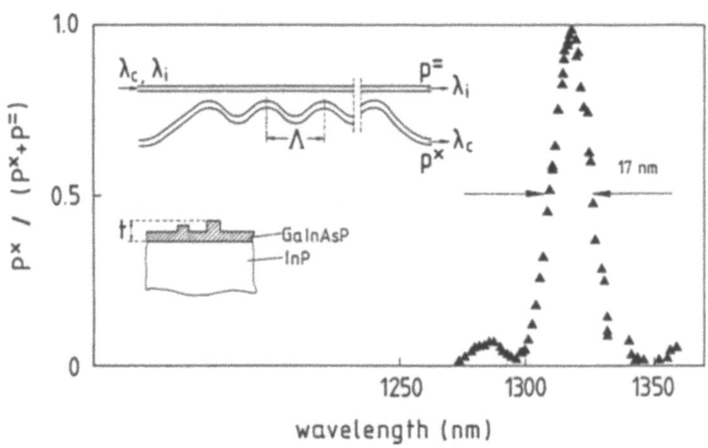

Fig. 3:

Transmission characteristics of meander measured for TE polarization (right side of figure), schematic top view (upper left) and cross section (lower left) of the device. For design data see text.

The transmission curve of the meander can still be improved by weighted coupling. This is the superposition of the periodic variation of κ already realized plus a slight and steady change of κ, which leads to a suppression of the sidelobes of the filter curve still present in Fig. 3.

Miscellaneous devices

All devices discussed so far are typical examples of structures, which are suited for monolithic integration, but several other concepts have also been realized and published in the literature. Among these are spectrometer type demultiplexers [13] , asymmetric Y-junctions for wavelength separation [14] and phase-controlled distributed feedback lasers serving as tunable wavelength filters [15]. Finally, a sequence of superlattice p-i-n photodetectors in a waveguide configuration, where applied voltages of different magnitude assure the wavelength selection due to the quantum-confined Stark effect [16] should also be mentioned. This device does not represent a multiplexer/demultiplexer in its rigorous sense, but it is equivalent to a demultiplexer/detector combination and it may replace a true demultiplexer in particular applications.

Conclusion

None of the devices developed so far is superior to the others. The width and the separation of the channels of an optical communication system are important determinants for the choice of a specific WDM structure, but the selection depends also strongly on the design of the OEIC which comprises the muldex as a central part, and on the technological processes available to realize the OEIC.

References

1. Ramaswamy, R.V.; Srivastava, R.: Ion-Exchanged Glass Waveguides: A Review. J. Lightwave Technol. **6** (1988) 984-1001.

2. Thylen, L.: Integrated Optics in $LiNbO_3$: Recent Developments in Devices for Telecommunications. J. Lightwave Technol. **6** (1988) 847-861.

3. Yi-Yan, A.; Carru, H.; Correc, P.; Chandouineau, J.P.; Dugrand, L.; Carre, M.: GaAs/GaAlAs rib waveguide grating filters for λ = 1.5 μm. Electron. Lett. **23** (1987) 622-624.

4. Cremer, C.; Heise, G.; März, R.; Riechert, H.; Schienle, M.: InGaAsP Y-branch grating demultiplexer. Electron. Lett. **23** (1987) 321-322.

5. Walker, R.G.; Bennion, I.; Carter, A.C.: Guided-Wave 1.3/1.55 μm Wavelength Duplexers in InGaAsP/InP. Proceedings of 5th European Conference on Integrated Optics, ECIO'89, paper 1/4/35.
   Walker, R. G.; Urquhart, J.; Bennion, I.; Carter, A. C.: 1.3/1.53 μm Mach-Zehnder wavelength duplexers for integrated optoelectronic transceiver modules. IEE Proceed. **137, Pt. J.** (1990) 33-38.

6. Müller, R.: Control of a reactive ion etching process for InP and related materials by in-situ ellipsometry in the near infra red. Proceed. of 2nd Internat. Conf. on InP and Related Materials, 23. - 25.4.90, Denver, Co. (USA).

7. Schienle, M.; Stoll, L.; Schulte-Roth, G; Müller-Nawrath, R; Mahlein, H.F; März, R.; Cremer, C.: GaInAsP/InP zero-gap directional couplers as compact optical WDM filters. Electron. Lett. **25** (1989) 1180-1181.

8. Bourbin, Y.; Enard, A.; Blondeau, R.; Rondi, D.; Glastre, G.; Papuchon, M.: Monolithically integrated multi/demultiplexer and access waveguides in InGaAsP/InP. Proceedings of 5th European Conference on Integrated Optics, ECIO'89, paper 1/4/37.

9. De Bernardi, C.; Morasca, S.; Rigo, C.; Sordo, B.; Stano, A.; Croston, I.R.; Young, T.P.: Two mode interference wavelength demultiplexer monolithically integrated on InP, for 1.5 or 1.3 μm operation. Proceedings of 5th European Conference on Integrated Optics, ECIO'89, paper 1/4/38.
De Bernardi, C.; Morasca, S.; Rigo, C.; Sordo, B.; Stano, A.; Croston, I. R.; Young, T. P.: Wavelength demultiplexer integrated on AlGaInAs/InP for 1.5 μm operation. Electron. Lett. **25** (1989) 1488-1489.

10. Yi-Yan, A.; Deri, R.J.; Seto, M.; Hawkins, R.J.: GaAs/GaAlAs asymmetric Mach-Zehnder demultiplexer with reduced polarization dependence. IEEE Photonics Technol. Lett. **1** (1989) 83-85.

11. Bornholdt, C.; Kappe, F.; Nolting, H.-P.; Stenzel, R.; Venghaus, H.; Weinert, C.M.: Wavelength multiplexer/demultiplexers based on codirectional couplers on GaInAsP/InP. Technical Digest on Integrated Photonics Research, 1990 (Optical Society of America, Washington, D.C., 1990), Vol. 5, p. 122.

12. März, R.; Nolting, H-P.: Spectral properties of asymmetrical optical directional couplers with periodic structures. Optical and Quant. Electron. **19** (1987) 273-287.

13. Gibbon, M.; Thompson, G.H.B.; Clements, S.J.; Moule, D.J.; Rogers, C.B.; Cureton, C.G.: Optical performance of integrated 1.5 μm grating wavelength-demultiplexer on InP-based waveguide. Electron. Lett. **25** (1989) 1441-1442.

14. Negami, T.; Haga, H.; Yamamoto, S.: Guided-wave optical wavelength demultiplexer using an asymmetric Y junction. Appl. Phys. Lett. **54** (1989) 1080-1082.

15. Numai, T.; Murata, S.; Mito, I.: 1.5 μm tunable wavelength filter with wide tuning range and high constant gain using a phase-controlled distributed feedback laser diode. Appl. Phys. Lett. **53** (1988) 1168-1169.

16. Larsson, A.; Andrekson, P.A.; Andersson, P.; Eng, S. T.; Salzman, J.; Yariv, A.: High-speed dual-wavelength demultiplexing and detection in a monolithic superlattice p-i-n waveguide detector array. Appl. Phys. Lett **49** (1986) 233-235.

Pin Photodiodes and Field-Effect Transistors for Monolithically Integrated InP/InGaAs Optoelectronic Circuits

H. Albrecht

Siemens Research Laboratories
Otto-Hahn-Ring 6, D-8000 München 83, Federal Republic of Germany

Summary

A monolithically integrated planar structure for an InP/InGaAs optoelectronic circuit consisting of a pin photodiode and a heterojunction field-effect transistor is described. The optimization for the FET has been obtained with a local buried p-n junction. The electrical and optical properties of pin, FET and pin-FET are presented.

Introduction

The integration of multiple optical and electronic functions on a single chip to achieve Opto-Electronic Integrated Circuits (OEIC), offers one approach to low cost high performance components to realize high-speed and economical transmitters and receivers for optical transmission systems [1]. In photoreceivers for direct and coherent detection the monolithic integration of pin photodiode and electronic amplifiers, using e.g. field-effect transistors (FET), reduces the stray capacitances and therefore has the potential of low equivalent input noise current and improved receiver sensitivity.

Device design

The monolithic integration of photodiode and FETs reveals the difficulties of OEIC device technology. Optimized pin photodiodes requires a thick absorption layer with low background doping concentration. In contrast FETs of high performance need a high doping concentration in a thin channel layer. To integrate both components with respect to their optimized device performances, different technologies have been developed on the basis of "horizontal" and "vertical" integrated structures [2]. In this paper we use a vertical integrated layer structure on a flat surface substrate. Fig. 1 shows a schematic cross section of our monolithically integrated planar InP/InGaAs pin-FET combination. The layer structure consists of (i) a thick n-doped InP buffer layer, which reduces the series resistance and transit time effects of

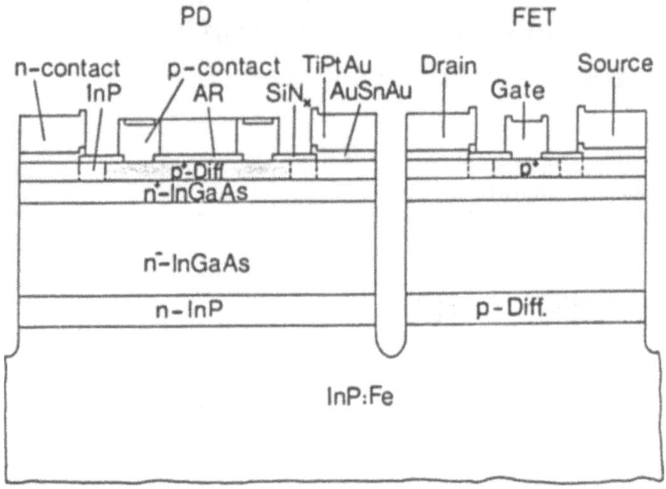

Fig.1. Schematic cross section of an InP/InGaAs pin-HJFET with buried p-layer

the pin, (ii) a thick undoped InGaAs absorption layer, (iii) a thin n-doped InGaAs channel layer and (iiii) a thin n-doped InP window layer, which also enables good passivation by the SiN_x coating. Beneath the region of the FET a buried p-layer has been incorporated. The space charge region of this p-n junction reaches through the undoped InGaAs absorption layer which acts as a buffer layer, and confines the lateral drain current to the thin n-doped channel layer. Besides this effect the buried p-layer in the FET structure improves the threshold voltage uniformity, suppresses drain conductance transients, alleviates short channel effects and reduces backgating effects [3]. To realize optimized performances for pin and FET the following conditions have to be fulfilled: (i) For a high transconductance of the FET the doping level of the channel layer should be high. However, the electric field strengh in InGaAs has to be kept low enough to avoid large tunnel currents in the gate diode and in the pin. (ii) In order to achieve high quantum efficiency in the pin, the absorption layer thickness should be large. But a limiting factor of the thickness is the transit time of photogenerated carriers through this layer, which affects the bandwidth of the pin. (iii) The sheet carrier concentration of the nominal undoped absorption layer should be low enough, so that in the region of the FET due to the buried p-n junction, this layer is depleted from charge carriers.

Device technology

The epitaxial layers were grown lattice matched onto a semi-insulating (100) InP:Fe substrate by low pressure MOVPE [4]. In a first step a 1 µm thick n-

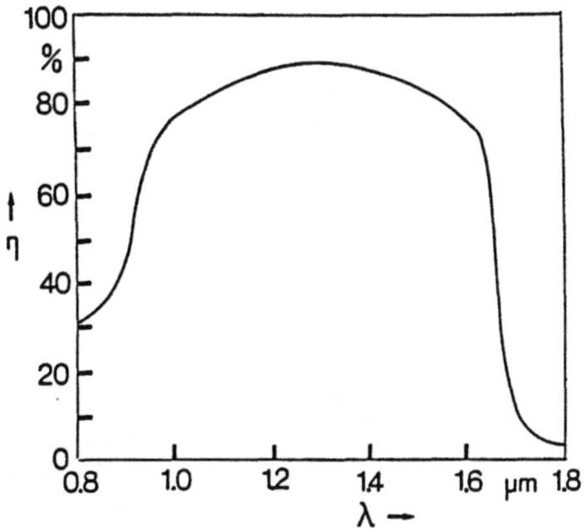

Fig.2. Spectral response of quantum efficiency of an InP/InGaAs pin diode

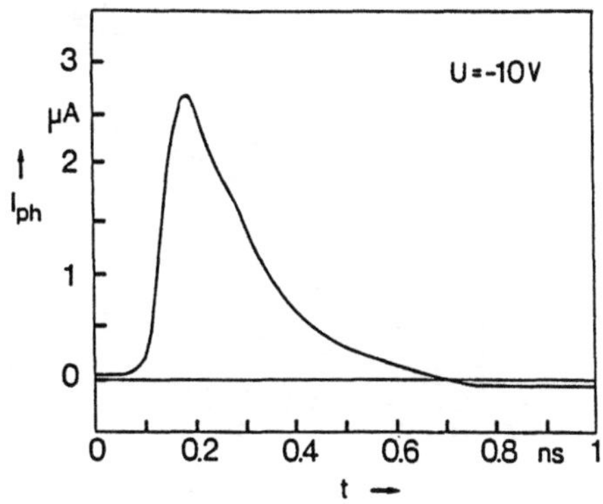

Fig.3. Time resolved photo-current of an InP/InGaAs pin Diode

doped InP buffer layer was grown. The epitaxial layer growth was interrupted and in the region of the FET a selective Zn diffusion was carried out using as a SiN_x mask. After removing the diffusion mask in a second growth step the undoped 1.5 μm thick InGaAs absorption layer, the 0.3 μm thick n-doped InGaAs channel layer (doping concentration is $8 \cdot 10^{16}$ cm^{-3}) and the 0.3 μm thick n-doped InP cap layer were deposited. Due to the low Schottky barrier height of metals on InP, which results in an undesired high gate leakage current, p-n junctions have been used for the pin and the gate of the FET. The p-region was fabricated by Zn diffusion with a SiN_x mask. The resulting planar p-n junction is 0.3 μm beneath the semiconductor surface very close to the InP/InGaAs heterojunction, providing a heterojunction field-effect transistor (HJFET). Details of further technological steps are described elsewhere [5]. The p-contact of the pin with a p-layer diameter of 80 μm is formed as a ring contact to permit surface illumination. Pin and HJFET were isolated by mesa etching down to the InP:Fe substrate.

<u>InP/InGaAs pin photodiode</u>

At a bias voltage of -5V a typical reverse bias characteristics of a pin show a dark current of 100nA and a capacitance of 0.35pF. For surface illumination of the pin the measured external quantum efficiency versus wavelength is shown in fig. 2. With the antireflection coating an external quantum efficiency of 90% at a 1.3 μm wavelength were obtained. Fig. 3 shows the time response of the pin measured at a reverse bias voltage of - 10V using short light pulses from a semiconductor laser with a 1.3 μm emission wavelength. The measured half-width value of the pin was less than 150 ps, corresponding

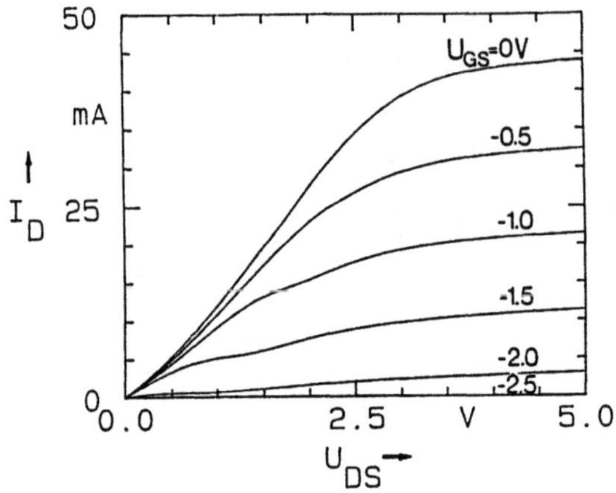

Fig.4. Typical output characteristic of an InP/InGaAs HJFET.

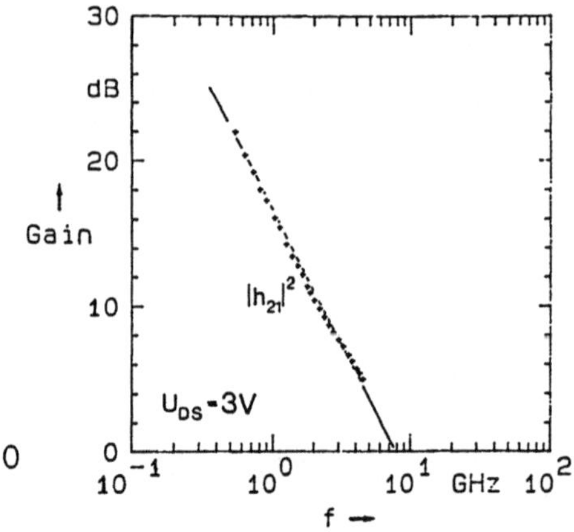

Fig.5. Current gain versus frequency of an InP/InGaAs HJFET.

to a 3db-bandwidth of 2.1 GHz. This value can be improved with a thicker and higher doped InP buffer layer.

<u>InP/InGaAs heterojunction field-effect transistor</u>

A typical output characteristic for an integrated InP/InGaAs HJFET with 1.6 µm gate length and 190 µm gate width is shown in fig. 4. The device exhibits an excellent pinch-off behaviour at a threshold voltage of -2.5V. At a gate-source voltage of U_{GS} = 0V the maximum drain current is 44mA or normalized to the gate width 232mA/mm, the external transconductance is 28mS or 147mS/mm and the output conductance is 1mS or 5.3mS/mm. Therefore the open loop voltage gain is $V_u = g_m/g_D = 28$. At low drain-source voltages the drain current shows a non-linear behaviour. This can be explained by non-ohmic drain/source contact due to a low doping concentation in the InP cap layer. This problem can be solved by a higher doped cap layer. At a gate-source voltage of -5V the measured gate leakage current is 100nA and the gate capacitance 0.4pF. The microwave S-parameter of an InP/InGaAs HJFET with buried p-n junction, wire-bonded to a microwave carrier, were measured at U_{GS} = 0V and U_{DS} = 3V in the frequency range from 0.5GHz to 5GHz using standard network analyzer technique with error correction. Based on the measured S-parameters, the current gain versus frequency was calculated as shown in fig. 5. An extrapolation of this curve to the frequency f_T, where the current gain is unity, gives f_T = 8GHz.

<u>Monolithically integrated InP/InGaAs pin-HJFET combination</u>

To determine the performance of the monolithically integrated pin-HJFET chip, signal and noise power measurements were performed in the frequency range

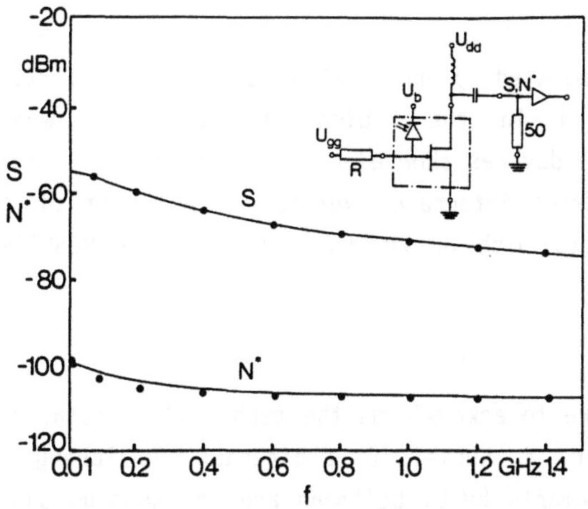

Fig.6. Signal S and noise power N* versus frequency at the output of the test circuit. Noise measurement bandwidth is 1MHz. The equivalent circuit is shown. measured: .... simulated: ———

from 15 MHz to 1.5 GHz with a network analyzer and a spectrum analyzer. For this purpose the device was mounted on a microwave carrier together with a photodiode load resistance of 1kΩ . The measured signal and noise characteristic versus frequency of modulated light at the output of the test circuit are shown in fig. 6 at a bias voltage of -6V for the pin and a drain voltage of 2V and a gate voltage of -1.1V for the HJFET. The measurement bandwidth for the noise investigation is 1MHz. Signal and noise measurements were analyzed with a PSPICE simulation program, which simulates the large signal and small signal characteristics of pin and HJFET [6]. The measured and simulated results deviate within ±3% (fig. 6). From the frequency response at the output of the test circuit an input capacitance of 0.73pF can be determined from fig. 6 with respect to the measured 3-dB-frequency of 200MHz in connection with the photodiode load resistance of 1kΩ. To describe the noise characteristic at the output of the test circuit following noise contributions were considered [7]: Thermal noise of the resistance, shot noise of the dark current, gate leakage current and signal current, 1/f-noise and the thermal channel noise.

With these measured curves the receiver sensitivity has been calculated according to the signal frequency response and the total output noise power value based on integrating the noise voltage spectrum within the bandwidth of the receiver. For a NRZ-signal with a transmission rate of 400MBit/s and a bit error rate of 10^{-9} a receiver sensitivity of -35dBm at 1.3μm wavelength without equalization can be determined from fig. 6. This result is in the range of sensitivity values, measured on monolithically integrated InP/InGaAs pin-preamplifier chips [8].

Conclusion

Monolithically integrated planar InP/InGaAs pin-HJFET combinations with buried p-n junction in the region of the HJFET were realized and characterized. These devices show electrical and optical properties which are comparable with hybrid integrated versions consisting of optimized single InP/InGaAs pin diodes with 80 μm diameter and GaAs MESFETs with 1μm gate length.

Acknowledgments

The author would like to acknowledge the technical assistance of R. Penz, Ch. Lauterbach, H. Huber, W. Kunkel, G. Ebbinghaus, T. Scherg, R. Strzoda, the signal noise measurements by L. Hoffmann and the helpful discussions with M. Plihal. This work has been supported by the Deutsche Bundespost of the Federal Republic of Germany.

References

1. Forrest, S.R.: Optoelectronic Integrated Circuits. Proc. IEEE 75 (1987) 1488-1497

2. Henning, I. D.: Materials for Optoelectronic and Photonic Integrated Circuits. Prog. Crystal and Charact. 19 (1989) 1-20

3. Heime, K.: InGaAs Field-Effect Transistors. New York: John Wiley & Sons Inc. 1989

4. Ebbinghaus, G., Strzoda, R., Scherg, T., Albrecht, H., Penz, R., Lauterbach, Ch.: Two-step MOVPE growth for planar InP/InGaAs/InP pin-FET combinations with locally diffused buffer layer. Submitted to J. Crystal Growth.

5. Albrecht, H., Penz, R., Lauterbach, Ch., Hoffmann, L., Ebbinghaus, G., Scherg, T., Strzoda, R.: Advanced Integrated Planar InP/InGaAs pin-FET Photoreceiver with Inhomogeneous Doped Layer Structure. Proc. EFOC/LAN (1990)

6. Albrecht, H., Hoffmann, L., Penz, R.: Signal-Rausch-Messungen an monolithisch integrierten InGaAs/InP:Fe-Photodiode-Feldeffekt-transistor-Kombination. VDE-Verlag. ITG-Fachbericht 112: Hetero-struktur-Bauelemente (1990) 177-182

7. Park, M.S., Shim, C.S., Kang, M.H.: Analysis of Sensitivity Degradation Caused by Flicker Noise of GaAs-MESFET in Fiber-Optic Receivers. J. Lightwave Techn. LT-6 (1988) 660-667

8. Lo, D.C., Forrest, S.R.: Performance of InGaAs and InP Junction Field-Effect Transistors for Optoelectronic Integrated Circuits. II. Optical Receiver Analysis. J. Lightwave Techn. LT-7 (1989) 966-971

Semiconductor Laser Amplifiers as Multifunctional Devices

H.G.Weber, G.Großkopf, R.Ludwig, R.Schnabel, N.Schunk
Heinrich-Hertz-Institut für Nachrichtentechnik Berlin GmbH,
Einsteinufer 37, D-1000 Berlin 10, FRG

Summary
Semiconductor laser optical amplifiers can be used as multifunctional devices in optical signal transmission, optical switching and optical signal processing. Examples described here are the linear optical amplifier, the all optical frequency converter and applications of the semiconductor laser amplifier in the electrooptic mode of operation for instance as optical switch, optical modulator or optoelectronic converter. We point out, that semiconductor laser amplifiers realized as multi-sections devices may be of special advantage, because they can be adjusted to the desired applications most appropriately.

1.Introduction

Optical amplifiers are likely to become an important component for optical signal transmission, optical switching and optical signal processing. Presently, there are two competing candidates for optical amplifiers: the semiconductor laser amplifier and the Er^{3+}-doped fibre amplifier. Both optical amplifiers were successfully operated in laboratories and in first field-trials and their properties were discussed in several recent review articles [1 - 6]. Both amplifiers offer a broad range of applications. As linear amplifiers they enhance the span in long-haul fibre optic transmission systems, they compensate for combining and splitting losses in distribution networks, and they improve the receiver sensitivity in high bit rate direct detection systems. Optical amplifiers can also be used in a nonlinear mode of operation and in an electrooptic mode of operation. In this paper we give a short outline of various applications of optical amplifiers focussed essentially on the semiconductor laser amplifier (SLA), because its range of applications is broader and better investigated today.

2. Properties of the Semiconductor Laser Amplifier (SLA)

The SLA shown schematically in Fig. 1 is a semiconductor laser diode having a typical length of 250 - 500 μm. Its facets exhibit reflectivities of commonly less than 10^{-3}. Low facet reflectivities eliminate the feedback required for laser oscillation. Thus an optical signal coupled into the active region of the laser diode is amplified by stimulated emission in a single pass gain (travelling wave amplifier). The device is biased by a DC-injection current. This current acts as a pump providing population inversion between conduction band and valence band in the active region. Today, in most research activities the laser diodes are InGaAsP devices operating in the 1.3 - 1.55 μm wavelength range because this is the wavelength window where optical fibres have minimum loss (1.55 μm) and zero dispersion (1.3 μm).

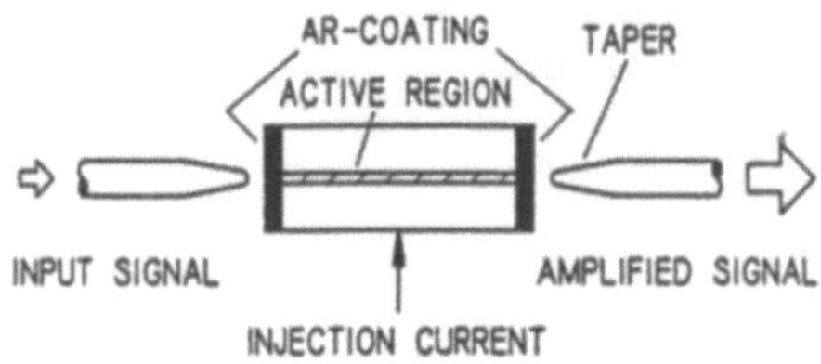

Fig. 1: Schematic of semiconductor laser amplifier (SLA)

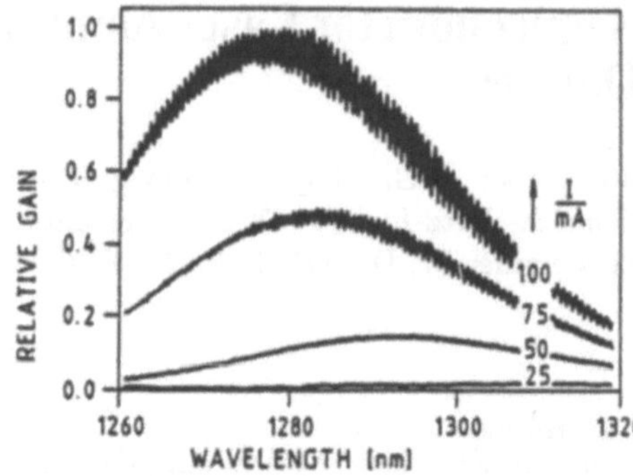

Fig. 2: TE-gain spectrum of the SLA for different injection currents I. The maximum gain for I = 100 mA is 24 dB

The reduction of facet reflectivity is one of the major problems of SLAs. The method most often used is deposition of a dielectric anti-reflection coating film on the facets. Optimum values obtained today are $R \approx 10^{-5}$. Other methods to reduce the facet reflectivity are tilted facets (angle $\approx 7°$) with respect to the waveguide direction [7] or an appropriately designed non-guiding region (window structure) between the facet and the waveguide region [8]. Both methods are usually combined with the deposition of an AR-coating.

The residual facet reflectivity R affects directly the gain spectrum of the SLA. Fig. 2 shows signal gain spectra for different injection currents I. With increasing injection current the maximum of the gain spectrum is shifted to shorter wavelengths (band filling effect) and the gain and the ripple of the gain curve increase. This ripple is the Fabry-Perot-resonance structure caused by the residual reflectivity $R \neq 0$. In the example shown in Fig. 2 the gain is 24 dB and the gain ripple is less than 2 dB at I = 100 mA. The 3 dB bandwidth of the SLA gain is about 4000 GHz. The bandwidth can be as large as 8000 GHz determined only by the material properties of the SLA if the gain ripple is further reduced.

The waveguide in a conventional SLA guides two perpendicularly polarized modes with different waveguide properties. This causes a dependence of the gain on the polarization of the optical waves. In general, light linearly polarized in the junction plane of the active region (TE) has a higher gain (for instance 5 dB for I = 100 mA in the example in Fig. 2) than light polarized perpendicularly to this plane (TM). Fig. 2 shows results for TE polarized waves. There are several ways to overcome this polarization problem [9, 10].

SLAs reveal saturation behaviour for large optical powers at the input or output. For instance, for the amplifier characterized by Fig. 2, the signal gain at I = 100 mA decreases by 3 dB at the output power +7 dBm (saturation output power). For MQW-structure-SLAs saturation output powers up to 20 dBm were already reported [11]. In optical amplifiers stimulated emission (gain) is always accompanied by spontaneous emission which is also amplified and which is the origin of amplifier noise. The total noise power of the optical amplifier, which is characterized by Fig. 2, is about -3 dBm at I = 100 mA. However, this noise power is distributed over a spectral range of about 8000 GHz. Consequently the spectral noise power is sufficiently low such, that this noise does not restrict the applications of SLAs seriously for instance in coherent optical transmission systems [1].

3. The SLA as linear optical amplifier

Typical applications in the linear mode of operation are: non-regenerative repeaters for long haul transmission systems [1], booster amplifiers for distribution networks [12], and low noise preamplifiers for high bit rate direct detection systems [13]. Here we give a short description of the application as non-regenerative repeater.

The transmission span of optical coherent communication systems is most often limited by fibre attenuation and not by fibre dispersion. Therefore a complete regeneration of the signals is not required at each repeater and a linear optical amplifier may be used to enhance the transmission span. Such in-line optical amplifiers have been demonstrated in direct and coherent optical transmission systems [e.g. 1, 6]. The main advantage of these amplifiers appears in multi-channel, wavelength division multiplexed (WDM) systems. As the SLAs exhibit gain bandwidths of several terahertz, they are able to amplify several optical signals simultaneously and independently of bit rate and data format. It is, therefore, expected, that the installation of SLAs will greatly simplify repeaters in multi-channel long haul transmission systems.

As an example we consider in the following the coherent optical multi-channel transmission system depicted schematically in Fig. 3. It consists of several transmitters [T] operating at the wavelength $\lambda_1 \ldots$ to λ_n, several in-line SLAs and finally heterodyne receivers [R], which may be tuned to one of the signal channels. Experimental and theoretical investigations on such a system led to the following conclusions:

The number of channels in a coherent optical transmission system with optical amplifiers seems to be limited not by the bandwidth of the amplifiers but by the amplifier saturation output power. An estimated value for the maximum number of channels is 100 if we assume the amplifier characteristic described in section 2. The channel spacing in the transmission system is in general not affected by the amplifiers. Crosstalk is in general no problem in coherent systems with FSK or PSK modulation. However, crosstalk affects multi-channel transmission in direct detection systems. The maximum number of cascaded optical amplifiers is determined by gain saturation due to accumulation of amplified spontaneous emission. The maximum number is about 20 if we use the optical amplifiers characterized in section 2, and if no narrow-band optical filters are inserted into the amplifier chain. An essential problem of cascaded optical amplifiers is the generation of echoes due to backward gain. Residual facet reflectivities of the amplifiers cause backward travelling waves which arrive at the receiver as echoes of the signal superimposed on the transmitted signal. Assuming an echo suppression by optical isolators and a 560 Mbit/s DPSK modulation scheme, we expect that 20 amplifiers inserted into a 100 channel transmission system give a net link loss improvement of 200 dB. This corresponds to an increased fibre length of 1000 km at 1.5 μm additionally to the link loss or fibre length without optical amplifiers. Long-haul transmission experiments with 10 in-line SLAs (516 km, 2.4 Gbit/s, direct detection [6]) have been reported.

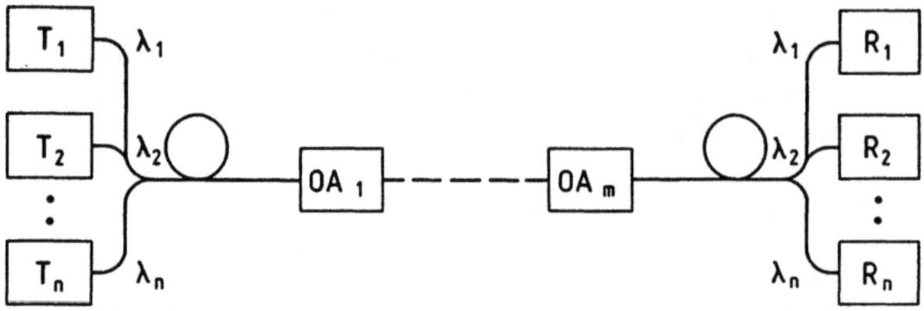

Fig. 3: Schematic of a coherent optical multi-channel transmission system with transmitters T, receiver R, and in-line optical amplifiers OA which are assumed to be SLAs here.

4. The SLA as all-optical frequency converter

SLAs exhibit nonlinear optical properties because of the interdependence of the complex index of refraction and the optical power in the active region. This nonlinear optical behaviour involves several interesting applications of the SLA, for example: optical bistability [14], all-optical switch [15], all-optical regenerative repeater [16] or all-optical frequency converter [17]. In the following we discuss the frequency converter shortly.

Optical frequency converters are expected to be important components for switching nodes for instance in coherent optical multi-channel systems. The function of the node is to switch a given data signal at the carrier frequency f_x to a desired carrier frequency f_y. This switching was demonstrated [17] all-optically with use of a SLA with a frequency conversion range $| f_x - f_y | \leq 4000$ GHz. The conversion efficiency was about -7 dB and the required optical input powers were less than 50 μW. The physical mechanism behind this frequency converter is Nearly Degenerate Four-Wave-Mixing (NDFWM) in the SLA together with homogeneous gain saturation of the SLA. NDFWM is a third order nonlinear optical effect ($\chi^{(3)}$). It results in a carrier density modulation at the beat frequency between two incident light waves, the signal wave with frequency f_x and a pump wave. This carrier density modulation affects homogeneously the entire gain spectrum of the SLA. If a second pump wave at frequency $\approx f_y$ is passed through the SLA, it experiences this carrier density modulation and generates two side bands. These side bands are frequency shifted replica of the input data signal. Turning on and off the second pump wave (it can be even more than only one) at a desired frequency enables switching of the data signal from a given carrier frequency f_x to one (or several) desired carrier frequency f_y.

5. The SLA in the electrooptic mode of operation

In the applications described before the SLA is biased by a constant DC-injection current. However, there are also applications of the SLA in which the injection current I has a controlling function. Switching on and off the current I changes the optical transmission of the device from 20 dB gain to 30 dB loss. These optical switches can be used to design photonic switching matrices which have the advantage of high fan out. If a DFB-laser is used as SLA one has a SLA with a narrow width (≈30 GHz) of the gain spectrum. Tuning the injection current I results in a tunable narrow band amplifier [18]. Modulation of the injection current causes carrier density modulation in the SLA. Consequently an optical wave travelling through the SLA experiences phase [19] and amplitude modulation [20]. Under suitable conditions the amplitude modulation can be much smaller than the phase modulation and the SLA can be used as phase modulator as experiments have demonstrated [21]. The SLA-phase modulator is very attractive, because it offers not only the possibility of optoelectronic integration on InP, but also an optical net gain. Moreover it requires lower modulation power than $LiNbO_3$ devices which are most often used as phase modulators, presently. Another application of SLAs is the use as optoelectronic converters because any modulation of the injected light intensity results in a corresponding modulation of the injection current. This application is of particular interest if the SLA is used as a general functional element in complex integrated optoelectronic circuits. As an example Fig. 4 shows an application of an SLA as a general functional device operating either as amplifier, as transmitter or as receiver [22]. A serious limitation of all electrooptic applications of the SLA is the restricted switching or modulation speed of this device. This speed is limited by the carrier lifetime which gives an optimistic upper limit of about 1 GHz.

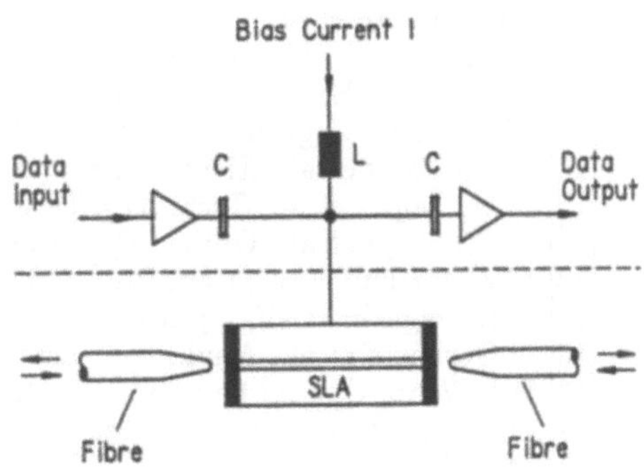

Fig. 4: The SLA as a functional device operating as optical amplifier, optical transmitter and optical receiver. The electronic part is above the broken line and the optic part below the broken line.

6. Conclusion

Optical amplifiers are likely to become key-components in future optical communication systems. We reported here mainly on semiconductor laser amplifiers (SLAs). The goal was to show that this device can be used as multifunctional device in many applications, for instance as linear amplifier, as all-optical frequency converter, as optical switch, as phase modulator etc. Presently, there are many research activities to improve this device with respect to high saturation output power, low spontaneous emission, low polarization sensitivity, and low residual reflectivity. The SLA considered here was the single-section SLA introduced in section 2. It has already been shown [23], that multi-sections SLAs will enable the construction of optical amplifiers as multi-functional devices, which are better adjusted to the specific applications than the conventional single-section devices. Some of the applications of the SLAs reported in this paper will eventually more favourably be performed by Er^{3+}-doped fibre amplifiers for instance the applications in the linear mode of operation. However, other applications will probably only be possible with SLAs.

Acknowledgement

The authors thank their colleagues in HHI for their support, particularly G.Heydt and E.Patzak for helpful discussions, R. Molt for antireflection-coating laser diodes and H. Knupke, B.Mikulski, H.J.Hensel, K.Peters and G.Reise for technical assistance. The work was supported by the Bundesminister für Forschung und Technologie of the Federal Republic of Germany under contract TK 0438. The authors alone are responsible for the contents.

References

[1] E.Dietrich et al., J.Lightw.Technol. Vol.7, No.12, Dec.1989 pp.1941-1955
[2] T.Saitoh, T.Mukai, J.Lightw.Technol. Vol.6, No.11, Nov.1988 pp.1656-1664
[3] M.J.O'Mahony, J.Lightw.Technol. Vol.6, No.4, Apr.1988 pp.531-544
[4] D.N.Payne, Tutorial at OFC'90 San Francisco (1990), USA, paper ThF
[5] S.Saito et al., Techn.Digest OFC'90, San Francisco (1990), USA, paper PD2
[6] S.Yamamoto et al., Techn.Digest IOOC'89, Kobe (1989), Japan, p.20 PDA9
[7] C.E.Zah et al., Electron.Lett., Vol.24, No. 20, Sept.1988 pp. 1275-1276
[8] I.Cha et al., Electron.Lett., Vol.25, No.18, Aug.1989 pp.1241-1242
[9] J.C.Simon, J.Lightw.Technol. Vol. LT-5, No.9, Sept.1987 pp. 1286-1295
[10] G.Großkopf et al., Electron.Lett., Vol. 23, No.25, Dec.1987 pp. 1387-1388
[11] D.M.Cooper et al., Techn.Dig.,OFC'90, San Francisco (1990), USA, p. PD32
[12] G.Großkopf et al., Optical Quantum Electron. in press
[13] I.W.Marshall et al., Conf.Publ.,P.1,ECOC'88,Brighton(1988),UK, pp 483-486
[14] H.Kawaguchi et al., Optics Lett., Vol.12, No.7, July 1987 pp.513-515
[15] G.Großkopf et al., Appl.Phys. Lett.B45, 1988, pp.171-174
[16] R.P.Webb, Optical Quantum Electron.19, 1987, pp.957-960
[17] G.Großkopf et al., Techn.Digest, IOOC'89, Kobe (1989), Japan, paper 19 C4-4
[18] M.P.Vecchi, Conf.Publ. P.1, ECOC'88, Brighton (1988), UK, pp.247-250
[19] G.Großkopf et al., Techn.Digest, IOOC'89, Kobe (1989), Japan, paper 21 B4-2
[20] M.Fujiwara et al., Trans.IECE, Vol. E71, No.10, Oct.1988 pp.972-974
[21] G.Großkopf et al., Electron.Lett., Vol.25, No.17, Aug.1989 pp.1188-1189
[22] U.Kohn and R.Ludwig, Patent application P3928 116.7-35, 1989
[23] R.Ludwig et al., to be published

Fibre Optical Amplifiers: Technology and System Applications

G. VEITH

SEL-ALCATEL Research Center

D-7000 Stuttgart 31, Germany

Summary

Erbium-doped fiber optical amplifiers (EDFAs) have undergone an enormous technological progress during recent years and are considered to be a key component for future broadband fiber communication networks. This paper reviews the state-of-the-art and potential future application areas of fiber optical amplifiers.

Introduction

Fiber communication systems in general are limited in performance by optical link and coupling losses and/or distribution losses. Optical amplifiers offer an attractive means to enhance the transmission capacity of long-haul trunk systems, to extend the repeater span of submarine links or to increase the potential number of subscribers in the subscriber systems. As basic approaches for optical amplification have been investigated in the past semiconductor-based and fiber-based amplifiers. Whereas semiconductor laser optical amplifiers (SCLOAs) have been studied over more than a decade with relatively modest progress /1/, optical amplifiers based on Erbium-doped fibers have been developed in a revolutionary way since its first publication in 1987 /2/. In the meantime Erbium-doped fiber amplifiers (EDFAs) are considered worldwide to be the most important approach for optical amplification.

Principles and technology

Fig. 1 shows the basic components of an Erbium-doped fiber amplifier (EDFA). The key element of an EDFA consists in an optical single-mode fiber with an Erbium-doped fiber core (typical Er^{3+}-concentration: 100 ppm). Erbium is a rare-earth material which can be incorporated efficiently into the core of silica based single mode fibers (e.g. SiO_2/GeO_2) by standard deposition-technologies (e.g. modified CVD, VAD).

Erbium exhibits optical absorption bands around 530 nm, 670 nm, 800 nm, 980 nm, 1470-1500 nm /2/. The most efficient pump wavelengths are 980 nm, and ≈ 1475 nm with reported efficiencies of 4.9 dB/mW /4/ and 2.6 dB/mW /5/, respectively (see fig. 2). Both pump bands are free of exited-state-absorption (ESA). The pump light can be coupled efficiently into the Erbium doped fiber by a wavelength-selective fused fiber coupler.

Due to the broad fluorescence band in the 1.5 µm range Erbium can be used as efficient amplifier medium for the third "transmission" window of fiber optic communication systems.

The gain profile of the EDFA ranges approximately from 1530-1570 nm and exhibits a distinct peak at 1535 nm and a second (less efficient) peak around 1550 nm. Fig. 3 shows the measured gain of an 980 nm pumped EDFA in the wavelength range 1530-1570 nm for two different input signal power levels (-32 dBm, -12 dBm).

The achievable gain is higher for lower signal power levels due to gain saturation, specifically in the 1535 nm band. Fig. 4 shows the gain saturation for increasing pump powers at two peak wavelengths (1536 nm, 1552 nm).

By increasing the pump power the gain and signal output power can be maximized. Up to date maximum gains of 46.5 dB and output saturation powers of +15.1 dBm (32.4 mW) /6/ have been reported.

In order to achieve these high performances the following parameters have to be optimized

- pump wavelength and coupling device efficiency
- Er^{3+} concentration and doping profile
- fiber core index profile and fiber length.

Erbium-doped fiber amplifiers exhibit the following advantages compared to 1.5 µm semiconductor based amplifiers:

- higher usable fiber-to-fiber gain (> 30 dB) due to lower insertion/coupling losses and high intrinsic gain in the 1.5 µm range,
- polarization independence,
- lower noise figure (3-5 dB)
- higher signal output power (> 10 dBm).

As pump sources can be used laser diodes which are already available (1480 nm) or which will be available in the next future (980 nm). This will allow the realization of very compact EDFA modules. First prototypes of EDFA modules have recently been presented /4/.

System experiments

The rapid progress of Erbium-doped-fiber amplifiers is reflected most spectacularly by the results of high speed laboratory transmission experiments, reported recently in international optical communication conferences /7/.

Fig. 5 shows the link length of high bitrate transmission experiments obtained with/without optical amplifiers including also results obtained with cascaded EDFAs. The longest link lengths up to now reported are 904 km in a 1.2 Gbit/s direct detection transmission experiment (12 inline EDFAs) /8/ and 2200 km in a 2.488 Gbit/s coherent FSK experiment (25 inline EDFAs) /9/.

Aside those "hero" results has been shown that the EDFA amplifiers are applicable at highest bitrates of 5 Gbit/s /10/ and $\geq$ 20 Gbit/s /11/ and can also be used for soliton amplification /12/.

It has been demonstrated that EDFAs can be used efficiently as booster amplifier, as in-line amplifier as well as pre-amplifier (i.e. before the optical frontend) which offers a wide range of potential application areas, e.g.:

- Submarine transmission systems (booster or preamplifier to enlarge the bridgeable unrepeatered link lengths).
- Multichannel (analogue or digital) TV-distribution systems (booster to compensate for the distribution losses and to increase the number of subscribers).
- Long-haul high bitrate digital trunk or feeder systems (booster or in-line amplifier to enhance the usable system gain and to increase the transmission capacity).
- Moreover Erbium-doped fiber amplifiers together with pulse compression techniques (e.g. chirped pulse compression, soliton transmission) provide the basic technologies to overcome conventional limitations of fiber optic systems (attenuation, dispersion).

Conclusion

During recent years has been demonstrated in numerous feasibility experiments the potential of Erbium-doped fiber amplifiers to enhance the performance of all types of fiber optical transmission systems.
Currently the emphasis of the activities in that field is mainly directed versus the optimization of subcomponents and the development of practical amplifier modules for field applications. First product prototypes have already been announced and many experts consider these devices to be key components for the introduction of future broadband communication systems.

Acknowledgement

The author would like to thank Th. Pfeiffer and B. Wedding (SEL-Alcatel Research Center Stuttgart) for the provision of laboratory results.

References

/1/ Simon J.C.: Semiconductor Optical Amplifiers, Proceed. OFC '90, Tutorial paper ThJ1, pp. 355-384, San Francisco, Cal. Jan. 1990.

/2/ Payne D.N.: Fiber Optical Amplifiers, Proceed. OFC '90, Tutorial paper ThF1, pp. 331-354, San Francisco, Cal., Jan. 1990.

/3/ Laming R.I., Reekie L., Payne D.N. et al.: Optimal pumping of Erbium-doped fiber optical amplifiers, Proceed. ECOC '88, pp. 25-28, Brighton, UK, Sept. 1988.

/4/ Shimizu M., Horiguchi M., Yamada M. et al.: Compact and highly efficient optical fiber amplifier module pumped by a 0.98 μm laser diode, Proceed. OFC '90, Postdeadline Paper RD 17, San Francisco, Cal., Jan. 1990.

/5/ Zyskind J.L., Giles C.R., Desurvire E., Simpson J.R.: Optimal pump wavelength in the I15/2-I13/2 absorption band for efficient Er-doped fiber amplifiers; IEEE Photon. Technol. Lett. 1 (1989), pp. 428-430.

/6/ Kimura Y., Suzuki K., Nakazawa M.: 46.5 dB gain in Er^{3+}-doped fibre amplifier pumped by 1.48 μm GaInAsP laser diodes, Electon. Lett. 25 (1989) pp. 1656-1657.

/7/ See e.g. papers presented at IOOC '89/Kobe, ECOC '89/Gotherburg and OFC '90/San Francisco.

/8/ Edagawa N., Yoshida Y., Taga H. et al.: 904 km, 1.2 Gbit/s nonregenrative optical transmission experiment using 12 Er-doped fibre amplifiers; Proceed ECOC '89, Postdealine Paper PDA-8, Gothenburg, Sweden, Sept. 1989.

/9/ Saito S., Imai T., Sugie T. et al.: An over 2200 km coherent transmission experiment at 2.5 Gbit/s using Erbium-doped fibre amplifiers, Proceed. OFC '90, Postdealine Paper PD2, San Francisco, USA, Jan. 1990.

/10/ Wedding, B., Pfeiffer, Th.: 5 Gbit/s transmission over 146 km using Erbium doped fibre amplifier; Proceed. ECOC '89, Vol. 1, pp. 86-89, Gothenburg, Sweden 1989.

/11/ Wedding, B., Pfeiffer, Th.: Submitted to ECOC '90.

/12/ Nakazawa M., Suzuki K., Kimura Y.: 20 Gb/s soliton amplification and transmission with an Er^{3+}-doped fiber; Proceed. IOOC '89, Postdealine Paper 20 PDA-3, Kobe, Japan, July 1989.

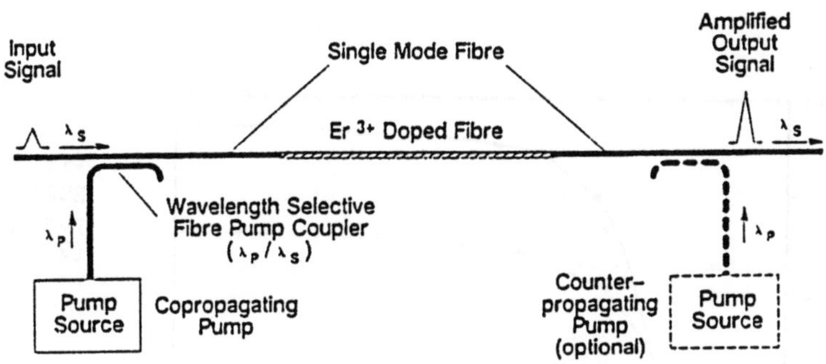

Fig. 1 Principle of Erbium-doped fibre amplifier (EDFA)

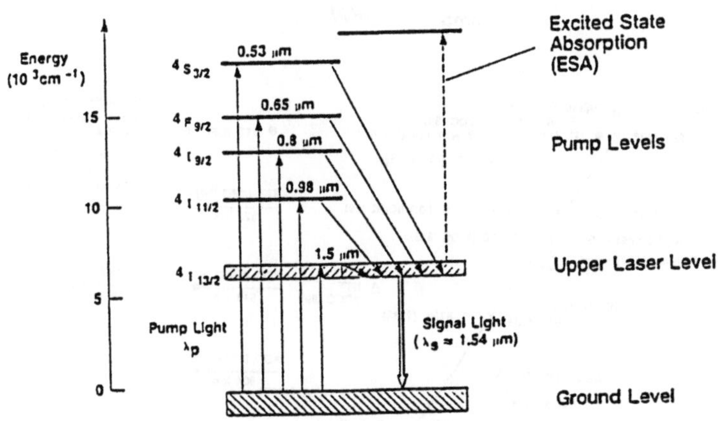

Fig. 2 Energy level diagram of Erbium (Er^{3+})

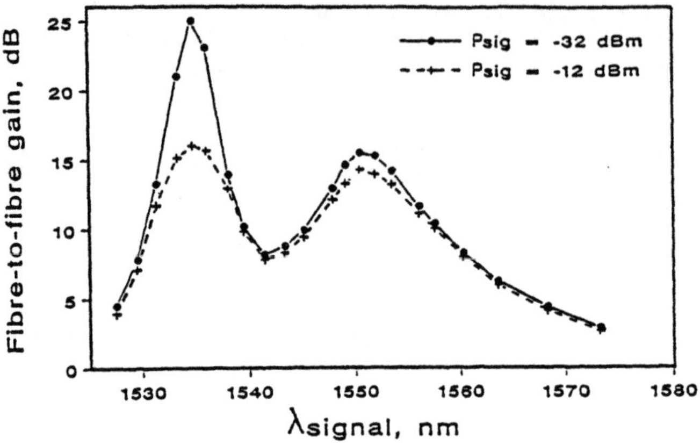

Fig. 3
EDFA gain vs. signal wavelength (980 nm pump, 21 mW pump power)

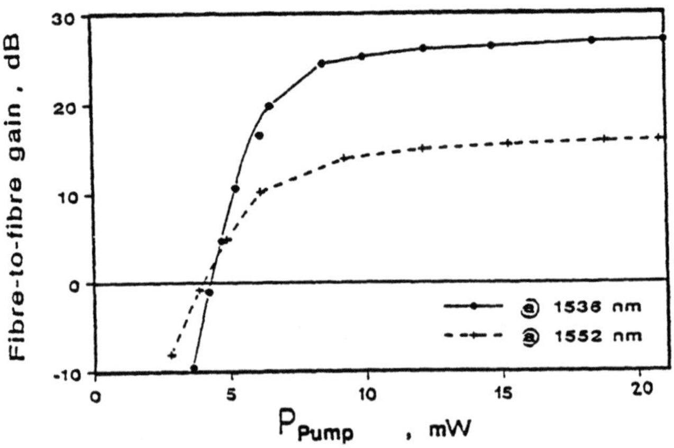

Fig. 4
EDFA gain vs. pump power (980 nm pump)

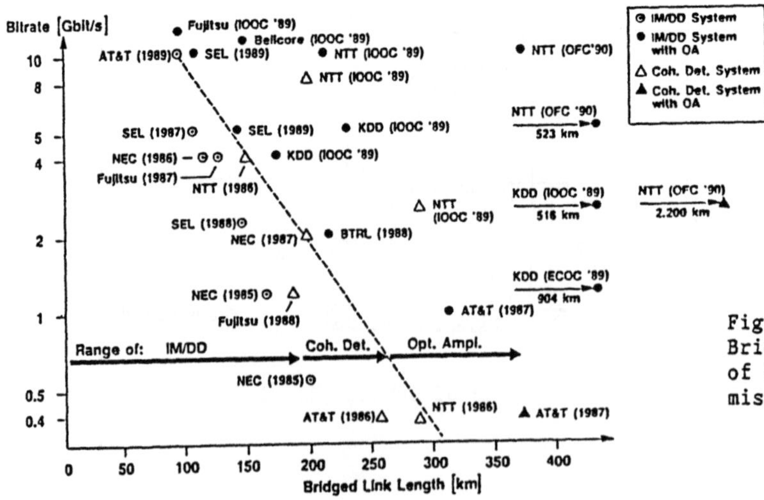

Fig. 5
Bridged link length of high speed trans mission experiments

Integrated Optic Polarimetric Refractometer

G. VOIRIN, P. DEBERGH, O. PARRIAUX, O. ZOGMAL

Centre Suisse d'Electronique et de Microtechnique S.A
CH-2007 Neuchâtel/Switzerland

Summary

One of the early practical applications of passive integrated optic circuits as sensors is the measurement of refractive indices. The usual transducing mechanism is the modal effective index change due to the index variation of the probe liquid placed as a waveguide overlay in the evanescent modal field. The readout is normally performed interferometrically in a Mach-Zehnder configuration; the problem is here the temperature stability of the reference waveguide branch.

We have demonstrated a new readout scheme using the interference between the two polarization modes of a single waveguide made by ion exchange in glass. Such waveguides exhibit a strong birefringence which is remarkably little temperature dependent. But the birefringence is very dependent on the refractive index of the waveguide overlay. A refractometer can be based on this effect in an integrated polarization interferometer configuration. The sensor structure simply involves a single sensing straight waveguide terminating as a symmetrical Y junction onto two detectors. The achievable sensitivity with a centimeter long K^+-Na^+ ion exchanged waveguide in BK7 is of the order of $9 \cdot 10^{-5}$.

Birefringence in ion exchanged waveguides

Ion exchange in glass produces an increase of the refractive index and is also accompanied by a tensile or compressive stress in the glass surface. Potassium-sodium exchange, which is used for glass strengthening is a well known example. The stress induces birefringence. In the case of a planar waveguide the stress is oriented in the plane of the ion exchanged surface. A typical birefringence figure is $n_{TM}-n_{TE} = 2.10^{-3}$ [1] where n_{TM}, n_{TE} are the effective indices of the TM and TE modes of a single mode waveguide. In the case of strip waveguides, as we will show hereafter, the stress can release in the other direction and the birefringence decreases to values between 2.10^{-4} to 2.10^{-5} depending on the width of the waveguide.

Measurement of the birefringence

The measurement of the birefringence was made by measuring the beatlength between the TM and TE modes:

$$Lb = \frac{\lambda}{(n_{TM}-n_{TE})}$$

This measurement is made by locally perturbing the polarisation state by a Faraday effect [2]. The magnetic field is created by a magnetic coil (tape recording head). The experimental set-up is presented on figure 1. The birefringence axes are parallel and normal to the glass surface. Laser light is coupled along the TE polarisation. Its state of polarisation is unchanged along the propagation direction if no field is applied. If, at one point in the waveguide a magnetic field is applied, an optical activity is induced, slightly changing the polarisation state.

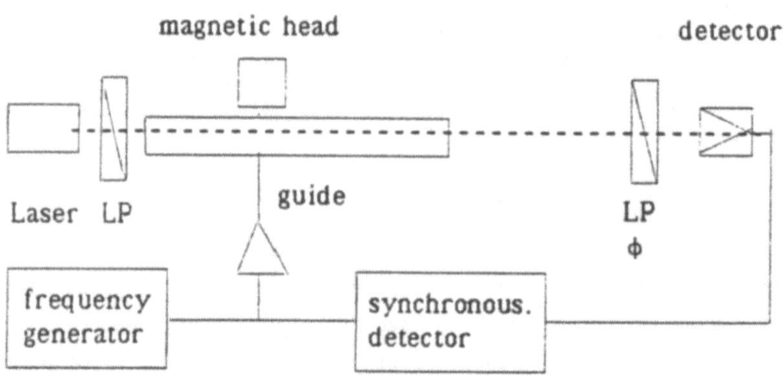

Figure 1: Experimental set up for beatlength measurement, LP: linear polarizer, laser: semiconductor laser at 820 nm.

A linear analyser is placed at the end of the waveguide, its axis is oriented in a direction slightly different from the TM axis. The signal delivered by the detector is synchronously detected. The magnetic head is scanned along the direction of propagation and a sinusoïdal signal is recorded versus the position. The recorded spatial period is equal to the beatlength between TE and TM modes of the guided wave (see Figure 2).

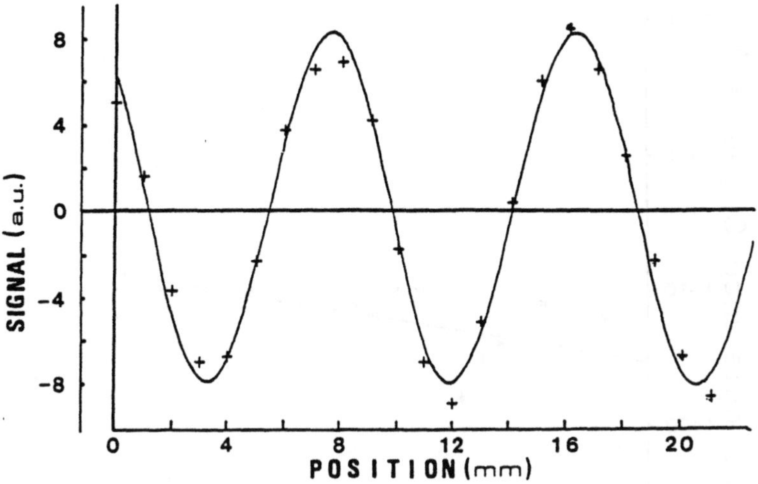

Figure 2: Measurement of the beatlength on a waveguide of 3µm nominal width at λ = 820 nm: the beatlength is obtained by a curve fitting and is 8.7 mm

Variation of the birefringence with the temperature

A linear polarisation is coupled at 45° relative to the axes of the waveguide and the polarisation state is analysed at the end of the waveguide either by means of a Soleil-Babinet compensator or by means of an analyser. As shown in figure 3, the variation of phaseshift between TE and TM modes versus temperature between -120°C and 80°C remains of the order of the measurement error. This remarkably low temperature dependence of the waveguide birefringence offers interesting possibilities for stable integrated polarisation elements as well as for temperature independent integrated polarimetric sensors.

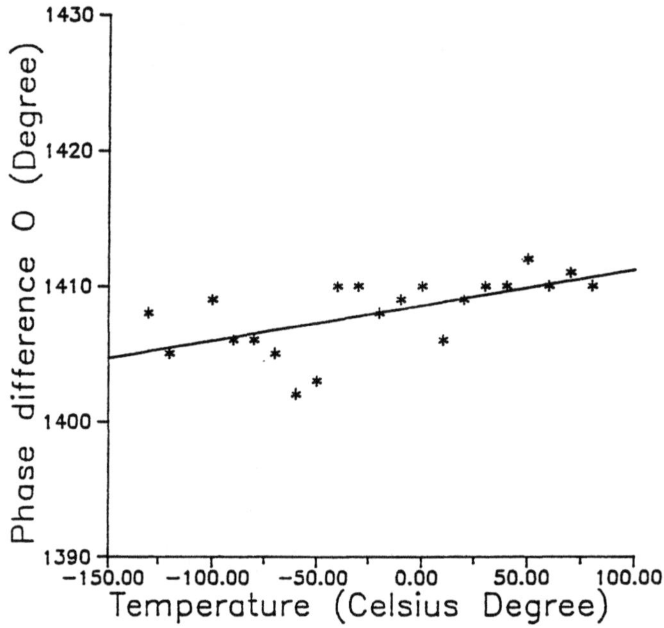

Figure 3: Variation of the phase difference Θ between TE and TM modes at the end of a 38 mm long K^+ ion exchanged waveguide versus the temperature.

<u>Variation of the beatlength with the refractive index of the overlay</u>

The beatlength was also measured with a deposited layer on the waveguide. The overlay was made of index liquid whose index was varied between 1.4 and 1.51 at room temperature. The beatlength varies from 9.8 mm in the case of a waveguide in air to 4.5 mm with a liquid of index 1.51. Figure 4 shows that the dependence of the birefringence on the overlay index is quite large, varying by a factor 2 between overlay index 1 and 1.51, the strongest variation occuring around 1.5.

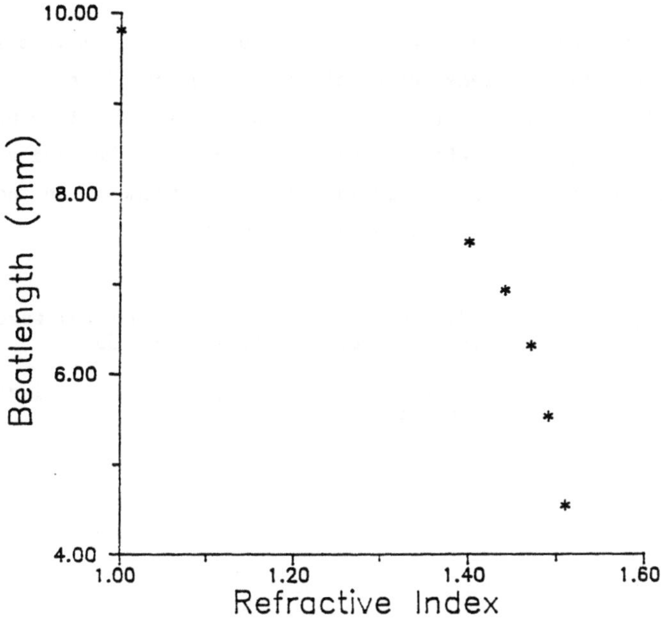

Figure 4: Variation of the beatlength with the index of the waveguide overlay (K^+ ion exchanged and 3 µm nominal width waveguide; λ = 820 nm).

Integrated polarisation interferometer configuration

The high sensitivity of the birefringence to the index of an overlay together with its low sensitivity to temperature can be used in a polarisation interferometer configuration. The refractometer is simply a single sensing waveguide terminating as a symmetrical Y junction. Each branch of the "Y" directs the same polarisation state onto two identical detectors with a polariser in between.

With this simple arrangement, the variation of the polarisation state around a given working point can be determined within an accuracy of ± 2°. Referring to figure 4, this means that a three centimeter long sensing waveguide gives a resolution on the index of 9.10^{-5}. Detailed results will be given at the conference.

Conclusion

A birefringent ion exchanged waveguide is used as a simple polarisation interferometer for the measurement of the refractive index of the waveguide overlay. The substantial practical advantage of this configuration is that the sensing and reference waves propagate in the same channel and that the temperature does not influence the measurement. An accuracy of 9.10^{-5} on the overlay index can be easily reached.

References

1. Yip, G.L.; Albert, J.: Characterization of planar optical waveguides by K^+-ion exchange in glass. Opt. Lett. 10 (1985) 151-153

2. Simon, A.; Ulrich, R.: Evolution of polarization along a singlemode fiber. Appl. Phys. Lett. 31 (1977) 517-520

Actuators

Integrated Fluid Control Systems on a Silicon Wafer

Shigeru Nakagawa, Shuichi Shoji and Masayoshi Esashi

Department of Precision Engineering
Tohoku University
Sendai 980, Japan

ABSTRACT

A micropump, a microvalve and a micro flow sensor were fabricated on silicon wafers. The micropump could control the flow of nl order and it could pump up the water to the 1.5m height as it was driven by 25Hz rectangular voltage of 100V amplitude. The microvalve had the open to close ratio of about 580 and 710 and the leakage from the closed valve could be reduced to 0.1μl/min using the nickel thin film as a gasket. The micro flow sensor was a kind of a differential pessure flow sensor and consisted of a one-way valve and a piezoresistive pressure sensor. It could measure the flow rate more than several tens μl/min.

INTRODUCTION

Recently, fluid control of a micro flow is required for VLSI fabrication processes, for medical applications and so on. In order to control the micro flow accurately, dead volume of the fluid control system have to be decreased. Micromachining technology can realize fluid controllers fabricated on silicon wafers by micromachining, e.g. micropumps and microvalves [1-4]. We have been studying on an integration of a fluid control system on a silicon wafer. A mass flow controller integrated on a silicon wafer has been already reported [1]. It consists of a normally close microvalve and a thermal flow sensor and it can control gas flow with fast response time of about 10msec. A flow control system for liquid flow are also integrated on a silicon wafer and it consists of micropumps, microvalves and micro flow sensors. These controllers have advantages that the dead volume inside them is negligible. Since they are integrated on a silicon wafer, a integrated fluid control system has a very small dead volume of about a few tens nl. Besides, its size is small enough to be placed near the other systems to be controlled and the dead volume between the components can be reduced. In the following, we will describe a micropump, a microvalve and a micro flow sensor for liquid flow fabricated on silicon wafers.

MICROPUMP

A micropump fabricated is a diaphragm pump and has a structure shown in Fig.1. Two Pyrex glass and one silicon wafer are anodically bonded to construct this pump. A diaphragm with a mesa is formed on the Pyrex glass and a piezoelectric actuator is put on the mesa. One-way valves of poly silicon are fabricated at the inlet and at the outlet of the pump using surface micromachining. This pump is driven by applying a periodic voltage to the actuator. The volume of the cavity under the diaphragm is very small and the liquid volume pushed out in one stroke of the actuator is also small, which is about 20nl. Since this micropump repeats pumping of such small volume quickly, liquid from it flows very smoothly and it can control flow rate accurately.

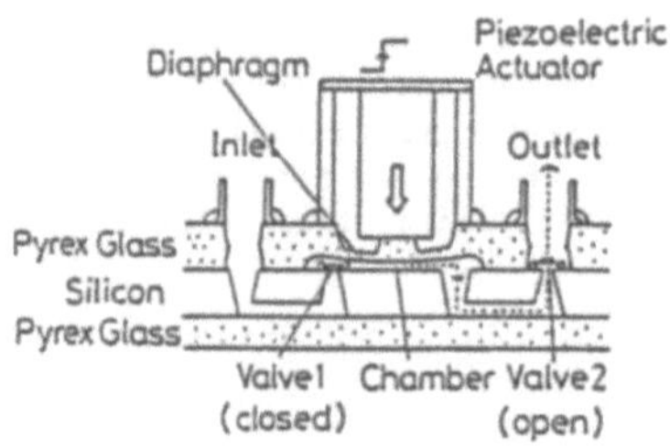

Fig.1 Structure of a micropump.

As the micropump was driven by applying a rectangular voltage, flow rate from it changed with the frequency as shown in Fig.2. The voltage amplitude was 100V and water was used as a working liquid. It could control the flow rate up to about 30μl/min and the flow rate became saturated at high frequency. The frequency characteristics of the micropump depends upon the flow resistance of the inlet one-way valve and the compliance of the diaphragm. Narrow flow channels fabricated on a silicon wafer have large flow resistance and decrease the flow rate from the micropump as they are connected to the micropump. The narrow flow channels of 5μm thickness which have various width, i.e. which have various flow resistance, were connected to the outlet of the micropump and the flow rate was measured. Fig.3 shows the flow rate as a function of the pressure drop at the flow channel. The pressure drop is caused between the inlet and the outlet of the channel. Flow rate decreased linearly with the increase of the pressure drop. As it is driven by the 25Hz rectangular voltage of which amplitude is 100V, its pumping pressure is approximately 1.5mH_2O.

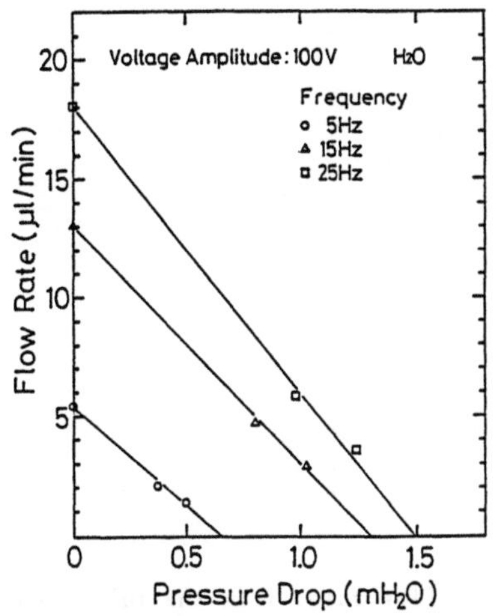

Fig.2 Relationship between a pumping rate of the micropump and a frequency of the applied rectangular voltage.

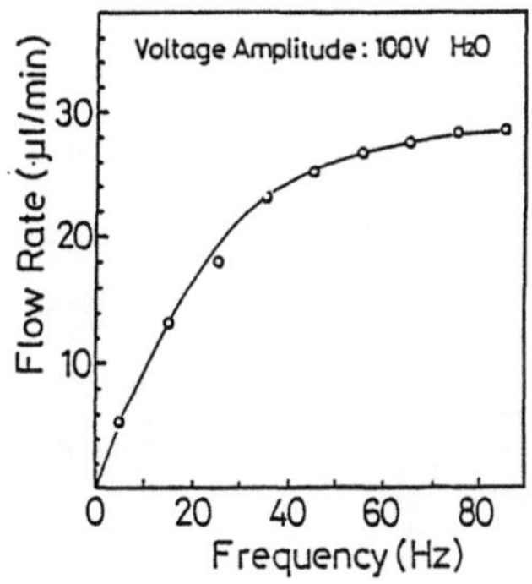

Fig.3 Relationship between the flow rate from the micropump and pressure drop at the flow channel connected to it.

MICROVALVE(THREE-WAY VALVE)

Fig.4 shows a top view and a cross section of a three-way valve. The three-way valve consists of two silicon wafers and one Pyrex glass which are anodically bonded. Inner and outer concentric circular mesas are formed on the silicon wafer and are supported by diaphragms. A piezoelectric actuator is put on the inner mesa and outer mesa has a nickel gasket. The valve opens the inlet1 or the inlet2 by changing a voltage applied to the actuator. Since the thermal expansion of the Pyrex glass and that of the silicon wafer differs, the substrates deform by thermal stress at the anodic bonding. A gap results between the Pyrex glass and the outer mesa to cause a leakage from the inlet1. A nickel gasket is formed on the outer mesa as shown in Fig.4 and it can compensate the gap. Further, the bias force due to the height of the gasket presses the outer mesa to the Pyrex glass and prevents a gap between the outer mesa and the Pyrex glass.

A voltage was applied to the actuator and the flow rates from the inlet1 and from the inlet2 were measured. Water was used as a working liquid. Fig.5 shows the flow rates as a function of the applied voltage at the water pressure of 220gf/cm². The open to close ratio at 0V and that at 200V were approximately 580 and 710 respectively. The leakage from the inlet1 and that from the inlet2 were both about 0.1μl/min. The leakage from the inlet1 were measured as a function of the water pressure. As shown in Fig.6, the leakage increased drastically when the water pressure exceeded 300gf/cm². As the water pressure gets higher than the bias force, a gap caused between the outside valve seat and the Pyrex glass increases the leakage.

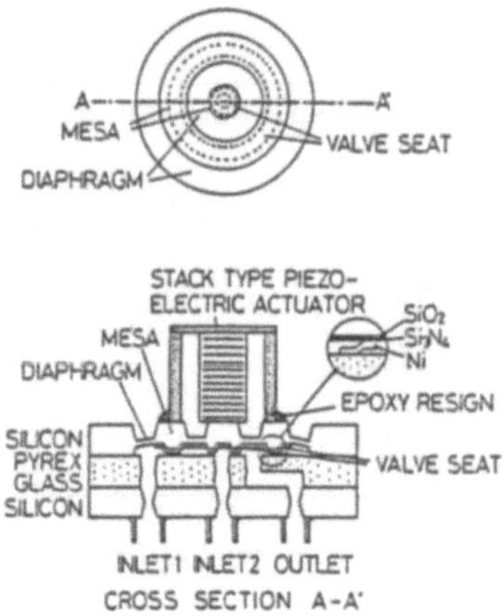

Fig.4 Top view and cross section of a three-way valve.

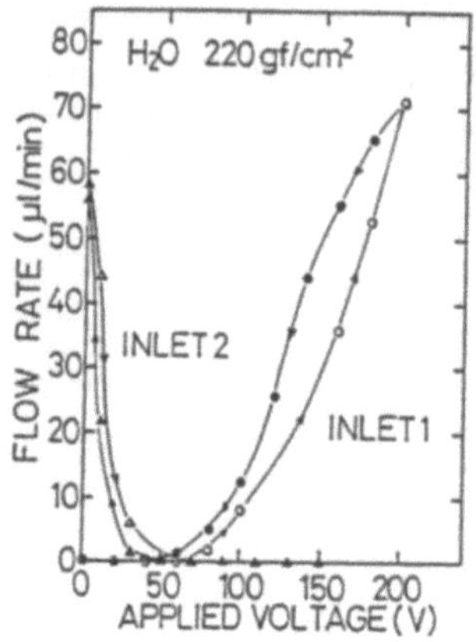

Fig.5 Flow rates from the inlet1 and inlet2 as a function of the voltage.

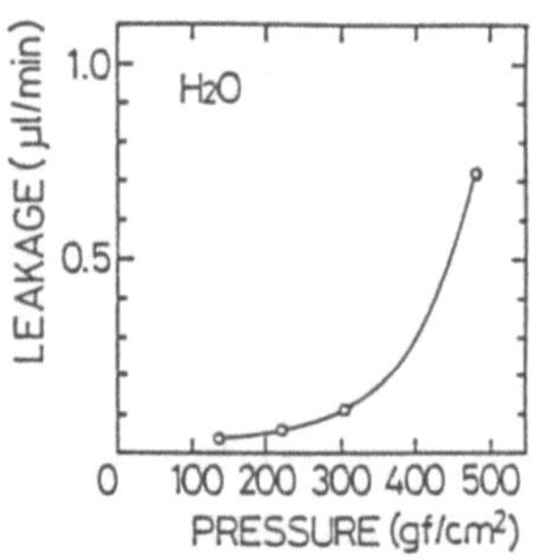

Fig.6 Leakage from inlet1 as a function of water pressure.

MICRO FLOW SENSOR

A principle of a micro flow sensor fabricated is that of a differential pressure flow sensor. The structure of the micro flow sensor is shown in Fig.7, which consists of a one-way valve of poly silicon and a piezoresistive pressure sensor. As the liquid flows through the one-way valve, it causes the pressure difference between the inlet and the outlet and the pressure difference is detected by the pressure sensor. In order to increase the sensitivity of the flow sensor, a large pressure drop caused at the one-way valve and a high sensitivity of the pressure sensor are required. Fig.8 shows the relationship between the flow rate through the one-way valve and the pressure difference across it. The one-way valve has the advantage that it causes relatively large pressure drop at small flow rate and as a result this micro flow sensor is suitable for the measurement of small flow rate.

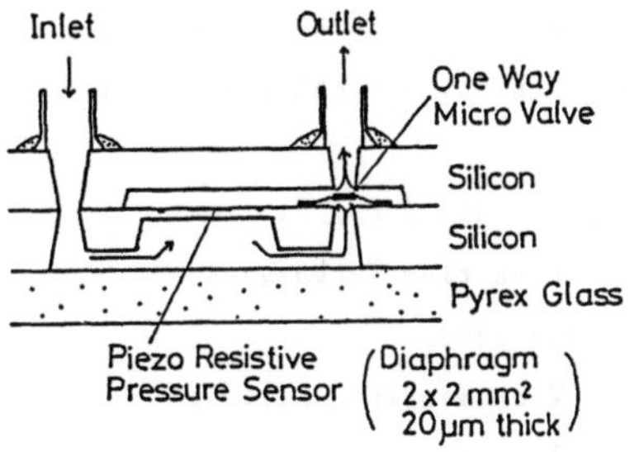

Fig.7 Structure of a micro flow sensor.

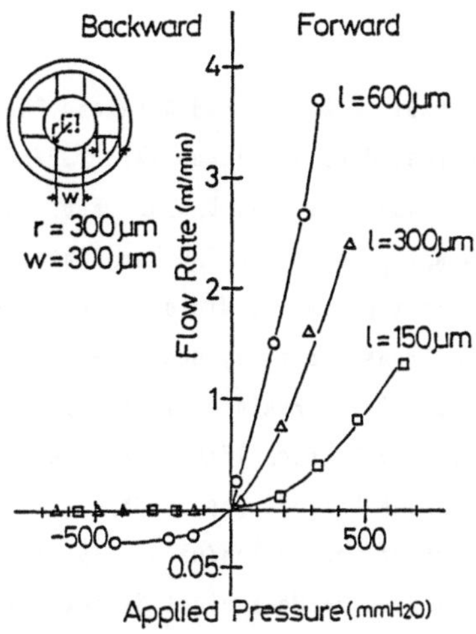

Fig.8 Relationship between flow rate through one-way valve and water pressure.

As water flew through the flow sensor, the output voltage of the pressure sensor increased with the flow rate of the water as shown in Fig.9. The sensitivity is not enough at small flow range. In order to measure the flow rate of the order of μ l/min, the sensitivity of the pressure sensor should be increased.

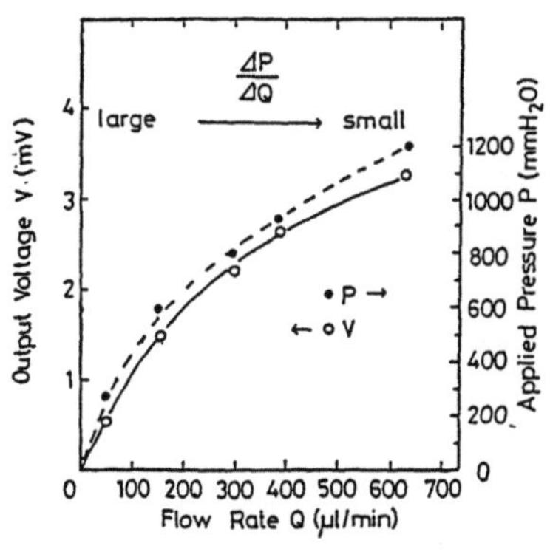

Fig.9 Output voltage of the pressure sensor as a function of the flow rate.

CONCLUSIONS

We have fabricated a micropump, a micro valve and a micro flow sensor on silicon wafers. The micropump and the micro valve use piezoelectric actuators and can control the flow rate of the order of μ l/min. The flow sensor can measure the flow rate more than several tens μ l/min. We intend to integrate a flow control system on a silicon wafer using the micropump, the micro valve and the flow sensor.

REFERENCES

[1] M.Esashi, S.Shoji, and A.Nakano,"Normally Close Microvalve and Micropump Fabricated on a Silicon Wafer",Proc. IEEE Micro Electro Mechanical Systems, Salt Lake City, UT, U.S.A., Feb. 1989, pp.29-34.

[2] F.C.M.Van de Pol, H.T.G.Van Lintel, M.Elwenspoek, and J.H.J.Fluitman, "A Thermopneumatic Micropump based on Micro Engineering Techniques", Proc.Transducers'89, Montreux, Switzerland, June 1989.

[3] T.Ohnstein, T.Fukiura, J.Ridley, and U.Bonne,"Micromachined Silicon Microvalve",Proc. IEEE Micro Electro Mechanical Systems, Napa Valley, CA, U.S.A., Feb. 1990, pp.95-98.

[4] A.Richter, and H.Sandmaier,"An Electrohydrodynamic Micropump",Proc. IEEE Micro Electro Mechanical Systems, Napa Valley, CA, U.S.A., Feb. 1990, pp.99-104.

Micro Liquid-Handling Devices - A Review

F.C.M. VAN DE POL

University of Twente, Faculty of Electrical Engineering, P.O. Box 217, 7500 AE Enschede, The Netherlands

J. BRANEBJERG

Technical University of Denmark, Lab for Semiconductor Technology, Building 348, DK 2800 Lyngby, Denmark

SUMMARY

An overview is given on research activities throughout the world, in the area of micro liquid-handling devices, such as pumps, valves and liquid-flow sensors. We also discuss ink-jet heads, which can be regarded as special types of micropumps, and complete systems for chemical analysis or medicine delivery. Some comments on future trends are given.

1. INTRODUCTION

Micro liquid-handling devices including pumps, valves and liquid-flow sensors are becoming more and more a "hot" research topic. In this context, "micro" stands for "micro-fabricated", i.e. made by applying all kinds of micro-techniques e.g. photolithography, micromachining of silicon and thin-film techniques, with silicon as a basic material.

Groups moving into this emerging field, usually aim at the development of complete systems for chemical analysis or medicine delivery. In general, these systems comprise parts for pumping, flow sensing and control.

2. MICROPUMPS

In the beginning of the eighties at the University of Twente, a micropump was realized applying bulk micromachining of silicon for the pump body, and anodic bonding of silicon to glass [1]. The pump comprises three active piezoelectric bimorph valves, and works in a peristaltic manner, see fig. 1.

Van Lintel *et al.* [2] reported on a pump with passive silicon check valves, and a piezoelectric monomorph glued onto a glass diaphragm as the driver, see fig. 2. In this reciprocating displacement type of pump the two functions (liquid displacement and valve action) are performed by separate components.

Smits [3] presented an intermediate type with active valves comprising piezo-electric monomorphs glued onto glass diaphragms and working peristaltically.

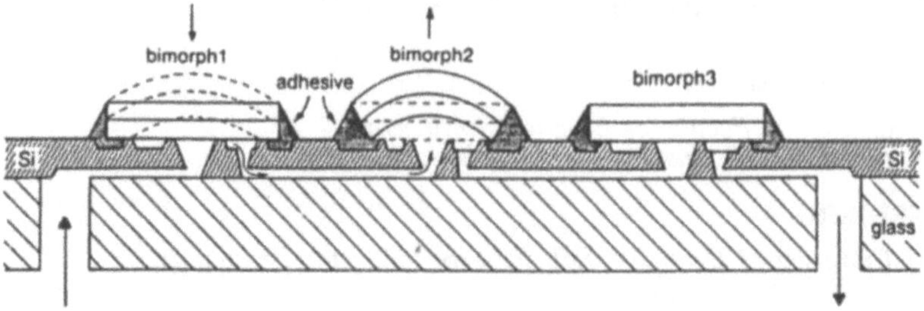

Fig. 1 *Piezoelectrically driven micropump with three active bimorph valves working peristaltically [1].*

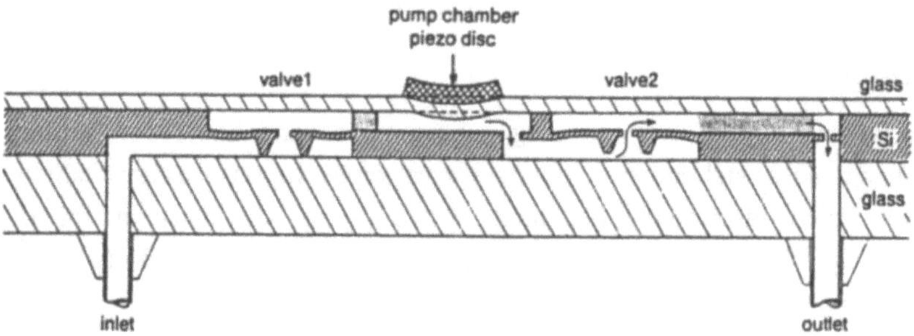

Fig. 2 *Piezoelectric, reciprocating displacement type of pump with passive silicon check valves [2].*

In the middle eighties at Tohoku University, Japan, Esashi and Shoji started research on microsystems for chemical analysis. They realized pumps with polysilicon valves fabricated by surface micromachining. These pumps are piezoelectrically driven by means of a stack of piezoelectric disks [4], see fig. 3.

Lindberg *et al.* [5] reported on a pump with electrochemically etched beams serving as valves. Van de Pol *et al.* [6] realized a reciprocating displacement type of micropump that is thermo-pneumatically driven. A small amount of air, contained in a cavity, is raised in temperature by resistive heating. The induced pressure is used to actuate a diaphragm which in turn displaces the liquid. A first attempt at constructing a electrohydrodynamic micropump for the displacement of charged liquids was described by Bart *et al.* [7].

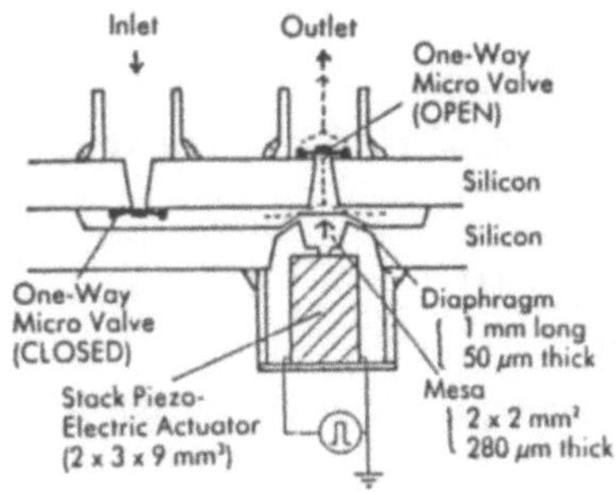

a. cross-sectional view

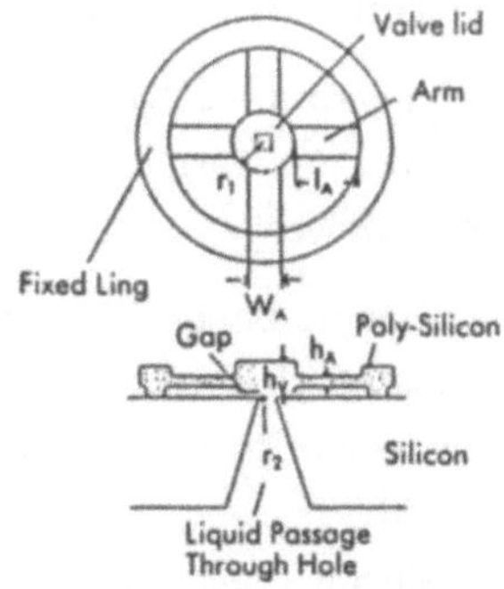

b. a valve in detail

Fig. 3 *Piezoelectric (stack), reciprocating displacement type of pump. The polysilicon check valves are fabricated by means of surface micromachining [4].*

3. INK-JET HEADS

Ink-jet nozzles based on micromachining of silicon were reported on by Bassous *et al.* [8]. Figure 4a shows a nozzle producing a stream of droplets as realized by Kuhn *et al.* [9]. A droplet can be deflected or allowed to pass to the paper, depending on whether or not is has been charged. In fig. 4b, a piezoelectrically driven printer head as described by Petersen [10] is shown. It comprises a nozzle and a periodically pressurized cavity filled with ink.

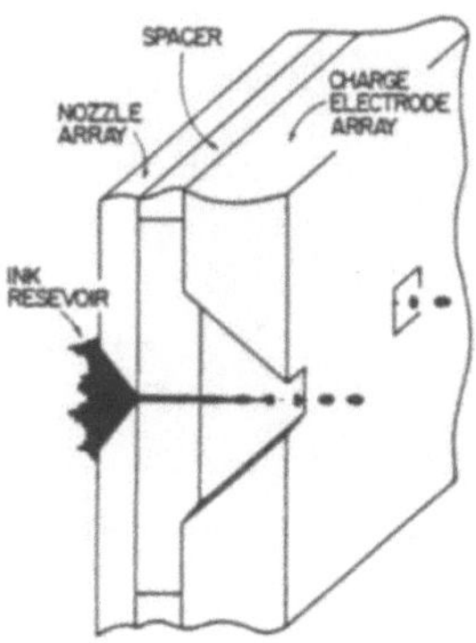

a.

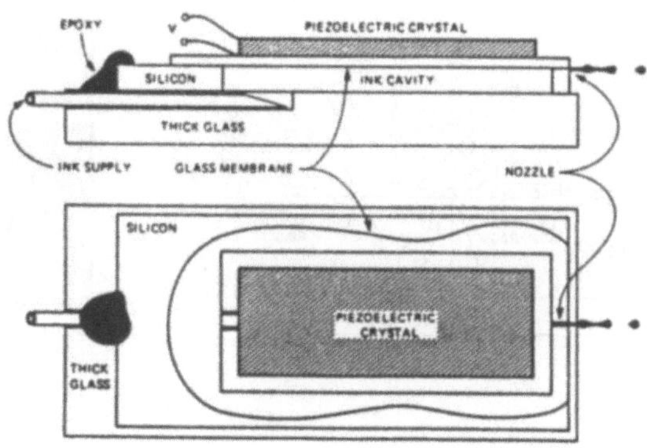

b.

Fig. 4 *Ink-jet printer heads based on micromachining of silicon:*
a. nozzle producing a stream of charged/uncharged droplets [9]
b. piezoelectrically driven ink-jet head [10].

4. MICROVALVES

Many types of microvalves have been realized; passive or active check valves, and control valves for various applications. A number of them have already been mentioned above:

- active piezoelectric bimorph valves, see fig. 1 [1]
- passive silicon check valves, see fig. 2 [2]
- active valves comprising piezoelectric monomorphs glued onto glass diaphragms [3]
- passive polysilicon check valves, see fig. 3 [4]
- electrochemically etched silicon beams as passive check valves [5].

Apart from these, a number of control valves based on micromachining of silicon are reported in the literature. Terry *et al.* [11,12] realized gas-pressure controlled valves with silicon diaphragms, and solenoid-actuated valves with nickel diaphragms, see fig. 5a, for use in a miniature gas chromatography system. Control valves actuated by piezoelectric stacks were described by Esashi *et al.* [13]. Shoji *et al.* [14] reported on a control valve actuated by a shape-memory-alloy spring for application in a miniature blood gas analysing system. Finally, we mention Zdeblick's work on different types of thermo-pneumatically actuated control valves, of which an example is shown in fig. 5b [15].

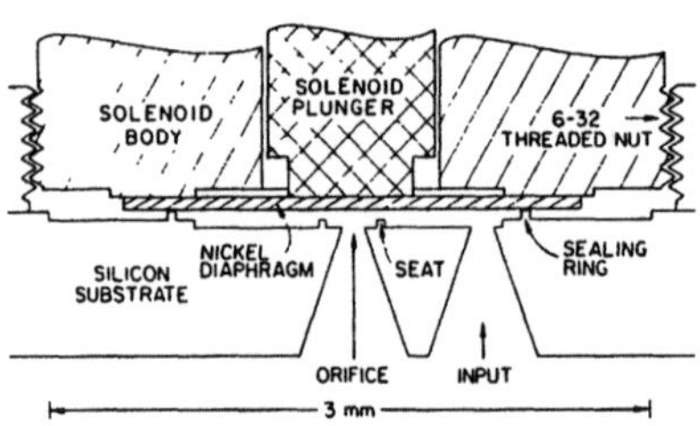

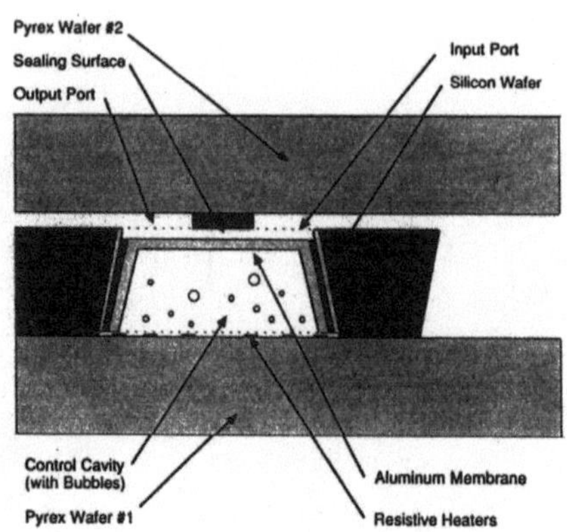

a. *b.*

Fig. 5 *Control valves based on micromachining of silicon:*
a. solenoid-actuated valve with a nickel diaphragm [12]
b. thermo-pneumatically actuated silicon control valve [15].

5. MICRO LIQUID-FLOW SENSORS

Many micro *gas*-flow sensors have been developed, most of them based on thermal working principles. A special type is the SAW gas-flow sensor described by Ahmad [16].

However, very few micro *liquid*-flow sensors have been reported. Suminto *et al.* [17] realized a venturi-tube differential-pressure liquid-flow sensor. It comprises a pyrex tube and two pressure sensors. Carr [18] worked on a fibre optic liquid-flow sensor, based on photon correlation spectroscopy. Fig. 6 shows a sketch of a liquid-flow sensor reported by Yang *et al.* [19]. The transit-time of a heat pulse, carried by the liquid, is measured.

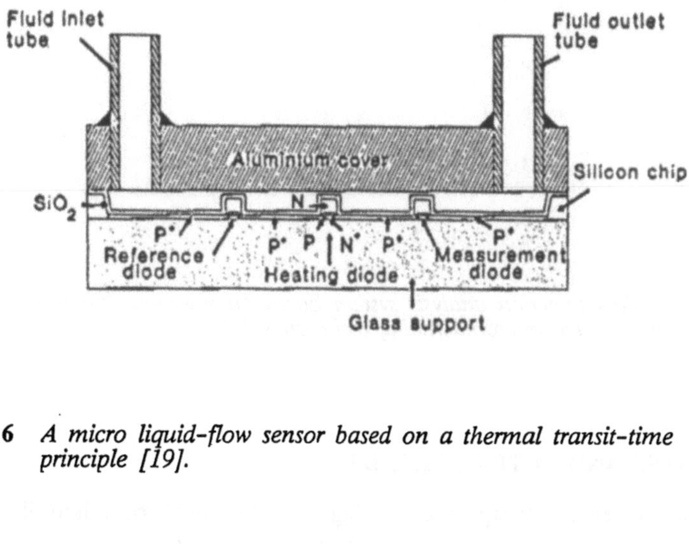

Fig. 6 *A micro liquid-flow sensor based on a thermal transit-time principle [19].*

6. MICRO LIQUID-HANDLING SYSTEMS

The gaschromatography system [11,12] and the blood gas analyzer [14] can be regarded as the first integrated silicon *gas*-handling systems. One of the first micro *liquid*-handling systems is the integrated flow injection analysis (FIA) system as worked on by Esashi and Shoji, see fig. 7 [20].

At present, work on a micro FIA system is carried out at the Technical University of Denmark, and there are a number of industrial research groups working on medicine delivery systems.

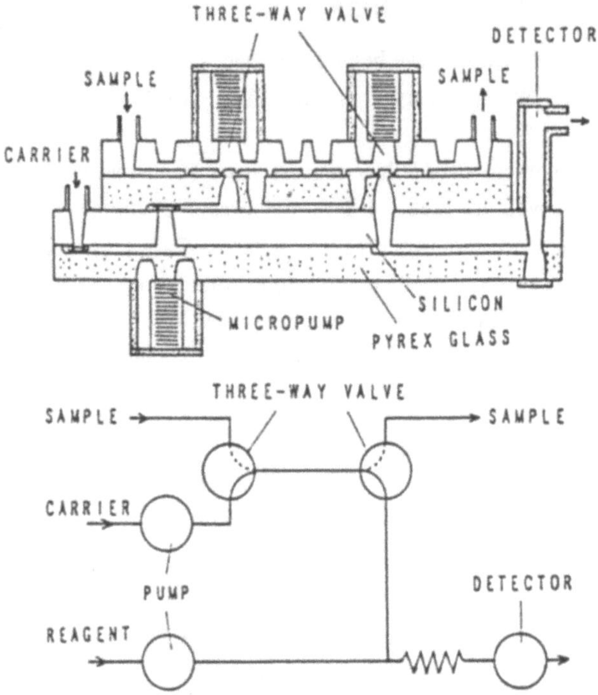

Fig. 7 *Integrated flow injection analysis system based on micromachining of silicon and surface micromachining techniques [20].*

7. CONCLUSIONS AND FUTURE TRENDS

More and more research groups are moving into the field of micro liquid-handling devices based on micromachining of silicon and surface micromachining techniques. Various types of components, such as pumps, valves and flow sensors have been realized, and a few complete systems have been reported.

Integrated microsystems, especially systems for chemical analysis and medicine delivery, are expected to be the main topics of research in the future. Main subjects thereby are design, physical system modelling, and technology development. Special attention should be paid to the theoretical aspects of flow in small channels etc., since it appears that there can be a significant deviation from the predictions from the Navier–Stokes equations [21].

The final goal is fully integrated microsystems for liquid-handling with on-chip signal processing.

REFERENCES

1. R.A. Groothengel, C. Hartevelt, J.G. Smits, *Verslag van een onderzoek aan een piezoelektrische micropomp*, internal report 060.0168, University of Twente, The Netherlands, 1984 (in Dutch)
2. H.T.G. van Lintel, F.C.M. van de Pol and S. Bouwstra, *A piezoelectric micropump based on micromachining of silicon*, Sensors and Actuators 15 (1988) 153-167
3. J.G. Smits, *Piezoelectric micropump with three valves working peristaltically*, Sensors and Actuators A21-A23 (1990) 203-206
4. S. Shoji and M. Esashi, *Micropump for integrated chemical analyzing systems*, Technical Digest of the 7th Sensor Symposium, Japan, 1988, pp.217-220
5. W. Lindberg, L. Tenerz, J. Tirèn and B. Hok, *Fabrication of microfluidic element by silicon technology*, Proceedings of the 13th Nordic Semiconductor Meeting, Stockholm, Sweden, June, 1988, pp. 456
6. F.C.M. van de Pol, H.T.G. van Lintel, M. Elwenspoek and J.H.J. Fluitman, *A thermopneumatic micropump based on micro-engineering techniques*, Sensors and Actuators A21-A23 (1990) 198-202
7. S.F. Bart, L.S. Tavrow, M. Mehregany and J.H. Lang, *Microfabricated electrohydrodynamic pumps*, Sensors and Actuators A21-A23 (1990) 193-197
8. E. Bassous, *Fabrication of novel three-dimensional microstructures by the anisotropic etching of (100) and (110) silicon*, IEEE Trans. on Electron Devices ED-25 (1978) 1178-1185
9. L. Kuhn, E. Bassous and R. Lane, *Silicon charge electrode array for ink jet printing*, IEEE Trans. Electron Devices ED-25 (1978) 1257-1260
10. K.E Petersen, *Fabrication of an integrated, planar silicon ink-jet structure*, IEEE Trans. on Electron Devices ED-26 (1979) 1918-1920
11. S.C. Terry, *A gas chromatography system fabricated on a silicon wafer using integrated circuit technology*, Technical report 4603-1, Stanford University USA, Ph.D. Thesis, 1975
12. S.C. Terry, J.H. Jerman and J.B. Angell, *A gas chromatographic air analyzer fabricated on a silicon wafer*, IEEE Trans. on Electron Devices ED-26 (1979) 1880-1886
13. M. Esashi e.a., *The fabrication of a micro valve by means of micromachining*, Proceedings of the 6th Sensor Symposium, Japan, 1986, pp. 269-272
14. S. Shoji, M. Esashi and T. Matsuo, *Prototype miniature blood gas analyzer fabricated on a silicon wafer*, Sensors and Actuators 14 (1988) 101-107
15. M.J. Zdeblick and J.B. Angell, *A microminiature electric-to-fluidic valve*, Proceedings Transducers '87 conference, Tokyo, Japan, June 2-5, 1987, pp. 827-829
16. N. Ahmad, *Surface acoustic wave flow sensor*, Proc. 1985 IEEE Ultrasonics Symposium, San Fransisco, USA, October 16-18, pp. 483-485
17. J.T. Suminto, G.-J. Yeh, T.M. Spear and W.H. Ko, *Silicon diaphragm capacitive sensor for pressure, flow, acceleration and attitude measurements*, Proceedings Transducers '87 conference, Tokyo, Japan, June 2-5, 1987, pp. 336-339
18. R.J.G. Carr, *Fibre optic sensors for the characterization of particle size and flow velocity*, Sensors and Actuators A21-A23 (1990) 1111-1117
19. C. Yang et al., *Monolithic flow sensor for milliliter per minute flow*, paper presented at the 14th Nordic Semiconductor Meeting, June, 1990
20. M Esashi, *Integrated micro flow control systems*, Proceedings Third Toyota Conference, Nissin, Japan, October 22-25, 1989, pp. 21-1 - 21-16
21. J. Pfahler, J. Harley, H. Bau and J. Zemel, *Liquid transport in micron and submicron channels*, Sensors and Actuators A21-A23 (1990) 431-434

Electrically-Activated, Micromachined Diaphragm Valves

J. H. JERMAN

IC SENSORS
Milpitas, CA 95035
USA

Abstract

Electrically-activated diaphragm valves have been fabricated using heated bimetallic structures to provide the operating force. These valves can be designed to provide fully proportional control of flows in the range of 0-300 cc/min at gauge input pressures from near zero to over 6 atm. The valves are batch fabricated using silicon micromachining techniques. By combining these valves with pressure or flow sensing elements, closed loop pressure or flow control is easily accomplished.

Introduction

By using advanced silicon micromachining techniques, high-performance valves have been developed with integral actuators to control gas pressure or flow by the application of an electrical signal. These devices provide an electrical-to-pressure function which complements the pressure-to-electrical transduction offered by the millions of monolithic silicon pressure sensors fabricated yearly.

Conventional valves for pressure and flow applications have typically used magnetic actuation in the form of discrete solenoids or motors to drive diaphragm or needle-type valves. In a miniature, monolithic silicon-based valve, however, magnetic actuation is not attractive due to the difficulties involved with providing sufficient force. Electrostatic valve actuation has been reported recently,[1] but the forces involved are large only for small electrode gaps. Additionally, only attractive force is feasible, and the nature of the force makes fully proportional control difficult.

Electrostatic actuation is possible using piezoelectric drivers, but the force and deflection properties of materials such as PZT produce either high forces with small deflection for button type actuators, or relatively low forces with high deflection for bimorph type drivers.

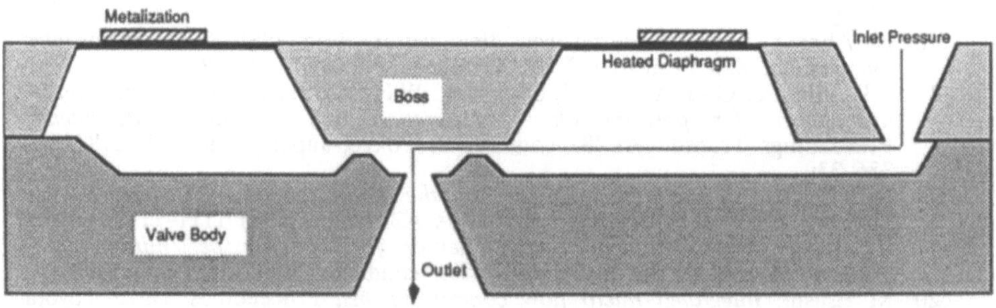

Fig. 1 Cross Section of a Micromachined, Bimetallically-Actuated Diaphragm Valve.

For very small structures, thermal actuation becomes more attractive due to the geometrical advantages of scaling. The amount of thermal mass decreases as the volume of material decreases, the thermal loss decreases as the thicknesses decrease, while the forces per unit area remains high. Bimetallic beams[2] have been shown to provide force and displacement as integrated actuation elements, however the previously described devices have shown insufficient force to use in most valve applications. The use of bimetallically-driven diaphragms increases considerably the force available from an integrated structure while simultaneously achieving large displacements and symmetric, vertical travel. Although a large variety of combinations of materials can be used, the use of a silicon diaphragm and an aluminum metallic layer is one of the most attractive.

Bimetallic Diaphragm Valve Structure

These valves, such as shown in Fig. 1, consist of a 3-mm diaphragm actuator with a central boss which mates to an etched silicon valve body. The actuator consists of a circular silicon diaphragm with integral diffused resistors, which acts as one element of the bimetallic structure. An annular aluminum region on that diaphragm forms the other element. By varying the electrical power dissipated in the resistors and thus the temperature of the silicon diaphragm, the displacement of the central boss can be controlled. The actuator chip and the valve body chip make a very compact valve structure without the large, external actuators common in previously developed valves.

In this design a nominally circular silicon diaphragm has been used, and this valve closes with increasing temperature of the diaphragm, making a normally-open valve. A photograph of such a diaphragm structure is shown in Fig. 2.

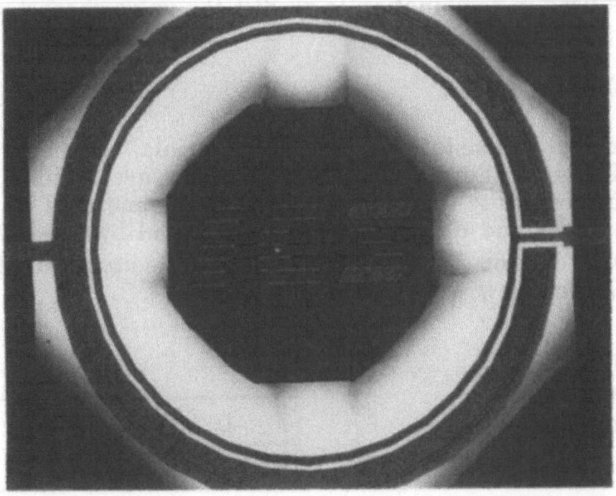

Figure 2. Photograph of typical bimetallic diaphragm with partial backside illumination.

An anisotropic silicon etch has been used to form the diaphragm, the extent of which can be seen from the light transmitted through the diaphragm. The corner compensation has been designed to provide an octagonal approximation to the outer diameter, while the inner diameter achieves a very close approximation to the desired circular shape. The 5-μm thick aluminum layer extends over the outer portion of the diaphragm. A diffused resistor is provided under the aluminum layer, and the aluminum is also used to provide electrical connection to the heater. The valve seat is on a mating silicon wafer, and is very similar to the original silicon valve structure developed by Terry at Stanford, now almost 18 years ago.[3]

The temperature of the bimetallic structure on the diaphragm is used to control the force applied to the central diaphragm boss. Thus, the gap between the valve seat and the bottom surface of the diaphragm boss can be varied, controlling the flow of gas through the valve. There is a gas feedthrough hole in the bimetallic diaphragm chip to allow gas flow through the structure and to equalize the pressure drop across the diaphragm. This minimizes the force necessary to operate the valve.

The flow characteristics of this valve are controlled by the diaphragm and metal thicknesses, the material properties of the metals, the outside diameter, boss ratio, and metal coverage on the diaphragm, and the geometry of the valve seat. The temperature of the bimetallic diaphragm is controlled by varying the power dissipated in heaters either on or in the diaphragm. Gas flows of more than 100 cc/min at pressures over 2 atm have been controlled with such a valve, with on/off flow ratios greater than 1000. Prototype valves have been cycled millions of times with no observable change in performance. These valves can be operated as either on-off valves or as proportional control valves by changing the flow geometry and dimensions of the valve elements. Other aspects of the valve operation, such as speed, power dissipation, and magnitude and direction of deflection can be controlled by adjusting the suspension characteristics of the diaphragms.

It is of crucial importance in the design and fabrication of these devices to adequately model the overall flow as a function of applied power to allow the design of this type of valve for a particular application. The analysis of the valves is complicated by the interaction between the gas flow, the diaphragm deflection, and the bimetallic force. Such complicated, linked analytical models can now be easily handled using computer programs which numerically solve the set of simultaneous non-linear equations. The individual models used in combination to predict the valve characteristics are described below.

Bimetallic Diaphragm Model

A bimetallic structure consists of a pair of materials, not necessarily metals, with different coefficients of thermal expansion, which are bonded together. As the temperature of the composite is changed, stresses are generated in the structure which can cause useful forces and deflections to be generated by the element. The original analytical model was developed by Timoshenko in a classic paper in 1925.[4] Schematically the structure is shown below:

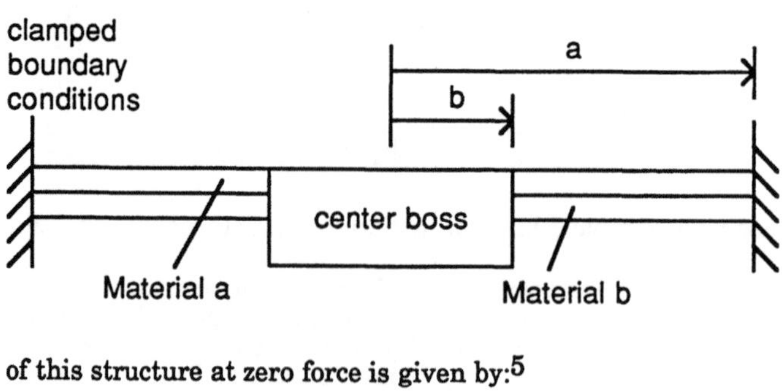

The deflection of this structure at zero force is given by:[5]

$$y_{(Q=0)} = \frac{K_y}{(1+v)}\,\Theta\, a^2$$

where a is the radius of the diaphragm and K_y is a constant which depends on the bossed ratio, b/a, and Θ is the temperature term:

$$\Theta = \frac{6(\gamma_b - \gamma_a)\Delta T(t_a + t_b)(1 + v_e)}{t_b^2 K_1}$$

where γ_a and γ_b are the thermal expansion coefficients of the two materials, t_a and t_b are the respective thicknesses, v_e is the effective Poisson ratio of the composite beam, and K_1 depends on the relative stiffness of the two components.

There are a number of different bimetal combinations possible for the structure, but one pair that is attractive is aluminum and silicon. The force and deflection are maximized for pairs which have a large difference in thermal expansion coefficients. It is often desirable to use materials with low thermal conductivities to limit power loss through conduction. Unfortunately both silicon and aluminum have high thermal conductivities, resulting in relatively large power dissipation when these materials are used with thick diaphragms.

For example, for a clamped diaphragm, 2.5 mm in diameter with a 9 μm silicon thickness, a 6 μm aluminum thickness, and a boss ratio (b/a) equal to 0.4, the deflection and force are maximized for aluminum which covers only about the outer half of the annular diaphragm. The theoretical maximum force is about 7100 dynes and the maximum deflection is 13.3 μm. This deflection is downward for the aluminum on the top surface of the device. Thinner diaphragms provide more deflection at lower force, while thicker diaphragms provide more force at smaller deflections. A silicon thickness between about 6 and 12 μm is optimum for most valve applications.

Flat silicon diaphragms are generally considered to have linear deflection vs. force characteristics. While this is true for the limited deflection of, for example, pressure transducers, in actuator structures the deflection is often a substantial fraction of the thickness of the diaphragm, where the non-linearities become significant.

Flow Model

The gas flow through the valve is determined by summing the appropriate forces on the valve diaphragm, which determines the deflection of the center of the diaphragm from the relation above. The flow through the valve is then assumed to be due to the restriction formed by the channel between the seating ring and the diaphragm boss. The volumetric, laminar flow through that channel is given by:

$$\text{flow} = \frac{\text{gap}^3\ w\,(P_{in}^2 - P_{out}^2)}{24\,\eta\, l\, P_{out}}$$

where w is the effective width of the seating ring, η is the gas viscosity, P_{in} and P_{out} are the absolute input and output pressures, and l the effective length of the seating ring.

This equation is only valid for laminar flow through the channel. The Reynold's number of the flow can be examined to determine if the flow is likely to be laminar. For a nitrogen flow velocity of 30,000 cm/s, nearly at the speed of sound, in a 5 μm channel, at a differential pressure of one atmosphere, the Reynold's number is only 79, well under the value of 2000 usually taken to be the onset of turbulent flow.

The average velocity at the output of the channel is found by dividing the volumetric flow given above by the area of the channel. The peak velocity is twice the average velocity for the parabolic flow profile found in laminar flow. For many combinations of high input pressure, low output pressure, and high flow, the average output flow velocity is found to exceed the speed of sound, using the laminar flow equation shown above. It is not possible, however, to obtain supersonic flow in a rectangular channel; that is only possible in a convergent-divergent nozzle. Flow through a sonic nozzle becomes only a function of the upstream pressure and the area of the nozzle, at constant temperature. Thus the flow characteristic of the valve changes from a cubic dependence on gap to a linear dependence on gap when sonic flow conditions are achieved.

Experimental Results

A normally open valve has been fabricated using the 2.5 mm diameter actuator shown in Fig. 2. It is mated to a valve body with an annular seat about 400 μm in diameter and 80 μm wide. The gas flow through this structure is typically from the topside of the structure, through a hole connecting the topside of the actuator diaphragm and the backside, across the valve seat, and through a hole in the valve body to the bottom of the valve assembly. By providing the inlet pressure to the topside of the actuator, the pressure tends to help close the valve, reducing the off-flow through the structure at high pressures. When closed, the valve will not suddenly open with inlet pressure pulses, as is the case with the electrostatic valve described by Ohnstein et. al.

The valve flow at a variety of inlet pressures is shown in Fig. 4. This valve has a 10 μm thick diaphragm, a 5 μm thick aluminum layer, and a nominal 4 μm initial gap when the actuator is at room temperature.

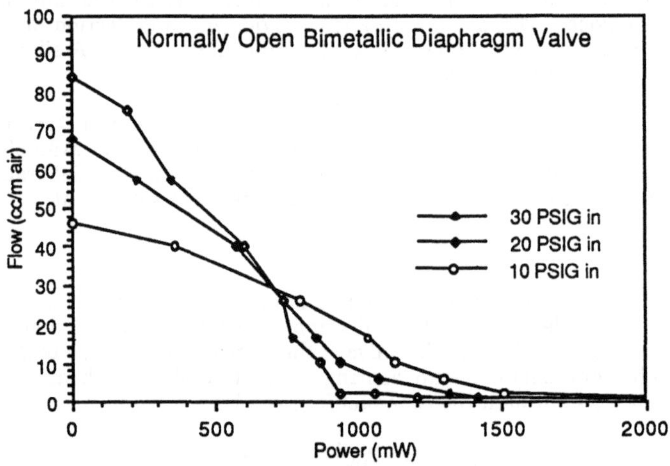

Figure 4. Experimentally determined flow characteristics of a valve with 2.5 mm diameter and 10 μm thick diaphragm.

Note that with 30 PSIG inputs, the valve actually closes earlier than with lower inlet pressures due to the action of the inlet pressure in helping to close the valve. The valves are fully proportional, with no obvious hysterysis. The leakage flow through this device at 20 PSIG input was about 0.045 cc/min, about a factor of 1,600 less than the full flow.

By using a combination of the models described earlier, the theoretical flow characteristics of such a valve structure can be calculated. The theoretical results are shown in Fig 5.

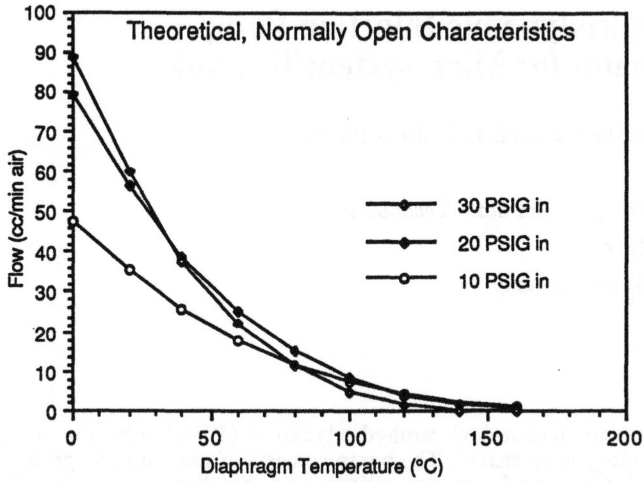

Figure 5. Theoretical flow characteristic of normally open valve structure.

The agreement with the experimental results is very encouraging. Here the initial gap has been selected to roughly match the room temperature 30 PSIG flow. The relatively thick diaphragm and small initial gap of this design results in a low flow, high pressure valve with relatively high power dissipation. By adjusting these parameters and the size of the diaphragms, a family of valves can be made with more optimum low pressure and high flow characteristics.

Conclusions

The use of bimetallic driven diaphragms has been shown to provide valve structures with flow and pressure ranges which are useful in a wide variety of applications. The characteristics of the valves can be greatly varied by changing the geometry of the structure, which can be used to optimize a valve for a particular application. The valves are batch fabricated and many hundreds of valves can be made on the same wafer, offering the possibility of a very low cost device. The theory, materials, and fabrication technologies are well understood, which shortens the development time for these devices. It is expected that these valves will greatly influence the types of integrated flow control systems available in the '90's.

References

1 T. Ohnstein, T. Fukiura, J. Ridley and U. Bonne, Micromachined Silicon Microvalve, Proc. IEEE Workshop on Micro-Electro Mech. Systems (MEMS), pp. 95-98, Feb. 1990.

2 W. Reithmuller and W. Benecke, Thermally Excited Silicon Microactuators, IEEE Trans. Elec. Dev., v 35 n 6, 1988

3 S.C. Terry, A Gas Chromatography System Fabricated on a Silicon Wafer Using Integrated Circuit Technology, Stanford ICL Technical Report #4603-1, May 1975, p. 41.

4 S.P. Timoshenko, Analysis of Bi-Metal Thermostats, J. Optical Soc. Am., V. 11, 1925, pp. 233-255.

5 R. Roark and W. Young, Formulas for Stress and Strain, McGraw-Hill, 1975, p.165.

An Electrohydrodynamic Injection Pump – A Novel Actuator for Microsystem Technology

Axel Richter, Hermann Sandmaier, Andreas Plettner

Fraunhofer–Institute for Solid State Technology
Paul–Gerhardt–Allee 42
D–8000 Munich; West Germany

Summary

A silicon micromachined electrohydrodynamic (EHD) injection pump for microsystem technology is presented. The pump consists of two opposed grids, which are micromachined from single crystal silicon and bonded together anodically. Pumping and static pressure generation was achieved with different polar fluids such as ethanole, propanole or acetone but most organic solvents can be pumped. With ethanole a maximum static pressure of 2.5 kPa (32 cm pumping head) and a pumping rate of 14 ml/min at a pressure of 420 Pa were achieved. For the realized pump the grid area measures 3 mm x 3 mm and the grid distance 350 μm. Pumping started at d.c. voltages as low as 40 V, wheras 700 V were needed for maximum throughput. Pumping was also observed with a.c. driving voltage. To increase pressure several of the devices can be stacked. This opens up new applications like micromachined cooling systems or microhydraulic actuators. The maximum flow rate and pressure head can be varied in a wide range by adjusting the dimensions of the pump. By further reducing the grid distance driving voltages below 100 V are expected.

INTRODUCTION

Microfabricated electromechanical systems (MEMS) composed of different active mechanical and electronic devices such as pumps, rotors, cantilevers, sensors and integrated electronics will lead to the realization of entirely new "machines" for which a whole range of new applications is expected. Micropumps and electrostatic or piezoelectrically excited mechanical actuators will be required as one of the main building blocks for future micro systems. Micropumps will be needed for integrated cooling devices, for the pumping and portioning of minute quantities of fluids and for microhydraulic actuators. Medical applications in the newly emerging field of chronobiology requires the controled pumping of drugs with adjustable flow rates and high reliability.

In principle electrohydrodynamic (EHD) micropumps offer the unique advantage that they require no moving parts and that the electric forces act *directly* on the fluid throughout the entire pumping volume. In contrast, micromachined pumps with thermally or piezoelectrically actuated moving membranes require more complicated fabrication processes and the assembling of different parts to realize mechanical structures in three dimensions, which possibly may also suffer from wear.

Electrostatic forces exerted upon dielectric fluids with low conductivity could lead to different hydrodynamic [1–3] effects such as electroconvection, mixing and pumping. The interaction of electric fields with free or bound charges in fluids leads to macroscopic motion of the fluid and can be used by a variety of electrohydrodynamic devices. Possible engineering applications include EHD pumps and generators, EHD diagnostics for fluids or EHD–induced augmentation of heat transfer. Microfabricated EHD pumps offer the further advantage of strongly reduced driving voltages compared to the known "large–scale" applications. The electrically-induced force density is a function of the applied electric field, which increases linearly when the voltage is fixed and the device is scaled down accordingly.

PRINCIPLE OF OPERATION

The ponderomotive force density $\vec{F}$ acting on a dielectric fluid with free space charge density q_f subjected to pressure p and electric field $\vec{E}$ is given by [4]

$$\vec{F} = -\nabla p + q_f\vec{E} + \vec{P}\cdot\nabla\,\vec{E} - \frac{1}{2}\,E^2\nabla\epsilon + \nabla\left[\frac{1}{2}\rho\,\frac{\delta\epsilon}{\delta\rho}\,E^2\right], \qquad (1)$$

where ϵ is the material permittivity and ρ the mass density of the fluid. The electric part of the terms on the right–hand side correspond to the Coulomb force on free space charge, the Kelvin polarization force, the dielectric or Korteweg–Helmholz force and the electrostrictive force respectively. The last term is relevant only for compressible media and corresponds to the gradient of a pressure. The polarization force density $\vec{P}\cdot\nabla\vec{E}$, sometimes also called dielectrophoretic force, can not lead to a permanent fluid motion for d.c. fields and is even small compared to coulomb forces for a.c. fields, so that it will not be considered further here.

In EHD injection pumps the motive force is the Coulomb force on ions injected from one or both electrodes into the fluid by electrochemical reactions [5–6]. EHD injection pumps require two permeable electrodes in direct contact with the fluid to be pumped. A pressure gradient will develop between the electrodes and leads to fluid motion between the emitter and collector. Thus the simplest arrangement consists of two electrically isolated grids within a tube separated by a distance d as used by

Pickard [7].Stuetzer [2] used a sharp point emitter with a collector ring to increase injection current in the inhomogenous field. Since the pumping performance of EHD pumps critically depends on electrode spacing and geometry silicon micromachining allows the realization of EHD injection pumps with grid separations in the micron size–scale range with optimized electrode geometry for field–enhanced injection of ions.

EHD pumps rely critically on the electrical properties of the fluid pumped, and the allowed values of permittivity ϵ and conductivity σ must be properly controlled. Typically the conductivity must lay in the range between 10^{-14} S/cm and 10^{-9} S/cm, whereas the permittivity indirectly determines the conductivity due to the increased ionizing power for ionic impurities with increased ϵ [5].

DEVICE FABRICATION

The realized EHD injection micropump mainly consists of two opposed grids, which are microfabricated from single crystal silicon wafers and bonded together anodically. Fig. 1 shows a SEM microphotograph of the cross section of the realized pump.

Fig. 1. Cross section of the EHD pump (SEM)

The two grids are almost identical, the grid area measuring 3 mm x 3 mm with 30 μm thickness. Two different sets with hole sizes of 70 μm and 140 μm respectively were fabricated. This structure is easily etched into <100> silicon using standard KOH etchant. Therefore the technology is quite simple and straightforeward, only 3 masking layers were needed for the shown devices.

A 1 μm thick SiO_2 layer is first thermally grown on n–type silicon wafers 380 μm thick with <100> orientation, which are polished on both sides. After removing the backside oxide and opening a substrate contact on the frontside a 5 μm thick pyrex layer is sputter deposited on the back of the upper electrodes. Then a nitride masking layer 0.4 μm thick is deposited by a PECVD–process on the front– and backside, which is subsequently structered on both sides after photolithgraphy with a double–sided mask aligner. Then the wafers are anisotropically etched with 33 % KOH at 80 °C as an orientation dependend etchant until the 30 μm deep pits are formed on the frontside. The etch depth is controlled *in situ* through a special test structure on the frontside. Now the nitride mask is removed from the frontside on which a new masking nitride is deposited. The wafers are again etched in KOH from the backside until the 30 μm grid is formed. Subsequently the grid area is metallized to prevent oxidation. A ohmic contact is formed on the backside of the lower and on the frontside of the upper grid. The wafers are diced and the individual grids bonded together anodically.

When the two grids are bonded together as described above, their separation distance is given by the wafer thickness. To decrease the required driving voltage further, the separation can be decreased by either thinning the upper wafer or etching back the upper grid from the frontside and mount it upside–down on the lower grid [8]. To increase the injection current, also a grid made from small silicon pyramids [8,9] is possible.

MEASUREMENTS

For the measurements the mounted pump is immersed upside down in the fluid to be pumped. To prevent leakage currents the bond wire and the lateral faces of the chip were encapsulated with epoxy. The pump was connected to a high voltage d.c. drive, voltage and current were measured. The static pressure was determined by the rise of the fluid within the tube after correcting for the capillary effect. The flow rate was determined by collecting the pumped fluid at a given tube hight for a fixed time and subsequent gravimetric measurement. Several organic solvents such as ethanole, methanole, acetone or propanole could be pumped with different yields, also depending on the purity of the fluid. The reported results were achieved with ethanole (99.8 %, p.a.) at a temperature of 22 °C with pumps having hole sizes of 140 μm. Different pumps showed slightly different characteristics, probably due to differences in the lateral adjustments of the grids.

Fig. 2a shows the pressure generated with zero flow rate as a function of the applied voltage. A maximum pressure of 2480 Pa is achieved with a driving voltage of 700 Volts. A quadratic variation of pressure with voltage is clearly seen.

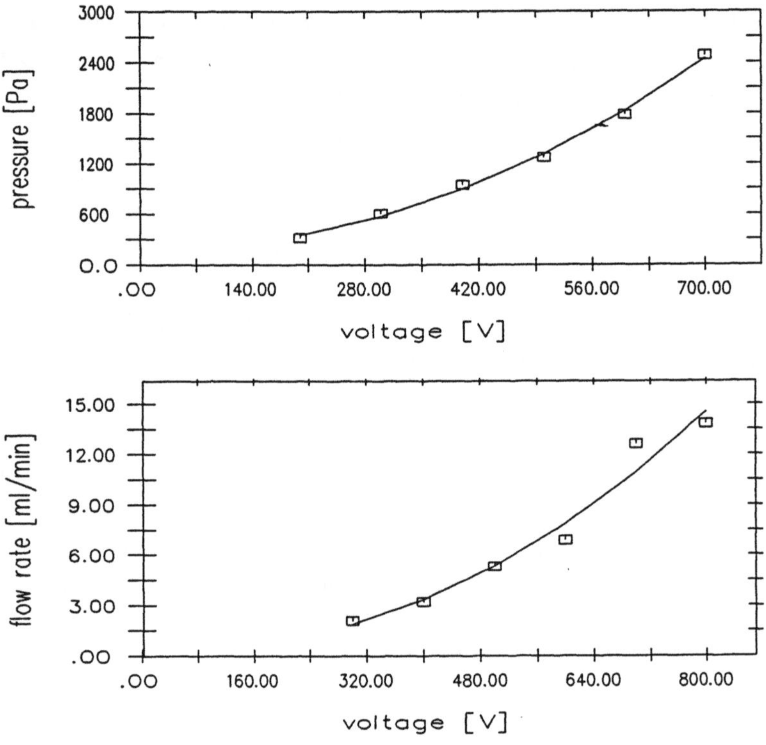

Fig. 2: a) static pressure vs. driving voltage b) flow rate vs. driving voltage

In Fig. 2b the flow rate is depicted as a function of the applied voltage at a constant pressure head of 420 Pa and shows a quadratic behavior too. A maximum flow rate of 14 ml/min was achieved. The pump was operated in this mode for periods of 1 h or more without degradation, albeit some fluctuations in the flow rate did apear. A slight temperature dependence is expected too [7], but was not measured yet.

CONCLUSIONS

A silicon micromachined electrohydrodynamic injection pump for relatively isolating dielectric fluids was presented. With an active area of 3 mm x 3 mm a flow rate almost three decades higher (ml insted of μl) than with piezoelectric or thermally excited micropumps [10,11] can be achieved, wheras the maximum pressure head is of the same order. The proposed pump offers the further advantage of simple design and fabrication since no moving parts are required. Through reduction of the grid distance the driving voltage can be further reduced, wheras the maximum flow rate is given by the operating conditions and the grid size, which can be adjusted depending on the application. The stacking of several devices is easily possible, so that even atmospheric pressures seems to be achievably. Aqueous solutions can not be pumped due to their high ionic conductivity, so that the use in medical or biological systems is restricted. In order to move such fluids the EHD pump can be used for the excitation of a micromechanical diaphragm pump.

ACKNOWLEDGEMENTS

The authors wish to thank Prof. I.Ruge for his encouragement for this work. They also acknowledge the assistance of S.Besson for device fabrication, D.Böhm for the SEM microphotographs and K.Marusczyk for the drawings.

REFERENCES

[1] A.C.Eringen, G.A.Maugin, "Electrodynamics of Continua Vol. II: Fluids and Complex Media", Springer, New York, 1990

[2] O.M.Stuetzer, "Ion Drag Pressure Generation", J. Appl. Phys, **30**, 7, pp.984–994, 1959

[3] A.Richter, H.Sandmaier, "An Electrohydrodynamic Micropump", Proc. 3rd IEEE Workshop on Micro Electro Mechanical Systems, Napa Valley, USA, pp. 99–104, February 1990

[4] J.R.Melcher, Continuum Electromechanics, MIT Press, Cambridge, USA, p. 3.13, 1981

[5] N.J.Felici, "D.C. Conduction in liquid dielectrics: A survey of recent progress (Part I)", Direct Current, **2**, 3, pp.90 –99, 1971

[6] N.J.Felici, "D.C. Conduction in liquid dielectrics: (Part II) electrohydrodynamic phenomena", Direct Current, **2**, 4, pp.147 –165, 1971

[7] W.F.Pickard, "Ion Drag Pumping. II. Experiment", J.Appl.Phys., **34**, 2, pp. 251–258, 1963

[8] A.Richter, H.Sandmaier, "Mikrominiaturisierte elektrostatische Pumpe", German Patent Application P 3925749.5

[9] H.H.Busta, R.R.Shadduck, W.J.Orvis, " Field Emission from Tungsten–Clad Silicon Pyramids", Trans. Electron Devices, **ED–36**, pp. 2679–2685, 1989

[10] S.Shoji, S.Nakagawa, M.Esashi, "Micropump and Sample–injector for Integrated Chemical Analyzing System", Proc. Transducer '89, Montreux, Switzerland

[11] F.C.M. Van De Pol, H.T.G. Van Lintel, M.Elwenspoek, J.H.J Fluitman, "A Thermopneumatic Micropump Based on Micro–engineering Techniques", Proc. Transducer '89, Montreux, Switzerland

Electrostatic and Superconducting Microactuators

Hiroyuki Fujita

Institute of Industrial Science
The University of Tokyo
Roppongi 7, Minato-ku, Tokyo 106 Japan

Summary

Rescent progress in MEMS (micro electro mechanical systems) fabricated by IC-based micromachining revealed a new possibility of making micro motion systems. This paper briefly discusses prospects and limitations of MEMS, which are considerably different from the simple miniaturization of macro machines. As the most critical element of MEMS, recent research on microactuators is reviewed with the stress on electrostatic and superconductive actuators.

Indroduction

Recent progress in silicon micromachining has attracted much interest in the field. Micromechanical elements [1,2] and electrostatic micromotors [3-5] fabricated by IC-compatible processes clearly showed the possibility to have micro electro mechanical systems (MEMS) on a silicon chip. On the other hand, somebody may have doubts about how we can construct three dimensional micro machines out of planar silicon structures. Some of the doubts come from the fact that the field is still very primitive and needs many technical breakthroughs. Others come from misunderstanding the purpose of the field; a typical example of such a misunderstanding is that the goal of the research is to make a miniaturized submarine, similar to the one which appeared in the SF movie "A Fantastic Voyage". This kind of goal is very easy to accept for non-technical people. It has, however, pitfalls and is very misleading. Criticisms such as "Planar IC processes cannot be applied to make real three-dimensional structures", "Silicon film is too thin to have enough strength to manipulate external objects", and "Electrostatic micomotors don't have enough output force to move something" are not wrong at all but miss the point of the discussion. Research on IC-based MEMS does not intend to make miniaturized version of macro machines.

From the viewpoint of MEMS research, the criticisms mentioned above are regarded as limitations of the present technology. The prospects of MEMS are being pursued within these limitations. The prospects can be summarized as follows;

(1) pre-assembly and batch fabrication capability of complicated mechanical systems,

(2) production of smart micro systems through the integration of sensors, logic, and actuators,

(3) creation of a decentralized system with many independent MEMS modules.

When these prospects and limitations are taken into account, the actual realization of MEMS will be different from the simple miniaturization of macro machines. Namely, they are not three-dimensionally assembled, but composed of planar units on which MEMS modules are tightly distributed. They may resemble a house of cards. The MEMS modules work cooperatively among each other in order to perform a task. Other possible realization is to have many micro mechanical elements on the narrow surface of a needle tip. This is useful to perform a task in very narrow spaces.

Since sensors and logic circuits on silicon have been fully established, the key missing element of MEMS is the microactuator. In the present paper, recent research on microactuators is reviewed with the stress on electrostatic and superconductive actuators. In the following, the word "micromechatronics" means "research on MEMS".

Microactuators

Breakthroughs in electrostatic microactuators have occurred in the last three years. One is the IC-processed electrostatic micromotors with diameters of 60-120 μm [3,4]. Figure 1 shows the "historical" picture by Tai, et al. [4] This success was based on the development of surface micromachining techniques, which are essential to make micro mechanical structures[1,2]. The technique resolved the problem of assembly and handling of micro structural elements by means of the pre-assembly process.

The reason why electrostatic force was chosen to drive the motor is that electrostatic force obeys a favorable scaling law when the characteristic dimention, L, is decreased. The force is proportional to L^2, while inertial force and electromagnetic force are proportial to L^3.

As compared to the magnetic field, the electric field can be produced between two separate electrodes and dosen't need windings and yokes. Therfore the structure of electrostatic motors are simple enough to be fabricated by IC-processes[3-5]. The energy density of the electric field is far less than that of the magnetic field in the macro world. The limiting factor is the weak breakdown strength of the air. When the characteristic dimension decreases, there are lower threshold voltages below which no electric discharge can take place in gases (Paschen's Law); the value is 330 V in air. We can expect the energy density of an electric field with 300 V across a 1 μm gap is comparable to that of a 1 T magnetic field [6].

The other breakthroughs in electrostatic microactuators were proposed prior to the micromotor, although they were not fully IC-compatible. Yano, et al. proposed an electrostatic linear motor based on the mechanism of human muscles [7]. The author and Omodaka fabricated the ESLAC (an electrostatic linear actuator) shown in Fig. 2 [6]. The former is important because it materialized the concept that "large force and displacement can be obtained by connecting many micro force generators in series and parallel by means of micro machining." The latter is interesting because the movable part of the ESLAC rotates on the stationary part to avoid large friction caused by sliding contacts. Frictional force, like electrostatic force, is dominant in small dimensions and considered to be one of the most crucial problems in micromechatronics. Many variations of microactuators utilizing rotational motion have been reported so far; cylindrical types [8,9], a conical type [10], and an IC-prrocessed type with side drive [5]. Figure 3 shows the conical type called the ESTOM.

Stimulated by these reports, many researchers proposed microactuators using micromachining techniques (some of them are fully micromachined and the others are partly). As shown in Table 1, microactuators are driven by many different forces.

In the first column is the electrostatic micromotor by Tai, et al [4]. It is called the side-drive type motor, since it utilizes the electrostatic force which acts between the edges of the rotor and the stator (Fig. 1). The rotor and stators are made of polysilicon film, 2 μms thick.

Stators are placed on a circle and connected in three phases. If the voltage is applied to each phase successively, the rotor rotates in synchronization. The voltage is up to 300 V across the 1-2 μm gap. The torque is estimated to be a few pNm. Rotational speed was reported to be on the order of 500 rpm. The speed is relatively low compared to the theoretical value[4,11]. The reason is the friction between the rotor and the shaft, although a silicon nitride film was deposited on the sliding surface to reduce the friction.

In the second column is the improved version of the side-drive micromotor reported by Mehregany, et al.[5]. Major points of improvement are as follows;

(1) clearance between the rotor and the shaft was reduced to be 0.3 μm rather than 1 μm,

(2) three dimples under the rotor served both as bushings and as electrical contacts,

(3) degradation of electrical contact between the rotor and the ground plane was prevented by a carefully-chosen rinse process and by operation in a nitrogen gas environment. (If the rotor is covered with an oxydization layer and is not effectively grounded, the charge accumulated on it produces a normal force which causes the rotor to stick to the substrate.) With these improvements, the motor was rotated at 15000 rpm and kept in operation for more than seven days. More than 7000 micromotors were fabricated on a single four-inch wafer— good proof of the mass-fabrication capability of micromachining.

Even for the improved micromotor, friction is a major problem. Using rotational motion is one solution to the problem (as was mentioned above). Other solutions are (1) elastic supports and (2) levitation of movable parts.

Four electrostatic actuators with elastic supports are shown in the third through the sixth columns. The author and co-workers reported a electrostatically deformed silicon membrane [12]. Closed-loop control using a capacitance-displacement sensor was applied. A position accuracy of 0.1 μm and a response time of 1 ms were obtained. Application to a varifocal mirror was examined. Next is an electrostatic resonator (Fig. 4) by Tang, et al. [13]. The resonator is supported by double-fold beams and actuated by comb-like structures. The teeth of the comb, attached to the moving part, overlap those fixed on the substrate. The force to increase the overlapping is generated when voltage is applied between the two combs. An alternating voltage of 10 V with a 40 V DC bias made the suspended part vibrate at resonance. The displacement was 10 μm and the resonant frequency was 18 kHz with 200μm-long supports.

Mahadevan, et al. [14] reported a linear actuator made by polyimide. The mover is a polyimide ladder-like structure sandwiched by two driving electrodes. The electrodes are also patterned in stripes which have the same pitch of the mover but are divided into some sections with different phase shifts. The mover is supported by four polyimide beams. Although the mover is not conductive, it is attracted in between electrodes which make up a parallel-plate-capacitor. The actuator is interesting because it utilizes the force acting on both sides of the mover rather than on the edge.

In the sixth column, an electrostatic valve is shown. A plate with one side fixed is driven electrostatically and seals an inlet orifice. The closure plate is composed of a metal electrode sandwiched by silicon nitride films. The valves are fabricated in a 5 by 5 aray, which results in larger flow rate and finer flow control just by closing some of the valves. It was possible to close the value against pressures of up to 110 mmHg with 30 V applied to the valve.

Some of the parameters of superconductive microactuator [16] are shown in the seventh column. The repulsive force between a permanent magnet and a superconductor is utilized to levitate a moving part. Since the repulsive force is caused by the Meissner effect, the actuator is named Meissnac. The lateral driving force is obtained by the Lorentz force. As shown in Fig. 5, the stator has a superconductive layer covered by a polyimide insulation layer. On the polyimide layer, copper lines (20 μm thick, 200 μm wide, 10mm long) were formed with spaces of 200 μm between the lines. If the stator is cooled by liquid nitrogen, the slider, composed of a permanent magnet (0.3x1x4mm^3, 8mg), was levitated about 1mm above the stator. The slider was moved by putting 0.4A through two adjacent lines in the alternate direction. The slider moved step by step (one step was 400 μm). With 0.4 A, the average velocity of the slider was 7mm/sec and the maximum acceleration was 3.5 m/sec^2. The driving force at 0.8 A was calculated to be 40 μN. The experimental value obtained from the acceleration and the weight was 30 μN. A drag force existed even though the slider had no contact with the stator. The reason is that the magnetic flux penetrates into the ceramic superconductor (YBaCuO) used in the actuator. As the slider moved, the flux also moved in the superconductor generating the drag force. The drag force was measured independently. The value was 10 μN, which accounts very well for difference between the calculated and observed driving forces.

Micro actuators which utilize other driving principles such as piezo electric[17], shape memory alloys[18-21] and thermal expansion [22-24] are all shown in Table 1 as a comparison. Each actuator in the table has advantages and disadvantages. The choice should be made according to the requirements of applications. A systematic design and analysis method is desired[23]. In general, the electrostatic actuator is suitable to perform tasks which can be completed within a chip (fluidic control, positioning of devices, light deflection, etc.), since it is easily integrated on a chip, easily controlled and requires little supplied power. On the contrary, the other types of actuators are more robust and suitable to perform external tasks (propulsion, to cut something, etc.).

Conclusion

Prospects and limitations of MEMS are discussed. Recent development of microactuators is also reviewed. Following conclusions are suggested.

(1) The actual realization of MEMS will be different from simple minaturized versions of macro machines. Namely, they are composed of planar units on which MEMS modules are tightly distributed. The MEMS modules work cooperatively among each other in order to perform a task. Other possible realization is to have many micro mechanical elements on the narrow surface of a needle tip.

(2) Many types of microactuators are under development. Elecrostatic microactuators are most promising as actuators which perform tasks within a chip. Other types of actuators are more robust and suitable to perform external tasks.

(3) In order to reduce friction, which is one of the most crucial problems of IC-processed micromotors, design, fabrication processes, and operational conditions were improved. Other methods to overcome friction were also proposed; elastic supports, rotational movement, and levitation. The Meissner effect of superconductors was applied to levitate micro movers.

References

(1)Fan, L. S., et al., "Integrated Movable Micromechanical Structures for Sensors and Actuators" IEEE Trans. Electron Devices, ED-35 (1988)724-730

(2)Mehregany, M., et al., "Integreated Fabrication of Polysilicon Mechanisms" ibid., 719-723

(3)Fan, L. S., et al., "IC-processed electrostatic micromotors" Sensors & Actuators, 20 (1989) 41-48

(4)Tai, Y. C., et al., "IC-processed electrostatic synchronous micromotors" ibid. 49-56

(5)Mehregany, M., et al., "Operation of Microfabricated Harmonic and Ordinary Side-Drive Motors" Proc. 3rd. IEEE MEMS Workshop, Napa Valley, Feb. (1990) 1-8.

(6)H. Fujita, et al., "Fabrication of an Electrostatic Linear Actuator" IEEE Trans. Electron Devices, ED-35 (1988) 731-734

(8)Jacobsen,S.C.,et al.,"A design overview of an eccentric-motion electrostatic microactuator" Sensors & Actuators, 20 (1989) 1-16
(9)Trimmer,W.S.N., et al.,"Harmonic Electrostatic Motors" Sensors & Actuators, 20 (1989) 17-24
(10)Sakata.M., et al.,"An Electrostatic Top Mptor and its Characteristics" Sensors and Actuators, A21-A23 (1990) 168-172
(11)Bart,S.F., et al.,"An Analysis of Electroquasistatic Induction Motors" Sensors and Actuators, 20 (1989) 97-106
(12)Fujita,H., et al., "An Integrated Micro Servosystem", IEEE Intn'l Workshop on Intelligent Robots and Systems (IROS'89) Tokyo, Japan, 15-20
(13)Tang,W.C., et al.,"Laterally driven polysilicon resonant microstructures" Sensors & Actuators, 20 (1989) 25-32
(14)Mahadevan,R.,et al.,"Application of Electric Microactuators to Micromechanics" Sensors and Actuators, A21-A23 (1990) 219-225
(15)Ohnstein, T., et al., "Miromachined Silicon Microvalve", Proc. 3rd IEEE Workshop on Micro Electro Mechanical Systems, Napa Valley, CA, Feb.(1990), 95.
(16)Kim,Y.K., et al.,"Fabrication and Testing of a Micro Superconductive Actuator Using Meissner Effect" Proc. 3rd. IEEE MEMS Workshop, Napa Valley, Feb.(1990) 61-66.
(17)Akamine,S., et al.,"A Planar Process for Microfabrication of Integrated Scanning Tunneling Microscopes" Sensors and Actuators, A21-A23 (1990) 964-970
(18)Walker,J.A.,et al."Thin-film Processing of TiNi Shape Memory Alloy" Sensors and Actuators, A21-A23 (1990) 243-246
(19)Ikuta,K. et al., "Crystallographic Analysis of TiNi Shape Memory Alloy Thin Film for Micro Actuator" Proc. 3rd. IEEE MEMS Workshop, Napa Valley, Feb.(1990), 38-39.
(20)Busch,J.D. et al., "Prototype Micro-Valve Actuator" ibid., pp.40-41.
(21)Kuribayashi,K., et al., "Reversible SMA Actuator for Micro Sized Robot", ibid., pp.217-221
(22)Zdeblick,M.J., et al., "A Microminiature Electric-to-Fluidic Valve", Transducers'87, June (1987) Tokyo, Japan, 827-829.
(23)Van De Pol,F.C.M., et al.,"A Thermo-pneumatic Actuation Priciple for a Microminiature Pump and Other Mechanical Devices" Sensors & Actuators, 17 (1989) 139-143
(24)Riethmuller,W., et al.,"Thermally Excited Silicon Microactuators" IEEE Trans on Electron Devices, ED-35 (1988) 758

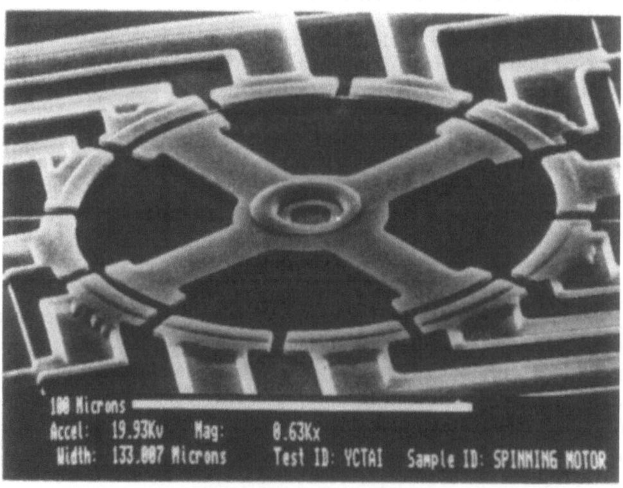

Fig. 1 IC-processed electrostatic micromotor[4]

Table 1. Micro Actuators (fully or partly IC-processed)

driving priciple	size	movement, application	support	speed, response	force, torque	material	input	ref. & authors
1. electrostatic	60～120μm (diameter)	rotation	sliding	500rpm	a few pNm	poly-Si	60～ 400V	[4] Y. C. Tai, et al.
2. electrostatic	100μm (diameter)	rotation	sliding	1500rpm	10pNm	poly-Si	50～ 300V	[5] M. Mehregany, et al.
3. electrostatic	0.05x10x10 mm	15μm (L. V.)	elastic	～3kHz	40mN	Si	0～300V	[12] M. Harada, et al.
4. electrostatic	5x100x100 μm^3	10μm (L. L.)	elastic	10～ 100kHz (resonace)	2μN	poly-Si	$40V_{DC}$ $+10V_{AC}$	[13] W. C. Tang, et al.
5. electrostatic	10x500x500 μm^3	2μm (L. L.)	elastic	N. A.	N. A.	poly-imide metal	200V	[14] R. Mahadevan, et al.
6. electrostatic	0.1x0.35 x0.39mm^3	on-off valve	elastic	N. A.	110mmHg	metal, Si_3N_4	30V	[15] T. Ohnstein, et al.
7. electrostatic	0.1x10x10 mm^3	5mm	levitation (Meissner effect)	20mm/s	30μN	YBaCuO, NdFeB	0.3～ 0.9A	[16] Y. K. Kim, et al.
8. piezoelectric	8μmx0.2mm x1mm	7μm (STM scan)	elastic	N. A.	23μN	ZnO	30V	[17] S. Akamine, et al.
9. shape memory alloy	2x30x2000 μm^3	a few μm	elastic	20Hz	N. A.	TiNi	2mA, /40V	[18] J. A. Walker, et al.
10. thermal	0.5x3x3mm^3	45μm (L. V.)	elastic	～5ms	0.6N	Si+ liquid	～200mV	[22] M. J. Zdeblik, et al.
11. thermal	0.5x8x8mm^3	23μm (L. V.)	elastic	>1Hz	0.1N	Si	13V	[23] F. C. van de Pol, et al.
12. thermal	6x100x500 μm^3	60μm (bending)	elastic	10Hz	N. A.	Si+Au	～150mA	[24] W. Riethmuller, et al.

*L. V.: linear motion in vertical direction, L. L.: linear motion in lateral direction.

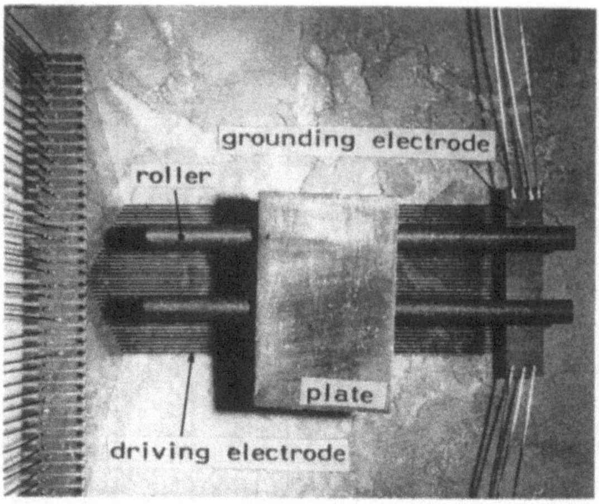

Fig. 2 Electrostatic linear microactuator (ESLAC)[6]

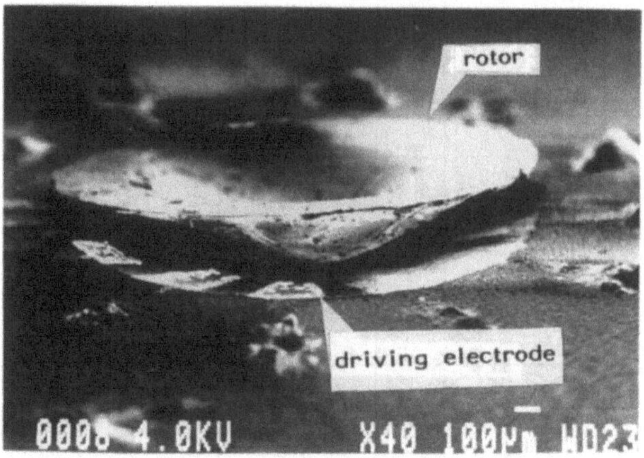

Fig. 3 Electrostatic top motor (ESTOM)[10]

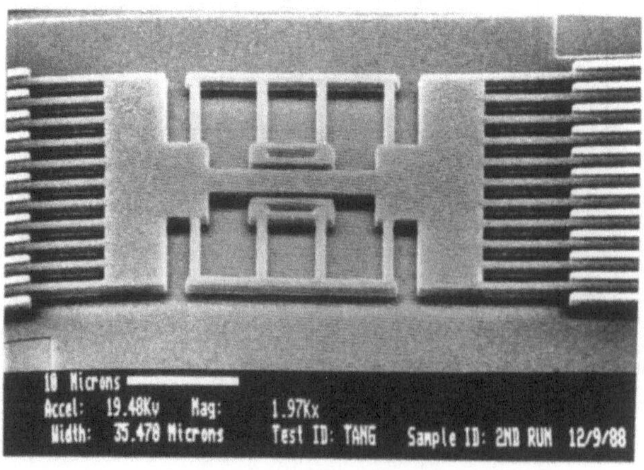

Fig. 4 Laterally driven poly-silicon resonator[13]

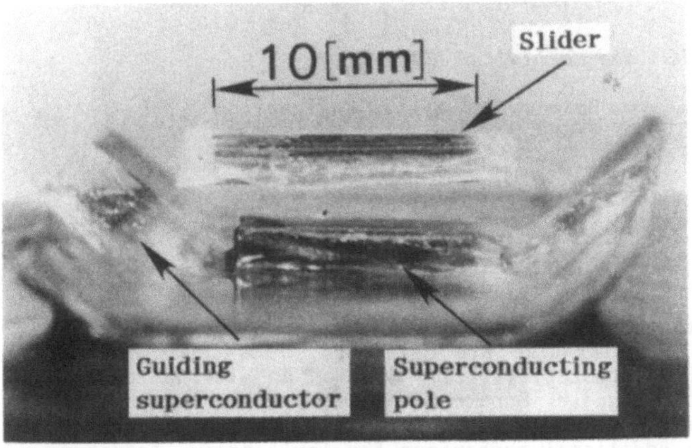

Fig. 5 Superconducting microactuator[16]

Levitational Bearing Systems by Meissner Effect

Tadashi KITAMURA and Mochimitsu KOMORI

College Of Computer Science And System Engineering
KYUSHU INSTITUTE OF TECHNOLOGY
Iizuka-City, Fukuoka, 820 JAPAN

Summary
A superconductive linear slide and a superconductive radial bearing of centimeter-size are developed. These bearings are characterized by stable levitation due to an alternating pole pattern of used magnets. Superconductors of these bearings are stably levitated by Meissner effect with neither mechanical contacts nor additional mechanisms for levitation. Actuators with the use of these bearings are also made. Static characteristics of the bearings are investigated experimentally and theoretically. Results of computer simulation and experiments show that an optimal size of magnets providing the maximum levitational force exists. The developed bearing mechanisms have a simple, miniaturizable structure feasible for microactuators which require frictionless motion.

1. LEVITATIONAL MECHANISM

Figure 1 shows the basic mechanism of levitation produced by Meissner effect and an alternating pole pattern of magnets. Magnetic fields between N poles and their adjacent S poles work as mechanical springs. Since these magnetic fields are uniformly arranged, a superconductor can be stably supported by Meissner effect. The advantage of this mechanism over the use of a single magnet is that the uniform levitational force is obtained everywhere on the magnets. A simple application of this mechanism provides a linear slide. Arranging magnets in the alternating pole pattern in a circle gives a radial bearing as shown in Figure 2.

2. DEVELOPED BEARINGS AND ACTUATORS

Figure 3 (a) shows a photograph of the developed linear slide. The magnetic guide includes ten pieces of bar magnets at the bottom for levitation and three pieces of the same dimensions (2mm×20mm×5mm) for its each side. The slider has disk-shaped nine superconductors (7mmϕ×1mm) on the base for levitation, square eight superconductors (5mm×5mm×1mm) on its each side for guide, and square superconductors on each end. All superconductors used in this study are made of YBaCuO in a standard fashion of sintering in the authors' laboratory. It is observed that the slider is stably levitated with no mechanical contacts with any side of the guide in the liquid nitrogen. A longitudinal view of this observation is shown in Figure 3(b).

A linear actuator with the use of this slide is made. A solenoid is placed facing the superconductors at each end of the slider to actuate the slider by Meissner effect. It is observed that the position of the slider is well controlled following a pulsatile referential curve of less than 0.5Hz by a fuzzy logic-based technique. The details of this actuator will be presented elsewhere.

Figure 4 shows photographs of a superconductive radial bearing. The magnetic housing of the inner diameter of 10 mm includes 10 pieces of bar magnets of the same dimensions. The width, thickness and length of the bar magnets are 3, 4 and 20 mm respectively. The radius and length of the columer shaft of YBaCuO are 6 mm and 15 mm respectively, weighing 2.5 g. Figure 4 (b) shows a photograph of the shaft being levitated stably in the liquid nitrogen.

3. STATIC CHARACTERISTICS OF THE BEARINGS

Static characteristics of the linear slide are measured using its simplified model with the experimental setup shown in Figure 5. The magnetic guide (20mm×5mm×~40mm) has the same alternating pole pattern as that used for the linear slide. A single YBaCuO disk of 23mm in diameter and 5mm in thickness is used. Some results of the measurements are as follows.

Relationships between the levitational force (F) and the width (d) of a bar magnet for various distances (h) between the magnets and the superconductor are shown in Figure 6. This figure suggests that the pole pattern of the guide can be optimized with respect to d providing the maximum force if h is fixed.

Computer simulation of the above experiments are carried out under assumptions of idealistic circumstances, such as, no critical magnetic field for the superconductor, complete reflected magnetic image in the superconductor. The magnetic field H is given by a standard formula:

$$dH = \frac{\rho}{4\pi\mu_0 r^2(x,y,z)} dxdydz \tag{1}$$

where μ_0 is the permeability of vacuum, and r is the distance between the magnetic surface and its reflected image, and ρ is the magnetization density of the used SmCo alloy. Levitational force is computed by integrating the force the reflected mgnetic image receives from dH.

4. RESULTS AND DISCUSSION

Figure 7 shows an idealistic counterpart of Figure 6. The shape of the curve for each distance (h) in Figure 7 is as a whole in a good agreement with that in Figure 6 except for the sharpness of the convexity of the curves. However the difference is that the theoretical value of the levitational force is about 200 times as large as that in Figure 6 for each distance (h). This is mainly because magnetic flux actually penetrates the superconductors under magnetic flux higher than H_{c1}, and this drastically weakens the levitational force.

The computer simulation also suggests that there is an optimal d providing the maximum stiffness of the slide for each h under idealistic circumstances. Further theoretical and experimental discussion is necessary for investigating dynamic characteristics of the slider and the radial bearing under flux penetration and pinning effect.

5. ACKNOWLEDGMENT

This study is in part supported by the Iwatani Naoji Foundation's Research Grant.

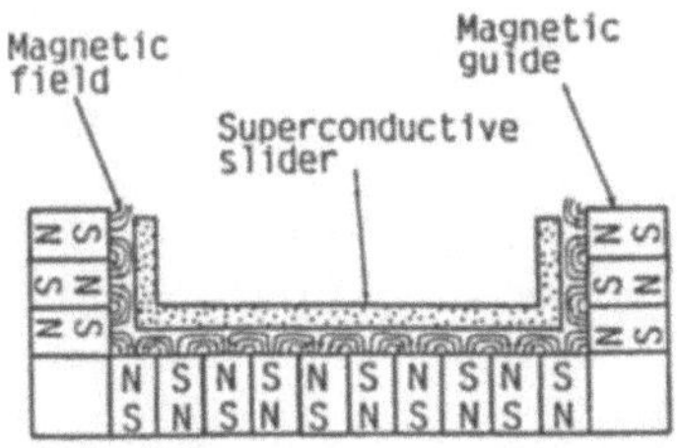

Fig.1 Levitational mechanism of a superconductive linear slide.

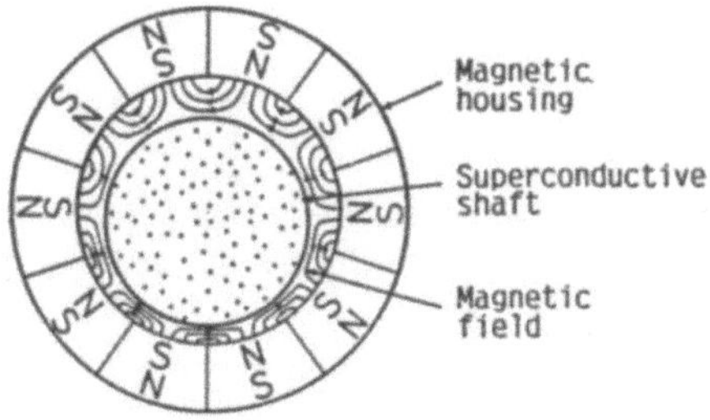

Fig.2 Levitational mechanism of a radial bearing

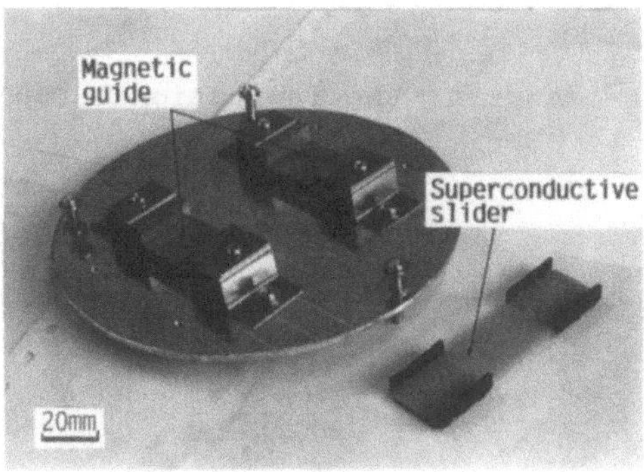

(a)

Fig.3 The superconductive linear slide.

(a): The magnetic guide and the superconductive slider.

(b)

(b): The slider levitated in the liquid nitrogen.

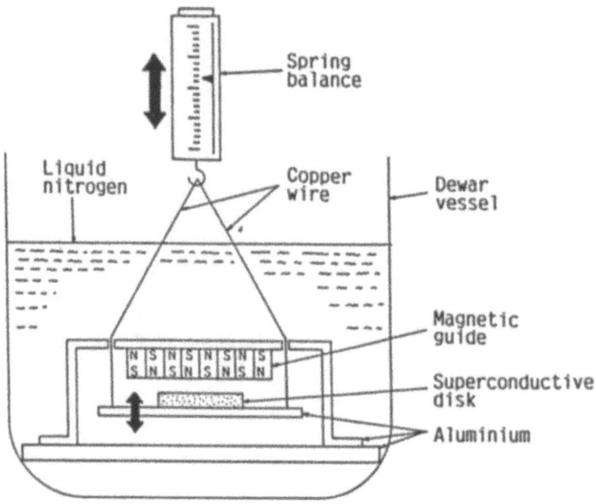

Fig.5 Experimental setup for measuring the levitational force

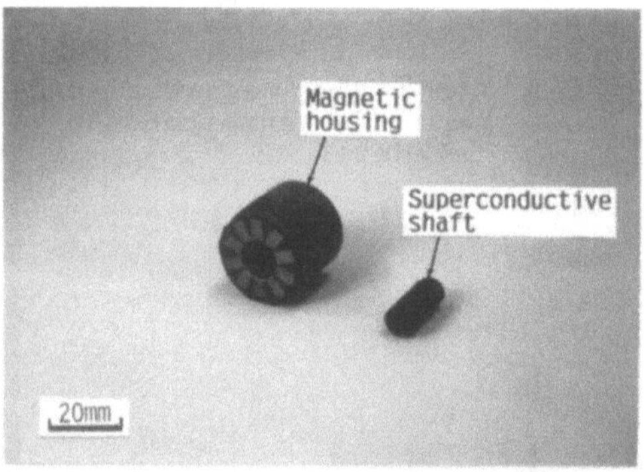

(a)

Fig.4 The superconductive radial bearing

(a) The columer shaft and its magnetic housing

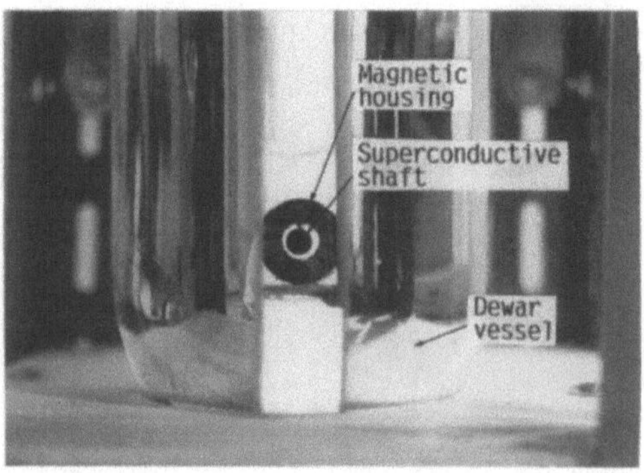

(b)

(b) The bearing levitated in the liquid nitrogen

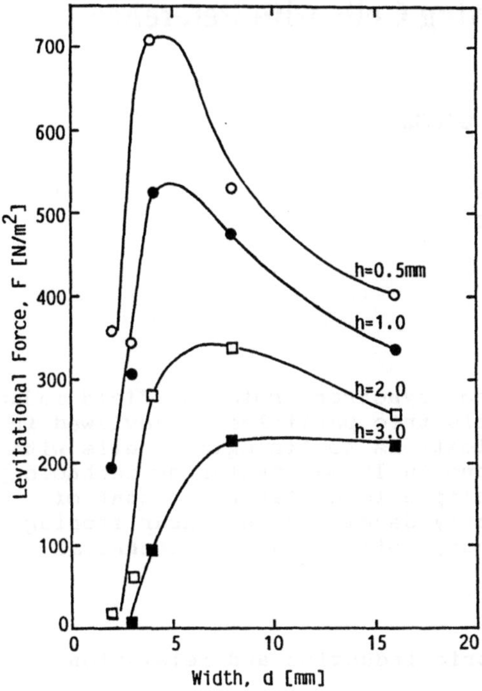

Fig.6 Levitational force (F) versus width (d) of bar magnets of the magnetic guide for distances (h).

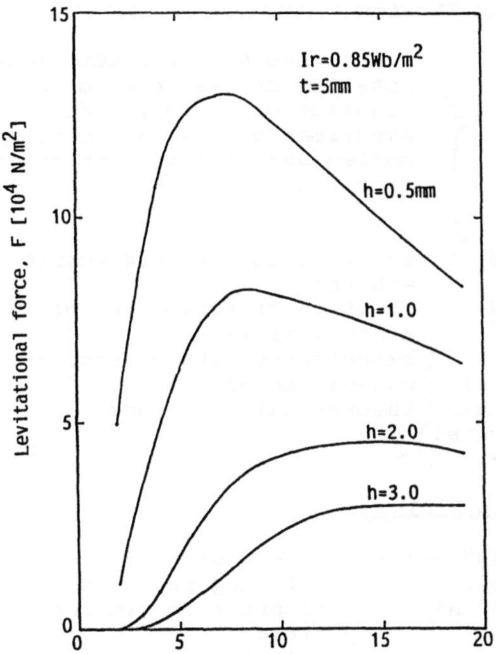

Fig.7 Theoretical relation between levitational force (F) and width (d) of bar magnets

Rotational Behaviour of Living Cells with Reference to Micro-Motors

G. FUHR, R. HAGEDORN, and J. GIMSA

Department of Biology
Humboldt University of Berlin
Invalidenstr. 42, Berlin 1040

Summary

In this paper the backround to asynchrone rotating field induced motion of liquid-immersed dielectric particles is reviewed in order to compare recent applications to biological cells with possible technical realizations in IC manufacturing technology. The here described motor principle is different to that of electrostatic motors as recently developed in planar etching technique [1-3]. History, theory, biological and technical applications are discussed.

History

Investigations of the dielectric induction and relaxation principle have a long history:

in Physics

[4]	1881	H. HERTZ	first theoretical predictions
[5]	1892	R. ARNO	rotation of dielectrics in
[6]	1896	G. QUINCKE	constant electric fields
[7]	1897	A. HEYDWEILLER	experiments with macroscopic
[8]	1906	A. LAMPA	bodies and torque calculations
[9]	1907	E.v.SCHWEIDLER	(rotating fields)
[10]	1920	M. BORN	
[11]	1921	P. LERTES	
[12]	1927	S.W. RICHARDSON	experiments with dielectric
[13]	1927	L.G. VEDY	spheres
[14]	1936	W. KRASNY-ERGEN	rotation of assamblies of microparticles
[15]	1956	I. SUMOTO	experiments with macroscopic
[16]	1968	P.E. SECKER et al.	Quincke-motors,
[17]	1969	J.R.MELCHER et al.	theoretical framework
[18]	1970	T. CODDINGTON et al.	
[19]	1971	P. SIMPSON et al.	

in Biology

[20]	1960	A.A.TEIXEIRA-PINTO et al.	observation of
[21]	1964	A.A.FüREDI et al.	spinning cells in
[22]	1981	U. ZIMMERMANN et al.	alternating electric
[23]	1982	H.A.POHL et al.	fields
[24]	1982	W.M.ARNOLD et al.	cell rotation induced by rotating fields

[25,26]	1983/84	G. FUHR et al.	development of rotating field spectroscopy, torque calculation of multi-layered particles
[27]	1985	F.A.SAUER et al.	
[28]	1985	V.Ph.PASTUSHENKO et al.	

in IC-Manufacturing Technology

[29]	1987	S.F.BART et al.	Comparison of IC-manufactured electrostatic motors with relaxation motors

Dielectric relaxation principle

If a dielectric body is exposed to electric fields traveling around, free electric image charges are induced at the interfaces of all rotor components. These image charges travel in synchronity with the electric field, but lag behind due to charge relaxation in the rotor and the surrounding medium. The resulting permanent displacment between field vector and image charges lead to a motive torque acting on the rotor. A maximal torque is reached, if the angular frequency of the external field is in the range of the dominant charge relaxation time ($\omega \approx 1/\tau$). At lower field frequencies the induced charges are only slightly displaced from the external field vector. Consequently, the resulting torque is not maximal. At higher

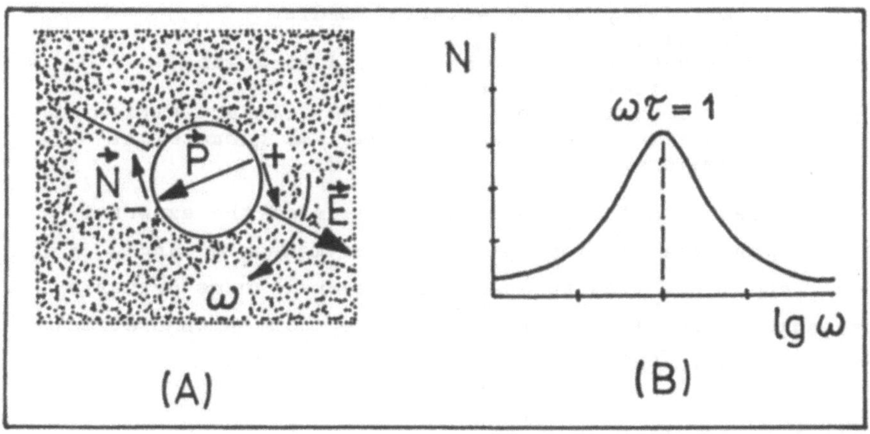

Fig. 1 (A) Dielectric particle driven by a rotating electric field (E - field vector, P - polarization vector, N - torque)
(B) Torque-frequency-characteristic of an isotropic sphere or cylinder immersed in a low conducting liquid.

frequencies very little image charges are induced and the torque is decreased too. A schematical cross section and typical torque-frequency-characteristic is given in Fig. 1.

Theory

Particles, such as biological cells, composed of several dielectrics show complex torque-frequency-characteristics due to a superposition of different charge relaxation processes. On the base of these spectra a new principle of dielectric spectroscopy of microscopical particles, called "electrorotation", could be developed [24-28], allowing characterization of individual living cells in suspension. Due to the frictional torque acting on liquid suspended particles cell rotation occurs slowly (0.1 - 10 rev/sec) and can therefore easily be measured. The angular velocity (ω_c), for example, of a multilayered cell (interior, membrane, cell wall ...), is approximately independent of the particle radius but a complex function of the angular field frequency (ω).

$$\omega_c = \frac{\varepsilon_0 \varepsilon_e}{2\eta} \cdot E^2 \, \frac{A_1 + A_2 \varepsilon_0 \omega + A_3 (\varepsilon_0 \omega)^2 + \ldots + A_n (\varepsilon_0 \omega)^{n-1}}{A_1' + A_2' \varepsilon_0 \omega + A_3' (\varepsilon_0 \omega)^2 + \ldots + A_n' (\varepsilon_0 \omega)^{n-1}}$$

where ϵ_e - permittivity of the external medium
η - viscousity of the liqiuid
E - amplitude of the external field strength
ω - angular field frequency
$A_1 \ldots A_n$, $A_1' \ldots A_n'$ - terms containing permittivities, conductivities of all particle components as well as geometrical parameters.

The rotor speed is a function of the square of the external field strength. the smallest particles we measured ranged between 1 and 5 µm. They rotated over several hours driven by 4 V rectangular pulses with nearly 100 rev/min. A theoretical review is given in [26].

Biological Application

The spin of individual cells is measured in four electrode chambers mounted on a glass slide as shown in Fig. 2. Examples for biological applications are summarized in the following table.

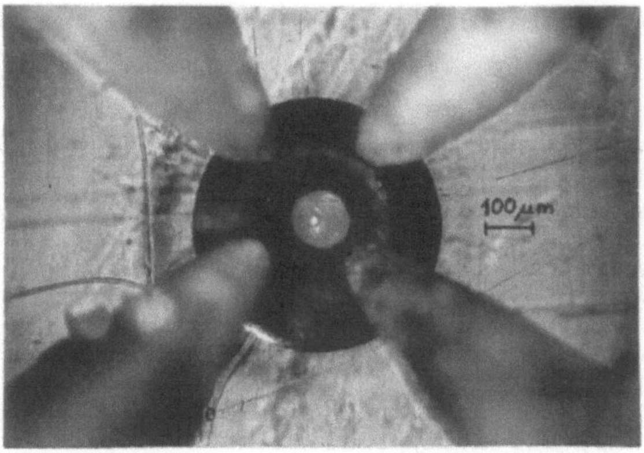

Fig. 2
Four-electrode chamber with liquid-immersed cells

Table 1
Reports on single cell applications of electrorotation (references are summarized in [30])

Investigated effect	Object
Characterization of the general physiological state of cells:	- plant cell organelles - plant protoplasts - Saccharomyces cerev. - Chlorella pyren. - hippocampus cells - mammalian eggs
Influence of ionophores:	- human erythrocytes - thylacoid vesicles - lymphocytes - platelets - plant protoplasts
Electric break down and subsequent resealing:	- human erythrocytes - plant protoplasts
Influence of heavy metals:	- Saccharomyces cerev.
Platelet activation:	- human platelets
Mitogenic stimulation:	- human lymphocytes
Fertilization:	- mouse oocytes
Virus-cell interaction:	- influenca virus human erythrocytes
Cell separation:	- plant cells

Technical application

To our knowledge up to now no relaxation motor was realized in IC manufactoring technique. We developed and investigated such motors in conventional technique having rotors (diameter 3 µm -

30 mm) composed of several dielectrics in sector-, sandwich- and shell design, respectively. The rotational behaviour of these motors can be characterized as followed:

1. the torque is independent on the angular position between rotor and stator electrodes
2. the torque-frequency characteristic can be well influenced via rotor architecture, the dielectrics used and the liquid surrounding the rotor
3. inhomogenities of dielectrics or asymmetry of rotors can be used to induce stepwise motor operation
4. the induced torque is comparable to that of electrostatic motors
5. the motor construction is simple and needs excitation voltages of nearly 5 V at typical motor dimensions of 500 µm in a frequency interval between 1 kHz and 100 MHz
6. well suited rotor materials are dielectrics such as semiconductors and isolators, e.g. silicon and its oxides.

Possible applications: micropumps, sensors (biosensors) and microactuators.

References

1. Tai, Y.-C., Fan, L.-S. and Muller, R.S., IC-processed electrostatic micro-motors, 1988 IEEE Int. Electron Devices Meeting, San Francisco, CA, U.S.A., Dec.11-14., 666-669
2. Jacobsen, S.C., Price, R.H., Wood, J.E., Rytting, T.H. and Rafaelof, M., A design overview of an eccentric-motion electrostatic microactuator (the wobblemotor), Sensors and Actuators 20, (1989) 1-16
3. Mehregany, M., Nagarkar, P., Senturia, S.D. and Lang, J.H., Operation of microfabricated harmonic and ordinary side-drive motors, IEEE (1990) 1-8
4. Hertz, H.R., über die Vertheilung der Electricität auf der Oberfläche bewegter Leiter, Wied. Ann. Physik,13 (1881) 266-275
5. Arnò, R., Sulla isteresa dielectrica viscosa, Atti Accad. Naz. Lincei Rend., 5 (5) (1896) 262-264
6. Quincke, G., Ueber Rotationen in constanten electrischen Felde. Wied. Ann. Physik, 59 (1896) 417-486
7. Heydweiller, A., Ueber Rotationen in constanten elecrischen Felde, Verh. Dtsch. Phys. Ges., 16 (1897) 32-36
8. Schweidler, E.v., über Rotationen im homogenen elektrischen Felde, Wien. Ber., 106 (1897) 526-532
9. Lampa, A., Ueber Rotationen im elektrostatischen Drehfelde, Wien. Ber. 2a, 115 (1906) 1659-1690
10. Born, M., über die Beweglichkeit der elektrolytischen Ionen, Z. Physik, 1 (1920) 221-249
11. Lertes, P., Untersuchungen über Rotationen von dielektrischen Flüssigkeiten im elektrostatischen Drehfeld, Z. Physik 4, (1921) 315-336
12. Richardson, S.W., Rotation of dielectric bodies in electrostatic fields, Nature, 119 (1927) 238
13. Vedy, L.G., Demonstration of the behaviour of bodies with non-conducting surfaces in electrostatic fields, Proc. Roy. Soc., 239 (1927) 169-170

14. Krasny-Ergen, W., Nicht-thermische Wirkungen elektrischer Schwingungen auf Kolloide, Hochfrequenztechn. Elektroakust., 48 (1936) 126-133
15. Sumoto, I., Rotary motion of dielectrics in a static electric field, Rep. Sci. Res. Inst., 32 (1956) 41-56
16. Secker, P.E. and Scialom, I.N., A simple liquid-immersed dielectric motor. J. Appl. Phys. 39, (1968) 2957-2961.
17. Melcher, J.R. and Taylor, G.I., Elektrohydrodynamics: A review of the role of interfacial shear stresses, Annu. Rev. Fluid Mech. 1, (1969) 111-146
18. Coddington, T. Pollard, A.F. and House, H., Operation of a dielectric motor with a low conductivity liquid. J. Phys. D.: Appl. Phys. 3 (1970) 1212-1218
19. Simpson, P. and Taylor, R.J., Characteristic rotor speed Variations of a dielectric motor with a low-conductivitiy liquid. J. Phys. D.: Appl. Phys. 4 (1971) 1893-1897
20. Teixeira-Pinto, A.A., Nejelski, L.L., Cutler, J.L. and Heller, J.H., The behavious of unicellular organism in an electromagnetic field, Exp. Cell. Res. 20 (1960) 548-564
21. Füredi, A.A. and Ohad, I., Behaviour of human erythrocytes in high-frequency electric fields and its relation to their age, Biochim. Biophys. Acta, 79 (1964) 1-8
22. Zimmermann, U., Vienken, J. and Pilwat, G., Rotation of cells in an alternating electric field: the occurrence of a resonance frequency, Z. Naturforsch. (C) 36 (1981) 173-177
23. Pohl, H.A. and Braden, T., Cellular spin resonance of aging yeast and mouse sarcoma cells, J. Biol. Phys.10 (1982) 17-30
24. Arnold, W.M. and Zimmermann, U., Rotating-field-induced rotation and measurement of the membrane capacitance of single mesophyll cells of Avens sativa, Z. Naturforsch.(C) 37 (1982) 908-915
25. Glaser, R., Fuhr, G. and Gimsa, J., Rotation of erythrocytes, plant cells, and protoplasts in an outside rotating electric field. Stud. Biophys., 96 (1983) 11-20
26. Fuhr, G., Gimsa, J. and Glaser, R., Interpretation of electrorotation of protoplasts. I. Theoretical considerations, Stud. Biophys. 108 (1985) 149-164
27. Sauer, F.A. and Schlögl, R.W., Torques exerted on cylinders and spheres by external electromagnetic fields. A contribution to the theory oef field induced cell rotation, in: A. Chiabrera, C. Nicolini and H.P. Schwan (Eds.), Interactions Between Electromagnetic Fields and Cells, Plenum, New York, NY, (1985) 203-251
28. Pastushenko, V.Ph, Kuzjimin, P,I. and Chizmadzhev, Yu. A., Dielectrophoresis and electrorotation - a unified theory of spherically symmetrical cells, Stud. Biophys. 110 (1985) 51-57
29. Bart, S.F. and Lang, J.H., An analysis of electrostatic induction micromotors. Sensors and Actuators, 20 (1989) 97-106
30. Arnold, W.M., and Zimmermann, U., Electro-rotation: development of a technique for dielectric measurements on individual cells and particles, J. Electrostatics 21 (1988) 151-191

Magnetically Driven Microactuator: Design Considerations

B. WAGNER, W. BENECKE

Fraunhofer-Institut für Mikrostrukturtechnik
Dillenburger Str. 53
D-1000 Berlin 33, Federal Republic of Germany

Summary

In this paper, design considerations for the use of magnetic forces in micromachined silicon structures are presented. Magnetic forces are of general interest for the fabrication of microactuators. The long range nature of the magnetic forces allows the design of microactuators with large deflection. It is proposed to mount a rare earth magnet on a movable micromachined silicon plate suspended by thin silicon bending beams, and further to integrate a planar coil on the chip. The basic configuration, with a moving permanent magnet and one static coil, can be used to realize a linear or a torsional actuator. A cubic permanent magnet, e.g. with a dimension of 1.5 mm and a ten-turn coil driven by a current of 300 mA generate a force of 1.5 mN. Various configurations of magnets and coils can be used to create actuators with optimum performance with respect to application.

Introduction

Silicon micromachining technologies have been used to produce a variety of microsensors. The use of silicon offers the unique advantage to integrate microelectronic components on the same chip [1]. For the future it will be a great challenge to fabricate sophisticated microsystems with integrated sensors, actuators and microelectronics [2,3]. Applications for actuators can be found among manipulators, motors, switches, shutters, pumps, valves etc.

Microactuators are based on effects such as electrostatic [4], thermomechanic [5], thermopneumatic [6], piezoelectric [7,8] and shape memory alloy [9] principles. Up to now electromagnetic forces, which dominate in the macro world, have been used scarcely. Magnetically driven micromanipulators have been fabricated using conventional fine mechanics and wire-wound solenoids [10,11]. On the other hand, silicon resonators have already been activated by electromagnetic forces with nearby external magnets [12-15].

The intent of this paper is to consider micromechanical magnetic actuators for static operation. The electromagnetic components should be integrated as far as possible on the chip.

Magnetic forces and micromechanics

Magnetic forces can be generated via the interaction of a magnetic field **H** with an electric current I or a magnetization **M** [16].

The Lorentz force

$$dF = \mu_0 \, I \, ds \times H \quad (1)$$

acts on a linear element ds of wire with current I ($\mu_0 = 4\pi \cdot 10^{-7}$ Vs/Am). On a permanent magnet with constant magnetization **M** and volume V, the field excerts the force

$$dF = \nabla(\mathbf{M} \cdot \mathbf{H})dV = \mathbf{M} \cdot \nabla \mathbf{H} \, dV \quad (2)$$

The force is only present in a nonhomogeneous field and points in the direction of the gradient of **H**. Moreover a torque

$$dT = \mathbf{M} \times \mathbf{H} \, dV \quad (3)$$

acts in order to align the magnetization parallel to the field. It should be mentioned that magnetic moments **M**dV can also be generated by soft magnetic materials or by current loops. These cases will not be considered in the following.

A basic advantage of magnetic forces is their long range. So, for example, the magnetic field of a wire decreases by 1/r, while electric fields decrease by $1/r^2$.

It is obvious that monolithically integrated coils can only have planar geometry. However, using the back of the chip or additional chips, several planes can be defined to produce, e. g., Helmholtz configurations. Thermal problems arising from the power dissipated in the coil favour solutions where the windings are placed on the substrate. Coils placed on the movable thin silicon part may generate bucking or bending of this bimorph structure. Therefore, the configurations using the Lorentz force will not be considered here.

Magnetic materials of various compositions can be deposited on the chip. However, the thickness is limited to several microns. On the other hand, the magnetic forces [Eqs. (2) and (3)] are proportional to the volume of the magnet. Therefore, using thin films, magnetic activation may only be useful for the stimulation of resonant motions.

However, for a static operation with large amplitudes one has to discuss hybrid integration of conventional discrete magnets. Therefore, the actuator described in this paper is based on a planar coil on the silicon frame and a moving permanent magnet mounted on a micromachined suspended silicon part. In this first approach, we focus on the most simple case of one coil and one magnet. This configuration allows the fabrication of either a vertical or a torsional actuator, depending on the orientation of the permanent magnet. When the magnetization has a vertical orientation (Fig. 1), the actuator is driven by the vertical force

$$dF_z = M_z \frac{d}{dz} H_z dV$$

When the center of the magnet is placed on the axis of the coil, all other components of force and torque vanish due to symmetry. The magnet can be fixed on top or on the back of the wafer, depending on application. Reversing the current of the coil reverses the direction of the force. In the configuration shown schematically in Fig. 2, the torque

$$dT_y = -M_x H_z dV$$

acts on the magnet with lateral magnetization. It is favourable to place the center of the magnet in the plane of the coil. Otherwise an additional lateral force would exist.

$$dF_x = M_x \frac{d}{dx} H_x \, dV$$

A lateral motion can also be suppressed by an appropriate spring design.

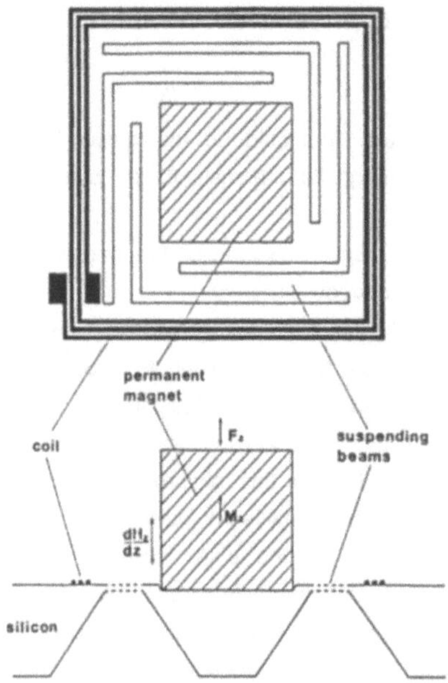

Fig. 1: Schematic view of a magnetic actuator for vertical motion

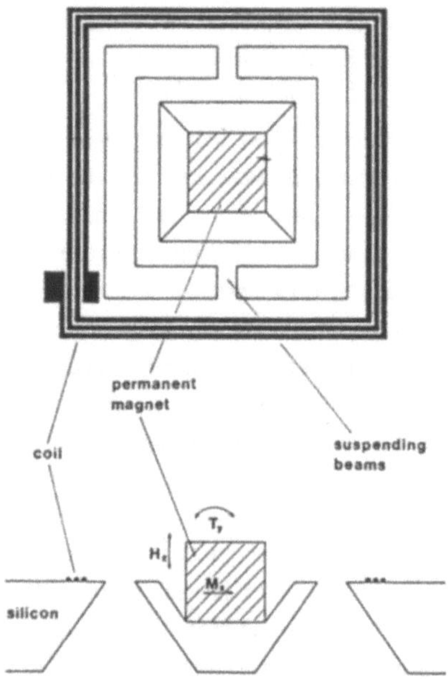

Fig. 2: Torsional actuator. The magnet is lowered to avoid lateral forces

<u>Fabrication process</u>

Micromachining of the silicon structure of the actuator is done by conventional anisotropic wet etching processes. The central plate is suspended by thin silicon beams with a thickness around 10 μm. The geometry of the suspending springs in Fig. 1 and Fig. 2 is exemplary and may be changed depending on the application. The thickness of the beams can be defined precisely by chemical or electrochemical etch-stop. Etched grooves in the central plate may be helpful for positioning of the magnet.

The planar coil may be fabricated using gold or aluminum. To reduce the resistivity of the coil, the thickness of the windings should be as high as possible. It has been shown that gold patterns with straight walls can be produced by conventional UV-lithography and electroplating processes [17] to a thickness of 20 μm with an aspect ratio of approx. 1. Even higher aspect ratios and thicker structures can be realized with X-ray lithography [18].

Permanent magnets with highest possible magnetization and coercivity are based on SmCo- or NdFeB-alloys [19]. With rectangular geometries they can be manufactured with dimensions less than 0.5 mm. Micromanipulators and glue dosing equipment, as used in SMT, may help in attaching the magnet to the chip.

Example

The magnetic field of a coil of arbitrary geometry can be computed by subdividing it into finite straight sections. The total field will then be the sum of the partial fields generated by each section. The magnetic field of a thin straight piece of wire (Fig. 3) can be calculated by integrating Biot-Savart's law to give [20]

$$H = \frac{I}{4\pi r}(\cos\alpha_1 + \cos\alpha_2)$$

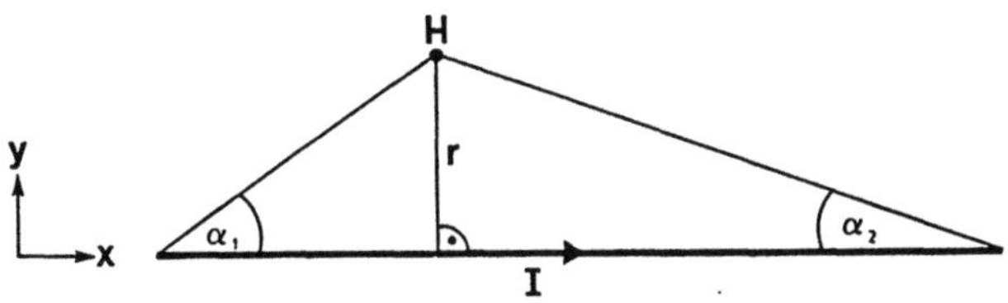

Fig. 3:

The components in a Cartesian coordinate system are

$$H_z = H \cdot y/r \qquad H_y = H \cdot z/r \qquad H_x = 0$$

Since the smallest permanent magnets are of rectangular shape, rectangular coils are also the most appropriate. We calculated the field for a quadratic ten-turn coil with a minimum to maximum edge length of 2.5 to 3.5 mm with a current of 300 mA. Using gold as metallization with a wire cross-section of 30x20 μm^2, the coil has a resistance of 4.5 Ω and dissipates a power of 0.4 W.

In Fig. 4, the H_z-component of the field is plotted for several z-values. In the center of the coil, an H-field of 920 A/m results, corresponding to a B-field of 1.16 mT.

To calculate the force for the vertical actuator, we need the gradient dH_z/dz, also shown in Fig. 4. The average gradient acting on a magnet with dimensions 1.5x1.5x1.5 mm^3 mounted as shown in Fig. 1 is approx. $4 \cdot 10^5$ A/m^2. NdFeB-magnets have a remanent magnetization of about 1.1 T and a mass density of 7.4 g/cm^3. This results in a magnetic force of 1.5 mN. In comparison, the gravitational force is about six times lower. Using a suspension with four silicon bending beams having dimensions of 1800 μm long, 250 μm wide and 10μm thick, the magnetic force leads to a vertical deflection of 50 μm.

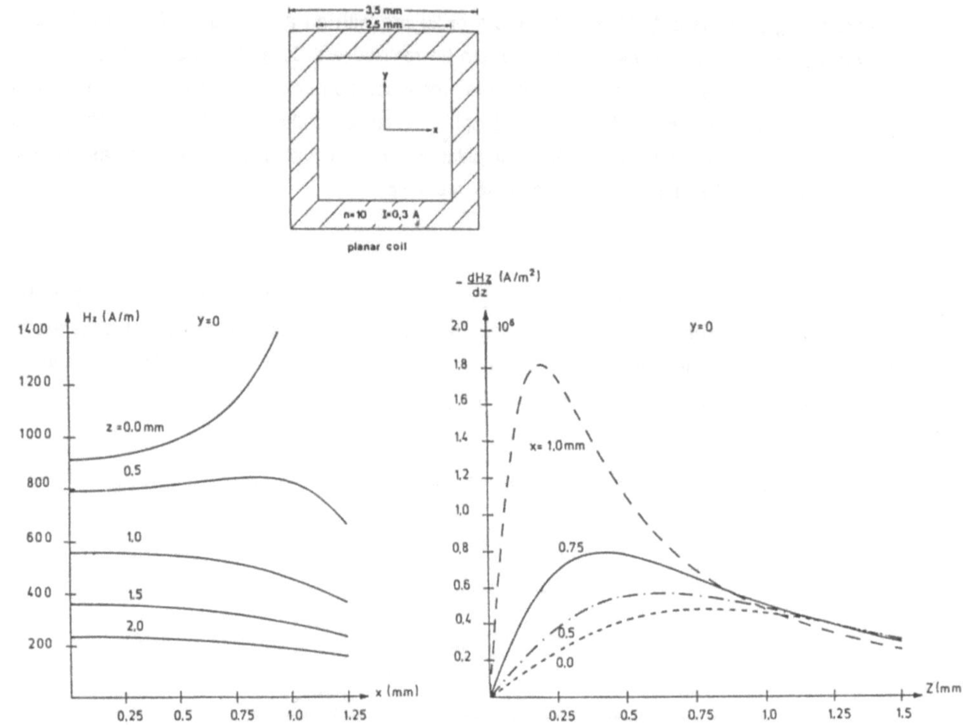

Fig. 4: Magnetic field and field gradient for the planar coil used in the example

Conclusion

Magnetic excitation of actuators is attractive because the long-range nature of the force can be used to achieve large deflections. This paper describes a microactuator design with an on-chip integrated planar coil and a hybrid mounted moving permanent magnet. The exact geometry of the mechanical microstructure and of the coil can be guaranteed using silicon micromachining and IC technologies. The precision of the magnet attachment, which is also important to achieve well-defined displacments, has to be proved in practice.

The presence of a magnet may be helpful in detecting the motion and building of an active controlled microactuator. Many other more complex configurations of magnets and coils are possible in creating actuators for various applications.

References

/1/ W. Riethmüller, Micromachined silicon devices integrated with standard IC processes, this conference

/2/ W.S.N. Trimmer, Microrobots and micromechanical systems, Sensors and Actuators 19 (1989) 267

/3/ R.S. Muller, Microdynamics, Sensors and Actuators A21-A23 (1990) 1

/4/ W.S.N. Trimmer, K.J. Gabriel, Design considerations for a practical electrostatic micromotor, Sensors and Actuators, 11 (1987) 189

/5/ W. Riethmüller, W. Benecke, Thermally excited microactuators, IEEE Trans. Electron Dev. ED-35 (1988) 758

/6/ F.C.M. Van de Pol, D.G.J. Wonnink, M. Elwenspoek, J.H.J. Fluitman, A thermo-pneumatic actuation principle for a microminiature pump and other micromechanical devices, Sensors and Actuators 17 (1989) 139

/7/ F.R. Blom, D.J. Yutema, F.C. M. Van de Pol, M. Elwenspoek, J.H.J. Fluitman, T.J.A. Popma, Thin-film ZnO as micromechanical actuator at low frequencies, Sensor and Actuators A21-A23 (1990) 226

/8/ M. Esashi, Integrated micro flow systems, Sensors and Actuators, A21-A23 (1990) 161

/9/ J.A. Walker, K.J. Gabriel, M. Mehregany, Thin film processing of TiNi shape memory alloy, Sensors and Actuators A21-A23 (1990) 243

/10/ R. Pelvine, I. Busch-Vishniac, Magnetically levitated micro-machines, IEEE Microrobots and teleoperators workshop, Hyannis 1987

/11/ D.P.E. Smith, S.A. Elrod, Magnetically driven micropositioners, Rev. Sci. Instr. 56 (1985) 1970

/12/ R.E. Hetrick, A vibrating cantilever magnetic-field sensor, Sensors and Actuators 16 (1989) 197

/13/ J.M. Naden, R.E. Jones, An optically-addressed silicon strain sensor, Sensors and Actuators 16 (1989) 135

/14/ R.A. Buser, N.F. de Rooij, Tuning forks in silicon, IEEE Micro Electro Mechanical Systems, Salt Lake City 1989, 94

/15/ K. Ikeda, H. Kuwayama, T. Kobayashi, T. Watanabe, T. Nishikawa, T. Yoshida, K. Harada, Silicon pressure sensor integrates resonant strain gauge on diaphragm, Sensors and Actuators A21-A23 (1990) 146

/16/ J.R. Melcher, Continuum Electromechanics, MIT Press 1981

/17/ U. Schnakenberg, W. Benecke, V. Wallendszus, K.P. Müller, A Heuberger, B. Lischke, Multi electron beam lithography: Fabrication of a control unit, Microelectronic Engineering 9 (1989) 205

/18/ E.W. Becker, W. Ehrfeld, D. Münchmeyer, H. Betz, A. Heuberger, S. Pongratz, W. Glashauser, H.J. Michel, R. v.Siemens, Production of separation-nozzle systems for uranium enrichment by a combination of X-ray lithography and galvanoplastics, Naturwissenschaften 69 (1982) 520.

/19/ W. Rodewald, F. Schmidt, Seltenerd-Dauermagnetwerkstoffe, Feinwerktechnik und Meßtechnik 97 (1989) 4

/20/ E. Philippow, Grundlagen der Elektrotechnik, VEB Verlag Technik, Berlin

Development of Micro Grippers

Y.ANDO*, H.SAWADA*, Y.OKAZAKI*, Y.ISHIKAWA*, T.KITAHARA*, Y.TATSUE* and K.FURUTA**

*Mechanical Engineering Laboratory
Namiki 1-2,Tsukuba,ibaraki,305(Japan)

**Seiko Instruments Inc.
Takatsukashinden 563,Matsudo,Chiba.,271(Japan)

Abstract

This paper describes the structures and basic performances of two micro-grippers, type I and type II, which were designed to hold a minute object by gripping. Each gripper has the same piezo-actuator but has different mechanism to activate the fingers. Type I, with spring joints and beams, indicated sufficient finger operation and holding force, but Type II, with gears, failed to activate the fingers.

1. Introduction

Conventional means to handle cells and small components mostly utilize suction force, and there have been very few tools which are designed to hold a object by gripping. Micro grippers, capable of holding a object by gripping, are absolutely needed to handle small machine elements of complex and indefinite forms. For this reason, Two micro gripper, type I and type II, were test manufactured. The each micro gripper has the same actuator, but it has different mechanism for transmitting and enlarging the displacement of the actuator.

To design such micro grippers, the following specifications were set.

(1) Overall size is approximately 20 mm.

(2) The size of the object to be gripped is 1 mm or less.

(3) Stroke of a finger point is 0.2mm or more.

(4) A force of 1mN is available at the finger point.

The structures and basic performances of the above micro grippers are reported in this paper.

2. Type I : Gripper with Spring Joints

2.1 Analysis by Finite Element Method (FEM)

The piezoelectric stack actuator, has the maximum elongation of only about 4μm even under a no-load condition, while the stroke of a finger is required 0.2mm or more. Accordingly, it was necessary to enlarge the displacement by approximately 50 times.

Initiatory, Type I was supposeed to be made of a plate in uniform thickness, and the principle of leverage and spring joints were used in designing the mechanism. The shape in mesh of the gripper is shown in Fig.1(a). Since the gripper is axisymmetic, only one side of the form is shown in the figure.

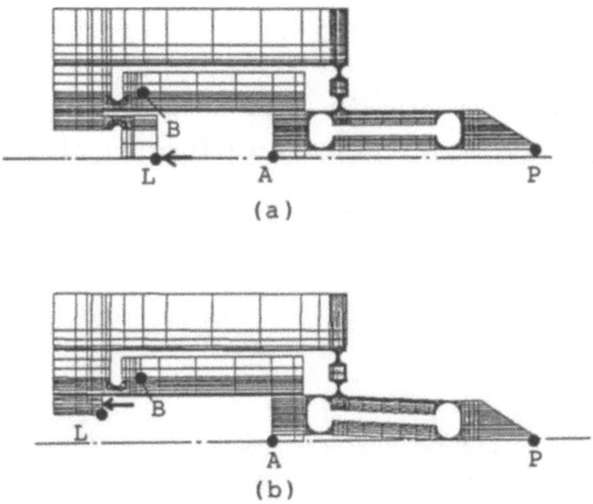

Fig.1. Shape in mesh of the gripper(type I)

FEM analysis was carried out for this form. Putting load on point L with point A fixed, the enlargement rate, that is to say, the displacement ratio of point L to P, was calculated, and the result found was only 5 times. With points A and B fixed, recalculation was conducted with the result that the value increased to 15 times, indicating that the parts except the spring joints exerted significant influence on the enlargement rate.

This value, however, was not large enough. To further increase the enlargement rate, the shape shown in Fig.1(b) was selected, and the value was calculated for both one-point and two-point fixing. The calculation results of the enlargement rate by FEM are summarized in Table 1.

Table 1. Calculation results of enlargement rate

	fig.1(a)	fig.1(b)
A fixed	5	9
A & B fixed	15	26

2.2 Structure

The type I eventually manufactured is shown in Fig.2. Type I consists of major plates, reiforcing plates and a piezo-actuator.

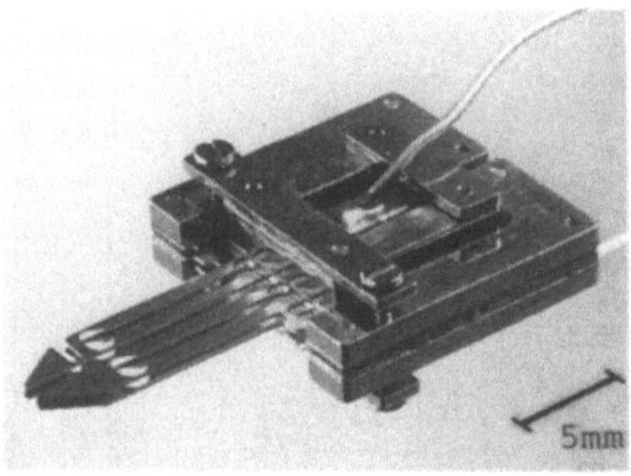

Fig.2. Photograph of micro gripper(type I)

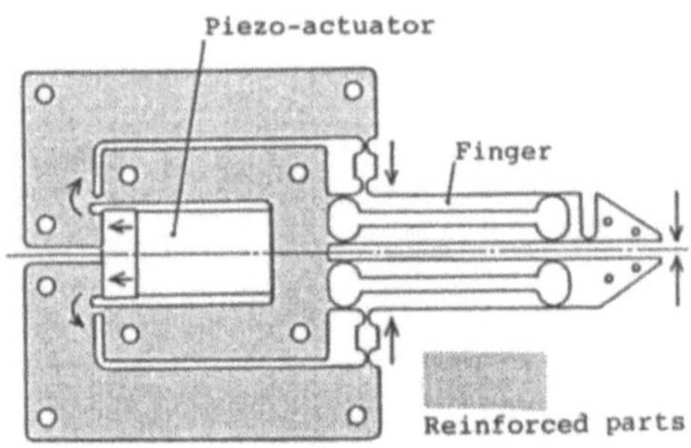

Fig.3. Enlarging mechanism of micro gripper(type I)

The major plate was formed by etching beryllium copper plate 0.15mm thick. The thinnest spring joint is approximately 0.07 mm. Two major plates were arranged parallel to each other with spacers of 0.3mm in between, and joined. Brass plates, for reinforcing, each 1.0 to 1.8mm thick, machine cut were placed around the major plates and joined to them as shown in Fig.3. The piezo-actuator, measuring 5 x 4 x 3 mm, was inserted and fixed at the center of the gripper.

2.3 Experimental Results

Fig.4 shows the characteristics of the impressed voltage-approach amount between the two finger points for the actually test manufactured type I. The approach amount was visually observed under an optical microscope by placing a scale of minimum scale value of 0.01mm. The ratio of the approach amount between the finger points to voltage was 1.6μm/V (0 to 100 V).

Fig.5 shows the impressed voltage-actuator displacement characteristics for the actuator actually mounted in the gripper. The elongation rate in this instance was 0.014μm/V (0 to 100V). From the above results, the enlargement rate for the mechanism was 57 times.

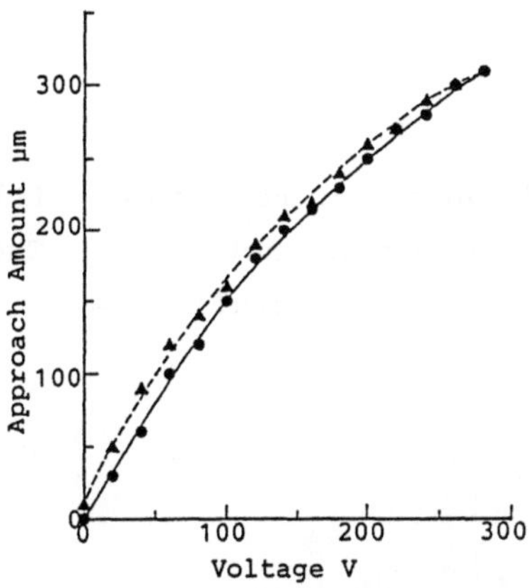

Fig.4. Approach amount of fingers versus voltage

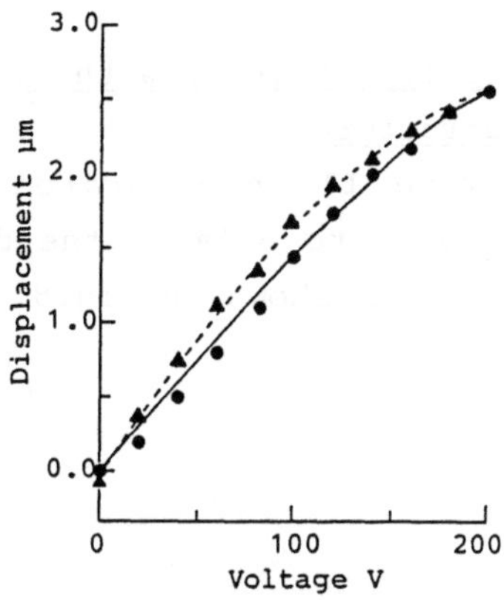

Fig.5. Actuator displacement versus voltage

An additional experiment, shown in Fig.6, was carried out to study what weight the finger could actually hold. A piece of plate 0.4mm thick was held by the finger and the voltage impressed on the actuator was varied in steps. For each test voltage, the force required to pull out the plate was measured.

Results of the measurements are shown in Fig.7. The voltage applied and the sliding force are proportional to each other, and the plate withstood a force of up to 8mN, indicating that the initially set target was realized with a good margin.

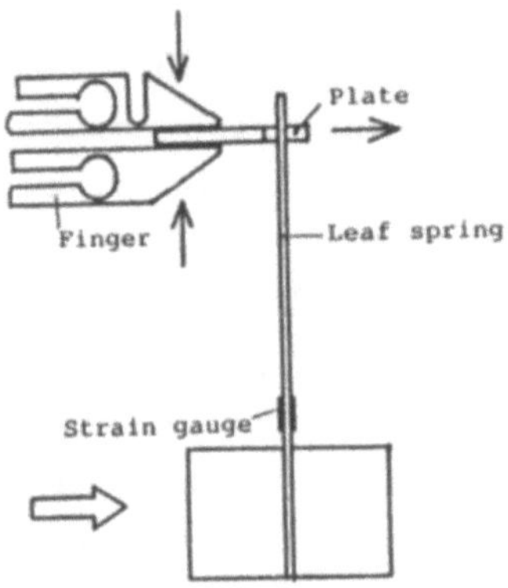

Fig.6. Schematic drawing on measurement of holding force

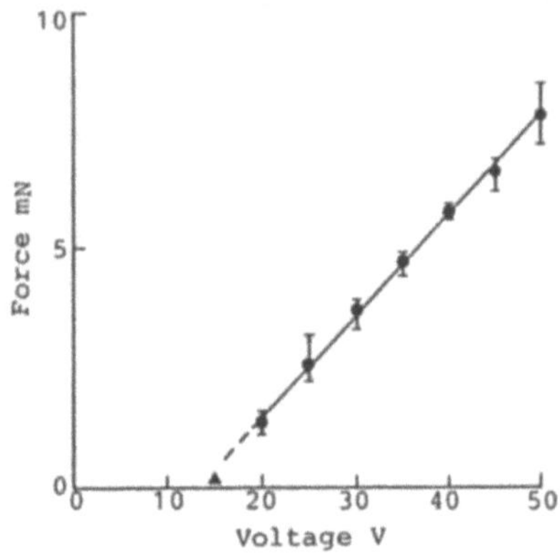

Fig.7. Holding force versus voltage

3. Type II : Gripper with Clock Gears

3.1 Structure

Type II incorporated a speed-increase mechanism with clock gears to enlarge the displacement of the actuator. The mechanism is shown in Fig.8. Fig.9 shows the overall view.

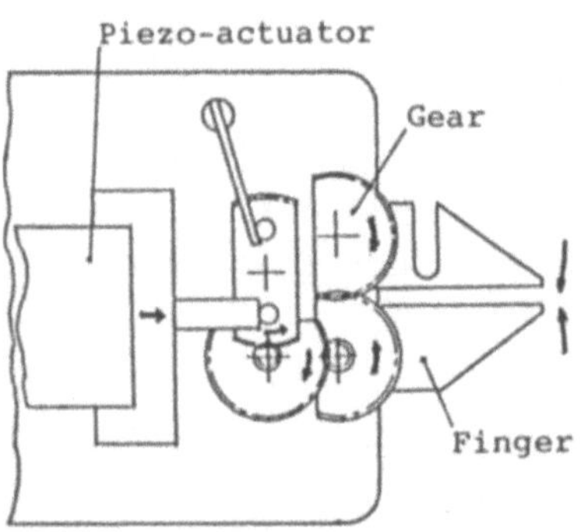

Fig.8. Enlarging mechanism of micro gripper(type II)

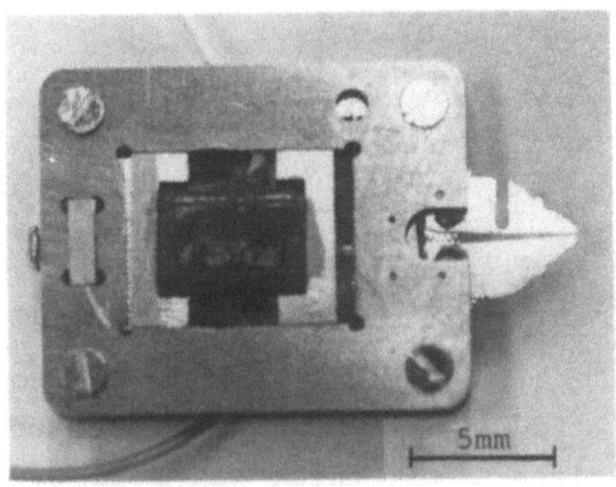

Fig.9. Photograph of micro gripper(type II)

Gear module is 0.07mm, and the displacement enlargement rate as calculated from the numbers of teeth is 75. Components, mostly made of brass, were manufactured by cutting and then assembled. A feed screw was arranged on the opposite side of the fingers to push the actuator, so that the positioning at setting could be done with ease.

3.2 Experimental Results

After completing the assembly of type II, voltage was applied to the piezo-actuator, with no resulting finger operations. This may be because the actuator displacement was completely absorbed by gear backlash. But the finger did not move even when the feed screw at the bottom of the actuator was turned, and for this reason, it was quite possible that friction or the shape precisions was the cause. Since, combination of a number of parts is required to enlarge the displacement and there are also many parts that undergoes friction as well.

There are also many problems to be solved such as the requirements for improved assembling precision and friction. It is difficult to design this system in a more compact form without the above improvements. However, if for example, the actuator is replaced by a small-sized motor, then a gear-type speed-reduction mechanism may be utilized, and it would be quite possible that the gear system is more advantageous.

4. Conclusions

In the case of the type I, spring joint system, the target performance was sufficiently realized by reinforcing the structural portions such as the beams with thick plates. This, in turn, confirmed the possibility of developing a smaller-sized higher performance micro gripper. However, it is necessary to study the optimum structure using FEM.

Type II, gear system, failed to activate the fingers. To make the gear system realistic, the frictions of the gears and their pins must be further studied and an actuator, which maximizes the gear characteristics, must be developed.

Micro-Actuator for Micro-Particles

Kunitoshi Nishimura

NTT LSI Laboratories
3-1, Morinosato Wakamiya, Atsugi-Shi
Kanagawa Pref., 243-01 Japan

Abstract

This paper describes a micro-actuator capable of carrying particles smaller than 0.5 mm in diameter. The actuator consists of a conductive guide with striped electrodes embedded in an insulation layer, and a driving circuit. The particles move along the guide when the voltage is switched from one electrode to the next. First, the feed performance is investigated using an enlarged model. Then, a device capable of carrying carbon particles is made on a silicon wafer.

1. Introduction

Semiconductor fabrication technology has rapidly been producing numerous applications for small mechanical systems. Micro-actuators based on the technology maintain a high potential in contrast to conventional machines. Several micro-actuators were reported by Trimmer and Gabriel.(1) The actuators are driven by electrostatic forces. This is because for smaller micro-mechanical systems, electrostatic forces are the most promising. Our aim is electrostatic micro-actuators able to handle particles smaller than 0.5mm in diameter.

The driving mechanism for a small roller was proposed by Fujita and Omodaka.(2) The driving force is the electrostatic force acting between a cylindrical electrode and striped electrodes embedded in an insulation layer. The roller is one of the electrodes. But, when a particle is expected as one of the electrodes, the reliability becomes low as the contacting area is very small. Accordingly, the particle is not an electrode in our actuator. When a particle is placed in a nonuniform electrostatic field, it is attracted towards high electric fields. We use this force in our micro-actuator.

2. Computer calculation of the electric field

Actuator design begins by estimating the force exerted by the electric field. A schematic diagram of the actuator is shown in Fig.1. Two-dimensional analysis is used to simplify the calculations. A roller replaces the particle. The two electrodes are separated by a distance L in one plane. The roller is parallel to the electrodes under the insulation layer. Since the roller is perfectly insulated, the total charge on the roller is always zero. The voltage applied between two electrodes forms an electric field. The equipotential lines, computed by the finite element method, is shown in Fig.2. The two-dimensional electric field analysis program for Fig.2 was developed by Saito et al..[3] The regions of high equipotential-line density indicate high electric fields. The forces attracting the roller can be obtained by the following procedure.

Electrostatic field $E(\theta)$ on the roller surface is the derivative of the potential ϕ with respect to the radius r.

$$E(\theta)=\partial\phi/\partial r$$

$E(\theta)$ is normal to the roller surface because the roller is conductive. The force f acting on a unit area of the surface is:

$$f=1/2\cdot\varepsilon\cdot E^2$$

where ε is the permittivity of the vacuum. The force f acts radially from the center of the roller. Integrating the components of f in the X-direction (parallel to the plane) and the Y-direction (normal to the plane) over the surface gives the driving force Fx and the sticking force Fy on a roller of length 2r. The relationship between the forces and the roller position is shown in Fig.3, for r = 2 mm and L = 3 mm. The maximum sticking force Fy is at the edge of the positive electrode. The driving force Fx has a steep slope intersecting the X-axis at the corner of the positive electrode.

Figure 4 shows the ratio Fx/(-Fy). When the weight of the roller is negligible and the coefficient of rolling friction is about 0.01, the figure shows that, when the roller is placed on the right or left side, it rotates, counterclockwise or clockwise to the positive electrode. The ratio hardly varies for insulation layer thicknesses ranging from 10 to 640 μm.

Figure 5 shows the relationship between the insulation layer thickness and the maximum values Fxmax, (-Fymax) of Fx, (-Fy). Thicker layers give smaller maximum forces, and the drop in force due to the increased thickness is smaller in Fx than Fy.

3. Micro-actuator configuration and performance

The configuration of the actuator for particles is shown in Fig.6. The electrodes are built in 3 phases to move the particles continuously when the voltage is switched from one electrode to the next. Biasing and phasing schemes are shown in Fig.7. In the 3-1 scheme, the voltage is pulsed on one of the three phases. In the 3-2 scheme, the voltage is pulsed on two phases. The relationship between the driving force and the particle position is shown in Fig.8, where the pitch of the electrodes is 2.4 mm and the particle is a steel ball 4 mm in diameter. The driving force in the 3-1 scheme is larger than in the 3-2. Assuming a rolling friction coefficient of 0.01, a 100 mg ball can be rotated by the 3-1 scheme. Strictly speaking, if the sticking force is 10 mg, the maximum weight of ball is reduced to 90 mg. A steel ball 4 mm in diameter cannot be driven because it weighs 260 mg.

If the scale of the actuator and the applied voltage are divided by ten, the electrostatic field intensity on the particle surface is the same, but the force on the particle reduces to one-hundredth because it is proportional to the particle surface area.

Several types of particles were tested, as shown in Table 1. The maximum driving forces are estimated from the experimental data for No. 1. The insulation layer thickness is not proportional to the particle diametér, but the deviation is neglected because the thickness has little effect on the force, as was shown in Fig.5.

For the particle to rotate, the driving force must overcome the frictional sticking force plus the gravitational force:

$$Fx > \mu [W + (-Fy)]$$

where the weight of the particle is W and the coefficient of rolling friction is μ.

The above equation can be rearranged as:

$$R = 1/ [W/Fx + (-Fy/Fx)] > \mu$$

Since the forces Fx and Fy change according to the particle position, R is estimated by the following equations to simplify the calculations.

$$(-Fy/Fx) \fallingdotseq 0.1$$
$$W/Fx \fallingdotseq W/Fxmax$$

Larger R gives better movement. The particle movement was experimentally confirmed for No. 2 to No. 4. The results indicate that the coefficient of friction is between 0.004 and 0.008.

Furthermore, the particle does not touch the guide on the surface of the insulation layer. This is because the conductive particle and guide close to each other have the same electric field potential, and thus repel each other.

The experiment for No. 5 is described in the next chapter.

4. Micro-actuator fabricated on silicon wafer

Electrodes with 3.2 μm pitch were made on a silicon wafer by semiconductor fabrication technology. The carbon particle is 5 μm in diameter. The electrodes are embedded in SiO_2 and the guides are made of aluminum.

The experiment showed the slight stepwise rotations took place at voltage ranging from 20 to 200 V just after the voltage was applied.

Figure 9 shows the cross-sectional SEM picture along the guide. The surface has a 0.3 μm roughness. This is a serious barrier to motion.

5. Conclusion

A micro-actuator capable of carrying particles smaller than 0.5 mm in diameter is proposed. Computer simulation shows that a roller placed near the electrodes rotates towards the positive electrode edge. The performance of a micro-actuator for micro-particles was confirmed by experiments on several types of the actuator. The roughness of the rolling surface is a serious barrier to rotation on the actuator made on a silicon wafer.

References

(1) W.S.Trimmer and K.J.Gabriel. Design Considerations for a Practical Electrostatic Micromotor. Sensors and Actuators. 11. pp:189-206 1987

(2) H.Fujita and A.Omodaka. The principle of an electrostatic linear actuator manufactured by micromachining. Transducer '87 pp:861-864 1987

(3) K.Saito, T.Okubo, K.Takamoto, Y.Uno and M.Kondo. Electron ray tracing with high accuracy. J. Vac. Sci. Technol. A4(4), Jul/Aug pp:1913-1917 1986

Table 1 Micro-actuator specification and performance

	Particle Diameter (mm)	Estimated Maximum Driving Force (gram)	Weight (gram)	Ratio R	Pitch of Electrodes (mm)	Thickness of Insulated Layer (μm)	Results
No 1	4.0	0.1×10^{-2} *	0.26	0.004	2.5	50 *	×
No 2	1.6	0.15×10^{-3}	0.16×10^{-1}	0.008	1.0	50 *	○
No 3	1.6	0.15×10^{-3}	0.62×10^{-2}	0.02	1.0	50 *	○
No 4	0.5	0.14×10^{-4}	0.1×10^{-3}	0.06	0.3	50 *	○
No 5	5 μm	0.4×10^{-8}	0.1×10^{-9}	0.1	3.2 μm	0.8 *	×

* Measured

○ Move
× Doesn't move

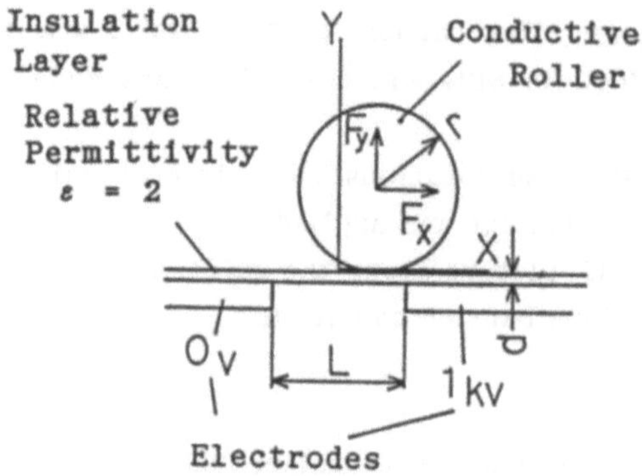

Fig.1 Basic micro-actuator structure

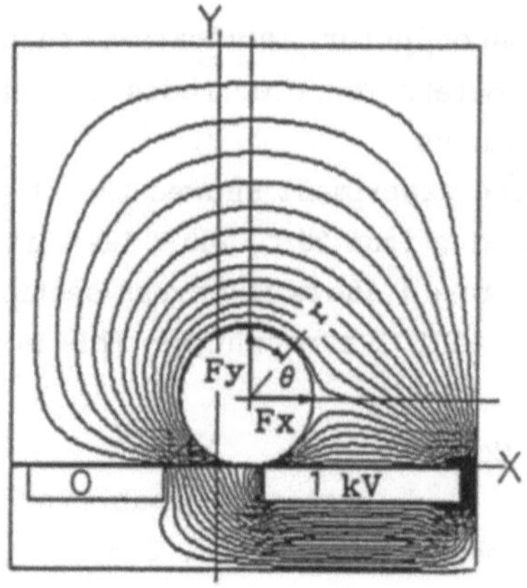

Fig.2 Equipotential lines from two-dimensional simulation

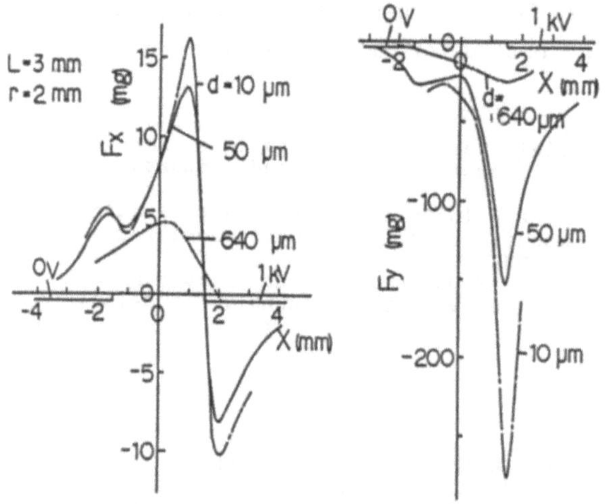

Fig.3 Forces versus roller position

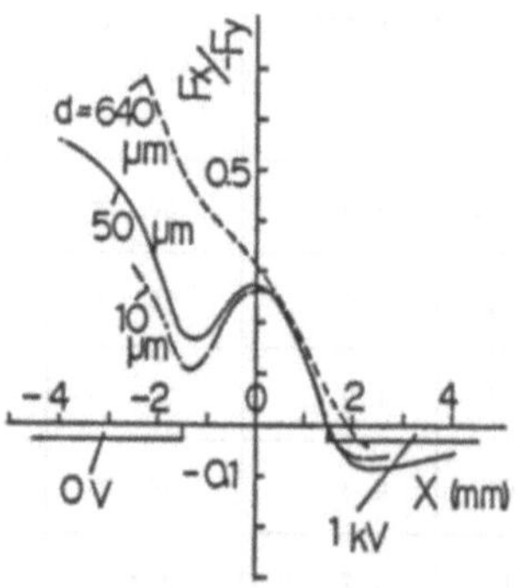

Fig.4 Ratio of forces versus insulation layer thickness

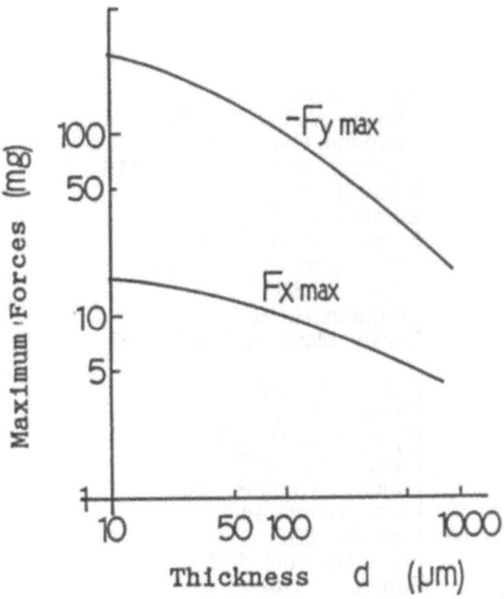

Fig.5 Maximum forces versus insulation layer thickness

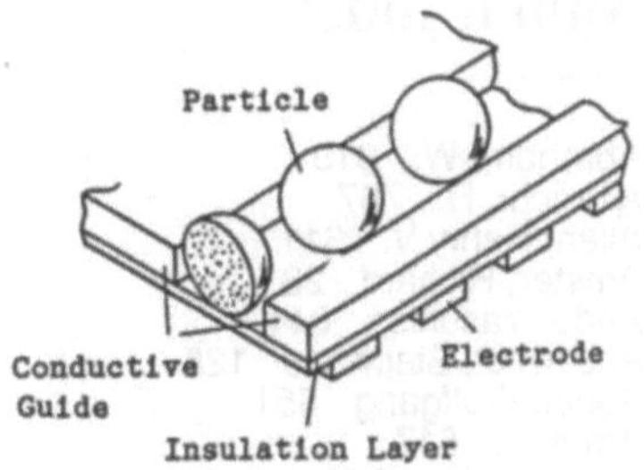

Fig.6 Micro-actuator for micro-particles

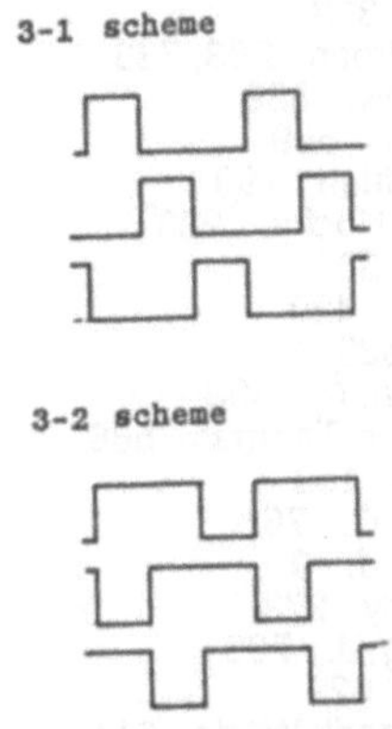

Fig.7 Biasing and phasing of a micro-actuator

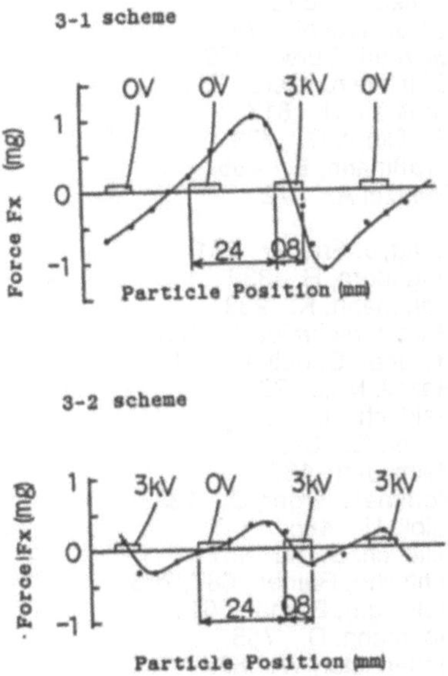

Fig.8 Driving forces versus particle position

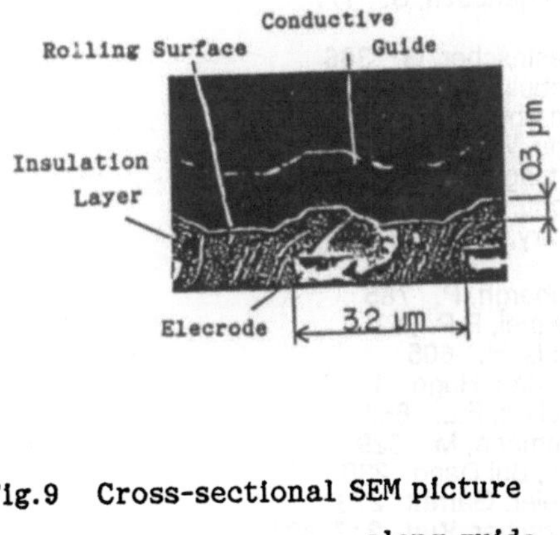

Fig.9 Cross-sectional SEM picture along guide

Author Index

MIX
Papier aus verantwortungsvollen Quellen
Paper from responsible sources
FSC® C105338
FSC
www.fsc.org

If you have any concerns about our products,
you can contact us on
ProductSafety@springernature.com

In case Publisher is established outside the EU,
the EU authorized representative is:
Springer Nature Customer Service Center GmbH
Europaplatz 3, 69115 Heidelberg, Germany

Printed by Libri Plureos GmbH
in Hamburg, Germany